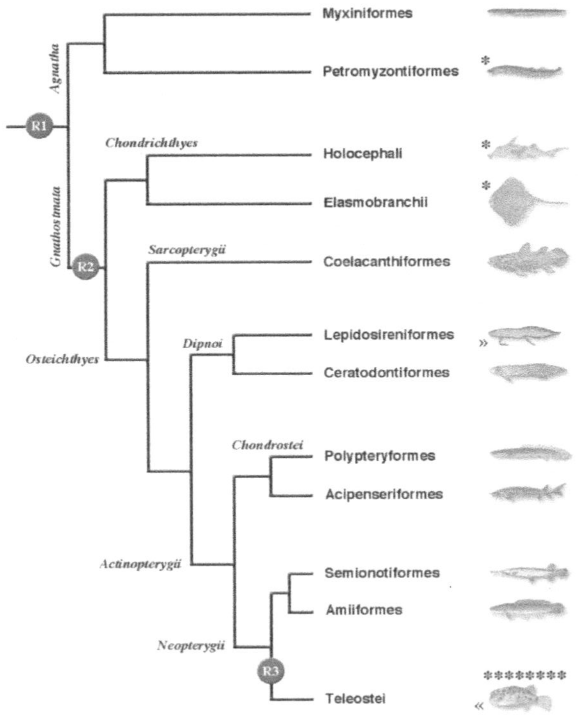

	yolk type	cleavage pattern	spawning environment FW	spawning environment SW	reproduction Ovp	reproduction Ovv	reproduction Vvp	egg type B	egg type P	fertilisation I	fertilisation E
Myxiniformes	T	meroblastic		+	+			+			+
Petromyzontiformes	M	holoblastic	+		+			+			+
Holocephali	T	meroblastic		+	+			+		+	
Elasmobranchii	T	meroblastic		+	+	+	+	+		+	
Coelacanthiformes	T	meroblastic		+			+			+	
Lepidosireniformes	M	semi-holoblastic	+		+			+			+
Ceratodontiformes	M	semi-holoblastic	+		+			+			+
Polypteryformes	M	semi-holoblastic	+		+			+			+
Acipenseriformes	M	semi-holoblastic	+		+			+			+
Semionotiformes	M	semi-holoblastic	+		+			+			+
Amiiformes	M	semi-holoblastic	+		+			+			+
Teleostei	T	meroblastic	+	+	+	+	+	+	+	+	+

Phylogenetic overview of extant fishes showing genome duplication events, type of yolk: telolecithal (T) or mesolecithal (M), cleavage pattern, spawning environment: freshwater (FW) or seawater (SW), mode of reproduction: oviparous (Ovp), ovoviviparous (Ovv), or viviparous (Vvp), type of egg spawned: benthic (B) or pelagic (P), and mode of fertilisation: internal (I) or external (E). * highlights current genome sequencing projects: Petromyzontiformes: sea lamprey (*Petromyzon marinus*); Holocephali: elephant shark (*Callorhinchus milii*); Elasmobranchii: Little Skate (*Raja erinacea*); Teleostei: zebrafish (*Danio rerio*); Atlantic salmon (*Salmo salar*); Atlantic cod (*Gadus morhua*); three-spined stickleback (*Gasterosteus aculeatus*); medaka (*Oryzias latipes*); Nile tilapia (*Oreochromis niloticus*); spotted green pufferfish (*Tetraodon nigroviridis*) and torafugu (*Takifugu rubripes*). » highlights Lepidosireniformes: marbled lungfish (*Protopterus aethiopicus*) as containing the largest animal genome (130 000 Mb) known to date; « highlights Tetraodontiformes: spotted green pufferfish (*Tetraodon nigroviridis*) as containing the smallest vertebrate genome (340 Mb) known to date.

FISH
LARVAL PHYSIOLOGY

FISH LARVAL PHYSIOLOGY

Editors

RN Finn

Department of Biology
University of Bergen
Bergen High Technology Centre
Thormøhlensgate 55
N-5008 Bergen, Norway
E-mail: nigel.finn@bio.uib.no

BG Kapoor

Formerly Professor of Zoology
The University of Jodhpur
Jodhpur, India
E-mail: bhagatgopal.kapoor@rediffmail.com

Science Publishers

Enfield (NH) Jersey Plymouth

Science Publishers www.scipub.net
234 May Street
Post Office Box 699
Enfield, New Hampshire 03748
United States of America

General enquiries : *info@scipub.net*
Editorial enquiries : *editor@scipub.net*
Sales enquiries : *sales@scipub.net*

Published by Science Publishers, Enfield, NH, USA
An imprint of Edenbridge Ltd., British Channel Islands
Printed in India

ISBN: 978-1-57808-388-6

Library of Congress Cataloging-in-Publication Data

Fish larval physiology/editors, R.N. Finn, B.G. Kapoor.
p. cm.
Includes bibliographical references and index.
ISBN 978-1-57808-388-6 (hardcover)
1. Fishes—Larvae—Physiology. I. Finn, R.N. II. Kapoor, B.G.

QL639.25.F567 2008
571.1'7—dc22

2007050712

Preface

To study physiology is to examine how organisms have evolved solutions to the business of living in an inanimate world. Our world is and has always been dominated by physical and chemical forces. A physicist might tell us that all things are physical, while a chemist is more concerned with the elementary nature of reactions. A physical chemist sees the bonds between these views, and a biochemist draws out the organic symphony of the vital pathways. A structural biologist adds shape to the chemical building blocks of life, while a molecular biologist tinkers with these structures. A cell biologist is closely related to a molecular biologist and cousin to an anatomist that is intrigued by the variety of phenotypes. Mapping the fabric between these fields is a developmental biologist that not only seeks to understand the mechanisms of cell signalling and differentiation, but place them in an evolutionary context. To understand how an organism functions, and adapts to changes in its external and internal environment, a physiologist must dabble in each of these fields. Consequently modern physiologists face a daunting task of acquiring a rich tapestry of knowledge and methods that will lead them from the whole organism to the depths of sub-cellular compartments and molecules responsible for life.

This book is intended as a resource for students and researchers interested in developmental biology and physiology and specifically addresses the larval stages of fish. Fish larvae (and fish embryos) are not small juveniles or adults. Rather they are transitionary organisms that bridge the critical gap between the single-celled egg and sexually immature juvenile. Fish larvae represent the stage of the life cycle that is used for differentiation, feeding and distribution. Like the juveniles and adults, however, they face the same physical and chemical challenges imposed by their aquatic environment, yet lack many of the cells, tissues or organs present in the more advanced stages to deal with such adversities. Many fish larvae, particularly those of the marine forms of teleosts, are small, in fact tiny, necessitating the use of magnifying devices just to see them. A common size of a newly hatched marine teleost is four mm or less, the same size as a one month old human embryo. As a consequence traditional physiological investigation of fish larvae has tended to be of the "black box" whole organism nature. With the advancement of molecular tools, and miniaturisation of manipulation instruments, investigators can now peer deep inside the developing systems and explain how they work. The genomic revolution might be regarded as the cavalry in this regard, where physiologists can not only investigate the ontogeny of expression, but modulate the genes through insertional mutations, or regulate them via hybridisation techniques. When coupled with more classical methods, an entirely new level of understanding emerges.

From a biodiversity perspective, fish champion the vertebrates, accounting for almost half of the known species of craniates. The most recent count is ~28 000 species spread among more than 500 families. The physiological diversity of fish larvae, in all of their myriad forms, is only beginning to be chartered. This book aims at providing a single-volume treatise that explains how fish larvae develop and differentiate, how they regulate salt, water and acid-base balance, how they transport and exchange gases, acquire and utilise energy, how they sense their environment, and move in their aquatic medium, how they control and defend themselves, and finally how they grow up. We hope that the reader learns as much from this text as we have editing it.

Roderick Nigel Finn
Bhagat Gopal Kapoor

Contents

Part 4: Sensory Physiology

Part 5: Movement

Part 6: Control and Defense

Part 7: Functional Changes in Form

List of Contributors

Bagatto Brian

Department of Biology, University of Akron, Akron, OH 44325, USA

E-mail: bagatto@uakron.edu

Björnsson Björn Th

Fish Endocrinology Laboratory, Department of Zoology/Zoophysiology, Göteborg University, Box 463, S40530, Göteborg, Sweden

E-mail: th.bjornsson@zool.gu.se

Brauner Colin J

Department of Zoology, University of British Columbia, 6270 University Blvd., Vancouver, BC, Canada, V6T 1Z4

E-mail: brauner@zoology.ubc.ca

Burggren Warren

Department of Biological Sciences, University of North Texas, Denton, TX 76205, USA

E-mail: burggren@unt.edu

Campinho MA

Comparative Molecular Endocrinology Group, Marine Science Centre (CCMAR), Universidade do Algarve, FERN, Campus de Gambelas, 8005-139 Faro, Portugal

E-mail: mcampinho@fm.ul.pt

Cortés Alfonso

Department of Cell Biology, Faculty of Biology, Complutense University, 28040 Madrid, Spain

E-mail: acortesp@pdi.ucm.es

Denizot Jean-Pierre
C.N.R.S., Unité de Neurosciences Intégratives et Computationnelles, 91198 Gif-sur-Yvette, Cedex, France
E-mail: denizot@inaf.cnrs-gif.fr

Døving Kjell B
Department of Molecular Bioscience, University of Oslo, PO Box 1041 Blindern, N-0316 Oslo, Norway
E-mail: k.b.doving@imbv.uio.no

Ebbesson Lars OE
Department of Biology, University of Bergen, Bergen High Technology Centre, N-5008 Bergen, Norway
E-mail: lars.ebbesson@bio.uib.no

Formicki Krzysztof
University of Agriculture in Szczecin, Department of Fish Anatomy and Embryology 4 K. Królewicza St. 71-550 Szczecin, Poland
E-mail: kformicki@fish.ar.szczecin.pl

Forward Jr Richard B
Duke University, Nicholas School of the Environment and Earth Sciences, Marine Laboratory, 135 Duke Marine Lab Road, Beaufort, NC 28516-9721, USA
E-mail: rforward@duke.edu

Govoni John J.
US Department of Commerce, National Oceanic and Atmospheric Administration, National Ocean Service, National Centers for Coastal Ocean Science, Center for Coastal Fisheries and Habitat Research, 101 Pivers Island Road, Beaufort, NC 28516-9727, USA
E-mail: jeff.govoni@noaa.gov

Hall Thomas E
Victor Chang Cardiac Research Institute, 384 Victoria Street, Darlinghurst, Sydney, NSW 2010, Australia
E-mail: t.hall@victorchang.unsw.edu.au

Hiroi Junya
Department of Anatomy, St. Marianna University School of Medicine, Miyamae, Kawasaki, Kanagawa 216-8511, Japan
E-mail: j-hiroi@marianna-u.ac.jp

Holmberg Anna
Department of Zoophysiology, University of Göteborg, Box 463, SE 405 30 Göteborg, Sweden
E-mail: anna.holmberg@gu.se

Holmgren Susanne
Department of Zoophysiology, University of Göteborg, Box 463, SE 405 30 Göteborg, Sweden
E-mail: s.holmgren@zool.gu.se

Kaneko Toyoji
Department of Aquatic Bioscience, Graduate School of Agricultural and Life Sciences, University of Tokyo, 1-1-1, Yayoi, Bunkyo, Tokyo 113-8657, Japan
E-mail: kaneko31@marine.fs.a.u-tokyo.ac.jp

Kasumyan Alexander O
Department of Ichthyology, Faculty of Biology, Moscow State University, R-119991, Russia
E-mail: alex_kasumyan@mail.ru

Kelsh Robert N
Centre for Regenerative Medicine, Department of Biology and Biochemistry, University of Bath, Claverton Down, Bath BA2 7AY, UK
E-mail: bssrnk@bath.ac.uk

Kirschbaum Frank
Leibniz-Institute of Freshwater Ecology and Inland Fisheries, Müggelseedamm 310, 12587 Berlin, Germany
Institute of Animal Sciences, Humboldt-University Berlin, Berlin, Germany
E-mail: fkirschb@igb–berlin.de

Loew Ellis R
Cornell University, Department of Biomedical Sciences, Ithaca, NY 14853, USA
E-mail: erl1@cornell.edu

McCormick Stephen D
USGS, Conte Anadromous Fish Research Center, Turners Falls, MA 01376, USA
E-mail: steve_mccormick@usgs.gov

Morais Sofia

Comparative Molecular Endocrinology Group, Marine Science Centre (CCMAR), Universidade do algarve, FERN, Campus de Gambelas, 8005-139 Faro, Portugal

E-mail: smorais@ualg.pt

Müller Ulrike K

Department of Biology, California State University Fresno, 2555 E San Ramon Avenue, Fresno CA 93740, USA

E-mail: umuller@csufresno.edu

Olsson Catharina

Department of Zoophysiology, University of Göteborg, Box 463, SE 405 30 Göteborg, Sweden

E-mail: c.olsson@zool.gu.se

Pankhurst Patricia M

School of Tropical and Marine Biology, Faculty of Science, Engineering and Information Technology, James Cook University, Townsville, Queensland 4811, Australia

E-mail: Tish.Pankhurst@jcu.edu.au

Parichy David M

Department of Biology and Institute for Stem Cell and Regenerative Medicine, University of Washington, Seattle WA 98195

E-mail: dparichy@u.washington.edu

Pelster Bernd

Institut für Zoologie, Universität Innsbruck, Technikerstraße 25, A-6020 Innsbruck, Austria

E-mail: bernd.pelster@uibk.ac.at

Power DM

Comparative Molecular Endocrinology Group, Marine Science Centre (CCMAR), Universidade do Algarve, FERN, Campus de Gambelas, 8005-139 Faro, Portugal

E-mail: dpower@ualg.pt

Rønnestad Ivar

Department of Biology, University of Bergen, Bergen High Technology Centre, N-5008 Bergen, Norway

E-mail: ivar.ronnestad@bio.uib.no

Silva N
Comparative Molecular Endocrinology Group, Marine Science Centre (CCMAR), Universidade do Algarve, FERN, Campus de Gambelas, 8005-139 Faro, Portugal
E-mail: nsilva@ualg.pt

Stefansson Sigurd O
Department of Biology, University of Bergen, Bergen High Technology Centre, N-5008 Bergen, Norway
E-mail: sigurd.stefansson@bio.uib.no

Suntsov Andrey **V**
Northwest Fisheries Science Center, National Oceanic and Atmospheric Administration, Newport OR 97365 USA
E-mail: Asuntsov@ucsd.edu

Sutton Tracey T
Harbor Branch Oceanographic Institution, Fort Pierce FL 34949 USA
E-mail: tsutton@hboi.edu

Terjesen Bendik F
AKVAFORSK, Institute of Aquaculture Research, N-6600, Sunndalsøra, Norway
E-mail: bendik.terjesen@akvaforsk.no

Wahl Christina M
Wells College, Department of Biological and Chemical Sciences, Aurora, NY 13026, USA
E-mail: cwahl@wells.edu

Widder Edith A
Ocean Research & Conservation Association, Fort Pierce FL 34949 USA
E-mail: ewidder@oceanrecon.org

Zapata Agustín G
Department of Cell Biology, Faculty of Biology, Complutense University, 28040 Madrid, Spain
E-mail: zapata@bio.ucm.es

PART 1

Ontogeny

CHAPTER

1

Pattern Formation

Thomas E Hall

INTRODUCTION

Pattern formation during development refers to the processes by which the structure of the organism is spatially and temporally organised. On a cellular level, this requires the management of cell activities within the embryo so that a well-ordered structure develops. These processes begin prior to fertilisation, during oogenesis, and eventually result in the differentiation of the tissue and organ systems essential for homeostasis in complex organisms.

The fishes are a large and diverse group containing approximately half of known craniate species (Nelson, 1994). Early ontogeny varies substantially between clades, although this diversity is partly artifactual, since fish are a paraphyletic taxon (Collazo *et al.*, 1994). As such, this chapter concentrates on the Actinopterygii (ray-finned fishes) and mainly on the group Teleostei, which benefits from a number of well-studied models such as the medaka (*Oryzias latipes*) and zebrafish (*Danio rerio*). In addition, the rise of intensive aquaculture in recent years has fuelled much developmental research on commercially important species.

Author's address: Victor Chang Cardiac Research Institute, 384 Victoria Street, Darlinghurst, Sydney, NSW 2010, Australia.
E-mail: t.hall@victorchang.unsw.edu.au

EARLY PATTERN FORMATION

Pre-cleavage Patterning

Typically, teleosts are externally fertilising (oviparous), and the entirety of embryonic development occurs outside the mother's body. The eggs are supplied with a large yolk volume to sustain early development in the absence of direct maternal nutrition. In the unfertilised egg, the yolk and cytoplasm are intermixed (Leung *et al.*, 2000). The first patterning information is apparent even at this early stage as the animal/vegetal axis can be predicted on the basis of the localisation of specific maternal mRNAs (Howley & Ho, 2000). Following sperm entry and extrusion of the second polar body, the cytoplasm and yolk begin to separate, forming a single blastomere at the animal pole, sitting atop a continuous acellular yolk mass. In many species (e.g. the zebrafish; Kimmel *et al.*, 1995) streaming of cytoplasmic constituents to the animal pole may be observed directly under the light microscope. An actin microfilament network drives this movement (Leung *et al.*, 2000).

Cleavage

All teleosts show a discoidal meroblastic cleavage pattern, where the large yolk volume restricts cell division to a small area at the animal pole. The early embryo forms as a small disc sitting atop the large yolk. Holoblastic cleavage, where division splits the entire mass, is thought to be the ancestral condition, and is common in many invertebrate chordate phyla, as well as the lampreys. However, it appears that most animals with telolecithal eggs (those with a high ratio of yolk:cytoplasm) have evolved the meroblastic "shortcut". A meroblastic cleavage pattern allows cell division to proceed rapidly, and avoids the need for synthesis of large amounts of new cell membrane. Indeed, meroblastic cleavage has evolved independently at least five times within the craniates (once in each of the lineages leading to the hagfish, elasmobranchs, coelacanths, teleosts and amniotes; Collazo *et al.*, 1994). Interestingly, present-day mammals show a reversion to holoblastic cleavage, possibly as a result of gaining nutrition directly from the mother, rather than having to lay down copious quantities of yolk.

The first divisions are vertical and there is no cytoplasmic growth during this period, resulting in a decrease in blastomere size with each successive division. Early cleavages tend to be incomplete, and the initial blastomeres maintain cytoplasmic bridges with the yolk cell. Cleavages occur synchronously between embryos and the first cell cycles are easy to follow within clutches. Mitotic spindles can often be seen as the disc cleaves to form two distinct cells (Kimmel *et al.*, 1995; Hall *et al.*, 2004). The second cleavage is also horizontal and results in a 2×2 array. Timing of the cell cycle during the first cleavage events is highly synchronous, with each cycle having approximately the same length (Marrable, 1965; Kimmel *et al.*, 1995). Around the 16 to 32 cell stage however, the mass of cells (the blastodisc), becomes stratified into more than one layer. In most species the blastomeres are regular in size and shape. However, inter- and intra-specific variation exists on the general pattern. In the Atlantic cod (*Gadus morhua*) horizontal stratification of the cell mass usually occurs at the sixth cleavage, between the 32 and the 64 cell stages, but is frequently observed at the previous or following cleavage (Hall *et al.*, 2004). In the medaka, the first horizontal cleavage is the fifth, occurring between the 16 and 32 cell stages (Iwamatsu, 1994) and in the ice goby (*Leucopsarion petersii*) it occurs even earlier, between the 4 and 8 cell stages (Nakatsuji *et al.*, 1997). Unusually within the teleosts, eggs of the wolffish (*Anarhichas lupus*), have unequal blastomere sizes during early divisions (Pavlov *et al.*, 1992). During later cleavages it is only the marginal blastomeres (sometimes called Wilson cells), positioned on the periphery of the cell mass (now called the blastodisc), which maintain their cytoplasmic bridges with the yolk.

The Early Blastula and the Yolk Syncytial Layer

Loss of cytoplasmic bridges coincides with the formation of the yolk syncytial layer (YSL; Trinkaus, 1992). Around the ninth and tenth cleavages, the marginal blastomeres collapse into a single large, multinucleate (or syncytial) cell (Kimmel & Law, 1985). The YSL is an organ unique to teleosts that no longer undergoes cytoplasmic cleavage. The nuclei, however, continue to divide mitotically, and spindles may be seen in the cytoplasm as the YSL spreads beneath the blastodisc, separating the true blastomeres from the yolk. Around the 12th cleavage the YSL nuclei become post-mitotic (Kane *et al.*, 1992). Three cellular layers are now visible in the blastula; an enveloping layer (EVL) constituting the most external cells, which is epithelial in nature and covers the blastoderm; a middle layer of deep cells which will become the embryo proper, and the YSL lying between the deep cells and the yolk. The YSL is continuous with the yolk cytoplasmic layer (YCL), a thin layer of anuclear cytoplasm surrounding the yolk. The YSL itself is often described as possessing two domains, the internal YSL, which spreads beneath the blastodisc, and the external YSL which lies around the circumference. Tight junctions between the YSL and the overlying blastomeres mark the boundary between the internal and external domains. The thinner internal YSL (beneath the blastodisc) is populated by nuclei derived from the external YSL (Kimmel & Law, 1985).

In the early development of many animals, the initial rapid cleavage stages are followed by a period where the cell cycle lengthens, the embryo begins to transcribe its own RNA, and the cells of the blastula become motile, in preparation for gastrulation. This is termed the mid-blastula transition (MBT; Kane & Kimmel, 1993). The teleost MBT begins around cell cycle eight-ten (cycle ten in the zebrafish and the Atlantic cod). From around the seventh cleavage, the cells begin to show metasynchrony in their divisions, and within a further two or three cell cycles, cleavage events cannot be distinguished. In contrast to the early cleavages when there was little net cell movement, cell divisions now often spread daughter cells more than a cell diameter apart, causing a mild dispersion of progeny and intermingling of distant cell lineages (Kimmel & Law, 1985). Pseudopodia are also present, consistent with the activation of cell motility. However, the fate map in the early blastula is non-determinative; the progeny of any blastomere may contribute to all tissues in the embryo. Cell movements at this stage are not "directed or coherent" (Kane & Adams, 2002). Interestingly, the onset of cell cycle lengthening at the MBT is correlated not with cleavage number, but with nucleocytoplasmic ratio (Kane & Kimmel, 1993).

EPIBOLY AND GASTRULATION

Gastrulation is effectively a way of organising the developing embryo into the three primordial tissue layers, moving the endoderm and the mesoderm inside the ectoderm (Kimelman & Schier, 2002). The ectoderm will form the epidermis and neural tissue, and organs such as the ears and eyes; the mesoderm will form organs such as the muscle, bone, heart, kidney and vasculature; and the endoderm will form the lining of the alimentary canal, the swim bladder and organs such as the liver and pancreas. Note that the EVL, considered an extra-embryonic tissue, will form the first skin of the embryo, the periderm (Kimmel *et al.*, 1990). Nuclei of the YSL, also considered an extra-embryonic tissue, have not been tracked beyond the end of gastrulation (D'Amico & Cooper, 2001).

The process of gastrulation in teleosts is associated with a cell movement called epiboly, where the blastoderm spreads over the yolk from the animal pole, eventually enveloping the entire mass. Strictly, epiboly begins before the onset of gastrulation, with the thinning of the blastodisc. The YSL appears to drive epiboly via a microtubule dependant process, and if the blastoderm is removed, continues to move autonomously (Trinkaus, 1951). The YSL undergoes a mitotic arrest shortly before epiboly in zebrafish and killifish (*Fundulus heteroclitus*), possibly because the simultaneous use of microtubules in both mitosis and motility is likely to be mutually exclusive (Kane & Adams, 2002).

Shortly after the onset of epiboly, and prior to gastrulation, the fate map becomes restricted at the tissue level. Cells within only a small number of cell diameters from the margin of the blastoderm will become endodermal and mesodermal tissues (the "mesendodermal" compartment). Cells close to the margin, but outside of this domain, will become purely mesodermal tissues. Cells further towards the animal pole, will become ectodermal tissues (Fig. 1a). The onset of gastrulation itself begins with involution. Involution is a movement within the deep cell layer, whereby cells on the periphery of the blastodisc "roll underneath" the blastoderm lip (which is advancing towards the vegetal pole), and move back in the direction of the animal pole. The resulting thickening around the edge of the blastodisc is known as the "germ ring". The involuted layer, termed the hypoblast, is specified as mesoderm and endoderm. By the end of gastrulation, the outside layer, the epiblast, is composed of ectodermal cells. There is some variability in the timing of the beginning of involution with respect to percent epiboly between species. In the Atlantic cod, medaka and common carp (*Cyprinus carpio*), hypoblast formation starts early and coincides with the thinning of the blastodisc at the very beginning of epiboly. In killifish, hypoblast formation begins at approximately 25% epiboly and in the zebrafish, does not start until 50% epiboly (Hall *et al.*, 2004).

Historically, there has been debate over the dynamics of hypoblast formation during teleost gastrulation. There are two models as to the manner in which the marginal cells internalise; ingression, where cells "sink" as individuals from the epiblast to the hypoblast, and involution, which specifically refers to the rolling of a sheet of cells around an edge. Early studies in the rainbow trout (*Oncorhynchus mykiss*), (formerly *Salmo gairdneri*), found no evidence of involution (Ballard, 1966). Similarly, in killifish, the classical process of involution appears to be absent (Trinkaus, 1984). However, hypoblast formation in the zebrafish (Warga & Kimmel, 1990), and the rosy barb (*Puntius conchonius*; Wood & Timmermans, 1988) has been clearly described as an involuting cellular sheet. A unifying model of these processes has since been proposed. In small-egged teleosts such as the zebrafish and rosy barb, where cell density is relatively high at the onset of gastrulation, cell movements to form the hypoblast do involve the internalisation of individual cells by a process of, or a process similar to, ingression. The passage of deep cells into the hypoblast layer is indeed confined to those cells at the blastoderm margin, and initially all cells appear to be moving together. However, as the cells begin to descend, they seem to be moving as individuals, often changing places with their neighbours and extending active protrusions. The close proximity of so many cells undergoing this movement within one or two cell diameters of the margin, gives the appearance of a flow of cells into the hypoblast (Kane & Adams, 2002).

Mesodermal and Endodermal Specification

The mesoderm and endodermal tissues of the hypoblast are first specified by inducing signals, released in a ring by the external edge of the YSL, affecting the overlying blastomeres (Mizuno *et al.*, 1996; Rodaway *et al.*, 1999). The nodal-related morphogen squint (Sqt) is one of the signals, inducing its own expression, and that of the related factor cyclops (Cyc) in the involuting cells. Cyc and Sqt show substantial functional redundancy, and are essential for the formation of all mesoderm and endoderm other than tail mesoderm, which is induced directly by an unknown TGF-β like signal from the YSL (reviewed by Schier, 2003).

In the zebrafish fate map (and almost certainly in those of other teleosts), there are relatively few endodermal progenitors compared to ectodermal and mesodermal progenitors (Warga & Stainier, 2002). The endodermal progenitors occupy the most marginal positions prior to involution. All endodermal cells are derived from clones situated within four cell diameters of the blastoderm lip, and are among the first to involute during gastrulation. However, in the early blastula, cells from this domain may also give rise to mesoderm, suggesting some indeterminacy in cell fate. Marginal cells situated further from the blastoderm lip give rise exclusively to mesoderm (Warga & Nusslein-Volhard, 1999). As the endodermal progenitors begin to involute, a number of endodermal genes become active, such as *casanova* (*cas*), *bonnie and clyde* (*bon*) and *faust* (*fau*/*gata5*; reviewed in Warga & Stainier, 2002). By the end of gastrulation, endodermal cells may be distinguished on the basis of morphology, as a monolayer of large flat cells with numerous filopodia lying against the yolk (Warga & Nusslein-Volhard, 1999). The mesodermal component of the hypoblast, in contrast, has limited contact with the yolk and shows less marked changes in cell morphology. There is increasing evidence that nodal signalling not only specifies the involuted cell compartment, but is also involved in patterning it. It is likely that there is a nodal-activity gradient along the animal-vegetal axis, with the highest levels at the blastoderm margin producing endoderm, prechordal plate and ventricular progenitors, lower levels producing notochord and other mesodermal fates, and the absence of nodal signalling allowing neural and tail specification (reviewed in Schier & Talbot, 2005).

Convergent-extension and the Fish "Organiser"

The third major cell movement that occurs during teleost gastrulation, (in addition to epiboly and involution/ingression), is convergent-extension. Convergent-extension breaks radial symmetry by the convergence of cells to the dorsal side of the embryo, and results in its animal/vegetal extension. The dorsal accumulation of cells is known as the embryonic shield, and defines the axis of the embryo. All three germ layers participate in the underlying cellular movements. As in the African clawed frog *(Xenopus laevis)*, a major driver of teleost convergent-extension is mediolateral intercalation of cells (Concha & Adams, 1998; Glickman *et al.*, 2003). However, active cell migration is also important in patterning the shield. Around mid-gastrulation, cells of the hypoblast, which are moving animal-ward on individual, meandering paths, acquire a directional bias towards the future dorsal side (Sepich & Solnica-Krezel, 2005; Sepich *et al.*, 2005). Depending on location along the animal-vegetal axis, cell migration occurs at different rates, resulting in a "fanning out" of cell trajectories, and the antero-posterior extension of the tissue layer. It is likely that convective or passive cell movements also contribute to convergent-extension, perhaps by adhesion to the epiblast or YSL (D'Amico & Cooper, 2001).

The concept of the organiser was introduced by Spemann (Spemann & Mangold, 1924; but see Gilbert & Raunio, 1997). The initial experiments, in amphibians, demonstrated that cells from the dorsal lip of the newt gastrula, when transplanted to a host of the same developmental stage, were able to elicit the formation of a secondary axis. The structures of the secondary axis, such as the somites, secondary notochord, and in some cases gut, were derived not from the grafted tissue but from host cells that normally would not participate in formation of these structures. These cells were subsequently referred to as the "Spemann organiser" (a functionally equivalent domain known as "Hensen's node" was later described in the chick; Hunt, 1929). There is some evidence that the embryonic shield in zebrafish is analogous to the Spemann organiser in amphibians. Transplantation of the shield can generate a complete secondary axis (Saude *et al.*, 2000). In addition, it appears that there are distinct cell populations within the shield, with grafts from deep regions of the shield preferentially inducing anterior structures, whilst the superficial region tends to induce posterior structures. Although the transplant experiments appear conclusive, the results of shield ablation experiments are more ambiguous. It appears that the shield does not define the organiser *per se*, but that the organiser forms earlier and over a wider region than the shield (reviewed by Hibi *et al.*, 2002).

DIFFERENTIATION OF THE GERM LAYERS

Organogenesis in the Mesoderm

The processes of organogenesis begin during late gastrulation or shortly after, when cells from the three tissue layers start to differentiate into specific structures. At this time, precursors of different cell types are arranged along the dorso-ventral and antero-posterior axes (Fig. 1b). It should be noted at this point that classical developmental periods such as gastrulation, segmentation, neurulation and organogenesis tend to show some overlap in teleosts, to a greater or lesser extent depending on species.

The mesoderm consists of three domains, the head, trunk and tail. Each of the three domains is specified early to produce different structures. Following gastrulation, the head domain (prechordal and head mesoderm) specifically expresses the marker *goosecoid* (*gsc*). Trunk and tail specification appear to be under the control of fibroblast growth factor (FGF) signalling, through the downstream genes *no tail* (*ntl*) and *spade tail* (*spt*). In the zebrafish, if FGFs are blocked, no trunk or tail mesoderm is formed (Griffin *et al.*, 1995). In the *ntl* and *spt* mutants, *ntl* has no tail mesoderm and *spt* has severely defective trunk mesoderm. The double mutant (*ntl/spt*), phenocopies the effect of blocking FGF signalling (reviewed by Sakaguchi *et al.*, 2002).

The first organ to delineate physical boundaries in the early embryo is the notochord. Two longitudinal furrows appear dividing the axial chorda mesoderm (which expresses the marker *floating head*, *flh*; Melby *et al.*, 1997) and the paraxial segmental plate mesoderm (which will become the somites). The process is driven by differing rates of convergent-extension between the two tissues. Cells in the presumptive notochord extend several times faster than those in the adjacent paraxial mesoderm, resulting in shearing along the boundary (Glickman *et al.*, 2003). As development proceeds, the furrows extend posteriorly down the axis of the embryo either side of the midline. The central cells of the notochord are known as chordoblasts, and each develops a large, fluid-filled vacuole. Vacuolisation is associated with the deposition of a surrounding thickened sheath of tissue which, together with the core, imparts the rigidity and axial support necessary for early swimming movements (Adams, 1990; Koehl *et al.*, 2000).

Shortly after the notochord furrows become visible, the first somite furrow appears perpendicularly in the paraxial mesoderm, marking the beginning of segmentation (so called because of the repeated nature of the somites). The process of somitogenesis is similar to that in amphibians, birds and mammals (Stickney *et al.*, 2000). It involves a mesenchymal-epithelial transition of cells within the paraxial mesoderm, to surround a central space, the somitocoele, which contains a loose meshwork of mesenchymal cells. In rainbow trout, the epithelial cells of the somite express collagen-1 (Rescan *et al.*, 2005). Somites are added posteriorly in a temporally linear, reiterative fashion until the final number is reached (for this reason somite number is the most useful staging tool in fish embryonic development). Final somite number is extremely variable within the teleosts, ranging from 18 in the ocean sunfish (*Mola mola*) (Brainerd *et al.*, 2001) to over 200 in some eels (Richardson *et al.*, 1998). As with the onset of gastrulation, there is some variability as to the beginning of segmentation. For instance, the first somite furrow in the Atlantic cod is formed at approx 45% epiboly. In contrast in the zebrafish, common carp, medaka and killifish, somitogenesis is not initiated until well after the completion of epiboly. Development of tissues from the onset of somitogenesis tends to proceed temporally in a rostral-caudal wave, with the most well developed structures towards the anterior, and the least developed towards the posterior. The process of vertebrate segmentation itself seems to be under the control of a segmentation clock, involving an intrinsic oscillator within the pre-somitic mesoderm, revealed by the expression of "cycling genes". The clock appears to function in parallel with a wavefront of differentiation involving FGF signalling, whose progression determines the correct position of somite boundaries (Dubrulle *et al.*, 2001; Sawada *et al.*, 2001; Freitas *et al.*, 2005).

Somites contain cells that will become a variety of mesodermal derivatives. The majority of the somite in fish becomes the myotome, and will form the body musculature. There are however, a number of other cellular compartments within the somite. The most ventral zone of the somite is the sclerotome, which will form the cartilage, and later the bone of the animal (Morin-Kensicki & Eisen, 1997). Fish, which exist within an aqueous medium, exert a lesser demand on the axial skeleton for maintenance and locomotion than terrestrial vertebrates, and possess an increased requirement for axial (swimming) muscle (Bone, 1966). As a result, the sclerotome in amniotes occupies a much larger area and is situated adjacent to the notochord. In teleosts, the sclerotome is a relatively small compartment, and is initially located ventrally within the somite. Only later in development do these cells undergo a dorsal-ward migration and position themselves adjacent to the notochord. The mechanism of vertebral formation is also fundamentally different. In amniotes the vertebral body rudiments are formed in cartilage and are derived from adjacent half-somites. In contrast, the development of teleost vertebrae involves the development of specialised notochordal structures called chordacentra, which first form as acellular calcified "rings" within the notochord sheath. The sclerotomally-derived bone is then deposited through direct ossification on the outside of the chordacentra and notochord, without the prior formation of cartilage (Grotmol *et al.*, 2003).

The structure of the body musculature in teleosts is peculiar within the vertebrates, in that the fast- and slow-twitch muscle fibres are spatially separated, with the slow muscle being confined to a relatively thin layer on the periphery of the myotome, and the fast fibres forming the bulk of the deep myotome (Bone, 1978). This peculiarity is reflected in myotomal ontogeny. In the undifferentiated somite, most cells are mesenchymal in nature. However, adjacent to the notochord lie a group of large cuboidal cells known as adaxial cells, so called because of their proximity to the midline (Felsenfeld *et al.*, 1991). Towards the end of somitogenesis, in response to the expression of hedgehog proteins from the notochord (Blagden *et al.*, 1997), these cells begin to express slow myosin heavy chain. They then perform an astounding migration, moving laterally through the overlying somitic cells, whilst elongating and fusing into the first myotubes (Devoto *et al.*, 1996). It has recently been shown that this migration is driven by cell sorting behaviour in response to laterally radiating waves of expression of the cell adhesion molecules n- and m-cadherin (Cortes *et al.*, 2003). A subset of these cells, which lie at the horizontal myoseptum, are known as the muscle pioneer cells (Felsenfeld *et al.*, 1991). They are distinct both morphologically and genetically, spreading laterally from the notochord to the myotome surface, and expressing the nuclear protein Engrailed (Roy *et al.*, 2001). Following adaxial cell migration, sclerotomal cells from the ventral somite migrate dorsally to occupy their position adjacent to the notochord, adopting the positional relationship seen in amniotes. Adaxial cell migration through the myotome also appears to be necessary to "prime" the myotome for invasion by the primary motor neurons (Zeller *et al.*, 2002). As the slow muscle cells reach the periphery of the myotome, the deeper, fast fibres begin to fuse and differentiate. Further fibre recruitment occurs from zones at the peripheries of the myotome, and at the horizontal myoseptum (reviewed by Johnston & Hall, 2004).

Other than the myotome and the sclerotome, cellular compartments within fish somites have not been extensively characterised. In amniotes, the dorsal epithelium of the somite is referred to as the dermomyotome, and is a multipotent tissue, able to give rise to dermis, vascular endothelia, trunk muscle fibres, and the muscles of the limbs and ventral body wall (intercostal muscles; Scaal & Christ, 2004). It has been shown that cells of the zebrafish somite can give rise to blood vessels (Morin-Kensicki & Eisen, 1997), and that muscle fibre recruitment occurs in the teleost myotome throughout life. In addition, the presence of so-called "external cells", a non-continuos layer of flattened mitochondrion-dense cells outside the slow muscle layer has been noted by many investigators (Waterman, 1969; van Raamsdonk *et al.*, 1978; Felsenfeld *et al.*, 1991) and this has been claimed represent the teleost dermomyotome (Devoto *et al.*, 2006). However, there is little functional evidence for this.

The heart, blood and kidney all form from more lateral non-somitic mesoderm (see Fig. 1c). Fate mapping studies have revealed that heart progenitors, giving rise to both myocardial and endocardial lineages, are clustered into two laterally located groups in the mesendodermal cell layer (reviewed by Stainier, 2001). By early somitogenesis, they are located close to the future hindbrain as two populations, either side of the embryonic midline. Within these populations the myocardial progenitors are located more laterally and the endocardial progenitors more medially. At this time, the pre-cardiac field is defined by the expression of the marker *nkx2.5*, which is conserved throughout the vertebrates (Schwartz & Olson, 1999). By mid-somitogenesis, the heart progenitors have migrated as far as the midline, and merge at the posterior to form a horseshoe-like structure, and then a shallow cone, with the apex being located more dorsally (Yelon *et al.*, 1999). During the latter period of somitogenesis, the apex of the cone tilts posteriorly and to the right, before its extension to form the heart tube. The tube consists of an atrial anterior end, to the left of the midline, and a posterior ventricular end, to the right, and is lined with endocardial cells. Subsequent morphogenetic events result in cardiac looping, and the formation of an organ with the ventricle positioned to the right of the atrium. It is noteworthy that the teleost heart tends to begin beating before the development of haemoglobin, and before the

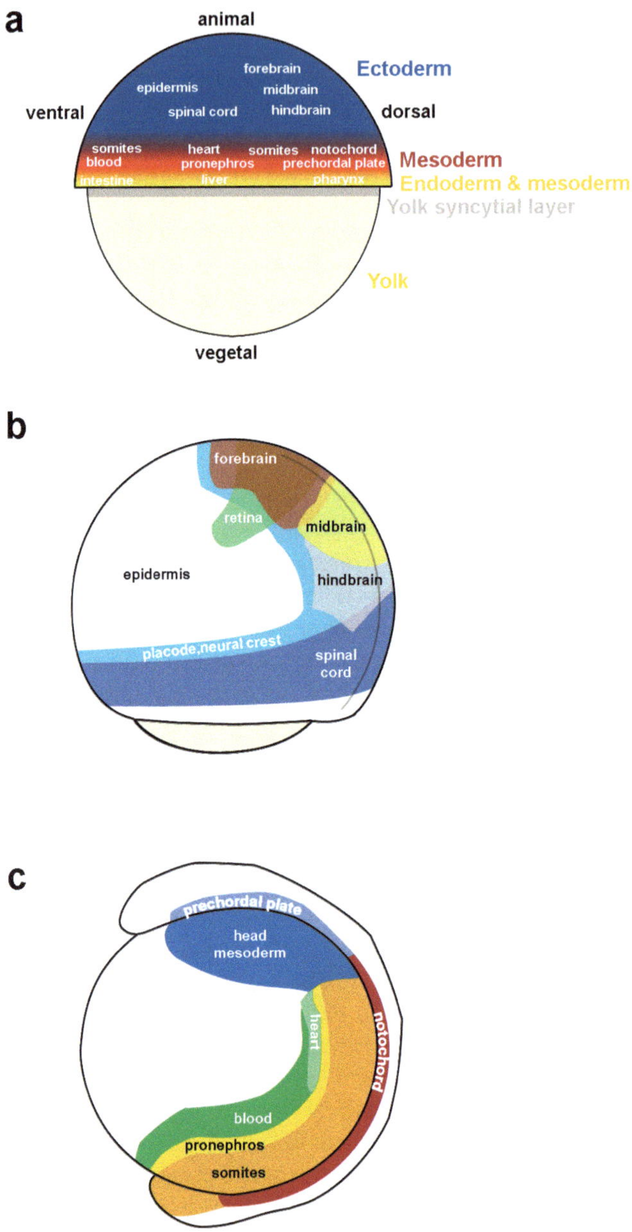

Fig. 1 Arrangement of presumptive tissue types in the zebrafish (*Danio rerio*) at the onset of gastrulation (a), shortly before the end of gastrulation (b), and early segmentation (c). Modified from Schier & Talbot (2005).

development of functional gills. It is likely that the function of this early circulation is nutritional, and for the distribution of the yolk substrates rather than gas exchange. Development of vasculature during the embryonic stages is highly variable between species and often reflects egg and larval size. For instance, in the salmonids, the entire yolk sac becomes vascularised early in development, in contrast to species with smaller embryos such as the Atlantic cod or zebrafish (Gorodilov, 1996; Hall *et al.*, 2004). It is thought that both vascular and haematopoietic cells originate from common progenitors known as haemangioblasts, present adjacent to the presumptive kidney at the end of gastrulation. All of the earliest forming blood vessels in teleosts appear to have direct homologues in other vertebrates, and the overall pattern changes little before first-feeding (Yelon *et al.*, 2002).

The embryonic kidney is the pronephros, which consists of two fused glomeruli situated ventral to the notochord at the approximate position of the first somites, sprouting a pair of laterally projecting pronephric tubules, each adjoining a pronephric duct running posteriorly above the alimentary canal (Serluca & Fishman, 2001). It arises from cells adjacent to the somitic mesoderm, and later in development, during the juvenile to adult transition, a more complex, mesonephric kidney arises (Drummond *et al.*, 1998).

Organogenesis in the Ectoderm

The ectodermal layer which, following gastrulation is the outermost tissue layer of the embryo, gives rise to the neural tissues, sensory organs, pigment cells, placodal tissues and the epidermis. Prior to organogenesis, the thickening of the ectoderm above the presumptive notochord and somites is known as the neural plate.

Towards the end of epiboly, neural plate cells above and dorso-lateral to the presumptive notochord extend ventrally, forming a deepened keel. This is quickly overridden by adjacent epidermal cells, and becomes a solid rod. A central lumen, the neurocoele, originates by cavitation, turning the neural rod into a tube. This mechanism, known as secondary neurulation, contrasts with the infolding mechanism (primary neurulation) observed in amniotes (Papan & Campos-Ortega, 1994).

Soon after the beginning of somite formation, the anterior neural tube swells and distends, but is subdivided by a number of constrictions (Kimmel, 1993). Initially, around ten swellings are apparent, and the rostral-most three of these become large and prominently sculpted. The rostral-most two swellings correspond to the subdivisions of the forebrain, the telencephalon and the diencephalon. The third makes up the midbrain. The seven smaller, more caudal bulges, termed "rhombomeres" make up the hindbrain (reviewed by Hjorth & Key, 2002). During their morphogenesis, a subset of cells located between the developing diencephalon and telencephalon, move laterally to form the optic primordia, which subsequently detach and invaginate (Easter & Malicki, 2002). The first neurons of the developing brain, known as primary neurons, arise in bilateral clusters, close to the centre of each rhombomere. The earliest axons extend between the clusters of cells in bundles, forming a stereotypically patterned scaffold. Primary neurons typically have larger soma, which extend axons long distances, and often pioneer pathways. The smaller secondary neurons develop later, utilising the initial scaffold laid down by the primary neurons.

The extent of the neural tube posterior to the brain differentiates into the spinal cord, and lies dorsal to the notochord. The first neurons in the spinal cord arise in a segmental pattern corresponding to the somites. In the zebrafish, each segment contains approximately 11 neurons; three primary motor neurons, three primary sensory Rohon-Beard neurons, and five interneurons (Kuwada & Bernhardt, 1990). The first neurons to develop are the dorsally located Rohon-Beard neurons, closely followed by the interneurons and the ventrally located motor neurons. The three primary motor neurons are located in stereotypical patterns relative to the somite boundaries, and are known as the rostral primary (RoP), middle primary (MiP) and caudal primary (CaP) neurons. The CaP axon pioneers the common pathway of motor axons to the horizontal myoseptum. All three axons extend to the muscle pioneer cells at the periphery, before growing along separate pathways to their final targets (Hjorth & Key, 2002).

One of the most fascinating cell types in the developing embryo is the neural crest. This population of relatively large cells is present at the border of the neural plate and the future epidermis, prior to formation of the neural tube (Thisse *et al.*, 1995). Neural crest cells are unusual in that they delaminate from the neuroepithelium during neurulation, and undergo long-range migrations throughout the embryo to form a variety of cell types, such as most of the peripheral nervous system, all epidermal pigment cells and much of the head skeleton. Cell types derived from vertebrate neural crest are as diverse as sensory, sympathetic and enteric neurons, glia, melanocytes, smooth muscle, dermis, connective tissue, cartilage and bone (reviewed by Baker & Bronner Fraser, 1997; Halloran & Berndt, 2003; see also Kelsh & Parichy, this volume). Cranial neural crest cells are those which emanate from the early brain to populate the anterior region of the head and pharyngeal arches. In the trunk region, individual trunk crest cells migrate ventrally, following either the "lateral pathway", between the lateral border of the somite and the overlying epidermis, or the "medial pathway", between the medial border of the somite and the neural tube. Cells that migrate along each pathway appear to have different fates. Medial pathway cells form sensory and sympathetic neurons, Schwann cells and pigment cells, whereas cells utilising the lateral pathway (which begin their migration later) make only pigment cells (Eisen & Weston, 1993). In birds and mammals, establishment of the medial migration pathway pattern in the trunk region is dependant upon sclerotomal cells, which occupy positions adjacent to the notochord. In teleosts however, the sclerotome is initially confined to the ventral aspect of each somite, and whilst sclerotomal cells do eventually migrate dorsally to meet crest cells at the dorso-ventral midline, this occurs later in development. Recent evidence suggests that the adaxial cells, which begin their movement shortly after the initiation of neural crest migration, are more important in the regulation of crest cell migration (Honjo & Eisen, 2005). In addition to cranial and trunk cells, cardiac crest cells derived from the caudal hindbrain and the pre-otic region of the neural tube, contribute to all segments of the myocardium (Sato & Yost, 2003).

Following neurulation, the placodes are first visible as thickenings of columnar epithelium in the non-neural ectoderm, which will go on to form the paired sense organs (olfactory, otic and lateral line placodes), the lens of the eye (lens placode), the anterior lobe of the pituitary gland (adenohypophyseal placode) and essential components of some ganglia (epibranchial and trigeminal placodes). It is believed that all of the seven vertebrate placodal tissues form from a common group of cells called pre-placodal ectoderm (PPE), located lateral to the neural crest and neural plate (see Fig. 1b). The PPE is characterised by the expression of *six* and *eya* genes, and is later regionalised by downstream transcription factors (reviewed by Baker & Bronner-Fraser, 2001; Schlosser, 2005). The lens placode is the only entirely non-neurogenic placode and gives rise to the lens of the eye.

The otic placode is first visible lateral to rhombomeres three-five, and approximately halfway between the eyes and the first somite in zebrafish (Kimmel *et al.*, 1995). The central fluid-filled chamber is formed by cavitation, rather than by invagination as in other vertebrates (Whitfield *et al.*, 2002). Now called the otic vesicle, this hollow ball of epithelium gives rise to all structures of the membranous labyrinth and the neurons of the statoacoustic ganglion. The dorsal part of the epithelium thins, whereas the ventral portion thickens, marking the future site of the sensory patches (maculae and cristae) which contain the hair cells. Cells of the statoacoustic ganglion are derived from the otic epithelium. These neuroblast cells undergo an epithelial to mesenchymal transition, and accumulate beneath the rostral half of the vesicle, where they differentiate into neurons (Haddon & Lewis, 1996). The adult ear contains three otoliths, one associated with the maculae of the utricle, saccule and lagena. Initially, only two otoliths form in the embryo, at the rostral and caudal ends of the vesicle. The third otolith forms much later, well into the larval period (Riley & Moorman, 2000).

Each olfactory placode is initially a homogeneous mass of cells. As the placode differentiates, the basal cells, sensory cells and support cells characteristic of the adult olfactory epithelium become apparent. However, before these develop, a class of pioneer neurons have been shown to extend path-finding axons to contact the telencephalon, before undergoing apoptosis (Whitlock & Westerfield, 1998). The sensory neurons of the olfactory placode then extend axons along the same path. Once within the CNS, the axons segregate into glomeruli, clustered within the developing olfactory bulb.

The first neuromasts of the lateral line system are also placodally derived. They are deposited by migrating primordia first detectable in the otic region through expression of *cadherin* 6 (Liu *et al.*, 2006). The neuromasts of the head form the anterior lateral line system (ALL) and are innervated by a ganglion located between the ear and eye. The neuromasts of the body and tail form the posterior lateral line system (PLL), whose ganglion is just posterior to the ear. Each neuromast is comprised of hair cells which resemble those of the inner ear, surrounded by support cells (see Pankhurst, this volume). In some species, through specialisation of the hair cell receptors, it has evolved into an electro-sensory system (Dambly-Chaudiere *et al.*, 2003; see Krischbaum & Denizot, this volume). The PLL primordium is the most extensively studied of the two, and in the zebrafish the extent of its migration is used as a staging tool (Kimmel *et al.*, 1995).

From its initial position posterior to the otic vesicle, the primordium migrates to the horizontal myoseptum. The most rostral of the cells become the innervating ganglion, whereas the most caudal cells migrate posteriorly, along the entire length of the horizontal myoseptum. The cells continue to divide as they migrate, to replace the periodically deposited neuromasts. It has recently been shown that migration of the PLL occurs along a pathway marked by the chemokine Sdf1a, and that its receptor Cxcr4 is expressed by the PLL cells. Furthermore, neuromast deposition appears to depend on HGF signalling from the myosepta, through the c-met tyrosine kinase receptor, also expressed by the PLL (Haines *et al.*, 2004). Despite major differences in the adult pattern of the lateral line system, it appears that the early migratory processes involved in migration and neuromast deposition are highly conserved, even between species such as the Mexican blind cave fish (*Astyanax fasciatus*) and the medaka (Sapede *et al.*, 2002).

Organogenesis in the Endoderm

After undergoing involution, the endodermal progenitors of the hypoblast sit beneath the mesodermal and ectodermal cells, on the surface of the yolk mass (Warga & Stainier 2002). The majority of the endodermal cells will contribute to structures of the digestive system; the alimentary canal, liver, gall bladder and pancreas. However, endoderm originating in the dorsoanterior region will also become the pharyngeal pouches of the head, giving rise to the segmental gill clefts, thymus, thyroid, parathyroid and parts of the inner ear. Confocal analysis in the zebrafish has shown that at early somite stages, endodermal cells constrict into a thin ribbon of midline cells, the intestinal rod, running posteriorly from the level of the first somite (Field *et al.*, 2003a). The liver, pancreas and swim bladder (which is not part of the digestive system) all develop by a process of budding from this central tract. Interestingly, the pancreas arises from two distinct anlagen, a dorsal posterior bud which gives rise only to endocrine cells, and a ventral anterior bud which gives rise to the exocrine cells, the pancreatic duct and a small proportion of endocrine cells (Field *et al.*, 2003b). The more anterior regions of the alimentary canal, the pharynx and oesophagus arise at mid to late somite stages from a seemingly separate population of precursor cells (Wallace & Pack, 2003). Further compartmentalisation of the alimentary canal occurs throughout the larval stages (Ng *et al.*, 2005; Rønnestad & Morais, this volume).

HOX GENES, SEGMENTAL PATTERNING AND THE EVOLUTION OF AXIAL COMPLEXITY

The homeotic or hox genes were first described in the fruitfly (*Drosophila melanogaster*) as key regulators of regional identity along the antero-posterior axis of the embryo. Eight linked genes were found to make up the homeotic complex (Lewis, 1978). Mutations in these transcription factors conferred dramatic phenotypes, whereby one body segment would take on the identity of another. For instance in the mutant *antennapedia*, legs grow on the head, in place of the antennae. *Hox* genes also exhibit "colinearity", meaning that their pattern of expression along the AP axis reflects the spatial organisation of genes, such that the most 3' genes are activated first, with sequential activation of more 5' sequences (Duboule, 1998). Complex interactions occur between adjacent hox genes such as sharing or competing for enhancer elements, which explains the maintenance of their tight clustering. Hox genes have since been described in diverse taxa within the bilateria (Balavoine *et al.*, 2002), but there is great variation in gene number. Interspecies variation reflects two types of gene duplication; the tandem duplication of genes within a cluster, and whole cluster duplication. Current models suggest that the single cluster arrangement found in the fruitfly originated by successive duplication from a single gene. The tetrapods (including mammals, birds, amphibians and reptiles) all have four hox clusters, consistent with two rounds of whole cluster duplication (Krumlauf, 1994). It is possible that this is a reflection of two rounds of whole genome duplication which occurred via polyploidisation during the origin of vertebrates; the so-called "2R" hypothesis (Sidow, 1996; Holland, 1999). A logical progression of this argument led to the suggestion that an increase in the number of hox genes, and/or an associated increase in genomic complexity was permissive for the evolution of more complex body plans, consistent with the long-standing theory that gene duplication is an important mechanism for generating genetic novelty in evolution (Ohno, 1970). Extensive phylogenetic studies have now been carried out to address this issue. Most surprisingly, an examination of hox genes in the zebrafish revealed the existence of seven hox clusters (Amores *et al.*, 1998), comprising 47 known genes (Kurosawa *et al.*, 2006). Furthermore, expanded hox clusters appear to be the common condition within the teleosts, with the medaka possessing 46 genes in seven clusters, and torafugu (*Fugu rubripes*) 48 genes in eight clusters. This reflects a further full-scale genome duplication, in the common ancestor of teleosts, elegantly demonstrated by Jaillon *et al.* (2004), using data from the spotted green pufferfish (*Tetraodon nigroviridis*) genome. They demonstrated firstly, that every chromosome was involved in large-scale duplication, and secondly, a striking pattern of double synteny, with one chromosomal region in humans matching two in spotted green pufferfish, across the whole genome. Consistent with these observations, the location of the eighth zebrafish hox cluster has been identified by synteny comparisons, but it contains no hox-related sequences, with the exception of a single microRNA (Woltering & Durston, 2006). The exact point within the actinopterygian lineage at which the duplication event occurred is unclear, although the available evidence suggests that it occurred shortly before the origin of the teleosts, since the non-teleost bichir, bowfin and paddlefish have hox clusters resembling those of the tetrapods (Crow *et al.*, 2006).

Hox genes across the vertebrates fall into 13 paralogue groups, although not all paralogues are represented in each cluster, due to gene loss since duplication. According to accepted nomenclature, hox gene loci in vertebrates are referred to as A, B, C and D clusters. Since the teleost loci are duplicated, they are referred to as hoxAa, hoxAb etc. (Kurosawa *et al.*, 2006; Finn & Kristoffersen, 2007). Classically, the retention of a duplicated gene is necessitated by the acquisition of novel functions (neofunctionalisation; Ohno, 1970), or by the independent loss of regulatory functions from each gene, resulting in paralogues with distinct but complementary properties (subfunctionalisation; Force *et al.*, 1999). As to whether duplication of hox genes allowed the evolution of more complex body plans, the jury is still out. It has been argued that invertebrates, which possess only a single hox cluster, exhibit by far greater diversity in terms of species richness and variety of body plans than do vertebrates. In addition, whilst most actinopterygians exhibit more hox clusters than sarcopterygians, and greater species diversity, it has been argued that "a zebrafish is not more complex than a mouse" (Bruce *et al.*, 2001, cited in Crow *et al.*, 2006).

The basic functions of hox genes between *Drosophila*, mice and teleosts are conserved; they act as selectors of regional identity along the primary body axis. Knockout studies in the mouse have demonstrated that hox paralogue groups are necessary for many diverse aspects of axial patterning including hindbrain specification, neural crest, pharyngeal and cranio-facial development. Particularly well characterised are effects on the axial skeleton and vertebrae, due to the ease of recognising morphological changes in vertebral structure (reviewed by Maconochie *et al.*, 1996). The hox genes are highly redundant in the mouse, with knockouts in single genes generally having mild phenotypes, due to compensation by their paralogues. For instance, the hox10 and 11 groups each contain three genes: *hoxa10*, *hoxc10*, *hoxd10*, and *hoxa11*, *hoxc11*, *hoxd11* respectively. In the absence of hox10 function, no lumbar vertebrae are formed. Instead, ribs project from all posterior vertebrae, extending caudally from the last thoracic vertebrae to beyond the sacral region. In the absence of hox11 function, sacral vertebrae are not formed and instead these vertebrae assume a lumbar identity. In each case, the redundancy among these paralogous family members is so great that this aspect of hox patterning is not apparent in mice that are mutant for five of the six paralogous alleles (Wellik & Capecchi, 2003). Having said this, for most paralogy groups a degree of subfunctionalisation is evident. In the zebrafish, hox genes have been shown to be activated in a colinear fashion (Prince *et al.*, 1998a). Genes from paralogue groups 1, 2, 3 and a subset of 4 are expressed within the developing hindbrain and show many similarities in pattern with those of the mouse (Prince *et al.*, 1998b). Genes from paralogue group 2 have also been shown to act to pattern the second pharyngeal arch (Hunter & Prince, 2002). It should be noted at this point that the axial organisation of tetrapod vertebrae is more complex than that of the zebrafish. Only two basic classes of teleost vertebrae are recognised: trunk vertebrae and tail vertebrae, in contrast to the tetrapods' five. Within the zebrafish trunk mesoderm however, many hox genes are expressed in a segmental manner. In addition, unlike in tetrapods, regionalised expression within the notochord has been described (Prince *et al.*, 1998c). The functional roles of these genes are currently unknown, but are likely to underpin more subtle aspects of regional identity.

FUTURE PERSPECTIVES

The process of early pattern formation in teleosts has been well studied historically. However, due to the sheer diversity of the lineage, many aspects of cell movement and differentiation, particularly during the later stages, are only documented from a single, or small number of species. The rise of intensive aquaculture is helping to address this shortcoming. Using the genetic models of the zebrafish and medaka, we are now beginning to understand the molecular basis of pattern formation. In addition, comparative genomics, utilising the zebrafish, torafugu and spotted green pufferfish genomes is allowing the examination of the very roots of morphological complexity, through the study of genome evolution.

Acknowledgments

I would like to thank Alison Rutstein and Abigail Gibson for comments on the manuscript.

References

Adams DS, Keller R, Koehl MA (1990) The mechanics of notochord elongation, straightening and stiffening in the embryo of *Xenopus laevis*. Development 110: 115-130

Amores A, Force A, Yan YL, Joly L, Amemiya C, Fritz A, Ho RK, Langeland J, Prince V, Wang YL, Westerfield M, Ekker M, Postlethwait JH (1998) Zebrafish hox clusters and vertebrate genome evolution. Science 282: 1711-1714

Baker CV, Bronner-Fraser M (1997) The origins of the neural crest. Part I: embryonic induction. Mech Dev 69: 3-11

Baker CV, Bronner-Fraser M (2001) Vertebrate cranial placodes I. Embryonic induction. Dev Biol 232: 1-61

Balavoine G, de Rosa R, Adoutte A (2002) Hox clusters and bilaterian phylogeny. Mol Phylogenet Evol 24: 366-373

Ballard WW (1966) Origin of the hypoblast in *Salmo* I. Does the blastodisc edge turn inward? J Exp Zool 161: 201-210

Blagden C, Currie P, Ingham P, Hughes S (1997) Notochord induction of zebrafish slow muscle mediated by sonic hedgehog. Genes Dev 11: 2163-2175

Bone Q (1966) On the function of two types of myotomal muscle fibre in elasmobranch fish. J Mar Biol Assoc UK 46: 321-349

Bone Q (1978) Locomotor muscle. In: Hoar WS, Randall DJ (eds) Fish Physiology, Vol 7. Academic Press, New York, p 361-424

Brainerd EL, Slutz SS, Hall EK, Phillis RW (2001) Patterns of genome size evolution in tetraodontiform fishes. Evolution 55: 2363-2368

Bruce AE, Oates AC, Prince VE, Ho RK (2001) Additional hox clusters in the zebrafish: divergent expression patterns belie equivalent activities of duplicate *hoxB5* genes. Evol Dev 3: 127-144

Collazo A, Bolker, JA, Keller, R (1994) A phylogenetic perspective on teleost gastrulation. Am Nat 144: 133-152

Concha ML, Adams RJ (1998) Oriented cell divisions and cellular morphogenesis in the zebrafish gastrula and neurula: a time-lapse analysis. Development 125: 983-994

Cortes F, Daggett D, Bryson-Richardson RJ, Neyt C, Maule J, Gautier P, Hollway GE, Keenan D, Currie PD (2003) Cadherin-mediated differential cell adhesion controls slow muscle cell migration in the developing zebrafish myotome. Dev Cell 5: 865-876

Crow KD, Stadler PF, Lynch VJ, Amemiya C, Wagner GP (2006) The "fish-specific" Hox cluster duplication is coincident with the origin of teleosts. Mol Biol Evol 23: 121-136

D'Amico LA, Cooper MS (2001) Morphogenetic domains in the yolk syncytial layer of axiating zebrafish embryos. Dev Dyn 222: 611-624

Dambly-Chaudière C, Sapède D, Soubiran F, Decorde K, Gompel N, Ghysen A (2003) The lateral line of zebrafish: a model system for the analysis of morphogenesis and neural development in vertebrates. Biol Cell 95: 579-587

Devoto SH, Melancon E, Eisen JS, Westerfield M (1996) Identification of separate slow and fast muscle precursor cells *in vivo*, prior to somite formation. Development 122: 3371-3380

Devoto SH, Stoiber W, Hammond CL, Steinbacher P, Haslett JR, Barresi MJ, Patterson SE, Adiarte EG, Hughes SM (2006) Generality of vertebrate developmental patterns: evidence for a dermomyotome in fish. Evol Dev 8: 101-110

Drummond IA, Majumdar A, Hentschel H, Elger M, Solnica-Krezel L, Schier AF, Neuhauss SC, Stemple DL, Zwartkruis F, Rangini Z, Driever W, Fishman MC (1998) Early development of the zebrafish pronephros and analysis of mutations affecting pronephric function. Development 125: 4655-4667

Duboule D (1998) Vertebrate hox gene regulation: clustering and/or colinearity? Curr Opin Genet Dev 8: 514-518

Dubrulle J, McGrew MJ, Pourquié O (2001) FGF signaling controls somite boundary position and regulates segmentation clock control of spatiotemporal Hox gene activation. Cell 106: 219-232

Easter SS, Jr., Malicki JJ (2002) The zebrafish eye: developmental and genetic analysis. Results Probl Cell Differ 40: 346-370

Eisen JS, Weston JA (1993) Development of the neural crest in the zebrafish. Dev Biol 159: 50-59

Felsenfeld AL, Curry M, Kimmel CB (1991) The *fub-1* mutation blocks initial myofibril formation in zebrafish muscle pioneer cells. Dev Biol 148: 23-30

Field HA, Ober EA, Roeser T, Stainier DY (2003a) Formation of the digestive system in zebrafish. I. Liver morphogenesis. Dev Biol 253: 279-290

Field HA, Dong PD, Beis D, Stainier DY (2003b) Formation of the digestive system in zebrafish. II. Pancreas morphogenesis. Dev Biol 261: 197-208

Finn RN, Kristoffersen BA (2007) Vertebrate vitellogenin gene duplication in relation to the "3R Hypothesis". Correlation to the pelagic egg and the oceanic radiation of teleosts. PLoS ONE 2(1): e169

Force A, Lynch M, Pickett FB, Amores A, Yan YL, Postlethwait J (1999) Preservation of duplicate genes by complementary, degenerative mutations. Genetics 151: 1531-1545

Freitas C, Rodrigues S, Saude L, Palmeirim I (2005) Running after the clock. Int J Dev Biol 49: 317-324

Gilbert SF, Raunio AM (1997) Embryology: Constructing the Organism. Sinauer Associates, Sunderland

Glickman NS, Kimmel CB, Jones MA, Adams RJ (2003) Shaping the zebrafish notochord. Development 130: 873-887

Gorodilov YN (1996) Description of the early ontogeny of the Atlantic salmon, *Salmo salar*, with a novel system of interval (state) identification. Environ Biol Fish 47: 109-127

Griffin K, Patient R, Holder N (1995) Analysis of FGF function in normal and no tail zebrafish embryos reveals separate mechanisms for formation of the trunk and the tail. Development 121: 2983-2994

Grotmol S, Kryvi H, Nordvik K, Totland GK (2003) Notochord segmentation may lay down the pathway for the development of the vertebral bodies in the Atlantic salmon. Anat Embryol (Berl) 207: 263-272

Haddon C, Lewis J (1996) Early ear development in the embryo of the zebrafish, *Danio rerio*. J Comp Neurol 365: 113-128

Haines L, Neyt C, Gautier P, Keenan DG, Bryson-Richardson RJ, Hollway GE, Cole NJ, Currie PD (2004) Met and Hgf signaling controls hypaxial muscle and lateral line development in the zebrafish. Development 131: 4857-4869

Hall TE, Smith P, Johnston IA (2004) Stages of embryonic development in the Atlantic cod *Gadus morhua*. J Morphol 259: 255-270

Halloran MC, Berndt JD (2003) Current progress in neural crest cell motility and migration and future prospects for the zebrafish model system. Dev Dyn 228: 497-513

Hibi M, Hirano T, Dawid IB (2002) Organizer formation and function. Results Probl Cell Differ 40: 48-71

Hjorth J, Key B (2002) Development of axon pathways in the zebrafish central nervous system. Int J Dev Biol 46: 609-619

Holland PW (1999) Gene duplication: past, present and future. Semin Cell Dev Biol 10: 541-547

Honjo Y, Eisen JS (2005) Slow muscle regulates the pattern of trunk neural crest migration in zebrafish. Development 132: 4461-4470

Howley C, Ho RK (2000) mRNA localization patterns in zebrafish oocytes. Mech Dev 92: 305-309

Hunt TE (1929) Hensen's node as an organizer in the formation of the chick embryo. Anat Rec 42: 22

Hunter MP, Prince VE (2002) Zebrafish hox paralogue group 2 genes function redundantly as selector genes to pattern the second pharyngeal arch. Dev Biol 247: 367-389

Iwamatsu T (1994) Stages of normal development in the medaka *Oryzias latipes*. Zool Sci 11: 825-839

Jaillon O, Aury JM, Brunet F, Petit JL, Stange-Thomann N, Mauceli E, Bouneau L, Fischer C, Ozouf-Costaz C, Bernot A *et al.*, (2004) Genome duplication in the teleost fish *Tetraodon nigroviridis* reveals the early vertebrate proto-karyotype. Nature 431: 946-957

Johnston IA, Hall TE (2004) Mechanisms of muscle development and responses to temperature change in fish larvae. In: Govoni JJ (ed) The development of form and function in fishes and the question of larval adaptation. American Fisheries Society Symp. 40, Bethesda, p 113-144

Kane DA, Adams R (2002) Life at the edge: epiboly and involution in the zebrafish. Results Probl Cell Differ 40: 117-135

Kane DA, Kimmel CB (1993) The zebrafish midblastula transition. Development 119: 447-456

Kane DA, Warga RM, Kimmel CB (1992) Mitotic domains in the early embryo of the zebrafish. Nature 360: 735-737

Kimelman D, Schier AF (2002) Mesoderm induction and patterning. Results Probl Cell Differ 40: 15-27

Kimmel CB (1993) Patterning the brain of the zebrafish embryo. Ann Rev Neurosci 16: 707-732

Kimmel CB, Ballard WW, Kimmel SR, Ullmann B, Schilling TF (1995) Stages of embryonic development of the zebrafish. Dev Dyn 203: 253-310

Kimmel CB, Law RD (1985) Cell lineage of zebrafish blastomeres. II. Formation of the yolk syncytial layer. Dev Biol 108: 86-93

Kimmel CB, Warga RM, Schilling TF (1990) Origin and organization of the zebrafish fate map. Development 108: 581-594

Koehl M, Quillin K, Pell C (2000) Mechanical design of fiber-wound hydraulic skeletons: The stiffening and straightening of embryonic notochords. Am Zool 40: 28-41

Krumlauf R (1994) Hox genes in vertebrate development. Cell 78: 191-201

Kurosawa G, Takamatsu N, Takahashi M, Sumitomo M, Sanaka E, Yamada K, Nishii K, Matsuda M, Asakawa S, Ishiguro H *et al.* (2006) Organization and structure of *hox* gene loci in medaka genome and comparison with those of pufferfish and zebrafish genomes. Gene 370: 75-82

Kuwada JY, Bernhardt RR (1990) Axonal outgrowth by identified neurons in the spinal cord of zebrafish embryos. Exp Neurol 109: 29-34

Leung CF, Webb SE, Miller AL (2000) On the mechanism of ooplasmic segregation in single-cell zebrafish embryos. Dev Growth Differ 42: 29-40

Lewis EB (1978) A gene complex controlling segmentation in *Drosophila*. Nature 276: 565-570

Liu Q, Liu B, Wilson AL, Rostedt J (2006) *Cadherin*-6 message expression in the nervous system of developing zebrafish. Dev Dyn 235: 272-278

Maconochie M, Nonchev S, Morrison A, Krumlauf R (1996) Paralogous Hox genes: function and regulation. Ann Rev Genet 30: 529-556

Marrable AW (1965) Variations in early cleavage of the zebrafish. Nature 184: 1160-1161

Melby AE, Kimelman D, Kimmel CB (1997) Spatial regulation of *floating head* expression in the developing notochord. Dev Dyn 209: 156-165

Mizuno T, Yamaha E, Wakahara M, Kuroiwa A, Takeda H (1996) Mesoderm induction in zebrafish. Nature 383: 131-132

Morin-Kensicki EM, Eisen JS (1997) Sclerotome development and peripheral nervous system segmentation in embryonic zebrafish. Development 124: 159-167

Nakatsuji T, Kitano, T, Akiyama, N, Nakatsuji, N (1997) Ice-goby (shiro-uo), *Leucopsarion petersii*, may be a useful material for studying teleostean embryogenesis. Zool Sci 14: 443-448

Nelson JS (1994) Fishes of the World. John Wiley and Sons Inc., New York

Ng AN, de Jong-Curtain TA, Mawdsley DJ, White SJ, Shin J, Appel B, Dong PD, Stainier DY, Heath JK (2005) Formation of the digestive system in zebrafish: III. Intestinal epithelium morphogenesis. Dev Biol 286: 114-135

Ohno S (1970) Evolution by Gene Duplication. Springer-Verlag, New York

Papan C, Campos-Ortega JA (1994) On the formation of the neural keel and neural tube in the zebrafish *Danio* (*Brachydanio*) *rerio*. Roux's Arch Dev Biol 203: 178-186

Pavlov D, Dzerzhinskiy, KF, Radzikhovskaya, EK. (1992) Assessing the quality of roe from White Sea wolffish (*Anarhichas lupus marisalbi* L.), obtained under experimental conditions. J Ichthyol 32: 88-104

Prince VE, Joly L, Ekker M, Ho RK (1998a) Zebrafish *hox* genes: genomic organization and modified colinear expression patterns in the trunk. Development 125: 407-420

Prince VE, Moens CB, Kimmel CB, Ho RK (1998b) Zebrafish *hox* genes: expression in the hindbrain region of wild-type and mutants of the segmentation gene, *valentino*. Development 125: 393-406

Prince VE, Price AL, Ho RK (1998c) *Hox* gene expression reveals regionalization along the anteroposterior axis of the zebrafish notochord. Dev Genes Evol 208: 517-522

Rescan PY, Ralliere C, Chauvigné F, Cauty C (2005) Expression patterns of collagen I (alpha1) encoding gene and muscle-specific genes reveal that the lateral domain of the fish somite forms a connective tissue surrounding the myotome. Dev Dyn 233: 605-611

Richardson MK, Allen SP, Wright GM, Raynaud A, Hanken J (1998) Somite number and vertebrate evolution. Development 125: 151-160

Riley BB, Moorman SJ (2000) Development of utricular otoliths, but not saccular otoliths, is necessary for vestibular function and survival in zebrafish. J Neurobiol 43: 329-337

Rodaway A, Takeda H, Koshida S, Broadbent J, Price B, Smith JC, Patient R, Holder N (1999) Induction of the mesendoderm in the zebrafish germ ring by yolk cell-derived TGF-beta family signals and discrimination of mesoderm and endoderm by FGF. Development 126: 3067-3078

Roy S, Wolff C, Ingham PW (2001) The *u-boot* mutation identifies a Hedgehog-regulated myogenic switch for fiber-type diversification in the zebrafish embryo. Genes Dev 15: 1563-1576

Sakaguchi T, Mizuno T, Takeda H (2002) Formation and patterning roles of the yolk syncytial layer. Results Probl Cell Differ 40: 1-14

Sapède D, Gompel N, Dambly-Chaudière C, Ghysen A (2002) Cell migration in the postembryonic development of the fish lateral line. Development 129: 605-615

Sato M, Yost HJ (2003) Cardiac neural crest contributes to cardiomyogenesis in zebrafish. Dev Biol 257: 127-139

Saude L, Woolley K, Martin P, Driever W, Stemple DL (2000) Axis-inducing activities and cell fates of the zebrafish organizer. Development 127: 3407-3417

Sawada A, Shinya M, Jiang YJ, Kawakami A, Kuroiwa A, Takeda H (2001) Fgf/MAPK signalling is a crucial positional cue in somite boundary formation. Development 128: 4873-4880

Scaal M, Christ B (2004) Formation and differentiation of the avian dermomyotome. Anat Embryol (Berl) 208: 411-424

Schier AF (2003) Nodal signaling in vertebrate development. Ann Rev Cell Dev Biol 19: 589-621

Schier AF, Talbot WS (2005) Molecular genetics of axis formation in zebrafish. Ann Rev Genet 39: 561-613

Schlosser G (2005) Evolutionary origins of vertebrate placodes: insights from developmental studies and from comparisons with other deuterostomes. J Exp Zool 304: 347-399

Schwartz RJ, Olson EN (1999) Building the heart piece by piece: modularity of cis-elements regulating *Nkx2-5* transcription. Development 126: 4187-4192

Sepich DS, Calmelet C, Kiskowski M, Solnica-Krezel L (2005) Initiation of convergence and extension movements of lateral mesoderm during zebrafish gastrulation. Dev Dyn 234: 279-292

Sepich DS, Solnica-Krezel L (2005) Analysis of cell movements in zebrafish embryos: morphometrics and measuring movement of labeled cell populations *in vivo*. Methods Mol Biol 294: 211-233

Serluca FC, Fishman MC (2001) Pre-pattern in the pronephric kidney field of zebrafish. Development 128: 2233-2241

Sidow A (1996) Gen(om)e duplications in the evolution of early vertebrates. Curr Opin Genet Dev 6: 715-722

Spemann H, Mangold H (1924) Uber induktion von embryonalanlagen durch implantation artfremder organisatoren. Roux Arch Entwick Org 100: 599-638

Stainier DY (2001) Zebrafish genetics and vertebrate heart formation. Nat Rev Genet 2: 39-48

Stickney HL, Barresi MJ, Devoto SH (2000) Somite development in zebrafish. Dev Dyn 219: 287-303

Thisse C, Thisse B, Postlethwait JH (1995) Expression of *snail2*, a second member of the zebrafish snail family, in cephalic mesendoderm and presumptive neural crest of wild-type and *spadetail* mutant embryos. Dev Biol 172: 86-99

Trinkaus JP (1951) A study of the mechanism of epiboly in the egg of *Fundulus heteroclitus*. J Exp Zool 118: 269-320

Trinkaus JP (1984) Mechanism of *Fundulus* epiboly—a current view. Am Zool 24: 673-688

Trinkaus JP (1992) The midblastula transition, the YSL transition and the onset of gastrulation in *Fundulus*. Dev Suppl 75-80

van Raamsdonk W, Pool CW, de Kronnie G (1978) Differentiation of muscle fiber types in the teleost *Brachydanio rerio*. Anat Embryol 153: 137-155

Wallace KN, Pack M (2003) Unique and conserved aspects of gut development in zebrafish. Dev Biol 255: 12-29

Warga RM, Kimmel CB (1990) Cell movements during epiboly and gastrulation in zebrafish. Development 108: 569-580

Warga RM, Nusslein-Volhard C (1999) Origin and development of the zebrafish endoderm. Development 126: 827-838

Warga RM, Stainier DY (2002) The guts of endoderm formation. Results Probl Cell Differ 40: 28-47

Waterman RE (1969) Development of the lateral musculature in the teleost, *Brachydanio rerio*: a fine structural study. Am J Anat 125: 457-493

Wellik DM, Capecchi MR (2003) Hox10 and Hox11 genes are required to globally pattern the mammalian skeleton. Science 301: 363-367

Whitfield TT, Riley BB, Chiang MY, Phillips B (2002) Development of the zebrafish inner ear. Dev Dyn 223: 427-458

Whitlock KE, Westerfield M (1998) A transient population of neurons pioneers the olfactory pathway in the zebrafish. J Neurosci 18: 8919-8927

Woltering JM, Durston AJ (2006) The zebrafish *hoxDb* cluster has been reduced to a single microRNA. Nature 38: 601-602

Wood A, Timmermans, LPM (1988) Teleost epiboly: a reassessment of deep cell movement in the germ ring. Development 102: 575-585

Yelon D, Horne SA, Stainier DY (1999) Restricted expression of cardiac myosin genes reveals regulated aspects of heart tube assembly in zebrafish. Dev Biol 214: 23-37

Yelon D, Weinstein BM, Fishman MC (2002) Cardiovascular system. Results Probl Cell Differ 40: 298-321

Zeller J, Schneider V, Malayaman S, Higashijima S, Okamoto H, Gui J, Lin S, Granato M (2002) Migration of zebrafish spinal motor nerves into the periphery requires multiple myotome-derived cues. Dev Biol 252: 241-256

CHAPTER

2

Pigmentation

Robert N Kelsh[1,*] and David M Parichy[2]

INTRODUCTION

Variations in pigmentation form one of the most striking of traits, readily apparent to even the most casual of observers. Pigmentation patterns in fish are as diverse as any in the animal kingdom (Lindholm *et al.*, 2004; Genner & Turner, 2005; Parichy, 2006). In general, fish species have two pigment patterns, since an early larval pattern, formed during the embryonic stages, is then replaced during metamorphosis with the adult pattern (Fig. 1). Larval patterns may have roles in camouflage and predator avoidance and tend to have conserved pattern elements. Adult pigment patterns are more variable, with functions ranging from camouflage to display to warning colouration.

Pigmentation is an important trait for several reasons. To geneticists, it has the tremendous advantage of being directly observable and thus easy to screen for variants. To developmental biologists, pigment cells are derived from the neural crest, a transient tissue likely to include cells with the properties of stem cells. These cells are marvellous models for understanding how multipotent cells adopt particular cell fates. To those in the fish farming industry, abnormal pigmentation will affect the market value of a fish.

Authors' addresses: [1]Centre for Regenerative Medicine, Department of Biology and Biochemistry, University of Bath, Claverton Down, Bath BA2 7AY, UK.

[2]Department of Biology and Institute for Stem Cell and Regenerative Medicine, University of Washington, Seattle WA 98195.

**Corresponding author:* E-mail: bssrnk@bath.ac.uk

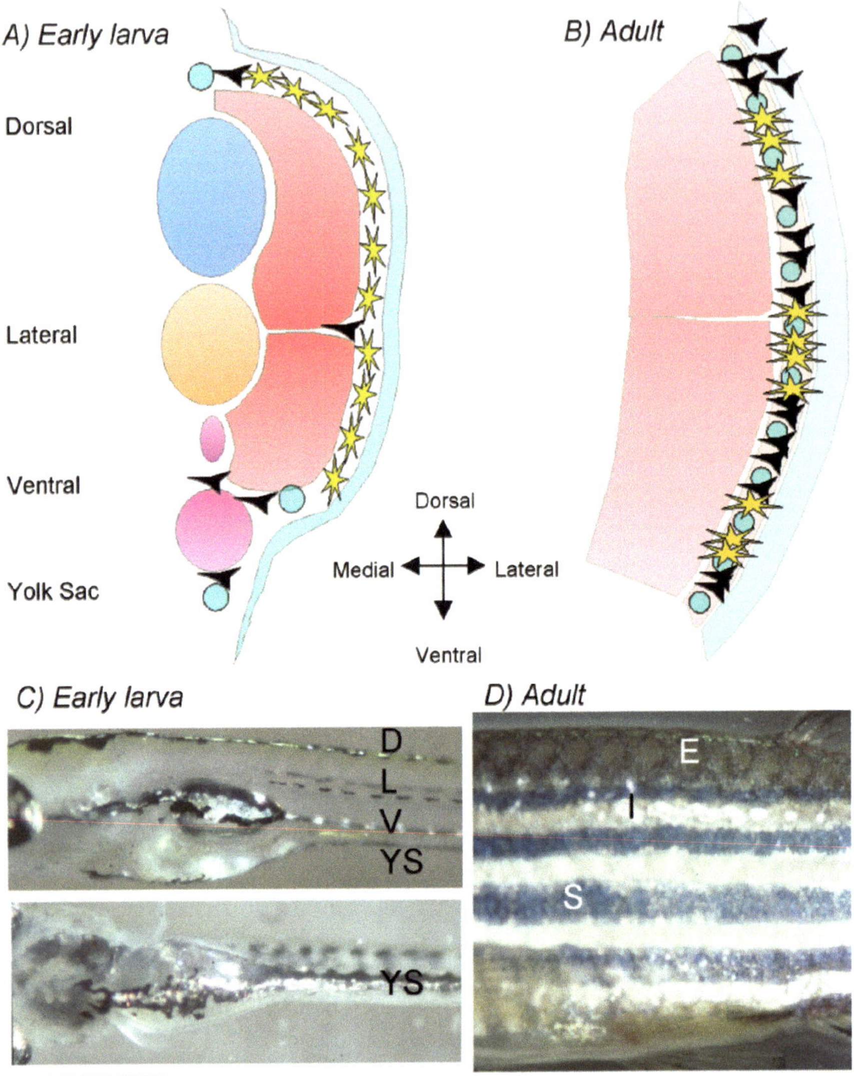

Fig. 1 Early larval and adult pigment patterns are distinct in zebrafish (*Danio rerio*). Schematics of transverse sections of early larva (A) and adult (B) to show position of melanophores (black), iridophores (bright blue circles) and xanthophores (yellow) with respect to epidermis (pale blue) and myotome (red). Neural tube shown in dark blue and notochord in orange. C) Lateral (C, upper) and ventral (C, lower) views of 5 days post-fertilisation zebrafish larva to show pigment cells. Note the four melanophore stripes (Dorsal, D; Lateral, L; Ventral, V; Yolk Sac, YS) and accompanying shiny iridophores (all except Lateral). Xanthophores are barely visible as a yellow cast, e.g. in anterior. D) Adult pigment pattern showing blue stripes (S) and silver interstripes (I); also epidermal melanophores dorsally (E). Adapted from Kelsh (2004).

To behavioural biologists, the contribution of pigmentation to signals in various behaviours such as shoaling and mate choice, and as an integral part of camouflage, make it a key focus. Finally, to evolutionary biologists, the diversity of patterns generates a multitude of questions regarding the molecular and cellular basis of selected traits, as well as roles for such patterns in speciation.

Pigments are contained within specialised pigment cells known as "chromatophores". Here we will focus on the diversity, origins, development and patterning of fish pigmentation, as generated by the numbers and arrangements of teleost chromatophores. An interesting additional feature of these cells is their ability to mobilise intracellular pigment granules in response to varied stimuli, and in so doing, lighten or darken the overall appearance of the fish. This phenomenon has been well studied and for more information, we refer the reader to several recent papers (Fujii, 2000; Kimler & Taylor, 2002; Kawauchi, 2006; Logan *et al.*, 2006).

CHROMATOPHORES

Fish pigment cells are diverse. Currently six different types are recognised (Table 1) although, since one of these was described only comparatively recently (reviewed in: Bagnara *et al.*, 2007), it is conceivable that others remain to be identified as the enormous diversity of fishes continues to be explored. The different cell-types and their relationships are nicely discussed by Bagnara (2006). Chromatophores are distinguished by their different colourations, due to different pigment organelles. In most cases pigmentation results from the colour of the pigment contained within the organelles, but for the reflecting platelets of iridophores and leucophores, colour results from the physical interaction of the guanine crystals with light and is regulated by their spacing and orientation (Kasukawa & Oshima 1987).

Table 1 Fish chromatophores

Chromatophore	*Colour*	*Pigment organelle*	*Pigment*
Melanophore	Black or brown	Melanosome	Melanin
Xanthophore	Yellow-orange	Pterinosome + carotenoid vesicles	Pteridines + carotenoids
Erythrophore	Red	Pterinosome + carotenoid vesicles	Pteridines + carotenoids
Iridophore	Shiny silver or blue	Reflecting platelet	Guanine crystals
Leucophore	Milky cream or white	Reflecting platelet	Guanine crystals
Cyanophore	Electric blue	Cyanosome	?

GENETIC AND GENOMIC RESOURCES FOR STUDYING FISH PIGMENTATION

Multiple fish species are now established as genetic models, including zebrafish (*Danio rerio*), medaka (*Oryzias latipes*), cichlids (e.g. *Oreochromis niloticus*), and three-spined stickleback (*Gasterosteus aculeatus*). Thus, it is no surprise that pigmentation mutants feature prominently in several collections of mutants. Embryonic zebrafish pigmentation mutants have been surveyed extensively in large and small-scale genetic screens (Streisinger *et al.*, 1981; Kelsh *et al.*, 1996; Amsterdam *et al.*, 1999; Lister *et al.*, 1999). These studies have defined approximately 100 embryonic pigmentation loci in zebrafish, although undoubtedly this list is incomplete. A growing number of these mutant loci have now been identified at the molecular level by positional cloning, candidate gene approaches or, most rapidly, by insertional mutagenesis (Table 2). Furthermore, reverse genetic approaches, especially the use of anti-sense modified oligonucleotides (morpholinos), have complemented forward genetic approaches and promise to identify a multitude of other genes critical for one or another aspect of chromatophore development (Pickart *et al.*, 2004). Beyond the embryo, numerous mutants also have been identified that affect adult pigmentation (Table 3). While some of these have defects limited to the formation of particular chromatophore lineages or chromatophore arrangements, others exhibit pigmentation changes as just one aspect of pleiotropic defects in multiple traits (http://protist.biology.washington.edu/dparichy/ImagesMutants.htm). Genetic resources in the form of mutants are supplemented in these and other model teleosts by extensive genomic resources (Table 4). As genome sequencing projects come to fruition, other fish genomes are likely to follow. Together these resources provide the background for a comprehensive molecular and cellular understanding of pigment cells, pigment cell patterning and pigment pattern evolution.

Table 2 Identified embryonic pigmentation loci in fish

Species	*Locus name*	*Gene*	*Pigment phenotype*	*Reference*
Zebrafish	*atp6ap2*	*atp6ap2*	Melanosome function	Amsterdam *et al.* (2004)
	atp6v0c	*atp6v0c*	Melanosome function	Golling *et al.* (2002)
	atp6v0d1	*atp6v0d1*	Melanosome function	Amsterdam *et al.* (2004)
	atp6v1e1	*atp6v1e1*	Melanosome function	Golling *et al.* (2002)
	atp6v1h	*atp6v1h*	Melanosome function	Golling *et al.* (2002)
	bleached *blond*	*atp6a1*	Melanosome function	Amsterdam *et al.* (1999)
	colourless	*sox10*	Specification of pigment cells from neural crest	Dutton *et al.* (2001); Elworthy *et al.* (2003)
	esrom	*mycb2*	Pteridine synthesis	Le Guyader *et al.* (2005)
	fading vision	*silva*	Melanosome biogenesis	Schonthaler *et al.* (2005)
	golden	*slc24a5*	Melanosome biogenesis	Lamason *et al.* (2005)
	kohtalo/ trapped	*trap230/* Med12	Coactivator of Sox9b-mediated transcription	Hong *et al.* (2005); Rau *et al.* (2006)
	mindbomb/ whitetail	*mib*	Specification of neural crest cells	Itoh *et al.* (2003)
	mont blanc/ lockjaw	*tfap2a*	Neural crest and/or chromatophore development	Barrallo-Gimeno *et al.* (2004); Knight *et al.* (2004); O'Brien *et al.* (2004)

Table 2 Contd. ...

Table 2 Contd. ...

	moonshine	*trim33*	Iridophore patterning	Ransom *et al.* (2004)
	nacre	*mitfa*	Melanophore specification & maintenance	Lister *et al.* (1999)
	narrowminded	*prdm1*	Specification of neural crest	Roy & Ng (2004); Hernandez-Lagunas *et al.* (2005)
	nicastrin	*ncstn*	Melanin synthesis	Amsterdam *et al.* (2005)
	paics	*paics*	Xanthophore differentiation	Amsterdam *et al.* (2004)
	panther/ salz/pfeffer	*csf1r/fms*	Xanthoblast dispersal	Parichy *et al.* (2000a)
	sandy	*tyrosinase*	Melanin synthesis	Page-McCaw *et al.* (2004)
	shady	*alk*	Iridophore specification?	Lopes, SL *et al.* (unpublished data)
	sox9b	*sox9b*	Induction & maintenance of neural crest	Yan *et al.* (2005)
	sparse	*kita*	Melanophore survival and migration	Parichy *et al.* (1999)
	sym1	*foxd3*	Induction & maintenance of neural crest?	Lister *et al.* (2006); Montero-Balaguer *et al.* (2006); Stewart *et al.* (2006)
	touchtone	*trpm7*	Melanophore survival	Elizondo *et al.* (2005)
	vps18	*vps18*	Melanosome trafficking	Maldonado *et al.* (2006)
Medaka	*B*	*aim-1*	Melanosome function	Fukamachi *et al.* (2001)
	ci	*somatolactin*	Melanophore number and leucophore differentiation?	Fukamachi *et al.* (2004b)
	i-3	*p*	Melanosome function	Fukamachi *et al.* (2004a)

Table 3 Identified pigmentation loci affecting adult pigment pattern in zebrafish

Locus	*Gene*	*Adult pigment phenotype*	*Reference*
jaguar/obelix	*kir7.1*	wider stripes	Iwashita *et al.* (2006)
hagoromo	*dactylin*	disrupted mel stripes	Kawakami *et al.* (2000)
leopard	*cx41.8*	fewer mel, spots	Watanabe *et al.* (2006)
panther/salz/pfeffer	*csf1r/fms*	xan absent, fewer mel, disrupted stripes	Parichy *et al.* (2000a)
picasso	*erbb3*	fewer mel, disrupted stripes	Lyons *et al.* (2005)
rose	*ednrb1*	irid absent, fewer melanophores	Parichy *et al.* (2000b)
shady	*ltk*	weak alleles irid absent; strong alleles lethal	Lopes, SL *et al.* (unpublished data)
sparse	*kita*	fewer stripe mel; scale mel absent	Parichy *et al.* (1999)

mel: melanophore; xan: xanthophore; irid: iridiphore

Table 4 Current genomic resources for fish model species

Species	*Resource*	*Resource website*	*Resources available*
Danio rerio	Zebrafish Information Network (ZFIN)	http://zfin.org	Genome project, genes, markers and clones, mutant and wild-type strains, gene expression patterns, zebrafish labs, etc.
Oryzias latipes	Medaka	http://biol1.bio.nagoya-u.ac.jp:8000/	Genome project, genes, markers and mutants
Fugu rubripes	Fugu Informatics Network Fugu Genome Project	http://fugu.biology.qmul.ac.uk http://www.fugu-sg.org/	Genome project, mapping resources, synteny viewer
Tetraodon nigroviridis	Tetraodon Genome Browser	http://www.genoscope.cns.fr/externe/tetraodon/	Genome project
Gasterosteus aculeatus	Stanford Genome Evolution Center	http://cegs.stanford.edu/index.jsp	Genome maps, markers
Xiphophorus	Xiphophorus	http://www.xiphophorus.org/	Species and strains, markers
Oreochromis niloticus and other cichlids	Cichlid Genome Consortium	http://hcgs.unh.edu/cichlid/	Genetic maps, markers, BAC and cDNA libraries

ORIGIN OF CHROMATOPHORES FROM THE NEURAL CREST

All fish chromatophores, except for the distinctively epithelial melanised cells of the pigmented retinal epithelium, are believed to derive from a transient embryonic tissue called the neural crest (Quigley & Parichy, 2002; Parichy, 2003; Kelsh, 2004; Parichy *et al.*, 2006). The neural crest is a set of cells that form at the junction between the nascent central nervous system and epidermis (i.e. the neural and non-neural ectoderm, see Hall, this volume). From this location, neural crest cells migrate along stereotypical pathways throughout the embryo, including a lateral (or dorsolateral) pathway, between the epidermis and somites, and a ventral pathway, between the neural tube/notochord and somites (Gilbert, 2006). In addition to pigment cells, neural crest cells generate craniofacial cartilage and bone, most cells of the peripheral nervous system and an assortment of other cell-types (Gilbert, 2006). While neural crest cells have been thought of as a vertebrate evolutionary innovation, a population of cells with neural crest-like properties also occurs in some species of ascidians, which are urochordates; these cells differentiate as chromatophores resembling the yellow-orange xanthophores or erythrophores of teleosts (Jeffery *et al.*, 2004).

Two of the most interesting features of neural crest cells and some of their immediate progeny are their multipotentiality and their ability to act as stem cells. While these qualities are sometimes equated with one another they are actually distinct. Cells that are multipotent can produce multiple derivative cell types (as distinguished from totipotent stem cells or embryonic stem cells that can produce all cell types of the embryo). By contrast, cells that function as stem cells are able to produce differentiated progeny (of multiple types, or of only one type) while also self-renewing. In amniotes, neural crest cells and their progeny often are multipotent with individual cells contributing to multiple derivatives; nevertheless some cells give rise to only one or a few derivatives and even multipotent cells typically undergo a series of progressive restrictions on their fates. Amniote neural crest-derived cells also have been found to have stem cell properties, in some instances maintaining their multipotentiality as well. It is currently unclear in fish whether neural crest cells and their immediate progeny have these properties, but given the highly conserved role of the neural crest in all vertebrates, it is very likely that at least some cells will show stem cell properties. Indeed, they do express at least one gene associated, in rodents, with maintenance of neural crest stem cell multipotentiality (Dutton *et al.*, 2001; Kim *et al.*, 2003).

Direct demonstration that pigment cells derive from the neural crest requires lineage labelling studies, in which neural crest cells are cleanly labelled in such a way that their progeny inherit the cellular marker. In zebrafish, such studies have confirmed that melanophores, xanthophores and iridophores are each neural crest-derived (Raible & Eisen, 1994; Schilling & Kimmel, 1994; Dutton *et al.*, 2001). A complementary approach is one of ablation of the neural crest and observing which cell-types are missing. Although in principal, this could be done genetically, that has not really been cleanly achieved in fish, although the few mutants that do strongly deplete neural crest cells in the embryo do have dramatic losses of pigment cells. Thus, mutants in the *alyron* gene show no pigment cells, as well as having profound depletion of other neural crest-derived cell-types (Cretekos & Grunwald, 1999). A strong *U-boot/narrowminded/prdm1* mutant results in severe depletion of the neural crest and significant reduction of melanophores at least (Artinger *et al.*, 1999).[1] Finally, *mindbomb* (also known as *whitetail*, which encodes a ubiquitin ligase required for notch signaling) mutants lack all melanophores, xanthophores and iridophores in the posterior trunk and tail (Kelsh *et al.*, 1996). A more direct approach, involving physical ablation of neural crest, constituted the first experimental indication that chromatophores were likely to be derived from the neural crest in amphibians, since embryos lacking neural crest also had corresponding pigment cell defects. Ablation experiments in zebrafish (Vaglia and Hall, 2000) and other species reveal analogous chromatophore defects in embryos and early larvae (reviewed in Hörstadius, 1950; Le Douarin & Kalcheim, 1999; Parichy *et al.*, 2006). The embryonic origin of leucophores and cyanophores has not, to our knowledge, been addressed definitively.

EMBRYONIC CHROMATOPHORES AND PIGMENT PATTERN FORMATION

Specification of Chromatophore Lineages

Understanding the genetic basis for pigment cell development involves analyses of several distinct steps in neural crest development. One of these concerns the lineage relationships of pigment cells to themselves and to other neural crest derivatives. Although lineage studies have been performed in zebrafish (see above), these have not revealed a clear pattern of relationship between different neural crest derivatives, in part because the *in vivo* clone sizes are so small. Phenotypic characterisation of *colourless/sox10* mutants led to the suggestion that precursors fated to ectomesenchymal derivatives (e.g. cartilage and fin mesenchyme) might segregate early from multipotent cells fated to non-ectomesenchymal derivatives (pigment cells, glia and peripheral neurons) (Kelsh & Eisen, 2000). This idea is consistent with the available cell lineage data, since mixed ectomesenchymal and non-ectomesenchymal clones have not yet been identified in zebrafish. The suggestion by Bagnara (1979) that all pigment cells shared an origin in a multipotent chromatoblast is attractive, but remains unproven.

Formation of neural crest cells is a prerequisite for formation of pigment cells. Multiple transcription factors expressed in early neural crest have been identified.

[1] Throughout this chapter we use the convention that genes are italicised whereas their protein products are not italicised. For most fish genes, names begin with lower case whereas their protein products begin with a capital letter; their mammalian counterparts are typically capitalised whether referring to gene or protein. Finally, many genes have been identified only after isolating them as a mutant phenotype. For mutants in which the underlying gene has been determined, we provide both names for easier reference to the literature, with the mutant name followed by the gene name. In some instances, a mutant has received more than one name, in which case we provide both mutant names followed by the gene name (e.g. *U-boot/narrowminded/prdm1*).

Some of these are likely to be important for the early specification or maintenance of the neural crest, including *sox9b, foxd3* and *tfap2a* (Barrallo-Gimeno *et al.*, 2004; Knight *et al.*, 2004; O'Brien *et al.*, 2004; Yan *et al.*, 2005; Lister *et al.*, 2006; Montero-Balaguer *et al.*, 2006; Stewart *et al.*, 2006). In the case of *tfap2a* mutants, numbers of all three chromatophore lineages are decreased, but so also are most neural crest derivatives examined, so it remains unclear whether *tfap2a* acts only in early neural crest or also later in the differentiation of each lineage. Nevertheless, characterisation of melanophore markers suggests that, in this lineage, *tfap2a* function may act in part through regulation of the melanophore survival factor *sparse/kita* (Barrallo-Gimeno *et al.*, 2004; O'Brien *et al.*, 2004). In contrast, disruption of *sox9b* function has unexpectedly demonstrated specific effects on individual derivatives, including pigment cells. Thus, numbers of iridophores, but not xanthophores nor melanophores, are strongly reduced in *sox9b* mutants (Yan *et al.*, 2005). It is thus plausible that *sox9b* has a late function in iridophore development, although explanations of the phenotype resulting from effects on early crest cell maintenance have not been eliminated. A similar iridophore-only phenotype has been described for *foxd3*, but immunofluorescence studies have not detected Foxd3 in differentiated iridophores, suggesting that defects result from a role in an earlier precursor (Lister *et al.*, 2006; Montero-Balaguer *et al.*, 2006; Stewart *et al.*, 2006).

In zebrafish, the basis for assignment (specification) of some cells to a melanophore fate has been the focus of significant attention and shows strong conservation with the mechanism in mammals. The distinctive phenotype of *nacre/mitfa* mutant embryos, in which melanoblasts are completely absent (Lister *et al.*, 1999), was consistent with the known pivotal role of *Mitf* in mammalian melanocyte development (Steingrimsson *et al.*, 1994). In zebrafish and mouse, neural crest cells were shown to respond to extracellular Wnt signals by transcriptional activation of *nacre/mitfa* (Dorsky *et al.*, 2000; Takeda *et al.*, 2000). The melanophore phenotype of *colourless/sox10* mutants is highly reminiscent of that in *nacre/mitfa* mutants. A combined *in vivo* and *in vitro* analysis showed that Sox10 is required intracellularly to activate *mitfa* transcription (Dutton *et al.*, 2001; Elworthy *et al.*, 2003), similar to the role shown in mouse (Bondurand *et al.*, 2000; Lee *et al.*, 2000; Potterf *et al.*, 2000). A common role for Sox10 in fate specification of pigment and neural cells from multipotent neural crest precursors has been proposed and is now supported by data for many lineages (Dutton *et al.*, 2001; Kelsh & Raible, 2002; Kelsh, 2006). Whether Sox10 might also function later, in the maintenance of melanophore differentiation, appears unlikely in the zebrafish (Elworthy *et al.*, 2003), although work in mammalian cell culture suggests there may be an ongoing role at least in initiation of expression of *tyrosinase*, a key gene in melanin synthesis (Hou *et al.*, 2006). Assignment of cells to other pigment cell fates is likely to depend upon Sox10 (Kelsh & Eisen, 2000; Dutton *et al.*, 2001), and may involve *shady* function (Kelsh *et al.*, 1996; SS Lopes, X Yang, J Müller & RN Kelsh, unpublished data). It is clear that this picture is far from complete, with multiple factors that act in combination with those noted above yet to be identified.

Subsequent development of chromatophores involves migration, active maintenance (survival) and differentiation. Although many relevant mutants exist (Kelsh *et al.*, 1996; Odenthal *et al.*, 1996), little is known about these processes in any chromatophore type, except melanophores. In this cell-type one study has begun to characterise genes involved in different stages of melanophore development, identifying genes regulating melanophore proliferation, migration, survival and differentiation in zebrafish (Kelsh *et al.*, 2000). This overview is being expanded and tested by detailed studies of individual gene phenotypes. In melanophores, *sparse/kita* function has been shown to be critical for migration of cells to ventral regions and for cell survival, although the receptor tyrosine kinase encoded by this locus is required only for a subset of these cells (Parichy *et al.*, 1999; Kelsh *et al.*, 2000). Also required for melanoblast survival is the *touchtone/nutria/trpm7* gene, which encodes a dual function cation channel and kinase (Arduini & Henion, 2004; Cornell *et al.*, 2004; Elizondo *et al.*, 2005). Trafficking of proteins to form melanosomes is critical for melanisation of melanophores and mutants affecting components of these pathways are now being identified (Maldonado *et al.*, 2006). Likewise, melanosome structural components and melanin synthetic enzyme genes underlie many of the albino zebrafish and medaka mutants (Fukamachi *et al.*, 2001; Fukamachi *et al.*, 2004a; Page-McCaw *et al.*, 2004; Lamason *et al.*, 2005; Schonthaler *et al.*, 2005). In the xanthophore lineage, an overview study classifies the roles of many xanthophore differentiation genes (Odenthal *et al.*, 1996). Again, further studies have provided a more detailed analysis of specific loci. Thus, *panther/csf1r*, encodes a receptor tyrosine kinase related to *kita* and is required for dispersal of xanthoblasts away from the dorsal neural tube (Parichy *et al.*, 2000a). The *esrom* locus encodes a Myc binding protein and functions in pteridine synthesis (Le Guyader *et al.*, 2005), which also requires the hormone somatolactin (Fukamachi *et al.*, 2004b). Finally, a recent medaka study indicates that shared components of vesicle-trafficking pathways may be required for pigment organelle formation in all chromatophores (Yu *et al.*, 2006). These studies show how revealing the analysis of individual mutant phenotypes can be and also indicate how much still remains to be learnt about pigment cell development in fish.

Mechanisms of Embryonic/Early Larval Pigment Pattern Development

Fish embryos have a distinctive pigment pattern, best described in zebrafish and medaka (Milos & Dingle, 1978a; Kimmel *et al.*, 1995; Kelsh *et al.*, 1996; McClure, 1999; Kelsh *et al.*, 2004; Lamoreux *et al.*, 2005). This pattern develops within the embryonic and early larval phases, being complete at hatching (medaka) or soon after hatching (zebrafish) (Figs. 1, 2). In both of these species, the melanophore pattern consists of a series of longitudinal stripes. These lie in the dorsalmost midline [i.e. above the brain and dorsal projections of the myotomes; Dorsal stripe in the nomenclature of Milos & Dingle (1978a), Kimmel *et al.*, (1995)], in the horizontal myoseptum (midway down the flanks; Lateral stripe), and ventral to the myotomes (and extending broadly across

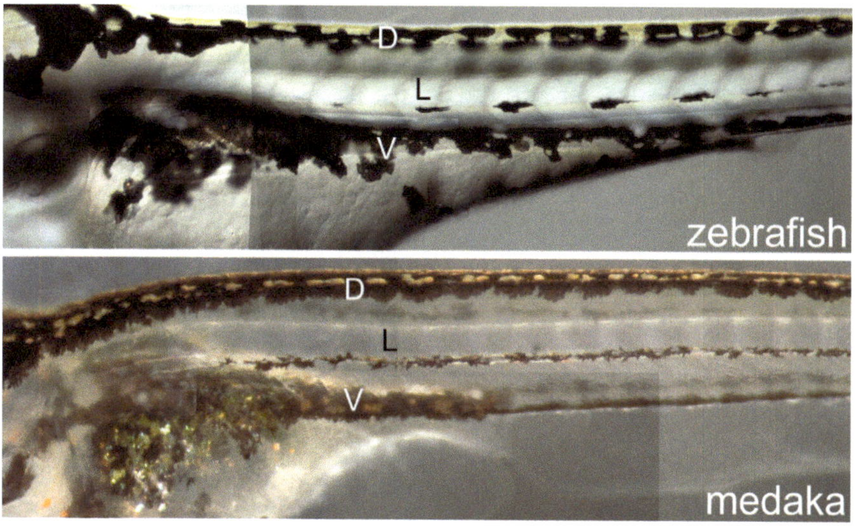

Fig. 2 Wild-type embryonic pigment patterns are strongly conserved in zebrafish (*Danio rerio*) and medaka (*Oryzias latipes*). 4 and 10 days post-fertilisation larvae are shown in dorsolateral view to illustrate the evident conservation of melanophore pattern in these distant taxa.

the viscera; Ventral stripe). In zebrafish the ventralmost midline has a fourth melanophore stripe (Yolk Sac stripe), but in medaka this is replaced by a Yolk Sac Cluster (Kelsh *et al.*, 2004). Xanthophores appear to fill in gaps between melanophores on the lateral migration pathway (i.e. under the epidermis). Iridophores, leucophores, or both decorate the Dorsal, Ventral and, in zebrafish, Yolk Sac stripes. The pattern appears to show many evolutionarily strongly conserved elements, although a wide-ranging, detailed phylogenetic analysis of these patterns is required. Thus, for example, the patterns in all *Danio* species examined are virtually indistinguishable from one another (McClure, 1999; Quigley *et al.*, 2004). Furthermore, even in medaka and common carp (*Cyprinus carpio*), three melanophore stripes are unambiguously recognisable, and the fourth may show an underlying homology (Fig. 1; RN Kelsh, unpublished data). Likewise, in these species, extensive flank xanthophores lying between the dermis and epidermis are the norm. Iridophores, and where present leucophores, are closely associated with the melanophore stripes and form distinctive patterns (Fig. 1). Careful study of the phylogenetic distribution of these cell types, which may be closely related (see Kelsh, 2004; Bagnara, 2005), is certainly required.

How the patterned distribution of any cell-type is determined is a classic problem in developmental biology. The process cannot be demonstrated in a more visual way than in the case of patterning of pigment cells. Nevertheless, the mechanisms guiding pigment pattern formation of embryonic pigment cells have been obscure. Two early studies of Lateral stripe patterning indicated that both pigmented melanophores and initially unpigmented cells, melanoblasts or multipotent precursors, together generated the pattern (Milos & Dingle, 1978a, b). Furthermore, these authors suggested that some sort of regulative process controlled the local density of melanophores in this stripe. A recent study identifies a chemokine, Sdf1a, as crucial in driving Lateral stripe formation. Thus, by taking a genetic approach to studying pigment pattern formation, Svétic and colleagues (Svétic *et al.*, 2007) analysed the cell-biological and molecular basis of two pigment patterning defects in *choker* mutant zebrafish. In these mutants, melanophores do not occur where the Lateral stripe should form, but ectopic melanophores are found in a collar in the anterior trunk. Time-lapse studies reveal that collar melanophores immigrate from the Dorsal and Ventral stripes, suggesting that in normal development the Dorsal and Ventral stripe melanophores are actively maintained by 'shutting off' the lateral migration pathway. In *choker* mutants, this shut-down either fails to occur, or is over-ridden. The molecular basis for Lateral stripe development in wild-type embryos and collar development in *choker* mutants lies with expression of the secreted chemokine, Sdf1a. In wild-type embryos, sdf1a is expressed in the vicinity of the horizontal myoseptum where it appears to act as a chemoattractant for melanophores to make the Lateral stripe. In *choker* mutants, however, sdf1a is absent from the horizontal myoseptum and is expressed instead in the anterior trunk, where it attracts dorsal and ventral melanophores to form the collar. Interestingly, melanophores themselves do not express detectable levels of the best characterised Sdf1 receptor, Cxcr4, so it remains to be seen whether Sdf1a influences pigment patterning directly, through an alternative receptor, or indirectly, via an intermediate cell type that is itself patterned directly by Sdf1a. The mechanisms directing melanophores to other stripes and maintaining their pattern, as well as mechanisms patterning other embryonic pigment cells, remain to be elucidated.

ADULT PIGMENT PATTERN FORMATION

Pigment patterns change markedly during the life cycles of many organisms including teleosts. In some species, such as the zebrafish, these changes are co-ordinated with others affecting a wide variety of traits at metamorphosis; in other species, formation of the definitive adult pigment pattern occurs subsequently, often at sexual maturity. Pigment pattern metamorphosis has been studied most extensively in zebrafish, in which the larval pattern of stripes begins to be remodelled at about two weeks post fertilisation (pf), with the completion of the juvenile pigment pattern by about four weeks pf (Johnson *et al.*, 1995; Parichy & Turner, 2003a). During the intervening time, new melanophores arise dispersed over the flank, outside of the early larval melanophore stripes. Some of these "early metamorphic" melanophores migrate short distances to the future sites of adult melanophore stripe formation. Subsequently, additional "late metamorphic" melanophores differentiate within the nascent adult stripes. As these events are occurring, many of the early larval melanophores die by apoptosis or other mechanisms, although some move short distances into the forming adult stripes. The result is a new pattern of two adult "primary" melanophore stripes bordering a light interstripe of xanthophores. Iridophores occur in both the stripes and the interstripe. During later growth, additional "secondary" melanophore stripes are added dorsally and ventrally.

Other fishes, and even other danios, develop a very different pigment pattern during metamorphosis. For example, medaka and pearl danio (*D. albolineatus*) develop uniform patterns. By contrast, *D. choprae* develop two horizontal stripes similar to zebrafish during metamorphosis, but develop a vertical barring pattern during later life. These examples highlight the diversity of adult pigment pattern formation and raise the dual questions of what mechanisms might be conserved among species, and how other mechanisms have changed evolutionarily. Answers to these questions are starting to be revealed, principally through studies of zebrafish and its relatives.

Genetic Requirements of Adult Chromatophore Development

Embryonic/early larval chromatophores and metamorphic chromatophores might be expected to have some commonalities, and some differences, in their genetic requirements. Studies of zebrafish have started to reveal these requirements. For example, several mutants lack particular chromatophore classes both following embryogenesis and after metamorphosis: *nacre/mitfa* (melanophores), *panther/csf1r* (xanthophores), and *shady/alk* (iridophores) (for examples, see Fig. 3). Other mutants identify loci required only by one or the other population. One such locus is *puma* for which mutants develop normal embryonic/early larval pigment patterns, but fail to develop most early and late metamorphic melanophores (Parichy & Turner, 2003a; Parichy *et al.*, 2003). The *puma* mutant phenotype is both temperature-sensitive and growth rate-dependent, exhibiting a spectrum of severity from complete absence of metamorphic melanophores at restrictive temperature or fast growth rates, to a wild-type complement of metamorphic melanophores and stripes at low temperature and slow growth rates. This temperature sensitivity allowed the identification of a critical period for *puma* activity during pigment pattern metamorphosis, and cell transplant studies revealed a role for *puma* that is intrinsic to the metamorphic melanophore lineage.

A second locus required specifically by metamorphic melanophores is *picasso*, which encodes a receptor tyrosine kinase, Erbb3 (Lyons *et al.*, 2005; E Budi & DM Parichy, unpublished data). Like *puma*, *picasso/erbb3* mutants develop normal embryonic/early larval pigment patterns then fail to develop most metamorphic melanophores (Quigley *et al.*, 2004). Despite the phenotypic similarity of *puma* and *picasso/erbb3* mutants, including additional shared defects in peripheral nervous system development, several lines of evidence point to different roles for the two loci in metamorphic melanophore development. For example, experiments with a pharmacological inhibitor show that *picasso/erbb3* is required during embryogenesis for adult pigment pattern formation, in contrast to the later, metamorphic requirement for *puma* (E Budi & DM Parichy, unpublished data). Thus, *picasso/erbb3* may be required to establish a population of precursor cells whereas *puma* contributes to recruiting these cells as metamorphic melanophores.

Fig. 3 Selected zebrafish (*Danio rerio*) adult pigment pattern mutants. Extraordinary variation in the numbers and arrangements of chromatophores provide a powerful resource for dissecting the mechanisms of adult pigment pattern formation. See text for details.

Beyond genes required by all metamorphic melanophores, additional studies have revealed genetically distinct subclasses within the metamorphic melanophore population. For example, *sparse/kita* and *rose/endothelin receptor b1* (*ednrb1*) mutants each develop about half the normal complement of metamorphic melanophores, but the timing of melanophore appearance differs. Genetic analyses show that *sparse/kit* is required specifically by early metamorphic melanophores that arise dispersed over the flank then migrate into stripes, whereas *rose/ednrb1* is required by specifically by late metamorphic melanophores that develop in stripes (Johnson *et al.*, 1995; Parichy *et al.*, 2000a). Genetically distinct subpopulations of metamorphic melanophores are found in at least one other danio, the pearl danio, and may be general for the group (Mills *et al.*, 2007). Conceivably, such populations have allowed additional flexibility in the sorts of patterns that can be generated evolutionarily.

Cellular and Physiological Mechanisms of Adult Pigment Pattern Development and Homeostasis

For a pigment pattern to form, not only must the appropriate chromatophores develop, the cells must be correctly organised. In zebrafish, this requires a complex interplay between the temporal and spatial differentiation of chromatophores, as well as interactions among the chromatophores themselves. Insights into these mechanisms have in part come from studies of *panther/csf1r* mutants (Parichy *et al.*, 2000b; Parichy

& Turner, 2003b). Adult *panther/csf1r* mutants have fewer metamorphic melanophores and these cells fail to migrate into well-defined stripes. These fish also completely lack xanthophores, and cell transplants show it is this absence of xanthophores that underlies the melanophore patterning defect. For example, when cells are transplanted from wild-type fish into *panther/csf1r* mutants, stripes of host melanophores form in regions where donor cells differentiate into xanthophores. The same result is seen when cells are transplanted between *nacre/mitfa* mutants (which lack melanophores but develop xanthophores), and *panther/csf1r* mutants (which lack xanthophores but develop melanophores). When the two cell types from different genetic backgrounds are juxtaposed, an organised stripe pattern forms.

Defects in adult patterning also are seen in *leopard/connexin41.8* mutants (Watanabe *et al.*, 2006) and *jaguar/obelix/kir7.1* mutants (Iwashita *et al.*, 2006). As suggested by its name, *leopard/connexin41.8* mutants develop melanophore spots rather than stripes, though an allelic series reveals a wide range of phenotypes from slightly irregular stripes to uniform patterns in which melanophores and xanthophores are intermingled (Asai *et al.*, 1999). In contrast, *jaguar/obelix/kir7.1* mutants develop only two broad melanophore stripes that are defective in containing xanthophores as well. Phenotypes of these two mutants have been interpreted as reflecting defects in stripe boundary formation, shape and integrity (Maderspacher & Nüsslein-Volhard, 2003). That *connexin41.8* encodes a gap junction protein whereas *kir7.1* encodes a potassium channel raises the possibility that gap junctional communication, perhaps modulating intracellular ionic concentrations, influences the morphogenesis or differentiation of these chromatophore lineages with the resulting pattern defects in mutants.

A final interesting, yet poorly understood, issue in adult pigment pattern formation concerns roles for neuronal inputs and hormones. While these factors have been most studied for mediating physiological colour change by the intracellular translocation of pigment granules, increasing evidence points to roles in determining the numbers and perhaps arrangements of chromatophores. Physiological colour change can occur rapidly, via neuronal input, and more gradually, via hormonal effects (Fujii, 2000). Among the neuronal mediators is norepinephrine, produced locally by sympathetic neurons that innervate melanophores, and causing rapid aggregation of melanin granules (melanosomes) to the cell center. Among the hormonal mediators are melanin-concentrating hormone (MCH) and α-melanocyte stimulating hormone (α-MSH), both secreted by the pituitary, which cause melanosome aggregation and dispersion, respectively. These and other effectors cause fish to become lighter overall, by melanosome aggregation, or darker overall, by melanosome dispersion, via mechanisms utilising both microtubules and the actin cytoskeleton (Logan *et al.*, 2005).

That a correlation exists between physiological colour change (affecting pigment granules) and morphological colour change (affecting cell numbers and arrangements) is illustrated nicely by studies of background adaptation both in wild-type and mutant zebrafish (for an excellent review, see Sugimoto, 2002). When fish are kept on a dark background they respond physiologically by dispersing melanosomes and over days to weeks this is followed by increases in melanophore numbers, possibly owing to effects of α-MSH. By contrast, when fish are kept on a light background they respond physiologically by aggregating melanosomes, initially in response to neuronal secretion of norepinephrine and over longer periods in response to increased MCH. Over days to weeks, melanophore numbers drop due to more frequent apopotosis (Sugimoto & Oshima, 1995; Sugimoto *et al.*, 2000, 2005). While these and similar studies have been conducted with wild-type fish, an intriguing genetic link between physiological and morphological change comes from the *jaguar/obelix/kir7.1* mutant zebrafish (Iwashita *et al.*, 2006). These fish exhibit not only defective melanophore stripes, but the melanophores within these stripes show an impaired response to signals that otherwise cause melanosome dispersion, raising the possibility that a common pathway is involved in both the physiological and morphological phenotypes.

Two final, potential intersections between pigment pattern development and organismal physiology are worthy of mention. The first of these also concerns α-MSH. In amniotes, α-MSH and its receptor melanocortin 1 receptor (MC1R) are well known for promoting melanisation and melanocyte development (reviewed by Slominski *et al.*, 2004). Furthermore the light ventrum of many amniotes results from local production of Agouti signalling protein, which antagonises MC1R signalling (Vrieling *et al.*, 1994). A homologous locus in teleosts is similarly expressed ventrally and similarly antagonises melanocortin receptors, raising the possibility that it functions in promoting the lighter ventrum in teleosts as well (Cerdá-Reverter *et al.*, 2005). The second interaction concerns roles for thyroid hormone signalling in pigment pattern formation. Thyroid hormones play a pivotal role in amphibian metamorphosis (reviewed by Shi, 2000) though their role in teleost metamorphosis remains understudied (see Power *et al.*, this volume). When thyroid hormone levels are manipulated pharmacologically, pigment cell differentiation and morphogenesis are perturbed (Brown, 1997; Liu & Chan, 2002; Hamre *et al.*, 2007; DM Parichy, unpublished data). Thyroid hormone receptors heterodimerise with retinoid-X receptors, and perhaps for this reason pharmacological manipulation of retinoids also can cause pigment cell defects. To date, however, it remains uncertain whether these effects are due to autonomous changes within the pigment cell lineages, local changes in the tissue environment encountered by pigment cells, or whole-animal changes in growth, development and physiology.

FUTURE PERSPECTIVES

This brief summary of the origins, differentiation and patterning of fish chromatophores makes it clear that there is much still to be learnt about these important and beautiful cells. One key question that remains is to establish definitively the lineage relationships between chromatophore precursors and, in particular, testing the chromatoblast concept of a shared progenitor for all pigment cell types. Further genetic dissection of neural crest development may provide a breakthrough here, but it seems likely that the use of clonal cell culture systems *in vitro* may be necessary to overcome the limitations of *in vivo* clone size. Clearly, detailed analysis of the gene regulatory networks underlying development of each pigment cell-type will be required and this may be furthered by the ability to sort cells of different types by Fluorescence Activated Cell Sorting from transgenic fish expressing fluorescent marker proteins (Carney *et al.*, 2006; Zou *et al.*, 2006). A major breakthrough would come from identification of the identity and localisation of pigment cell stem cells and for this the development of a powerful chemical tool to ablate melanophores may well prove helpful (Yang & Johnson, 2006). From the perspective of evolutionary developmental biology, the growth in the number and accuracy of completely sequenced fish genomes, as well as mechanisms of pigment pattern development in model species, will provide developmental biologists and geneticists with unparalleled resources for understanding the factors underlying pigment pattern diversification.

References

Amsterdam A, Burgess S, Golling G, Chen W, Sun Z, Townsend K, Farrington S, Haldi M, Hopkins N (1999) A large-scale insertional mutagenesis screen in zebrafish. Genes Dev 13: 2713-2724

Amsterdam A, Nissen RM, Sun Z, Swindell EC, Farrington S, Hopkins N (2004) Identification of 315 genes essential for early zebrafish development. Proc Nat Acad Sci USA 101: 12792-12797

Amsterdam A, Nissen RM, Sun Z, Swindell E, Farrington S, Hopkins N (2005) Submission and curation of data from an insertional mutagenesis screen. ZFIN Direct Data Submission. ZFIN Direct Data Submission

Arduini BL, Henion PD (2004) Melanophore sublineage-specific requirement for zebrafish touchtone during neural crest development. Mech Dev 121: 1353-1364

Artinger KB, Chitnis AB, Mercola M, Driever W (1999) Zebrafish *narrowminded* suggests a genetic link between formation of neural crest and primary sensory neurons. Development 126: 3969-3979

Asai R, Taguchi E, Kume Y, Saito M, Kondo S (1999) Zebrafish *leopard* gene as a component of the putative reaction-diffusion system. Mech Dev 89: 87-92

Bagnara JT, Matsumoto J, Ferris W, Frost SK, Turner WA, Tchen TT, Taylor JD (1979) Common origin of pigment cells. Science 203: 410-415

Bagnara JT, Matsumoto J (2006) Comparative anatomy and physiology of pigment cells in nonmammalian tissues. In: Nordlund JJ, Boissy RE, Hearing VJ, King RA, Oetting WS, Ortonne J-P (eds) The Pigmentary System: Physiology and Pathophysiology, 2nd Edition. Oxford University Press, New York, p 11-59

Bagnara JT, Fujii PJ, Fujii R (2007) On the blue coloration of vertebrates. Pigment Cell Res 20: 14-26

Barrallo-Gimeno A, Holzschuh J, Driever W, Knapik EW (2004) Neural crest survival and differentiation in zebrafish depends on *mont blanc/tfap2a* gene function. Development 131: 1463-1477

Bondurand N, Pingault V, Goerich DE, Lemort N, Sock E, Caignec CL, Wegner M, Goossens M (2000) Interaction among SOX10, PAX3 and MITF, three genes altered in Waardenburg syndrome. Hum Mol Genet 9: 1907-1917

Brown DD (1997) The role of thyroid hormone in zebrafish and axolotl development. Proc Natl Acad Sci USA 94: 13011-13016

Carney TC, Dutton KA, Greenhill E, Delfino-Machin M, Dufourq P, Blader P, Kelsh RN (2006) A direct role for Sox10 in specification of neural crest-derived sensory neurons. Development 133: 4619-4630

Cornell RA, Yemm E, Bonde G, Li W, d'Alencon C, Wegman L, Eisen J, Zahs A (2004) *Touchtone* promotes survival of embryonic melanophores in zebrafish. Mech Dev 121: 1365-1376

Cretekos CJ, Grunwald DJ (1999) *alyron*, an insertional mutation affecting early neural crest development in zebrafish. Dev Biol 210: 322-338

Dorsky RI, Raible DW, Moon RT (2000) Direct regulation of *nacre*, a zebrafish *MITF* homolog required for pigment cell formation, by the Wnt pathway. Genes Dev 14: 158-162

Dutton KA, Pauliny A, Lopes SS, Elworthy S, Carney TJ, Rauch J, Geisler R, Haffter P, Kelsh RN (2001) Zebrafish *colourless* encodes *sox10* and specifies non-ectomesenchymal neural crest fates. Development 128: 4113-4125

Elizondo MR, Arduini BL, Paulsen J, MacDonald EL, Sabel JL, Henion PD, Cornell RA, Parichy DM (2005) Defective skeletogenesis with kidney stone formation in dwarf zebrafish mutant for *trpm7*. Curr Biol 15: 667-671

Elworthy S, Lister JA, Carney TJ, Raible DW, Kelsh RN (2003) Transcriptional regulation of *mitfa* accounts for the *sox10* requirement in zebrafish melanophore development. Development 130: 2809-2818

Fujii R (2000) The regulation of motile activity in fish Chromatophores. Pigment Cell Res 13: 300-319

Fukamachi S, Shimada A, Shima A (2001) Mutations in the gene encoding B, a novel transporter protein, reduce melanin content in medaka. Nat Genet 28: 381-385

Fukamachi S, Asakawa S, Wakamatsu Y, Shimizu N, Mitani H, Shima A (2004a) Conserved function of medaka *pink-eyed dilution* in melanin synthesis and its divergent transcriptional regulation in gonads among vertebrates. Genetics 168: 1519-1527

Fukamachi S, Sugimoto M, Mitani H, Shima A (2004b) Somatolactin selectively regulates proliferation and morphogenesis of neural-crest derived pigment cells in medaka. Proc Nat Acad Sci USA 101: 10661-10666

Genner MJ, Turner GF (2005) The mbuna cichlids of Lake Malawi: a model for rapid speciation and adaptive radiation. Fish and Fisheries 6: 1-34

Gilbert SF (2006) Developmental Biology, 8th Edition. Sinauer, Sunderland MA

Golling G, Amsterdam A, Sun Z, Antonelli M, Maldonado E, Chen W, Burgess S, Haldi M, Artzt K, Farrington S, Lin SY, Nissen RM, Hopkins N (2002) Insertional mutagenesis in zebrafish rapidly identifies genes essential for early vertebrate development. Nat Genet 31: 135-140

Hamre K, Holen E, Moren M (2007) Pigmentation and eye migration in Atlantic halibut (*Hippoglossus hippoglossus* L.) larvae: new findings and hypotheses. Aquaculture Nutrition 13: 65-80

Hernandez-Lagunas L, Choi IF, Kaji T, Simpson P, Hershey C, Zhou Y, Zon L, Mercola M, Artinger KB (2005) Zebrafish *narrowminded* disrupts the transcription factor *prdm1* and is required for neural crest and sensory neuron specification. Dev Biol 278: 347-357

Hong SK, Haldin CE, Lawson ND, Weinstein BM, Dawid IB, Hukriede NA (2005) The zebrafish kohtalo/trap230 gene is required for the development of the brain, neural crest, and pronephric kidney. Proc Nat Acad Sci USA 102: 18473-18478

Hörstadius S (1950) The Neural Crest. Oxford University Press, London

Hou L, Arnheiter H, Pavan WJ (2006) Interspecies difference in the regulation of melanocyte development by SOX10 and MITF. Proc Nat Acad Sci USA 103: 9081-9085

Itoh M, Kim CH, Palardy G, Oda T, Jiang YJ, Maust D, Yeo SY, Lorick K, Wright GJ, Ariza-McNaughton L, Weissman AM, Lewis J, Chandrasekharappa SC, Chitnis AB (2003) Mind bomb is a ubiquitin ligase that is essential for efficient activation of Notch signaling by Delta. Dev Cell 4: 67-82

Iwashita M, Watanabe M, Ishii M, Chen T, Johnson SL, Kurachi Y, Okada N, Kondo S (2006) Pigment pattern in *jaguar/obelix* zebrafish is caused by a Kir7.1 mutation: Implications for the regulation of melanosome movement. PLoS Genetics 2: 1861-1870

Jeffery WR, Strickler AG, Yamamoto Y (2004) Migratory neural crest-like cells form body pigmentation in a urochordate embryo. Nature 431: 696-699

Johnson SL, Africa D, Walker C, Weston JA (1995) Genetic control of adult pigment stripe development in zebrafish. Dev. Biol. 167: 27-33

Kasukawa H, Oshima N (1987) Divisionistic generation of skin hue and the change of shade in the scalycheek damselfish, *Pomacentrus lepidogenys*. Pigment Cell Res 1: 152-157

Kawakami K, Amsterdam A, Shimoda N, Becker T, Mugg J, Shima A, Hopkins N (2000) Proviral insertions in the zebrafish *hagoromo* gene, encoding an F-box/WD40-repeat protein, cause stripe pattern anomalies. Curr Biol 10: 436-466

Kawauchi A (2006) Functions of melanin-concentrating hormone in fish. J Exp Zool 305A: 751-760

Kelsh RN (2004) Genetics and evolution of pigment patterns in fish. Pigment Cell Res 17: 326-336

Kelsh RN (2006) Sorting out *Sox10* functions in neural crest. Bioessays 28: 788-798

Kelsh RN, Eisen JS (2000) The zebrafish *colourless* gene regulates development of non-ectomesenchymal neural crest derivatives. Development 127: 515-525

Kelsh RN, Raible DW (2002) Specification of zebrafish neural crest. In: Solnicka-Kresel L (ed) Pattern Formation in Zebrafish. Springer-Verlag, New York, p 216-236

Kelsh RN, Brand M, Jiang YJ, Heisenberg CP, Lin S, Haffter P, Odenthal J, Mullins MC, van Eeden FJ, Furutani-Seiki M, Granato M, Hammerschmidt M, Kane DA, Warga RM, Beuchle D, Vogelsang L, Nüsslein-Volhard C (1996) Zebrafish pigmentation mutations and the processes of neural crest development. Development 123: 369-389

Kelsh RN, Schmid B, Eisen JS (2000) Genetic analysis of melanophore development in zebrafish embryos. Dev Biol 225: 277-293

Kelsh RN, Inoue C, Momoi A, Kondoh H, Furutani-Seiki M, Ozato K, Wakamatsu Y (2004) The Tomita collection of medaka pigmentation mutants as a resource for understanding neural crest cell development. Mech Dev 121: 841-859

Kim J, Lo L, Dormand E, Anderson DJ (2003) SOX10 maintains multipotency and inhibits neuronal differentiation of neural crest stem cells. Neuron 38: 17-31

Kimler VA, Taylor JD (2002) Morphological studies on the mechanisms of pigmentary organelle transport in fish xanthophores and melanophores. Microsc Res Tech 58: 470-480

Kimmel CB, Ballard WW, Kimmel SR, Ullmann B, Schilling TF (1995) Stages of embryonic development of the zebrafish. Dev Dyn 203: 253-310

Knight RD, Javidan Y, Nelson S, Zhang T, Schilling T (2004) Skeletal and pigment cell defects in the *lockjaw* mutant reveal multiple roles for zebrafish *tfap2a* in neural crest development. Dev Dyn 229: 87-98

Lamason RL, Mohideen MA, Mest JR, Wong AC, Norton HL, Aros MC, Jurynec MJ, Mao X, Humphreville VR, Humbert JE, Sinha S, Moore JL, Jagadeeswaran P, Zhao W, Ning G, Makalowska I, McKeigue PM, O'Donnell D, Kittles R, Parra EJ, Mangini NJ, Grunwald DJ, Shriver MD, Canfield VA, Cheng KC (2005) SLC24A5, a putative cation exchanger, affects pigmentation in zebrafish and humans. Science 310: 1782-1786

Lamoreux LM, Kelsh RN, Wakamatsu Y, Ozato K (2005) Pigment pattern formation in the medaka embryo. Pigment Cell Res 18: 64-73

Le Douarin NM, Kalcheim C (1999) The Neural Crest. Cambridge University Press, Cambridge

Le Guyader S, Maier J, Jesuthasan S (2005) *Esrom*, an ortholog of PAM (protein associated with *c-myc*), regulates pteridine synthesis in the zebrafish. Dev Biol 277: 378-386

Lee M, Goodall J, Verastegui C, Ballotti R, Goding CR (2000) Direct regulation of the microphthalmia promoter by Sox10 links Waardenburg-Shah syndrome (WS4)-associated hypopigmentation and deafness to WS2. J Biol Chem 275: 37978-37983

Lindholm AK, Brooks R, Breden F (2004) Extreme polymorphism in a Y-linked sexually selected trait. Heredity 92: 156-162

Lister JA, Cooper C, Nguyen K, Modrell M, Grant K, Raible DW (2006) Zebrafish Foxd3 is required for development of a subset of neural crest derivatives. Dev Biol 290: 92-104

Lister JA, Robertson CP, Lepage T, Johnson SL, Raible DW (1999) *nacre* encodes a zebrafish microphthalmia-related protein that regulates neural-crest-derived pigment cell fate. Development 126: 3757-3767

Liu Y-W, Chan W-K (2002) Thyroid hormones are important for embryonic to larval transitory phase in zebrafish. Differentiation 70: 36-45

Logan DW, Burn SF, Jackson IJ (2006) Regulation of pigmentation in zebrafish melanophores. Pigment Cell Res 19: 206-213

Lyons, DA, Pogoda HM, Voas MG, Woods IG, Diamond B, Nix R, Arana N, Jacobs J, Talbot WS (2005) erbb3 and erbb2 are essential for schwann cell migration and myelination in zebrafish. Curr Biol 15: 513-524

Maderspacher, F, Nüsslein-Volhard C (2003) Formation of the adult pigment pattern in zebrafish requires *leopard* and *obelix* dependent cell interactions. Development 130: 3447-3457

Maldonado E, Hernandez F, Lozano C, Castro ME, Navarro RE (2006) The zebrafish mutant *vps18* as a model for vesicle-traffic related hypopigmentation diseases. Pigment Cell Res 19: 315-326

McClure M (1999) Development and evolution of melanophore patterns in fishes of the genus *Danio* (Teleostei: Cyprinidae) J Morphol 241: 83-105

Mills MG, Nuckels RJ, Parichy DM (2007) Deconstructing evolution of adult phenotypes: genetic analyses of *kit* reveal homology and evolutionary novelty during adult pigment pattern development of *Danio* fishes. Development 134: 1081-1090

Milos N, Dingle AD (1978a) Dynamics of pigment pattern formation in the zebrafish, *Brachydanio rerio*. I. Establishment and regulation of the lateral line melanophore stripe during the first eight days of development. J Exp Zool 205: 205-216

Milos N, Dingle AD (1978b) Dynamics of pigment pattern formation in the zebrafish, *Brachydanio rerio*. II. Lability of lateral line stripe formation and regulation of pattern defects. J Exp Zool 205: 217-224

Montero-Balaguer M, Lang MR, Sachdev SW, Knappmeyer C, Stewart RA, La Guardia A, Hatzopoulos AK, Knapik EW (2006) The *mother superior* mutation ablates *foxd3* activity in neural crest progenitor cells and depletes neural crest derivatives in zebrafish. Dev Dyn 235: 3199-3212

O'Brien EK, d'Alencon C, Bonde G, Li W, Schoenebeck J, Allende ML, Gelb BD, Yelon D, Eisen JS, Cornell RA (2004) Transcription factor Ap-2α is necessary for development of embryonic melanophores, autonomic neurons and pharyngeal skeleton in zebrafish. Dev Biol 265: 246-261

Odenthal J, Rossnagel K, Haffter P, Kelsh R, Vogelsang E, Brand M, van Eeden F, Furutani-Seiki M, Granato M, Hammerschmidt M, Heisenberg C, Jiang Y, Kane D, Mullins M, Nüsslein-Volhard C (1996) Mutations affecting xanthophore pigmentation in the zebrafish, *Danio rerio*. Development 123: 391-398

Page-McCaw PS, Chung SC, Muto A, Roeser T, Staub W, Finger-Baier KC, Korenbrot JI, Baier H (2004) Retinal network adaptation to bright light requires tyrosinase. Nat Neurosci 7: 1329-1336

Parichy DM (2003) Pigment patterns: fish in stripes and spots. Curr Biol 13: R947-950

Parichy DM (2006) Evolution of danio pigment pattern development. Heredity 97: 200-210

Parichy DM, Turner JM (2003a) Zebrafish *puma* mutant decouples pigment pattern and somatic metamorphosis. Dev Biol 256: 242-257

Parichy DM, Turner JM (2003b) Temporal and cellular requirements for Fms signaling during zebrafish adult pigment pattern development. Development 130: 817-833

Parichy DM, Rawls JF, Pratt SJ, Whitfield TT, Johnson SL (1999) Zebrafish *sparse* corresponds to an orthologue of c-*kit* and is required for the morphogenesis of a subpopulation of melanocytes, but is not essential for hematopoiesis or primordial germ cell development. Development 126: 3425-3436

Parichy DM, Ransom DG, Paw B, Zon LI, Johnson SL (2000a) An orthologue of the *kit*-related gene *fms* is required for development of neural crest-derived xanthophores and a subpopulation of adult melanocytes in the zebrafish, *Danio rerio*. Development 127: 3031-3044

Parichy DM, Mellgren EM, Rawls JF, Lopes SS, Kelsh RN, Johnson SL (2000b) Mutational analysis of *endothelin receptor b1* (*rose*) during neural crest and pigment pattern development in the zebrafish *Danio rerio*. Dev Biol 227: 294-306

Parichy DM, Turner JM, Parker NB (2003) Essential role for *puma* in development of postembryonic neural crest-derived cell lineages in zebrafish. Dev Biol 256: 221-241

Parichy DM, Reedy MV, Erickson CA (2006) Regulation of melanoblast migration and differentiation. In: Nordlund JJ, Boissy RE Hearing VJ, King RA, Ortonne JP (eds) The Pigmentary System: Physiology and Pathophysiology, 2nd Edition. Oxford University Press, New York p 108-139

Pickart MA, Sivasubbu S, Nielsen AL, Shriram S, King RA, Ekker SC (2004) Functional genomics tools for the analysis of zebrafish pigment. Pigment Cell Res 17: 461-470

Potterf SB, Furumura M, Dunn KJ, Arnheiter H, Pavan WJ (2000) Transcription factor hierarchy in Waardenburg syndrome: regulation of MITF expression by SOX10 and PAX3. Hum Genet 107: 1-6

Quigley IK, Parichy DM (2002) Pigment pattern formation in zebrafish: a model for developmental genetics and the evolution of form. Microsc Res Tech 58: 442-455

Quigley IK, Turner JM, Nuckels RJ, Manuel JL, Budi EH, Macdonald EL, Parichy DM (2004) Pigment pattern evolution by differential deployment of neural crest and post-embryonic melanophore lineages in *Danio* fishes. Development 131: 6053-6069

Quigley IK, Manuel JL, Roberts RA, Nuckels RJ, Herrington ER, Macdonald EL, Parichy DM (2005) Evolutionary diversification of pigment pattern in *Danio* fishes: differential *fms* dependence and stripe loss in *D. albolineatus*. Development 132: 89-104

Raible DW, Eisen JS (1994) Restriction of neural crest cell fate in the trunk of the embryonic zebrafish. Development 120: 495-503

Ransom DG, Bahary N, Niss K, Traver D, Burns C, Trede NS, Paffett-Lugassy N, Saganic WJ, Lim CA, Hersey C, Zhou Y, Barut BA, Lin S, Kingsley PD, Palis J, Orkin SH, Zon LI (2004) The zebrafish *moonshine* gene encodes transcriptional intermediary factor 1γ, an essential regulator of hematopoiesis. PLoS Biol 2: E237

Rau MJ, Fischer S, Neumann CJ (2006) Zebrafish Trap230/Med12 is required as a coactivator for Sox9-dependent neural crest, cartilage and ear development. Dev Biol 296: 83-93

Roy S, Ng T (2004) Blimp-1 specifies neural crest and sensory neuron progenitors in the zebrafish embryo. Curr Biol 14: 1772-1777

Schilling TF, Kimmel CB (1994) Segment and cell type lineage restrictions during pharyngeal arch development in the zebrafish embryo. Development 120: 483-494

Schonthaler HB, Lampert JM, von Lintig J, Schwarz H, Geisler R, Neuhauss SC (2005) A mutation in the silver gene leads to defects in melanosome biogenesis and alterations in the visual system in the zebrafish mutant *fading vision*. Dev Biol 284: 421-436

Shi Y-B (2000) Amphibian Metamorphosis: From Morphology to Molecular Biology. Wiley-Liss, New York

Slominski A, Tobin DJ, Shibahara S, Wortsman J (2004) Melanin pigmentation in mammalian skin and its hormonal regulation. Physiol Rev 84: 1155-1228

Steingrimsson E, Moore KJ, Lamoreux ML, Ferre-D'Amare AR, Burley SK, Zimring DC, Skow LC, Hodgkinson CA, Arnheiter H, Copeland NG, *et al.* (1994) Molecular basis of mouse *microphthalmia* (*mi*) mutations helps explain their developmental and phenotypic consequences. Nat Genet 8: 256-263

Stewart RA, Arduini BL, Berghmans S, George RE, Kanki JP, Henion PD, Look AT (2006) Zebrafish *foxd3* is selectively required for neural crest specification, migration and survival. Dev Biol 292: 174-188

Streisinger G, Walker C, Dower N, Knauber D, Singer F (1981) Production of clones of homozygous diploid zebrafish (*Brachydanio rerio*) Nature 291: 293-296

Sugimoto M (2002) Morphological color changes in fish: regulation of pigment cell density and morphology. Microsc Res Tech 58: 496-503

Sugimoto M, Oshima N (1995) Changes in adrenergic innervation to chromatophores during prolonged background adaptation in the medaka, *Oryzias latipes*. Pigment Cell Res 8: 37-45

Sugimoto M, Uchida N, Hatayama M (2000) Apoptosis in skin pigment cells of the medaka, *Oryzias latipes* (Teleostei), during long-term chromatic adaptation: the role of sympathetic innervation. Cell Tissue Res 301: 205-216

Sugimoto M, Yuki M, Miyakoshi T, Maruko K (2005) The influence of long-term chromatic adaptation on pigment cells and striped pigment patterns in the skin of the zebrafish, *Danio rerio*. J Exp Zool 303A: 430-440

Takeda K, Yasumoto K, Takada R, Takada S, Watanabe K, Udono T, Saito H, Takahashi K, Shibahara S (2000) Induction of melanocyte-specific microphthalmia-associated transcription factor by Wnt-3a. J Biol Chem 275: 14013-14016

Vaglia JL, Hall BK (2000) Patterns of migration and regulation of trunk neural crest cells in zebrafish (*Danio rerio*). Int J Dev Biol 44: 867-881

Vrieling H, Duhl DM, Millar SE, Miller KA, Barsh GS (1994) Differences in dorsal and ventral pigmentation result from regional expression of the mouse *agouti* gene. Proc Nat Acad Sci USA 91: 5667-5671

Watanabe M, Iwashita M, Ishii M, Kurachi Y, Kawakami A, Kondo S, Okada N (2006) Spot pattern of *leopard Danio* is caused by mutation in the zebrafish *connexin41.8* gene. EMBO Rep 7: 893-897

Yan YL, Willoughby J, Liu D, Crump JG, Wilson C, Miller CT, Singer A, Kimmel C, Westerfield M, Postlethwait JH (2005) A pair of Sox: distinct and overlapping functions of zebrafish *sox9* co-orthologs in craniofacial and pectoral fin development. Development 132: 1069-1083

Yang CT and Johnson SJ (2006) Small molecule-induced ablation and subsequent regeneration of larval zebrafish melanocytes. Development 133: 3563-3573

Yu JF, Fukumachi S, Mitani H, Hori H, Kanamori A (2006) Reduced expression of *vps11* causes less pigmentation in medaka, *Oryzias latipes*. Pigment Cell Res 19: 628-634

Zou J, Beermann F, Wang J, Kawakami K, Wei X (2006) The Fugu *tyrp1* promoter directs specific GFP expression in zebrafish: tools to study the RPE and the neural crest-derived melanophores. Pigment Cell Res 19: 615-627

CHAPTER

3

Bioluminescence

Andrey V Suntsov[1], Edith A Widder[2,*] and Tracey T Sutton[3]

INTRODUCTION

The production of living light – bioluminescence, is a widespread but very unevenly distributed phenomenon among living organisms. As presently known, some five terrestrial and 14 marine phyla and more than 700 genera are known to have luminous species (Herring, 1987; Herring & Widder, 2001). The morphological diversity of luminous organs, physiology and taxonomic occurrence of bioluminescence in fishes is higher than in any other group of organisms and confined exclusively to marine habitats.

A significant body of information on fish bioluminescence is available from several detailed reviews (Herring & Morin, 1978; Herring, 1982; Haygood, 1993). Historically, the characterisation of fish bioluminescence was almost exclusively confined to adults. As emphasis on early ontogeny of fishes has increased in recent decades, the development of many species with bioluminescent properties has become increasingly well documented. According to these studies, bioluminescence often arises early in ontogeny. To date, however, very few studies have focused upon the early ontogeny of

Authors' addresses: [1]Northwest Fisheries Science Center, National Oceanic and Atmospheric Administration, Newport OR 97365 USA.
[2]Ocean Research & Conservation Association, Fort Pierce FL 34949 USA.
[3]Harbor Branch Oceanographic Institution, Fort Pierce FL 34949 USA.
**Corresponding author:* E-mail: ewidder@oceanrecon.org

bioluminescence in the majority of marine fish groups. Ontogenetic aspects of bioluminescence are clearly needed for a better understanding of the mechanisms of light production in fishes, the different roles played by bioluminescence during a particular period of life, and the evolution of bioluminescence in particular groups and in fishes in general.

The purpose of this review is to synthesise the available information on various aspects of the development of bioluminescence during the early life history of marine fishes, to examine the functional significance and physiological basis for light production during early ontogeny, and to suggest directions for future studies.

PHYSIOLOGICAL BASIS FOR LIGHT PRODUCTION

Bioluminescence is an oxidative chemical reaction where the enzyme, known as luciferase and the substrate, known as luciferin, are synthesised by living organisms. Luciferin and luciferase are generic terms (derived from Lucifer, the light bearer) and require a taxon prefix to differentiate chemicals that have been isolated from different organisms. Chemically the luciferins and luciferases produced by different organisms are so dissimilar (Fig. 1) that it is apparent that bioluminescence has arisen independently multiple times during evolutionary history (Hastings, 1983). Substrates range from the simple, such as the aldehyde luciferin found in earthworms to the complex tetrapyrrole luciferin of dinoflagellates. The only commonality is that enzymes, although structurally dissimilar, are all oxygenases that require molecular oxygen to create the excited electron state that precedes light emission. Because of the reactivity of luciferins with reactive oxygen species, it has been proposed that these substrates originally evolved as detoxifiers of deleterious oxygen derivatives (Rees *et al.*, 1998). According to this theory it was only after selection for antioxidative defense mechanisms lessened, possibly because of migration into deeper waters where oxidative stress is reduced, that transitions to light-emitting function occurred. Additionally, Seliger (1993) has proposed that the oxidative breakdown of pigment molecules could produce electronically excited states capable of emitting visible light. Therefore, as fish colonised the deeper, darker depths of the ocean any pigment spots that were used for visual signalling would be enhanced by a mutation that resulted in light emission and immediately selected for. Although the colours of bioluminescence cover the entire visible spectrum, the colour most commonly found in marine organisms is the blue (~475 nm) that penetrates farthest through seawater and is therefore capable of enhancing communication over the greatest distances (Widder *et al.*, 1983).

Methods of controlling light emission are even more diverse than the light emitting chemicals. Bacteria are unusual in that their light emission is continuous rather than transient, apparently as a consequence of the light emitting reaction being a shunt of the respiratory pathway (Hastings, 1983). As a consequence, symbioses between animals and bacteria often involve the use of mechanical shutters to control the light emission. In non-symbiotic emitters, where the light emitting chemicals are intrinsic to the organism, light emission may be under either neural or hormonal control. Often a

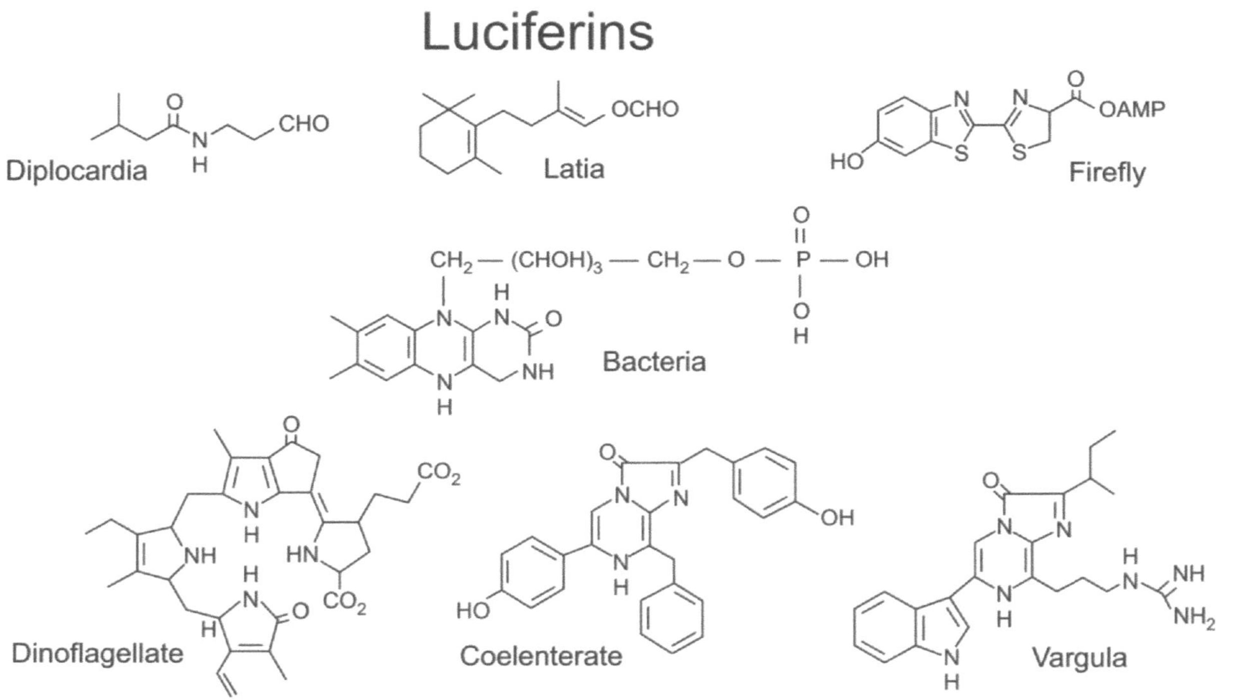

Fig. 1 Chemical structures of different luciferins redrawn from Hastings & Morin (1991). The simplest chemical structure is that of the earthworm, *Diplocardia*, while the most complex is the tetrapyrrolic luciferin extracted from photosynthetic dinoflagellates. Freshwater examples of bioluminescence are virtually unknown except for the limpet, *Latia*. Coelenterate luciferin, also known as coelenterazine because it was originally extracted from an anthozoan, is known to occur in at least seven different phyla. The ostracod, *Vargula*, releases its luciferin and luciferase into the water, while fishes exhibiting a dietary dependence on ostracod luciferin retain their light emitting chemicals internally.

co-factor is required in addition to the luciferin and luciferase, such as ATP in firefly luminescence or Ca^{2+} in coelenterates and in some cases another protein may serve as a secondary emitter that shifts the emission to longer wavelengths, such as the green fluorescent protein (GFP) extracted from some coelenterates.

Two major types of bioluminescence are known in marine fishes: 1) light production due to intrinsic luminescence, and 2) emission of light due to luminous bacteria harboured in special organs.

In the case of intrinsic luminescence the organism may either synthesise the enzyme and substrate *de novo* from basic biochemical building blocks or it may require an exogenous source of luciferin. When an exogenous source is required it is passed on maternally via the yolk or it is acquired by direct feeding later in the maturation process. Once an exogenous source of luciferin is acquired it can be recycled if the appropriate enzymes are present or if those enzymes are induced by the presence of the substrate. In such instances the exogenous source does not need to be replenished. However, if enzymes for recycling the substrate are not present, then the substrate must constantly be replenished through the diet. No examples of exogenously acquired luciferases are known, presumably because they are broken down by proteolytic enzymes during digestion.

The best documented example of dietary dependence of luminescence in fishes is in the midshipman (*Porichthys notatus*). Two factors have contributed to the extensive research that has been done on this species. The first is its accessibility. Female midshipman deposit their eggs subtidally, attached to abalone shells or in rock cavities. Males fertilise the eggs and then guard the nest throughout the approximately 40 days it takes for the juveniles to develop and become free-swimming (Anctil, 1977). This behaviour makes the nests relatively easy to find and transfer to a laboratory setting. The second is that the oxyluciferin produced by the light reaction is strongly fluorescent, a characteristic that is not shared by all bioluminescent reactions. Because fluorescence is much brighter and less transient than bioluminescence, it is easier to observe, thereby simplifying the process of monitoring dietary acquisition of the substrate.

Although near-shore bioluminescent fishes generally depend on luminous bacteria for light production (Morin, 1983), the midshipman is unusual in that its luminescence is intrinsic. Light originates from rows of hundreds of photophores distributed ventrally and ventro-laterally, along the head and trunk, and light emission is dependent on an exogenous source of luciferin derived from feeding on bioluminescent ostracods such as *Vargula hilgendorfii* and *V. tsujii* (Herring & Morin, 1978). The substrate, an imidazolopyrazine, is initially passed on maternally through the yolk, while the luciferase is synthesised *de novo* 28 days after the eggs are deposited (Tsuji *et al.*, 1972). Luciferin reserves, present when juveniles detach from the nest, can be depleted within three weeks following repeated stimulation of bioluminescence (Mensinger & Case, 1991) indicating that the enzymes needed for recycling luciferin are not present. However, when members of a non-luminescent population of midshipman were fed a single dose of *V. hilgendorfii* luciferin, they became luminescent and remained so for two years, producing more photons than could be accounted for by the amount of luciferin that was administered (Thompson *et al.*, 1987). This was interpreted as being due to the exogenously supplied luciferin inducing the synthesis of enzymes required for recycling. These investigators inferred that the recycling mechanisms were not initially present because injection of oxyluciferin, i.e. spent luciferin, to dark fish did not result in luminescence, therefore it was not recycled back to luciferin. These results may be explained either by a developmental shift in the capacity for recycling that occurs sometime after the first three weeks following nest detachment or by a different metabolic pathway for dietary compared to maternally acquired luciferin.

A dietary dependence on *Vargula*-type luciferin may also exist in shallow-water apogonid fishes of the Far East as well as several species of the genera *Parapriacanthus* and *Pempheris*, all of which lack photophores, but have light organs associated with the digestive tract and elaborate accessory structures that function as reflectors, diffusers and light guides to produce a diffuse ventral glow. As in midshipman, their luciferin and luciferase extracts cross react with *Vargula* luciferin and luciferase, suggesting that they too may require this ostracod in their diet in order to remain luminescent (Herring & Morin, 1978).

Another imidazolopyrazine luciferin, known as coelenterazine, because it was originally extracted from an anthozoan, is found in fishes as well as squids, radiolarians, a chaetognath and some crustaceans (Shimomura *et al.*, 1980; Campbell & Herring, 1990; Haddock & Case, 1994). Although similar in structure to *Vargula*-luciferin, coelenterazine does not cross react with *Vargula* luciferase, nor does *Vargula*-luciferin cross react with coelenterate luciferase. Additionally these luciferins appear to be derived from different metabolic pathways, since *Vargula* luciferin is formed from the cyclisation of the three amino acids, tryptophan, isoleucine, and arginine, while coelenterazine is assembled from two tyrosines and a phenylalanine (Herring & Gruner, 2004). A dietary dependence on coelenterazine has been documented in the deep-sea lophogastrid shrimp (*Gnathophausia ingens*) (Frank *et al.*, 1984). However, whether or not the coelenterazine that is found in fishes such as silver hatchetfish (*Argyropelecus hemigymnus*), viperfish (*Chauliodus sloani*), spotted lanternfish (*Myctophum punctatum*), slender lightfish (*Vinciguerria attenuata*) and Brauer's bristlemouth (*Cyclothone braueri*) (Mallafet & Shimomura, 1995) is acquired exogenously has yet to be established. Ironically, it was recently shown that a hydromedusa from which coelenterazine was extracted, *Aequorea victoria* (Shimomura *et al.*, 1974), is itself dependent on an exogenous source of coelenterazine, raising the possibility that this substrate is misnamed (Haddock *et al.*, 2001). Since the best evidence for *de novo* synthesis of coelenterazine is in the embryos of the decapod shrimp, *Systellaspis debilis* (Thomson *et al.*, 1995), it is possible that, like the imidazolopyrazine required by midshipman, coelenterazine is actually of crustacean origin.

Fishes that depend on bacterial symbionts for light production are less common than intrinsic light emitters, but include such well known groups as the flashlight fishes (Beryciformes), ponyfishes (Perciformes), deep-sea anglerfishes (Lophiiformes) and rattails (Gadiformes). There are three culturable bacterial symbionts that have been isolated from specialised organs of fishes; *Vibrio fischeri* and *Photobacterium leiognathi*, that grow best at warmer temperatures and have been isolated from shallow-water species, and *Photobacterium phosphoreum*, which thrives in cooler waters and has been isolated from deep-sea species. The taxonomy of luminescent bacteria has been revised often with *Vibrio* species, alternatively designated as *Photobacterium*, *Achromobacter*, *Lucibacterium*, *Neisseria* and *Beneckea* (Herring, 2002). The most recent contribution to the name game is the suggestion that *P. phosphoreum* in deep-sea fishes is an incorrect identification and should be replaced by *P. kishitanii* (Ast & Dunlap, 2005).

The primary means by which symbionts are transmitted between generations are either horizontal (environmental) or vertical (transovarian) (Nyholm & McFall-Ngai, 2004). Vertical transmission is known to occur in some insect species, but is generally rare compared to horizontal transmission, which includes the enteric microbiota found in all animal digestive tracts. Also known as cyclic transmission, because the symbionts must be reacquired from the environment with each new generation, horizontal transmission includes both direct and indirect transfer from the adult host. Direct transfer is relatively rare and requires the adult host to facilitate the transmission as, for example, when cows inoculate their calves with rumen microbiota while grooming them. Indirect transfer, where inoculation of the young is from the general environment, is the most common means of transmission, but one that requires a remarkable degree of co-ordination in order to insure that the host selects only desirable bacteria from the environment, rejects undesirable species and establishes growth conditions that favour the symbiont without allowing runaway growth (McFall-Ngai, 1998). The best documented case of environmental transmission of bioluminescent bacteria is in the Hawaiian bobtail squid (*Euprymna scolopes*) that systematically eliminates environmental intruders in favour of the bacterial symbiont, *Vibrio fischeri*, which then triggers developmental changes in the host, resulting in a functional light organ (Nyholm & McFall-Ngai, 2004). Inoculation of the juvenile light organ is facilitated by the daily expulsion into the water column of more than 90% of the adult light-organ symbionts, thereby augmenting their local concentration (Boettcher *et al.*, 1996). A similar situation has been demonstrated in the ponyfish (*Leiognathus nuchalis*), which has a circumoesophageal light organ populated with the symbiont, *Photobacterium leiognathi* (Wada *et al.*, 1999). Light emission is ventral through a complex series of reflectors and diffusers and apparently functions in counter-illumination (Hastings, 1971). In a study by Wada *et al.* (1999) it was found that juveniles that were reared separately from adults failed to become luminescent, indicating that bacteria were not transferred vertically. However, juveniles placed with adults or with homogenate from adult light organs became luminescent within 48 hours. During normal development juveniles became luminescent 45 days after hatching, apparently selecting the appropriate bacterial symbiont from the local environment, which is enriched with *Photobacterium leiognathi* expelled from adult light organs.

Similar examples of horizontal transmission of symbionts have been demonstrated in the monocentrid pineconefish (*Monocentris japonicus*), which has two small light organs located on the underside of the lower jaw that open to ambient seawater, and the sea urchin cardinalfish (*Siphamia versicolor*), which has a ventral light organ that opens to the intestine. The symbiont in the monocentrids has been identified as *Vibrio fischeri* (Ruby & Nealson, 1976) while in sea urchin cardinalfish it appears to be *Photobacterium leiognathi* (Leis & Bullock 1986). Although the bacterial symbionts from the five species of flashlightfish, *Anomalops katoptron, Photoblepharon palpebratus, P. steinitzi, Kryptophanaron alfredi* and *Phthanophaneron harveyi*, have never been cultured or identified, there is evidence that they are acquired anew with each generation, via ducts that connect to the surrounding seawater. Interestingly, it appears that the flashlightfish symbionts vary between species and may have co-evolved with their hosts (Wolfe & Haygood, 1991). A similar situation may exist in deep-sea anglerfishes (Lophiiformes: Ceratioidea), which also have unculturable species-specific symbionts and a duct that connects the interior of the escal light organ with ambient seawater (Haygood & Distel, 1993). However, because deep-sea anglerfish larvae reside in surface waters and only descend into the depths at the start of metamorphosis, the question arises as to how the larvae can acquire the appropriate symbiont when there are no adults in proximity to provide the inoculum. One possible solution proposed by Haygood (1993) is that bacteria discharged from adult female escae during spawning might adhere to the eggs and then grow in the mucus layer that sheathes anglerfish larvae. During metamorphosis the esca develops as an ingrowth of epidermal cells that could therefore provide the required inoculum via the mucus sheath (Munk, 1999).

A TAXONOMIC SURVEY OF LARVAL BIOLUMINESCENCE

In order to simplify our survey, we will review the early development of fishes with luminescent properties by large taxonomic grouping (orders). The distribution of both bacterial and intrinsic bioluminescence in modern teleost fishes is presented in Figure 2 and follows the latest classification (Nelson, 2006).

Order Saccopharyngiformes

Among eels, two aberrant deep-sea families, Saccopharyngidae (swallower eels) and Eurypharingidae (gulpers or pelican eels), have been reported to have bioluminescent properties (Herring & Morin, 1978). However, the evidence of light production in these two families is still equivocal and is mainly based on indirect evidence.

For nine species of swallower eels, Nielsen & Bertelsen (1985) found it likely that the terminal caudal organ could serve as a bioluminescent lure for prey based on a single observation of a live specimen (Beebe, 1932) and the morphology of the organ itself, which bears numerous filaments hypothesised to aid in the luring process. Histological examinations of caudal organs in swallower eels have yielded no bacteria or tubular glands characteristic of bacterial light organs; thus, if caudal organs in the genus *Saccopharynx* are indeed photogenic, then they are likely non-bacterial. In addition, *Saccopharynx* species possess so-called "white line" organs, also called nuchal troughs, which run dorsally from just behind the head down the length of the body and tail (Herring & Morin, 1978; Nielsen & Bertelsen, 1985). Beebe (1932) described

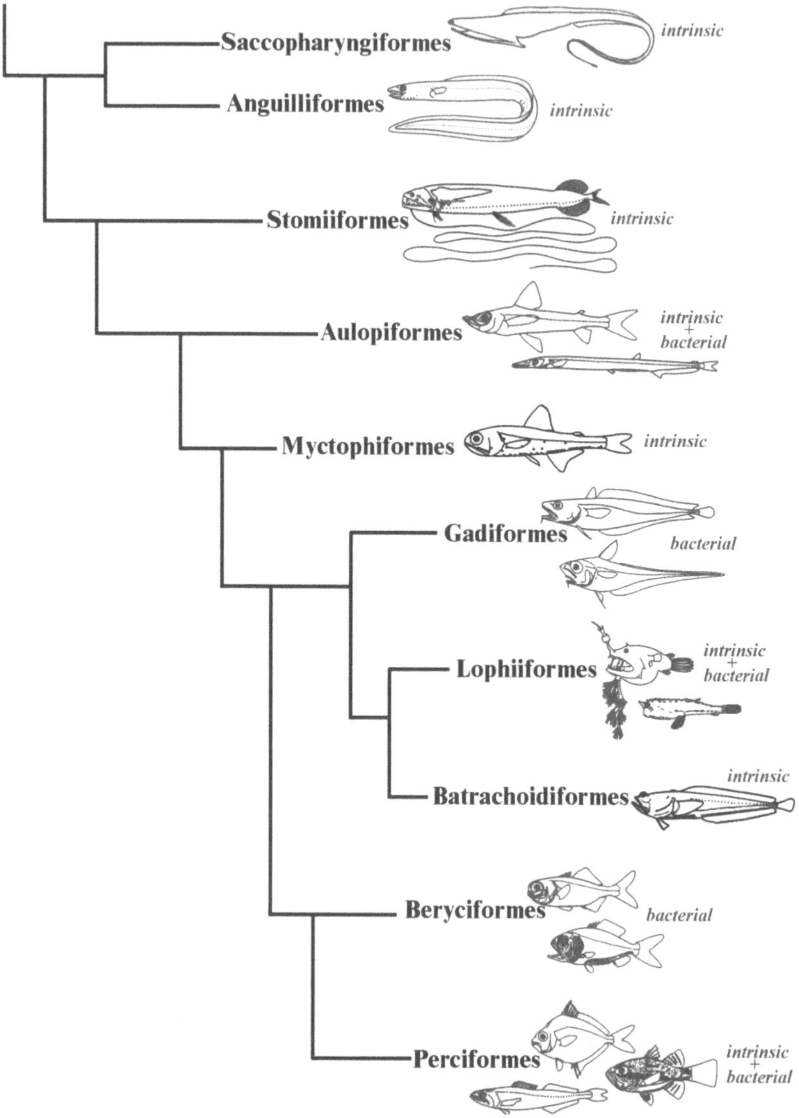

Fig. 2 Cladogram showing distribution of intrinsic and bacterial light production among various orders of teleostean fishes. Phylogeny after Nelson (2006).

these lines, running along both sides of the body below the dorsal midline as "filled with bluish luminous substance," although later studies failed to confirm bioluminescence of these structures (Nielsen & Bertelsen, 1985). The bioluminescent properties of the caudal organ of the pelican eel (*Eurypharynx pelecanoides*) are even less evident. It is not equipped with filaments and observations on luminescence from this organ (Owre & Bayer, 1970) have not been confirmed (Nielsen *et al.*, 1989). However, observations and video recordings of luminescence from the "white line" organs in the pelican eel have revealed them to be capable of brilliant luminescent flashes of such short duration that they are probably under neural control (E. Widder, unpublished data).

Similar to other elopomorph fishes, all members of the Order Saccopharyngiformes pass through a unique leptocephalus larval stage and undergo extreme metamorphic changes prior to becoming juveniles. The leptocephali of *Saccopharynx* and *Eurypharynx* are known and possess no structures that can be attributed to incipient light organs.

Order Anguilliformes

The total absence of bioluminescent representatives among deep-sea anguilliform eels is perplexing, considering that a number of species thrive in the meso- and bathypelagic zones of the World Ocean. At present, no early life history information is available on the shallow-living, benthic congrid eel (*Lumiconger arafura*), the only species that has been shown to have bacterially induced bioluminescence (Castle & Paxton, 1984). We suspect, however, that the unique leptocephalus stage and the radical morphological changes taking place during metamorphosis preclude early bioluminescent capabilities in this species. Lack of incipient light organs in known larval saccopharyngiforms provides some credence to such views.

Order Clupeiformes

The only known bioluminescent species – the goldspotted grenadier anchovy (*Coilia dussumieri*, Engraulidae), with several dozen photophores along the isthmus, behind the eye and in four to six rows along the ventral and lateral surfaces of the body, is found from coastal India to Java (Whitehead *et al.*, 1988). The bioluminescence of the goldspotted grenadier anchovy is apparently endogenous, but no information is available on the early ontogeny of photophores in this apparently unique engraulid species.

Order Argentiniformes

Suborder Argentinoidei In the highly specialised deep-sea spookfishes of the genera *Opisthoproctus*, *Winteria* and *Rhynchohyalus*, light organs are of the open type and are a diverticulum of the gut, hosting luminous bacteria (*Photobacterium phosphoreum*) (Bertelsen & Munk, 1964; Bertelsen *et al.*, 1965; Herring, 1975). In *Opisthoproctus* larvae (10.0-14.5 mm standard length, SL), the unusual light reflecting structure of the abdomen (sole) already shows precocious development and differentiation of parts (rectal bulb and parts of the reflector organ; Fig. 3a, b) found in adults (Bertelsen & Munk, 1964). The only record of a larval barreleye (*Winteria telescopa*) is that of a tentatively identified 12.0-mm specimen (Belyanina, 1982). Although no description was provided, an illustration shows an intensely pigmented terminal part of the intestine, the place for the developing rectal light organ (Fig. 3c). A similarly pigmented rectal

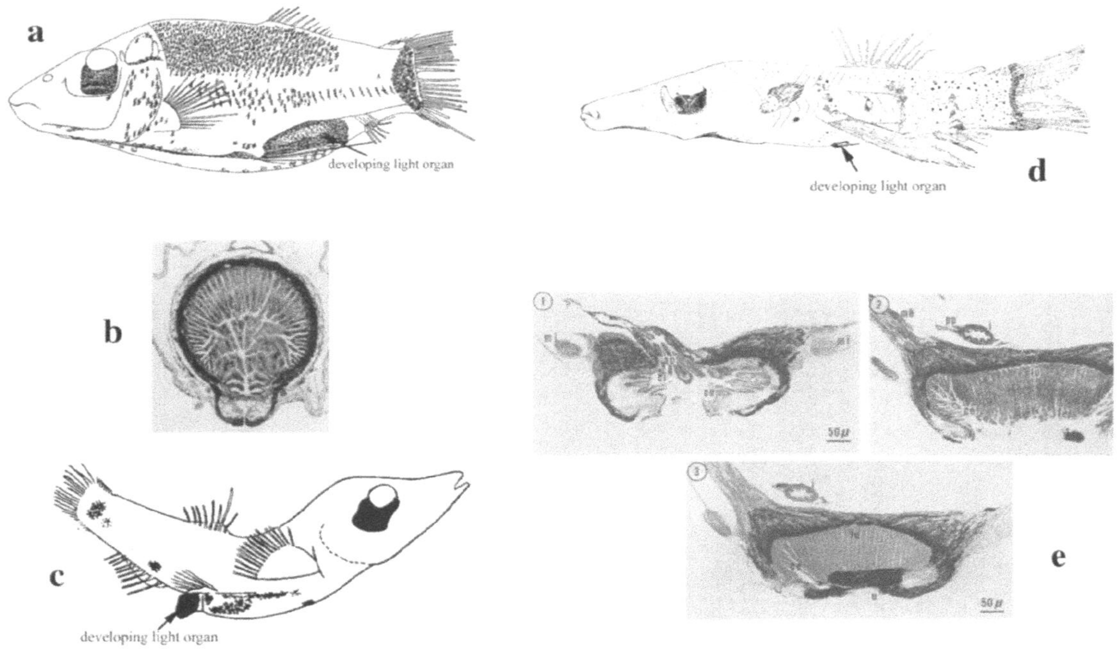

Fig. 3 The light organ development in Argentiniformes (suborder Argentinoidei). **a** – larva of barrel-eye (*Opisthoproctus soleatus*, Opisthoproctidae) 16.2 mm SL after Evseenko & Sunstov (1995); **b** – transverse section through rectal bulb of larval mirrorbelly (*Opisthoproctus grimaldii*, Opisthoproctidae) 10.0 mm SL after Bertelsen & Munk (1964); **c** – larva tentatively identified as *Winteria telescopa* (Opisthoproctidae) 12.0 mm SL with possible precocious development of a rectal bulb after Belyanina (1982); **d** – postlarva of *Rhynchohyalus natalensis* (Opisthoproctidae) 23.0 mm SL after Bertelsen *et al.* (1965); **e** – (1) transverse section through anterior, (2) middle and (3) posterior parts of the rectal bulb of *R. natalensis* 23.0 mm SL.

part of the intestine has also been noted for larval *O. grimaldii* and *O. soleatus* (Schmidt, 1918; Bertelsen & Munk, 1964; Evseenko & Suntsov, 1995). Thus, a precocious development of the rectal light organ in *Winteria* is also likely the case for *Opisthoproctus*.

An examination of a small post-larval specimen of the glasshead barreleye (*Rhynchohyalus natalensis*), still displaying a number of larval features (Fig. 3d, e), revealed the presence of a light organ (Bertelsen *et al.*, 1965), a simple rectal bulb without lens cells or a reflector layer. Unlike *Opisthoproctus* and *Winteria*, which emit light indirectly, the light organ of *Rhynchohyalus* appears to emit light directly through the anus (Bertelsen *et al.*, 1965; Herring, 1975).

Although information on the early stages of luminous spookfishes (Opistoproctidae) is very limited, it appears that their development is more or less gradual, similar to gradual development in related argentinoids (Ahlstrom *et al.*, 1984). No actual bioluminescence in larval spookfishes has been observed, but precocious development of light organs could facilitate the initiation of the ventral camouflage, suggesting a deeper distribution than (shallower) larvae utilising transparency as primary camouflage. The two species of *Opisthoproctus* possess an elaborate abdominal reflecting structure, putatively assisting in bioluminescent counter-illumination, while *Winteria* and *Rhynchohyalus* develop less elaborate light organs.

Suborder Alepocephaloidei

Four genera of Alepocephalidae – *Xenodermichthys*, *Photostylus*, *Rouleina* and *Microphotolepis*, apparently forming a natural group, have been reported to have photophores (Herring & Morin, 1978). Subsequently, Sazonov (1995) mentioned the presence of photophores on the chin of the bathypelagic genera *Bathyprion* and *Mirognathus*. However, one of us (TT Sutton) has examined the latter species and did not detect the presence of photophores. Alepocephalid photophores show significant structural diversity, being sessile (*Xenodermichthys*, *Rouleina*), borne on stalks (*Photostylus*) or covered with scales (*Microphotolepis*) and the bioluminescence is endogenous.

For still unclear reasons, alepocephalid larvae and juveniles are rarely collected, and this naturally limits the available information on development of their light producing organs. At present, early ontogeny of photophores is known for the bluntsnout smooth-head (*Xenodermichthys copei*) and Schmidt's slickhead (*Microphotolepis schmidti*) (Badcock & Larcombe, 1980; Sazonov, 1995). In the bluntsnout smooth-head (Fig. 4a), first to appear are the primary photophores on the head (OpV, PO, OpD and Md) and ventral regions of the body (in series PV and VA), the latter being present in 9.0 mm SL larvae. The pattern of body photophore development is fairly constant and the first photophore in any given series is invariably ventral-most, with subsequent photophores added in a ventro-dorsal direction. Badcock & Larcombe (1980) found that the postflexion larva of another bioluminescent alepocephalid, the starry smooth-head (*Photostylus pycnopterus*) (11.8 mm SL), lacked photophores completely. Development of primary photophores in the smallest juveniles of *M. schmidti*

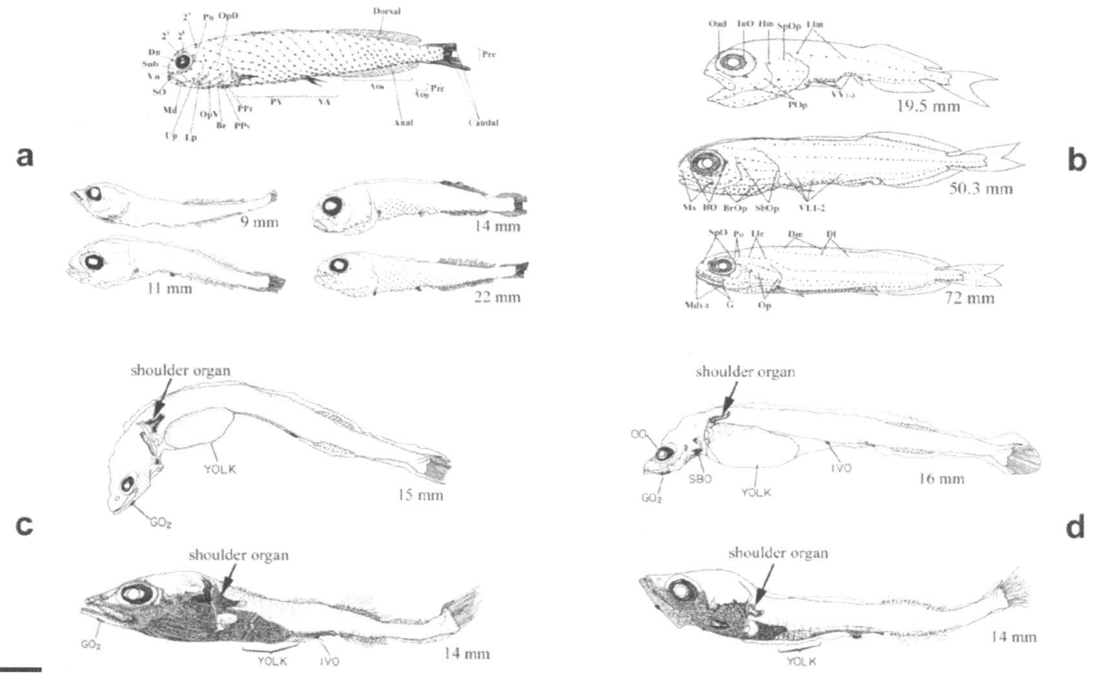

Fig. 4 The light organ development in Argentiniformes (suborder Alepocephaloidei). **a** – photophore terminology and sequence of photophore development in bluntsnout smooth-head (*Xenodermichthys copei*, Alepocephalidae), photophore acronyms: AOa – anterior anal; AOp – posterior anal; BR – branchiostegal; Dn – dorsonasal; LP – lower preopercular; Md – mandibular; OpD – latero-dorsal opercular; OpV – latero-ventral opercular; PO – postorbital; Prc – precaudal; PV – pectoral to pelvic series; SO – symphysial; Sub – suborbital; UP – upper preopercular; VA – pelvic to anal series; Vn – ventro-nasal; after Badcock & Larcombe (1980); **b** – sequence of photophore development in Schmidt's slickhead (*Microphotolepis schmidti*, Alepocephalidae), photophore acronyms: BrOp – branchiostegal; Dm – dorsomedian; DL – dorsolateral; G – gular; Hm – hyomandibular; IfO – infraorbital; InO – infrarorbital ring; Llc – continuation of mediolateral series on head; Llm – mediolateral; $Md_{1\text{-}2}$ – mandibular; Mx – maxillary; SpO – supraorbital; SpOp – supraopercular; Oad – antorbital; Op – opercular; PO – postorbital; Pop – preopercular; SbOp – subopercular; $Vl_{1\text{-}2}$ – ventrolateral; $Vl_{1\text{-}3}$ – three ventropelvic series. After Sazonov (1995); **c** – larval development in streaklight tubeshoulder (*Holtbyrnia latifrons*, Platytroctidae), GO_2 – posterior gular organ, IVO intraventral organ; **d** – larval development in shining tubeshoulder (*Sagamichthys abei*, Platytroctidae), OO – orbital organ, SBO – subopercular organ. After Matsui (1991).

(~ 19.5 mm SL) parallels that of the bluntsnout smooth-head (Fig. 4b), with the first appearance on the lower parts of the body and head. Secondary photophores appear later in ontogeny, at 35-40 mm SL (Sazonov, 1995).

All genera of tubeshoulders (Platytroctidae) are characterised by a unique shoulder organ, containing blue-green luminous cells that can be ejected into the water (Herring, 1972). In addition to shoulder organs, most species of platytroctids have variously developed body photophores, primarily found on the ventral surface of the body (Matsui & Rosenblatt, 1987).

Similar to alepocephalids, early stages of platytroctids are uncommon in plankton collections. Earlier studies noted distinct ontogenetic changes in the orientation of photophores in platytroctids, where larvae and juveniles are characterised by horizontally oriented light organs, changing to ventrally directed in adults (Matsui & Rosenblatt, 1971). Detailed development from yolk-sac larvae to juveniles was subsequently described for two platytroctids – the streaklight tubeshoulder (*Holtbyrnia latifrons*) and the shining tubeshoulder (*Sagamichthys abei*) (Matsui, 1991). Shoulder organ and head photophores in both species show extremely precocious development (Fig. 4c, d), being already distinct in yolk-sac larvae (15-16 mm SL). Photophores in the ventral series on the body develop in late larvae and juveniles (Matsui, 1991). The description of another, tentatively identified platytroctid species, the palegold tubeshoulder (*Maulisia argipalla*), also reports the precocious development of the shoulder organ (John *et al.*, 2000).

Order Stomiiformes

Fishes in this order (4-10 families depending on the classification scheme), along with lanternfishes (Order Myctophiformes), form the vast majority of luminous pelagic ichthyofauna of the oceanic midwaters. All representatives of this primarily deep-water assemblage are self-bioluminescent. Larvae of numerous stomiiform families are quite common in plankton collections, but complete development is documented mostly for zooplanktivorous families.

Zooplanktivorous stomiiforms (Diplophidae, Gonostomatidae, Phosichthyidae and Sternoptychidae) mostly possess large "primary" photophores on the head and ventral surface of the body, with infrequent occurrence of smaller secondary photophores in some species. Primary photophores can be separate or in groups.

In stomiiforms in general, light organs form during a sometimes prolonged metamorphosis and are laid down as "white photophores," which are apparently non-functional. Transformation length and patterns of light organ development vary significantly in different stomiiform families and are detailed by Moser (1996). Ventral photophores of diplophids, gonostomatids and phosichthyids develop rapidly and simultaneously during a "white photophore" stage, while lateral photophores form later and more gradually. In *Cyclothone* spp. (Gonostomatidae), first to form are photophores of the BR and VAV series, while in *Gonostoma* and *Sigmops* spp. (Gonostomatidae) and *Diplophos* spp. (Diplophidae) the OP_3 is first (Ahlstrom, 1973; Watson, 1996a).

Species of Sternoptychidae have photophores within at least some of the photophore groups united in common glands and are characterised by gradual acquisition of both photophore groups and photophores within each group (Fig. 5b, e) during metamorphosis (Watson, 1996b).

Light organ diversity is higher in barbeled stomiiforms (Stomiidae, *sensu* Fink, 1985). In addition to the regular series of ventral and lateral photophores, most piscivorous stomiiforms have bioluminescent barbels and minute secondary photophores scattered over the body. Also present in a few species are pre-, sub-, and postorbital photophores which can be sexually dimorphic, as well as species-specific luminous patches or spherules on the body and fins (O'Day, 1973; Jorgensen & Munk, 1979; Parin & Borodulina, 2003). In addition, three stomiid genera (*Aristostomias*, *Malacosteus*, and *Pachystomias*) have developed extremely rare long-wavelength (red) bioluminescence (Widder *et al.*, 1984; Herring & Cope, 2005).

The reference to larval length at the time of development of light organs in predatory stomiiforms can often be misleading, since their larvae are known to undergo significant shrinkage during metamorphosis. Thus, less advanced larvae with "white photophores" are significantly larger than more developed transitional specimens (30-40 mm SL) with pigmented and probably functional photophores. Similarly, larvae with "white photophores" may be smaller than early larvae with no photophores.

The early development of predatory stomiiforms has been documented in detail in very few cases, precluding broad generalisations on time and order of photophore formation. In the viperfish (*Chauliodus sloani*), photophores appear at 33-37 mm SL after significant shrinkage of up to 30% of its previous size. The shrinkage is complete at 25-27 mm SL, by which time a barbel with a terminal photophore is present. Small secondary photophores subsequently appear on the head and body (Belyanina, 1977), and the barbel later regresses to form a vestigial nub. In the longfin dragonfish (*Tactostoma macropus*), photophores in the ventral series appear at 47-49 mm SL and development of the primary photophores precedes that of the smaller secondary photophores (Kawaguchi & Moser, 1993). In another stomiid fish (*Astronesthes spatulifer*), branchiostegal photophores (BR series) are first to form, appearing at 40-41 mm SL, followed by almost simultaneous development of "white photophores" in other ventral series (PV, VAV, AC) at 46-47 mm SL (Suntsov, 1999). A review of the published information on the larvae of barbeled stomiiforms indicates that specialised luminous organs and structures appear much later in ontogeny (Kawaguchi & Moser, 1984; Moser, 1996).

Order Aulopiformes

Bioluminescence is not widespread in this diverse group of primarily deep-sea fishes (15 families), of which only four families – the benthopelagic greeneyes (Chlorophthalmidae), and the pelagic pearleyes (Scopelarchidae), barracudinas (Paralepididae) and sabertooth fishes (Evermannellidae) have few bioluminescent representatives. Rarity and morphological disparity of light organs in this group suggest

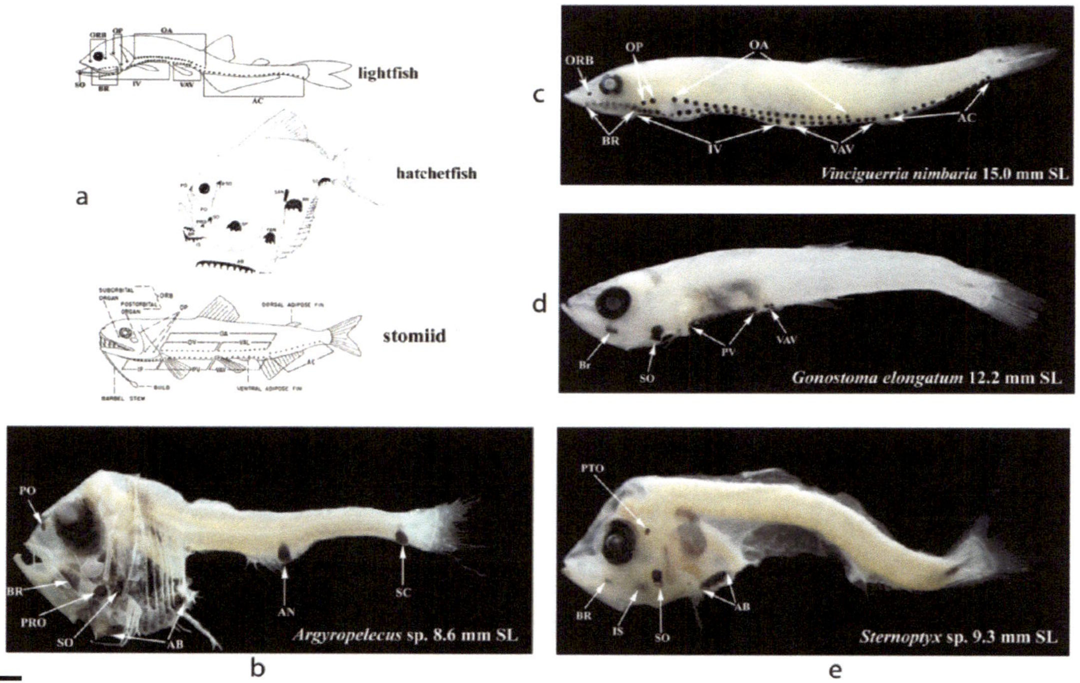

Fig. 5 Development of light organs in the order Stomiiformes. **a** – photophore terminology for different stomiiform groups. After Moser & Watson (1996); **b**, **e** (Sternoptychidae), **c** (Phosichthyidae), **d** (Gonostomatidae) – representatives of larval stomiiforms showing precocious development of light organs. AB – abdominal photophores; AC – photophores from the anal fin base to caudal fin base; AN – anal; Br – branchiostegal; Is – isthmal; IV – isthmus to pelvic series; PAN – preanal; OA – lateral series from the opercle to anal fin base; ORB – orbital; OP – opercular; PO - preorbital; PTO – postorbital; PRO – preopercular; SAN – supraanal; SC – subcaudal; SO – subopercular; SP – suprapectoral; VAV – pelvic to anal fin base series.

that the ability to produce light may have evolved multiple times due to similar selection pressure in the oceanic midwaters.

Species of *Chlorophthalmus* spp. were reported to have a bacterial light organ associated with a peri-anal groove (Somiya, 1977). Available information on the early ontogeny of *Chlorophthalmus* indicates that a light organ is not present in the larvae (Okiyama, 1984). Among pearleyes (Scopelarchidae), only Zugmayer's pearleye (*Benthalbella infans*) and *Scopelarchoides kreffti* are known to have ventral luminescent organs (four and two, respectively) and are considered self-bioluminescent (Merrett *et al.*, 1973). The photogenic tissue of Zugmayer's pearleye is unique among teleosts because of its derivation from muscle cells (Johnston & Herring, 1985). Light organs are barely visible as whitish areas in 20-mm postlarvae of Zugmayer's pearleye and are fully developed only in adults (Merrett *et al.*, 1973). The ontogeny of most scopelarchids is well known, but no information on the larvae of *S. kreffti* is available and the possibility of early development of light organs remains open (Johnson, 1984a).

Of the three genera of sabertooth fishes (Evermannellidae), only the Atlantic sabertooth *(Coccorella atrata)* and possibly *C. atlantica* have light organs as part of the gut wall of the intestine, the rectum and the anterior pyloric caecae (Herring, 1977). Light organs do not appear early in the development of *Coccorella* species (Johnson, 1984b). Two genera of barracudinas (Paralepididae) – *Lestrolepis* and *Lestidium*, are self-bioluminescent, having either single (*Lestidium*) or double (*Lestrolepis*) abdominal bands of luminous tissue. In addition, some species are reported to have ventral photophores (Herring & Morin, 1978). The early development of barracudinas is still poorly known and no incipient light organs are evident in published accounts of paralepidid ontogeny (Rofen, 1966; Okiyama, 1984).

Order Myctophiformes

Bioluminescence is found in both families of this rather morphologically homogenous group, the blackchins (Neoscopelidae) and the lanternfishes (Myctophidae). Of the blackchins, which includes three genera and six species, three species of *Neoscopelus* have large primary photophores in horizontal rows on the body and tongue. No additional luminous organs are present. Available information on the development of *Neoscopelus* suggests that no photophores are present in the larvae (Okiyama, 1984).

All 235 currently known species of lanternfishes are luminous. In addition to large primary photophores arranged in distinct groups on the head and body (Fig. 6a), various species of myctophids possess small secondary photophores on the head and body, supra- and infracaudal glands, photophores associated with the eyes, and a variety of luminous patches (Paxton & Hulley, 1999). Unlike neoscopelids, all species of myctophids show relatively early development of photophores (Fig. 6b-f). Early myctophid ontogeny is well documented and the sequence of photophore development in different genera can be significantly different (Table 1). However, the initial light organs to form are always the middle pair (Br_2) of branchiostegal photophores. The only exception to this rule is peculiar larvae of *Diaphus* from the western Pacific,

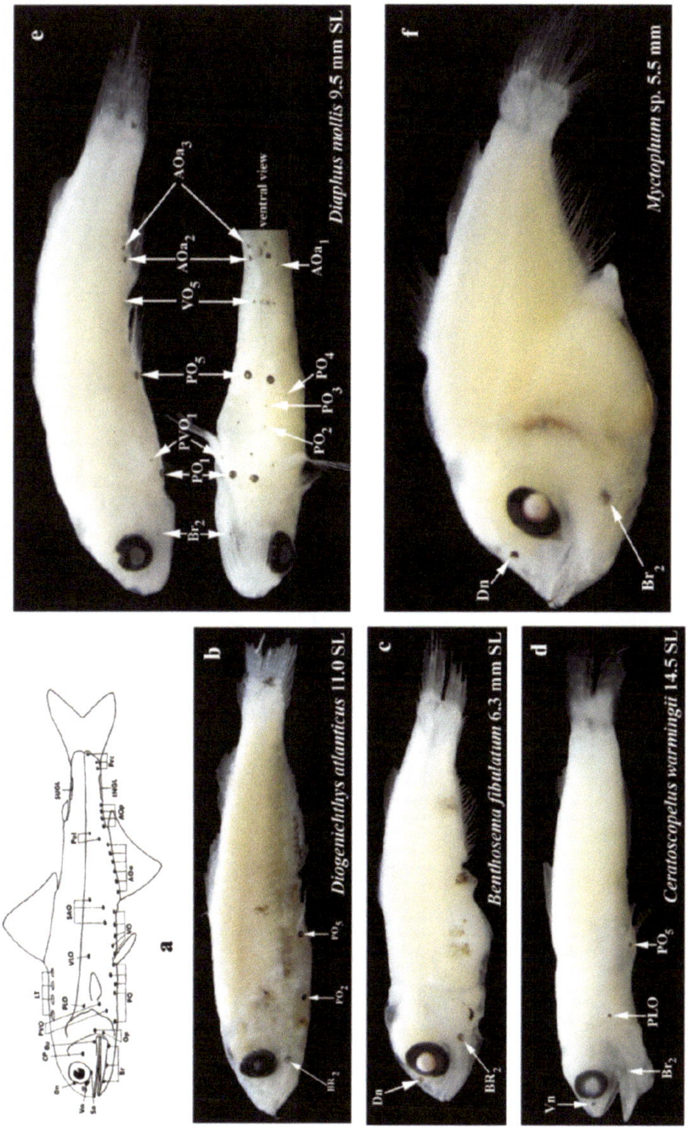

Fig. 6 **a** – The general distribution and terminology of the luminous organs (photophores) in the family Myctophidae. After Paxton (1972); **b-f** – representative larval myctophids showing early appearing photophores. Photophore acronyms: AOa – anterior anal; AOp – posterior anal; Br – branchiostegal; Bu – buccal; Cp – cheek; Dn – dorsonasal; INGL – infracaudal luminous gland; Lt – patches of luminous tissue; Op – opercular; PLO – suprapectoral; PO – thoracic or pectoral; Pol – posterolateral; Prc – precaudal; PVO – subpectoral; SAO – supraanal; So – suborbital; SUGL – supracaudal luminous gland; VLO - supraventral; Vn – ventronasal; VO – ventral.

Table 1 Sequence of formation of photophores in selected genera of Myctophidae. The Br appear first in all genera listed. Parentheses indicate late appearing photophores after Moser *et al.* (1984).

	Br_1	Br_3	Dn	Vn	OP_2	PO_1	PO_2	PO_3	PO_4	PO_5	PVO_1	PVO_2	PLO	VLO	VO_1	VO_5	AOa_1	AOa_2
Benthosema																		
suborbitale	2	2	—	—	2	1	1	3	3	3	—	—	—	—	—	—	3	3
glaciale	—	—	—	—	(1)	(1)	(1)	(1)	(1)	(1)	—	—	—	—	—	—	—	—
pterota	—	—	1	—	4	6	—	—	—	2	3	5	—	—	5	—	6	—
fibulatum	—	—	1	—	—	3	5	—	—	2	—	—	—	6	—	—	4	6
Diogenichthys																		
laternatus	—	—	—	—	—	—	—	—	—	1	—	—	—	—	—	—	—	—
atlanticus	—	—	—	—	—	—	1	—	—	2	—	—	—	—	—	—	3	—
Myctophum																		
spinosum	—	—	1	—	—	—	—	—	—	—	—	—	2	—	—	—	—	—
lychnobium	—	—	1	—	—	—	—	—	—	—	—	—	2	—	—	—	—	—
asperum	—	—	1	—	—	—	—	—	—	—	—	—	2	—	—	—	—	—
brachygnathum	—	—	1	—	—	2	—	—	—	—	—	—	2	—	—	—	—	—
obtusirostre	—	—	1	—	—	3	—	—	—	—	—	—	2	—	—	—	—	—
selenops	—	—	—	—	—	3	—	—	—	—	—	—	2	—	—	—	—	—
Lobianchia	—	—	—	—	—	1	—	—	—	2	3	4	—	—	—	—	—	—
Diaphus																		
theta	—	—	—	—	(5)	2	(4)	(7)	(8)	1	—	—	—	(9)	(3)	(6)	—	—
pacificus	—	—	—	—	—	2	(3)	(5)	—	1	(4)	—	—	—	(6)	—	—	—
Gymnoscopelus	—	—	—	—	—	2	4	—	—	1	—	—	—	—	3	—	—	—
Lampanyctodes	—	—	—	1	—	4	—	—	—	2	—	—	3	—	—	—	—	—
Scopelopsis	—	—	—	2	—	—	—	—	—	1	—	—	—	3	—	—	—	—
Lampichthys	—	—	—	2	—	4	—	—	—	1	—	—	3	—	—	—	—	—
Notoscopelus	—	—	—	2	—	—	—	—	—	1	—	—	3	—	—	—	—	—
Lampadena	—	—	—	3	—	3	—	—	—	2	—	—	1	—	—	—	—	—
Ceratoscopelus	—	—	—	1	—	—	—	—	—	3	—	—	2	—	—	—	—	—
Lepidophanes	—	—	—	1	—	—	—	—	—	1	—	—	1	—	—	—	—	—
Bolinichthys	—	—	—	(1)	—	—	—	—	—	(1)	—	—	(1)	—	—	—	—	—

designated as *Diaphus* sp. XII, which never develop Br_2 photophores even during metamorphosis (Ozawa, 1986). In the multispot lanternfish (*Scopelopsis multipunctatus*), Br_2 are visible at 5.4 mm SL and are well formed in 10.8-mm larvae. In the same species, the posterior-most pair of PO photophores is next to form, followed by Vn and VLO photophores (Moser & Ahlstrom, 1972). The functional state of the larval photophores in myctophids has not been investigated to date.

Order Gadiformes

Bioluminescence is widespread in this primarily benthopelagic group and is most commonly found in fishes inhabiting continental slopes between 100-500 m (Marshall & Cohen, 1973). Among the grenadiers (Macrouridae), all species in 10 (out of 27 total) genera have a ventral light organ in front of the anus, and the light is produced by luminous bacteria (Herring & Morin, 1978). Although certain developmental stages in several bioluminescent genera (e.g. *Coelorhynchus*, *Malacocephalus*, *Nezumia*, *Ventrifossa*) are described, no special attention is usually given to development of luminous organs (Fahay & Markle, 1984; Merrett, 1989; Ambrose, 1996). A macrourine alevin, with a head length (HL) of 19 mm, was found with a light organ with two separate dermal windows (Fig. 7a) between the pelvic bases (Merrett, 1989). Pelagic larvae of Sagami grenadier (*Ventrifossa garmani*) (5.1 mm HL, 90+ mm total length, TL) appear to initiate the development of its two dermal windows (Fig. 7b) as part of the anal light organ (Fukui & Tsuchiya, 2005).

Species of *Hymenocephalus* have a long, tubular, abdominal light organ, with two (anterior and posterior) light glands (Fig. 7c) equipped with lenses and connected by a common duct (Haneda, 1951). The light organ of *Hymenocephalus* is accompanied by fine skin striae ventrally and laterally, probably serving as indirect diffusers of light (Marshall & Iwamoto, 1973). Larvae of *Hymenocephalus* sp. (3.6 mm HL, 20.5 mm TL) display a light organ similar to that of the adult, with anterior and posterior lenses connected by a duct, but with no apparent striations on the ventral portion of the abdomen (Endo *et al.*, 1992). Light organs in the species of *Coelorhynchus* display significant structural variation, ranging from simple bulbous light glands near the anus to tubular organs similar to that of *Hymenocephalus* (Okamura, 1970). There is some evidence that in some species of *Coelorhynchus* bioluminescence may be restricted or better developed in juveniles than adults (Haneda, 1951)

The only bioluminescent merlucciid species, the luminous hake (*Steindachneria argentea*, Merlucciidae) (Fig. 7d), has a doughnut-shaped light gland surrounding the anus, with bacterially produced light emitted through the striated skin and hypaxial abdominal musculature (Cohen, 1964). Although no special attention has been given to the exact time of the development of bioluminescent structures, published illustrations of early development in luminous hake show that the light organ system is probably developed by 24.0 mm SL (Fahay, 1989).

Several genera of morid cods (Moridae), *Physiculus*, *Brosmiculus*, *Gadella*, *Tripterophycis*, and perhaps *Antimora* and *Lotella*, are known to have bioluminescent

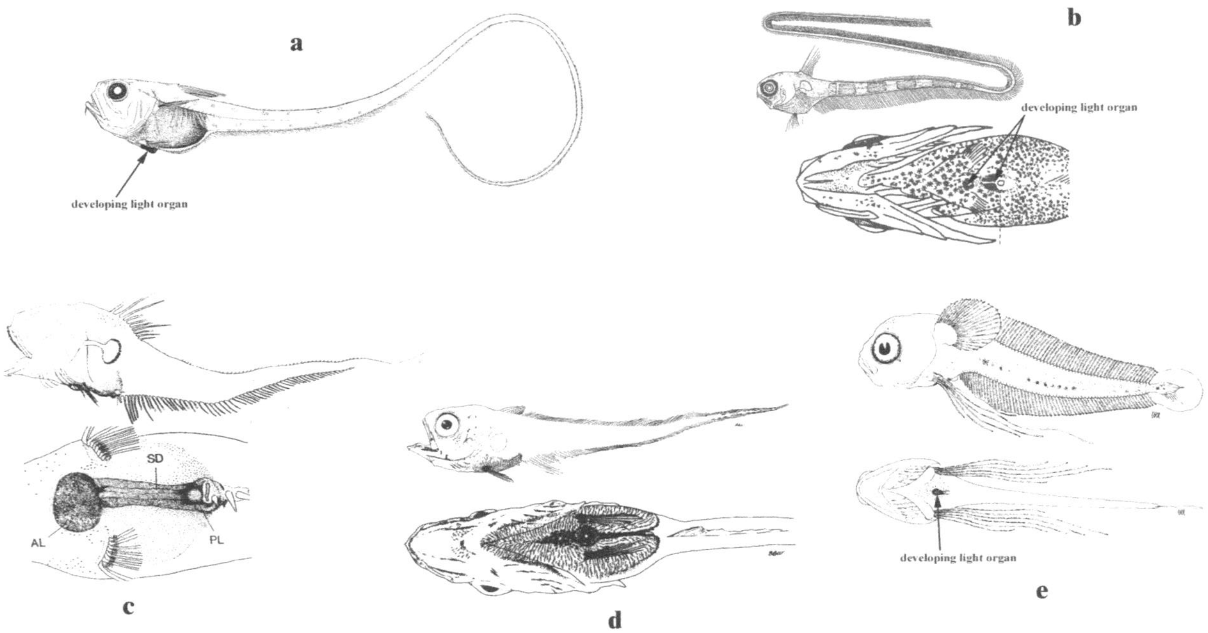

Fig. 7 Development of light organs in the order Gadiformes. **a** – unidentified macrourid alevin (19 mm HL) with a developing light organ (after Merrett, 1989); **b** – pelagic larva of Sagami grenadier (*Ventrifossa garmani*, Macrouridae) 5.1 mm HL (90 + mm TL). Lateral view and ventral view of the abdomen showing incipient light organ modified after Fukui & Tsuchiya (2005); **c** – larval *Hymenocephalus* sp. (Macrouridae) 20.5 mm TL, lateral and ventral views of developing light organs after Endo *et al.* (1992); **d** – late larva of luminous hake (*Steindachneria argentea*, Merluciidae) 24.0 mm SL, showing developing light organ and striated abdominal musculature after Fahay & Markle (1984); **e** – larva of charcoal mora (*Physiculus nematopus*, Moridae) 9.2 mm SL and ventral view of larva 14.1 mm SL after Fahay & Markle (1984). Abbreviations: SL – standard length, TL – total length, HL – head length, AL – anterior lens, PL – posterior lens, SD – secondary duct.

organs (Marshall & Cohen, 1973). Early life history information on morid species is scarce, but similar to other bioluminescent gadiform representatives, the published larval descriptions (e.g. *Physiculus*, Fig. 7e) indicate rather early development of luminous structures (Fahay & Markle, 1984).

Order Batrachoidiformes

Within this order, bioluminescence occurs only in the genus *Porichthys*, with 14 species found in coastal waters of North and South America. The midshipman has served as a model organism in a variety of bioluminescence studies (Nicol, 1957; Baguet & Case, 1971). In this species, numerous photophores (up to 850) in rows are found along the lateral lines and light production is endogenous (Herring & Morin, 1978).

During early ontogeny, photophores are first externally visible in larvae 12-13 mm SL and display an asynchronous development, spreading from head to tail. First to appear are the mandibular photophores, followed by differentiation of body photophores. There are two distinct stages in the functional development of midshipman photophores. At around 18-21 mm SL (28 days after hatching), when the yolk is nearly or fully resorbed, larvae emit a green fluorescence from most photophores under ultraviolet light and exhibit strong and fluctuating luminescence upon application of hydrogen peroxide. A luciferin and luciferase are detected in larvae at this stage (Tsuji *et al.*, 1972). However, the endogenous control of the bioluminescence, i.e. when luminescence is induced by electrical stimulation or norepinephrine injection, is observed somewhat later, at 32-33 days post hatching. Detailed anatomical studies indicate that the photophore anlage originates as an outgrowth from the basal cell layer of the epidermis and thus is of ectodermal origin. Conversely, the reflector is of mesodermal origin, developing from the fibroblast layer enveloping the dermal aspect of the developing photophore (Anctil, 1977).

Order Lophiiformes

The remarkable assemblage of deep-sea anglerfishes (superfamily Ceratioidea), currently with ~158 species in 11 families, represents the most diverse group of bioluminescent bathypelagic fishes. In addition to extreme sexual dimorphism in size (females greatly larger than males) and some other unique morphological trends, only female ceratioids (except *Neoceratias* and *Caulophryne*) are known to produce light (Bertelsen, 1951; Pietsch, 2005). The most common and widespread type of bioluminescence in ceratioids is due to non-culturable luminous bacteria hosted in the esca – the terminal bulbous part of the ilicium (a modified dorsal ray). Some studies suggest that each ceratioid species may have its own species of luminous bacterium (Haygood & Distel, 1993). Furthermore, females of the family Ceratiidae possess caruncles – additional bacterial light organs in front of the soft dorsal fin (Bertelsen, 1951). Females of the genus *Linophryne* (Linophrynidae) display a dual bioluminescent system, with both a bacterial light organ within the esca as well as a self-luminescent hyoid (chin) barbel (Hansen & Herring, 1977).

The larvae of ceratioids, similar to many deep-sea fish groups, primarily live in the productive epipelagic zone and descend rapidly to deeper levels at the start of their metamorphosis. A club-shaped, rudimentary ilicium is formed early in ontogeny and is present in numerous larval female ceratioid species (Bertelsen, 1984). The early development of escal light glands was studied in larval *Melanocetus* spp. and triplewart seadevil (*Cryptopsaras couesii*), and metamorphosing bulbous dreamer (*Oneirodes eschrichtii*), soft leafvent angler (*Haplophryne mollis*), and Murray's abyssal anglerfish (*Melanocetus murrayi*) specimens. No or few luminous bacteria were found in the larvae and light glands of recently metamorphosed specimens (Munk and Herring, 1996; Munk *et al.*, 1998). Thus, the functionality of the luminous glands appears to be postponed until the end of metamorphosis, at which time morphological differentiation of intraglandular lumina occurs and a duct to the exterior is formed (Munk, 1999). Furthermore, metamorphosis is associated with an accelerated morphological development of the esca (Munk & Herring, 1996).

The acquisition of the right type of luminous symbiotic bacteria apparently occurs around the time of metamorphosis, a view corroborated by the presence of partly-colonised light glands in juvenile soft leafvent angler (Herring and Munk, 1994). A similar situation was described for the caruncles found in ceratiid species (Munk & Herring, 1996). The self-luminescence found in the hyoid barbels of the genus *Linophryne*, where 10-mm larvae possess only rudimentary thickening of the skin in the place of the barbel, is also apparently confined to adults (Bertelsen, 1984).

The only other lophiiform representative known to have bioluminescent properties is the batfish (*Dibranchus atlanticus*, Ogcocephalidae), reported to have dorsal skin luminescence (Crane, 1968). No further information on bioluminescence in this species is available.

Order Beryciformes

Three families of this order have bioluminescent representatives – the slimeheads (Trachichthyidae), the flashlightfishes (Anomalopidae) and the pinecone or knight fishes (Monocentridae). Bioluminescence is of bacterial origin and light organs can be variously placed near the anus, on the lower jaw, or beneath the eyes (Herring & Morin, 1978).

Among trachichthyids, only species of *Aulotrachichthys* (formerly a subgenus of *Paratrachichthys*) are bioluminescent. However, the presence of characteristic striated areas in other representatives (e.g. *Sorosichthys*) can be an indication of a more frequent occurrence of bioluminescence in these fishes (Kotlyar, 1996). In *Aulotrachichthys* sp., a light organ surrounding the anus first appears in 3.6-mm larvae and is well-developed and heavily pigmented by 4.9 mm SL (Fig. 8b). Striated tissue, probably serving as a light guiding structure similar to that found in some gadiform fishes, is developed somewhat later, by 7.9 mm SL (Jordan & Bruce, 1993). Similarly, Konishi & Okiyama (1997) reported developed light organs in a 4.6-mm larva and the development of striated tissue in 7.4-mm larva of this species.

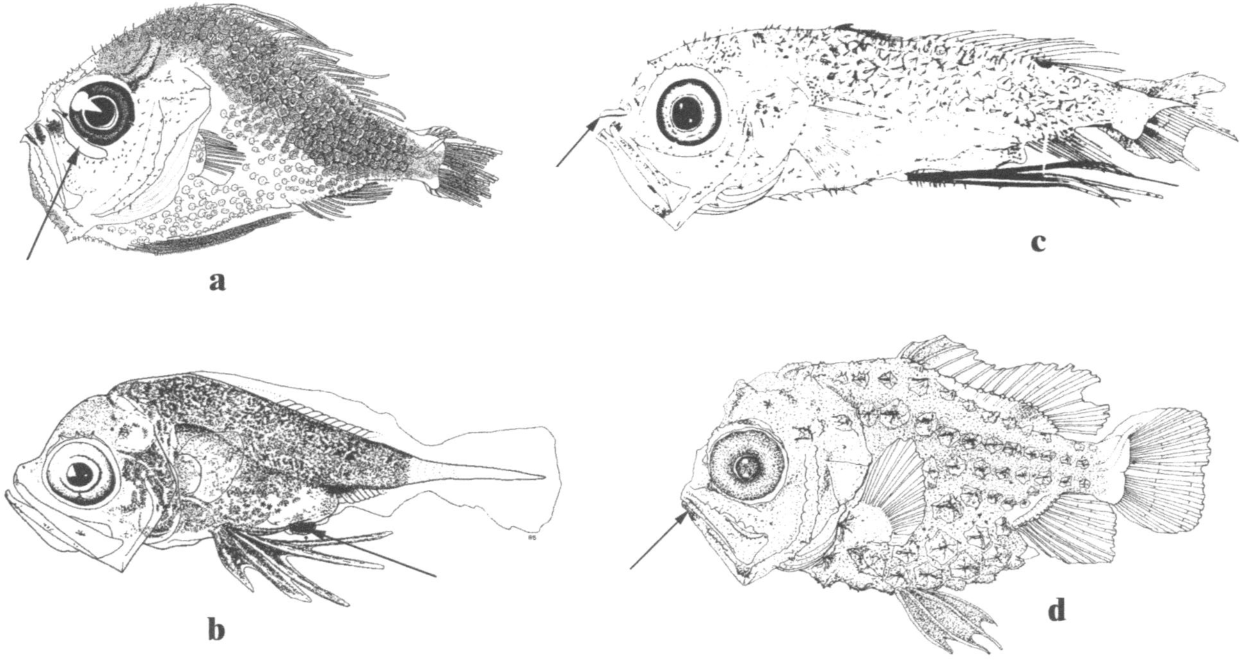

Fig. 8 Development of light organs in the order Beryciformes. **a** – larva of splitfin flashlightfish (*Anomalops katoptron*, Anomalopidae) 5.8 mm SL after Konishi & Okiyama (1997); **b** – larva of *Aulotrachichthys* sp. (Trachichthyidae) 4.4 mm SL after Jordan & Bruce (1993); **c** – larva of Atlantic flashlightfish (*Kryptophanaron alfredi*, Anomalopidae) 6.2 mm SL after Baldwin & Johnson (1995); **d** – larva of pineconefish (*Monocentris japonicus*, Monocentridae) 6.5 mm SL after Okiyama (1988). Arrows indicate developing light organs.

All five genera and six species of anomalopids are bioluminescent, bearing prominent subocular light organs packed with luminous bacteria (Kotlyar, 1996). In splitfin flashlightfish (*Anomalops katoptron*) larvae, incipient light organs appear as rod-shaped projections on the ascending process of the premaxilla in flexion-stage larvae (Fig. 8a), subsequently becoming crescent-shaped and migrating below the eye (Konishi & Colin, 2000). The exact state of bacterial inoculation of the developing light organ in postflexion larva of splitfin flashlightfish 5.8 mm SL is not clear, since it was not examined histologically, but the developing light organ was referred to as "incipient" (Konishi & Okiyama, 1997). In the only known larva of Atlantic flashlightfish (*Kryptophanaron alfredi*) (6.2 mm notochord length), the light organ develops as an anteriorly directed, rod-like projection on the snout (Fig. 8c), with internal invaginations similar to adult light organs. At this stage, no luminous bacteria are present (Baldwin & Johnson, 1995).

Three species of nocturnal, bottom-living monocentrids are unusual in having luminous organs located on the lower jaw. The incipient black-coloured light organ in pineconefish appears on the symphysis of the lower jaw (Fig. 8d) by 6.5 mm SL (Konishi, 2000). No information on the early development of the western Australian pineapplefish (*Cleidopus gloria-maris*) is available.

Order Perciformes

Bioluminescence is a rare phenomenon in this largest assemblage of modern teleost fishes (160 families). Only six distantly related families contain bioluminescent representatives, with both endogenous (some Apogonidae, Pempheridae, Sciaenidae, Chiasmodontidae) and bacterially induced light production (some Apogonidae, Acropomatidae, Leiognathidae). Nocturnal lifestyles (Apogonidae, Pempheridae), as well as certain environmental conditions such as deep-sea darkness (Chiasmodontidae, some Acropomatidae) or turbid estuarine or coastal waters (Leiognathidae, Sciaenidae) have made light emitting properties in certain perciform groups selectively advantageous.

The bacterial luminescence in cardinalfishes (Apogonidae) is apparently restricted to the genus *Siphamia* (Herring & Morin, 1978). The small luminous gland of *Siphamia* is found ventrally in the anterior part of the abdominal cavity and is connected to the gut (Iwai, 1971). An additional orobranchial luminous organ in the same genus was described recently by Fishelson *et al.* (2005). Three other genera of luminous apogonids (*Apogon*, *Archamia* and *Rhabdamia*), with endogenous (*Cypridina*-like) bioluminescence, show diversity in the morphology of their internal light organs, ranging from pouch-like protrusions of the intestine, rectum or even a modified pyloric caecum (Haneda *et al.*, 1969).

To date, only the larvae of the sea urchin cardinalfish *(Siphamia versicolor)* have been studied closely in terms of the early ontogeny of luminous organs (Leis and Bullock, 1986). In this species, the light gland develops as a pigmented gut diverticulum at 2.4 mm SL and moves to a more anterior position near the cleithrum with larval growth (Fig. 9a). The final positioning of the light gland, reached by 2.8 mm SL, is accompanied by the development of associated light diffusers, which gradually spread

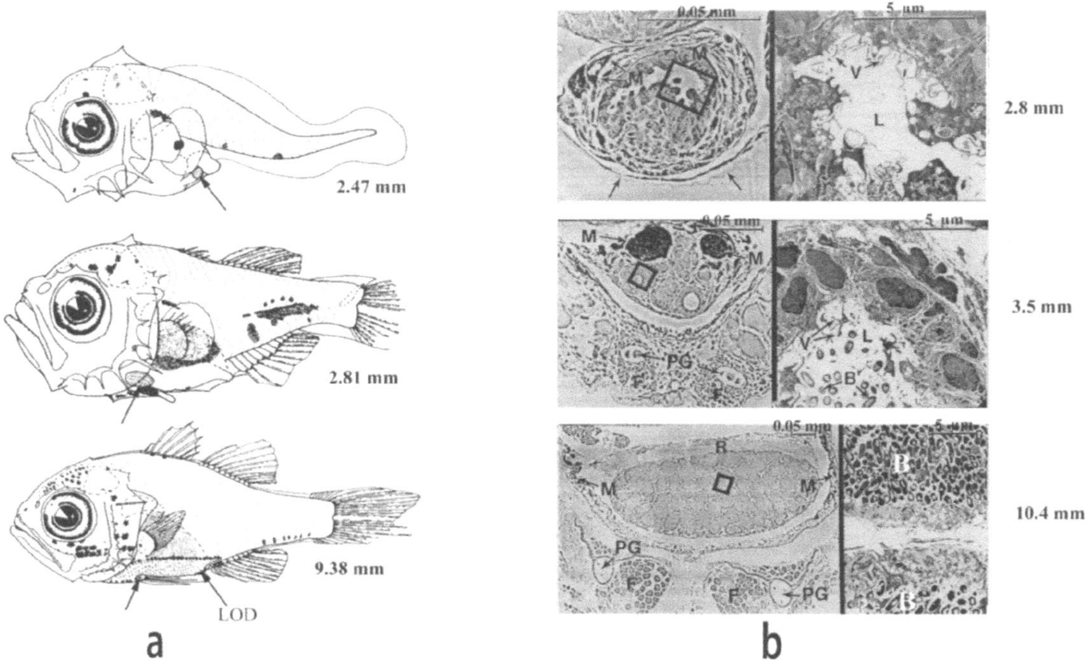

Fig. 9 Development of light organ in sea urchin cardinalfish (*Siphamia versicolor*, Apogonidae, Perciformes) modified after Leis & Bullock (1986). **a** – Migration of the light organ in developing larvae of sea urchin cardinalfish. Arrows point to developing light gland. LOD – light organ diffuser. **b** – phase contrast photographs of transverse sections of the light organ in larval sea urchin cardinalfish 2.8 mm, 3.5 mm and 10.4 mm SL. Regions similar to those shown within boxes are enlarged on right. Scale 0.05 mm. Muscle fibres (F) of the light diffuser surround the developing bones of pelvic girdle (PG). R – dorsal layer of fibrous tissue. The light organ contains many densely packed vesicles packed with bacteria. V – microvilli, M - melanin bodies, L – lumen, B – bacteria.

anterior and posterior to the light organ itself. Symbiotic bacteria (*Photobacterium leiognathi*) were observed in the larval light organs starting at 3.5 mm SL and the gross morphology of the light organ in 10.4-mm larva is basically identical to that of the adults (Leis & Bullock, 1986). This study suggests that the light organ of sea urchin cardinalfish should be functional before reaching 10.4 mm SL, when associated reflector and diffusers are already fully developed and while the larvae are still pelagic. Luminous bacteria are apparently acquired *de novo* from the environment by developing larvae, but when this happens is still a matter of conjecture (Leis and Bullock, 1986). The presence of a recently described oral light organ in a second species of *Siphamia* (Fishelson *et al.*, 2005), as well as its early development, remains to be confirmed. Although a number of other forms of apogonid larvae have been described, few have been ascribed to particular species and no developing light organs are evident from available illustrations and descriptions (Leis & Rennis, 2000a).

The early stages of development are known for at least some species in the remaining bioluminescent perciforms – e.g. Acropomatidae (*Acropoma*), Leiognathidae (*Leiognathus*), Pempheridae (*Pempheris*, *Parapriacanthus*) – but no references to developing light organs are made in available descriptions and no precocious light organs are evident from published illustrations (Haque & Ozawa, 1995; Sandknop & Watson, 1996; Leis & Rennis, 2000b; Trnski & Leis, 2000). Thus, more detailed studies of larval ontogeny, including histological investigations, are needed to confirm the presence or absence of light organs in the larvae of perciform fishes with internal light organs.

The self-luminescent perciform genera *Pseudoscopelus* (Chiasmodontidae) and *Collichthys* (Sciaenidae), unlike perciforms with internal light glands, possess numerous small photophores on lower parts of the body (Herring & Morin, 1978). The early ontogeny of *Pseudoscopelus* species is barely known, but the absence of light organs in available illustrations suggests that development of photophores is postponed until probably the juvenile stage (Watson & Sandknop, 1996). No early life history information is available on luminous species of *Collichthys*, but late development of body photophores is most likely the case in this genus as well.

FUNCTIONAL SIGNIFICANCE OF BIOLUMINESCENCE IN LARVAL FISHES

Although a significant body of information has accumulated on the early development of teleost fishes with bioluminescent properties as adults, numerous aspects of the early ontogeny of light production, such as the exact timing when photophores become functional, possible changes in physical properties of the emitted light during growth and maturation, and/or ontogenetic variations in light organ development between different ecological groups remain unknown. In addition, although a significant number of observations of live bioluminescence is available for adult fishes, such data for larval stages are almost completely lacking. In fact, only the larvae of coastal midshipman have been closely studied to date in this respect (Anctil, 1977).

Despite the numerous gaps in our knowledge of early light production in fishes, certain clues to its possible function and biological significance can be deduced from the accumulated ecological and developmental information on marine fish larvae. This is especially true for the most diverse and abundant bioluminescent teleost representatives, the midwater ichthyofauna, where in some cases 97% of the individuals and 75% of the fish species can be bioluminescent (Badcock & Merrett, 1976). Due to the abundance and ubiquity of mesopelagic fish larvae in various oceanic regions, more information on their development and habitat preferences is available than for any other bioluminescent fish group.

Unlike shallow-living and coastal species, where adult and larval lives are spent in relatively similar depth strata, most deep-sea pelagic fishes initially develop in the productive surface waters, forming a significant portion of the zooplankton assemblage (the ichthyoplankton). With growth and approaching the juvenile stage, these larvae initiate their descent to deeper levels common for adults, a phenomenon known as ontogenetic vertical migration. Most oceanic ichthyoplankton are found within the upper 200 m and the bulk is composed of larvae of light-producing fishes such as stomiiforms and myctophids (Loeb, 1979; Suntsov, 2000; Sassa *et al.*, 2002).

The available information on the development of deep-sea fishes reveals a significant variation in developmental timing of particular photophores or photophore groups. It is not clear, however, whether these developing photophores actually produce light, and to date no live bioluminescence at these early stages has been reported. In adult deep-sea fishes, the uses of bioluminescence are manifold and often several functions can be attributable to a single light organ. Among most commonly cited functions of light production in fishes are ventral counter-illumination to match the downwelling light, attraction or illumination of prey, intra- and inter-specific communication, and deterrence of predators (Herring, 1982; Widder, 1999). One can hypothesise a number of similar uses in fish larvae, but at least some known functions (e.g. attracting a mate) can be excluded from early ontogeny with confidence. In addition, the full diversity of adult light organs is not acquired until the completion of metamorphosis, making the larval bioluminescent repertoire intrinsically more limited.

Ventral photophores are the first to develop in certain midwater fishes and the use of bioluminescence for counter-illumination seems to be the most plausible explanation. According to experimental data, the effective upper depth limit for counter-illumination is approximately 400 m (Young *et al.*, 1980). Thus, during daytime, light production is of limited value in the well-lit epipelagic zone, because no light organ can match the high intensity of downwelling irradiance.

In oceanic lanternfishes, whose larvae develop primarily at depths less than 100 m (Loeb, 1979), first to develop are the Br_2 photophores, followed by some other photophores on the head (Fig. 6). The acquisition of photophores on the ventral surface of the body is postponed to later stages and is gradual. This particular feature, as well as the occurrence of lanternfish larvae in the upper levels of the water column, suggests that ventral counter-illumination is likely not to occur during early development of myctophids. An apparently similar situation exists in the larvae of some other luminous fishes that develop in well-lit levels of the upper water column. For example, only non-functional, white photophores are found in the larvae of *Cyclothone* spp. occurring above 200 m, with photophore completion and metamorphosis occurring at depths greater than 350 m (Ahlstrom, 1973; Loeb, 1979).

The early appearance of some light organs in myctophid larvae could potentially be an indication of other uses of bioluminescence at low light levels, such as at night. For example, early forming head photophores (Br_2, Dn, Vn and others) could offer a certain selective advantage (if functional) and be used for attraction of prey organisms such as copepods (Moser, 1981), most of which are known to have positive phototaxis. Interestingly enough, development of light organs on the head (e.g. Br, So) in larval stomiiforms, also precedes the appearance of body photophores (Watson, 1996a, b; Suntsov, 1999), possibly indicating a common ecological cause in both groups. However, the production of light has its own metabolic costs and for such a strategy to be effective the advantages should outweigh the possible losses. There is also the possibility that certain patterns of light organ development could be a reflection of specific phylogenetic patterns and have little or no functional significance in early ontogeny.

Interestingly, the timing in photophore development seems to have a positive correlation with different ecological preferences in different groups of myctophid larvae. Two currently recognised subfamilies of oceanic lanternfishes, Myctophinae and Lampanyctinae, have markedly different larval morphologies and also vary in their habitat preferences, with larvae of the former subfamily usually found deeper in the water column, while most larval lampanyctines inhabit shallower depths (Ahlstrom, 1959; Loeb, 1979; Sassa *et al.*, 2002). Available information on the development of deeper dwelling lanternfish larvae (subfamily Myctophinae) indicates that more photophores develop early and have more advanced development in this subfamily than in the other group with shallow dwelling larvae (subfamily Lampanyctinae) (Moser & Ahlstrom, 1996). It is possible that lower irradiance as well as lower prey abundance at deeper strata would render precocious development in light organs and their early functional state more useful and advantageous for early life.

The larvae of oceanic hatchetfishes (Sternoptychidae) display strikingly precocious development of photophores in the ventral series. The adult hatchetfishes, with their extremely flattened bodies, silvery flanks, and clusters of abdominal photophores, are exemplary deep-sea fishes utilising ventral counter-illumination in oceanic midwaters (Denton *et al.*, 1972). Larvae of *Argyropelecus* and *Sternoptyx* (and possibly some other genera) are unusual among other types of oceanic ichthyoplankton in developing at markedly deeper levels of the water column. Larvae of *Sternoptyx* are found between 500-1000 m and *Argyropelecus* between 100-500 m in the eastern Atlantic (Badcock & Merrett, 1976). In the central Pacific the situation is somewhat different with larval *Sternoptyx* occurring at 100-350 m and *Argyropelecus* at 350-600 m, but this still reflects deeper dwelling preferences (Loeb, 1979). It seems plausible that early developing clusters of ventral photophores in hatchetfish larvae could well be functional and used for counter-illumination at low irradiance levels in deeper water. At the opposite end of the vertical distribution spectrum, the larvae of *Diplophos* (Diplophidae), whose adults have well-formed abdominal photophores, develop at very shallow depths (0-25 m; Loeb, 1979). The formation of photophores in this genus is not apparent during the larval stage and evidently quite delayed.

Data on the early development of another deep-pelagic fish family, the Platytroctidae, suggest a similar trend in light organ development, i.e. deeper water levels - early forming photophores. A number of platytroctid species with documented early ontogeny display very precocious development of certain body photophores as well as the tube shoulder organ (Fig. 4). This is in agreement with present knowledge on their vertical distribution, where larvae occur at deeper strata – 300-1000 m (Matsui, 1991), than the majority of larval fishes in the open ocean. Although the functional state of these early forming light organs is also a matter of conjecture, dim downwelling light of the mesopelagic zone could be a factor, making the early emission of light from larval photophores selectively advantageous.

One point to consider is that if light production is indeed present in the larvae of some bioluminescent mesopelagic fishes, this light could be quite different from that of adults. Fully developed photophores can be quite complex structures, equipped with reflectors, screens, filters, diffusers or light conducting channels, which modulate the spectrum and angular distribution of the emissions (Herring, 1982; Widder *et al.*, 1983). Thus, the final state of a particular light organ is probably achieved only after metamorphosis, and before that, the emission of light may not be tuned as in adults.

It is apparent from previous accounts of fish bioluminescence (Herring & Morin, 1978; Herring, 1982) that the multiple origins of light production in unrelated teleost groups indicate a great degree of convergence in evolution of both bacterial and endogenous light organs due to common selection pressures. Intrinsic luminescence is usually confined to single cutaneous light organs or photophore groups and is found in both ancient deep-water fishes (*sensu* Andriashev, 1953) such as stomiiforms, myctophiforms and alepocephaloids as well as more phylogenetically recent deep-sea families, e.g. swallowers (Chiasmodontidae). Although both "primary" and "secondary" deep-sea fishes possess multiple body photophores as adults, only ancient deep-water forms display early the appearance of light organs in larvae. Evidently, longer evolutionary periods under deep-sea conditions have not only produced an array of unique morphological and physiological adaptations in primitive teleosts, but also influenced their early ontogeny, warranting early development of light organs in larvae.

FUTURE PERSPECTIVES

It is apparent that bioluminescence has played a significant, if not pivotal, role in the evolution of certain teleost groups, facilitating a number of adaptive radiations in a diversity of habitats ranging from deep-sea to coastal waters and coral reefs. To date, the research emphasis in studies of fish bioluminescence has been on the adult stage. Ideally, however, every adaptive feature or function should be investigated in its entirety, from earliest appearance to fully developed state, if we are to gain a true understanding of its biological significance. Considering the still nascent state of larval bioluminescence research, the prospects for incorporating data from early ontogeny in the studies of light production in fishes are very promising and this largely unexplored field will no doubt yield many important insights and discoveries in the future.

One of the primary goals of future research should be to document the functional ontogeny of light producing organs or structures. This information, coupled with available data on various developmental patterns in different teleost lineages, could supply valuable insight into environmental adaptations, developmental constraints and/or ecological and behavioural factors relevant to light production during early life stages. A taxonomically broad approach in such studies will also likely provide insight into convergent evolution of bioluminescent phenomena.

Differential rates of development of photophores, changes in intensity and spectral composition of light during development would also be important and interesting aspects to address in future studies. For example, heterochrony has been a common evolutionary mechanism in the long history of teleost fishes. It is intriguing to consider the potential role of changes in the order and timing of developmental events in reference to developing photophores and other specialised light organs. Another important and equally demanding component of larval bioluminescence research will be to understand the exact pathways and origins of luciferins in self-luminescent groups and also mechanisms and timing of bacterial acquisition in species with bacterial bioluminescence.

Detailed morphological, physiological and histological studies of the early ontogeny in a few bioluminescent "oddballs" – e.g. rare species in otherwise non-bioluminescent groups, could be particularly promising in gaining understanding of the independent evolution of light production, possible limiting factors precluding its development, as well as mechanisms or morphologies which overcome such restrictions. Similarly, detailed comparison of light producing structures in phylogenetically "old" and "modern" teleost groups would be valuable in deciphering the degree of convergence of bioluminescent phenomena.

Several of these avenues of research would clearly require a great deal of experimental field work and laboratory study. Such research would be particularly interesting for pelagic deep-sea fishes considering their outstanding diversity. However, deep-sea fishes are rarely maintained in captivity because of their strict physiological and ecological demands. In this regard, experimental work on larvae might prove more realistic, considering their often less strict physiological requirements.

In conclusion, teleost larvae are incredibly diverse morphologically, behaviorally and ecologically and the body of knowledge in this area is impressive. Capitalising on this wealth of information to understand the early ontogeny of bioluminescence is a challenging task which should prove rewarding for any researcher ready to take on the challenge.

Acknowledgements

The authors would like to thank the American Society of Ichthyologists and Herpetologists and many individual researchers for granting permission to reproduce some of their published illustrations. During the preparation of this manuscript A. Suntsov received financial support from Harbor Branch Oceanographic Institution (USA), Institute of Aquatic Resources of the North (Russia) and National Research Council (USA). E. Widder received support from the Office of Naval Research Grant #N00014-02-1-0949 and T. Sutton received support from Harbor Branch Oceanographic Institution, the MAR-ECO project of the Census of Marine Life, and the National Science Foundation Ocean Sciences Division – Biological Oceanography Program Grant #OCE 0623551.

References

Ahlstrom EH (1959) Vertical distribution of pelagic fish eggs and larvae off California and Baja California. Fish Bull 60: 107-146

Ahlstrom EH (1973) The diverse patterns of metamorphosis in gonostomatid fishes – an aid to classification. In: Blaxter JHS (ed) The Early Life History of Fish. Springer-Verlag, Berlin, Germany, p 659-674

Ahlstrom EH, Moser HG, Cohen DM (1984) Argentinoidei: development and relationships. In: Moser HG, Richards WJ, Cohen DM, Fahay MP, Kendall AW Jr, Richardson SL (eds) Ontogeny and Systematics of Fishes. American Society of Ichthyologists and Herpetologists, Special Publication no 1, La Jolla, California, p 155-169

Ambrose DM (1996) Macrouridae: Grenadiers. In: Moser HG (ed) The Early Stages of Fishes in the California Current Region. La Jolla, California: CalCOFI Atlas no 33, p 483-499

Andriashev AP (1953) Ancient deep-water and secondary deep-water fishes and their importance in a zoogeographical analysis. In: Lindberg GU (ed) Essays on Special Problems in Ichthyology. IzdatelstvoAkademii Nauk SSSR, Moscow-Leningrad, p 58-64

Anctil M (1977) Development of bioluminescence and photophores in the Midshipman fish, *Porichthys notatus*. J Morphol 151: 363-396

Ast JC, Dunlap PV (2005) Phylogenetic resolution and habitat specificity of members of the *Photobacterium phosphoreum* species group. Environ Microbiol 7: 1641-1654

Badcock JR, Larcombe RA (1980) The sequence of photophore development in *Xenodermichthys copei* (Pisces: Alepocephalidae). J Mar Biol Assoc UK 60: 277-294

Badcock JR and Merrett NR (1976) Midwater fishes in the eastern North Atlantic - I. Vertical distribution and associated biology in 30°N, 23°W, with developmental notes on certain myctophids. Prog Oceanog 7: 3-58

Baguet F and Case JF (1971) Luminescence control in *Porichthys* (Teleostei): excitation of isolated photophores. Biol Bull 140: 15-27

Baldwin CC, Johnson GD (1995) A larva of the Atlantic flashlight fish, *Kryptophanaron alfredi* (Beryciformes : Anomalopidae), with a comparison of beryciform and stephanoberyciform larvae. Bull Mar Sci 56: 1-24

Beebe W (1932) Nineteen new species and four post-larval deep-sea fish. Zoologica NY, 13: 47-107

Belyanina TN (1977) Materials on development of *Chauliodus* species (Chauliodontidae, Pisces). Trudy Instituta Okeanologii AN SSSR 109: 113-132

Belyanina TN (1982) Larvae of midwater fishes in the western tropical Pacific Ocean and the seas of the Indo-Australian Archipelago. Trudy Instituta Okeanologii AN SSSR 118: 5-42

Bertelsen E (1951) The ceratioid fishes. Ontogeny, taxonomy, distribution and biology. Dana Reports 39: 1-276

Bertelsen E (1984) Ceratioidei: development and relationships. In: Moser HG, Richards WJ, Cohen DM, Fahay MP, Kendall AW Jr, Richardson SL (eds) Ontogeny and Systematics of Fishes. American Society of Ichthyologists and Herpetologists, Special Publication no 1, La Jolla, California, p 325-334

Bertelsen E, Munk O (1964). Rectal light organs in the argentinoid fishes *Opisthoproctus* and *Winteria.* Dana Reports 62: 1-17

Bertelsen E, Theisen B, Munk O (1965) On a post-larval specimen, anal light organ, and tubular eyes of the argentinoid fish *Rhynchohyalus natalensis* (Gilchrist and von Bonde). *Vidensk Meddr Jansk naturh Foren* 128: 357-371

Boettcher KJ, Ruby EG, McFall-Ngai, MJ (1996) Bioluminescence in the symbiotic squid *Euprymna scolopes* is controlled by a daily biological rhythm. J Comp Physiol 179: 65-73

Campbell AK, Herring PJ (1990) Imidazolopyrazine bioluminescence in copepods and other marine organisms. Mar Biol 104: 219-225

Castle PHJ, Paxton JR (1984) A new genus and species of luminescent eel (Pisces, Congridae) from the Arafura Sea, northern Australia. Copeia 1984: 72-81

Cohen DM (1964) Bioluminescence in the Gulf of Mexico anacanthine fish *Steindachneria argentea*. Copeia 1964: 406-409

Crane JM (1968) Bioluminescence in the batfish *Dibranchus atlanticus*. Copeia 1968: 410-411

Denton EJ, Gilpin-Brown JB, Wright PG (1972) The angular distribution of the light produced by some mesopelagic fish in relation to their camouflage. Proc R Soc Lond B 182: 145-158

Endo H, Yabe M, Amaoka, K (1992) A rare macrourid alevin of the genus *Hymenocephalus* from the Pacific Ocean. Jap J Ichthyol 39: 265-267

Evseenko SA, Suntsov AV (1995) Early stages of development of rare deep water fishes from notal waters of the southwestern Pacific. J Ichthyol 35: 111-122

Fahay MP (1989) The ontogeny of *Steindachneria argentea* Goode and Bean with comments on its relationships. In: Cohen DM (ed) Papers on Systematics of Gadiform Fishes. Los Angeles County Natural History Museum, Science Series 32: 143-158

Fahay MP, Markle DF (1984) Gadiformes: development and relationships. In: Moser HG, Richards WJ, Cohen DM, Fahay MP, Kendall AW Jr, Richardson SL (eds) Ontogeny and Systematics of Fishes. American Society of Ichthyologists and Herpetologists, Special Publication no 1, La Jolla, California, p 265-283

Fink WL (1985) Phylogenetic interrelationships of the stomiid fishes (Teleostei: Stomiiformes). Misc Publ Mus Zool Univ Mich No 171, p i-vii + 1-127

Fishelson L, Gon O, Goren M, Ben-David-Zaslow R (2005) The oral cavity and bioluminescent organs of the cardinal fish species *Siphamia permutata* and *S. cephalotes* (Perciformes, Apogonidae). Mar Biol 147: 603-609

Frank TM, Widder EA, Latz MI, Case JF (1984) Dietary maintenance of bioluminescence in a deep-sea mysid. J Exp Biol 109: 385-389

Fukui A, Tsuchiya T (2005) Pelagic larvae of *Ventrifossa garmani* (Gadiformes: Macrouridae) from Suruga Bay and offshore waters of Japan. Ichthyol Res 52: 311-315

Haddock SHD, Case JF (1994) A bioluminscent chaetognath. Nature 367: 225-226

Haddock SHD, Rivers TJ, Robison BH (2001) Can coelenterates make coelenterazine? Dietary requirement for luciferin in cnidarian bioluminescence. Proc Nat Acad Sci USA 98: 11148-11151

Haneda Y (1951) The luminescence of some deep-sea fishes of the families Gadidae and Macrouridae. Pacific Science 5: 372-378

Haneda Y, Tsuji FI, Sugiyama N (1969) Luminescent systems in apogonid fishes from the Philippines. Science 165: 188-190

Hansen K, Herring PJ (1977) Dual bioluminescent systems in the anglerfish genus *Linophryne* (Pisces. Ceratioidea). J Zool 182: 103-124

Haque MM, Ozawa T (1995) Ontogenic larval characters of 3 leiognathid species in Kagoshima Bay, southern Japan. Jap J Ichthyol 42: 137-146

Hastings JW (1971) Light to hide by: ventral luminescence to camouflage the silhouette. Science 173: 1016-1017

Hastings JW (1983) Diversity, chemistry and evolution of bioluminescence. J Mol Evol 19: 309-321

Hastings JW, Morin JG (1991) Bioluminescence. In: Prosser CL (ed) Comparative Animal Physiology. Neural and Integrative Animal Physiology, 4th Edition, Wiley-Liss, New York, p 131-170

Haygood MG (1993) Light organ symbioses in fishes. Critical Rev Microbiol 19: 191-216

Haygood MG, Distel DL (1993) Bioluminescent symbionts of flashlight fishes and deep-sea anglerfishes form unique lineages related to the genus *Vibrio*. Nature 363: 154-156

Herring PJ (1972) Bioluminescence of searsiid fishes. J Mar Biol Assoc UK 52: 879-887

Herring PJ (1975) Bacterial bioluminescence in some argentinoid fishes. In: Barnes H (ed) Proc of 9th Eur Mar Biol Symp, Aberdeen University Press, Aberdeen, p 563-572

Herring PJ (1977) Bioluminescence in an evermannellid fish. J Zool 181: 297-307

Herring PJ (1982) Aspects of bioluminescence of fishes. Oceanogr Mar Biol Ann Rev 20: 415-470

Herring PJ (1987) Systematic distribution of bioluminescence in living organisms. J Biolum Chemilum 1: 147-163

Herring PJ (2002) Marine microlights: the luminous marine bacteria. Microbiol Today 29: 174-176

Herring PJ, Cope C (2005) Red bioluminescence in fishes: on the suborbital photophores of *Malacosteus, Pachystomias* and *Aristostomias*. Mar Biol 148: 383-394

Herring PJ, Gruner HE (2004) Luminous organs and luminescence. In: Forest J, von Vaupel Klein JC, Schram FR (eds.) Treatise on Zoology, Anatomy, Taxonomy, Biology, Vol. 1. The Crustacea. EJ Brill, Leiden, The Netherlands, p 381-413

Herring PJ, Morin JG (1978) Bioluminescence in fishes. In: Herring PJ (ed) Bioluminescence in Action. Academic Press, New York, p 273-329

Herring PJ, Munk O (1994) The escal light gland of the deep-sea anglerfish *Haplophryne mollis* (Pisces: Ceratioidei) with observations on luminescence control. J Mar Biol Assoc UK 74: 747-763

Herring PJ, Widder EA (2001) Bioluminescence in plankton and nekton. In: Steele JH, Thorpe SA, Turekian KK (eds) Encyclopedia of Ocean Science, Vol 1. Academic Press, San Diego, p 308-317

Iwai T (1971) Structure of luminescent organs of apogonid fish, *Siphamia versicolor*. Jap J Ichthyol 18: 125-127

John H-Ch, Zelk C, Burkel D (2000) Description of a larval tubeshoulder, tentatively assigned to *Maulisia argipalla* Matsui and Rosenblatt, 1979 (Teleostei, Platytroctidae). Mitt Hamb Zool Mus Inst 97: 117-122

Johnson RK (1984a) Scopelarchidae: development and relationships. In: Moser HG, Richards WJ, Cohen DM, Fahay MP, Kendall AW Jr, Richardson SL (eds) Ontogeny and Systematics of Fishes. American Society of Ichthyologists and Herpetologists, Special Publication no 1, La Jolla, California, p 245-250

Johnson RK (1984b) Evermannellidae: development and relationships. In: Moser HG, Richards WJ, Cohen DM, Fahay MP, Kendall AW Jr, Richardson SL (eds) Ontogeny and Systematics of Fishes. American Society of Ichthyologists and Herpetologists, Special Publication no 1, La Jolla, California, p 250-254

Johnston IA, Herring PJ (1985) The transformation of muscle into bioluminescent tissue in the fish *Benthalbella infans* Zugmayer. Proc R Soc Lond B 255: 213-218

Jordan AR, Bruce BD (1993) Larval development of three roughy species complexes (Pisces: Trachichthyidae) from southern Australian waters, with comments on the occurrence of orange roughy *Hoplosthetus atlanticus*. Fish Bull 91: 76-86

Jorgensen JM, Munk O (1979) Photophores and presumably luminous chin barbel and pectoral fin ray filaments of *Thysanactis dentex* (Pisces, Stomiatoidea). Acta Zool 60: 33-42

Kawaguchi K, Moser HG (1984) Stomiatoidea: Development. In: Moser HG, Richards WJ, Cohen DM, Fahay MP, Kendall AW Jr, Richardson SL (eds) Ontogeny and Systematics of Fishes. American Society of Ichthyologists and Herpetologists, Special Publication no 1, La Jolla, California, p 169-181

Kawaguchi K, Moser HG (1993) Development and distribution of the early-life history stages of the mesopelagic fish *Tactostoma macropus* (Stomiidae) in the transitional waters of the North Pacific. Jap J Ichthyol 40: 161-172

Konishi Y (2000) Monocentridae (pineconefishes). In: Leis JM, Carson-Ewart BM (eds) The Larvae of Indo-Pacific Coastal Fishes. An Identification Guide to Marine Fish Larvae, EJ Brill, Leiden, The Netherlands, p 182-185

Konishi Y, Okiyama M (1997) Morphological development of four trachichthyoid larvae (Pisces: Beryciformes), with comments on trachichthyoid relationships. Bull Mar Sci 60: 66-88

Konishi Y, Colin PL (2000) Anomalopidae (Lanterneyes, Flashlightfishes). In: Leis JM, Carson-Ewart BM (eds) The Larvae of Indo-Pacific Coastal Fishes. An Identification Guide to Marine Fish Larvae, EJ Brill, Leiden, The Netherlands, p 165-169

Kotlyar AN (1996) Beryciform Fishes of the World Ocean. VNIRO Publishing, Moscow

Leis JM, Bullock S (1986) The luminous cardinalfish *Siphamia* (Pisces, Apogonidae): Development of larvae and the luminous organ. In: Uyeno T, Arai R, Taniuchi T, Matsuura K (eds) Proc 2nd Int Conf Indo-Pacific Fishes, Ichthyol Soc Japan, Tokyo, p 703-714

Leis JM, Rennis DS (2000a) Apogonidae (Cardinalfishes). In: Leis JM, Carson-Ewart BM (eds) The Larvae of Indo-Pacific Coastal Fishes. An Identification Guide to Marine Fish Larvae. EJ Brill, Leiden, The Netherlands, p 237-279

Leis JM, Rennis DS (2000b) Pempherididae (Sweepers, bullseyes) In: Leis JM, Carson-Ewart BM (eds) The Larvae of Indo-Pacific Coastal Fishes. An Identification Guide to Marine Fish Larvae. EJ Brill, Leiden, The Netherlands, p 425-429

Loeb VJ (1979) Vertical distribution and development of larval fishes in the North Pacific Central Gyre during summer. Fish Bull 77: 777-793

Mallefet J, Shimomura O (1995) Presence of coelenterazine in mesopelagic fishes from the Strait of Messina. Mar Biol 124: 381-385

Marshall NB, Cohen DM (1973) Order Anacanthini (Gadiformes), characters and synopsis of families. In: Cohen DM (ed) Fishes of the Western North Atlantic, Mem Sears Found Mar Res No 1, Pt 6, p 479-495

Marshall NB, Iwamoto T (1973) Family Macrouridae. In: Cohen DM (ed) Fishes of the Western North Atlantic. Mem Sears Found Mar Res No 1, Pt 6, p 496-665

Matsui T (1991) Description of young of the mesopelagic platytroctids *Holtbyrnia latifrons* and *Sagamichthys abei* (Pisces, Alepocephaloidea) from the Northeastern Pacific Ocean. Fish Bull 89: 209-219

Matsui T, Rosenblatt RH (1971) Ontogenetic changes in patterns of light organs in searsiids and the taxonomy of *Sagamichthys* and *Persparsia*. Copeia 1971: 440-448

Matsui T, Rosenblatt RH (1987) Review of the deep-sea fish family Platytroctidae (Pisces: Salmoniformes). Bull Scripps Inst Oceanogr Univ Calif v 26, p vii + 159

McFall-Ngai MJ (1998) The development of cooperative associations between animals and bacteria: Establishing détente among domains. Am Zool 38: 593-608

Mensinger AF, Case JF (1991) Bioluminescence maintenance in juvenile *Porichthys notatus*. Biol Bull 181: 181-188

Merrett NR (1989) The elusive macrourid alevin and its seeming lack of potential in contributing to intrafamilial systematics. In: Cohen DM (ed) Papers on Systematics of Gadiform Fishes. Los Angeles County Natural History Museum, Science Series 32: 175-185

Merrett NR, Badcock J, Herring PJ (1973) The status of *Benthalbella infans* (Pisces: Myctophoidei), its development, bioluminescence, general biology and distribution in the eastern North Atlantic. J Zool 170: 1-48

Morin JG (1983) Coastal bioluminescence: patterns and functions. Bull Mar Sci 33: 787-817

Moser HG (1981) Morphological and functional aspects of marine fish larvae. In: Lasker R (ed), Marine Fish Larvae – Morphology, Ecology and Relation to Fisheries. Washington Sea Grant Program, Seattle and London, p 89-131

Moser HG (ed) (1996) The Early Stages of Fishes in the California Current Region. CalCOFI Atlas 33

Moser HG, Ahlstrom EH (1972) Development of the lanternfish, *Scopelopsis multipunctatus* Brauer 1906, with a discussion of its phylogenetic position in the family Myctophidae and its role in a proposed mechanism for the evolution of photophores patterns in lanternfishes. Fish Bull 70: 541-564

Moser HG, Ahlstrom EH (1996) Myctophidae: lanternfishes. In: Moser HG (ed) The Early Stages of Fishes in the California Current Region. CalCOFI Atlas 33, p 387-475

Moser HG, Watson W (1996) Stomiiformes. In: Moser HG (ed) The Early Stages of Fishes in the California Current Region. CalCOFI Atlas 33, p 244-246

Moser HG, Ahlstrom EH, Paxton JR (1984) Myctophidae: development. In: Moser HG, Richards WJ, Cohen DM, Fahay MP, Kendall AW Jr, Richardson SL (eds) Ontogeny and Systematics of Fishes. American Society of Ichthyologists and Herpetologists, Special Publication no 1, La Jolla, California, p 218-239

Munk O (1999) The escal photophore or ceratioids (Pisces; Ceratioidei) – a review of structure and function. Acta Zool 80: 265-284

Munk O, Herring PJ (1996) An early stage in development of escae and caruncles in the deep-sea anglerfish *Cryptopsaras couesi* (Pisces; Ceratioidei). J Mar Biol Assoc UK 76: 517-527

Munk O, Hansen K, Herring PJ (1998) On the development and structure of the escal light organ of some melanocetid deep-sea anglerfishes (Pisces; Ceratioidei). J Mar Biol Assoc UK 78: 1321-1335

Nafpaktitis BG (1977) Family Neoscopelidae. In: Gibbs RH Jr (ed) Fishes of the Western North Atlantic, Memoirs of Sears Foundation for Marine Research, New Haven, Connecticut vol 7, p 1-12

Nelson JS (2006) Fishes of the World, Fourth Edition. John Wiley & Sons, Hoboken.

Nicol JAC (1957) Observations on photophores and luminescence in the teleost *Porichthys*. Q J Microscop Sci 98: 179-188

Nielsen JG, Bertelsen E (1985) The gulper-eel family Saccopharyngidae (Pisces, Anguilliformes). Steenstrupia 11: 157-206

Nielsen JG, Bertelsen E, Jespersen A (1989) The biology of *Eurypharynx pelecanoides* (Pisces, Eurypharyngidae). Acta Zool 70: 187-197

Nyholm SV, McFall-Ngai MJ (2004) The winnowing: Establishing the squid-*Vibrio* symbiosis. Nature Rev Microbiol 2: 632-642

O'Day WT (1973) Luminescent silhouetting in stomiatoid fishes. Contributions to Science, Los Angeles Natural History Museum no 246: 1-8

Okamura O (1970) Studies on the macrourid fishes of Japan – morphology, ecology and phylogeny. Reports of Usa Marine Biological Station, Kochi University 17: 1-179

Okiyama M (1984) Myctophiformes: development. In: Moser HG, Richards WJ, Cohen DM, Fahay MP, Kendall AW Jr, Richardson SL (eds) Ontogeny and Systematics of Fishes. American Society of Ichthyologists and Herpetologists, Special Publication no 1, La Jolla, California, p 206-218

Okiyama M (1988) Monocentridae In: Okiyama M (ed) An Atlas of the Early Stages of Fishes in Japan. Tokai University Press, Tokyo, p 358-359

Owre HB, Bayer FM (1970) The deep-sea gulper *Eurypharynx pelecanoides* Vaillant, 1982 (order Lyomeri) from the Hispaniola basin. Bull Mar Sci 20: 186-192

Ozawa T (1986) Early life history of the family Myctophidae in the ocean off Southern Japan. In: Ozawa T (ed) Studies on the Oceanic Ichthyoplankton in the Western North Pacific. Kyushu University Press, p 114-188

Parin NV, Borodulina OD (2003) Phylogeny, systematics, and zoogeography of the mesopelagic genus *Astronesthes* (Astronesthidae, Stomiiformes). J Ichthyol 43: 557-576

Paxton JR (1972) Osteology and relationships of the lanternfishes (family Myctophidae). Bulletin of Natural History Museum, Los Angeles County no 13: 1-81

Paxton JR, Hulley PA (1999) Myctophidae. In: Carpenter KE, Niem VH (eds) FAO Species Identification Guide for Fishery Purposes. The Living Marine Resources of the Western Central Pacific. Volume 3. Batoid fishes, chimaeras and bony fishes part 1 (Elopidae to Linophrynidae). Rome, FAO, p 1957-1965

Pietch TW (2005) Dimorphism, parasitism, and sex revisited: modes of reproduction among deep-sea ceratioid anglerfishes (Teleostei: Lophiiformes). Ichthyol Res 52: 207-236

Rees J-F, Wergifosse BD, Noiset O, Dubuisson M, Janssens B, Thompson EM (1998) The origins of marine bioluminescence: Turning oxygen defence mechanisms into deep-sea communication tools. J Exp Biol 201: 1211-1221

Rofen RR (1966) Family Paralepididae. In: Mead GW (ed) Fishes of the western North Atlantic. Memoirs of Sears Foundation for Marine Research, New Haven, Connecticut, no 1, pt 5, p 205-461

Ruby EG, Nealson KH (1976) Symbiotic association of *Photobacterium fischeri* with the marine luminous fish *Monocentris japonica*: a model of symbiosis based on bacterial studies. Biol Bull 151: 574-586

Sandknop EM, Watson W (1996) Howellidae: Pelagic basslets. In: Moser HG (ed) The Early Stages of Fishes in the California Current Region. La Jolla, California, CalCOFI Atlas 33, p 1072-1077

Sassa C, Moser HG, Kawaguchi K (2002) Horizontal and vertical distribution patterns of larval myctophid fishes in the Kuroshio Current region. Fish Oceanogr 11: 1-20

Sazonov YI (1995) Morphology and significance of the luminous organs in alepocephaloid fishes. Int Symp Deep Sea Extreme Shallow-Water Habitats, Vienna (Austria), Oct 1995

Sazonov YI (1995) Three rare species of slickheads (Alepocephalidae), found for the first time in the Indian Ocean, with remarks about the status of *Microphotolepis schmidti*. J Ichthyol 35: 130-139

Schmidt J (1918) Argentinidae, Microstomidae, Opisthoproctidae, Mediterranean Odontostomidae, Reports of Dana Oceanographic Expedition, 1908-1910 Mediterranean and Adjacent Seas, Biology 2: 1-40

Seliger HH (1993) Bioluminescence: Excited states under cover of darkness. Naval Res Rev XLV: 5-11

Shimomura O, Johnson FH, Morise H (1974) Mechanism of the luminescent intramolecular reaction of aequorin. Biochemistry 13: 3278-3286

Shimomura O, Inoue S, Johnson FH, Haneda Y (1980) Widespread occurrence of coelenterazine in marine bioluminescence. Comp Biochem Physiol 65B: 435-437

Somiya H (1977) Bacterial bioluminescence in chlorophthalmid deep-sea fish: a possible interrelationship between the light organ and the eyes. Experientia 33: 906-909

Suntsov AV (1999) Early developmental stages of *Astronesthes spatulifer* from the subtropical convergence region in the southwestern Pacific. J Ichthyol 39: 762-771

Suntsov AV (2000) Ichthyoplankton assemblages in the North Pacific Central Gyre. J Ichthyol, Suppl 1 40: S152-S162

Thompson EM, Nafpaktitis BG, Tsuji FI (1987) Induction of bioluminescence in the marine fish, *Porichthys*, by *Vargula* (crustacean) luciferin. Evidence for *de novo* synthesis or recycling of luciferin. Photochem Photobiol 45: 529-533

Thomson CM, Herring PJ, Campbell AK (1995) Evidence for *de novo* biosynthesis of coelenterazine in the bioluminescent midwater shrimp, *Systellaspis debilis*. J Mar Biol Assoc UK 75: 165-171

Trnski T, Leis JM (2000) Acropomatidae (Glowbellies, Splitfins). In: Leis JM, Carson-Ewart BM (eds) The larvae of Indo-Pacific Coastal Fishes. An Identification Guide to Marine Fish Larvae. EJ Brill, Leiden, The Netherlands, p 263-268

Tsuji FI, Barnes AT, Case JF (1972) Bioluminescence in the marine teleost, *Porichthys notatus*, and its induction in a non-luminous form by *Cypridina* luciferin. Nature 237: 515-516

Wada M, Azuma N, Mizuno N, Kurokura H (1999) Transfer of symbiotic luminous bacteria from parental *Leiognathus muchalis* to their offspring. Mar Biol 135: 683-687

Watson W (1996a) Gonostomatidae: Lightfishes. In: Moser HG (ed) The Early Stages of Fishes in the California Current Region. La Jolla, California, CalCOFI Atlas 33, p 247-267

Watson W (1996b) Sternoptychidae: Hatchetfishes. In: Moser HG (ed) The Early Stages of Fishes in the California Current Region. La Jolla, California: CalCOFI Atlas 33, p 268-283

Watson W, Sandknop EM (1996) Chiasmodontidae: swallowers. In: Moser HG (ed) The Early Stages of Fishes in the California Current Region. La Jolla, California: CalCOFI Atlas 33, p 1131-1137

Whitehead PJP, Nelson GJ, Wongratana T (1988) Clupeoid fishes of the world (Suborder Clupeoidei). An annotated and illustrated catalogue of the herrings, sardines, pilchards, sprats, anchovies and wolf-herrings. Part 2. Engraulididae. FAO species catalogue, vol 1, part 2, p 305-579

Widder EA (1999) Bioluminescence. In: Archer SN, Djamgoz MBA, Loew E, Partridge JC, Vallerga S (eds) Adaptive Mechanisms in the Ecology of Vision. Kluwer Academic Publishers, Dordrecht, The Netherlands, p 555-581

Widder EA, Latz MI, Case JF (1983) Marine bioluminescence spectra measured with an optical multichannel detection system. Biol Bull 165: 791-810

Widder EA, Latz MI, Herring PJ, Case JF (1984) Far-red bioluminescence from two deep-sea fishes. Science 225: 512-514

Wolfe CJ, Haygood MG (1991) Restriction fragment length polymorphism analysis reveals high levels of genetic divergence among the light organ symbionts of flashlight fish. Biol Bull 181: 135-143

Young RE, Kampa EM, Maynard SD, Mencher FM, Roper CFE (1980) Counterillumination and the upper depth limits of midwater animals. Deep-Sea Res 27A: 671-691

PART 2

Respiration & Homeostasis

CHAPTER

4

Gas Exchange

Bernd Pelster[1,2,*]

INTRODUCTION

The form and function of animal respiratory and circulatory systems have attracted considerable attention for several hundred years. While the main organs of gas exchange (JO_2, JCO_2, JNH_3) in adult fishes are the gills, they are also responsible for hydromineral and acid-base balance (Evans *et al.* 1999, 2005), nitrogenous waste excretion (Randall *et al.* 1999; Terjesen, this volume), hormone production (Zaccone *et al.* 1996, 2006), and activation or inactivation of circulating metabolites (Olson, 1998). Although the circulatory system is typically the first functioning organ during early embryonic development (Rombough, 1997; Pelster, 1999, 2002, Hall, this volume, Burggren & Bagatto, this volume), gills develop much later (Rombough, 2002, 2004). Gas exchange, however, is required right from the start of development of the fertilised egg. This implies that gas exchange initially must occur through the cellular integument of the developing embryo, having traversed the boundary layer, chorion and perivitelline fluids (PVF) of the embryo, or the surface epithelium of the hatched larva. The gills only supersede this function once they have developed. In addition, the autonomic

Author's addresses: [1]Institut für Zoologie, Universität Innsbruck, Technikerstraße 25, A-6020 Innsbruck, Austria.
[2]Center of Molecular Biosciences, Leopold Franzens Universität Innsbruck, Technikerstraße 25, A-6020 Innsbruck, Austria.
**Address for Correspondence:* Institut für Zoologie, Leopold-Franzens-Universität Innsbruck, A-6020 Innsbruck, Austria.
E-mail: bernd.pelster@uibk.ac.at

nervous system, which controls cardiac activity and other organs, only becomes fully functional during later stages of development (Protas & Leontieva, 1992; Jacobsson & Fritsche, 1999; Schwerte *et al.*, 2006; Holmberg *et al.*, this volume). If this is also true for the control of ventilatory activity, the early embryonic or larval control of ventilation may be quite different from the system known for adult fish.

This chapter does not repeat earlier treatise (Rombough, 1988a, 2004), rather it summarises recent advances in our understanding of gas exchange during early fish ontogeny. Starting with the basic principles of gas exchange, which also apply to developing embryos and larvae, the contribution of the chorion and PVF to gas exchange resistance are discussed. Of special interest is the switch from bulk diffusion through the body surface to gill respiration. The influence of oxygen deficiency and also of temperature on gas exchange is discussed. Finally, the development of ventilatory control and the response to hypoxic conditions are considered.

THE PRINCIPLES OF GAS EXCHANGE

Aerobic energy metabolism is dependent on an adequate supply of oxygen, which is reduced to water by mitochondrial cytochrome aa3 activity, and on the removal of CO_2, which is produced mainly in the citric acid cycle. Accordingly, respiratory gas exchange comprises the transport of oxygen and CO_2 between the respiratory medium, i.e. the water, and the mitochondria. It involves convectional transport of gas, but also the diffusion of gas molecules. Starting from the respiratory medium, the first step is convective transport in the ventilatory water current. Uptake from the water into the organism at the gas exchange areas requires diffusional transport, which is ultimately driven by the partial pressure gradients. Once inside the blood stream convectional transport is responsible for the transport between the tissues and the gas exchange area. Diffusion again is required for the transport between the blood and tissue mitochondria. This transport chain can be visualised as a cascade (Dejours, 1981). Typically this cascade depicts the oxygen gradient between the environmental water and the mitochondria. Especially for water breathing animals, oxygen transport indeed appears to be much more critical than CO_2 transport, due to the formation of bicarbonate in water and the resulting high transport capacity for aqueous CO_2, which is indeed almost similar to the CO_2 transport capacity of air (Dejours, 1981). Accordingly, oxygen exchange is much more critical for water breathers than CO_2 exchange. Most studies focusing on gas exchange therefore discuss oxygen exchange, while very little attention is paid to CO_2 exchange.

Considering the decrease in pO_2 along the transport cascade, it becomes apparent that diffusion is the critical bottleneck. The diffusive flux of gas across a gas exchanger (either the gas exchange area or the connection between blood and tissue) is described by the Fick equation

$$J = K \cdot \Delta p \cdot A/L$$

where J is the rate of gas transfer, K represents Krogh's diffusion constant, A is the area available for the gas transfer, and L is the thickness of the gas exchanger, i.e. the length of the pathway, that has to be surpassed by diffusion, Δp is the partial pressure gradient across the diffusional pathway.

Krogh's diffusion constant comprises the diffusion coefficient (D) for the gas analysed, and also the gas solubility (α), or, if chemical binding is involved as for CO_2 and O_2, the transport capacity of the gas (ß) in water or body fluids

$$K = D \cdot ß$$

Krogh's diffusion constant is dependent on temperature, because an increase in temperature increases Brownian motion and thus the diffusion coefficient. Increasing temperature, however, decreases the solubility of gases in aqueous solutions. As a consequence, Krogh's diffusion constant increases only by a factor of 1.1 for a temperature increase of 10°C (Dejours, 1981).

While the value for Krogh's diffusion constant is similar in fish gills and in the skin, K_{O_2} for water is about 2.5 times higher than K_{O_2} for tissues (Rombough, 2004). Kranenbarg *et al.* (2000) assume that the diffusion coefficient in tissues is about 33% of the value measured in water. The morphology of the gas exchanger, i.e. its thickness and the area available for the exchange of gases are crucial for its effectiveness. A large surface area combined with a short diffusion distance is characteristic for an effective gas exchanger, with the Δp providing the driving force.

Convective transport of gases is achieved by ventilation and by blood flow. In quantitative terms the amount of gas taken up from the ventilatory water current must be transported by the blood stream. Ideally, ventilation-perfusion ratios are balanced, so that the amount of gas brought to one side of the exchanger is carried away on the other side. Due to the difference in the binding capacity for CO_2 and O_2 in water this ideal situation cannot be achieved for both gases at the same time.

Considering embryonic and larval gas exchange, convective gas transport in the blood stream can only occur after the onset of cardiac activity, establishing blood flow in the circulatory system. As we will see later, the circulatory system in most larvae becomes important for gas transport only in later developmental stages, so that bulk diffusion is mainly responsible for gas transport during early development (Kranenbarg *et al.*, 2000; Territo & Altimiras, 2001).

EGGS

Egg Size and Egg Position in a Clutch

A critical parameter determining the effectiveness of diffusional gas transport is the distance that has to be surpassed. Accordingly, for eggs incubated in water it has been suggested that for embryos of large eggs it would be more difficult to obtain the oxygen needed to meet their metabolic demands. With increasing diameter the surface to volume ratio will become smaller, so that the increasing oxygen requirements of the larger egg can no longer be met by diffusion. Thus, in oxygen-limited aquatic environments small eggs should be advantagous, while after hatching, survival should increase with increasing size of the larvae (Hendry *et al.*, 2001).

The idea that larger eggs are disadvantagous has been tested for brown trout (*Salmo trutta*). Survival rate of small and large sibling eggs of brown trout under normoxic conditions was similar, but under hypoxic conditions (72 μmol $O_2 \cdot L^{-1}$ = 2.3 $mg \cdot L^{-1}$ = 18% saturation) survival of larger eggs was significantly higher than survival of smaller eggs (Einum *et al.*, 2002). A possible explanation for this result was obtained in additional experiments with Atlantic salmon (*Salmo salar*). Oxygen consumption increased relatively slowly with increasing body mass (scaling factor 0.44), which may be due to a variation in yolk mass, rather than in body mass in eggs of different stages. The authors concluded that prehatching selection may also influence egg size evolution (Einum *et al.*, 2002), and the results are also of particular relevance to the observation that egg size commonly increases with female body size (Roff, 1992). While these results are not in line with the assumed negative effect of egg size on egg survival, this also cannot substantiate a statement such as "the larger the better". Clearly, diffusion does set limitations to the size increase of an egg. In a theoretical study Kranenbarg *et al.* (2000) calculated the optimal size of an egg as $R_{max} = 3V/A$ (where V = volume; A = surface area) when modeled as spheres in stagnant water. Application of this analysis to 8 different teleost eggs revealed that the actual size of the eggs was below the predicted maximum size for the particular species, suggesting a well balanced oxygen supply to the egg, and no oxygen shortage due to the size of the egg.

Normal development of an egg will hardly occur in totally stagnant water, so that no oxygen shortage can be expected under normal oxygen conditions. In addition, embryonic muscle contractions will mix the PVF and thus reduce the effect of unstirred boundary layers (UBL) (see below). In egg clutches parental fin-fanning can be observed, which also improves oxygen supply to the eggs. In eggs of fire clownfish (*Amphiprion melanopus*; Pomacentridae), dissolved oxygen increases immediately after fanning. The fanning activity of the parent fishes changes with the daily cycle, being highest when ambient pO_2 is low, and lowest when ambient pO_2 is high, i.e. during midday (Green & McCormick, 2005).

While egg size correlates with female body size (Roff, 1992) the size of newly hatched larvae may be dependent on the position of the egg within a clutch (Green *et al.*, 2006). In fire clownfish, newly hatched larvae from the periphery of an egg clutch (5 mm from edge) were smaller in standard length, cranial depth, eye diameter and body area (7%, 8%, 4% and 11%, respectively) than larvae from the interior positions within clutches. The eggs from the periphery also had a 33% lower rate of oxygen consumption than did embryos from the clutch interior, throughout development. The relationships between position within a clutch and egg size, oxygen consumption and larval size imply that size variation in larval fish at hatching is partly generated during early embryogenesis, either from maternal endowment or maternal nest design, and was amplified throughout development (Green *et al.*, 2006).

The Chorion as a Diffusion Barrier

The egg capsule or chorion represents a mechanical barrier and provides protection for the developing egg from external environmental conditions. It is mainly composed of protein and glycoprotein with a thickness of between 15 and 70 µm in salmonids (Cotelli *et al.*, 1988; Groot & Alderdice, 1985). The question, whether this robust structure also represents a diffusion barrier to oxygen, has been discussed intensively (Rombough, 1988b). Several studies reported a significant decrease in critical pO_2 after removal of the chorion (Rombough, 1986, 1988a). Early estimates of the diffusion coefficient for the chorion were as low as $1/10^{th}$ of the diffusion coefficient for water (Rombough, 1988b). For the egg case of the spotted dogfish (*Scyliorhinus canicula)* a much higher value of 0.285×10^{-5} $cm^2 \cdot s^{-1}$ has been determined (Diez & Davenport, 1987), but the egg case of elasmobranchs has a completely different structure than the chorion of a teleost fish egg. Measurements of the oxygen profile across the chorion of eggs of the weatherfish (*Misgurnus fossilis*) revealed only a small drop in pO_2 across the egg capsule, suggesting that the diffusion coefficient could not be as low as initially assumed (Berezovsky *et al.*, 1979). Alderdice *et al.* (1984) observed that the hydrostatic pressure exerted on the egg capsule of steelhead trout (*Oncorhynchus mykiss*) was lower than the osmotic pressure of the PVF. Therefore, an osmotic inflow of water must exist, which should also bring dissolved oxygen into the egg. This would suggest that the water pores of the egg might represent an important pathway for oxygen uptake.

That the chorion represents a barrier to oxygen diffusion was also shown by the analysis of growth and muscle cellularity of the Atlantic salmon (Matschak *et al.*, 1995, 1997). Oxygen availability also modified the cellularity of developing rainbow trout (*Oncorhynchus mykiss*) muscle tissue (Matschak *et al.*, 1998) and oxygen deficiency caused vertebral deformities in red seabream (*Pagrus major*) (Hattori *et al.*, 2004). Dechorionated embryos of the Atlantic salmon grew longer and also had a greater white muscle cross sectional area than chorionated siblings. Fibre hyperplasia was reduced at higher temperatures only in the presence of the egg capsule, while fibre hypertrophy was increased at higher temperatures. The differences in the effect of temperature on muscle cellularity in embryonic Atlantic salmon grown with or without the chorion were explained by a higher oxygen demand combined with restricted oxygen availability at the higher temperature (Matschak *et al.*, 1995). As stated earlier, the chorion represents an important mechanical barrier for the developing egg. It could therefore also be that removal of the barrier has an impact on muscular activity of the embryo. Removal of the chorion in rainbow trout eggs 3 to 6 days before the natural time of hatching resulted in 36 times more movements of the embryos (Ninness *et al.*, 2006). By 10 hours post hatch (ph) the number of larval body movements of these previously dechorionated embryos was no longer elevated compared to control larvae which hatched naturally. Muscular activity may in turn have an influence on muscle cellularity, as shown for zebrafish (*Danio rerio*) larvae exposed to a constant water current (Pelster *et al.*, 2003). The recent study on chorionated and dechorionated rainbow trout also revealed an acceleration of growth after removal of the chorion, but this effect disappeared again with proceeding development. At first feeding the larvae from dechorionated and chorionated eggs had the same body mass (Ninness *et al.*, 2006). Therefore after hatching, the growth rate of the previously dechorionated larvae must have been retarded compared to the control animals, which is not easy to explain or to interpret. If growth rate is mostly under genetic control (Blaxter, 1988), one could assume that growth is designed to achieve an optimum body size at first feeding.

Recent studies addressing this question revealed that in rainbow trout, the presence of the chorion did not modify the influence of oxygen availability on growth of the embryos. Hypoxia resulted in retarded growth rates of the embryos, but this was similar in the presence and absence of the chorion (Ciuhandu *et al.*, 2005). In a subsequent study of this group the oxygen gradient across the egg capsule was measured with an oxygen microelectrode, revealing a drop in pO_2 across the chorion (Ciuhandu *et al.*, 2007). Analysis of the impact of hypoxia on cardiac activity of embryos of whitefish (*Coregonus lavaretus*) and of vendace (*C. albula*) in whole eggs and in excised (dechorionated) eggs revealed that hypoxic bradycardia was consistently observed at higher pO_2 values in whole eggs than in excised eggs (Czerkies *et al.*, 2002). From the pO_2 differences at the onset of the bradycardia the authors calculated a pO_2 gradient of about 11-13 Torr (1.46-1.73 kPa) over the egg envelope (chorion) at 4°C. At higher temperatures (8°C) this gradient increased significantly, possibly due to the increase in metabolic rate. Taken together these data clearly indicate that the chorion can represent a diffusion barrier to oxygen under hypoxic conditions, but apparently it is not as large as initially expected. Under normoxic conditions this barrier does not appear to impair oxygen supply to the developing embryo.

Unstirred Layers

Measurements of pO_2 in the vicinity of aquatic animals revealed that boundary layers exist near the surface of the body, which are not well mixed and therefore represent semistagnant regions (Feder & Pinder, 1988; Pinder & Feder, 1990; Feder & Booth, 1992). In these so-called "unstirred layers" a depletion of oxygen is observed due to the removal of oxygen by the uptake at the body surface. This oxygen has to be replenished by diffusion from outer layers. These unstirred layers affect not only oxygen, but also metabolic waste products or ions that are taken up or removed from the body.

Using microelectrodes intravascular pO_2 was measured at various sites in anaesthetised rainbow trout larvae in still and in moving water in order to identify possible regions that are especially important for cutaneous gas exchange (see Terjesen, this volume). Intravascular pO_2 values were in the range of 0.4-1.7 kPa in still water, significantly lower than pO_2 values recorded at the body surface of rainbow trout larvae (Rombough, 1992). In flowing water intravascular pO_2 values were significantly elevated, demonstrating that the unstirred boundary layer represented the main resistance to oxygen uptake. It was calculated that 95% of the resistance to oxygen uptake was due to the boundary layer in still water, while in flowing water the boundary layer contributed 82% to total resistance (Rombough, 1992). The low pO_2 values recorded in this study were quite surprising, but it can be expected that the pO_2 values in free-living larvae and in free-flowing water are higher than those recorded in Rombough's (1992) study.

In a follow up study, the pO_2 gradient in the diffusive boundary layer adjacent to seven positions on the skin surface (one on the head, two on the yolk-sac, two on the trunk, one at the base of the dorsal fin-fold, and one on the proximal portion of the caudal fin-fold) of rainbow trout larvae in still water was measured with microelectrodes (Rombough, 1998). pO_2 levels in the boundary layer increased fairly slowly with distance at all sites tested and typically did not approach free-stream values until 3-5 mm from the body surface (Fig. 1). According to the 'pipes in a wall' model (Malvin, 1988), differences in capillary density should have little impact on uptake efficiency if the total diffusion distance is long compared with the distance between capillaries. This appears to be the case for trout larvae.

In still water the boundary layer is thicker than in flowing water. Pinder & Feder (1990) analysed the effect of water velocity on boundary layer thickness in adult bullfrogs and reported that the proportion of the total resistance to oxygen uptake represented by the boundary layer increased from 35% at a water velocity of 5 $cm \cdot s^{-1}$ to over 90% at 0.1 $cm \cdot s^{-1}$. At water velocities below 1 $cm \cdot s^{-1}$ oxygen uptake was limited by the resistance of the boundary layer, and at 0.1 $cm \cdot s^{-1}$, the pO_2 immediately adjacent to the skin was only 2 kPa. For an immobilised frog this pO_2 was tantamount to placing

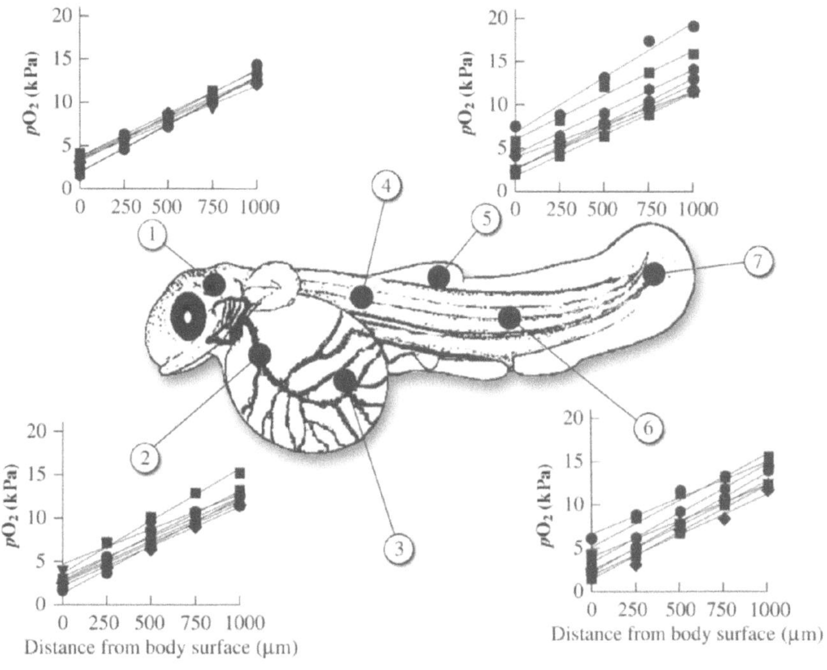

Fig. 1 Microelectrode measurements of boundary layer pO_2 near various positions of the skin of rainbow trout (*Oncorhynchus mykiss*) larvae. The various symbols indicate results obtained using different individuals after Rombough (1998). Reproduced with permission of the Company of Biologists.

it in anoxic water. Boundary layers were efficiently disrupted by body movements. The reduction in the thickness of the UBL with increasing water flow rate was also demonstrated for rainbow trout eggs and larvae. A decrease in flow rate from 0.6 $cm \cdot s^{-1}$ to 0.2 $cm \cdot s^{-1}$ caused a 32% increase in the oxygen gradient across the chorion (Ciuhandu *et al.*, 2007). An interesting observation was that prolonged hypoxia caused a decrease in the oxygen gradient across the chorion.

The PVF may also add to the resistance to gas exchange (Rombough, 1988b). The oxygen diffusion coefficient is probably similar to the value determined for other body fluids or water, i.e. it is probably in the range of $\sim 2.5 \times 10^{-5}$ $cm^2 \cdot s^{-1}$ (Dejours, 1981). The distance across the perivitelline space can be as large as 500 μm (Rombough, 1988b), depending on the size of the egg. Rombough (1988a) interpreted the increase in critical pO_2 observed in rainbow trout as an impact of the perivitelline fluid. The critical pO_2 of the embryos in relation to metabolic rate increased rapidly at the time when the embryos started to move and thus to stir the PVF. The flexing movement of the embryo stirring the PVF reduces boundary layers within the egg. Thereafter, critical pO_2 values increased more slowly and in proportion to the increase in metabolic rate (see below).

THE SITE OF GAS EXCHANGE

Gas exchange of the egg and the developing embryo is via cutaneous gas exchange. Gills, and especially the secondary lamellae, the main gas exchanger of adult fish, can only contribute to gas exchange once they have developed. Gills typically form quite late during development, and are initially considered to primarily function as ionoregulatory organs (Rombough, 2002). In Mozambique tilapia (*Oreochromis mossambicus*) (Li *et al.*, 1995; Van der Heijden *et al.*, 1999), Japanese flounder (*Paralichthys olivaceous*) (Hiroi *et al.*, 1999), brown trout (Rojo *et al.*, 1997; Pisam *et al.*, 2000), and rainbow trout (Rombough, 1999), large numbers of mitochondria-rich cells appear on the gills well before there is any indication of secondary lamellae, the definitive adult gas-exchange structure.

This of course implies a switch in the gas exchange area from solely cutaneous gas exchange to gas exchange largely or perhaps exclusively located at the gills. A significant portion of the early developing embryo is the yolk, and the yolk-sac has been discussed as a possible site of gas exchange. In chinook salmon (*Oncorhynchus tshawytscha*) for example, at developmental stage 37.4 (accumulated thermal units (ATU); = tissue mass ~ 0.04 g; = approximate hatching time) the surface of the yolk-sac accounts for almost 60% of the total surface area, while the gills contribute only 4% (Rombough & Moroz, 1990). In Atlantic salmon at a comparable developmental stage the yolk-sac contributes almost 50% to total surface area (Wells & Pinder, 1996a). For larvae of the air-breathing fish, the swamp eel (*Monopterus albus*), the well-developed capillary network on the yolk-sac surface appears to be an important structure for gas exchange (Liem, 1981). These larvae are frequently exposed to hypoxic water, and the well developed pectoral fins generate water currents, that flow countercurrent to the main direction of blood flow on the yolk surface. A reversal of the flow direction from countercurrent to concurrent significantly reduced the percentage of oxygen removed from the water. In larvae reared under hypoxic conditions an increase in the vascularisation of the yolk-sac has been observed (Garside, 1959; Balon, 1975; Brooke & Colby, 1980), which also suggests that the yolk-sac may contribute to gas exchange.

In rainbow trout larvae, however, blood vessel density of the yolk-sac had no significant impact on gas exchange of this organ (Rombough, 1997). Using microelectrodes, intravascular pO_2 has been measured at various sites of rainbow trout larvae in order to detect possible important sites for cutaneous gas exchange, but there was no region within the body, where particularly high internal pO_2 values have been recorded (Rombough, 1992). Especially high internal pO_2 values would be indicative for a very good oxygen supply. Similarly, direct measurements of the thickness of the UBL at various sites of the body surface including the surface of the yolk-sac of rainbow trout larvae failed to identify a special region where the magnitude of the boundary layer or the pO_2 gradient within the boundary layer indicated an especially effective oxygen uptake (Fig. 1) (Rombough, 1998). In Atlantic salmon, the yolk-sac accounts for almost 50% of the total surface area, but contributes only 33% to total oxygen uptake, indicating that the surface of the yolk-sac is a less efficient gas exchanger than the body skin or the gills (Wells & Pinder, 1996b).

In contrast to swamp eel larvae, larvae of salmonid fishes typically develop in well oxygenated water. It therefore seems that the contribution of the yolk-sac surface to cutaneous oxygen uptake is dependent on the species, and species vary considerably in the extent to which the yolk is vascularised (Rombough, 1997). Nevertheless, there appears to be little correlation between the efficiency of cutaneous gas exchange and the number of capillaries in the skin or yolk-sac, because diffusion through the UBL represents the main limitation for oxygen exchange (Malvin, 1988; Rombough, 1997). This is also in line with the observation that in quantitative terms cutaneous gas exchange per unit surface area of adult fish is in the same range as the oxygen uptake of larval fish (Kirsch & Nonnotte 1977; Rombough & Ure, 1991; Wells & Pinder, 1996a; Rombough, 2004).

For fins the situation may be different, since they move quickly and show oscillatory movements, effectively breaking up, or at least limiting the size of the UBL. Besides well vascularised fins, some species develop special structures to assist embryonic or larval respiration, including, for example, external gills, which are rarely observed in actinopterygians (Rauther, 1937; Balon, 1975; Rombough, 1988b), but present in several elasmobranchs (Pelster & Bemis, 1992).

An interesting question for the time of cutaneous gas exchange is the internal pO_2, especially as the UBL represents the main resistance to oxygen uptake at this stage. Intravascular pO_2 values have been shown to be surprisingly low in rainbow trout larvae (Rombough, 1992). Considering the intravascular pO_2 values, however, it has to be taken into account that in these early larval stages the circulatory system and the blood do not significantly contribute to oxygen transport to tissues. Thus, the driving force for oxygen diffusion to the tissues in cutaneously respiring larvae is not the pO_2 gradient between the blood and the tissue, but between the body surface and the tissues. In zebrafish larvae intravascular haemoglobin oxygen saturation has been used to determine pO_2 in venous blood (Grillitsch *et al.*, 2005). In zebrafish larvae with a body mass about 10-times smaller than that of rainbow trout, much higher intravascular pO_2 values have been determined, but even in these small larvae haemoglobin was only partially saturated in certain developmental stages, also when incubated under normoxic pO_2 values.

With development of the gills, the overall contribution of cutaneous gas exchange to oxygen uptake decreases, and the gills become the predominant site for oxygen uptake and CO_2 release. Due to the sequential development of gill filaments and gill lamellae, with the lamellae appearing much later then the filaments, the gill filaments initially appear to be important for gas exchange. With proceeding development and growth, the lamellae gradually take over and in adult fish become the definitive site of gas exchange (Fig. 2). In chinook salmon, at the time of hatching (37.4 ATU) lamellae are too small to be measured, but at 103.4 ATU, they contribute about 48% to total gill surface area (Rombough & Moroz, 1990).

Scaling the potential importance of cutaneous and branchial surfaces in respiratory gas exchange in developing chinook salmon revealed that initially 96% of the oxygen uptake occurs via the skin (Rombough & Moroz, 1990). As the body mass of the embryo and larvae increases, gill surfaces expand much more rapidly than cutaneous

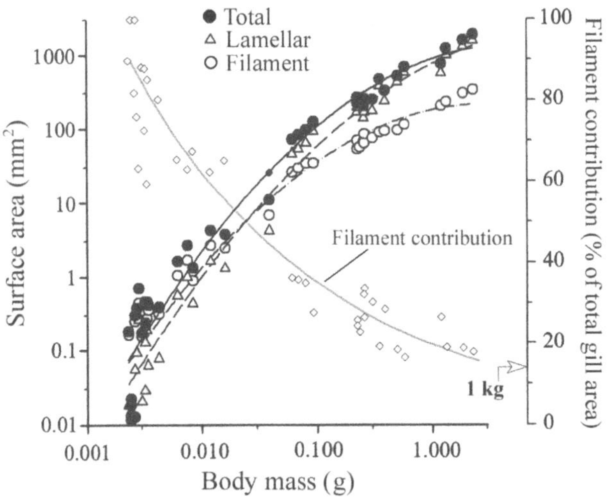

Fig. 2 Increase of the surface area of gill filaments and lamellae as well as of total surface area of lamellae plus filaments in developing walleye (*Stizostedion vitreum*) as functions of body mass. Dotted line shows the relative contribution of the filaments to total gill area. Lines are best-fitted second-order polynomials after Rombough & Moroz (1997). Reproduced with permission of the Company of Biologists.

surfaces. Accordingly, the contribution of the gills to gas exchange increases, but it was not before a body mass of 2.5-4.0 g that branchial gas exchange exceeded cutaneous gas exchange in chinook salmon. The rapid expansion of the gill surface during early development also exceeds the normal surface to volume ratio where the volume of an animal varies with the cube of its linear dimension, while the surface varies with the square of the dimensions. As oxygen consumption typically is related to body mass, this implies that at this time of development the gill surface area expands more rapidly than metabolic rate.

In juvenile and adult fish, in which branchial gas exchange is clearly predominating, the aerobic metabolic capacity is correlated to the expansion of gill surface area (Rombough & Moroz, 1990). In line with this notion in chinook salmon larvae following yolk-sac absorption, the mass exponent for total surface area with a value of 0.85 was not significantly different from the mass exponent for metabolic rate (0.8-0.9; Hughes, 1984; Duthie & Hughes, 1987; Rombough & Moroz, 1990). In yolk-sac larvae, however, the mass exponent for total surface area was significantly smaller, indicating a change in the relative efficiencies of the surface areas contributing to gas exchange.

Measurements of oxygen consumption in yolk-sac larvae (alevins) and fry of this species using a two-chambered respirometer supported this conclusion (Rombough & Ure, 1991). Over the body mass range from 0.1-1.2 g total oxygen uptake scaled with body mass with an exponent of 0.89. While the cutaneous exponent was as low as 0.48, the branchial exponent was much higher with a value of 1.33. Accordingly, the relative importance of the skin decreased with development, while the relative importance of the gills increased significantly with development. In other words, the area-specific oxygen uptake of the skin increased only slightly during the observation period, while the area-specific oxygen uptake of the gills increased almost six-fold over the same developmental time period. In conclusion, the increase in branchial oxygen uptake observed with continued development does not only reflect an increase in the branchial surface area, it also reflects an increase in the efficiency of the gills as a respiratory organ.

The increase in branchial gas exchange efficiency can at least in part be attributed to a significant decrease in the diffusion distances across the gill filaments and lamellae. Wells & Pinder (1996a) conducted a similar study on gill development of Atlantic salmon. As in chinook salmon, relative cutaneous surface area decreased with further development, while branchial surface area significantly increased. While the harmonic mean diffusion distance through the skin increased from 20 μm at hatching to 70 μm in 11.2 g fish, filament diffusion distance decreased from 11 μm to 3.7 μm, and lamellar diffusion distance from 11 μm to 2.4 μm, respectively. The anatomical diffusion factor (ADF), calculated as the mass-specific surface area per unit diffusion distance continuously decreased with development of the skin, while branchial ADF increased several fold until the end of yolk absorption, and remained constant thereafter (Fig. 3) (Wells & Pinder, 1996a).

As in chinook salmon, measurements of oxygen consumption by Atlantic salmon in a two-chambered respirometer supported the morphometric data. With further

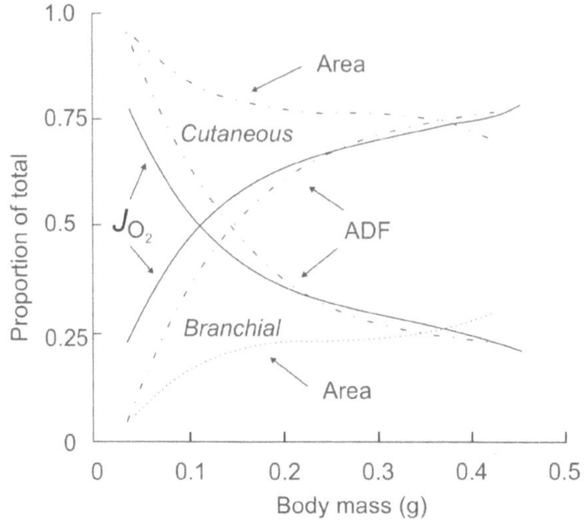

Fig. 3 Change in the contribution of cutaneous and branchial surface area to total gas exchange area, in the contribution of cutaneous and branchial oxygen uptake to total oxygen uptake, and in the anatomical diffusion factor (ADF), calculated as the mass-specific surface area per unit diffusion distance, of developing Atlantic salmon (*Salmo salar*) after Wells & Pinder (1996b). Reproduced with permission of the Company of Biologists.

development the proportion of branchial oxygen uptake to total oxygen uptake increases, while the proportion of cutaneous oxygen uptake decreases (Wells & Pinder, 1996b). In addition, the area-specific rate of oxygen uptake of the gills increases with development, while cutaneous area-specific oxygen uptake decreases, confirming that the efficiency of the gills as a gas exchange organ increases with development, while the efficiency of cutaneous gas exchange decreases. A comparison of the ADF normalised to body mass between early larval stages and adult fish reveals that the branchial ADF is lower in adult fish. This indicates that, compared to adult fish, the anatomical arrangement of the newly developed gills is more than adequate to meet the oxygen uptake requirements.

While a poor correlation between total respiratory surface area and metabolic demand was observed in chinook salmon larvae (Rombough & Moroz, 1990), the mass specific exponents for metabolism and respiratory surface area were similar in larvae of Atlantic salmon (Wells & Pinder, 1996a, b). The correlation between metabolic activity and respiratory surface area was also analysed in the walleye (*Stizostedion vitreum*) (Rombough & Moroz, 1997). The larvae of walleye are pelagic and much smaller than a typical salmonid larva, and at the time of hatching the yolk-sac accounts for only 10-15% of total surface area. The results of this study and also a literature survey revealed a mismatch between respiratory surface area and metabolic oxygen demand, as described for chinook salmon (Rombough & Moroz, 1997). During early development the effective respiratory surface area declines in most larvae, indicating that initially there is an excess in the gas exchange capacity (Fig. 4). Compared to adult fish, the effective surface area for gas exchange in small larvae is between 2.5 and 6 times larger. The point at which changes in respiratory surface area match with metabolic oxygen demand is reached only some time after hatching, or in the juvenile state.

An interesting question in this respect is the timing of the switch from primarily cutaneous respiration to branchial respiration. Gill development is quite different between different species. In the walleye gill, development begins as early as 2 days PH at a body mass of 2.5 mg, while salmonids typically are several times larger when the gills begin to form. By comparing relative lamellar surface area with the body mass of larvae from different families Rombough & Moroz (1997) concluded that with the body mass of below 100 mg, larvae do not require gills for oxygen uptake, i.e. the transition to branchial respiration must occur at a body mass of about 100 mg or not much beyond (Fig. 4).

OXYGEN UPTAKE

Energy metabolism of developing embryos and yolk-sac larvae of the turbot (*Scophthalmus maximus*) using calorespirometry demonstrated that until 12 days post

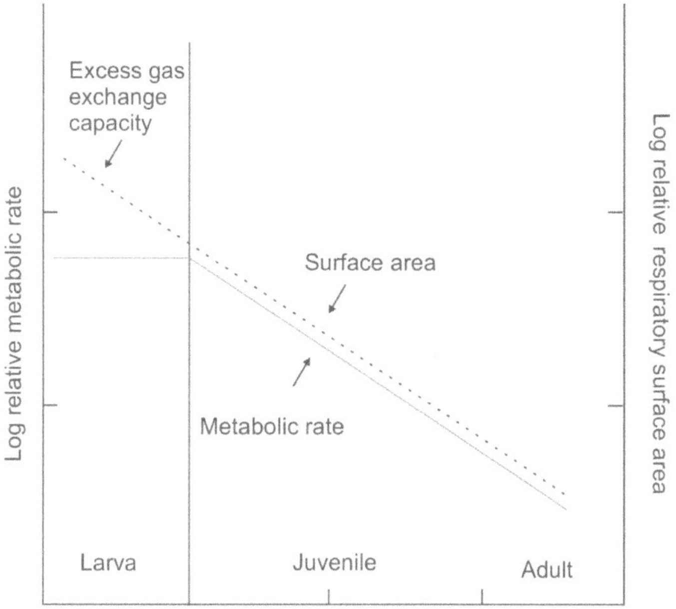

Fig. 4 Schematic diagram illustrating differences in scaling patterns for metabolic rate and relative respiratory surface area in larvae, in juvenile and in adult fish. Larvae appear to have surplus exchange capacity compared with older fish after Rombough & Moroz (1997). Reproduced with permission of the Company of Biologists.

fertilisation (pf) aerobic metabolism fully accounts for total energy turn over, under resting conditions there is no contribution of anaerobic metabolism (Finn *et al.*, 1995a). Similarly, in Atlantic halibut (*Hippoglossus hippoglossus*) metabolism of developing yolk-sac larvae was almost exclusively aerobic (>99%) (Finn *et al.*, 1995b), and during endogenous development of Atlantic cod (*Gadus morhua*) larvae anaerobic metabolism did not contribute significantly to energy dissipation (Finn *et al.*, 1995c). Nevertheless, even early larval stages have been shown to be capable of anaerobic energy dissipation and lactate accumulation during forced exercise (Wieser, 1985; Wieser *et al.*, 1985). Under normoxic conditions without excessive muscular activity, however, aerobic metabolism based on the uptake of oxygen appears to be dominating. In zebrafish larvae, even hypoxemic conditions under normoxia did not evoke any increase in whole body lactate levels (Jacob *et al.*, 2002). A recent study revealed that metabolic rate of eyed-stage eggs of Arctic charr (*Salvelinus alpinus*) varied significantly between families, suggesting that family and/or genetic effects have an influence on embryonic metabolism (Pakkasmaa *et al.*, 2006).

For marine fish such as Atlantic cod that do not contain oil globules, and turbot, that do contain oil globules, it has been shown that during the first stages of development (prior to gastrulation and epiboly) glycogen is the sole substrate for energy metabolism (Finn *et al.*, 1995a, c, d). After formation of the syncytium, however, free amino acids and also free fatty acids represent a major fuel of the catabolic metabolism during endogenous feeding (Finn *et al.*, 1995c, 1996). Calorimetric studies combined with physiological studies revealed that in Atlantic cod larvae 42% of the total yolk enthalpy is actually dissipated for energy metabolism, while 53% are used for growth, fueling growth rates of more that 25% d^{-1} (Finn *et al.*, 1995d, 2002).

For a comprehensive review on oxygen uptake of various fish embryos and larvae, including aspects of aerobic metabolism in relation to development, body mass or activity, see Rombough (1988b). Crucial for the amount of oxygen taken up is the mass of respiring tissue. Although intuitively this is expected to apply not only to adult animals, but also to developing stages, several studies indicate that this is not true for all developmental stages. For a significant fraction of larval development, mass-specific oxygen uptake appears to be independent of body mass, i.e. in the allometric equation the mass-specific exponent is close to 0.0, i.e. metabolic rates scale isometrically (Rombough, 1988b; Giguère *et al.*, 1988; Post & Lee, 1996; Rombough & Moroz, 1997). Bochdansky & Legget (2001) reported a curvilinear relationship between oxygen uptake and body mass during ontogeny by analysing a data set of 25 species. On the other hand, there are also species in which routine metabolic rate scales allometrically with body mass, as in Atlantic cod, for example (Finn *et al.*, 2002). Accordingly, the relation between metabolic rate and body mass in fish is species-specific, and a generalised statement does not appear appropriate. It could well be that a specific ontogenetic reorganisation of tissues, which contribute to body mass but are more or less metabolically active, contributes to these differences in metabolic scaling. This could be changes in the relative contribution of red and white muscle tissue, for example, (Galloway *et al.*, 1999; Finn *et al.*, 2002) or the accumulation of large ametabolic energy stores as described for the *Leptocephalus* larvae (Bishop *et al.*, 2000).

Initially the rate of oxygen consumption is low, but increases on average by the factor of about 20 between fertilisation and hatching, although there appear to be severe differences between species. During hatching, mass-specific oxygen uptake is elevated, which may be related to the removal of the chorion, and the PVF as an oxygen diffusion barrier (Davenport, 1983; Gruber & Wieser, 1983).

Mass-specific oxygen uptake of embryos and larvae is quite variable, but on average it appears to be higher than for adult fish (Rombough, 1988b). Measurement of oxygen uptake of the zebrafish starting with the egg and over the first 100 days of development showed a more than 10-fold increase in oxygen uptake with hatching and initial growth until 10 days PF ($3.6\ \mu L \cdot g^{-1} \cdot h^{-1}$ to $39.4\ \mu L \cdot g^{-1} \cdot h^{-1}$). Oxygen uptake then decreased to $5.8\ \mu L \cdot g^{-1} \cdot h^{-1}$ until 60 days PF, and showed little additional change until 100 days PF (Barrionuevo & Burggren, 1999).

In salmonids, a similar peak in JO_2 has been described (Rombough, 1988a, 1988b; Rombough & Ure, 1991). The high metabolic rate may be linked to organogenesis and the conversion of yolk into metabolising biomass (Barrionuevo & Burggren, 1999). In several species mass-specific oxygen consumption does not start to decline immediately after hatching, but continues to increase until well into the feeding period, before it starts to decline (Rombough, 1988b).

Oxygen uptake of embryos within a clutch of the demersal tropical fire clownfish has been shown to be dependent on the positions of the egg within the clutch (Green *et al.*, 2006). Although the oxygen supply can be expected to be best at the periphery of the clutch, eggs located there had a 33% lower oxygen uptake rate than the eggs from the interior of the clutch. In addition, the hatched larvae from the periphery were significantly smaller than the hatched larvae from the center of the clutch. The authors speculate that the higher risk of predation for the eggs at the peripheral region of the clutch may contribute to these differences in metabolic activity and development. Although the yolk content of all eggs was similar, females may produce smaller eggs at the periphery of the clutch, but the lower metabolic rate of peripheral eggs appears to increase the size differences between peripheral and central hatched larvae.

The Influence of Hypoxia

Oxygen uptake often decreases in response to hypoxia. For animals that, in the face of decreasing oxygen tensions, are initially able to keep the rate of oxygen uptake fairly constant or respond with very moderate declines in oxygen uptake (so called "oxygen regulators"), a critical pO_2 (p_C) is defined, below which JO_2 declines rapidly. Below the p_C, anaerobic metabolic pathways (such as lactate production) significantly contribute to the synthesis of ATP. The p_C of an egg is dependent on the rate of oxygen uptake, the diffusion coefficient and the geometry of the egg and can be calculated as $p_C = JO_2 \cdot r^2/6\ D$ (where r = radius; D = diffusion coefficient) (Hamdorf, 1961). After the onset of convective oxygen transport, analysis of p_C becomes more complicated since the diffusion distance between the blood and the surface must be taken into account. In addition, a reduced oxygen availability may also result in a more or less severe reduction in metabolic activity (Guppy *et al.*, 1994; Guppy, 2004). Accordingly, the capability of the embryo or larvae to reduce oxygen demand by reducing metabolic activity will also be of importance for the definition of the p_C, and this may even change with development (Hastings & Burggren, 1995). Nevertheless, the p_C is dependent on JO_2, and will therefore be affected by changes in mass-specific JO_2.

While adult fish typically are so called "oxyregulators", embryos and larvae quite often appear to be oxyconformers (Davenport, 1983; Gruber & Wieser, 1983; Pelster, 1999). This appears especially the case for large eggs, and therefore may be related to the chorion and the PVF as a diffusion barrier (Rombough, 1988b).

The p_C typically increases with development of the embryo. In rainbow trout the slope of the increase in p_C decreases after the embryos start muscular contractions, which results in a mixing of the PVF. This suggests that the unstirred PVF indeed adds to gas exchange resistance (Rombough, 1988a, 1997). In addition, the highest p_C often is observed at the time of hatching. After hatching p_C decreases significantly, levelling off to a plateau, or it may increase again (Rombough, 1988a, b). In zebrafish larvae the capability to cope with hypoxic conditions increases with development, and the lowest p_C has been reported for 100 day pf animals (Barrionuevo & Burggren, 1999).

The onset of blood flow may also reduce the p_C, because internal oxygen transport will be much more efficient and effective. This contention, however, is not apparent from the current literature. In zebrafish the circulatory system becomes important for gas exchange at about 12-14 days pf, but there appears to be no special reduction in p_C (Barrionuevo & Burggren, 1999). Another important parameter contributing to a decrease in p_C may be the formation of gill lamellae (De Silva & Tytler, 1973; Spoor, 1984; Rombough, 2004). Quite often, however, no changes in p_C are observed (Barrionuevo & Burggren, 1999; Rombough, 2004). Considering the fact that the circulatory system initially is not required for oxygen transport, and that the gills initially may primarily serve ionoregulatory functions (Rombough, 2002), this is perhaps not surprising.

In several species of fish, the eggs have been shown to be quite resistant to severe hypoxia or even anoxia, but as development proceeds the embryos become more and more sensitive to a decrease in environmental oxygen tension (Hayes *et al.*, 1951; Alderdice *et al.*, 1958; Rombough, 1986; Czerkies *et al.*, 2001). Embryos and larvae clearly have the capacity for aerobic and anaerobic metabolism, and the rainbow trout heart has a considerable tolerance of hypoxia during early life. The p_C for rainbow trout heart is in the range of 0.1 kPa or even below (Rombough, 1997). Embryos of Arctic charr, at about 80% of embryonic development, survive 8 h of anoxia. Direct calorimetry revealed that during anoxia heat dissipation decreased to a few percent of control, indicating a severe metabolic depression in the test embryos (Pelster, 1999). Soon after fertilisation, zebrafish eggs are surprisingly resistant towards a lack of oxygen. Zebrafish embryos survive 24 h of anoxia in a state of suspended animation, in which cardiac activity ceases and mitotic activity of the blastomeres is arrested (Padilla & Roth, 2001).

It could well be that the metabolic depression observed under these conditions is mainly achieved at the cellular level and does not include central humoral or nervous signalling. The decrease in heart rate (bradycardia) and in cardiac output observed in very early stages of fishes and amphibians during periods of oxygen deficiency, for example, has been attributed to a direct effect of the lack of oxygen on the metabolism of cardiomyocytes (Orlando & Pinder, 1995; Fritsche & Burggren, 1996; Pelster, 1999).

In later embryonic stages hypoxia may stimulate hatching. In embryos of whitefish and vendace at developmental stage 13 (eye movement visible) or 14, hypoxic exposure induces precocious hatching (Czerkies *et al.*, 2001). The rate of precocious hatching increased with the degree of hypoxia. Below the p_C aerobic metabolism can no longer be sustained and secretion of chorionase ensues, followed by precocious hatching. In whitefish the p_C increased with development.

Another set of experiments provides useful information in this context. Hypoxia or anoxia can experimentally be arranged by decreasing or eradicating the aqueous pO_2. It can also be achieved by chemical means. The addition of cyanide blocks the respiratory chain, while the application of CO or phenylhydrazine, blocks or chemically destroys haemoglobin, thus reducing the oxygen carrying capacity of the blood in order to reduce convectional oxygen transport in the blood stream. The zebrafish can be raised in an atmosphere containing 1-2% CO, which should be sufficient to essentially inhibit haemoglobin-mediated oxygen transport (Pelster & Burggren, 1996; Jacob *et al.*, 2002). These studies demonstrated that oxygen uptake of CO-incubated zebrafish was not different from control during the first days of embryonic and larval development. This clearly shows that hypoxemia, i.e. a reduction in the oxygen carrying capacity of the blood, does not provoke a hypoxic response in zebrafish larvae. Accordingly, carrier-mediated oxygen transport is not required at this stage of development.

Hypoxia and Ventilation

Although convective oxygen transport is not required in early developmental stages of fish, sensors to detect environmental oxygen are present in early stages, and the larvae respond to changes in environmental pO_2. Even in very early larvae a decrease in environmental oxygen availability usually provokes a stimulation of ventilatory activity, but also cardiac activity may be enhanced (Jacob *et al.*, 2002; Jonz & Nurse, 2005). In newly hatched Arctic charr, ventilatory activity and the co-ordination of ventilatory movements increases during hypoxic exposure (McDonald & McMahon, 1977). An increase in swimming activity under hypoxic conditions has also been reported for several fish larvae such as speckled trout (*Salvelinus fontinalis*) (Shepard, 1955), smallmouth bass (*Micropterus dolomieui*) and largemouth bass (*M. salmoides*) (Spoor, 1977, 1984) or northern anchovy (*Engraulis mordax*) (Weihs, 1980).

A detailed analysis of the ventilatory response of zebrafish larvae revealed that in the earliest stages ventilatory frequency is quite irregular and increases until 5 dpf (Turesson *et al.*, 2006). With further development ventilation rate becomes more regular. Beyond day 5-6 pf, ventilation rate decreases, reaching only a few ventilatory movements per minute at 10 dpf (Jonz & Nurse, 2005; Turesson *et al.*, 2006). The decreasing frequency increases the scope for stimulation of ventilation. Hypoxic exposure enhances ventilation rate starting at 2 dpf (Jonz & Nurse, 2005), and the maximum ventilatory response increases with proceeding development (Turesson *et al.*, 2006). The maturation of ventilatory activity and the hypoxic response may reflect a maturation of the nervous system controlling ventilation.

Oxygen receptors necessary to detect hypoxic conditions in the environmental water are located in neuroepithelial cells (NEC) of the zebrafish gill filaments. NECs were first detected in gill filament primordia of larvae at day 5 pf and were fully innervated by day 7 pf. Accordingly, this pathway cannot account for the hypoxic stimulation of ventilation as early as day 2 pf. In these stages oxygen-sensing must occur through another pathway. It has been proposed that a new type of serotonin-(5-HT)-positive NECs of the gill arches may account for this earlier hypoxic response (Jonz & Nurse, 2005).

Typically, glutamatergic *N*-methyl-D-aspartate (NMDA) receptors are involved in ventilatory control (Sundin *et al.*, 2003; Turesson & Sundin, 2003). In zebrafish larvae, maturation of the glutamatergic system started at about 8 dpf, as indicated by the partial inhibition of the ventilatory response by inhibition of the NMDA receptors. But only 13 dpf zebrafish larvae and older ones completely rely upon a glutamatergic component for the mediation of the ventilatory response to hypoxia (Turesson *et al.*, 2006). As pointed out earlier, hypoxic stimulation of ventilation has been observed in zebrafish as early as 2 dpf and therefore cannot be explained by the activity of the glutamatergic system. Two non-NMDA receptors are the ionotropic α-amino-3-hydroxy-5-methyl-4-isoxazolepropionate (AMPA) and the γ-aminobutyric acid (GABA)-A receptors, and both may be involved in the control of ventilation in very early stages (Turesson *et al.*, 2006).

Hypoxia and Haemoglobin Function

In developing vertebrates haematopoiesis is divided into two separate waves (see Burggren & Bagatto, this volume; Zapata & Cortès, this volume). In zebrafish the first proerythroblasts are formed in the intermediate cell mass (ICM), located in the trunk ventral to the notochord. These cells of the first or primitive wave enter circulation at about 24 hours pf (de Jong & Zon, 2005). This is also the time when the heart starts to beat. In addition, anterior mesoderm, namely the rostral blood island, marks another haematopoietic area. At about 5 dpf the primitive erythrocytes begin to be replaced by cells produced in the second wave of erythropoiesis. Cells of the second wave originate from the aorta-gonad mesonephros. Unlimited self-renewal and the potential of differentiation to all mature haematopoietic lineages are characteristic to the definitive wave. At about this stage erythropoiesis also starts in the kidney marrow, which together with the spleen, represents the lifelong location of red blood cell formation. This change in the site of erythropoiesis is combined with a switch in the expression of different haemoglobin isoforms, from the embryonic haemoglobin $\alpha\beta$ to the adult haemoglobin $\alpha_2\beta_2$ isoforms. It appears quite likely that these different isoforms also have different oxygen affinities.

Hypoxic conditions have been shown to induce a change in the oxygen affinity of haemoglobin in lower vertebrates. Compared to adult haemoglobin, the haemoglobin of viviparous embryonic vertebrates is characterised by a higher oxygen affinity, and this appears to be true for higher as well as for lower vertebrates. In viviparous fishes and elasmobranchs the lower p_{50} is in part due to a higher intrinsic oxygen affinity of the pigment. In part the higher affinity can be attributed to changes in the concentration of organo-phosphates within the red blood cells (Ingermann & Terwilleger, 1984; Ingermann *et al.*, 1984; Hartvig & Weber, 1984; Weber & Hartvig, 1984; Korsgaard & Weber, 1989).

A modification of haemoglobin oxygen affinity during hypoxia will only be beneficial if convective oxygen transport contributes to the oxygen supply of the tissues. The modification of the affinity of the respiratory pigment in response to hypoxia therefore may well reflect differences in the importance of convective gas transport for the metabolism of the developing embryo or larvae.

TEMPERATURE EFFECTS

A comparison of Q_{10} values for oxygen consumption from more than 20 fish larvae revealed an enormous variation, with a number of species with a remarkable temperature sensitivity and Q_{10} values higher than 4, while there are also several species, in which metabolism remains quite constant in the face of varying temperatures, as indicated by a Q_{10} lower than 2 (Rombough, 1988b). In a carefully conducted study on minnow embryos and free swimming larvae the Q_{10} for oxygen consumption in animals raised at 15°C and at 25°C was 1.7 for the egg and of 2.4 for the larvae at the time of swimbladder inflation (Schönweger *et al.*, 2000). For developing embryos and yolk-sac larvae of the turbot a Q_{10} of 2.6 has been reported (Finn & Rønnestad, 2003). Besides temperature, light was also identified as an important factor influencing aerobic metabolism of the turbot. Between 9 and 12 dpf yolk-sac larvae under constant light had a significantly higher oxygen consumption compared to larvae reared in the dark (Finn & Rønnestad, 2003).

A high temperature sensitivity of the early stages has been demonstrated for zebrafish embryos. For the temperature range of 25-28°C and for 28-31°C Q_{10} values in the range of 4 to 5 were calculated, while between 20 and 60 dpf the Q_{10} dropped to about 2 to 3. In later stages up to 100 dpf the Q_{10} increased again, reaching values of between 4 and 5 (Barrionuevo & Burggren, 1999). The high Q_{10} values reported especially for the earliest stages indicate stenothermic metabolism. Early embryonic development apparently is designed to function optimally at a narrow temperature range (Finn &

Rønnestad, 2003). Deviations from this range result in thermal stress, and a significant activation of metabolic activity. Overall, the narrow windows of thermal tolerance in the larvae may be important in reducing energy expenditure in larval fish (Pörtner *et al.*, 2006).

Given that many adult fish have a Q_{10} for metabolism between 2 and 3, the very high Q_{10} values recorded for 100 day old zebrafish larvae was somewhat surprising. The authors suggest that adult zebrafish are quite stenothermic animals (Barrionuevo & Burggren, 1999), and reproduction of zebrafish indeed is severely impaired at temperatures below 25°C, for example. The high Q_{10} values may also indicate that, at this stage, the extrinsic mechanisms responsible for temperature compensation in adult fish hearts are not yet functional.

A common observation with respect to temperature effects on metabolism and on cardiac activity, i.e. heart rate, stroke volume or cardiac output (see Burggren & Bagatto, this volume), is that there is no match between the stimulation of metabolism by an increase in temperature and the stimulation of cardiac activity with increasing temperature. Typically, temperature related stimulation of metabolism far exceeds the activation of cardiac activity (Barrionuevo & Burggren, 1999; Schönweger *et al.*, 2000). In adult vertebrates cardiac function is typically viewed as tightly coupled to metabolic activity, but this coupling is not yet established in embryonic or larval fish (Pelster, 1999, 2002; Pörtner *et al.*, 2006). Larval respiration is largely independent from convective oxygen transport in the blood. For the zebrafish, it has been shown that blood circulation and gill function become essential for aerobic metabolism at about 12-14 days PF (Jacob *et al.*, 2002; Rombough, 2002, 2004).

FUTURE PERSPECTIVES

Available physiological data for the early developmental stages of fishes clearly demonstrates that embryos and larvae of lower vertebrates are not just small adults. In particular, hormonal control mechanisms appear to dominate in early stages, while neuronal control is established only in later stages. With respect to metabolism, metabolic scaling and thermal sensitivity of the early stages follow their own rules. Quite surprising is the observation that embryos and even larval stages of several fishes are quite resistant towards a transient lack of oxygen. Considerable work is still required to understand and explain such differences between adults and early developmental stages. This work necessarily will also require studies at the molecular level, in order to identify molecular pathways and the molecular background for these physiological observations. Of particular interest will be studies on how the environment influences gene expression and gas exchange in fish embryos.

Acknowledgments

Parts of the study were financially supported by the Fonds zur Förderung der wissenschaftlichen Forschung (FWF, P12571-BIO and P14976-BIO).

References

Alderdice DF, Jensen JOT, Velsen FPJ (1984) Measurement of hydrostatic pressure in salmonid eggs. Can J Zool 62: 1977-1987

Alderdice DF, Wickett WP, Brett JR (1958) Some effects of temporary exposure to low dissolved oxygen levels on Pacific salmon eggs. J Fish Res Bd Can 15: 229-249

Balon EK (1975) Reproductive guilds of fishes: A proposal and definition. J Fish Res Bd Can 32: 821-864

Barrionuevo WR and Burggren WW (1999) O_2 consumption and heart rate in developing zebrafish (*Danio rerio*): influence of temperature and ambient O_2. Am J Physiol 276: R505-R513

Berezovsky VA, Goida EA, Mukalov IO, Sushko BS (1979) Experimental study of oxygen distribution in *Misgurnus fossilis* eggs. Fiziol Zh 25: 379-389

Bishop RE, Torres JJ, Crabtree RE (2000) Chemical composition and growth indices in leptocephalus larvae. Mar Biol 137: 205-214

Blaxter JHS (1988) Patterns and variety in development. In: Hoar WS, Randall DJ (eds) Fish Physiology, Vol VIII, Academic Press, New York, p 1-58

Bochdansky AB, Leggett WC (2001) Winberg revisited: convergence of routine metabolism in larval and juvenile fish. Can J Fish Aquat Sci 58: 220-230

Brooke LT, Colby PJ (1980) Development and survival of embryos of lake herring at different constant oxygen concentrations and temperatures. Progressive Fish-Culturist 42: 3-9

Ciuhandu CS, Stevens ED, Wright PA (2005) The effect of oxygen on the growth of *Oncorhynchus mykiss* embryos with and without a chorion. J Fish Biol 67: 1544-1551

Ciuhandu CS, Stevens ED, Goldberg JI, Wright PA (2007) Parameters influencing the dissolved oxygen in the boundary layer of rainbow trout (*Oncorhynchus mykiss*) embryos and larvae. J Exp Biol 210: 1435-1445

Cotelli F, Andronico F, Brivio M, Lamia CL (1988) Structure and composition of the fish egg chorion (*Carassius auratus*). J Ultrastruct Mol Struct Res 99: 70-78

Czerkies P, Brzuzan P, Kordalski K, Luczynski M (2001) Critical partial pressures of oxygen causing precocious hatching in *Coregonus lavaretus* and *C. albula* embryos. Aquaculture 196: 151-158

Czerkies P, Kordalski K, Golas T, Krysinski D, Luczynski M (2002) Oxygen requirements of whitefish and vendace (Coregoninae) embryos at final stages of their development. Aquaculture 211: 375-385

Davenport J (1983) Oxygen and the developing eggs and larvae of the lumpfish, *Cyclopterus lumpus*. J Mar Biol Assoc UK 63: 633-640

de Jong JLO, Zon LI (2005) Use of the zebrafish system to study primitive and definitive hematopoiesis. Annu Rev Gen 39: 481-501

De Silva C, Tytler P (1973) The influence of reduced environmental oxygen on the metabolism and survival of herring and plaice larvae. Neth J Sea Res 7: 345-362

Dejours P (1981) Principles of Comparative Respiratory Physiology. Elsevier, Amsterdam

Diez JM, Davenport J (1987) Embryonic respiration in the dogfish (*Scyliorhinus canicula* L.). J Mar Biol Assoc UK 67: 249-261

Duthie GG, Hughes GM (1987) The effects of reduced gill area and hypoxia on the oxygen consumption and swimming speed of rainbow trout. J Exp Biol 127: 349-354

Einum S, Hendry AP, Fleming IA (2002) Egg-size evolution in aquatic environments: does oxygen availability constrain size? Proc R Soc Lond B 269: 2325-2330

Evans DH, Piermarini PM, Choe KP (2005) The multifunctional fish gill: Dominant site of gas exchange, osmoregulation, acid-base regulation, and excretion of nitrogenous waste? Physiol Rev 85: 97-177

Evans DH, Piermarini PM, Potts WTW (1999) Ionic Transport in the Fish Gill Epithelium. J Exp Zool 283: 641-652

Feder ME, Booth DT (1992) Hypoxic boundary layers surrounding skin-breathing aquatic amphibians: Occurrence, consequences and organismal responses. J Exp Biol 166: 237-251

Feder ME, Pinder AW (1988) Ventilation and its effect on "infinite pool" exchangers. Am Zool 28: 973-983

Finn RN, Widdows J, Fyhn HJ (1995a) Calorespirometry of developing embryos and yolk-sac larvae of turbot (*Scophthalmus maximus*). Mar Biol 122: 157-163

Finn RN, Ronnestad I, Fyhn HJ (1995b) Respiration, nitrogen and energy metabolism of developing yolk-sac larvae of Atlantic halibut (*Hippoglossus hippoglossus* L.). Comp Biochem Physiol 111A: 647-671

Finn RN, Fyhn HJ, Evjen MS (1995c) Physiological energetics of developing embryos and yolk-sac larvae of Atlantic cod (*Gadus morhua*). I. Respiration and nitrogen metabolism. Mar Biol 124: 355-369

Finn RN, Henderson JR, Fyhn HJ (1995d) Physiological energetics of developing embryos and yolk-sac larvae of Atlantic cod (*Gadus morhua*). II. Lipid metabolism and enthalpy balance. Mar Biol 124: 371-379

Finn RN, Fyhn HJ, Henderson RJ, Evjen MS (1996) The sequence of catabolic substrate oxidation and enthalpy balance of developing embryos and yolk-sac larvae of turbot (*Scophthalmus maximus*L.). Comp Biochem Physiol 115A: 133-151

Finn RN, Ronnestad I, van der Meeren T, Fyhn HJ (2002) Fuel and metabolic scaling during the early life stages of Atlantic cod *Gadus morhua*. Mar Ecol Prog Ser 243: 217-234

Finn RN, Ronnestad I (2003) The effect of acute changes in temperature and light on the aerobic metabolism of embryos and yolk-sac larvae of turbot (*Scophthalmus maximus*). Can J Fish Aquat Sci 60: 1324-1331

Fritsche R, Burggren WW (1996) Development of cardiovascular responses to hypoxia in larvae of the frog *Xenopus laevis*. Amer J Physiol 271: R912-R917

Galloway TF, Kjorsvik E, Kryvi H (1999) Muscle growth and development in Atlantic cod larvae (*Gadus morhua* L.), related to different somatic growth rates. J Exp Biol 202: 2111-2120

Garside ET (1959) Some effects of oxygen in relation to temperature on the development of lake trout embryos. Can J Zool 37: 689-698

Giguere LA, Cote B, St-Pierre JF (1988) Metabolic rates scale isometrically in larval fishes. Mar Ecol Prog Ser 50: 13-19

Green BS, Anthony KRN, McCormick MI (2006) Position of egg within a clutch is linked to size at hatching in a demersal tropical fish. J Exp Mar Biol Ecol 329:144-152

Green BS, McCormick MI (2005) O_2 replenishment to fish nests: males adjust brood care to ambient conditions and brood development. Behav Ecol 16: 389-397

Grillitsch S, Medgyesy N, Schwerte T, Pelster B (2005) The influence of environmental PO_2 on hemoglobin oxygen saturation in developing zebrafish *Danio rerio*. J Exp Biol 208: 309-316

Groot EP, Alderdice DF (1985) Fine structure of the external egg membrane of five species of Pacific salmon and steelhead trout. Can J Zool 63: 553-566

Gruber K, Wieser W (1983) Energetics of development of the Alpine charr, *Salvelinus alpinus*, in relation to temperature and oxygen. J Comp Physiol 149: 485-493

Guppy M, Fuery CJ, Flanigan JE (1994) Biochemical principles of metabolic depression. Comp Biochem Physiol 109B: 175-189

Guppy M (2004) The biochemistry of metabolic depression: a history of perceptions. Comp Biochem Physiol 139B: 435-442

Hamdorf K (1961) Die Beeinflussung der Embryonal- und Larvalentwicklung der Regenbogenforelle (*Salmo irideus* Gibb.) durch die Umweltfaktoren, O_2-Partialdruck und Temperatur. Z vergl Physiol 44: 523-549

Hartvig M, Weber RE (1984) Blood adaptations for maternal-fetal oxygen transfer in the viviparous teleost, *Zoarces viviparus* L. In: Seymour RS (ed) Respiration and Metabolism of Embryonic Vertebrates. Dr W Junk Publishers, Dordrecht, p 17-30

Hastings D, Burggren WW (1995) Developmental changes in oxygen consumption in larvae of the South African clawed frog *Xenopus laevis*. J Exp Biol 198: 2465-2475

Hattori M, Sawada Y, Kurata M, Yamamoto S, Kato K, Kumai H (2004) Oxygen deficiency during somitogenesis causes centrum defects in red sea bream, *Pagrus major* (Temminck et Schlegel). Aquac Res 35: 850-858

Hayes FR, Wilmot IR, Livingstone DA (1951) The oxygen consumption of the salmon egg in relation to development and activity. J Exp Zool 116: 377-395

Hendry AP, Day T, Cooper AB (2001) Optimal size and number of propagules: allowance for discrete stages and effects of maternal size on reproductive output and offspring fitness. Am Nat 157: 387-407

Hiroi J, Kaneko T, Tanaka M (1999) *In vivo* sequential changes in chloride cell morphology in the yolk-sac membrane of Mozambique tilapia (*Oreochromis mossambicus*) embryos and larvae during seawater adaptation. J Exp Biol 202: 3485-3495

Hughes GM (1984) Measurements of gill area in fishes: practises and problems. J Mar Biol Assoc UK 68: 489-498

Ingermann RL, Terwilliger RC (1984) Facilitation of maternal-fetal oxygen transfer in fishes: Anatomical and molecular specializations. In: Seymour RS (ed) Respiration and Metabolism of Embryonic Vertebrates. Dr W Junk Publishers: Dordrecht, p 1-15

Ingermann RL, Terwilliger RC, Roberts MS (1984) Foetal and adult blood oxygen affinities of the viviparous seaperch, *Embiotoca lateralis*. J Exp Biol 108: 453-458

Jacob E, Drexel M, Schwerte T, Pelster B (2002) The influence of hypoxia and of hypoxemia on the development of cardiac activity in zebrafish larvae. Am J Physiol 283: R911-R917

Jacobsson A, Fritsche R (1999) Development of adrenergic and cholinergic cardiac control in larvae of the African clawed frog *Xenopus laevis*. Physiol Biochem Zool 72: 328-338

Jonz MG, Nurse CA (2005) Development of oxygen sensing in the gills of zebrafish. J Exp Biol 208: 1537-1549

Kirsch R, Nonnotte G (1977) Cutaneous respiration in three freshwater teleosts. Resp Physiol 29: 339-354

Korsgaard B, Weber RE (1989) Maternal-fetal trophic and respiratory relationships in viviparous ectothermic vertebrates. Adv Comp Environm Physiol 5: 209-233

Kranenbarg S, Muller M, Gielen JLW, Verhagen JHG (2000) Physical constraints on body size in teleost embryos. J Theor Biol 204: 113-133

Li J, Eygensteyn J, Lock R, Verbost P, Heijden A, Bonga S, Flik G (1995) Branchial chloride cells in larvae and juveniles of freshwater tilapia *Oreochromis mossambicus*. J Exp Biol 198: 2177-2184

Liem KF (1981) Larvae of air-breathing fishes as countercurrent flow devices in hypoxic environments. Science 211: 1177-1179

Malvin GM (1988) Microvascular regulation of cutaneous gas exchange in amphibians. Am Zool 28: 171-177

Matschak TW, Hopcroft T, Mason PS, Crook AR, Stickland NC (1998) Temperature and oxygen tension influence the development of muscle cellularity in embryonic rainbow trout. J Fish Biol 53: 581-590

Matschak TW, Stickland NC, Mason PS, Crook AR (1997) Oxygen availability and temperature affect embryonic muscle development in Atlantic salmon (*Salmo salar* L.). Differentiation 61: 229-235

Matschak TW, Stickland NC, Crook AR, Hopcroft T (1995) Is physiological hypoxia the driving force behind temperature effects on muscle development in embryonic Atlantic salmon (*Salmo salar* L.)? Differentiation 59: 71-77

McDonald DG, McMahon BR (1977) Respiratory development in Arctic char *Salvelinus alpinus* under conditions of normoxia and chronic hypoxia. Can J Zool 55: 1461-1467

Ninness MM, Stevens ED, Wright PA (2006) Removal of the chorion before hatching results in increased movement and accelerated growth in rainbow trout (*Oncorhynchus mykiss*) embryos. J Exp Biol 209: 1874-1882

Olson KR (1998) Hormone metabolism by the fish gill. Comp Biochem Physiol 119A: 55-65

Orlando K, Pinder AW (1995) Larval cardiorespiratory ontogeny and allometry in *Xenopus laevis*. Physiol Zool 68: 63-75

Padilla PA, Roth MB (2001) Oxygen deprivation causes suspended animation in the zebrafish embryo. Proc Natl Acad Sci USA 98: 7331-7335

Pakkasmaa S, Penttinen OP, Piironen J (2006) Metabolic rate of Arctic charr eggs depends on their parentage. J Comp Physiol 176B: 387-391

Pelster B (1999) Environmental influences on the development of the cardiac system in fish and amphibians. Comp Biochem Physiol 124A: 407-412

Pelster B (2002) Developmental plasticity in the cardiovascular system of fish, with special reference to the zebra fish. Comp Biochem Physiol 133A: 547-553

Pelster B, Bemis WE (1992) Structure and function of the external gill filaments of embryonic skates (*Raja erinacea*). Resp Physiol 89: 1-13

Pelster B, Burggren WW (1996) Disruption of hemoglobin oxygen transport does not impact oxygen- dependent physiological processes in developing embryos of zebra fish (*Danio rerio*). Circ Res 79: 358-362

Pelster B, Sänger AM, Siegele M, Schwerte T (2003) Influence of swim training on cardiac activity, tissue capillarization, and mitochondrial density in muscle tissue of zebrafish larvae. Amer J Physiol 285: R339-R347

Pinder AW, Feder ME (1990) Effect of boundary layers on cutaneous gas exchange. J Exp Biol 154: 67-80

Pisam M, Massa F, Jemmet C, Prunet P (2000) Chronology of the appearance of ß, A, and α mitochondria-rich cells in the gill epithelium during ontogenesis of the brown trout (*Salmo trutta*). Anat Rec 259: 301-311

Pörtner HO, Bennett AF, Bozinovic F, Clarke A, Lardies MA, Lucassen M, Pelster B, Schiemer F, Stillman JH (2006) Trade-offs in thermal adaptation: The need for a molecular ecological integration. Physiol Biochem Zool 79: 295-313

Post JR, Lee JA (1996) Metabolic ontogeny of teleost fishes. Can J Fish Aquat Sci 53: 910-923

Protas LL, Leontieva GR (1992) Ontogeny of cholinergic and adrenergic mechanisms in the frog (*Rana temporaria*) heart. Amer J Physiol 262: R150-R161

Randall DJ, Wilson JM, Peng KW, Kok TWK, Kuah SSL, Chew SF, Lam TJ, Ip YK (1999) The mudskipper, *Periophthalmodon schlosseri*, actively transports NH_4^+. Amer J Physiol 277C: R1562-R1567

Rauther M (1937) Kiemen der animnier - Kiemendarmderivate der cyclostomen fische. In: Bolke L, Göppert E, Kallius E, Lubosch W (eds) Hanbuch der Vergleichenden Anatomie der Wirbeltiere. Urban & Schwarzenberg, Berlin, p 211-278

Roff AD (1992) The Evolution of Life Histories; Theories and Analysis. Chapman & Hall, New York

Rojo MC, Blánquez MJ, González ME (1997) Ultrastructural evidence for apoptosis of pavement cells, chloride cells, and hatching gland cells in the developing branchial area of the trout *Salmo trutta*. J Zool Lond 243: 637-651

Rombough PJ (1986) Mathematical model for predicting the dissolved oxygen requirements of steelhead (*Salmo gairdneri*) embryos and alevins in hatchery incubators. Aquacult 59: 119-137

Rombough PJ (1988a) Growth, aerobic metabolism, and dissolved oxygen requirements of embryos and alevins of steelhead, *Salmo gairdneri*. Can J Zool 66: 651-660

Rombough PJ (1988b) Respiratory gas exchange, aerobic metabolism and effects of hypoxia during early life. In: Hoar WS and DJ Randall (eds) Fish Physiology, The Physiology of developing fish, Vol. XI, Academic Press, San Diego New York, p 59-161

Rombough PJ (1992) Intravascular oxygen tensions in cutaneously respiring rainbow trout (*Oncorhynchus mykiss*) larvae. Comp Biochem Physiol 101A: 23-27

Rombough PJ (1997) Piscine cardiovascular development. In: Burggren WW, Keller BB, (eds) Development of Cardiovascular Systems: Molecules to Organisms, University Press: Cambridge, p 145-165

Rombough PJ (1998) Partitioning of oxygen uptake between the gills and skin in fish larvae: a novel method for estimating cutaneous oxygen uptake. J Exp Biol 201: 1763-1769

Rombough PJ (1999) The gill of fish larvae. Is it primarily a respiratory or an ionoregulatory structure? J Fish Biol 55 Suppl A: 186-204

Rombough PJ (2002) Gills are needed for ionoregulation before they are needed for O_2 uptake in developing zebrafish, *Danio rerio*. J Exp Biol 205: 1787-1794

Rombough PJ (2004) Gas Exchange, ionoregulation, and the functional development of the teleost. In: Govoni JJ (ed) The Development of Form and Function in fishes and the Question of Larval Adaptation. Am Fish Soc 40: 47-83

Rombough PJ, Moroz BM (1990) The scaling and potential importance of cutaneous and branchial surfaces in respiratory gas exchange in young chinook salmon (*Oncorhynchus tshawytscha*). J Exp Biol 154: 1-12

Rombough PJ, Moroz BM (1997) The scaling and potential importance of cutaneous and branchial surfaces in respiratory gas exchange in larval and juvenile walleye *Stizostedion vitreum*. J Exp Biol 200: 2459-2468

Rombough PJ, Ure D (1991) Partitioning of oxygen uptake between cutaneous and branchial surfaces in larval and young juvenile chinook salmon *Oncorhynchus tshawytscha*. Physiol Zool 64: 717-727

Schönweger G, Schwerte T, Pelster B (2000) Temperature-dependent development of cardiac activity in unrestrained larvae of the minnow *Phoxinus phoxinus*. Am J Physiol 279: R1634-R1640

Schwerte T, Prem C, Mairosl A, Pelster B (2006) Development of the sympatho-vagal balance in the cardiovascular system in zebrafish (*Danio rerio*) characterized by power spectrum and classical signal analysis. J Exp Biol 209: 1093-1100

Shepard MP (1955) Resistance and tolerance of young speckled trout (*Salvelinus fontinalis*) to oxygen lack, with special reference to low oxygen acclimation. J Fish Res Bd Canada 12: 387-433

Spoor WA (1977) Oxygen requirements of embryos and larvae of the largemouth bass, *Micropterus salmoides* (Lacépède). J Fish Biol 11: 77-86

Spoor WA (1984) Oxygen requirements of larvae of the smallmouth bass, *Micropterus dolomieui* Lacépède. J Fish Biol 25: 587-592

Sundin L, Turesson J, Burleson M (2003) Identification of central mechanisms vital for breathing in the channel catfish, *Ictalurus punctatus*. Resp Physiol Neurobiol 138: 77-86

Territo PR, Altimiras J (2001) Morphometry and estimated bulk oxygen diffusion in larvae of *Xenopus laevis* under chronic carbon monoxide exposure. J Comp Physiol 171B: 145-153

Turesson J, Schwerte T, Sundin L (2006) Late onset of NMDA receptor-mediated ventilatory control during early development in zebrafish (*Danio rerio*). Comp Biochem Physiol 143A: 332-339

Turesson J, Sundin L (2003) N-methyl-D-aspartate receptors mediate chemoreflexes in the shorthorn sculpin *Myoxocephalus scorpius*. J Exp Biol 206: 1251-1259

Van der Heijden AJH, van der Meij JCA, Flik G, Wendelaar Bonga SE (1999) Ultrastructure and distribution dynamics of chloride cells in tilapia larvae in fresh water and sea water. Cell Tiss Res 297: 119-130

Weber RE, Hartvig M (1984) Specific fetal hemoglobin underlies the fetal-maternal shift in blood oxygen affinity in a viviparous teleost. Mol Physiol 6: 27-32

Weihs D (1980) Respiration and depth control as possible reasons for swimming of northern anchovy, *Engraulis mordax*, yolk-sac larvae. Fish Bull 78: 109-117

Wells PR, Pinder AW (1996a) The respiratory development of Atlantic salmon. I. Morphometry of gills, yolk sac and body surface. J Exp Biol 199: 2725-2736

Wells PR, Pinder AW (1996b) The respiratory development of Atlantic salmon. II. Partitioning of oxygen uptake among gills, yolk sac and body surfaces. J Exp Biol 199: 2737-2744

Wieser W (1985) Developmental and Metabolic Constraints of the Scope for Activity in Young Rainbow Trout (*Salmo gairdneri*). J Exp Biol 118: 133-142

Wieser W, Platzer U, Hinterleitner S (1985) Anaerobic and aerobic energy production of young rainbow trout (*Salmo gairdneri*) during and after bursts of activity. J Comp Physiol 155B: 485-492

Zaccone G, Mauceri A, Fasulo S (2006) Neuropeptides and nitric oxide synthase in the gill and the air-breathing organs of fishes. J Exp Zool 305A: 428-439

Zaccone G, Mauceri A, Fasulo S, Ainis L, Lo Cascio P, Ricca MB (1996) Localization of immunoreactive endothelin in the neuroendocrine cells of fish gill. Neuropeptides 30: 53-57

CHAPTER

5

Cardiovascular Anatomy and Physiology

Warren W Burggren[1,*] and Brian Bagatto[2]

INTRODUCTION

Larval fishes face a number of challenges to survival. Indeed, mortality rates can exceed 99% in the larvae of pelagic fishes e.g. Atlantic herring (*Clupea harengus*), Atlantic mackerel (*Scomber scombrus*). Some of the challenges are common to all near-microscopic organisms and relate to their small size and the physical nature of their environment, while others are unique to larval fishes. In many instances these challenges have acted as selection pressures that, through evolution, have shaped cardiovascular function in larval fishes as well as the adult forms. Before beginning an exploration of the larval cardiovascular system, let us briefly consider these shaping forces.

Environmental Challenges to Survival

While considerable recent research has focused on the mechanisms by which larval fishes acquire prey (e.g. Gahtan *et al.*, 2005; McElligott & O'Malley, 2005), larval fishes are also prey themselves. Rates of predation on larval fishes by other fish species, by amphibians, and by aquatic predatory insects can be very high, especially in the absence of paternal care. Larval fishes are especially vulnerable to predation not only because of

Authors' addresses: [1]Department of Biological Sciences, University of North Texas, Denton, TX 76205, USA.

[2]Department of Biology, University of Akron, Akron, OH 44325, USA.

**Corresponding author:* E-mail: burggren@unt.edu

their small size and immature locomotory system, but also because of water's high kinematic viscosity (the ratio of the dynamic viscosity of a fluid to its density), which limits their locomotion (see Müller, this volume). In larval herring, for example, viscosity has major effects on swimming speeds when larvae have body lengths of less than 1 cm (Fuiman & Batty, 1997). Indeed, in these smaller larvae, temperature-related viscosity effects have more influence over locomotion than temperature-related metabolic effects. The combination of these factors make evasive actions more difficult for larval fishes, aiding especially larger prey.

Abiotic challenges to larval fish survival result from potentially sharp temperature and dissolved oxygen gradients in aquatic microhabitats. Movement of only a few millimeters can result in significant decreases in oxygen and increases in temperature in shallow freshwater habitats (accentuated in unique habitats such as intertidal pools). While being small affords the exploitation of microhabitats (e.g. respiration from the well-oxygenated surface zone of water or retreat from sun-heated substrate), the physico-chemical characteristics of such microhabitats can also be highly unstable. Yet, larval fish must successfully cope at organismal through molecular levels with these environmental challenges.

Anatomical/Physiological Challenges to Survival

Irrespective of their environment, the very make-up of larval fishes also results in great challenges to their survival. Larval fishes experience phenomenal rates of growth through the dual mechanisms of hyperplasia and hypertrophy (Rowlerson *et al.*, 1995; Paturno *et al.*, 1998; Johnston, 2006). Rapid somatic growth presents escalating challenges in terms of ensuring adequate nutrient and oxygen delivery as well as CO_2 and waste removal. Critical in this delivery is the surface to volume ratio of not only growing cells, but the growing embryo/larva, itself. Younger larval fish have a very high surface area to volume ratio, which translates into a very large surface area across which materials can be gained or lost through the body wall by diffusion or osmosis. On the one hand, gas exchange (both gain of O_2 and loss of CO_2) is achieved with ease, largely across the non-specialised body wall rather than the developing gills (see below). On the other hand, osmoregulation is likely to be an especially great challenge, especially in freshwater fishes, because of the strong proclivity towards inward water movement and outward salt loss (see Kaneko & Hiroi, this volume).

Another developmental challenge to survival involves the numerous transitions of functions from one system or site to another. Many key physiological functions are "handed off" in full or in part during development. For example, the transition from yolk subsistence to active feeding requires the successful activation of the foregut, coordination of mouth motions for feeding with motions for breathing, etc. Another example involves the transition of osmoregulatory processes from general skin surface to the emerging gills (e.g. chloride cells at the base of each gill lamella) and kidneys. Each of these major transitions requires major structural, physiological and even behavioural processes, and is a phase in the life cycle marked by disproportionately large mortality.

A final example of a challenge to survival in larvae of the great number of oviparous fish species results from their relatively precocial release into the environment well in advance of having regulated physiological systems, and certainly well before their cardio-vascular system is regulated. Yet, as free-living animals, they experience many of the same demands and challenges of more mature individuals.

The Changing Role of the Circulation during the Embryonic to Larval Transition

Based heavily on adult studies, our traditional view of the vertebrate circulation is that of a system for internal transport of respiratory gases, metabolic fuels and wastes, pressure and heat. While the circulation of the adult fish may be involved in most if not all of these functions, an argument can be made that the circulation of larval fishes shows distinct transitions in primary function, with emphasis on regulated internal transport of gases and wastes only developing during intermediate and later stages of larval development. Indeed, several studies have shown that the early fish larva can acquire necessary O_2 and eliminate CO_2 across the non-specialised body surface, apparently substantially without the aid of circulation through gills or with functional haemoglobin (Pelster & Burggren, 1996; Wells & Pinder, 1996a; Rombough & Moroz, 1997; Rombough, 1998; Jacob *et al.*, 2002). In part, the efficiency of the skin as an organ for gas exchange results from the very high surface to volume ratio of larval fishes (see Pelster, this volume). Rombough & Moroz (1997) estimate for the walleye (*Stizostedion vitreum*) that the surface area does not become limiting to gas exchange until body mass exceeds 100 mg. Related to this, Mirkovic & Rombough (1998), investigated the linkage between metabolism and cardiac output in larvae of the rainbow trout (*Oncorhynchus mykiss*) and found that O_2 delivery to the tissues depended less upon convection by the cardiovascular system and more upon direct oxygen diffusion across the body wall in the first few days of development. Similarly, Kopp *et al.* (2005), studying larvae of the minnow (*Phoxinus phoxinus*), concluded that the early larval circulation does not serve a major function in oxygen transport – this occurring instead by diffusion.

The independence of gas exchange from circulatory function in early developmental stages may be a broad general characteristic, having been documented not only in fishes, but also in amphibians and birds (Burggren, 2004, 2005). Collectively, these data suggest that the heart and associated cardiovascular system develops and begins convective blood flow before existing as a functional necessity – a phenomenon called prosynchronotropy (Burggren & Territo, 1995; Territo & Burggren, 1998; Burggren, 2005). Various non-respiratory hypotheses have been proposed for why convective blood flow and pressure is generated in early developmental stages. One hypothesis describes the function of the early embryonic heart as a pressure generator that produces shear-strain within the interior of endothelial-lined vessels (Burggren, 2005). These forces promote hormonal stimulation of angiogenesis, especially in the periphery (Burggren, 2004). Alternative (or complementary) hypotheses ascribe an osmoregulatory function of the early circulation (Rombough, 2002), which is supported by the anecdotal observation that larval fish or amphibians that are cardiovascular mutants with impaired circulation often show pathological accumulation of pericardial and/or peritoneal fluid.

Eventually, the larval fish's diminishing surface area to volume ratio, combined with the increasing distances for diffusion from the larva's general body surface to its interior, result in the development of elaborate, heavily perfused gills responsible for the majority of gas exchange. Concurrently, larvae become more active, and consequently may encounter a variety of microhabitats with respect to temperature and oxygen. Thus, as larval fishes grow they will require an increasingly complex cardiovascular system that is regulated by regional and systemic needs for transport of oxygen, carbon dioxide, nutrients and metabolic rates.

CARDIOVASCULAR DEVELOPMENT

There have been many publications on cardiovascular development in fish, though each typically focuses on a very specific anatomical or physiological phenomenon in a specific species. Rombough (1997) provided a comprehensive compilation of data on fish cardiovascular development up to that time. It is not our intention to revisit what he has so eloquently written, especially the valuable historical contexts. However, some review is necessary here to set the context for the new information published since 1997. In this regard, the current chapter can be used in complement with Rombough's (1997) chapter.

Heart Formation

The cardiovascular system is the first organ system to function in vertebrates and the heart is the first organ to become functional within this system. The heart anlage originates from two lateral masses of splanchnic mesoderm (primordial) located on either side of the ventral axis that marks the embryo's centre (Laale, 1984; Driever, 2000; Hall, this volume). These two masses of tissue move towards the dorsal midline and eventually merge and form a tube located near the midline, ventral to the otic anlage (Rombough, 1997). Movements are directed by the *Mil* (miles apart) gene and sphingosine-1-phosphate, a lysosphingolipid acting as a ligand for *Mil* (Kupperman *et al.*, 2000). The heart begins contracting in an irregular and weak and peristaltic fashion, propelling what blood cells have already formed through a rudimentary circuit (Rombough, 1997). As beating continues, the contractions become stronger and more regular. During this time, the heart moves to a more anterior-posterior position and the linear tube shape compresses slightly into an S-shaped curve (Laale, 1984; Stainier & Fishman, 1992). Constriction then occurs at the boundary between the atrium and ventricle while the orientation of the heart continues to migrate until the atrium is anterior to the ventricle (Fig. 1). Peristaltic contraction, the primary mechanism propelling blood in a one-way direction, gradually changes into a chamberised contraction after which valves form to prevent backflow of blood. Apart from onset of

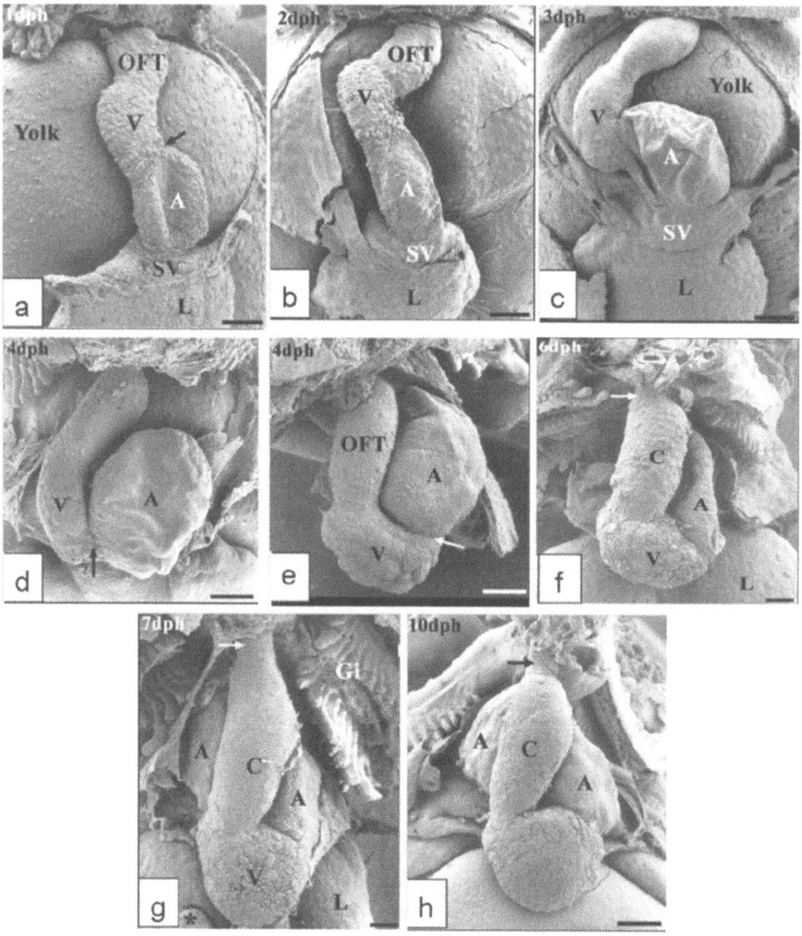

Fig. 1 Scanning electron micrograph composite revealing a ventral view of the development of the heart of the Adriatic sturgeon (*Acipenser naccarii*). Developmental age is given in days post-hatching (dph). A, atrium; C, myocardial conus arteriosus; Gi, gills; L, liver; OFT, outflow tract; SV, sinus venosus; V, ventricle; *, gall bladder. **a.** On 1 dph the heart is largely represented by a straight tube, with the four chambers of the heart arranged in a craniocaudal sequence. Arrow indicates atrioventricular segment in panel a, as well as b-d. **b.** On 2 dph outflow tract ventricle migrates towards the right, but the sinus venosus remains along the midline, attached to the developing liver. **c.** On 3 dph rapid atrial growth begins. The ventricle and its outflow tract have developed their maximum bending to the right. **d** and **e.** Different views of the heart 4 dph. Rapid atrial growth continues as the ventricle rotates caudally in a counter-clockwise direction. **f.** At 6 dph the outflow tract has divided into a proximal, myocardial conus arteriosus and a distal, bulbus segment. White arrow indicates the distal (bulbus) segment of the outflow tract. **g.** The outflow tract now obscures the atrium, while the trapezoid-shaped ventricle is caudally located. **h.** By 10 dph the heart exhibits the adult configuration after Icardo *et al.* (2004).

physiological regulation, there is little morphological change in the heart other than growth in size.

The events described above are the general events of heart development as documented in most fish species. There are, however, variations in some of the sequences and the morphology and full detailed accounts have been documented in at least 16 species (Rombough, 1997; Icardo *et al.*, 2004). Table 1 shows a summary of the major cardiovascular developmental events in the zebrafish (*Danio rerio*) larva. While these data are likely to be generally representative of teleost fishes, there are likely to be interesting and important species-specific variations worthy of description.

The cells of the heart (cardiac myocytes) undergo several distinct transitions during the formation and subsequent growth of the heart. Cardiac myocytes display two developmentally programmed forms of growth, hyperplasia and hypertrophy. During the process of heart formation described above, myocytes proliferate rapidly and increase in total cell numbers via hyperplasia. Upon development of separate atrial and ventricular chambers, the cell-cycle ceases and existing myocytes begin to increase in size – i.e. hypertrophy. Hypertrophic cardiac growth usually parallels body mass increase, although there have been data in some species showing a marked cardiac hypertrophy accompanying sexual maturation (Clark & Rodnick, 1998). During hypertrophic heart growth, there are also increases in mitochondrial and capillary density, as well as a transition from uninucleate to multi-nucleate myocardiocytes (MacLellan & Schneider, 2000).

Conduction System of the Heart

The conduction system in fish hearts, while functionally similar to mammalian hearts, shows distinct anatomical differences. The cardiac pacemaker in adult fish is found in the sinus venosus and serves to regulate the rate of cellular depolarisation (Arbel *et al.*, 1977). Within the sinus venosus of elasmobranchs, marine teleosts and freshwater teleosts, these pacemaker potentials have been localised to the sinoatrial valve (Saito, 1969). In some fish, however, the atrioventricular funnel and the bulboventricular junction contain the pacemaker sites (Saito, 1973). The atrioventricular myocardium creates a delay between atrial and ventricular contraction (much like the mammalian atrioventricular node) and prevents atrial tachyarrhythmias from reaching the ventricle (Irisawa, 1978). The heart is also able to achieve the characteristic contraction from apex to base, not with a differentiated His-Purkinje system, but rather via ventricular trabeculae (Sedmera *et al.*, 2003). These trabecular bands form direct myocardial continuity between the atrioventricular canal and the apex of the ventricle (Sedmera *et al.*, 2003). Pacemaker cells of all fish studied also have the fundamental property discovered in all animal hearts studied to date – stretch induced tachycardia (Irisawa, 1978).

The electrocardiogram in adult fishes follows the vertebrate paradigm, having the characteristic P (atrial depolarisation), R (ventricular depolarisation), and T (ventricular repolarisation) waves (Randall, 1970). The cardiac cycle time intervals of P-R and

Table 1 Major morphological and physiological events in the development of the zebrafish (*Danio rerio*) cardiovascular system

Chronological time (hpf at 28.5C)	*Heart morphology and physiology*	*Vessel morphology and physiology*	*Blood cells*	*References*
12			Intermediate cell mass (ICM) precursors adopt either a haematopoietic stem cell fate or vascular fate	Zon (1995)
14-15			Cells above differentiate into pro-erythroblasts or endothelial cells of the trunk vasculature	Davidson & Zon (2004)
15-16		Endothelial cells migrate to site of vessel formation	Macrophage precursors from the rostral blood island migrate toward the head midline	Ward *et al.* (2003)
18		Major vasculature established		Isogai *et al.* (2003)
24-25	Brief irregular beating followed by regular peristaltic beating[1]	Blood flows through a simple circulatory loop – completion of axial patterning[2]	ICM comprised about 300 proerythroblasts[3]	[1]Isogai *et al.* (2001) [2]Nye *et al.* (2006) [3]Long *et al.* (1997)
26-30	Heart tube starts to loop[1]		Functional macrophage activity[2]	[1]Stainier & Fishman (1992) [2]Herbomel *et al.* (1999)
36	Chamber boundaries are evident			Stainier & Fishman (1992)
48			Neutrophillic granulocytes detected in circulation[1] Thrombocytes present[2]	[1]Willet *et al.* (1999) [1]Lieschke *et al.* (2001) [2]Lin *et al.* (2001) [2]Gregory & Jagadeeswaran (2002)

Table 1 Contd. ...

Table 1 Contd. ...

60-65	Cardiac valves present[1]		Thymi are populated with immature lymphoblasts[2]	[1]Weinstein & Fishman (1996) [2]Willett *et al.* (1999)
72 (3 dpf)		Dorsoventrally aligned intersegmental vessels Dorsal longitudinal anastomotic vessels Longitudinal parachordal vessels		Isogai *et al.* (2001) Nye *et al.* (2006)
4dpf	Heart responds chronotropically to cholinergic agonists[1]		Immunoglobin gene rearrangements detected[2]	[1]Bagatto (2005) [2]Danilova & Steiner (2002)
5 dpf	Functional haemoglobin necessary[1]		All proerythroblasts have matured into flattened elliptical erythrocytes[2]	[1]Liao *et al.* (2002) [1]Pelster & Burggren (1996) [2]Weinstein *et al.* (1996)
6 dpf	Heart responds to adrenergic agonists	Vessels respond to vasodilators		Bagatto (2005)
7 dpf			Erythroblasts, myeloblasts, neutrophilic granulocytes observed within the pronephros Transcripts for the membrane form of IgM	Danilova & Steiner (2002)
8 dpf		Vessels respond to vasoconstrictors		Bagatto (2005)
15 dpf	Presence of cholinergic tone			Bagatto (2001)
19 dpf			Lymphoblasts observed within the pronephros	Danilova & Steiner (2002)

R-T in the eel ammocoete (*Lethenteron japonicum*) are 20% and 54%, respectively of the total cardiac cycle (Prosheva, 2005), and are similar to those in the silver carp (*Hypophthalmichthys molitrix*) (Chuiko *et al.*, 1977) and gilthead seabream (*Sparus aurata*) (Aissaoui *et al.*, 1998). Additionally, Aissaoui *et al.* (1998) have reported a clear relationship between R-T interval and fish mass in adult gilthead seabream. The change in T-P interval is the main component influencing heart rate changes and the modulation of the P-R and R-T intervals by neural autonomic influence on the heart (Aissaoui *et al.*, 1998). Perhaps due to the technical difficulties of accurately measuring electrocardiograms (ECG) in larval fish of only a few milligrams, the developmental profile of electrocardiology in fishes remains to be described.

Vasculogenesis and Angiogenesis

Vasculogenesis, *de novo* formation of blood vessels through the combined growth of endothelial cells, often involves significant migration of endothelial cells to the site of vessel formation (Lawson & Weinstein, 2002). Although vessel formation in fish is a highly dynamic process, it follows a stereotypic vertebrate program. The overall form of the vasculature is established early, before the initiation of circulation, and the pattern of vessel tracts is reproducible from embryo to embryo (Isogai *et al.*, 2003). In the zebrafish, cells from the dorsal margin of the lateral plate migrate medially to form the ICM (Weinstein & Fishman, 1996). This mass gives rise to both the endothelia of the major trunk vessels and the first cohort of blood cells (Childs *et al.*, 2002). Even with this predictability of vessel formation, many early vasculogenic vessels first appear as a network or plexus of smaller vessels, with little apparent pattern or differentiated identity (Weinstein, 1999). The final pattern and identity of vessels emerges only upon angiogenic remodeling of the primary networks, and the particular final pattern formed is plastic and easily altered by changes in the local environment (Weinstein & Fishman, 1996; Yelon, 2001; Rossant & Hirashima, 2003).

Angiogenesis, in contrast to vasculogenesis, is the formation of new blood vessels from pre-existing ones and is often associated with cell division of endothelial cells that contribute to the growing blood vessel (Lawson & Weinstein, 2002). In the last few years, there have been major breakthroughs in the understanding of the genetic control of the normal processes of angiogenesis and how this process contributes to overall vascular development. In addition to assessing the many vascular mutants generated in mice, the zebrafish has quickly become a powerful genetic model for studying vascular cell biology *in vivo* (Weinstein, 2002; Ny *et al.*, 2006). Embryonic phenotypes that fail to develop different phases of the normal vasculature have been reported for the vascular endothelial growth factor (VEGF)/VEGF receptor (VEGFR) (Habeck *et al.*, 2002), Neuropilin-1 (NRP-1, a membrane glycoprotein) (Lee *et al.*, 2002), and angiopoietin/Tie families of vascular-specific signalling molecules (Coultas *et al.*, 2005). Similar results have been shown for even more widely used signalling pathways such as platelet-derived growth factor (PDGF), transforming growth factor-β (TGF-β), ephrin, Notch and pathways thought to be primarily involved in axon guidance (such as semaphorins, netrins and Robo) (Bikfalvi & Bicknell, 2002). The steps in forming a functional circulation from endothelial progenitors have been detailed in Coultas *et al.* (2005) and Ny *et al.* (2006).

While the zebrafish has served as an extremely useful model for the study of angiogenesis, little is known of this process in other fishes, including more phyletically primitive species.

Vascular Structure and Patterns

Vasculogenesis takes place during the first 24 hours post fertilisation (hpf) in the zebrafish developing at 28.5C. The following is a summary of major vascular developmental events in the zebrafish at 28.5C as described in extensive detail by Isogai *et al.* (2001) and Ny *et al.* (2006) (for complete vascular anatomy online, search "zebrafish vascular atlas"). Briefly, at 24 hpf, blood flows through a simple circulatory loop indicating the completion of axial vessel patterning. An evolutionarily conserved network of secondary branching vessels is then formed in the trunk region via angiogenesis. This network includes the dorsoventrally aligned intersegmental vessels, two separate dorsal longitudinal anastomotic vessels, and longitudinal parachordal vessels. This network is typically functional by 72 hpf. Some vessels then continue to form branching networks as tissue mass increases, while other networks become simpler, forming single plexiform vessels. This is evident in the dorsal artery and vein as multiple paths of blood flow become integrated into a coherent vessel, signified by a decreasing cross sectional area during early development (Fig. 2).

Zebrafish offer a number of advantages that allow imaging of the cardiovascular system during development. Development proceeds rapidly compared to larger salmonids and the embryo is transparent. Additionally, the synthesis of pigments that normally occurs later in development can be blocked by a variety of techniques, including exposure of the embryos to 0.2 mM 1-phenyl-2-thio-urea (Karlsson *et al.*, 2001). Pigment elimination allows imaging over a very large time window of development compared to many other well studied fish species such as salmonids and bass. Ample techniques are available for imaging the heart and vascular system of transparent embryos using conventional light microscopy and associated imaging software (Burggren & Fritsche, 1995; Schwerte & Pelster, 2000; Schwerte & Fritsche, 2003; Bagatto & Burggren, 2006). Light microscopy techniques also allow measurement of cardiovascular parameters *in vivo*, which has permitted more complex experiments in developmental physiology. Three-dimensional imaging of the cardiovascular system in zebrafish uses confocal microangiography (Weinstein *et al.*, 1995), which allows high resolution imaging of vessels. Even more advanced techniques are being developed to image small particles injected into the vascular system. For example, digital particle image velocimetry allows for even more complex analysis of the fluid dynamics of red cells and plasma in the developing cardiovascular system (Hove *et al.*, 2003; Hove, 2006).

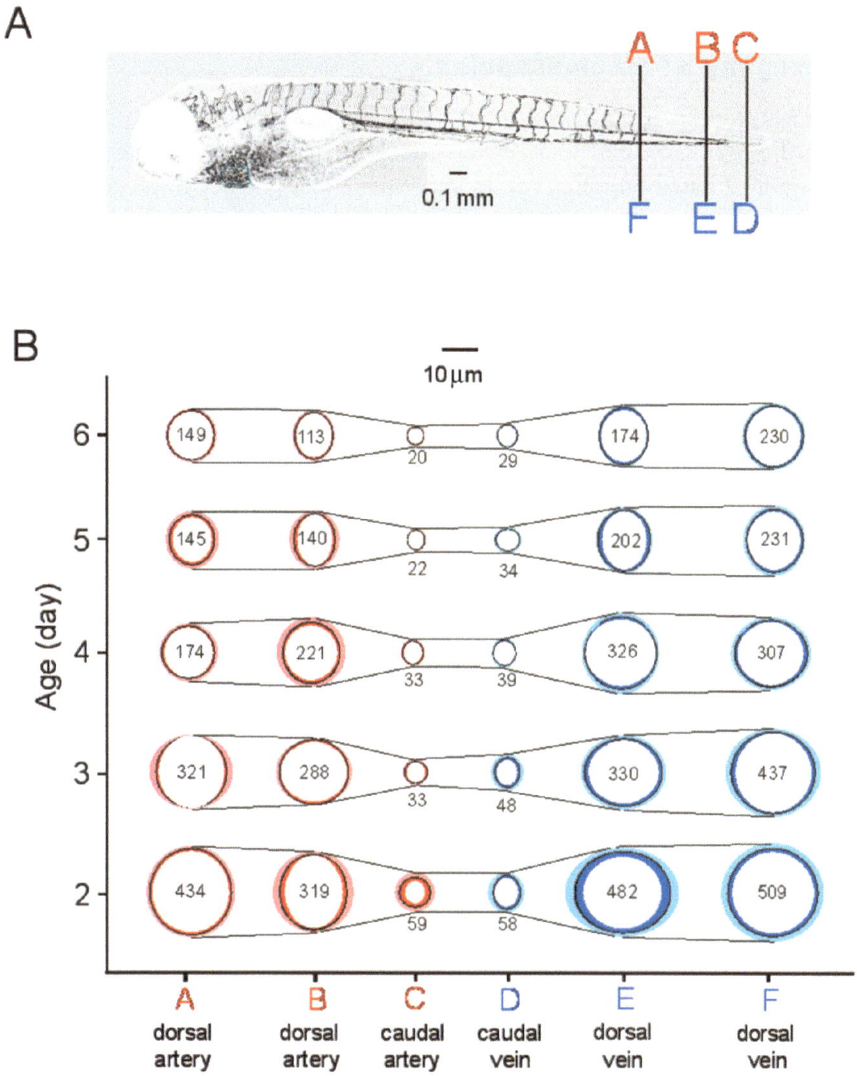

Fig. 2 (A) Cross-sectional area of major arteries and veins in the posterior trunk of the zebrafish (*Danio rerio*) larva during early development. The lines A-F, B-E, and C-E indicate vessel measurements at transects between the 24th and 25th intersegmental vessels, between the 28th and 29th intersegmental vessels, and at a vessel midway in the caudal loop, respectively. (B) Images of the mean cross-sectional area of major arteries and veins during early development according to the letter schematic previously mentioned. The black outline indicates the mean vessel cross section while the colour shading indicates the standard error of the mean. Numerical values indicate the corresponding vessel cross-sectional area (µm^2). $N = 8$ for each angle of each vessel on each day. After Bagatto & Burggren (2006).

Ontogeny of Haematopoiesis

The basic process of haematopoiesis occurs in a similar fashion across all vertebrates. Two distinct phases occur in haematopoiesis, primitive and definitive, and these are determined based on difference in site, timing and potency of progenitor cell differentiation, morphology of differentiated blood cells, and on the molecular species of expressed globins (Zon, 1995; see also Zapata & Cortés, this volume). Primitive haematopoiesis occurs intra-embryonically in the ICM. The ICM precursors adopt either a haematopoietic stem cells (HSC) fate or a vascular fate by 12 hpf at 28.5C. These cells migrate from the bilateral stripes of cells in the posterior mesoderm to the trunk midline and differentiate into pro-erythroblasts or endothelial cells of the trunk vasculature in the zebrafish (Davidson & Zon, 2004). During this migration, there is a sequential activation of genes important for erythroid cell function (Davidson & Zon, 2004). Blood cell differentiation proceeds in a rostral to caudal direction. By the time blood circulation begins at 25 hpf in the zebrafish, for example, the ICM comprised about 300 pro-erythroblasts (Long *et al.*, 1997), each expressing the genes necessary for haeme and globin synthesis, iron utilisation, and membrane stability (Wingert & Zon, 2003). Over the next four days, the circulating pro-erythroblasts synchronously mature into flattened elliptical erythrocytes (Weinstein *et al.*, 1996).

A second site of haematopoiesis begins to develop in the anterior mesoderm of the embryonic zebrafish, referred to as the rostral blood island (RBI) (Davidson & Zon, 2004). In a similar manner to the ICM, the cells in the RBI acquire either a haematopoietic (myeloid) or a vascular cell fate. At 15-16 hpf, macrophage precursors from the RBI region migrate toward the head midline and form a partially covered mass of cells between the eye and the heart primordium (Ward *et al.*, 2003). These cells then migrate laterally, scattering into single cells across the yolk sac. Some of these macrophage precursors enter circulation via the ducts of Cuvier, the yolk sac blood channels that connect the trunk vasculature to the heart, while others migrate into the head (Herbomel *et al.*, 1999). Functional macrophage activity in the ducts of Cuvier in zebrafish has been observed as early as 26 hpf. It is not known whether the macrophages originating from the RBI represent a transient wave of primitive macrophages or if they are permanently retained into adulthood.

Teleosts all have at least two types of granulocytes: neutrophilic granulocytes and eosinophilic granulocytes. In developing zebrafish, neutrophilic granulocytes can be morphologically detected in the circulatory system and in loose connective tissue as early as 48 hpf (Willett *et al.*, 1999; Lieschke *et al.*, 2001). However, eosinophilic granulocytes have not been detected in embryos up to 5 dpf (Lieschke *et al.*, 2001). Additionally, no eosinophil-specific molecular markers have been isolated to date. Therefore, the origin and developmental stage at which eosinphil granulocytes appear during embryogenesis remains unknown (Davidson & Zon, 2004).

As definitive haematopoiesis begins, the ability to produce additional blood cell types likely develops. Teleosts possess a fully functioning immune system including T cells, B cells, antigen-presenting cells, and natural killer-like cells (Traver *et al.*,

2003; Zapata & Cortés, this volume). Ultrastructural analysis in zebrafish reveals that the thymi are the first to become populated with immature lymphoblasts at around 65 hpf (Willett *et al.*, 1999), although the origin of the lymphoid progenitors is not known definitively. An absence of transcripts for *Pax5*, a gene important for B-cell differentiation, suggests that the lymphoid cells in the thymus at this stage are responsible for the T-cell lineage (Trede *et al.*, 2001). Lymphopoiesis occurs relatively late in the kidney and is not detected until 3 weeks pf (Willett *et al.*, 1999). Relative to T lymphopoiesis, B-cell formation is initiated later in zebrafish development. Immunoglobin gene rearrangements are first detected at 4 dpf, whereas transcripts for the membrane form of IgM appear at 7 dpf (Danilova & Steiner, 2002). Additionally, transcripts for *Rag1* have been found in the region of the pancreas primordium indicating that this organ is the initial site of B lymphopoiesis (Danilova & Steiner, 2002).

Little is known about the origin and ontogeny of the first thrombocytes in any teleost. Current literature indicates the presence of presumptive thrombocytes at 48 hpf as supported by two different methodologies (Lin *et al.*, 2001; Gregory & Jagadeeswaran, 2002). Zebrafish thrombocytes appear as small round nucleated cells that are morphologically similar to T lymphocytes (Jagadeeswaran & Sheehan, 1999). Based on their adhesion and aggregation properties, it appears that teleost thrombocytes are functional homologues to mammalian platelets (Gregory & Jagadeeswaran, 2002). This suggests that factors involved in the formation of these cell types may be conserved.

The kidney and spleen are sites for haematopoiesis in both developing and mature teleosts. In the zebrafish, haematopoietic cells appear in the pronephros at 4dpf and become organised into cellular cords that surround the blood vessels between the renal tubules and glomeruli (Willett *et al.*, 1999). By 7 dpf, erythroblasts, myeloblasts, and neutrophilic granulocytes are readily observed within the pronephros, whereas lymphoblasts are not observed until 19 dpf (Danilova & Steiner, 2002). Thus, with the exception of T lymphocytes (educated in the thymus), all mature blood cell lineages in zebrafish are derived from the adult kidney, with the spleen storing mainly mature erythrocytes (Traver *et al.*, 2003). This is in contrast to brown trout (*Salmo trutta*) and perch (*Perca fluviatilis*), where the spleen is an additional or the sole source of adult haematopoiesis, respectively (Catton, 1951).

Ontogeny of Haemoglobin and Respiratory Gas Transport

Haemoglobin switching (the transition from embryonic to adult forms of haemoglobin) occurs during the transition from primitive to definitive haematopoiesis. Among teleosts, the switch from embryonic to adult haemoglobin has been well investigated in rainbow trout (Iuchi, 1985). Haemoglobin of embryos immediately after hatching is distinctly different from that of 2-year-old rainbow trout in molecular properties such as stability, electrophoretic mobility and antigenicity, as well as physiological properties such as oxygen affinity (Iuchi & Yamagami, 1969; Iuchi, 1973, 1982). Embryonic trout haemoglobin consists of eight globins with different retention times from the eight globins constituting the adult haemoglobin, with the two α and two β globins clearly distinguishable from their adult counterparts (Maruyama *et al.*, 1999). Embryonic β globin cDNA has also been indentified in zebrafish (Quinkertz & Campos-Ortega, 1999). Many regard the expression of embryonic globins to be closely related to the primitive haematopoiesis, while that of adult globins is related to definitive haematopoiesis. However, it is difficult to blanket this statement for all teleosts, given that the Medaka (*Oryzias latipes*) does not show this pattern of haemoglobin switching to match changes in gene expression (Maruyama *et al.*, 2002).

The concept of "prosynchronotropy", mentioned earlier, also applies to the red blood cells forming and circulating before the absolute need for convective oxygen delivery in fish larvae. The bloodless (*bls*) mutant in the zebrafish has a defect in ICM haematopoiesis and is able to survive for 5 days until it starts to recover circulating erythroid cells (Liao *et al.*, 2002). When blood oxygen binding is completely inhibited by using carbon monoxide, or red blood cells are pharmaceutically lysed, zebrafish are able to survive until at least 5 dpf (Pelster & Burggren, 1996). These and other similar studies indicate that normal tissue growth and metabolism can proceed independent of tissue perfusion in the early fish embryo, with effective oxygen, nutrient, carbon dioxide, and nitrogenous waste exchange proceeding via diffusion.

ONTOGENY OF CARDIOVASCULAR REGULATION

In the previous section we reviewed our knowledge of the onset, development, and maturation of the cardiovascular system in embryonic and larval fishes. Of course, any organ system is only as functional as is its ability to be regulated, so we now turn to the ontogeny of regulatory mechanisms in larval fishes.

Ontogeny of Heart Regulation

There are essentially two phases of cardiovascular performance during development: an early *metronomic* phase, and a later regulated phase.

Metronomic phase of larval heart/Cardiovascular function

The heart of the embryo/early fish larva first starts to beat in a rather irregular fashion, but within hours typically settles into a more constant metronomic rhythm that is characterised by little spontaneous variation, and little or no changes associated with hypoxia or locomotory activity. Although functional skeletal muscle produces sometimes intense movements even in egg-bound embryos, such intermittent bouts of activity do not appear to be consistently accompanied by increases in heart rate in larval fishes. In fact, extended training (swimming) bouts during the first four weeks of development have little effect on heart rate, ventricular volumes, and cardiac output in zebrafish (Pelster *et al.*, 2003). Though metronomic, these early heart rates are both highly species-specific and stage dependent.

Even in this metronomic phase, heart rates are of course highly influenced by temperature, exhibiting a Q_{10} of approximately 2 in freshwater telesosts such as the zebrafish (Barrionuevo & Burggren, 1999; Schonweger *et al.*, 2000; Kopp *et al.*, 2005) (Fig. 3). However, to our knowledge, temperature effects on cardiovascular function have not been examined in thermally interesting larval fishes, such as those that inhabit either eurythermal environments such as tidepools or highly stenothermal environments.

Regulated phase of larval heart function

The circulation of vertebrates is typically controlled through a combination of regulatory mechanisms: Extrinsic cardiac regulation, intrinsic cardiac regulation, peripheral adjustments in vascular tone, and autoregulation at the local vessel level. Extrinsic regulation involves the influences of sympathetic stimulatory and parasympathetic inhibitory innervation, along with the stimulatory effects of circulating catecholamines. The cardiac pacemaker(s) respond with changes in rate of depolarisation and consequently in heart rate, while cardiac muscle responds with changes in rate and strength of contraction. Intrinsic regulation involves the classic Frank-Starling relationship, in which dynamic extension of the myocytes result in a subsequently stronger contraction, over the mid-range of myocyte extension. Practically, this translates into stronger cardiac contractions (larger stroke volume) stimulated by increased venous output, thereby keeping circulating blood from pooling. Peripheral vasodilation-vasoconstriction of arterial and venous vasculature modifies peripheral resistance, and exerts a powerful control over both the overall level of cardiac output, as well as its distribution between various vascular beds. Finally, autoregulation involves localised responses of the microvasculature (arterioles, pre-capillary sphincters) to changes in microenvironment (e.g. hypoxia, temperature change, inflammation, etc.), mediated by locally released vasoactive agents (e.g. NO, histamine).

The interacting roles of these various cardiovascular regulatory responses are generally well defined for adult vertebrates and more specifically, regulation of the circulatory system of adult fishes has been extensively studied (Randall, 1970; Johansen & Burggren, 1980; Satchell, 1991; Burggren *et al.*, 1996; Farrell, 2002). Clearly, resting heart rate in adult fish is variable and the pacemaker cells of the sinus venosus can be modified by either autonomic control and/or humoral control (Randall, 1970; Burggren *et al.*, 1996). In primitive fishes such as lampreys or hagfish (cyclostomes), there has been confusion in even recent literature as to the nature and extent of heart innervation. This confusion is mostly mediated by the polyphyletic grouping of hagfish and lamprey into discussions of the Cyclostomes in general, when it is clear that their hearts are quite different. Intact hagfish hearts do not respond to acetylcholine or adrenaline (Jensen, 1965; Labat *et al.*, 1968) and there is no published evidence to date to suggest any direct cardiac innervation whatsoever. However, despite an apparent opposite role of acetylcholine on the lamprey heart (Itina & Balanov, 1964), there is clear evidence that some species of lamprey, i.e. Arctic lamprey (*Lampetra japonica*), have two sets of

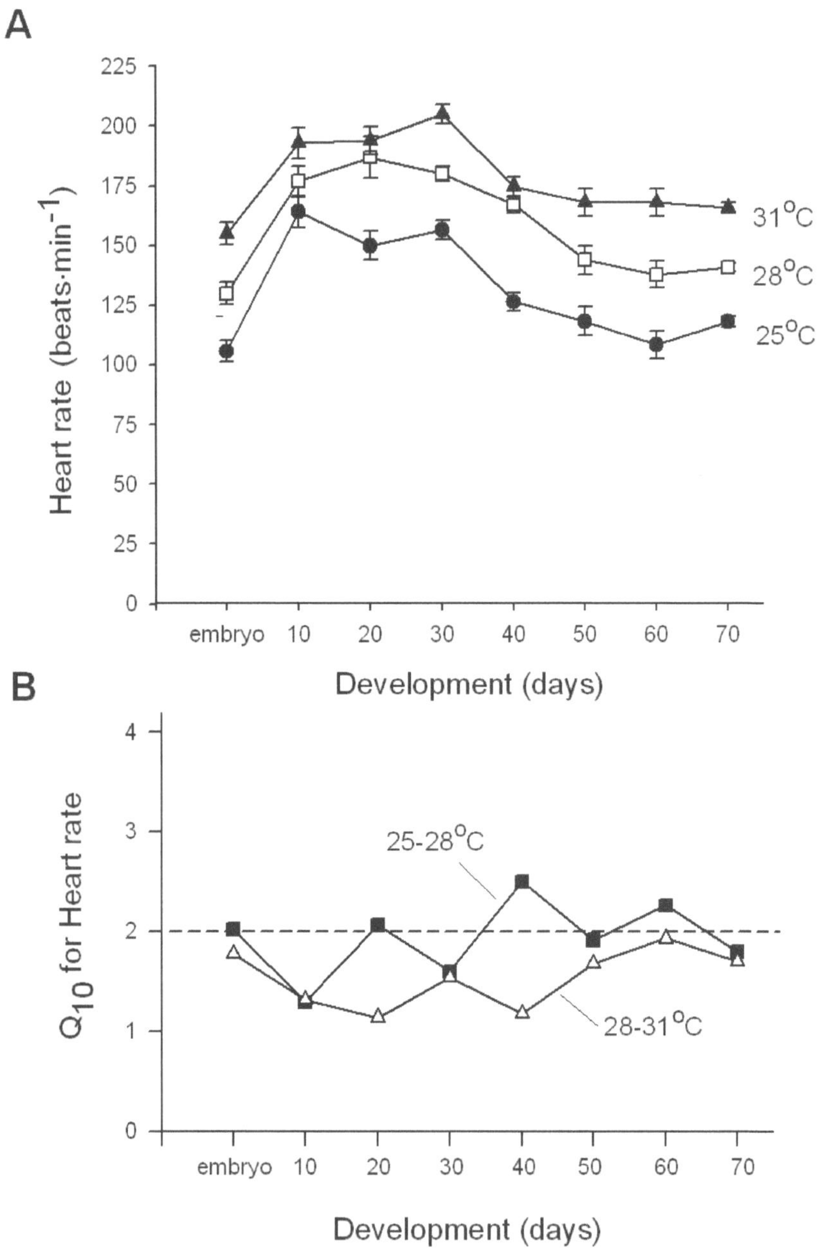

Fig. 3 Temperature effects on heart rate in the zebrafish *Danio rerio*. (A) Heart rate as a function of temperature and development from embryo to 70 days. (B) Q_{10} values for heart rate (from A). After Barrionuevo & Burggren (1999).

nerve fibres entering the heart (Nakao *et al.*, 1981). The developmental onset of cardiac innervation in any species of lamprey is unknown, but ammocoete larvae of European brook lamprey (*Lampetra planeri*) do respond to the application of exogenous sympathetic drugs, indicating the occurrence of β-adrenoreceptors in the developing myocardium (Lignon, 1979).

In adult teleost hearts, both sympathetic and parasympathetic components of the autonomic nervous system innervate the myocardium and the pacemaker region and exhibit significant influence on cardiac function. Inhibitory parasympathetic or vagal action on the heart is a feature of all vertebrates (with the exception of the cyclostomes noted above). Modulation of the vagal action or tonus on the teleost heart is a major regulatory mechanism for altering both heart rate and possibly atrial contraction (Laurent *et al.*, 1983; Nilsson, 1994). Sympathetic nerve fibres are present in holosteans and teleosts, and allow a direct, rapid, and restricted increase in cardiac activity in adults. In addition, sympathetic neurons have been shown to surround and control transmission in the vagal pathway at the intracardiac ganglion (Holmgren, 1977; Nilsson, 1994). As for the mammalian gut, these adrenergic fibres may act by a presynaptic α-adrenoreceptor mediated inhibition of acetylcholine release from the preganglionic cholinergic nerve endings (Gershon *et al.*, 1979).

While these characteristics of the adult fish heart innervation and its control are well documented, once again we know relatively little about the ontogeny of innervation of the fish heart and the extent of central control. Most information is gleaned in pieces from investigations reporting the first appearance of physiological responses (typically changes in gill ventilation rate and heart rate) to acute or chronic hypoxia/hyperoxia or activity (e.g. Holeton, 1971; McDonald & McMahon, 1977; Liem, 1981; Barrionuevo & Burggren, 1999; Jacob *et al.*, 2002; Pelster *et al.*, 2003; Vulesevic *et al.*, 2006). No clear patterns emerge from these observations apart from the obvious, namely that there is a very early metronomic period in the embryonic-to-larval transition period when coordinated physiological regulatory responses are lacking, and then a species-specific point during subsequent development when physiological responses first appear and then grow in intensity. This general pattern is actually mirrored in the embryos and larvae of other vertebrates, which bear mention simply because of the evolutionarily conservative nature of developmental processes. In the avian embryo – likely the most studied of any developing vertebrate model – there is a species-specific pattern of regulatory development that typically begins with functional receptors for blood-borne, circulating catecholamines and ends with vagal control appearing somewhat later in development (Burggren & Crossley, 2002; Moriya *et al.*, 2002; Crossley *et al.*, 2003a, b). A similar pattern of staggered onset of individual control elements has been reported for amphibian larvae (Burggren & Just, 1992; Fritsche & Burggren, 1996; Jacobsson & Fritsche, 1999; Kloberg & Fritsche, 2002). In larvae of the African clawed frog (*Xenopus laevis*) the early response of the developing heart to hypoxia is reduced stroke volume and cardiac output, likely reflecting a direct depression of cardiac tissue (Fritsche & Burggren, 1996). With additional development, however, the heart responds to hypoxia with an increase in stroke volume and apparently a reflex-initiated bradycardia. Similar ontogenetic patterns could well occur in larval fishes.

Very few studies of larval fishes have attempted pharmacological probes, histological study, or molecular studies of putatative receptors involved in cardiovascular control. Some of the most complete pharmacological investigations of the control of heart rate of larval fishes to date involve injection or ambient exposure to cholinergic and adrenergic agonists and antagonists, with subsequent observation of heart rate (Bagatto, 2001; 2005). In zebrafish up to 4 dpf, carbamylcholine chloride caused no significant decrease in heart rate, but on day 4 and thereafter carbamylcholine chloride injection induced bradycardia. Further in development, exposure to atropine resulted in a tachycardia, suggesting the establishment of a cholinergic tone by day 15. Epinephrine, on the other hand, induced tachycardia by as early as day 6. Thus, there are differences in the timing of the onset of cholingergic and adrenergic control of the larval zebrafish heart.

Some limited information on the sensory reflexes and motor responses has been determined for larval fishes (but again, nothing to the same extent as for bird embryos). The time course of development, innervation and function for neuroepithelial cells (NECs), which form an important leg of the oxygen-sensing system in teleost fish, has been a recent focus, especially in zebrafish larvae (Jonz *et al.*, 2004; Jonz & Nurse, 2005, 2006). NO has been shown to play a role in cardiovascular development (e.g. angiogenesis) (Pelster *et al.*, 2005). NO is also involved in acute vascular tone adjustments in larval zebrafish before functional innervation of the peripheral vascular system (Fritsche *et al.*, 2000). The developmental trajectory (progressive onset from 4 dpf through 16 dpf) for *N*-methyl-D-aspartate (NMDA) receptors and their role in the hypoxic ventilatory response has been described for teleost fishes (Turesson *et al.*, 2006), but whether they play a role in cardiovascular regulation remains to be determined.

Continuing the litany of unresolved issues for cardiac regulation in larval fishes, we know little about whether environmental perturbation can alter the timing regulatory onset in individual animals – a phenomenon called heterokairy, as distinct from heterochrony, which denotes such changes on an evolutionary scale (Spicer & Burggren, 2003). Conceptually, we know that the regulatory elements (receptors, effectors, integrators) of the vertebrate cardiovascular system come into operation over time, and that these elements do not begin functioning concurrently (Burggren & Crossley, 2002). Indeed, Spicer & Burggren (2003) have suggested that the onset of cardiovascular regulation involves multiple control elements, each becoming functional at its own rate (Fig. 4).

Evidence for heterokairy in larval fishes emerges from recent studies on zebrafish. Chronic hypoxia exposure not only delays the initial onset of cardiovascular regulation in zebrafish larvae, but additionally shifts the onset of regulation relative to timing of onset of other morphological features (Bagatto, 2005). Although delayed in chronological time, Fig. 5 does not show the considerable acceleration of the response to adrenergic agonists induced by hypoxia when compared to morphological time. Interestingly, family (sibship) differences interjects considerable variability of the

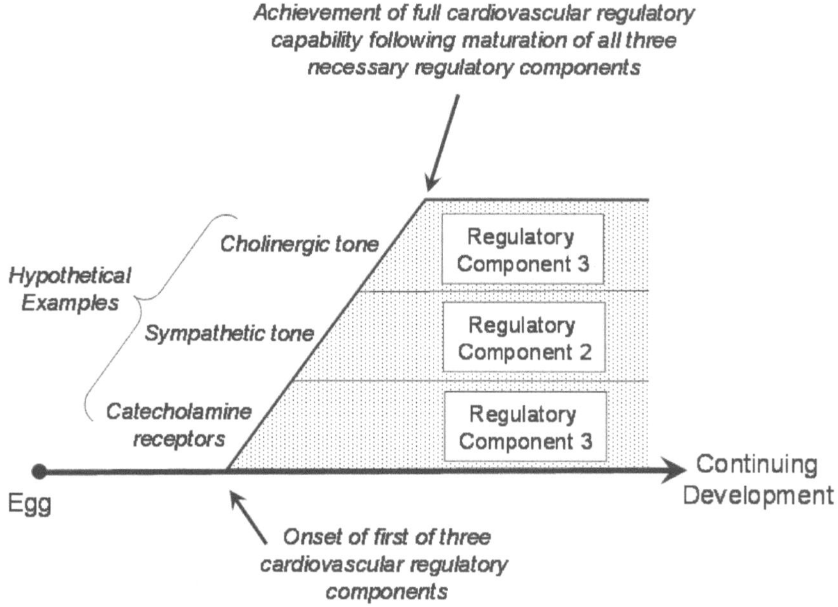

Fig. 4 The ontogeny of cardiovascular regulation can be depicted as a series of discrete but interacting regulatory components, each of which first begins to operate and then matures at a different time from the others. In the hypothetical, simplified regulatory system depicted in this figure, three components are required for full cardiovascular regulation. Receptors for circulating catecholamines begin to develop before the onset of development of sympathetic tone, which itself appears before vagal tone. Full cardiovascular regulation is not achieved until all components have each matured and begin to operate in concert. Of course, actual cardiovascular regulatory systems in vertebrates will have many more processes and structures, each maturing according to the developmental plan of the animal. After Spicer & Burggren (2003).

cardiovascular response to hypoxia in zebrafish, suggesting that cardiovascular developmental plasticity is itself a heritable trait (Moore *et al.*, 2006).

Finally, it warrants mention that heart rate (as well as respiration rate), has long been a "bio-indicator" of environmental toxicity in larval fishes, and consequently there is a great deal of heart rate data published for both larval freshwater and marine fishes showing heart rate changes during toxicological stress (for recent examples, see Vines *et al.*, 2000; Latif *et al.*, 2001; Svecevicius *et al.*, 2005; Vosyliene *et al.*, 2005; Fraysse *et al.*, 2006). Though valuable, these data are somewhat ancillary to the purpose of understanding the larval fish circulation, in that such heart rates are typically measured and reported as a byproduct of viability assessment, rather than as part of a structured assessment of cardiovascular performance and how it changes during growth.

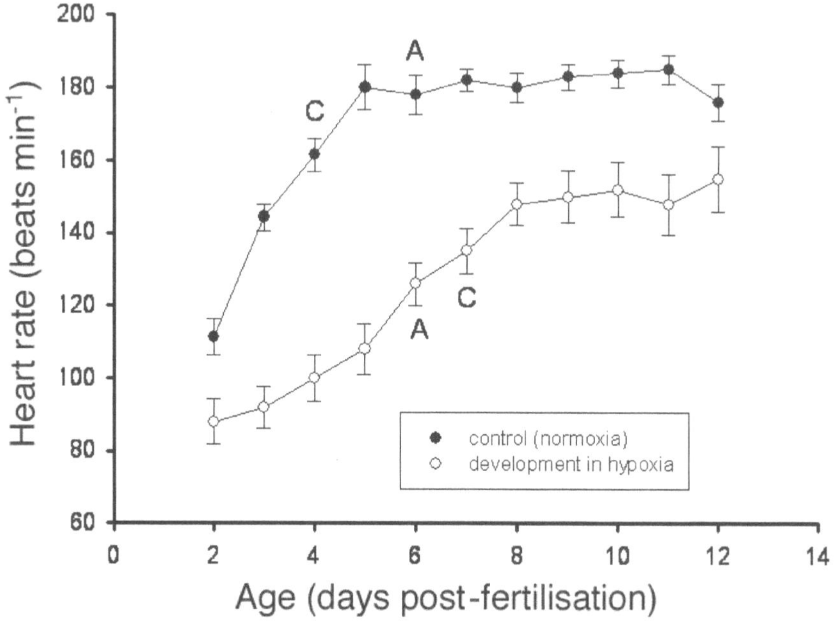

Fig. 5 Mean resting heart rate during early larval development in the zebrafish (*Danio rerio*) at 25°C in normoxia and hypoxia. The letter C indicates the first significant negative chronotropic response to cholinergic agonists during development and the letter A indicates the first significant positive chronotropic response to adrenergic agonists. All hypoxic data points are significantly different from their corresponding controls. Data shown are means ± S.E.M. After Bagatto (2005).

Peripheral Vascular Regulation

Our understanding of the changing nature of the peripheral circulation and its performance has been greatly advanced in recent years by software intensive visual imaging techniques that capitalise upon the considerable transparency of larval fishes, especially in early stages prior to heavy melanin synthesis and deposition (Burggren & Fritsche, 1995; Schwerte & Fritsche, 2003; Schwerte *et al.*, 2003; Forouhar *et al.*, 2006; Hove, 2006). Using frame-to-frame comparison of video images of the microcirculation, fairly complete regional vascular maps can be constructed (Schwerte & Fritsche, 2003; Bagatto & Burggren, 2006). Figure 2 shows vascular cross-sections in larval zebrafish, as well as the changing dimensions of the major categories of vessels during normal growth.

Unfortunately, these techniques have yet to be applied to *in vivo* studies of vasoconstriction and vasodilation in the peripheral circulation of the developing larval fish. Short-term change in peripheral resistance is, of course, a powerful method for regulating regional blood flow. Since vasomotor activity can arise through either neural stimulation or through circulating hormones, it is likely that modification of peripheral resistance and regional distribution of blood flow develops relatively early in larval fishes.

Microcirculatory Autoregulation

For completeness sake, we mention here the process of autoregulation of the circulation at the microcirculatory level in larval fish, involving intrinsic, metabolic, and neurohumoral factors. Almost certainly, such autoregulation develops during larval ontogeny, as such localised responses of the microcirculation require neither the development of innervation or secretory capacity by endocrine glands. At least in a few species, full autoregulation is complete after the full chamberisation of the heart has been achieved (Balashov & Soltitskij, 1991; Balashov *et al.*, 1991; Bagatto, 2005). Generally, however, development of autoregulation in fishes – indeed in any vertebrate – is poorly understood.

CARDIOVASCULAR PERFORMANCE OF LARVAL FISHES

A key feature to the survival of larval fishes is the ability to achieve effective, appropriate levels of cardiac output not only under resting conditions, but also when environmental challenges loom. Here we consider available data on normal cardiovascular performance and how it may be modified by stressors.

Baseline Function

Heart rate

When the embryonic fish heart first begins to beat, contractions are weak and irregular (see above). Contractions become more regular and then increase in rate throughout embryogenesis, reaching a maximum rate around the time of hatching (Laale, 1984; Barrionuevo & Burggren, 1999). This increase in heart rate is due to an increase in spontaneous electrical activity, as regulatory systems have not yet developed (Laale, 1984). Heart rate begins to decline after hatching, but the onset and the rate of decline varies widely between species. For example, in walleye larvae, heart rate does not decline until after yolk absorption, which is well after hatching (McElman & Balon, 1979). Other larvae including the rainbow trout (Holeton, 1971), Arctic charr (*Salvelinus alpinus*) (McDonald & McMahon, 1977), and zebrafish (Bagatto, 2005), begin to decrease very soon following hatching. This pattern is similar in elasmobranchs (for example the little skate (*Raja erinacea*)) except that the maximum heart rate and subsequent decrease occurs well before hatching (Pelster & Bemis, 1991). It remains unclear as to whether this pattern is consistent throughout all elasmobranchs or whether this pattern is due to the fact that the little skate is much larger in size compared to all other fish measured to date. The overall decrease in heart rate, regardless of the timing, is believed to be due to the onset of vagal tone, rather than a direct modification of the intrinsic electrical activity of the cardiac myocytes. Interestingly, an early pattern of initial increase in heart rate even as body mass increases (i.e. a counterintuitive change based on allometric considerations), followed by a subsequent decline more conventionally correlated with increasing body mass, seems to be a general feature of vertebrate development (Burggren & Warburton, 1994).

Blood pressure

As with all vertebrate hearts, the pressure generated upon the initiation of contraction is very low, in part because the myocardium is only a few cells thick and incapable of generating high pressures. In contrast to heart rate patterning during fish development, ventricular pressure begins and continues to rise along with increasing body mass. In elasmobranchs, ventricular pressures are around 0.13 kPa for a 10 mg little skate embryo (Pelster & Bemis, 1991). Ventricular pressure in this elasmobranch continues to rise with increasing body mass, but it was found that pressure at hatching did not differ significantly from adult pressure in little skates, which peaked at around 1.8 kPa (Pelster & Bemis, 1991). Another elasmobranch, the Epaulette shark (*Hemiscyllium ocellatum*) had a mean systemic blood pressure of 2.5 kPa for a 659 g animal (Soderstrom *et al.*, 1999). In teleosts, the only larval blood pressure that has been recorded is for zebrafish, at approximately 0.5 kPa at 96 hpf (Pelster & Burggren, 1996). There is no reason not to assume that blood pressure for developing teleosts also increases with growth. An unresolved issue remains in comparing teleost species of vastly different sizes, while larger adult fish may tend to have higher blood pressures compared to smaller versions within each species, most species documented have ventral aortic pressures between 3 and 4 kPa as adults, regardless of differences in mass e.g. common carp (*Cyprinus carpio*) 3-4 kPa (Stecyk & Farrell, 2006); rainbow trout, 4 kPa (Perry *et al.*, 1999); bald notothen (*Pagothenia borchgrevinki*) 3 kPa (Sundin *et al.*, 1999); Japanese eel (*Anguilla japonica*) 3-4 kPa (Takei & Tsukada, 2001). There is also unpublished evidence that ventricular pressures of adult zebrafish are approximately 4 kPa (B Bagatto & J Lucitti, unpublished data), as opposed to the far lower published values of 0.5 kPa (Hu *et al.*, 2000). This recent result likely indicates that blood pressure varies more according to variables such as physical activity and temperature, as opposed to fish mass. With a recent technique for estimating blood pressure in non-living fishes ("necrophysiology"), comparing many species has become a much easier task (Jones *et al.*, 2005). This, along with other scaling literature, has confirmed that while heart mass, heart rate, and cardiac output scales with body mass, blood pressure is rather constant across a large body mass range.

Stroke volume and cardiac output

Because of the difficulty in obtaining long term measurements of stroke volume and cardiac output (especially in larger larval species where optical techniques are less feasible), the only reported measurements for stroke volume and cardiac output during development are for the zebrafish. At 2 dpf, zebrafish have a stroke volume of around 0.15 nL and a cardiac output of around 18 $nL{\cdot}min^{-1}$ (Bagatto & Burggren, 2006).

Stroke volume doubles within 4 days and, because of an increasing heart rate, cardiac output increases by over 175% within the same time period (Bagatto & Burggren, 2006). Over the longer term as heart rate begins to show its characteristic decline, it is outmatched by the increase in stroke volume as the heart grows. Thus, in a 30-day-old fish, the stroke volume has increased to about 6 nL and cardiac output has increased to about 800 nL·min^{-1} (Fig. 6).

Responses to Environmental Challenges

Temperature

Temperature is arguably the most important abiotic factor affecting development of fishes. The eggs and larvae of fish can experience wide temperature fluctuations in their surrounding micro-environments (Pelster, 1997), profoundly affecting the rate and overall success of development. In fish, lower temperatures may delay hatching or even prevent embryogenesis (Schirone & Gross, 1968), and although higher temperatures will accelerate development, an upper limit is signified by increased morphological abnormalities and increased mortality. In the zebrafish, regular successive cleavages and a distinct continuity of morphological features proceed normally between 23°C and 34°C. Embryos kept at temperatures outside these limits immediately following fertilisation may not complete embryogenesis (Schirone & Gross, 1968). Superimposed on thermal tolerance is a significant increase in the viable temperature range within the first few days of embryo development in zebrafish (Rombough, 1996). Thus, thermal tolerance for normal embryonic development is more restricted than in adult fish (Rombough, 1996; Hochachka & Somero, 2002). These thermal concepts apply to the formation of any organ system, including the cardiovascular system.

Vascular responses to sodium nitroprusside (a potent peripheral vasodilator) and phenylephrine (an α-adrenergic receptor agonist) are delayed by development at a lower temperature. Lower temperature also reduces heart rate both acutely and chronically (Barrionuevo & Burggren, 1999; Bagatto, 2005). Additionally, developmental onset of chronotropic responses (both cholinergic and adrenergic) is significantly delayed at 20°C (Bagatto, 2005). It is likely that these physiological delays are mediated by delaying the chronological time in developmental programming – heterokairy – as described by Spicer & Burggren (2003). Temperature has a significant effect on not only the physiology, but also the morphological development of the cardiovascular system during development in zebrafish. A reduction in developmental temperature from 25°C to 20°C delays the formation of distinct vessel boundaries and delays vessel growth (Bagatto, 2005).

Water current (acute and chronic swim training)

Many fish species experience a broad range of water current velocity either over developmental time or from location to location at the same developmental stage. This common environmental challenge has the potential of inducing plasticity in fish both during development (Bagatto *et al.*, 2001; Pelster *et al.*, 2003), and during adult stages

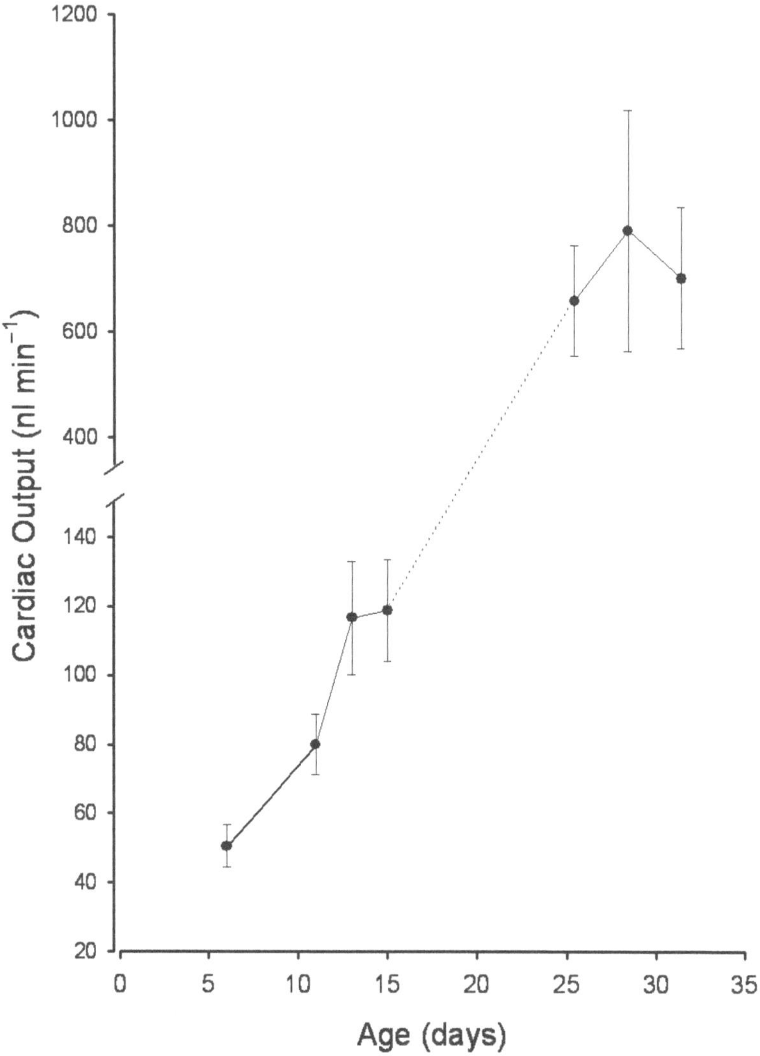

Fig. 6 Mean cardiac output of the zebrafish (*Danio rerio*) during early larval development. After Bagatto (2001).

(for review, see Davison, 1997; Kieffer, 2000). For example, swim training zebrafish larvae enhances swimming efficiency, increases mitochondrial density in red and intermediate muscle, increases capillarisation, and increases tolerance to hypoxia (Bagatto, 2001; Pelster *et al.*, 2003). In adult salmonids, such training leads to increases in growth rate and food-conversion efficiency (Davison, 1997).

Clearly water current can cause a vast array of phenotypically plastic responses but it is unclear what effect, if any, this developmental plasticity has on the onset and function of cardiovascular regulation. Swimming during development likely has more of a profound physiological than anatomical effect. Chronic swimming in a constant current does not alter vessel cross-sectional area in zebrafish (Bagatto, 2005). Pelster *et al.* (2003) report a significant increase in muscle capillarisation in trained zebrafish, suggesting that microcirculatory changes might be the main plastic response to swim training rather than larger level cross-sectional changes in individual main vessels. Developmental onset of vascular responses to sodium nitroprusside and phenylephrine does not change during exposure to constant swimming (Bagatto, 2005), further suggesting that constant swimming is an unlikely factor in inducing major changes in developmental program for cardiovascular anatomy.

Zebrafish developing in water current have significantly higher resting heart rates after 6 dpf compared to control zebrafish (Bagatto, 2005). Bagatto *et al.* (2001) also show an increased routine metabolic rate in some groups of trained zebrafish of the same age, a response also seen in the routine metabolic rate of zebrafish larvae from parents with high swimming performance (M Gore & W Burggren, unpublished data). Additionally, Bagatto (2001) documents increased cardiac output in continuously trained zebrafish of the same age, which was contributed to both by slightly increased heart rate and stroke volumes. Contrary to this, Pelster *et al.* (2003) showed no differences in heart rate, stroke volume and cardiac output in trained zebrafish at 15 dpf. However, this difference could be attributed to diurnal rather than constant exercise in that study. The effects of chronic swim training have not been followed into adulthood, and would provide interesting insights into the "fetal programming" in fishes.

Hypoxia

Hypoxia is a common stressor in many fish populations, and plays a key role in a whole suite of organismal, cellular, and genetic responses in fish (see Pelster, this volume). Hypoxia has a profound effect on fish behaviour (Marks *et al.*, 2005), cardiovascular physiology (Jacob *et al.*, 2002; Moore *et al.*, 2006), most ionic and substrate membrane transport systems (for review, see (Nikinmaa, 2002)), intracellular signalling systems (Salama & Nikinmaa, 1990), and gene expression in fish (Gracey *et al.*, 2001). Because of its multiple effects on fish physiology, morphology, biochemistry and molecular biology, it is not surprising that we are discovering complex interactions of these factors that influence development in addition to the "typical" effect of hypoxia decreasing the overall developmental program.

Zebrafish experiencing development in hypoxia have significantly smaller vessel cross-sectional areas relative to normoxia-reared zebrafish (Bagatto, 2005). However, relative to morphological development, the first vessel response to phenylephrine appears significantly earlier in developmental hypoxia compared to normoxia. Development in chronic hypoxia also significantly lowers heart rate throughout development compared to controls (Bagatto, 2005). Zebrafish have a mild tachycardia by 5 dpf when developing in a pO_2 of 10 kPa (Jacob *et al.*, 2002). At lower pO_2's, such as 3 kPa (Bagatto, 2005), the resulting bradycardia is likely due to the limited oxygen supply.

In zebrafish, developmental onset of chronotropic responses (both adrenergic and cholinergic) is significantly delayed by development in severe hypoxia (Bagatto, 2005). Although a positive chronotropic response was recorded on 6 dpf for normoxic and hypoxic treatments, the hypoxic zebrafish were not as morphologically developed compared to normoxic zebrafish. Therefore, the adrenergic response was shifted to earlier in development in chronically hypoxic zebrafish. This result adds to a small body of literature showing that although hypoxia may delay the overall developmental program, it can selectively accelerate certain aspects of development, including respiratory regulation in brine shrimp (Spicer & El-Gamal, 1999), for example.

Clearly, the developmental environment can affect more than just developmental rate. Hypoxia has a profound affect on the onset of cardiovascular regulation in that it can accelerate the onset of adrenergic and vasoconstrictive responses relative to the developmental program. It will take many more studies to elucidate the molecular mechanisms by which hypoxia exerts its effects.

CARDIOVASCULAR DIVERSITY AMONG LARVAL FISHES

Extant adult fishes show an extremely wide diversity of form and function, and in many instances this diversity is supported by similarly large variations in cardiovascular anatomy and physiology. Yet, in embryonic form, all fishes (indeed, all vertebrates) exhibit a very similar, basic body plan (see Burggren, 2000) and the same basic processes for cardiovascular development, described above in the section on Cardiovascular Development, will hold for all species. Obviously, growth and development requires increasing vascularisation as organogenesis proceeds, and somatic growth accelerates, irrespective of species. Thus, there is a potentially interesting developmental story in the vascularisation of every organ and in the creation of every vascular bed.

Here, we consider two major categories of cardiovascular diversity involving species-specific differences relating to numbers of gill arches and the evolution of air-breathing.

Diversity in Branchial Structure

The branchial circulation of adult fish is elaborate and highly regulated (for reviews see Satchell, 1991; Olson, 2002; Perry & Gilmour, 2002; Sundin & Nilsson, 2002; Evans *et al.*, 2005), and during larval development this same degree of sophistication of the growing branchial circulation must also unfold. Considerable differences in branchial circulatory patterns (primarily related to numbers of gill arches) exist among extant adult fishes, and these patterns must develop as the larval fish grows. Let us first consider the variation in branchial arch morphology among adult fishes.

Most bony fishes (and almost all teleosts) have four functional gill arches, each receiving deoxygenated blood via an afferent artery from the ventral aorta, and sending deoxygenated blood in an efferent artery to the dorsal aorta. Lungfishes have gill modifications that vary with species (see below), but generally have four arches (with a hyoidean hemibranch evident in *Neoceratodus*). Elasmobranchs typically have 9, 11, or 13 hemibranchs (an odd number because of the presence of a posterior hemibranch associated with the hyoid arch) associated with 5, 6 or 7 gill slits. The primitive fishes show yet further variety. Lampreys have eight gill arches, while hagfishes have 5-15 pairs of gills, or more correctly, gill pouches. In some species, a vascularised pseudobranch (a remnant of the primitive gill of the mandibula arch) is exposed to water in the branchial chamber (or, in elasmobranchs and sturgeons, via a spiracle). Given the pseudobranch's small surface area yet significant vascularisation, this structure is thought either to serve a sensory function (Reid & Perry, 2003; Jonz & Nurse, 2006) or to provide highly oxygenated blood directly to the optic choroid and retina (Bridges *et al.*, 1998). Interestingly, it has been speculated that the pseudobranch may have a more active role in larval fishes prior to full development of the gill arches.

Despite these considerable differences in vascular patterns in adult fish, the in-depth and comprehensive biochemical, morphological and physiological studies of the branchial circulation have in relatively few cases been extended back to larval forms. Even, here, studies are more related to the functions of the gill arches such as gas exchange and osmoregulation (Wells & Pinder, 1996a; Rombough, 2002; Varsamos *et al.*, 2005) than the actual cardiovascular physiology and anatomy supporting these critical functions. Certainly, there are numerous studies indicating branchial morphological changes associated with development e.g. sea lamprey (*Petromyzon marinus*) (Youson & Freeman, 1976); Arctic charr (McDonald & McMahon, 1977); banded gourami (*Colisa fasciatus*) (Prasad, 1989); little skate (Pelster & Bemis, 1992), but few characterise the developmental waxing (or waning, in the case of larval external gills) of branchial structures in terms of quantitative vascularisation and blood flow. Even fewer take an experimental approach examining the plasticity of the developmental plan for the branchial circulation of larval fishes.

One caveat to the above involves the recent explosion of developmental studies using piscine developmental models such as the zebrafish, or, to a lesser extent, the common mummichog (*Fundulus heteroclitus*). Many important molecular and biochemical characterisations of embryonic gill structure are emerging (Hogan *et al.*, 2004; Nakamura *et al.*, 2005; Wang *et al.*, 2006). These studies are of great value in understanding fundamental developmental mechanisms, but are often not conducted from a comparative physiological, ecological or environmental perspective[1]. Full exploitation of model systems is best achieved with full awareness of the animal's natural history (Burggren, 2000).

Diversity in Respiration–Circulatory Support for Air Breathing

Air breathing has independently evolved in at least 49 different families of fishes (Graham, 1997). Not surprisingly, the number of air breathing organs and supporting

[1] Monte Westerfield's (1993) now classic *The Zebrafish Book* has introduced many investigators to this wonderful fish, which hails from the Ganges River in India. Interestingly, it deals with "natural history" in the very first line of the book, thusly..... *Zebrafish are available at pet stores throughout the world*".

cardiovascular arrangements are manifold. For a comprehensive description of known adult cardiovascular patterns, the reader is directed to any of several reviews describing the cardiovascular systems of air breathing fishes (Johansen, 1970; Johansen & Burggren, 1980; Randall *et al.*, 1981; Little, 1983; Burggren *et al.*, 1986, 1996; Graham, 1997). Briefly, however, the major cardiovascular specialisations revolve around a combination of:

1) delivering primarily deoxygenated blood to the air breathing organ (swim bladder, modified branchial cavity, gut, skin or actual lung);
2) ensuring that the newly oxygenated blood is not lost from the gills to surrounding water (which is often hypoxic, serving as a major selection pressure for air breathing in the first place) and;
3) delivering the newly oxygenated blood primarily to the systemic tissues without recirculation through the air breathing organs.

Not surprisingly, some of the circulatory modifications ensuring these conditions involve extensive modifications to the basic piscine circulatory pattern, as well as the evolution of new structures. For example, the most extreme situation is likely that of the lungfishes (*Protopterus, Neoceratodus, Lepidosiren*), which have evolved a true lung (opening derived from the ventral oesophagus). Although there are differences between genera, the lungfishes have a pulmonary artery derived from the 6th branchial arch, a pulmonary artery draining the lung, and separate left and right cardiac atrial chambers with the left atrium receiving pulmonary venous blood. Although there remains a single ventricle, it is highly trabeculate, which helps preserve separation of left and right atrial blood streams. The *conus arteriosus* contains an elaborate spiral valve running longitudinally along its length that helps channel primarily oxygenated blood from the left atrium into the branchial circulation, and primarily deoxygenated blood from the right atrium towards the pulmonary circulation and on to the lungs. Finally, there is considerable dimorphism between anterior and posterior gill arches. The posterior arches are more typical of aquatic fishes, and allow for the uptake of oxygen and elimination of carbon dioxide. The anterior arches, however, bear far fewer and smaller lamellae, and correspondingly have a much lower surface area. Lungfish can differentially shunt blood between these arches, resulting in the more oxygenated blood originating from the pulmonary venous drainage being preferentially directed through the nearly denuded anterior arches, thus reducing the threat of loss of blood-borne oxygen across the gills and into hypoxic water.

While the lungfish have the most cardiovascular modifications and specialisations of all air breathing fishes, there are many other complex patterns in extant fishes that do involve modifications other than a pulmonary circulation (see Randall *et al.*, 1981; Little, 1983; Graham, 1997; Perry *et al.*, 2001). A wide variety of physostome fishes use a modified swim bladder to assist in gas exchange e.g. garfishes (*Lepisosteus* sp.), aimara (*Hoplerythrinus unitaeniatus*), and bowfin (*Amia calva*). The labrynthidontid fishes such as *Betta* sp. and *Trichogaster* sp. use an elaborate, branchially-derived organ (the labyrinth) residing in an enlarged epibranchial chamber to breathe air. Many fishes have evolved the capacity to swallow air and absorb oxygen through the gut wall e.g. Corydoras (*Corydoras* sp.) and weatherfish (*Misgurnus fossilis*). Still others rely to varying degrees on a lightly or non-scaled general body surface for gas exchange e.g. swamp eels (*Monopterus* sp. and *Symbranchus* sp.), silurid catfishes (*Clarias* sp.), and mudskippers (*Periophthalmus* sp.).

Irrespective of the cardiovascular format in support of these air breathing mechanisms in these fishes, their larvae must undergo an early developmental transition away from the basic piscine circulatory plan (ventricle > gills > systemic tissues > single atrium) to the more complex pattern required by air breathing. This transition can be quite rapid. For example, the three spot gourami (*Trichogaster trichopterus*) begins air breathing 18-20 dpf (Hisaoka & Firlit, 1962), while banded gourami and climbing perches (*Anabas* sp.) begin breathing 18-20 and 20-22 dpf, respectively (Prasad, 1989). Concurrent with the onset of air breathing must be the appearance of regulatory mechanisms that integrate control of blood flow with aquatic and aerial respiration – essentially, building additional regulatory components into the conceptual scheme shown in Fig. 4.

Unfortunately, after this significant build-up to what must be a series of very interesting developmental stories, we know of very few studies of the circulatory development of air-breathing fish. Respiratory development in air breathing fishes has been described sporadically at various developmental times in various air breathing fish larvae (e.g. Das, 1927; Bader, 1937; Mishra & Singh, 1979; Prasad & Singh, 1984; Hughes *et al.*, 1986; Prasad, 1989; Ahmad & Hasnain, 2005). There have also been occasional observations of the developmental changes in blood respiratory properties (Babiker, 1984). Cardiovascular development in air breathing fishes also remains enigmatic. Typically, observations are limited to the onset and rate of heart rate – e.g. snakehead (*Channa punctatus*) at 22 hpf begins heart beat at 166-175 beats min^{-1} (Ahmad & Hasnain, 2005), while heart beat and circulation begin in three spot gourami at 20 hpf (Hisaoka & Firlit, 1962). Only a very few longitudinal studies have characterised the onset of the cardiovascular (and respiratory) function over time. Table 2 presents data for climbing perch (*Anabas testudineus*), an anabantid air-breathing fish from India (Hughes *et al.*, 1986; Munshi & Hughes, 1992). This table should be viewed as a template for future studies that extend into detailed physiological and morphological observations of the circulation, which not only indicates the onset of function, but also tracks how functions change both quantitatively and qualitatively with subsequent development.

Studies integrating form and function of the circulation of larval air breathing fishes are also almost non-existent. In one of the few such studies, Liem (1981) reported that larvae of swamp eel have a number of modifications apparently driven by their occupation of water that is frequently very hypoxic. Larval swamp eel have dense capillary networks in the skin and a relatively short small blood-water barrier across their skin. These modifications are not unexpected, since morphological changes are seen in other larval aquatic vertebrates developing under hypoxia, e.g. bullfrog (*Rana*

Table 2 Cardio-respiratory development in the air-breathing fish, climbing perch (*Anabas testudineus*) after Munshi & Hughes (1992)

Organ system	*Organ/Tissue*	*Time to differentiation*	*Time to first function*
Blood vascular system	Red blood cells	10 h	15 h
	Heart	15 h	60 h
	Arteries	17 h	60 h
	Veins	18 h	60 h
Respiratory system (Gills and Labyrinth)	Skin	15 h	15 h
	Scales	Day 11	Day 13
	Gill arches	15-60 h	15-60 h
	Gill filaments	15-60 h	15-60 h
	Gill ventilation	-	15-60 h
	Secondary lamellae	15-60 h	15-60 h
	Suprabranchial chamber	15-60 h	Day 14
	Labyrinth plates	15-60 h	Day 14
	Air breaths	-	Day 14

catesbeiana) (for references see Burggren & Just, 1992). However, Liem (1981) describes an apparently unique pectoral fin mechanism that pushes oxygen-rich surface water posteriorly along the length of the body, even as blood is directed primarily anteriorly in the capillary network of the skin. This highly efficient counter-current flow of water and blood will doubtlessly aid oxygen uptake from hypoxic water. The modifications required for turning the larva's body into what Liem (1981) calls the functional analogue of fish gill lamellae are relatively simple, involving large pectoral fins directing water posteriorly, and a re-orientation of the skin capillaries parallel to the longitudinal axis of the larva's body. Whether such a counter current arrangement occurs in other air-breathing fishes, or even in aquatic fishes developing in hypoxic water, remains to be determined.

Doubtlessly, the investigation of the ontogeny of air-breathing fishes will be instructive not only in how the piscine cardiovascular system and its regulation processes develop, but additionally how hypoxia and access to aerial respiration have evolutionarily sculpted the basic plan for circulatory development in fishes.

MEASURING CARDIOVASCULAR DEVELOPMENT IN LARVAL FISHES

Typically, developmental rate in fish is measured either by recording the rate of cleavage in early embryonic phases, or by recording the progression of morphological characteristics in later stages. Measuring cleavage rate has been a successful technique in studying the thermal tolerances of early development in many fish species. Not surprisingly, developmental rate, its temperature dependence, and the range of temperature over which development is possible, are all species-specific. For example, the common carp is a very eurythermic species completing temperature dependent cleavage divisions over a range of 10–26°C (Ignat'eva, 1974). In contrast, the peled (*Coregonus peled*) is able to have successful cleavage divisions as low as 0.5°C (Ignat'eva, 1974). Not only are these cleavage rate temperature profiles species-specific, they are also determined by the spawning temperature of that particular clutch. Although spawning temperature in the rainbow trout can range from 3–13°C, spawn produced at 3°C have a higher Q_{10} for cleavage time (i.e. increased sensitivity to temperature increases) at lower temperatures compared to spawn produced at 13°C (Lebedeva & Meshkov, 1969). This research has really set the stage for molecular investigation into cellular clocks regulating developmental processes. Counting early cell cycles in specific tissues has just begun in fish (Pourquie, 1998).

Because temperature has a profound effect on the rate of development, there is an inherent danger in making comparisons at the same chronological age (as opposed to morphological age) of populations reared at different temperatures. Ideally, morphological characteristics (i.e. their appearance, absence, or description) should be matched with an animal's physical measurements throughout the developmental period, but this is generally not possible once a fish reaches a juvenile stage as outward gross morphological changes are typically minimal or absent. Apart from the rainbow trout (Vernier, 1969), Atlantic salmon (*Salmo salar*) (Gorodilov, 1996), Atlantic cod (*Gadus morhua*) (Hall *et al.*, 2004), zebrafish (Kimmel *et al.*, 1995), some centrarchids (Meyer, 1970), and the Japanese anchovy (*Engraulis japonicus*) (Takahashi & Watanabe, 2004), there is little information on developmental staging on a broad comparative level. In the case of zebrafish, the most widely used staging scheme was determined at 28.5°C (Kimmel *et al.*, 1995), which in many cases has corralled subsequent research into using this temperature so that precise developmental staging can be performed. This led to the idea that development can be expressed in terms of accumulated thermal units (ATUs) (Rombough, 1988; Wells & Pinder, 1996b), which relate development to time spent at specific temperatures. The concept of degree days also has similar connotations (Highley *et al.*, 1986; Hamel *et al.*, 1997). While the notion of degree days may be useful in comparing developmental differences within a species, it assumes development increases linearly with temperature, which certainly creates major obstacles in inter-species comparisons given the species-specific Q_{10}s for development. Even within species, the thermal history of the parents and spawning environment can have significant effects on the Q_{10} for development. Other temperature based models have been published and used successfully, however, each have their limitations (reviewed by Hamel *et al.*, 1997).

With mainly external morphology remaining as the metric for early developmental rate, there are still significant gaps in our ability to measure and compare developmental rates. By using morphological indicators, once a particular structure forms and remains unchanged in terms of shape (it may continue to grow in size), it is no longer useful as a morphological indicator. Bass and trout metrics rely on external morphology as the embryo is quite large and internal structures are obscured by pigments and many layers of muscle and epithelia. This is the reason why the zebrafish has become the standard in fish development as it affords a longer time window during which to view development. Whether by employing phenyl-thio-urea for inhibiting pigment formation or by using albino strains, the zebrafish larvae can remain transparent and allow both morphological and physiological measurements well beyond 30 dpf at 28°C, which approaches the morphology of a juvenile (Bagatto, 2001). Even with this tool, our understanding of how internal organs develop and respond to changing environmental circumstances during development is in its infancy.

Using a relatively unstable scaffolding of morphological indicators to measure development, it is even more foreboding to begin the task of quantifying the development of physiological characteristics, such as those in the cardiovascular system, which inherently show relatively large variability as well as plasticity. As mentioned earlier, the heart of the zebrafish attains its adult shape by 6 dpf at 28.5°C (Stainier *et al.*, 1993). Interestingly, although the morphological action occurs before 6 dpf, the bulk of the onset of physiological regulation does not occur until after this time (Bagatto, 2005). Thus, there is no gross external morphological change in the heart – except for increase in overall size – while significant physiological regulation is coming online. Although morphological development can be increased or decreased, especially by increased and decreased temperature, respectively, this is not the case with physiological development. In hypoxia, although morphological development is slowed, the physiological onset of adrenergic regulation is upregulated at an earlier time relative to morphology and therefore shows heterokairy (Bagatto, 2005). Although the data are scant, they are beginning to reveal an enormous complexity in the response of physiological development to changes in the environment.

Over the life of a typical teleost, the fraction of time in which we can measure development is quite small compared to the period of growth. Stated another way, there is a large portion of time over which development may be occurring and which we are only able to portray this in terms of mass and length. As is the case in many ectotherms, fish tend to grow slower, but over the long term, end up larger than their warmer water counterparts (Angilletta & Dunham, 2003). The advantages and disadvantages for each strategy are discussed in detail (Angilletta *et al.*, 2003) but the fact remains that measuring size (mass, length, or body mass index) does not indicate stage of development or provide reliable rate information. This holds not only for general body features, but for specific cardiovascular tissues such as the ventricle. Unless temperature, oxygen, and other abiotic conditions are completely controlled, developmental comparisons are highly imprecise. When investigating environmental effects on the development of any system, including the cardiovascular system, the existing framework is extremely weak. Therefore it will be up to future techniques and experiments to provide support for this framework.

FUTURE PERSPECTIVES

Larval fish undergo complex and dynamic patterns of change during their development, and the growth and elaboration of the cardiovascular system is crucial to larval ontogeny.

Particularly intriguing are the changing roles of the circulation as larval surface-to-volume ratios change, and as water-breathing (and in some cases, air-breathing) supplants simple diffusion as a way to achieve gas, nutrient and/or waste exchange. Larval fishes also provide valuable insights into vertebrate cardiovascular development in general because they have high experimental utility, in large part because they exist independently in the environment (as opposed to enclosed in an egg shell or protected by a uterus), are relatively transparent (aiding in imaging techniques), and can be often acquired in large numbers.

Whether as models for studying vertebrate development or as a highly divergent class worthy of study in their own right, larval fishes – and especially the characteristics of their circulation – have not been as extensively investigated as amphibian larvae or avian embryos. Consequently, there are numerous areas that we can identify as benefiting from additional research activity, including:

- The time course of the ontogeny of cardiovascular regulation in larval fishes, and the various molecular, cellular and organismal components of these regulatory systems
- The plasticity of cardiovascular developmental programs, especially the extent to which common environmental factors such as temperature and oxygen are influential in modifying morphological and physiological ontogeny in larval fishes.
- The role of non-genomic factors (e.g. maternal effects) on cardiovascular development in larval fishes.
- The developmental transitions occurring in non-model fish species (i.e. the vast majority of species) that are associated with phyletic differences, air breathing, long distance larval migration, etc.

Then, in terms of techniques, research on development of the cardiovascular system will be advanced by:

- Recognition of a universal (or standardised) developmental measurement system that can be used to make inter- and intra-specific comparison – not only for the cardiovascular system, but for all aspects of development in larval fishes.
- Creation of microtechniques for cardiovascular research (especially in chronic as opposed to acute measurements), so that *in vivo* research will continue to expand

Finally, at the conceptual level, the study of the cardiovascular system of larval fishes (indeed, the study of any system), will continue to benefit from:

- aggressive pursuit of the "(X)omics", such as "genomics", 'metabolomics', "physiomics", to provide a hierarchical linkage between molecular biology, biochemistry and physiology, creating an integrated view of cardiovascular development.

References

Ahmad R, Hasnain AU (2005) Ontogenetic changes and developmental adjustments in lactate dehydrogenase isozymes of an obligate air-breathing fish *Channa punctatus* during deprivation of air access. Comp Biochem Physiol 140B: 271-278

Aissaoui A, Altimiras J, Tort L (1998) Cardiac conduction times in *Sparus auratus* at different heart rates. Influence of body weight. J Fish Biol 52: 1154-1164

Angilletta MJ, Dunham AE (2003) The temperature-size rule in ectotherms: Simple evolutionary explanations may not be general. Am Nat 162: 332-342

Angilletta MJ, Wilson RS, Navas CA, James RS (2003) Tradeoffs and the evolution of thermal reaction norms. Trends Ecol Evol 18: 234-240

Arbel ER, Liberthson R, Langendorf R, Pick A, Lev M, Fishman AP (1977) Electrophysiological and anatomical observations on the heart of the African lungfish. Am J Physiol 232: H24-H34

Babiker MM (1984) Adaptive respiratory significance of organophosphates (ATP and GTP) in air-breathing fishes. Hydrobiologia 110: 339-349

Bader R (1937) Bau, Entwicklung und Funktion des akzessorischen Atmungsorgans der Labyrinthfishe. Z wiss Zool 149: 323-401

Bagatto B (2001) The developmental physiology of the zebrafish: influence of environment on metabolic and cardiovascular attributes. PhD Dissertation, University of North Texas, Denton, Texas, USA

Bagatto B (2005) Ontogeny of cardiovascular control in zebrafish (*Danio rerio*): Effects of developmental environment. Comp Biochem Physiol A141: 391-400

Bagatto B, Burggren W (2006) A three-dimensional functional assessment of heart and vessel development in the larva of the zebrafish (*Danio rerio*). Physiol Biochem Zool 79: 194-201

Bagatto B, Pelster B, Burggren WW (2001) Growth and metabolism of larval zebrafish: effects of swim training. J Exp Biol 204: 4335-4343

Balashov NV, Soltitskij VV (1991) On the development of cardiac muscle reactivity in Black Sea mullets at early stages of ontogenesis. In: Kulikova NG (ed) Mullet Culture in the Azov and Black Sea Basins. VNIRO, Moscow, p 51-59

Balashov NV, Soltitskii VV, Makukhina LI (1991) The effect of noradrenaline and acetylcholine on cardiac muscle of the golden mullett *Liza auratus* at early stages of ontogenesis. Z Evol Biokhim Fiziol 27: 255-257

Barrionuevo WR, Burggren WW (1999) O_2 consumption and heart rate in developing zebrafish (*Danio rerio*): influence of temperature and ambient O_2. Am J Physiol 276: R505-R513

Bikfalvi A, Bicknell R (2002) Recent advances in angiogenesis, anti-angiogenesis and vascular targeting. Trends Pharm Sci 23: 576-582

Bridges CR, Berenbrink M, Muller R, Waser W (1998) Physiology and biochemistry of the pseudobranch: An unanswered question? Comp Biochem Physiol 119A: 67-77

Burggren W, Fritsche R (1995) Cardiovascular Measurements in Animals in the Milligram Range. Braz J Med Biol Res 28: 1291-1305

Burggren WW (2000) Developmental physiology, animal models, and the August Krogh principle. Zool Anal Complex Syst 102: 148-156

Burggren WW (2004) What is the purpose of the embryonic heart beat? or how facts can ultimately prevail over physiological dogma. Physiol Biochem Zool 77: 333-345

Burggren WW (2005) Developing animals flout prominent assumptions of ecological physiology. Comp Biochem Physiol 141A: 430-439

Burggren WW, Just JJ (1992) Developmental changes in amphibian physiological systems. In: Feder ME, Burggren WW (eds) Environmental Physiology of the Amphibia. University of Chicago Press, Chicago

Burggren WW, Warburton SJ (1994) Patterns of form and function in developing hearts – contributions from nonmammalian vertebrates. Cardioscience 5: 183-191

Burggren WW, Territo P (1995) Early Development of blood oxygen transport. In: Houston J, Coates J (eds) Hypoxia and Brain, Vol 5 Queen City Printer, Burlington, Vermont, p 45-56

Burggren W, Crossley DA (2002) Comparative cardiovascular development: improving the conceptual framework. Comp Biochem Physiol 132A: 661-674

Burggren WW, Johansen K, McMahon BR (1986) Respiration in primitive fishes. In: Foreman RE, Gorbman A, Dodd JM, Olsson R (eds) The Biology of Primitive Fishes. Plenum, New York, p 217-252

Burggren WW, Farrell AP, Lillywhite HB (1996) Vertebrate cardiovascular systems. In: Dantzler W (ed) Handbook of Comparative Physiology. Oxford University Press, Oxford, p 215-308

Catton WT (1951) Blood cell formation in certain teleost fishes. Blood 6: 39-60

Childs S, Chen JN, Garrity DM, Fishman MC (2002) Patterning of angiogenesis in the zebrafish embryo. Development 129: 973-982

Chuiko VA, Ganzii TV, Lemeshko VV, Kal'ko AI (1977) Electrical activity of the hearts of *Hypophthalmicthys molitrix* fish during its cooling and warming. Zh Evol Biokhim Fiziol 13: 91-94

Clark RJ, Rodnick KJ (1998) Morphometric and biochemical characteristics of ventricular hypertrophy in male rainbow trout (*Oncorhynchus mykiss*). J Exp Biol 201: 1541-1552

Coultas L, Chawengsaksophak K, Rossant J (2005) Endothelial cells and VEGF in vascular development. Nature 438: 937-945

Crossley DA, Burggren WW, Altimiras J (2003a) Cardiovascular regulation during hypoxia in embryos of the domestic chicken *Gallus gallus*. Am J Physiol 284: R219-R226

Crossley DA, Bagatto BR, Dzialowski EM, Burggren WW (2003b) Maturation of cardiovascular control mechanisms in the embryonic emu (*Dromiceius novaehollandiae*). J Exp Biol 206: 2703-2710

Danilova N, Steiner LA (2002) B cells develop in the zebrafish pancreas. Proc Natl Acad Sci USA 99: 13711-13716

Das BK (1927) The bionomics of certain air-breathing fishes of India, together with an account of the development of their air-breathing organs. Phil Trans R Soc Lond B216: 183-217

Davidson AJ, Zon LI (2004) The 'definitive' (and 'primitive') guide to zebrafish hematopoiesis. Oncogene 23: 7233-7246

Davison W (1997) The effects of exercise training on teleost fish, a review of recent literature. Comp Biochem Physiol 117A: 67-75

Driever W (2000) Developmental biology – Bringing two hearts together. Nature 406: 141-142

Evans DH, Piermarini PM, Choe KP (2005) The multifunctional fish gill: Dominant site of gas exchange, osmoregulation, acid-base regulation, and excretion of nitrogenous waste. Physiol Rev 85: 97-177

Farrell AP (2002) Cardiorespiratory performance in salmonids during exercise at high temperature: insights into cardiovascular design limitations in fishes. Comp Biochem Physiol 132A: 797-810

Forouhar AS, Liebling M, Hickerson A, Nasiraei-Moghaddam A, Tsai HJ, Hove JR, Fraser SE, Dickinson ME, Gharib M (2006) The embryonic vertebrate heart tube is a dynamic suction pump. Science 312: 751-753

Fraysse B, Mons R, Garric J (2006) Development of a zebrafish 4-day toxicity of embryo-larval bioassay to assess chemicals. Ecotoxicol Environ Saf 63: 253-267

Fritsche R, Burggren WW (1996) Development of cardiovascular responses to hypoxia in larvae of the frog *Xenopus laevis*. Am J Physiol 271: R912-R917

Fritsche R, Schwerte T, Pelster B (2000) Nitric oxide and vascular reactivity in developing zebrafish, *Danio rerio*. Am J Physiol 279: R2200-R2207

Fuiman LA, Batty RS (1997) What a drag it is getting cold: Partitioning the physical and physiological effects of temperature on fish swimming. J Exp Biol 200: 1745-1755

Gahtan E, Tanger P, Baier H (2005) Visual prey capture in larval zebrafish is controlled by identified reticulospinal neurons downstream of the tectum. J Neurosci 25: 9294-9303

Gershon MD, Dreyfus CF, Rothman TP (1979) The mammalian enteric nervous system: A third autonomic division. In: Kalsner S (ed) Trends Autonom Pharmacol, Vol 1 Urban and Schwarzenberg, Baltimore, p 59-101

Gorodilov YN (1996) Description of the early ontogeny of the Atlantic salmon, *Salmo salar*, with a novel system of interval (state) identification. Environ Biol Fish 47: 109-127

Gracey AY, Troll JV, Somero GN (2001) Hypoxia-induced gene expression profiling in the euryoxic fish *Gillichthys mirabilis*. Proc Natl Acad Sci USA 98: 1993-1998

Graham JB (1997) Air-Breathing Fishes, Academic Press, New York

Gregory M, Jagadeeswaran P (2002) Selective labeling of zebrafish thrombocytes: Quantitation of thrombocyte function and detection during development. Blood Cell Mol Dis 28: 418-427

Habeck H, Odenthal J, Walderich B, Maischien HM, Schulte-Merker S (2002) Analysis of a zebrafish VEGF receptor mutant reveals specific disruption of angiogenesis. Curr Biol 12: 1405-1412

Hall TE, Smith P, Johnston IA (2004) Stages of embryonic development in the Atlantic cod *Gadus morhua*. J Morphol 259: 255-270

Hamel P, Magnan P, East P, Lapointe M, Laurendeau P (1997) Comparison of different models to predict the in situ embryonic developmental rate of fish, with special reference to white sucker (*Catostomus commersoni*). Can J Fish Aquat Sci 54: 190-197

Herbomel P, Thisse B, Thisse C (1999) Ontogeny and behaviour of early macrophages in the zebrafish embryo. Development 126: 3735-3745

Highley LG, Pedigo LP, Ostlie KR (1986) DEGDAY: A program for calculating degree-days, and assumptions behind the degree-day approach. Environ Entomol 15: 999-1016

Hisaoka KK, Firlit CF (1962) The embryology of the blue gourami, *Trichogaster trichopterus* (Pallus). J Morphol 111: 239-253

Hochachka PW, Somero GN (2002) Biochemical Adaptation: Mechanism and Process in Physiological Evolution, Oxford University Press, New York

Hogan BM, Hunter MP, Oates AC, Crowhurst MO, Hall NE, Heath JK, Prince VE, Lieschke GJ (2004) Zebrafish *qcm2* is required for gill filament budding from pharyngeal ectoderm. Dev Biol 276: 508-522

Holeton GF (1971) Respiratory and circulatory responses of rainbow trout larvae to carbon monoxide and to hypoxia. J Exp Biol 55: 683-694.

Holmgren S (1977) Regulation of the heart of a teleost, *Gadus morhua*, by autonomic nerves and circulating catecholamines. Acta Physiol Scand 99: 62-74

Hove JR (2006) Quantifying cardiovascular flow dynamics during early development. Pediatr Res 60: 6-13

Hove JR, Koster RW, Forouhar AS, Acevedo-Bolton G, Fraser SE, Gharib M (2003) Intracardiac fluid forces are an essential epigenetic factor for embryonic cardiogenesis. Nature 421: 172-177

Hu N, Sedmera D, Yost HJ, Clark EB (2000) Structure and function of the developing zebrafish heart. Anat Rec 260: 148-157

Hughes GM, Munshi JSD, Ojha J (1986) Post-embryonic development of water and air-breathing organs of *Anabas testudineus* (Bloch). J Fish Biol 29: 443-450

Icardo JM, Guerrero A, Duran AC, Domezain A, Colvee E, Sans-Coma V (2004) The development of the sturgeon heart. Anat Embryol 208: 439-449

Ignat'eva GM (1974) Temperature dependence of cleavage rates in carp, pike, and whitefish. Ontogenez 5: 27-32

Irisawa H (1978) Comparative physiology of the cardiac pacemaker mechanism. Physiol Rev 58: 461-498

Isogai S, Horiguchi M, Weinstein BM (2001) The vascular anatomy of the developing zebrafish: An atlas of embryonic and early larval development. Dev Biol 230: 278-301

Isogai S, Lawson ND, Torrealday S, Horiguchi M, Weinstein BM (2003) Angiogenic network formation in the developing vertebrate trunk. Dev Dis 130: 5281-5290

Itina NA, Balanov LY (1964) Study of the lamprey ECG in relation with the issue of the role of acetylcholine. Fiziol Zh SSSR im IM Sechenova 50: 335-364

Iuchi I (1973) Chemical and physical properties of the larval and the adult hemoglobins in rainbow trout, *Salmo gairdnerii irideus*. Comp Biochem Physiol 44B: 1087-1101

Iuchi I (1982) Characterization of rabbit antisera against rainbow trout (*Salmo gardnerii irideus*) adult hemoglobins and preparation of their specific antibodies. J Fac Sci Univ Tokyo Sec IV 15: 156-171

Iuchi I (1985) Cellular and molecularbases of the larval-adult shift of hemoglobins in fish. Zool Sci 2: 1-23

Iuchi I, Yamagami K (1969) Electrophoretic pattern of larval hemoglobins of the salmonid fish, *Salmo gairdnerii irideus*. Comp Biochem Physiol 28: 977-979

Jacob E, Drexel M, Schwerte T, Pelster B (2002) Influence of hypoxia and of hypoxemia on the development of cardiac activity in zebrafish larvae. Am J Physiol 283: R911-R917

Jacobsson A, Fritsche R (1999) Development of adrenergic and cholinergic cardiac central in larvae of the African clawed frog *Xenopus laevis*. Physiol Biochem Zool 72: 328-338

Jagadeeswaran P, Sheehan JP (1999) Analysis of blood coagulation in the zebrafish. Blood Cell Mol Dis 25: 239-249

Jensen D (1965) The anural heart of the hagfish. Ann New York Acad Sci 127: 443-458

Johansen K (1970) Air breathing in fishes. In: Hoar WE, Randall DJ (eds) Fish Physiology, Vol IV. Academic Press, New York, p 361-411

Johansen K, Burggren WW (1980) Cardiovascular function in lower vertebrates. In: Bourne GH (ed) Hearts and Heart-like Organs. Academic Press, New York, p 61-117

Johnston IA (2006) Environment and plasticity of myogenesis in teleost fish. J Exp Biol 209: 2249-2264

Jones DR, Perbhoo K, Braun MH (2005) Necrophysiological determination of blood pressure in fishes. Naturwissenschaften 92: 582-585

Jonz MG, Nurse CA (2005) Development of oxygen sensing in the gills of zebrafish. J Exp Biol 208: 1537-1549

Jonz MG, Nurse CA (2006) Ontogenesis of oxygen chemoreception in aquatic vertebrates. Resp Physiol Neurobiol 154: 139-152

Jonz MG, Fearon IM, Nurse CA (2004) Neuroepithelial oxygen chemoreceptors of the zebrafish gill. J Physiol 560: 737-752

Karlsson J, von Hofsten J, Olsson PE (2001) Generating transparent zebrafish: A refined method to improve detection of gene expression during embryonic development. Mar Biotechnol 3: 522-527

Kieffer JD (2000) Limits to exhaustive exercise in fish. Comp Biochem Physiol A126: 161-179

Kimmel CB, Ballard WW, Kimmel SR, Ullmann B, Schilling TF (1995) Stages of embryonic development of the zebrafish. Dev Dyn 203: 253-310

Kloberg AJ, Fritsche R (2002) Catecholamines are present in larval *Xenopus laevis*: A potential source for cardiac control. J Exp Zool 292: 293-303

Kopp R, Schwerte T, Pelster B (2005) Cardiac performance in the zebrafish breakdance mutant. J Exp Biol 208: 2123-2134

Kupperman E, An SZ, Osborne N, Waldron S, Stainier DYR (2000) A sphingosine-1-phosphate receptor regulates cell migration during vertebrate heart development. Nature 406: 192-195

Laale HW (1984) Fish embryo culture – cardiac monolayers and contractile activity in embryo explants from the zebrafish, *Brachydaniorerio*. Can J Zool 62: 878-885

Labat R, Demer J, Laffont J (1968) Recherches sur la regulation cardiaque chez quelques vertebres inferieurs. Bull Soc Histoire Natur Toulouse 104: 69-81

Latif MA, Bodaly RA, Johnston TA, Fudge RJ (2001) Effects of environmental and maternally derived methylmercury on the embryonic and larval stages of walleye (*Stizostedion vitreum*). Environ Pollut 111: 139-148

Laurent P, Holmgren S, Nilsson S (1983) Nervous and humoral control of the fish heart: structure and function. Comp Biochem Physiol 76A: 525-542

Lawson ND, Weinstein BM (2002) Arteries and veins: Making a difference with zebrafish. Nature Reviews Genetics 3: 674-682

Lebedeva OA, Meshkov MM (1969) Change in the times of organ formation and duration of embryogenesis in the rainbow trout (*Salmo irideus* Gibb.) in relation to temperature. Izv GosNIORKh 68: 136-155

Lee P, Goishi K, Davidson AJ, Mannix R, Zon L, Klagsbrun M (2002) Neuropilin-1 is required for vascular development and is a mediator of VEGF-dependent angiogenesis in zebrafish. Proc Natl Acad Sci USA 99: 10470-10475

Liao EC, Trede NS, Ransom D, Zapata A, Kieran M, Zon LI (2002) Non-cell autonomous requirement for the bloodless gene in primitive hematopoiesis of zebrafish. Development 129: 649-659

Liem KF (1981) Larvae of air-breathing fishes as countercurrent – flow devices in hypoxic environments. Science 211: 1177-1179

Lieschke GJ, Oates AC, Crowhurst MO, Ward AC, Layton JE (2001) Morphologic and functional characterization of granulocytes and macrophages in embryonic and adult zebrafish. Blood 98: 3087-3096

Lignon JM (1979) Responses to sympathetic drugs in the ammocoete heart: probable influence of the small intensely fluorescent (SIF) cells. J Mol Cell Cardiol 11: 447-465

Lin HF, Paw BH, Gregory M, Jagadeeswaran P, Handin RI (2001) Production and characterization of transgenic zebrafish (*Danio rerio*) with fluorescent thrombocytes and thrombocyte precursors. Blood 98: 514A-514A

Little C (1983) The Colonisation of Land, Cambridge University Press, Cambridge

Long QM, Meng AM, Wang H, Jessen JR, Farrell MJ, Lin S (1997) GATA-1 expression pattern can be recapitulated in living transgenic zebrafish using GFP reporter gene. Development 124: 4105-4111

MacLellan WR, Schneider MD (2000) Genetic dissection of cardiac growth control pathways. Annu Rev Physiol 62: 289-319

Marks C, West TN, Bagatto B, Moore FB-G (2005) Developmental environment alters conditional aggression in zebrafish. Copeia 2005: 900-907

Maruyama K, Yasumasu S, Iuchi I (1999) Characterization and expression of embryonic globin in the rainbow trout, *Oncorhynchus mykiss*: intra-embryonic initiation of erythropoiesis. Dev Growth Differ 41: 589-599

Maruyama K, Yasumasu S, Iuchi I (2002) Characterization and expression of embryonic and adult globins of the teleost *Oryzias latipes* (Medaka). J Biochem 132: 581-589

McDonald DG, McMahon BR (1977) Respiratory development in Arctic char *Salvelinus alpinus* under conditions of normoxia and chronic hypoxia. Can J Zool 55: 1461-1467

McElligott MB, O'Malley DM (2005) Prey tracking by larval zebrafish: Axial kinematics and visual control. Brain Behav Evol 66: 177-196

McElman JF, Balon EK (1979) Early ontogeny of walleye, *Stizostedion vitreum*, with steps of saltatory development. Environ Biol Fish 4: 309-348

Meyer FA (1970) Development of some larval centrarchids. Prog Fish Cult 32: 130-136

Mirkovic T, Rombough P (1998) The effect of body mass and temperature on the heart rate, stroke volume, and cardiac output of larvae of the rainbow trout, *Oncorhynchus mykiss*. Physiol Zool 71: 191-197

Mishra AP, Singh BR (1979) Oxygen uptake through water during early life of *Anabas testuidensuI* (Bloch). Hydrobiologia 66: 129-133

Moore FB-G, Hosey M, Bagatto B (2006) Cardiovascular system in larval zebrafish responds to developmental hypoxia in a family specific manner. Front Zool 3: 4

Moriya K, Kato K, Matsumura M, Dzialowski E, Burggren WW, Tazawa H (2002) Cardiac rhythms in developing emu hatchlings. Comp Biochem Physiol A131: 787-795

Munshi JSD, Hughes GM (1992) Air-breathing Fishes of India: Their Structure, Function, and Life History, AA Balkema, The Netherlands

Nakamura T, Amanuma K, Aoki Y (2005) Frameshift mutations induced by the acridine mustard ICR-191 in embryos and in the adult gill and hepatopancreas of rpsL transgenic zebrafish. Mutat Res-Fund Mol M 578: 272-283

Nakao T, Suzuki S, Saito M (1981) An electron-microscopic study of the cardiac innervation in larval lamprey. Anat Rec 199: 555-563

Nikinmaa M (2002) Oxygen-dependent cellular functions – why fishes and their aquatic environment are a prime choice of study. Comp Biochem Physiol A133: 1-16

Nilsson S (1994) Evidence for adrenergic nervous control of blood pressure in teleost fish. Physiol Zool 67: 1347-1359

Ny A, Autiero M, Carmeliet P (2006) Zebrafish and *Xenopus* tadpoles: Small animal models to study angiogenesis and lymphangiogenesis. Exp Cell Res 312: 684-693

Olson KR (2002) Gill circulation: Regulation of perfusion distribution and metabolism of regulatory molecules. J Exp Zool 293: 320-335

Patruno M, Radaelli G, Mascarello F, Carnevali MDC (1998) Muscle growth in response to changing demands of functions in the teleost *Sparus aurata* (L.) during development from hatching to juvenile. Anat Embryol 198: 487-504

Pelster B (1997) Oxygen, temperature and pH influences on the development of non-mammalian embryos and larvae. In: Burggren W, Keller BB (eds) Development of Cardiovascular Systems. Cambridge University Press, Cambridge, p 227-239

Pelster B, Bemis WE (1991) Ontogeny of heart function in the little skate (*Raja erinaceai*). J Exp Biol 156: 387-398

Pelster B, Bemis WE (1992) Structure and function of the external gill filaments of embryonic skates (*Raja erinacea*). Resp Physiol 89: 1-13

Pelster B, Burggren WW (1996) Disruption of hemoglobin oxygen transport does not impact oxygen-dependent physiological processes in developing embryos of zebra fish (*Danio rerio*). Circ Res 79: 358-362

Pelster B, Grillitsch S, Schwerte T (2005) NO as a mediator during the early development of the cardiovascular system in the zebrafish. Comp Biochem Physiol 142A: 215-220

Pelster B, Sanger AM, Siegele M, Schwerte T (2003) Influence of swim training on cardiac activity, tissue capillarization, and mitochondrial density in muscle tissue of zebrafish larvae. Am J Physiol 285: R339-R347

Perry SF, Gilmour KM (2002) Sensing and transfer of respiratory gases at the fish gill. J Exp Biol 293: 249-263

Perry SF, Fritsche R, Hoagland TM, Duff DW, Olson KR (1999) The control of blood pressure during external hypercapnia in the rainbow trout (*Oncorhynchus mykiss*). J Exp Biol 202: 2177-2190

Perry SF, Wilson RJA, Straus C, Harris MB, Remmers JE (2001) Which came first, the lung or the breath? Comp Biochem Physiol 129A: 37-47

Pourquie O (1998) Clocks regulating developmental processes. Curr Opin Neurobiol 8: 665-670

Prasad MS (1989) Relationship between oxygen uptake and water-blood diffusion barrier at the secondary gill lamellae in two anabantid fish during early life. Proc Indian Natn Sci Acad B55: 239-244

Prasad MS, Singh BR (1984) Oxygen uptake through water during early life in *Colisa (trichogaster) fasciatus*. (Bl. and Schn.). Pol Arch Hydrobiol 31: 153-159

Prosheva VI (2005) Morphofunctional characteristics of the heart of larva of the Arctic brook lamprey *Lethenteron japonicum*. J Evol Biochem Physiol 41: 73-77

Quinkertz A, Campos-Ortega JA (1999) A new beta-globin gene from the zebrafish, beta E1, and its pattern of transcription during embryogenesis. Dev Genes Evol 209: 126-131

Randall DJ (1970) The circulatory system. In: Hoar WE, Randall DJ (eds) Fish Physiology, Vol IV Academic Press, New York, p 133-172

Randall DJ, Burggren WW, Haswell MS, Farrell AP (1981) The Evolution of Air Breathing in Vertebrates, Cambridge University Press, Cambridge

Reid SG, Perry SF (2003) Peripheral O_2 chemoreceptors mediate humoral catecholamine secretion from fish chromaffin cells. Am J Physiol 284: R990-R999

Rombough P (1996) The effects of temperature on embryonic and larval development. Soc Exp Biol Semin Ser 61: 177-223

Rombough P (2002) Gills are needed for ionoregulation before they are needed for O_2 uptake in developing zebrafish, *Danio rerio*. J Exp Biol 205: 1787-1794

Rombough PJ (1988) Respiratory gas exchange, aerobic metabolism, and effetcs of hypoxia during early life. In: Hoar WS, Randall DJ (eds) Fish Physiology, Vol XI, Part A, Academic Press, San Diego, p. 59-161

Rombough PJ (1997) Piscine cardiovascular development. In: Burggren WW, Keller BB (eds) Development of Cardiovascular Systems. Cambridge University Press, Cambridge

Rombough PJ (1998) Partitioning of oxygen uptake between the gills and skin in fish larvae: A novel method for estimating cutaneous oxygen uptake. J Exp Biol 201: 1763-1769

Rombough PJ, Moroz BM (1997) The scaling and potential importance of cutaneous and branchial surfaces in respiratory gas exchange in larval and juvenile walleye *Stizostedion vitreum*. J Exp Biol 200: 2459-2468

Rossant J, Hirashima M (2003) Vascular development and patterning: making the right choices. Curr Opin Genet Dev 13: 408-412

Rowlerson A, Mascarello F, Radaelli G, Veggetti A (1995) Differentiation and growth of muscle in the fish *Sparus aurata* (L). 2. Hyperplastic and hypertrophic growth of lateral muscle from hatching to adult. J Muscle Res Cell M 16: 223-236

Saito T (1969) Electrophysiological studies on the pacemaker of several fish hearts. Zool Mag 78: 291-296

Saito T (1973) Effect of vagal stimulation on the pacemaker action potential of carp heart. Comp Biochem Physiol 44A: 191-199

Salama A, Nikinmaa M (1990) Effect of oxygen tension on catecholamine-induced formation of Camp and on swelling of carp red blood cells. Am J Physiol 259: C723-C726

Satchell GH (1991) Physiology and Form of Fish Circulation, Cambridge University Press, Cambridge

Schirone RC, Gross L (1968) Effect of temperature on early embryological development of the zebra fish, *Brachydanio rerio*. J Exp Zool 169: 43-52

Schonweger G, Schwerte T, Pelster B (2000) Temperature-dependent development of cardiac activity in unrestrained larvae of the minnow *Phoxinus phoxinus*. Am J Physiol 279: R1634-R1640

Schwerte T, Pelster B (2000) Digital motion analysis as a tool for analysing the shape and performance of the circulatory system in transparent animals. J Exp Biol 203: 1659-1669

Schwerte T, Fritsche R (2003) Understanding cardiovascular physiology in zebrafish and *Xenopus* larvae: the use of microtechniques. Comp Biochem Physiol 135A: 131-145

Schwerte T, Überbacher D, Pelster B (2003) Non-invasive imaging of blood cell concentration and blood distribution in zebrafish *Danio rerio* incubated in hypoic conditions *in vivo*. J Exp Biol 206: 1299-1307

Sedmera D, Reckova M, deAlmeida A, Sedmerova M, Biermann M, Volejnik J, Sarre A, Raddatz E, McCarthy RA, Gourdie RG, Thompson RP (2003) Functional and morphological evidence for a ventricular conduction system in zebrafish and *Xenopus* hearts. Am J Physiol 284: H1152-H1160

Soderstrom V, Renshaw GMC, Nilsson GE (1999) Brain blood flow and blood pressure during hypoxia in the epaulette shark *Hemiscyllium ocellatum*, a hypoxia-tolerant elasmobranch. J Exp Biol 202: 829-835

Spicer JI, El-Gamal MM (1999) Hypoxia accelerates the development of respiratory regulation in brine shrimp – But at a cost. J Exp Biol 202: 3637-3646

Spicer JI, Burggren WW (2003) Development of physiological regulatory systems: altering the timing of crucial events. Zoology 106: 91-99

Stainier DYR, Fishman MC (1992) Patterning the zebrafish heart tube – acquisition of anteroposterior polarity. Dev Biol 153: 91-101

Stainier DYR, Lee RK, Fishman MC (1993) Cardiovascular development in the Zebrafish. 1. Myocardial fate map and heart tube formation. Development 119: 31-40

Stecyk JAW, Farrell AP (2006) Regulation of the cardiorespiratory system of common carp (*Cyprinus carpio*) during severe hypoxia at three seasonal acclimation temperatures. Physiol Biochem Zool 79: 614-627

Sundin L, Nilsson S (2002) Branchial innervation. J Exp Zool 293: 232-248

Sundin L, Axelsson M, Davison W, Forster ME (1999) Cardiovascular responses to adenosine in the Antarctic fish *Pagothenia borchgrevinki*. J Exp Biol 202: 2259-2267

Svecevicius G, Syvokiene J, Stasiunaite P, Mickeniene L (2005) Acute and chronic toxicity of chlorine dioxide (ClO_2) and chlorite (ClO_2^-) to rainbow trout (*Oncorhynchus mykiss*). Environ Sci Pollut R 12: 302-305

Takahashi M, Watanabe Y (2004) Developmental and growth rates of Japanese anchovy *Engraulis japonicus* during metamorphosis in the Kuroshio-Oyashio transitional waters. Mar Ecol-Prog Ser 282: 253-260

Takei Y, Tsukada T (2001) Ambient temperature regulates drinking and arterial pressure in eels. Zool Sci 18: 963-967

Territo PR, Burggren WW (1998) Cardio-respiratory ontogeny during chronic carbon monoxide exposure in the clawed frog *Xenopus laevis*. J Exp Biol 201: 1461-1472

Traver D, Herbomel P, Patton EE, Murphey RD, Yoder JA, Litman GW, Catic A, Amemiya CT, Zon LI, Trede NS (2003) The zebrafish as a model organism to study development of the immune system. Adv Immunol 81: 254-330

Trede NS, Zapata A, Zon LI (2001) Fishing for lymphoid genes. Trends Immunol 22: 302-307

Turesson J, Schwerte T, Sundin L (2006) Late onset of NMDA receptor-mediated ventilatory control during early development in zebrafish (*Danio rerio*). Comp Biochem Physiol 143A: 332-339

Varsamos S, Nebel C, Charmantier G (2005) Ontogeny of osmoregulation in postembryonic fish: A review. Comp Biochem Physiol 141A: 401-429

Vernier JM (1969) Table chronologique du développement embryonnaire de la truite arc-enciel, *Salmo gardneri* Rich 1836. Anat Embryol 4: 495-520

Vines CA, Robbins T, Griffin FJ, Cherr GN (2000) The effects of diffusible creosote-derived compounds on development in Pacific herring (*Clupea pallasi*). Aquat Toxicol 51: 225-239

Vosyliene MZ, Kazlauskiene N, Joksas K (2005) Toxic effects of crude oil combined with oil cleaner simple green on yolk-sac larvae and adult rainbow trout *Oncorhynchus mykiss*. Environ Sci Pollut R 12: 136-139

Vulesevic B, McNeill B, Perry SF (2006) Chemoreceptor plasticity and respiratory acclimation in the zebrafish *Danio rerio*. J Exp Biol 209: 1261-1273

Wang L, Scheffler BE, Willett KL (2006) CYP1C1 mRNA expression is inducible by Benzo[a]pyrene in *Fundulus heteroclitus* embryos and adults. Toxicol Sci 93: 331-340

Ward AC, McPhee DO, Condron MM, Varma S, Cody SH, Onnebo SMN, Paw BH, Zon LI, Lieschke GJ (2003) The zebrafish *spi1* promoter drives myeloid-specific expression in stable transgenic fish. Blood 102: 3238-3240

Weinstein B, Stemple DL, Driever W, Fishman MC (1995) *Gridlock*, a localized heritable vascular patterning defect in zebrafish. Nature Med 11: 1143-1147

Weinstein BM (1999) What guides early embryonic blood vessel formation? Dev Dyn 215: 2-11

Weinstein BM (2002) Vascular cell biology in vivo: A new piscine paradigm? Trends Cell Biol 12: 439-445

Weinstein BM, Fishman MC (1996) Cardiovascular morphogenesis in zebrafish. Cardiovasc Res 31 Spec No: E17-24

Weinstein BM, Schier AF, Abdelilah S, Malicki J, Solnica-Krezel L, Stemple DL, Stainier DY, Zwartkruis F, Driever W, Fishman MC (1996) Hematopoietic mutations in the zebrafish. Development 123: 303-309

Wells PR, Pinder AW (1996a) The respiratory development of Atlantic salmon II. Partitioning of oxygen uptake among gills, yolk sac and body surfaces. J Exp Biol 199: 2737-2744

Wells PR, Pinder AW (1996b) The respiratory development of Atlantic salmon I. Morphometry of gills, yolk sac and body surface. J Exp Biol 199: 2725-2736

Westerfield M (1993) The Zebrafish Book: A Guide for the Laboratory Use of Zebrafish (*Danio rerio*). University of Oregon Press, Eugene

Willett CE, Cortes A, Zuasti A, Zapata AG (1999) Early hematopoiesis and developing lymphoid organs in the zebrafish. Dev Dyn 214: 323-336

Wingert RA, Zon LI (2003) Hematopoietic Stem Cells, Landes Bioscience, Georgetown

Yelon D (2001) Cardiac patterning and morphogenesis in zebrafish. Dev Dyn 222: 552-563

Youson JH, Freeman PA (1976) Morphology of the gills of larval and parasitic adult sea lamprey, *Petromyzon marinus*. J Morphol 149: 73-103

Zon LI (1995) Developmental biology of hematopoiesis. Blood 86: 2876-2891

CHAPTER

6

Osmo- and Ionoregulation

Toyoji Kaneko[1,*] and Junya Hiroi[2]

INTRODUCTION

It is an essential prerequisite for most vertebrate species to maintain various chemical and physical conditions of the body fluids (blood and extracellular fluids) within certain physiological ranges in order to ensure normal activities of the cells that constitute the body. Among the important factors prescribing the characteristics of the body fluids is osmotic pressure, or osmolality, which is determined by the total concentration of the solutes, mostly inorganic electrolytes. Since Na^+ and Cl^- are the major ions in the body fluids, accounting for more than 90% of inorganic electrolytes, the regulation of both Na^+ and Cl^- is critical for osmoregulation. Another important factor is regulation of water balance, since water serves as solvent for osmolytes. Vertebrates apparently regulate their body fluid osmolality by controlling ion concentration and water content.

Teleost species that inhabit aquatic environments face osmotic problems, which differ from those in terrestrial vertebrates. The aquatic animals possess the gill, which is permeable to both ions and water to a large extent, as their respiratory organ. Ion and

Authors' addresses: [1]Department of Aquatic Bioscience, Graduate School of Agricultural and Life Sciences, University of Tokyo, 1-1-1, Yayoi, Bunkyo, Tokyo 113-8657, Japan.
[2]Department of Anatomy, St. Marianna University School of Medicine, Miyamae, Kawasaki, Kanagawa 216-8511, Japan.
**Corresponding author:* E-mail: kaneko31@marine.fs.a.u-tokyo.ac.jp

water movements occur according to concentration and osmotic gradients, respectively, between the body fluid and external aquatic environments. This is not the case with terrestrial vertebrates, in which gas exchanges take place in the air through the lung. To maintain ion and water balance in a wide variety of environmental salinity, teleosts have developed superior osmoregulatory mechanisms.

As in higher vertebrates, blood osmolality in teleosts is generally maintained at about one third that of seawater osmolality both in freshwater and in seawater. Osmoregulation in adult teleosts is largely the result of integrated ion and water transport activities of the gills, kidney and intestine (Evans, 1993; Bentley, 2002). Freshwater teleosts face water load and salt loss through their permeable body surfaces, the most of which is occupied by the gill epithelia. To meet these problems, they discharge excess water by producing dilute urine in the kidney, and uptake ions through the gill epithelia. On the other hand, marine teleosts must deal with passive water loss and salt load. The loss of water is compensated for by drinking seawater and absorbing water across the intestine, while the excess ions are actively excreted from the gills and kidney. Thus, in adult fish, the gills, kidney and intestine are important osmoregulatory organs, creating ionic and osmotic gradients between the body fluid and external environments. During early life stages, however, those osmoregulatory organs in adult fish are not yet developed or are not fully functional. One may wonder whether fish embryos and larvae actually regulate body fluid osmolality, and, if so, how osmoregulation can be achieved without adult-type osmoregulatory organs.

OSMOREGULATION DURING EARLY LIFE STAGES

Development of Osmoregulatory Organs during Early Life Stages of Teleosts

Adult teleosts are usually equipped with four pairs of gills. Each gill consists of a gill arch, filaments and lamellae, where the two rows of gill filaments radiate posteroventrally from the arch. The gill lamellae extend dorsally and ventrally from both sides of the filament. The gill epithelium is comprised of pavement, chloride and mucous cells. While respiratory pavement cells, which occupy most of the gill surface area, are in charge of gas exchanges, chloride cells, also known as as mitochondria-rich cells (MRC) or ionocytes, are responsible for active ion transport across the gill epithelium. Chloride cells are the major site of ion secretion and absorption, thus being important in seawater and freshwater adaptation, respectively.

In most cases, gills seemingly start functioning as a respiratory and osmoregulatory organ after hatching, although the precise timing may differ in different species. In Japanese flounder (*Paralichthys olivaceus*), the gills are not distinguishable in newly-hatched larvae (Hiroi *et al.*, 1998). Four pairs of gill arches develop 4 days post hatching (ph), but neither gill filaments, lamellae nor chloride cells are yet differentiated at this early life stage. Gill filaments sprout from the gill arches in premetamorphoic larvae on day 8 ph, when chloride cells first appear in the filament epithelium. In premetamorphoic larvae on day 18 ph, the filaments are further developed, lamellae are extended from the filaments, and chloride cells are extensively distributed in the filaments. In common mummichog (*Fundulus heteroclitus*), immature chloride cells first appear in the gill filaments on day –1 ph, but it is after hatching that morphologically functional chloride cells become evident in the gills, in which lamellae are not yet formed (Katoh *et al.*, 2000). Similarly, the gill chloride cells are still underdeveloped in eyed-stage embryos of chum salmon (*Oncorhynchus keta*) (Kaneko *et al.*, 1995).

Marine teleosts imbibe the surrounding seawater to balance osmotic water loss across the gills; in contrast, freshwater teleosts produce a large amount of urine to balance osmotic water uptake. Mozambique tilapia (*Oreochromis mossambicus*) and medaka (*Oryzias latipes*) larvae have been shown to drink more ambient water in seawater than in freshwater, suggesting that the function of the gastrointestinal tract as an osmoregulatory organ has already started at the larval stages (Miyazaki *et al.*, 1998; Kaneko & Hasegawa, 1999). In Mozambique tilapia, however, drinking takes place only after hatching, but they do not ingest water during embryonic stages before hatching, suggesting lack of a water-absorbing mechanism in fish embryos (Miyazaki *et al.*, 1998). The kidney, which produces dilute urine to extrude excess water with monovalent ions retained in freshwater-adapted teleosts, appears to be underdeveloped during embryonic stages. During the late embryonic stages of chum salmon, although a pair of the nephric ducts are developed, neither glomeruli nor convoluted tubules are formed, and thus the kidney is not yet functional (Kaneko *et al.*, 1992). All these observations indicate that adult-type osmoregulatory organs are not able to contribute to osmoregulation at least during embryonic stages.

Osmoregulatory Ability of Teleost Embryos

In spite of lacking adult-type osmoregulatory organs, fish embryos definitely possess osmoregulatory ability (Alderdice, 1988; Kaneko *et al.*, 2002), which has been clearly demonstrated in chum salmon embryos (Kaneko *et al.*, 1995). In freshwater where chum salmon embryos develop, blood osmolality is maintained at 360-370 mOsm kg^{-1} H_2O. Although the value is slightly higher than that in adult fish (typically 300 mOsm kg^{-1} H_2O), it shows that hyperosmotic regulation is already functioning and an osmotic gradient is created between the body fluid and external freshwater environment. Eyed-stage embryos of chum salmon already develop seawater adaptability to some extent, although they stay in freshwater until migration down to the sea in the fry stage. Transfer of eyed-stage embryos (4-5 weeks post fertilisation) of chum salmon from freshwater to seawater does not cause any significant change in the whole egg weight. However, embryos shrink when the eggs are transferred to seawater, whereas the perivitelline space between the embryo and egg shell, filled with the perivitelline fluid, increases at the expense of the embryo. Osmolality of the perivitelline fluid increases rapidly to reach an environmental level within 3 h after transfer (Fig. 1). This indicates that the chorion (egg shell) is permeable to both ions and water, and lacks osmoregulatory ability. Thus, the embryos are substantially exposed to environmental

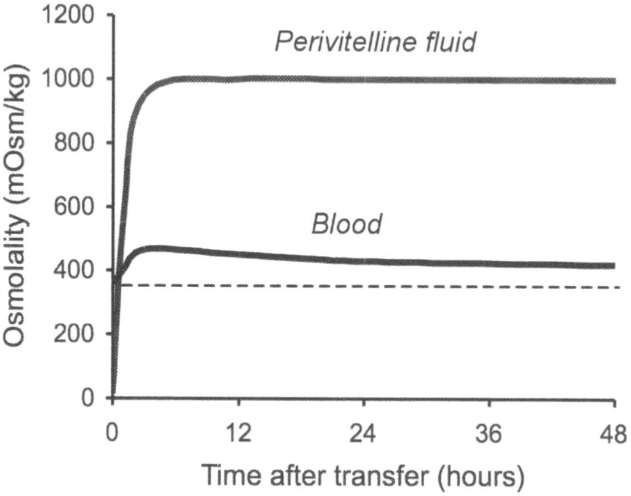

Fig. 1 Changes in perivitelline fluid and blood osmolality following transfer from freshwater to seawater in eyed-stage embryos of chum salmon (*Oncorhynchus keta*). A dotted line indicates the initial level of blood osmolality. After Kaneko *et al.* (1995).

media in spite of the presence of the chorion. In contrast with perivitelline fluid osmolality, blood osmolality is apparently regulated to cope with environmental salinity changes: blood osmolality increases shortly after transfer to seawater, but then decreases gradually toward the initial level (Fig. 1). Such osmoregulatory ability in embryos is essentially the same as that in chum salmon fry ready to migrate down to the sea (Hasegawa *et al.*, 1987). These observations indicate that eyed-stage embryos of chum salmon have already acquired a hypo-osmoregulatory mechanism, which acts toward reducing blood osmolality in fish exposed to hyperosmotic environments.

Extrabranchial Chloride Cells in Embryos and Larvae

In chum salmon embryos, despite the absence of functional gills, chloride cells have been shown to be present in the yolk-sac membrane, which surrounds the yolk sac and occupies a large proportion of the body surface during the late embryonic stage (Kaneko *et al.*, 1995). These extrabranchial chloride cells are associated with hypo-osmoregulation during the early life stages of chum salmon. The chloride cells in the yolk-sac membrane become activated following seawater transfer, as shown by increased cell size and frequent appearance of their apical openings in the outer surface of the yolk-sac membrane.

The occurrence of extrabranchial chloride cells in the yolk-sac membrane and other body surfaces during early life stages is not restricted to chum salmon, but is commonly observed in fish embryos and larvae (Shen & Leatherland, 1978; Guggino,

1980; Hwang, 1989). Hiroi *et al.* (1998) examined the developmental sequence of chloride cells in the body skin and gills of Japanese flounder larvae, and described the occurrence of chloride cells in the body skin in premetamorphic larvae. Similarly, the shift of chloride cell distribution was examined during early life stages of common mummichog (Katoh *et al.*, 2000). Chloride cells first appear in the yolk-sac membrane in the early embryonic stage, followed by their appearance in the body skin in the late embryonic stage. The principal site for chloride cell distribution shifts from the yolk-sac membrane and body skin during embryonic stages to the gills and opercular membrane in larval and later developmental stages. In Japanese eel (*Anguilla japonica*) embryos and larvae, chloride cells are also detected in the yolk-sac membrane and body skin (Sasai *et al.*, 1998). Figure 2 depicts the distribution of extrabranchial chloride cells in an eel larva 10 days post hatching (dph). Chloride cells are scattered all over the body surface in this early stage, in which the yolk has been absorbed but the gills are still underdeveloped. It could be thus speculated that extrabranchial chloride cells play a central role in osmoregulation during early developmental stages in place of gill chloride cells.

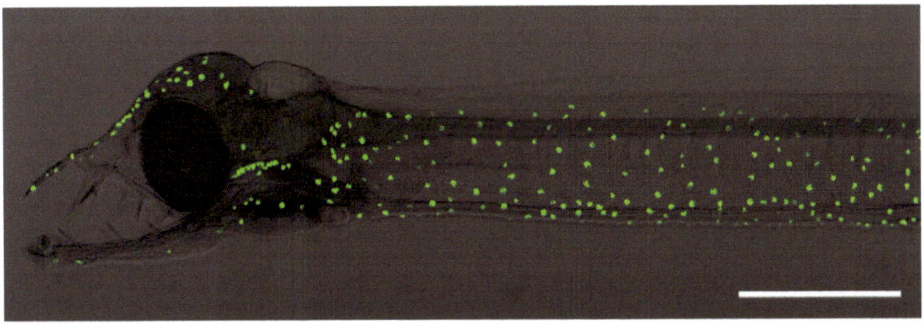

Fig. 2 Distribution of extrabranchial chloride cells in the body skin of a 10 dph Japanese eel (*Anguilla japonica*) larva. Chloride cells (green) were detected immunocytochemically with anti-Na^+/K^+-ATPase, and observed by confocal laser scanning microscopy. Scale bar, 0.5 mm.

CHLORIDE CELLS IN THE YOLK-SAC MEMBRANE OF MOZAMBIQUE TILAPIA

Mozambique tilapia is a suitable fish species for studies on osmoregulation, especially during early developmental stages. They can breed both in freshwater and in seawater, and embryos develop normally in respective media. Mozambique tilapia embryos are able to survive direct transfer from freshwater to seawater and *vice versa*. Their strong euryhalinity enables us to investigate the difference in osmoregulatory mechanisms between freshwater- and seawater-adapted embryos and larvae.

Detection of Chloride Cells

In general, chloride cells are characterised by the presence of numerous mitochondria and an extensive tubular system continuous with the basolateral membrane (Philpott, 1980; Pisam & Rambourg, 1991). The cells are readily identified with

2-(4-dimethylaminostyryl)-1-ethylpyridinium iodide (DASPEI), a vital fluorescent dye specific for mitochondria, under a fluorescent microscope. Immunocytochemistry with a specific antiserum for Na^+, K^+-ATPase (NKA) is also applicable to the detection of chloride cells, since this enzyme is intensively located in the tubular system. The extensive distribution of the tubular system in the cytoplasm results in labelling over the cell with the nucleus unstained.

Ayson *et al.* (1994) examined the development and changes in the activity of chloride cells in the yolk-sac membrane of Mozambique tilapia embryos and larvae, which spawned and hatched in freshwater and seawater. According to their observations, DASPEI-positive chloride cells are detected in the yolk-sac membrane in both freshwater and seawater embryos, starting from 2 days post fertilisation (dpf) (3 days before hatching) at 25°C. The size of chloride cells in embryos and larvae in seawater is always significantly larger than that in freshwater. The density of chloride cells in both freshwater and seawater increases towards hatching and thereafter decreases as the yolk is gradually resorbed. They also investigated effects of transfer between freshwater and seawater: when newly-hatched larvae in freshwater are transferred directly to seawater, the chloride cell size increases significantly, while the cell size does not change in those retained in freshwater. Conversely, the cell size decreases in newly-hatched larvae transferred from seawater to freshwater, whereas there is no difference in those retained in seawater. These findings indicate that chloride cells are consistently larger in fish adapted to seawater than in those adapted to freshwater, suggesting a significant role of chloride cells particularly in seawater environments. According to scanning electron microscopic observations (Shiraishi *et al.*, 1997), apical openings of chloride cells in contact with the external environment are located at the boundary of the pavement cells possessing arrays of microridges. Pavement cells cover a large part of chloride cells other than the apical opening. The apical openings are generally smaller in freshwater than in seawater, and become larger after transfer from freshwater to seawater. These observations also imply physiological significance of chloride cells in seawater adaptation.

Although ion-transporting functions of chloride cells in the yolk-sac membrane of embryos and the branchial epithelium of adult fish are considered to be identical to each other, one notable morphological difference between them is in the shape of the cells. Chloride cells in the yolk-sac membrane are flattened compared to the elongated or spherical shape of chloride cells in the gills. The chloride cell height and consequently the shape appear to be limited by the thickness of the epidermis, in which they are located.

Freshwater- and Seawater-type Chloride Cells

In addition to the difference in chloride cell size between freshwater- and seawater-adapted embryos and larvae, transmission electron microscopic observations revealed more distinct differences in their fine structures (Shiraishi *et al.*, 1997). In freshwater, the apical membrane forms a shallow pit, possessing villous cytoplasmic projections. In contrast, the apical membrane in seawater is deeply invaginated to form an enlarged apical pit. In the larvae transferred to seawater, mitochondria are more numerous, and the tubular system forms a denser network of anastomosed tubules than in the freshwater larvae. Furthermore, two or three cells frequently congregate and share a common apical pit in seawater, forming a multicellular complex. The complex typically consists of one main chloride cell and one or two accessory cells, which are attached to the shoulder of the main cell. They interdigitate with the main cell, and extend their cytoplasmic processes to its apex. Cytoplasmic processes of the main and accessory cells are arranged alternately on a horizontal section cut through an apical pit. Such structures of the complex may be advantageous to Na^+ extrusion, since Na^+ is probably secreted through a paracellular pathway down its electrical gradient in seawater. The accessory cells may also reflect enhanced turnover of the chloride cells in seawater, as accessory cells are considered to be a population of partially differentiated chloride cells. From these observations, chloride cells in the yolk-sac membrane can be classified into two distinct types; that is, a single freshwater type and a multicellular seawater type (Fig. 3). It should be noted, however, that single freshwater-type cells are also present in seawater-adapted fish, whereas multicellular seawater-type cells are only rarely observed in freshwater. The multicellular complex of chloride cells has also been described in the gills of adult Mozambique tilapia and other species (Sardet *et al.*, 1979; Hootman & Philpott, 1980; Hwang, 1987; Uchida *et al.*, 2000).

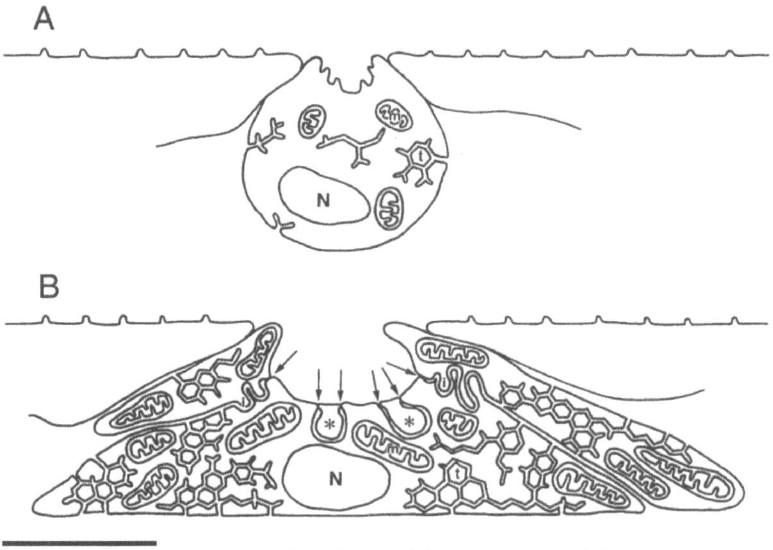

Fig. 3 Schematic drawings of freshwater (A)- and seawater (B)-type chloride cells in the yolk-sac membrane of Mozambique tilapia (*Oreochromis mossambicus*) embryos and larvae. Whereas chloride cells exist individually in freshwater, cells form multicellular complexes together with accessory cells in seawater. The chloride and accessory cells share a common apical pit. The accessory cells interdigitate with the main chloride cell and extend the cytoplasmic processes (asterisks) toward the apical pit. Arrows indicate multiple shallow junctions between the main chloride cell and accessory cells. N, nucleus.

Ion-secreting and Absorbing Functions of Chloride Cells

The seawater-type chloride cells are considered to secrete excess Na^+ and Cl^- in hyperosmotic environments, whereas freshwater-type cells presumably uptake ions in hypo-osmotic environments. Chloride cells in seawater-adapted fish have been well established as the Cl^--secretory cells; using the vibrating probe technique, Foskett & Scheffey (1982) have clearly demonstrated Cl^--secreting activity of chloride cells in the opercular membrane of seawater-adapted Mozambique tilapia.

In view of the possible significance of multicellular complexes of chloride cells in the yolk-sac membrane, we attempted to localise Cl^--secreting sites in the yolk-sac membrane of seawater-adapted Mozambique tilapia larvae using the chloride test (Kaneko & Shiraishi, 2001). This technique is based on the chemical reaction that Cl^- reacts with Ag^+ to form photosensitive AgCl, which is reduced to Ag deposition after sunlight exposure. Under a microscope equipped with a Nomarski's differential interference contrast device, chloride cells in the yolk-sac membrane are readily identified as large, round cells, when focused on the sub-surface plane. The dark deposition, representing the presence of Cl^-, is detected in the apical pits of seawater-type chloride cells, whereas such deposition is not associated with freshwater-type cells. These findings indicate that Cl^- is secreted from the apical pit of seawater-type chloride cells. This is also confirmed by X-ray microanalysis, showing that the Ag distribution is confined to the apical pit of the chloride cell and its adjacent area (Fig. 4). This indicates that Ag is the reaction product of the chloride test, and that multicellular seawater-type chloride cells definitely secrete Cl^- through the apical pit in seawater-adapted Mozambique tilapia larvae.

On the other hand, although ion-absorbing mechanisms are less understood and direct evidence for this is still lacking, it has been shown that increased Cl^- uptake is closely associated with enlarged apical surfaces of chloride cells (Lin & Hwang, 2001). When ambient Cl^- levels are extremely reduced, chloride cells enlarge their apical area, probably to upregulate their Cl^- uptake capacity; conversely, when ambient Cl^- levels are increased, chloride cells are inactivated by constriction of their apical openings and become totally covered by adjacent pavement cells (Lin & Hwang, 2004).

Functional Differentiation of Chloride Cells

The strong euryhalinity of Mozambique tilapia embryos and larvae could be attributed to the extrabranchial chloride cells. From both physiological and morphological points of view, it is of great interest to determine whether chloride cells are replaced by newly-differentiated cells of a different function after transfer from freshwater to seawater, or whether the same chloride cells function in both freshwater and seawater. The most effective way to answer this question would be to examine sequential changes in individual chloride cells during adaptation to a different salinity. However, conventional

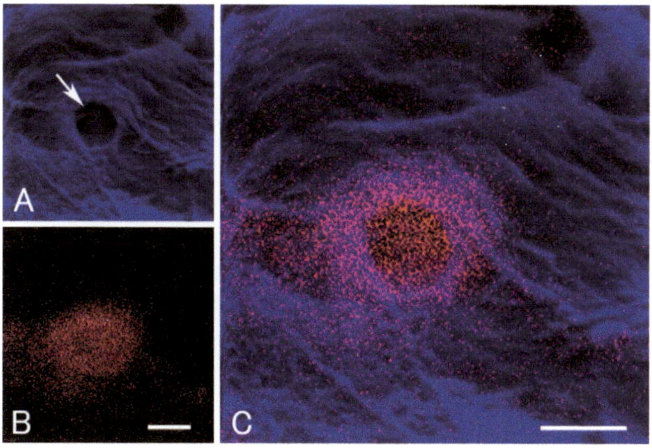

Fig. 4 Scanning electron microscopic image of the yolk-sac membrane of a seawater-adapted Mozambique tilapia (*Oreochromis mossambicus*) larva (A: blue), the distribution image of Ag detected by the X-ray characteristic of Ag (B: red), and their overlapped image (C). X-ray signals, representing the presence of Cl^-, are confined to the apical pit (arrow) of the chloride cell and its adjacent area. Scale bars, 5 μm. After Kaneko & Shiraishi (2001).

methods do not allow us to follow time-course changes in the same chloride cells in the gills due to their complex, three-dimensional structures.

Using Mozambique tilapia embryos, Hiroi *et al.* (1999) succeeded in observing *in vitro* sequential changes in chloride cell morphology in the yolk-sac membrane: individual chloride cells, vital-stained with DASPEI, were observed continuously under a confocal laser scanning microscope. According to the sequential observations, freshwater-type chloride cells become larger to form multicellular complexes in fish transferred from freshwater to seawater (Fig. 5), while the cell size does not change in those maintained in freshwater. During the formation of seawater-type, multicellular complexes, accessory cells develop from undifferentiated cells attaching to the main chloride cells and gradually intrude into the main cells (Fig. 6). These findings clearly indicate that freshwater-type single chloride cells are transformed into seawater-type multicellular complexes during seawater adaptation, suggesting plasticity in altering ion-transporting functions of chloride cells.

Functional Classification of Chloride Cells

Recently, more detailed functional classification of chloride cells has been introduced on the basis of distribution patterns of ion-transporting proteins involved in ion secretion and absorption (Hiroi *et al.*, 2005a). A currently accepted model for NaCl secretion by chloride cells consists of the cooperative action of three major ion transporters: NKA, $Na^+/K^+/2Cl^-$ cotransporter (NKCC) and Cl^- channel (Silva *et al.*, 1977; Marshall, 1995). A NKA, which is localised to the basolateral membrane of chloride cells, creates

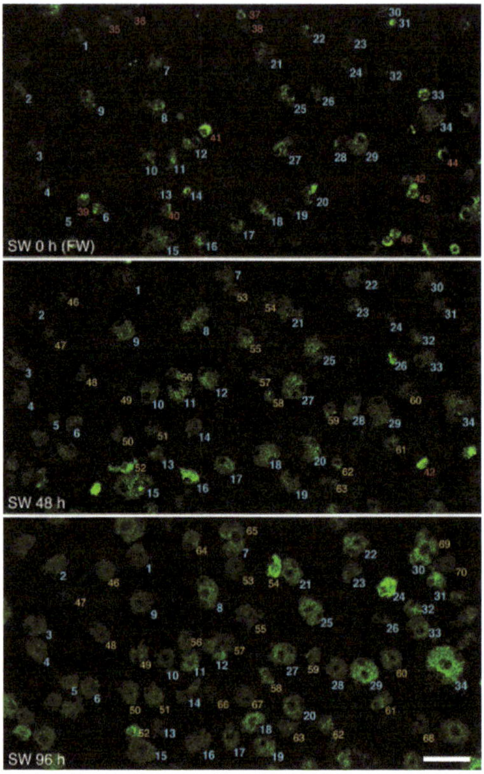

Fig. 5 Sequential confocal images of DASPEI-stained chloride cells in the yolk-sac membrane of a tilapia embryo transferred from freshwater (FW) to seawater (SW). Scale bar: 50 μm. After Hiroi *et al.* (1999).

low intracellular Na^+ and a highly negative charge within the cell. The Na^+ gradient is used to transport Na^+, K^+ and $2Cl^-$ into the cell through a basolateral NKCC. Cl^- then leaves the cells down on an electrical gradient through an apical Cl^- channel, which is homologous to human cystic fibrosis transmembrane conductance regulator (CFTR). Na^+ is transported back outside the cells via NKA, and then secreted by a paracellular pathway between chloride and accessory cells. K^+ is considered to be recycled by a basolateral K^+ channel (Suzuki *et al.*, 1999). Based on distribution patterns of those ion transporters determined by triple-colour whole-mount immunofluorescence staining, chloride cells in Mozambique tilapia embryos have been classified into four distinct types: type I, showing only basolateral NKA staining; type II, basolateral NKA and apical NKCC; type III, basolateral NKA and basolateral NKCC; and type IV, basolateral NKA, basolateral NKCC and apical CFTR (Fig. 7).

The relative frequency of those four types of chloride cells has been shown to change greatly in fish transferred from freshwater to seawater and *vice versa* (Fig. 8).

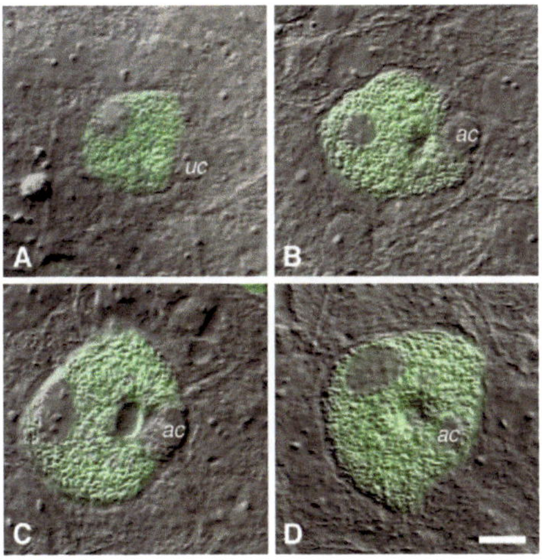

Fig. 6 A series of differential interference contrast (DIC) and immunofluorescence images arranged in the presumed developmental stages from a single freshwater-type chloride cell to a multicellular seawater-type cell. (A) An undifferentiated cell (*uc*) is attached to a single, small chloride cell (stained green). (B, C) The chloride cell becomes larger, indented by an accessory cell (*ac*) originating from the undifferentiated cell. (D) The accessory cell is finally enveloped by the chloride cell, so that the boundary between these cells becomes indistinct. Scale bar, 10 μm. After Hiroi *et al.* (1999).

Among the four types, the basolateral distributions of NKA and NKCC and a distinct apical CFTR distribution in type-IV cells are completely consistent with the currently accepted model for ion secretion by chloride cells, indicating that type-IV cells are seawater-type, ion-secretory cells. This is also supported by the fact that type-IV cells are the predominant cell-type in seawater-adapted embryos but absent from freshwater-adapted ones. Both type-III and type-IV cells show basolateral staining of NKA and NKCC, but type-III cells lack apical CFTR staining. When Mozambique tilapia embryos are transferred from freshwater to seawater, freshwater-predominant type-III cells decrease in number and concomitantly seawater-specific type-IV cells increase (Fig. 8). It is thus considered that type-III cells synthesise CFTR *de novo* and arrange it at the apical membrane to be transformed into the type IV. Transfer from seawater to freshwater shows these changes in reverse: the numbers of type-III cells increase, whereas type-IV cells decrease. It is most likely that type-IV cells lose apical CFTR and stop their ion secretory function to change into type-III cells. From these observations, it can be concluded that type-III cells are a proto- or dormant-type of type-IV cells that become active in ion secretion after placement of CFTR in the apical membrane.

Type-II cells possess apical NKCC, basolateral NKA and show freshwater-specific changes in number during transfer experiments: type-II cells were absent in seawater, appeared following transfer from seawater to freshwater, and tended to decrease following transfer from freshwater to seawater (Fig. 8). These responses to salinity

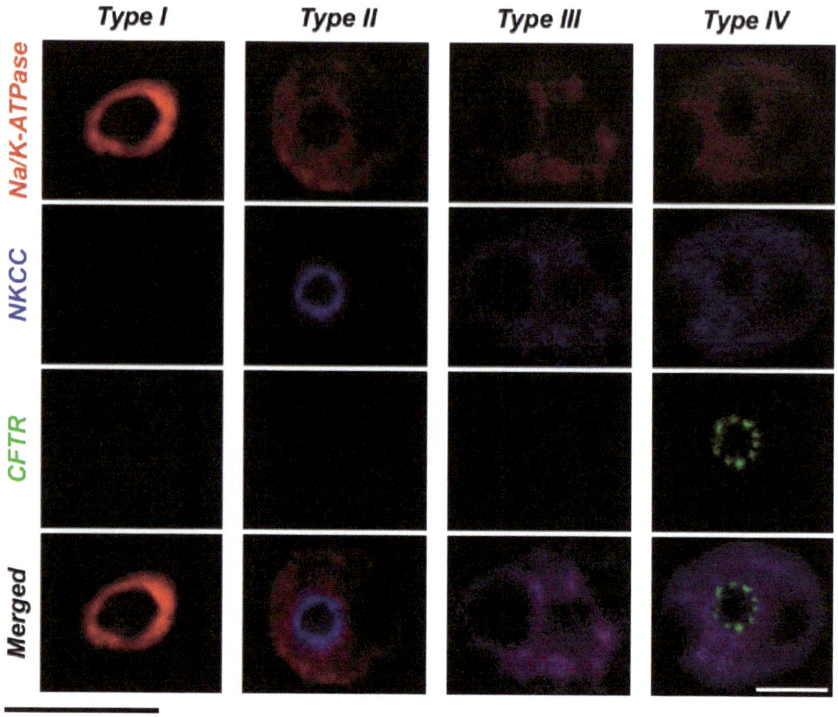

Fig. 7 Functional classification of chloride cells in the yolk-sac membrane of Mozambique tilapia (*Oreochromis mossambicus*) embryos into four types (I-IV), on the basis of distribution patterns of ion-transporting proteins: Na^+/K^+-ATPase (NKA), $Na^+/K^+/2Cl^-$ cotransporter (NKCC) and Cl^- channel (CFTR). After Hiroi *et al.* (2005a).

changes provide indirect evidence that type-II cells possess an ion transport function favourable for freshwater acclimation, presumably ion absorption. However, the apical localisation of NKCC in chloride cells is somewhat surprising. A number of previous reviews on ion transport mechanisms of teleost gills have suggested Na^+ absorption by H^+-ATPase and a Na^+ channel, or by a Na^+/H^+-exchanger, and a Cl^- absorption by Cl^-/HCO_3^- exchanger, but have not referred to apical NKCC (Marshall, 2002; Hirose *et al.*, 2003; Perry *et al.*, 2003; Evans *et al.*, 2005). An active NaCl uptake mechanism with apical NKCC has been proposed in crabs such as green crab (*Carcinus maenas*) and burrowing crab (*Chasmagnathus granulatus*) in brackish water (Riestenpatt *et al.*, 1996; Onken *et al.*, 2003), and only Kirschner (2004) has proposed that the mechanism may exist in estuarine fish. In other euryhaline teleost species examined so far, the presence of apical NKCC has not been reported: in Atlantic salmon (*Salmo salar*) and Hawaiian goby (*Stenogobius hawaiiensis*), NKCC has a basolateral localisation in chloride cells in both freshwater and seawater conditions (Pelis *et al.*, 2001; McCormick *et al.*, 2003); in common mummichog, NKCC is condensed and localised in lateral parts of chloride cell complexes in freshwater, and apparently not in the apical membrane area

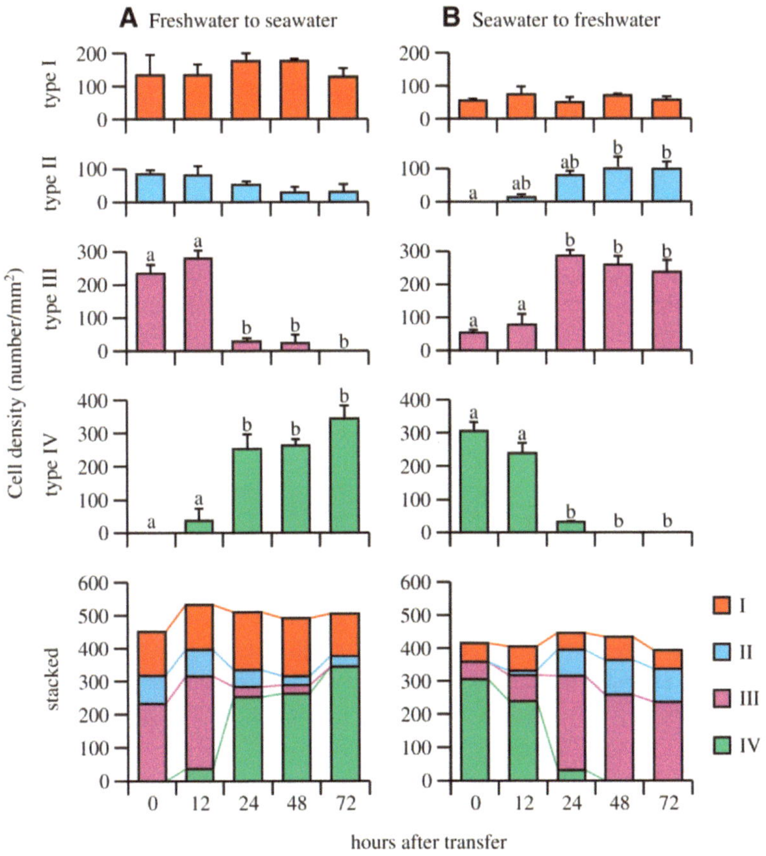

Fig. 8 Changes in the density of the four chloride cell types following transfer from freshwater to seawater (A) and *vice versa* (B) in the yolk-sac membrane of Mozambique tilapia (*Oreochromis mossambicus*) embryos. The data are expressed as the mean ± standard error of the mean. Columns with different letters are significantly different (P<0.0125). *red*, type-I cell; *cyan*, type-II cell; *magenta*, type-III cell; *green*, type-IV cell. After Hiroi *et al.* (2005a).

of chloride cells (Marshall *et al.*, 2002). Apical NKCC staining is currently detectable only in chloride cells in the gills (Wu *et al.*, 2003) and embryonic yolk-sac membrane (Hiroi *et al.*, 2005a) of freshwater-acclimated Mozambique tilapia.

Meanwhile, type-I cells are characterised by their relatively small size, indistinct apical opening and possessing only NKA immunoreactivity, indicating they are probably immature chloride cells. Immature type-I cells may develop into other cell types: types II, III and IV. Type-I cells exist consistently in both freshwater and seawater conditions, suggesting that the recruitment of new chloride cells occurs irrespective of environmental salinity.

The characteristics of the four chloride cell types and their presumptive interrelationships are summarised in Figure 9. Immature type-I cells are considered to develop into type-II ion-absorptive cells (arrow A) and type-III cells (arrow B) in freshwater, and into type-IV ion-secretory cells in seawater (Arrow C). Following transfer from freshwater to seawater, it seems likely that most type-III cells are transformed into type-IV cells (arrow D), while a certain number of type-I cells may also develop into type-IV cells (arrow C). When transferred from seawater back to freshwater, type-IV cells seem to change into type-III cells to stop ion secretion (arrow E), and type-I cells develop into type-II cells to absorb ions (arrow A). The euryhalinity of Mozambique tilapia embryos seems to depend on the rapid and reversible changes between type-III cells and type-IV cells, as well as the continual presence of immature type-I cells, which can develop into the other cell types.

"YOLK BALL" INCUBATION SYSTEM

Although many studies have focused in the past on the effects of environmental salinity and hormones, their direct effects on chloride cell differentiation are not yet clear. This may be partly due to the lack of suitable *in vitro* experimental models. Under these circumstances, Shiraishi *et al.* (2001) have established a unique *in vitro* experimental model named "yolk balls", in which the yolk-sac was separated from the embryonic body of Mozambique tilapia and subjected to *in vitro* incubation (Fig. 10A-D). The yolk-ball preparation consists of the yolk and the covering yolk-sac membrane, which contains a rich population of chloride cells. After appropriate cutting, the incision

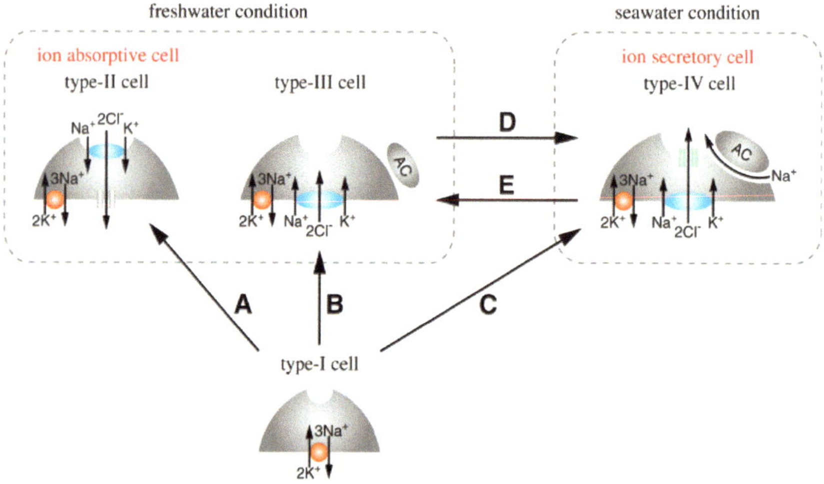

Fig. 9 Schematic diagrams of the four chloride cell types, showing intracellular localization of Na^+/K^+-ATPase (red), $Na^+/K^+/2Cl^-$ cotransporter (NKCC, cyan) and cystic fibrosis transmembrane conductance regulator (CFTR, green). The presumed interrelationship among the four cell types are indicated by arrows A-E. After Hiroi *et al.* (2005a).

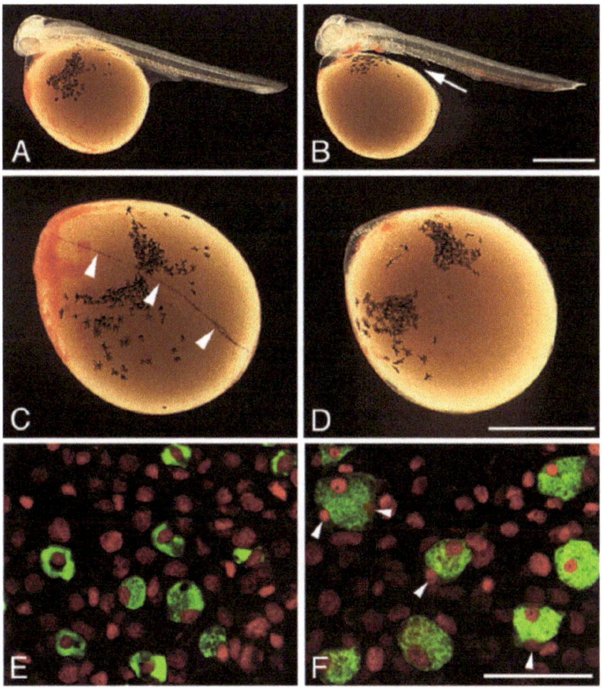

Fig. 10. (A, B) An intact Mozambique tilapia (*Oreochromis mossambicus*) embryo at 2 days before hatching (A) and an embryo whose yolk-sac was nearly cut off from the embryonic body (B). An arrow indicates the incision. (C, D) Incised wounds on the yolk-ball preparations at 0 h (C) and 3 h (D) after surgical operation, stained with Trypan Blue. The wound (arrowheads) has healed after 3 h incubation in balanced salt solution. (E, F) Chloride cells in the yolk-sac membrane of the yolk-ball preparations incubated in freshwater (E) and seawater (F) for 48 h, double-stained with FITC-labeled anti-Na^+, K^+-ATPase (green) and propidium iodide (PI, red). Chloride cells in seawater form multicellular complexes together with accessory cells (arrowheads), as evidenced by the presence of more than one PI-labelled nucleus. Scale bars: A, B, 1 mm; C, D, 1 mm; E, F, 50 µm. After Shiraishi *et al.* (2001).

on the yolk ball was healed during incubation in balanced salt solution for 3 h, so that the yolk-sac membrane completely enclosed the yolk.

The yolk-ball incubation system has great advantages over conventional *in vitro* experimental models for chloride cells. Primary cultures of isolated cells from gill epithelia have been utilised in physiological studies; however, dispersed chloride cells do not survive under *in vitro* conditions. Another problem of primary cultures is the loss of the cellular polarity, since the location of respective ion-transporting proteins in the plasma membrane is critical for ion-transporting mechanisms of chloride cells. Although the opercular and yolk-sac membrane preparations maintains the cellular polarity, chloride cells in those preparations can neither survive long nor differentiate under *in vitro* conditions. Moreover, the yolk balls prepared in balanced salt solution are able to survive transfer to freshwater and seawater. The Cl^- turnover rate measured by whole-body influx of $^{36}Cl^-$ was about 60 times higher in yolk balls in seawater than in freshwater, while there was no essential difference in Cl^- content between them (Hiroi *et al.*, 2005b). This is exactly what happens in intact embryos, thus indicating that the yolk-sac membrane of the yolk balls maintain the ion-transporting properties of the intact membrane.

Shiraishi *et al.* (2001) examined whether or not functional differentiation of chloride cells in the yolk-sac membrane occurred in the yolk balls without embryonic control. The yolk balls prepared from freshwater embryos were incubated either in freshwater or in seawater up to 48 h to elucidate the morphological alteration of chloride cells in response to the environmental salinity (Fig. 10E, F). The chloride cells in the yolk balls incubated in seawater become larger than those in freshwater and form multicellular complexes, characteristic of seawater-type chloride cells. In those transferred to freshwater, on the other hand, the cells remain small and rarely form a complex. More precisely, types I, II and III chloride cells are identified in freshwater yolk balls (Hiroi *et al.*, 2005b). It is likely that freshwater yolk balls preserve a chloride cell population of the intact embryos in freshwater, being favourable for freshwater adaptation. The yolk balls incubated in seawater, on the other hand, are characterised by the predominant occurrence of type-IV cells, which can account for active ion secretion in seawater yolk balls. Such responses of chloride cells are identical to those observed in intact embryos. These findings indicate that the chloride cells in the yolk-sac membrane are equipped with an autonomous mechanism of functional differentiation, independent of embryonic endocrine and nerve systems, although it is generally believed in adult fish that chloride cell activities are under hormonal control.

POSSIBLE OSMORECEPTION BY CHLORIDE CELLS

The functional differentiation of chloride cells independent of embryonic endocrine and nerve systems implies existence of a mechanism that chloride cells detect environmental salinity changes. It is most probable that, in response to transfer from freshwater to seawater, freshwater-type cells can sense salinity changes in a direct or indirect manner, and this triggers transformation into the seawater type. One possible way is to detect a fluctuation in the external salinity *via* the apical membrane of chloride cells facing the external medium. It is more likely, however, that increased environmental salinity results in a slight increase in internal osmolality, which could be detected by the basolateral membrane of chloride cells. In fact, changes in osmolality on the basolateral side have been shown to affect the rate of Cl^- secretion by opercular epithelia with a rich population of chloride cells in common mummichog (Zadunaisky *et al.*, 1995; Marshall *et al.*, 2000).

Meanwhile, we have recently identified aquaporin 3 (AQP3), a water-channel molecule, from the Mozambique tilapia gills, and demonstrated that AQP3 is intensively located in the basolateral membrane of chloride cells (Watanabe *et al.*, 2005). Based on this finding, it is hypothesised that basolateral AQP3 is involved in osmoreception by chloride cells. It is expected that the AQP3-rich basolateral membrane is more permeable to water than the other membranes. This is supported by the finding that the oocytes of African clawed frog (*Xenopus laevis*) expressing tilapia AQP3 exhibited higher water permeability than the control oocytes, resulting in oocyte swelling in the hypo-osmotic medium. Since the surface area of the basolateral membrane is enlarged due to intensive infoldings of the tubular system, the ratio of the surface area to the cell volume is supposed to be much higher in chloride cells than in any other cell type in the gills or yolk-sac membrane. Such structural characteristics may further enhance the possible osmosensitivity of chloride cells. Transfer of fish from seawater to freshwater, for example, may induce a slight decrease in blood osmolality, which creates an osmotic gradient between the intracellular fluid and blood. Subsequently, water moves into chloride cells through basolateral AQP3 according to the osmotic gradient, resulting in cell swelling. Conversely, transfer from freshwater to seawater may leads to shrinkage of chloride cells. Although a more detailed mechanism of the osmoreception is not yet known, AQP3 expressed in chloride cells seems to be closely related not only to autonomous mechanisms of their functional differentiation, but also to euryhalinity of teleost fishes.

FUTURE PERSPECTIVES

The molecular mechanisms of ion uptake in chloride cells are still controversial in both adult and early life stages. The current ion uptake models involve H^+-ATPase and/or NKA as major driving forces for the operation of the relevant ion transporters in chloride cells (Marshall, 2002; Hirose *et al.*, 2003; Perry *et al.*, 2003; Evans *et al.*, 2005). H^+-ATPase has been considered to be coupled to Na^+ uptake in chloride cells: electrogenic H^+ efflux via the apical H^+-ATPase generates a negative intracellular potential, which in turn acts as a driving force to allow Na^+ entry through the apical Na^+ channel. However, convincing molecular and cellular data are very few and fragmentary, partly because genetic information is limited in the euryhaline species that have been widely used as experimental models. Meanwhile, zebrafish (*Danio rerio*) is increasingly being recognised as a powerful model to elucidate ion uptake mechanisms, as well as organogenesis, in fish.

Using a non-invasive electrophysiological technique, Hwang and his coworkers have recently succeeded in demonstrating H^+ secretion in the skin of zebrafish larvae (Lin *et al.*, 2005). Strong signals of H^+ secretion are detected at the yolk-sac membrane and pericardial surface, which contain a rich population of chloride cells. Surprisingly, H^+ secretion is attributed to novel H^+-secreting "chloride cells", which are H^+-ATPase rich and distinct from the traditional mitochondria-rich chloride cells, or MR cells. Thus, two different types of ionocytes (rather than chloride cells) present in the skin of zebrafish larvae are proposed to be responsible for the uptake of different ions. Further studies with zebrafish will surely provide new insights to the current models of ion transport.

In this chapter, we have reviewed recent advances in studies on osmoregulation during early life stages of teleost fishes, with special reference to extrabranchial chloride cells. Successful osmoregulation is achieved by maintaining ion and water balance of the body fluid. It is now clear that chloride cells in the yolk-sac membrane and body skin are critical for ionic regulation during early life stages. Those extrabranchial chloride cells function as ion-secreting sites in seawater and as ion-absorbing sites in freshwater. This may explain ion regulation during early developmental stages of fish, which lack functional gills. On the other hand, an active mechanism for maintaining water balance is lacking in fish embryos and larvae until the onset of drinking and urine production. Before the intestine and kidney become functional, diffusional water permeability of the epidermis is low to prevent water load in freshwater and water loss in seawater (Miyazaki *et al.*, 1998). The yolk-ball incubation system introduced here will serve as a suitable material to investigate water, as well as ion, permeability of the yolk-sac membrane. A better understanding of ion and water regulation during early developmental stages will greatly contribute to future advances in the field of fish osmoregulation in general.

References

Alderdice DF (1988) Osmotic and ionic regulation in teleost eggs and larvae. In: Hoar WS, Randall DJ (eds) Fish Physiology, Vol XI, Part A, Academic Press, San Diego, p 163-251

Ayson FG, Kaneko T, Hasegawa S, Hirano T (1994) Development of mitochondrion-rich cells in the yolk-sac membrane of embryos and larvae of tilapia, *Oreochromis mossambicus*, in fresh water and seawater. J Exp Zool 270: 129-135

Bentley PJ (2002) Endocrines and osmoregulation: A Comparative Account in Vertebrates. Springer-Verlag, Berlin

Evans DH (1993) Osmotic and ionic regulation. In: Evans DH (ed) The Physiology of Fishes. CRC Press, Boca Raton, p 315-341

Evans DH, Piermarini PM, Choe KP (2005) The multifunctional fish gill: dominant site of gas exchange, osmoregulation, acid-base regulation, and excretion of nitrogenous waste. Physiol Rev 85: 97-177

Finn RN (2007) The maturational disassembly and differential proteolysis of paralogous vitellogenins in a marine pelagophil teleost: A conserved mechanism of oocyte hydration. Biol Reprod 76: 936-948

Finn RN, Østby G, Norberg B, Fyhn HJ (2002) *In vivo* oocyte hydration in Atlantic halibut (*Hippoglossus hippoglossus*). Proteolytic liberation of free amino acids, and ion transport are driving forces for osmotic water influx. J Exp Biol 205: 211-224

Foskett KJ, Scheffey C (1982) The chloride cell: definitive identification as the salt secreting cell in teleosts. Science 215: 164-166

Guggino WB (1980) Salt balance in embryos of *Fundulus heteroclitus* and *F. bermudae* adapted to seawater. Am J Physiol 238: R36-R41

Hasegawa S, Hirano T, Ogasawara T, Iwata M, Akiyama T, Arai S (1987) Osmoregulatory ability of chum salmon, *Oncorhynchus keta*, reared in fresh water for prolonged periods. Fish Physiol Biochem 4: 101-110

Hiroi J, Kaneko T, Seikai T, Tanaka M (1998) Developmental sequence of chloride cells in the body skin and gills of Japanese flounder (*Paralichthys olivaceus*) larvae. Zool Sci 15: 455-460

Hiroi J, Kaneko T, Tanaka M (1999) *In vitro* sequential changes in chloride cell morphology in the yolk-sac membrane of Mozambique tilapia (*Oreochromis mossambicus*) embryos and larvae during seawater adaptation. J Exp Biol 202: 3485-3495

Hiroi J, McCormick SD, Ohtani-Kaneko R, Kaneko T (2005a) Functional classification of mitochondrion-rich cells in euryhaline Mozambique tilapia (*Oreochromis mossambicus*) embryos, by means of triple immunofluorescence staining for Na^+/K^+-ATPase, Na^+/K^+2Cl^- cotransporter and CFTR anion channel. J Exp Biol 208: 2023-2036

Hiroi J, Miyazaki H, Katoh F, Ohtani-Kaneko R, Kaneko T (2005b) Chloride turnover and ion-transporting activities of 'yolk balls', yolk-sac preparations separated from Mozambique tilapia embryos, incubated in freshwater and seawater. J Exp Biol 208: 3851-3858

Hirose S, Kaneko T, Naito N, Takei Y (2003) Molecular biology of major components of chloride cells. Comp Biochem Physiol 136B: 593-620

Hootman SR, Philpott CW. (1980) Accessory cells in teleost branchial epithelium. Am J Physiol 238: R199-R206

Hwang PP (1987) Tolerance and ultrastructural responses of branchial chloride cells to salinity changes in the euryhaline teleost *Oreochromis mossambicus*. Mar Biol 94: 643-649

Hwang PP (1989) Distribution of chloride cells in teleost larvae. J Morphol 200: 1-8

Kaneko T, Hasegawa S (1999) Application of laser scanning microscopy to morphological observations on drinking in freshwater medaka larvae and those exposed to 80% seawater. Fish Sci 65: 492-493

Kaneko T, Shiraishi K (2001) Evidence for chloride secretion from chloride cells in the yolk-sac membrane of Mozambique tilapia larvae adapted to seawater. Fish Sci 67: 541-543

Kaneko T, Hasegawa S, Hirano T (1992) Embryonic origin and development of the corpuscles of Stannius in chum salmon (*Oncorhynchus keta*). Cell Tissue Res 268: 65-70

Kaneko T, Hasegawa S, Takagi Y, Tagawa M, Hirano T (1995) Hypoosmoregulatory ability of eyed-stage embryos of chum salmon. Mar Biol 122: 165-170

Kaneko T, Shiraishi K, Katoh F, Hasegawa S, Hiroi J (2002) Chloride cells in early life stages of fish and their functional differentiation. Fish Sci 68: 1-9

Katoh F, Shimizu A, Uchida K, Kaneko T (2000) Shift of chloride cell distribution during early life stages in seawater-adapted killifish, *Fundulus heteroclitus*. Zool Sci 17: 11-18

Kirschner LB (2004) The mechanism of sodium chloride uptake in hyperregulating aquatic animals. J Exp Biol 207: 1439-1452

Lin LY, Horng JL, Kunkel JG, Hwang PP (2005) Proton pump-rich cell secretes acid in skin of zebrafish (*Danio rerio*) larvae. Am J Physiol, Cell Physiol 290: C371-C378

Lin LY, Hwang PP (2001) Modification of morphology and function of integument mitochondria-rich cells in tilapia larvae (*Oreochromis mossambicus*) acclimated to ambient chloride levels. Physiol Biochem Zool 74: 469-476

Lin LY, Hwang PP (2004) Mitochondria-rich cell activity in the yolk-sac membrane of tilapia (*Oreochromis mossambicus*) larvae acclimated to different ambient chloride levels. J Exp Biol 207: 1335-1344

McCormick SD, Sundell K, Björnsson BT, Brown CL, Hiroi J (2003) Influence of salinity on the localization of Na^+/K^+-ATPase, $Na^+/K^+/2Cl^-$ cotransporter (NKCC) and CFTR anion channel in chloride cells of the Hawaiian goby (*Stenogobius hawaiiensis*). J Exp Biol 206: 4575-4583

Marshall WS (1995) Transport processes in isolated teleost epithelia: opercular epithelium and urinary bladder. In: Wood CM, Shuttleworth TJ (eds) Fish Physiology, Vol 14, Academic Press, San Diego, p 1-23

Marshall WS (2002) Na^+, Cl^-, Ca^{2+} and Zn^{2+} transport by fish gills: retrospective review and prospective synthesis. J Exp Zool 293: 264-283

Marshall WS, Bryson SE, Luby T (2000) Control of epithelial Cl^- secretion by basolateral osmolality in the euryhaline teleost *Fundulus heteroclitus*. J Exp Biol 203: 1897-1905

Marshall MS, Lynch EM, Cozzi RR (2002) Redistribution of immunofluorescence of CFTR anion channel and NKCC cotransporter in chloride cells during adaptation of the killifish *Fundulus heteroclitus* to sea water. J Exp Biol 205: 1265-1273

Miyazaki H, Kaneko T, Hasegawa S, Hirano T (1998) Developmental changes in drinking rate and ion and water permeability during early life stages of euryhaline tilapia, *Oreochromis mossambicus*, reared in fresh water and seawater. Fish Physiol Biochem 18: 277-284

Onken H, Tresguerres M, Luquet, CM (2003) Active NaCl absorption across posterior gills of hyperosmoregulating *Chasmagnathus granulatus*. J Exp Biol 206: 1017-1023

Philpott CW (1980) Tubular system membranes of teleost chloride cells: osmotic response and transport sites. Am J Physiol 238: R171-R184

Pelis RM, Zydlewski J, McCormick SD (2001) Gill Na^+-K^+-$2Cl^-$ cotransporter abundance and location in Atlantic salmon: effects of seawater and smolting. Am J Physiol 280: R1844-R1852

Perry SF, Shahsavarani A, Georgalis T, Bayaa M, Furimsky M, Thomas SL (2003) Channels, pumps, and exchangers in the gill and kidney of freshwater fishes: their role in ionic and acid-base regulation. J Exp Zool A 300: 53-62

Pisam M, Rambourg A (1991) Mitochondria-rich cells in the gill epithelium of teleost fishes: an ultrastructural approach. Int Rev Cytol 130: 191-232

Riestenpatt S, Onken H, Siebers D (1996) Active absorption of Na^+ and Cl^- across the gill epithelium of the shore crab *Carcinus maenas*: voltage-clamp and ion-flux studies. J Exp Biol 199: 1545-1554

Sardet C, Pisam M, Maetz J (1979) The surface epithelium of teleostean fish gills. J Cell Biol 80: 96-117

Sasai S, Kaneko T, Tsukamoto K (1998) Extrabranchial chloride cells in early life stages of the Japanese eel, *Anguilla japonica*. Ichthyol Res 45: 95-98

Shen ACY, Leatherland JF (1978) Structure of the yolksac epithelium and gills in the early developmental stages of rainbow trout (*Salmo gairdneri*) maintained in different ambient salinities. Environ Biol Fish 3: 345-354

Shiraishi K, Hiroi J, Kaneko T, Matsuda M, Hirano T, Mori T (2001) *In vitro* effects of environmental salinity and cortisol on chloride cell differentiation in embryos of Mozambique tilapia, *Oreochromis mossambicus*, measured using a newly developed "yolk-ball" incubation system. J Exp Biol 204: 1883-1888

Shiraishi K, Kaneko T, Hasegawa S, Hirano T (1997) Development of multicellular complexes of chloride cells in the yolk-sac membrane of tilapia (*Oreochromis mossambicus*) embryos and larvae in seawater. Cell Tissue Res 288: 583-590

Silva P, Solomon R, Spokes K, Epstein, F (1977) Ouabain inhibition of gill Na-K-ATPase: relationship to active chloride transport. J Exp Zool 199: 419-426

Suzuki Y, Itakura M, Kashiwagi M, Nakamura N, Matsuki T, Sakuta H, Naito N, Takano K, Fujita T, Hirose S (1999) Identification by differential display of a hypertonicity-inducible inward rectifier potassium channel highly expressed in chloride cells. J Biol Chem 274: 11376-11382

Uchida K, Kaneko T, Miyazaki H, Hasegawa S, Hirano T (2000) Excellent salinity tolerance of Mozambique tilapia (*Oreochromis mossambicus*): elevated chloride cell activity in the branchial and opercular epithelia of the fish adapted to concentrated seawater. Zool Sci 17: 149-160

Watanabe S, Kaneko T, Aida K (2005) Aquaporin-3 expressed in the basolateral membrane of gill chloride cells in Mozambique tilapia *Oreochromis mossambicus* adapted to freshwater and seawater. J Exp Biol 208: 2673-2682

Wu YC, Lin LY, Lee TH (2003) Na^+,K^+,$2Cl^-$-cotransporter: a novel marker for identifying freshwater- and seawater-type mitochondria-rich cells in gills of the euryhaline tilapia, *Oreochromis mossambicus*. Zool Studies 42: 186-192

Zadunaisky JA, Cardona S, Au L, Roberts DM, Fisher E, Lowenstein B, Cragoe EJ Jr, Spring KR (1995) Chloride transport activity by plasma osmolarity during rapid adaptation to high salinity of *Fundulus heteroclitus*. J Membrane Biol 143: 207-217

CHAPTER

7

Acid-base Balance

Colin J Brauner

INTRODUCTION

Regulation of intra- and extra-cellular pH (pHi and pHe, respectively) is crucial to the normal functioning of all vertebrates, including fish. Deviations in pH alter the local charge on proteins affecting enzyme function and membrane channel properties, which ultimately impact upon cell-cell signalling, cell-volume regulation, gene expression, contractility of the heart and skeletal muscles, and whole animal metabolism (reviewed by Putnam & Roos, 1997). Consequently pH regulation is one of the most tightly controlled processes and not surprisingly a great deal is known about the mechanisms involved in acid-base regulation in fish (Heisler, 1984, 1993; Claiborne *et al.*, 2002; Evans *et al.*, 2005). Most of this knowledge, however, comes from juvenile and adult fish and our understanding of the mechanisms and patterns of acid-base regulation during development in fish is very limited. This chapter will briefly review acid-base regulation in juvenile and adult fish, discuss the limited literature on acid-base regulation in developing fish, and provide a framework for future studies required to advance our understanding of acid-base regulation in developing fish; a field ripe for investigation.

Author's address: Department of Zoology, University of British Columbia, 6270 University Blvd., Vancouver, BC, Canada, V6T 1Z4.

E-mail: brauner@zoology.ubc.ca

MECHANISMS OF ACID-BASE REGULATION IN JUVENILE AND ADULT FISH

Under control steady-state conditions, most fish exhibit a small whole body net acid excretion, as is observed in mammals, which is required for pH homeostasis. Acid-base status in fish can be influenced by a wide variety of conditions, including changes in the external environment, such as gas tensions (CO_2, NH_3, and/or O_2 levels), temperature, salinity and water pH, and internally induced challenges associated with processes such as feeding and exercise (predator avoidance, prey capture, migration). Acid-base disturbances can loosely be grouped into two categories. The first is respiratory in origin, where changes in pH are associated with changes in blood CO_2 levels. The second is metabolic in origin, and is a result of non-CO_2 related changes in pH, as occurs during anaerobic metabolism. In many instances, acid-base disturbances may be both respiratory and metabolic in origin. For example, exposure to hypoxia results in hyperventilation, inducing a respiratory alkalosis that may be combined with a metabolic acidosis due to inadequate O_2 delivery to the tissues and subsequent anaerobic metabolism. The majority of acid-base disturbances experienced in fish involve an acidosis, and consequently compensation for an acidosis will represent the main emphasis in this review. Experimentally, exposure to elevated environmental CO_2 (hypercapnia) is one of the most common means to induce an acidosis through which the pattern and process of pH recovery can be followed. This is because CO_2 can easily be maintained at a pre-determined, constant environmental level over the exposure duration, and it diffuses rapidly from the water into the fish to induce an acidosis. During an acid-base disturbance, pH changes can be buffered or actively restored through a number of pathways as described below.

Whole Animal Acid-base Regulation

In vertebrates in general, acid-base disturbances can be compensated by: i) physicochemical buffering with bicarbonate and non-bicarbonate buffers, ii) changes in ventilation to alter pCO_2 and thus pH, via the CO_2-HCO_3^- buffer system, or iii) net transport of acid-base equivalents between the body and the environment. In contrast to air-breathers, the partial pressure of CO_2 in the blood of water breathing fish is low due to the high ventilatory requirements to secure O_2 uptake from water (Dejours, 1988), and thus compensatory strategies i) and ii) have relatively little effect on blood pH regulation. Consequently, disturbances in blood pH in fish are generally compensated for by transepithelial transfer of acid-base relevant ions between the animal and the aquatic medium. The gills, intestine and kidney have all been implicated to some degree in acid-base regulation in fish, however, it is generally accepted that the gills are the dominant surface (about 90%) across which net acid-base transfer occurs (Heisler, 1984, 1993). The gills represent a large surface area that is in intimate contact with the aquatic medium. The gill epithelium is provided with high water flow rates, and the aquatic medium can act as an unlimited source and sink for acid-base equivalents and respective counter ions, provided that the composition of the aquatic medium is appropriate (i.e. not too low in ionic composition or acidic). In general, the rate and degree of pH compensation is influenced by ionic composition of water, where fish in hard freshwater (FW) compensate more completely and rapidly than fish in soft FW (Larsen & Jensen, 1997), and those in seawater (SW) more completely and rapidly than those in FW (Toews *et al.*, 1983).

In general, recovery of blood pH following an acid-base disturbance results from adjustments in blood HCO_3^- levels (Evans *et al.*, 2005). In studies measuring net acid excretion during an acidosis, it is impossible to distinguish between net H^+ excretion and HCO_3^- uptake (Heisler, 1984). In most fish investigated to date, plasma HCO_3^- elevation during an acidosis is associated with a 1:1 reduction in plasma Cl^-, implicating net brachial Cl^-/HCO_3^- exchange (Toews *et al.*, 1983; Cameron & Iwama, 1989; Larsen & Jensen, 1997) as the compensatory mechanism. Some species appear to utilise net branchial Na^+/H^+ exchange (Hyde & Perry, 1989; Hirata *et al.*, 2003) to compensate for acidosis, however, this pathway appears to be much less common. In a recent study on two species of marine teleost fishes, Japanese flounder (*Paralichthys olivaceus*) and yellow tail (*Seriola quinqueradiata*), during exposure to 1% CO_2, the elevation in plasma HCO_3^- was matched with a similar reduction in plasma Cl^-, however at higher CO_2 tensions of 3-5% CO_2, the rise in HCO_3^- exceeded the fall in Cl^-. Interestingly, under these conditions there is an increase in plasma Na^+ concentration that makes up most of the increase in plasma HCO_3^- not accounted for by a reduction in Cl^-. Thus, in some marine fish, there appears to be a biphasic response likely involving net branchial Cl^-/HCO_3^- exchange at low levels of hypercapnia, coupled with net branchial Na^+/H^+ exchange at higher CO_2 levels (Hayashi *et al.*, 2004).

While most fish are capable of regulating blood pH back to close to pre-disturbance levels during an acid-base disturbance within hours to days (Ultsch, 1996), provided the acidosis is not too severe, a few species possess a very blunted extracellular acid-base regulatory response (Heisler, 1982; Brauner *et al.*, 2004; Sanchez *et al.*, 2005). During exposure to environmental hypercapnia in the facultative air-breathing sailfin catfish (*Liposarcus pardalis*), blood pH may be reduced by up to 1 pH unit, with minimal pH compensation over the following 96 h, despite 100% survival. In both the sailfin catfish and another facultative air-breathing fish, the marbled swamp eel (*Synbranchus marmoratus*), the lack of compensation for a reduction in plasma pH is associated with complete preferential pHi regulation of heart, liver and muscle (Heisler, 1982; Brauner *et al.*, 2004). In most fish species investigated to date, pHi is regulated back to pre-disturbance levels; however, this usually requires significant recovery in plasma pH for this to be accomplished (Brauner *et al.*, 2004). The mechanisms responsible for complete preferential pHi regulation in the absence of pHe compensation, and the ubiquity of this pattern of acid-base regulation among fish species remains to be investigated. It is not known whether larval stages of these species exhibit this same strategy in acid-base regulation.

Cellular Processes and Transporters Associated with Acid-base Regulation

Earlier studies on acid-base regulation primarily focused on whole animal responses and net acid-base status following alterations in primarily extracellular, but also intracellular pH. While it is now well established that net adjustments in plasma HCO_3^- during an acid-base disturbance are associated with Na^+ and/or Cl^- linked acid-base transport in both FW and SW fish (Evans *et al.*, 2005), there are differences in both the cellular and sub-cellular pathways involved at the gills, and in the types of transport proteins used in the gills of different fishes. These differences may be species specific, and may be affected by the ambient environment, such as water pH and salinity. While recently there has been considerable interest in elucidating the cellular, sub-cellular and molecular pathways through which acid-base regulation is accomplished (reviewed by Claiborne *et al.*, 2002 and Evans *et al.*, 2005), there remain large gaps in our understanding of the processes involved in pH regulation in juvenile and adult fish. Not surprisingly, much less is known in this regard for larval fishes.

In general, the mitochondrial rich cell (MRC, also commonly referred to as the chloride cell, see Kaneko & Hiroi, this volume) of both FW and SW fishes is thought to be the primary cell associated with acid-base regulation. However, there is also evidence for the pavement cells in FW fish playing a role. Two distinct sub-types of MRCs exist in freshwater fish, an α sub-type, and a β sub-type (Pisam *et al.*, 1987). One is thought to be a net acid excreting cell and the other a net base secreting cell (Galvez *et al.*, 2002; Evans *et al.*, 2005) as is observed in the mammalian kidney.

There are two general models for net Cl^-/HCO_3^- exchange across the gills during an acidosis associated with hypercapnia in freshwater. The first incorporates direct uptake of HCO_3^- from the water in exchange for Cl^- (Claiborne & Heisler, 1986; Larsen & Jensen, 1997), presumably across a branchial anion exchanger. This model is supported by the observation that an elevation in water HCO_3^- at constant pCO_2 results in a more rapid and complete compensation in blood pH (Larsen & Jensen, 1997) than in waters with lower HCO_3^- levels. The second model is based upon morphological changes in the gill filament associated with hypercapnia exposure. Within several hours to days following exposure to environmental CO_2, there is a physical covering of MRCs by pavement cells, reducing the MRC fractional area that is correlated with a reduction in branchial Cl^- influx (Goss *et al.*, 1992). These morphological changes are associated with an increase in net acid excretion and retention of plasma HCO_3^-. The net acid excretion is hypothesised to occur either by removal of Cl^-/HCO_3^- exchangers from the MRCs and/or insertion of proton pumps into the pavement cells (Goss *et al.*, 1995). These two models for net HCO_3^- accumulation in the plasma during hypercapnia are not mutually exclusive and it may be that both operate under different environmental conditions or developmental stages within a species, or possibly even simultaneously within an individual.

Characterisation of the pathways responsible for acid-base regulation in the gills of fish at the level of protein transporters, has been accomplished through the use of immunohistochemistry, pharmacology and molecular biology which have been the focus of a few recent reviews (Claiborne *et al.*, 2002; Evans *et al.*, 2005). Given that acid-base relevant ion transfer is associated with Na^+ and/or Cl^- exchange, acid-base regulation is integrally coupled with ionoregulation. Consequently, models of ionoregulation in FW and SW (Perry, 1997; Marshall & Bryson, 1998; Marshall, 2002; Evans *et al.*, 2005) are crucial to understand the molecular and cellular basis for acid-base regulation. The emerging consensus is that there are two acid secreting mechanisms in fish that are linked with Na^+ transport. The first appears to be restricted to FW fish and relies upon apical V-type ATPase that generates an electrical gradient to drive Na^+ uptake through an apical sodium channel (based upon studies largely limited to salmonids; Perry, 1997). The second is an apical electroneutral Na^+/H^+ exchanger that operates in some FW (Hirata *et al.*, 2003) and likely most SW teleost fish (Claiborne *et al.*, 2002). Processes related to base excretion, and thus net branchial Cl^-/HCO_3^- exchange are less well understood, which is interesting given that under most acid-base disturbances, this is thought to be the primary route through which acid-base disturbances are corrected. In general it is thought that Cl^-/HCO_3^- exchange occurs across an apical anion exchanger (AE1) or pendrin (Piermarini *et al.*, 2002), but clearly much remains to be investigated in juvenile and adult fish, let alone developing fish.

MECHANISMS OF ACID-BASE REGULATION IN EARLY LIFE STAGES

Mass-specific metabolic rate scales inversely with body size in vertebrates (Schmidt-Nielsen, 1975). Given the net acid production associated with metabolism, smaller animals must have high mass specific acid excretion. Furthermore, considering the dramatic developmental changes that occur, many of which are likely pH sensitive, acid-base homeostatic processes must be operational in early life stages. However, very few studies have focused upon the ontogeny of acid-base regulation in fish, and surprisingly little is known from the period following fertilisation through to larvae.

Whole Animal Acid-base Regulation Shortly Following Fertilisation

In one of the few studies investigating acid-base regulation shortly after fertilisation, Molich and Heisler (2005) observed that in zebrafish (*Danio rerio*) embryos, exposure to hypercapnia resulted in a rapid acidification of the embryo. Exposure to a 10 fold elevation in water pCO_2, from 0.097 to 0.99 kPa (1% CO_2), or from 0.32 to 3.2 kPa (3.5% CO_2), acidified both the interstitial and intracellular fluid, with maximal reductions in pH achieved within about 10 minutes. As surface area to volume ratios decrease with growth, the time at which maximal pH reduction occurs during exposure to hypercapnia may increase, but because of the high diffusivity of CO_2 in water and biological membranes, equilibrium between the environment and intracellular fluid would always be expected in the order of minutes rather than hours in water-breathing vertebrates. In the 10 to 256 cell zebrafish embryo preparation used by Molich and Heisler (2005), pH disturbances in both the interstitial and intracellular compartments associated with exposure to 1% CO_2 were largely compensated for within 2h, presumably due to active pH regulatory processes associated with net HCO_3^- accumulation in both compartments. HCO_3^- accumulation in these compartments must be associated with acid-base relevant ion transfer between the embryo and the environment as in adult fish, but appears to occur at a much higher rate. Nothing is known of the specific mechanisms involved. It should be mentioned that in this study, the bathing medium in which the embryo was held was similar in composition to the interstitial fluid which differs markedly from FW, and this may have influenced the dynamics of acid-base regulation in that experimental setup. At higher pCO_2 levels (3.5%), pH compensation also occurred, but to a greater extent intracellularly than interstitially. From this study, it is clear that embryos are capable of regulating cellular and interstitial pH quite tightly very early in development. Rearing embyros at high pCO_2 (3.3 vs 0.3 kPa) for 24h shortly following fertilisation, had no affect on mortality or growth (Molich & Heisler, 2005). Longer exposure durations were not conducted in this study, but would be very interesting given that chronically elevated CO_2 levels of this magnitude have been shown to reduce growth and feed conversion over longer term exposure in juvenile spotted wolffish (*Anarhichas minor*) (Foss *et al.*, 2003), Atlantic salmon (*Salmo salar*) (Fivelstad *et al.*, 1998), and white sturgeon (*Acipenser transmontanus*) (Crocker & Cech, 1996).

Whole Animal Acid-base Regulation Following Hatch

Although very few physiological studies have investigated the ontogeny of acid-base regulation in fish, there have been a number of studies investigating the ontogeny of ionoregulation (see Kaneko & Hiroi, this volume). These studies may shed important insights into the development of acid-base regulatory processes because of the tight interaction between pH and ion regulation. Active uptake of Na^+ from the water (unidirectional Na^+ uptake, $J_{in}Na^+$) in developing salmonid embryo's is very low up to hatch (Rudy & Potts, 1969; Eddy & Talbot, 1985; Brauner & Wood, 2002a, b), but then increases almost exponentially following hatch which correlates with a large increase in whole body Na^+,K^+-ATPase (NKA) activity (Fig. 1; Brauner & Wood, 2002a). During yolk-sac absorption, $J_{in}Na^+$ levels far exceed that of juvenile and adult fish until swim-up, at which time $J_{in}Na^+$ have been reduced to adult levels (Brauner *et al.*, 2003). During development, there is a progressive increase in whole body Na^+ content (a three-fold increase between hatch and swim-up), indicating that there is significant net Na^+ uptake from the environment. Interestingly, the same relationship is not seen for whole animal Cl^- dynamics, where $J_{in}Cl^-$ is much lower than $J_{in}Na^+$ (Misiaszek, 1996) and between hatch and swim-up there is no significant change in whole body Cl^- content. During development, the yolk represents a considerable store for both Na^+ and Cl^-, but appears to be more significant for Cl^- as yolk concentrations

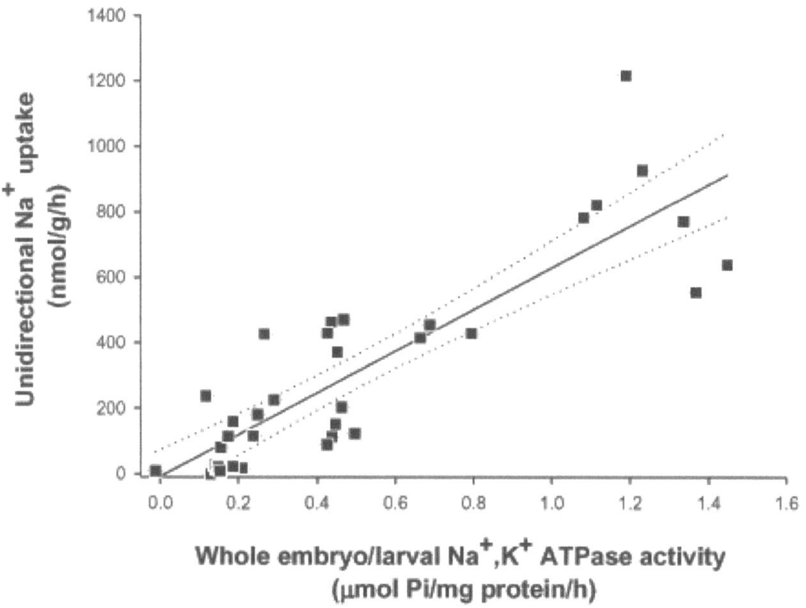

Fig. 1 Relationship between unidirectional Na^+ uptake ($J_{in}Na^+$), and whole body Na^+,K^+-ATPase activity in developing rainbow trout (*Oncorhynchus mykiss*) embryos and larvae (up to swim-up). Squares represent the mean of 10 animals, where standard errors of the mean are contained within each square (least squares regression r^2=0.75; Dotted lines about regression indicate 95% confidence limits. After Brauner *et al.* (2003).

are initially 4-fold higher for Cl^- than Na^+. Assuming these data indicate a limited ability for Cl^- exchange between the developing organism and the environment, this has very interesting implications for acid-base regulation during development given that in most FW fish, acid-base compensation in the blood compartment is associated with a net Cl^-/HCO_3^- exchange. While $J_{in}Na^+$ at the gills is much greater than $J_{in}Cl^-$ during development, it is also far in excess of that required for homeostasis (Misiaszek, 1996). Taken together these data may indicate that acid-base regulation at this stage of development may be more dependent upon net Na^+/H^+ exchange than Cl^-/HCO_3^- exchange, and then switches to the latter some time later in development. Another possibility is that rather than a net acid-base exchange between the organism and the environment to correct for an acid-base disturbance, there is net exchange between the developing fish and yolk-sac contents. Clearly more work is required to investigate these possibilities.

There are also very interesting dynamics in relation to ammonia production and accumulation in developing embryos that may be of importance as a temporal buffer system. During development, there appears to be a progressive increase in embryonic and yolk total ammonia up to hatch, following which ammonia levels quickly decrease (reviewed by Wright and Fyhn, 2001). NH_3 is produced as an end product of protein and amino acid catabolism. Up to and following hatch, developing fish have an ammonia quotient (mol ammonia excretion/ mol oxygen consumption) of between 0.10 and 0.30. The pK_a for ammonia is between 9.0 and 9.5, and thus at the pH of various body compartments, at least 95% of the total ammonia will be as NH_4^+. Consequently, as total ammonia accumulates, it will bind and store protons. Whether ammonia accumulation in the body and yolk is important for acid-base balance in embryonic and larval fish is not known, but worthy of further study.

Cellular Processes and Transporters Associated with Acid-base Regulation

In adult fish, the gills are clearly the predominant structure for acid-base regulation (Heisler, 1984, 1993). However, early in development, functional gills do not exist and therefore cannot be involved. From the discussion above based upon a single study, it appears that fish embryos have the ability to regulate acid-base status very effectively shortly after fertilisation. Given that acid-base regulation is tightly linked with ionoregulation, and at the cellular level, acid-base relevant ion exchange is largely associated with MRCs and NKA, studies documenting the relative distribution of MRCs and NKA over the body surface throughout development may yield important insight into the ontogeny of acid-base regulation.

It is well documented that MRCs exist on both the skin and yolk-sac epithelium of developing fishes (Alderdice, 1988; Hwang, 1989; Hwang *et al.*, 1994, 1999; Chang *et al.*, 2003; Varsamos *et al.*, 2005; Kaneko & Hiroi, this volume), and MRC density on these structures increase with development prior to gill formation (Rombough, 1999). It is hypothesised that skin and yolk-sac epithelium MRCs play an important role in osmoregulation until gill MRCs become functional (Kaneko *et al.*, 2002), but this is largely based upon morphology, where there appears to be considerable homology between skin and yolk-sac MRCs, and those of the gills in juvenile and adult fish. Recent studies have also demonstrated the functional importance of these cells for Cl^- excretion in seawater (see Kaneko & Hiroi, this volume). It is conceivable and likely that skin and yolk-sac MRCs also play a vital role in acid-base regulation; however this remains to be investigated.

An *in vitro* "yolk ball' preparation has recently been developed (Shiraishi *et al.*, 2001; see also kaneko & Hiroi, this volume), where the yolk-sac epithelium is separated from the developing larvae, and then incubated in a balanced salt solution so that the yolk-sac membrane, containing a rich population of MRCs, seals around the yolk completely. This preparation lends itself nicely to physiological studies to investigate the function of these MRCs in an environment that is very similar to *in vivo* conditions. To date, some studies have directly investigated the ionoregulatory but not acid-base regulatory role of MRCs using the "yolk ball" preparation.

Traditionally, the gills have been thought to develop for gas exchange in embryonic and larval fishes to compensate for a reduction in total surface area to volume that occurs with growth. However, this dogma has recently been challenged by Rombough

(1999), who has proposed that the gills more likely develop for ionoregulation and/or acid-base regulation. This hypothesis is based upon numerous observations that MRCs develop on the gill lamellae long before the gills are required for gas exchange. For example, in rainbow trout (*Oncorhynchus mykiss*), MRCs appear on the gill filament a few days prior to hatch and long before the lamellae form. At hatch, the MRC density on the gill filament is already similar to that seen in adult fish (Perry *et al.*, 1992), and the filament possesses 22% of larval MRCs despite the gills accounting for only 7% of total surface area (Rombough, 1999, 2004). Following yolk-sac absorption in rainbow trout, about 75% of the MRCs are found on the gills, while total gill surface area only accounts for 37% of the total body surface. There have been no studies conducted on the relative role of the gills, yolk-sac or skin surface for whole animal ionoregulation or acid-base regulation throughout early development and so this field remains a fruitful area for investigation.

In a SW teleost, the Japanese flounder, exposure to 1 and 5% CO_2 results in elevated gill NKA, and causes an increase in MRC size in larval Japanese sillago (*Sillago japonica*) and red seabream (*Pagrus major*), further supporting the role of the MRC in acid-base regulation during development in SW fishes (Ishimatsu *et al.*, 2004). In FW, exposure to hypercapnia also affects MRC morphology as described above. Thus, MRCs are likely associated with acid-base regulation in larval fishes, at least during exposure to environmental hypercapnia, and changes in MRC characteristics are presumably indicative of processes associated with acid-base compensation during hypercapnia.

ACID-BASE REGULATION IN RELATION TO THE ENVIRONMENT IN EARLY LIFE STAGES

Many studies have investigated the effect of water pH on growth and survival of early life stages of fish (Eddy & Talbot, 1985; Alderdice, 1988), largely in relation to concerns with acid rain and environmental acidification. Most of the negative effects of a low pH environment in juvenile and adult fish are thought to be more related to disruption of ionoregulation, than to acid-base disturbances *per se* (Wood, 1989). In some cases the ionoregulatory disturbances in juvenile and adult fish are associated with a change in gill permeability associated with low environmental pH; in others it is related to a limitation of the V-type H^+-ATPase to generate an electrical gradient to drive Na^+ uptake (Lin & Randall, 1991). In the latter case, compensation for the effect of low water pH may be informative on development of acid-base regulation for fish species that rely on net Na^+/H^+ exchange for acid-base regulation, but in general, these changes are more likely to be associated with adjustments to the ionoregulatory system.

Environmental hypercapnia is a useful tool to investigate the acid-base regulatory capacity as described above, but it is also of applied interest as global CO_2 levels continue to rise. Projected increases in atmospheric CO_2 levels are hypothesised to elevate surface water CO_2 levels and reduce the pH of these waters by up to 0.7 pH units (Caldeira & Wickett, 2003). CO_2 injection into the world's oceans has been proposed as a means through which an elevation in atmospheric CO_2 levels may be minimised despite continued industrial CO_2 production. The consequences of this practice on marine organisms have recently been questioned (Seibel & Walsh, 2001), as organisms will have to tolerate high local CO_2 concentrations at the site of disposal. While elevated CO_2 levels will reduce pH, there is more interest in CO_2 effects because of the high permeability of gaseous CO_2 relative to hydrogen ions. At constant levels of SW acidification, effects on embryos and larvae of marine fish were much greater if acidification was accomplished through hypercapnia than with acid addition (Kikkawa *et al.*, 2004).

Recently, a number of studies have investigated acute CO_2 tolerance in marine fish species at different early developmental stages to address issues related to increases in global CO_2. The approach taken in these studies was more toxicological in design, where 6h and 24 h LC_{50} values (lethal concentration resulting in 50% mortality) could be calculated at different developmental stages. The basis for differences in CO_2 LC_{50} values among species at similar developmental stages, or within species at different developmental stages, are likely related to: i) differences in tolerance to a given reduction in pH, ii) differences in buffer values that alter the magnitude of the acidosis at a given pCO_2, or iii) differences in active pH regulatory capacity up until the point that pH changes affect survival. From the time of fertilisation, through to larval and juvenile stages, there appears to be tremendous variability in CO_2 sensitivity, with 6 h LC_{50}s ranging from 1 to 15% CO_2 (summarised in Ishimatsu *et al.*, 2005) among species. There appear to be interesting patterns of CO_2 sensitivity with development. In two species of fish, Japanese sillago and red seabream, the median lethal CO_2 at 6 h was between 1-3 kPa (1 and 3% CO_2) following cleavage, then increased to 6-7 kPa (6-7% CO_2) at the pre-flexion stage (Kikkawa *et al.*, 2003). Following this, acute tolerance to CO_2 progressively decreased at the post-flexion and juvenile stage (3 kPa, 3% CO_2) to values similar to those of adult fish. The authors of this study attribute the reduction in acute CO_2 tolerance at the post-flexion and juvenile stages to gill development which initiates at the pre-flexion stage in these species. Perhaps the more intriguing question is what is responsible for the greatly elevated CO_2 tolerance at the pre-flexion stage, a time when great developmental changes are occurring, yet CO_2 sensitivity is dramatically reduced.

FUTURE PERSPECTIVES

Acid-base regulation is one of the most tightly controlled physiological processes in the body, due to the degree to which changes in pH can affect protein charge, enzyme function, metabolism and finally, whole animal performance. Most of the studies investigating the mechanisms and pathways involved in acid-base regulation in fish have been conducted in juvenile and adult life stages, and it is surprising how much remains unknown. In early life stages of fish, the ontogeny of acid-base regulation and mechanisms involved at the whole animal, tissue, cellular and protein level, remain virtually unstudied. Understanding the development of any system within an animal, and the influence that changes in the environment have on developmental trajectories is currently topical (Burggren & Warburton, 2005). Early life stages in fish are often considered the most sensitive to environmental perturbations, and the more we understand about the basic physiological mechanisms operating and their thresholds for perturbation the better able we are to make predictions about the organisms tolerance to changing environments, such as global changes in CO_2 levels. Investigations into the ontogeny of acid-base regulation in early life stages of fishes are sorely needed, and the possibilities endless. The following are particular areas in need of examination in relation to the material discussed above:

1) What is the relative role of the different surfaces (yolk-sac epithelium, body surface, gills) in acid-base regulation, and how do they change during development under different environmental conditions?
2) Are the same processes operating in developing fish as in juveniles and adults? For example, are some species, such as salmonids reliant upon net Na^+/H^+ exchange early in development, and then Cl^-/HCO_3^- later in development?
3) What are the predominant acid-base regulatory mechanisms involved at the whole animal, tissue, cellular and protein level in developing fish?
4) Are MRCs the site of net acid/base exchanges between the developing fish and environment? How do changes in MRCs and expression of specific protein transporters correlate with whole animal performance during acid-base disturbances?
5) What are the implications of water hardness and salinity on rates of acid-base regulation in developing fishes?
6) What is the metabolic cost of acid-base regulation during development? Do elevated and/or fluctuating CO_2 levels associated with global warming or culture conditions influence growth and development?
7) Does ammonia accumulation in the whole body and yolk during development act as a temporal buffer system?

Acknowledgements

CJB was supported by an NSERC Discovery grant and thanks Daniel Baker and Louise Kuchel for comments on an earlier version of this manuscript.

References

Alderdice DF (1988) Osmotic and ionic regulation in teleost eggs and larvae. In: Hoar WS, Randall DJ (eds) Fish Physiology, Vol XI, Part A, Academic Press, New York, p 163-251

Brauner CJ, Wood CM (2002a) Ionoregulatory development and the effect of chronic silver exposure on growth, survival, and sublethal indicators of toxicity in early life stages of rainbow trout (*Oncorhynchus mykiss*). J Comp Physiol B172: 153-162

Brauner CJ, Wood CM (2002b) Effect of long-term silver exposure on survival and ionoregulatory development in rainbow trout (*Oncorhynchus mykiss*) embryos and larvae, in the presence and absence of added dissolved organic matter. Comp Biochem Physiol 133C: 161-173

Brauner CJ, Wilson J, Kamunde C, Wood CM (2003) Water chloride provides partial protection during chronic exposure to waterborne silver in rainbow trout (*Oncorhynchus mykiss*) embryos and larvae. Physiol Biochem Zool 76: 803-815

Brauner CJ, Wang T, Wang Y, Richards JG, Gonzalez RJ, Bernier NJ, Xi W, Patrick M, Val AL (2004) Limited extracellular but complete intracellular acid-base regulation during short-term environmental hypercapnia in the armoured catfish, *Liposarcus pardalis*. J Exp Biol 207: 3381-3390

Burggren W, Warburton S (2005) Comparative developmental physiology: An interdisciplinary convergence. Ann Rev Physiol 67: 203-223

Caldeira K, Wickett ME (2003) Anthropogenic carbon and ocean pH. Nature 425: 365

Cameron JN, Iwama GK (1989) Compromises between ionic regulation and acid-base regulation in aquatic animals. Can J Zool 67: 3078-3084

Chang I, Wei Y, Chou F, Hwang P (2003) Stimulation of Cl^- uptake and morphological changes in gill mitochondria-rich cells in freshwater tilapia (*Oreochromis mossambicus*). Physiol Biochem Zool 76: 544-552

Claiborne J, Heisler N (1986) Acid-base regulation and ion transfers in the carp (*Cyprinus carpio*): pH compensation during graded long- and short-term environmental hypercapnia, and the effect of bicarbonate infusion. J Exp Biol 126: 41-61

Claiborne J, Edwards S, Morrison-Shetlar A (2002) Acid-base regulation in fishes: cellular and molecular mechanisms. J Exp Zool 293: 302-319

Crocker CE, Cech JJ (1996) The effects of hypercapnia on the growth of juvenile white sturgeon, *Acipenser transmontanus*. Aquaculture 147: 293-299

Dejours P (1988) Respiration in Water and Air: Adaptations, Regulation, Evolution. Elsevier, Amsterdam, p179

Eddy FB, Talbot C (1985) Sodium balance in eggs and dechorionated embryos of the Atlantic salmon (*Salmo salar*) L. exposed to zinc, alumunium and acid waters. Comp Biochem Physiol C 81: 259-266

Evans DH, Piermarini PM, Choe KP (2005) The multifunctional fish gill: Dominant site of gas exchange, osmoregulation, acid-base regulation, and excretion of nitrogenous waste. Physiol Rev 85: 97-177

Fivelstad S, Haavik H, Løvik G, Olsen AB (1998) Sublethal effects and safe levels of carbon dioxide in seawater for Atlantic salmon postsmolts (*Salmo salar* L.): ion regulation and growth. Aquaculture 160: 305-316

Foss A, Rosnes BA, Øiestad V (2003) Graded environmental hypercapnia in juvenile spotted wolffish (*Anarhichas minor* Olafsen): Effects on growth, food conversion efficiency and nephrocalcinosis. Aquaculture 220: 607-617

Galvez F, Reid SD, Hawkings G, Goss GG (2002) Isolation and characterization of mitochondria-rich cell types from the gill of freshwater rainbow trout. Am J Physiol 282: R658-668

Goss GG, Perry SF, Wood CM, Laurent P (1992) Mechanisms of ion and acid-base regulation at the gills of freshwater fish. J Exp Zool 263: 143-159

Goss G, Perry S, Laurent P (1995) Ultrastructural and morphometric studies on ion and acid-base transport processes in freshwater fish. In: Wood CM, Shuttleworth TJ (eds) Fish Physiology: Cellular and Molecular Approaches to Fish Ionic Regulation, Academic Press, New York, p 257-284

Hayashi M, Kita J, Ishimatsu A (2004) Acid-base responses to lethal aquatic hypercapnia in three marine fishes. Mar Biol 144: 153-160

Heisler N (1982) Intracellular and extracellular acid-base regulation in the tropical fresh-water teleost fish *Synbranchus marmoratus* in response to the transition from water breathing to air breathing. J Exp Biol 99: 9-28

Heisler N (1984) Acid-base regulation in fishes. In: Hoar WS, Randall DJ (eds) Fish Physiology, Vol X, Academic Press, New York, p 315-401

Heisler N (1993) Acid-base regulation. In: Evans DH (ed) The Physiology of Fishes, CRC Press, Boca Raton, p 343-378

Hirata T, Kaneko T, Ono T, Nakazato T, Furukawa N, Hasegawa S, Wakabayashi S, Shigekawa M, Chang M-H, Romero MF, Hirose S (2003) Mechanism of acid adaptation of a fish living in a pH 3.5 lake. Am J Physiol 284: R1199-1212

Hwang PP (1989) Distribution of chloride cells in teleost larvae. J Morph Suppl 200: 1-8

Hwang PP, Tsai YN, Tung YC (1994) Calcium balance in embryos and larvae of the freshwater-adapted teleost, *Oreochromis mossambicus*. Fish Physiol Biochem 13: 325-333

Hwang P, Lee TH, Weng CF, Fang MJ, Cho GY (1999) Presence of Na-K-ATPase in mitochondria-rich cells in the yolk-sac epithelium of larvae of the teleost *Oreochromis mossambicus*. Physiol Zool 72: 138-144

Hyde DA, Perry SF (1989) Differential approaches to blood acid-base regulation during exposure to prolonged hypercapnia in two freshwater teleosts the rainbow trout (*Salmo gairdneri*) and the American eel (*Anguilla rostrata*). Physiol Zool 62: 1164-1186

Ishimatsu A, Kikkawa T, Hayashi M, Lee K-S, Kita J (2004) Effects of CO_2 on marine fish: larvae and adults. J Oceanogr 60: 731-741

Ishimatsu A, Hayashi M, Lee KS, Kikkawa T, Kita J (2005) Physiological effects on fishes in a high-CO_2 world. J Geophys Res 110: C09S09.

Kaneko T, Shiraishi K, Katoh F, Hasegawa S, Hiroi J (2002) Chloride cells during early life stages of fish and their functional differentiation. Fish Sci 68: 1-9

Kikkawa T, Ishimatsu A, Kita J (2003) Acute CO_2 tolerance during the early developmental stages of four marine teleosts. Environ Toxicol 18: 375-382

Kikkawa T, Kita J, Ishimatsu A (2004) Comparison of the lethal effect of CO_2 and acidification on red sea bream (*Pagrus major*) during the early developmental stages. Mar Pollut Bulletin 48: 108-110

Larsen BK, Jensen FB (1997) Influence of ionic composition on acid-base regulation in rainbow trout (*Oncorhynchus mykiss*) exposed to environmental hypercapnia. Fish Physiol Biochem 16: 157-170

Lin H, Randall DJ (1991) Evidence for the presence of an electrogenic proton pump on the trout gill epithelium. J Exp Biol 161: 119-134

Misiaszek CM (1996) The development of ion regulation in larval rainbow trout (*Oncorhynchus mykiss*). MSc thesis, Department of Biology, McMaster University, Hamilton, Ontario, Canada.

Marshall WS (2002) Na^+, Cl^-, Ca^{2+} and Zn^{2+} transport by fish gills: retrospective review and prospective synthesis. J Exp Zool 293: 264-283

Marshall WS, Bryson SE (1998) Transport mechanisms of seawater teleost chloride cells: An inclusive model of a multifunctional cell. Comp Biochem Physiol 119A: 97-106

Molich A, Heisler N (2005) Determination of pH by microfluorometry: Intracellular and interstitial pH regulation in developing early-stage fish embryos (*Danio rerio*). J Exp Biol 208: 4137-4149

Perry SF (1997) The chloride cell: Structure and function in the gills of freshwater fishes. Ann Rev Physiol 59: 325-347

Perry SF, Goss GG, Laurent P (1992) The interrelationships between gill chloride cell morphology and ionic uptake in four freshwater teleosts. Can J Zool 70: 1775-1786

Piermarini PM, Verlander JW, Royaux IE, Evans DH (2002) Pendrin immunoreactivity in the gill epithelium of a euryhaline elasmobranch. Am J Physiol 283: R983-992

Pisam M, Caroff A, Rambourg A (1987) Two types of chloride cells in the gill epithelium of a freshwater-adapted euryhaline fish: *Lebistes reticulatus*; their modifications during adaptation to saltwater. Am J Anat 179: 40-50

Putnam R, Roos A (1997) Intracellular pH. In: Hoffman J, Jamieson J (eds) Handbook of Physiology, Vol 14, Cell Physiology, Oxford University Press, Oxford, p 389-440

Rombough PJ (1999) The gill of larvae. Is it primarily a respiratory or an ionoregulatory structure? J Fish Biol 55: 186-204

Rombough PJ (2004) Structural and functional development of the larval gill. In: Govoni JJ (ed), The Development of Form and Function in Fishes and the Question of Larval Adaptation, American Fisheries Society Symposium 40, p 47-83

Rudy PP, Potts WTW (1969) Sodium balance in the eggs of the Atlantic salmon (*Salmo salar*). J Exp Biol 50: 239-246

Sanchez AP, Giusti H, Bassi M, Glass ML (2005) Acid-Base regulation in the South American lungfish *Lepidosiren paradoxa*: Effects of prolonged hypercarbia on blood gases and pulmonary ventilation. Physiol Biochem Zool 78: 908-915

Schmidt-Nielsen K (1975) Scaling in biology the consequences of size. J Exp Zool 194: 287-307

Seibel BA, Walsh PJ (2001) Potential impacts of CO_2 injection on deep-sea biota. Science 294: 319-320

Shiraishi K, Hiroi J, Kaneko T, Matsuda M, Hirano T, Mori T (2001) *In vitro* effects of environmental salinity and cortisol on chloride cell differentiation in embryos of Mozambique tilapia, *Oreochromis mossambicus*, measured using a newly developed "yolk-ball" incubation system. J Exp Biol 204: 1883-1888

Toews DP, Holeton GF, Heisler N (1983) Regulation of the acid-base status during environmental hypercapnia in the marine teleost fish *Conger conger*. J Exp Biol 107: 9-20

Ultsch GR (1996) Gas exchange, hypercarbia and acid-base balance, paleoecology, and the evolutionary transition from water-breathing to air-breathing among vertebrates Palaeogeogr Palaeocl Palaeoecol 123: 1-27

Varsamos S, Nebel C, Charmantier G (2005) Ontogeny of osmoregulation in postembryonic fish: A review. Comp Biochem Physiol 141A: 401-429

Wood CM (1989) The physiological problems of fish in acid waters. In: Morris R, Taylor EW, Brown DJ, Brown JA (eds), Acid-base Regulation in Fishes: Biological Responses and Limitations. Cambridge University Press, Cambridge, p 125-152

Wright PA, Fyhn HJ (2001) Ontogeny of nitrogen metabolism and excretion. In: Wright PA, Anderson PM (eds) Nitrogen Excretion. Fish Physiology, Vol 20, Academic Press, San Diego, p 149-200

PART 3

Nutrition and Energy

CHAPTER

8

Digestion

Ivar Rønnestad[1,*] and Sofia Morais[2]

INTRODUCTION

The major function of the digestive system is to supply dietary nutrients to the body tissues. The efficiency of this process is one of the key factors that determines the supply of substrates to the rapidly growing larval tissues, particularly during the first critical stages when the larvae start exogenous feeding. The larval intestine appears very simple, in some species being a straight tube but, when given the proper feedstuff, it has a remarkable processing capacity and efficiency. This chapter will deal mainly with teleosts. The digestive physiology of larval stages of these species has received much attention in recent years due to the large interest in mariculture of these species world-wide, combined with problems of survival and quality that are apparently related to the performance of the digestive system during ontogeny. Most of the knowledge accumulated regarding the larval digestive processes over the last decades has been related to lipid and fatty acids (FA), protein and amino acids (AA), while little is still known regarding their handling of carbohydrates. This chapter focuses mainly on the physiology of digestion and intestinal absorption but some basic points will be made on the functional development and gross anatomical differentiation of the digestive tract,

Authors' addresses: [1]Department of Biology, University of Bergen, Bergen High Technology Centre, N-5008 Bergen, Norway.

[2]Comparative Molecular Endocrinology Group, Marine Science Centre (CCMAR), Universidade do Algarve, FERN, Campus de Gambelas, 8005-139 Faro, Portugal.

**Corresponding author:* E-mail: ivar.ronnestad@bio.uib.no

including associated organs such as the liver, bile and exocrine pancreas, in order to better understand the larval digestive physiological processes.

DEVELOPMENTAL PATTERNS – MORPHOLOGY AND HISTOLOGY

The Digestive Tract

The ontogeny of the digestive system varies between species. Balon (1975) distinguishes *precocial* species that develop directly from embryos to juveniles with a transitional stage (e.g. salmonids) from *altricial* species that develop from a distinct larval stage with temporal structures that change during metamorphosis, e.g. Japanese flounder, (*Paralichthys olivaceus*), Fig. 1, and ayu (*Plecoglossus altivelis*), Fig. 2, or gilthead seabream, (*Sparus aurata*). A typical feature of altricial species with a distinct larval stage is that they lack a functional gastric stomach. Other altricial species, such as zebrafish (*Danio rerio*), go through a larval stage but never develop a stomach. Although there are considerable differences in terms of the timing of the developmental events relative to each other (Fig. 3), the basic ontogenetic patterns appear to be similar (Falk-Petersen, 2005). A thorough description of the differentiation and growth of the intestine is outside the scope of this chapter but readers can consult recent work on zebrafish by Wallace & Pack (2003) and Wallace *et al.* (2005).

Intestine

In general, the alimentary tract of first-feeding larvae is a simple, relatively undifferentiated tube with only one sphincter (intestinal-rectal valve) before the anal opening. Some species have a rotated gut at the onset of exogenous feeding (Fig. 1), while other species have a straight gut (Fig. 2). As development proceeds, the subsequent stages of ontogenetic development of the intestine are mainly quantitative changes, such as an increase in length and mucosal surface area, as well as an intensification of the activities of brush border enzymes (Segner *et al.*, 1994). The area will increase as a combined result of increased length (sometimes with increased number of rotated intestinal loops) and an increase in the epithelial folding (Fig. 4). The folding of the intestinal wall appears in some species as a regular pattern (e.g. ayu and Atlantic herring (*Clupea harengus*); Fig. 4B) while in other species (e.g. Atlantic halibut (*Hippoglossus hippoglossus*), and zebrafish) the folds appear more irregular (Kamisaka, 2005; Wallace *et al.*, 2005; Fig. 4C). The intestinal tube consists of a simple epithelium surrounded by connective tissue, smooth muscle and enteric nerves (Wallace *et al.*, 2005; Holmberg *et al.*, this volume).

There are different cell types that can be identified in the epithelial layer (Fig. 5, A-C). These include enterocytes (the most numerous of all cell types, responsible for nutrient absorption), single enteroendocrine cells (produce and secrete peptide hormones) and goblet cells (mucus producing cells). In addition, there also appears to be special enterocytes identified as "antigen presenting cells" (Wallace *et al.*, 2005).

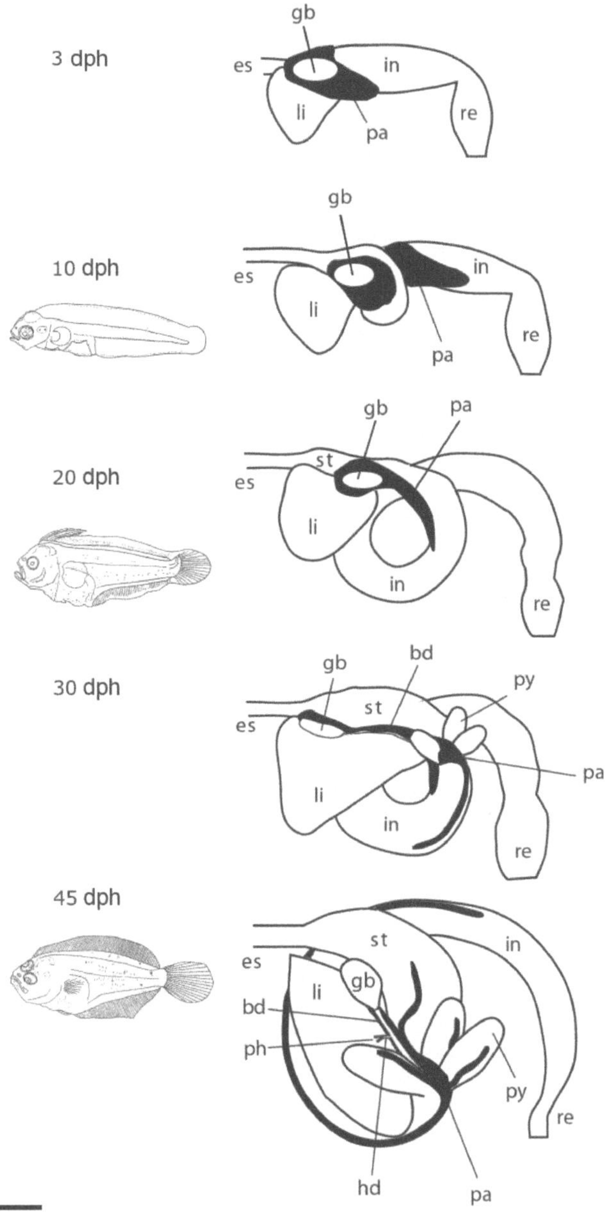

Fig. 1 Development of the digestive system in Japanese flounder *(Paralichthys olivaceous)* a species with a rotated gut at the onset of exogenous feeding, at different ages (days post hatch, dph). bd: bile duct; es: oesophagus; gb: gall bladder; hd: hepatic duct; in: intestine; li: liver; pa: pancreas; ph: porta hepatis; py: pyloric caeca; re: rectum: st: stomach. After Kurokawa *et al.* (1996). Reprinted with permission from Elsevier. Drawings of fish by Dr. O. Fukuhara (pers. comm.).

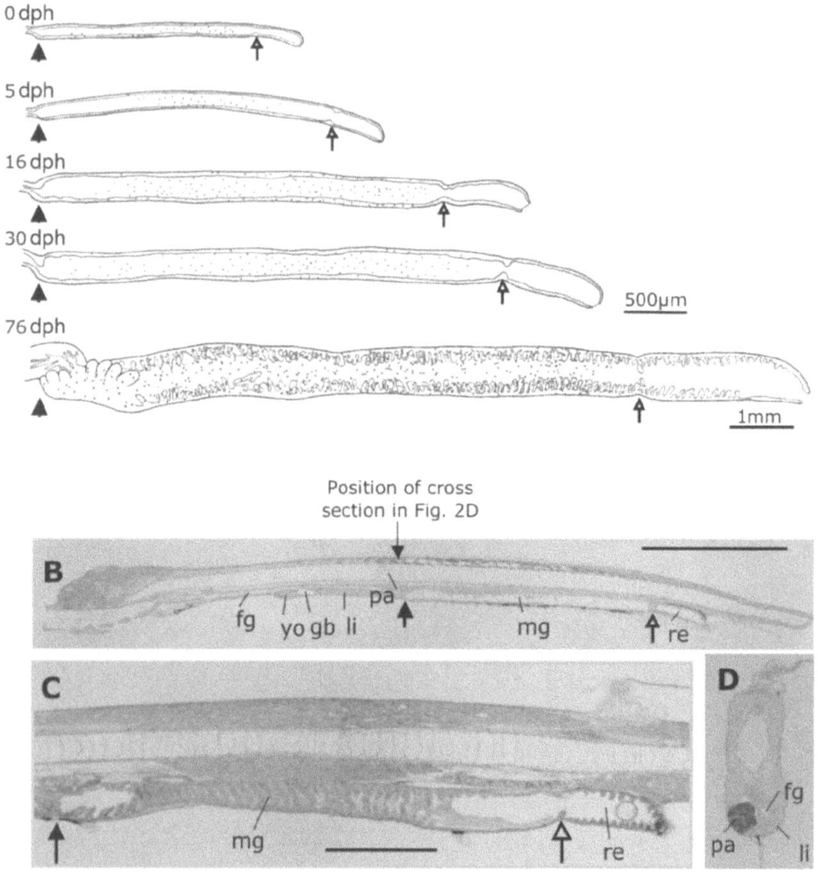

Fig. 2 Development of the digestive system in ayu *(Plecoglossus altivelis)* a species with a straight gut at the onset of exogenous feeding. (A) drawings of midgut and rectum with individual CCK-immunoreactive cells localised (dots). Plates: Photomicrographs of gut in ayu larvae. (B) gross morphology of 0 days post hatch (dph) larva, (C) 30 dph larva from the foregut to the rectum, and (D) cross section of 0 dph larva with immunoreaction to anti-trypsinogen. Solid arrow identifies the pyloric sphincter and open arrow identifies the intestinal-rectal sphincter. Scale bars in plates indicate 1 mm. fg: foregut; gb: gall bladder; li: liver; mg: midgut; pa: pancreas; re: rectum; yo: yolk. After Kamisaka *et al.* (2003) and Kamisaka (2005). Reprinted with permission from Elsevier and the author, respectively.

Goblet cells can be identified in all intestinal regions (Wallace *et al.*, 2005), while the distribution of the enteroendocrine cells varies between larval types. For instance, Kamisaka *et al.* (2005) have described clear differences in the distribution pattern of cholecystokinin (CCK)-producing cells and in the timing of their appearance in the gut among several studied teleost species. In species having a straight gut, strongly-stained CCK-producing cells were present from hatching and were widely distributed

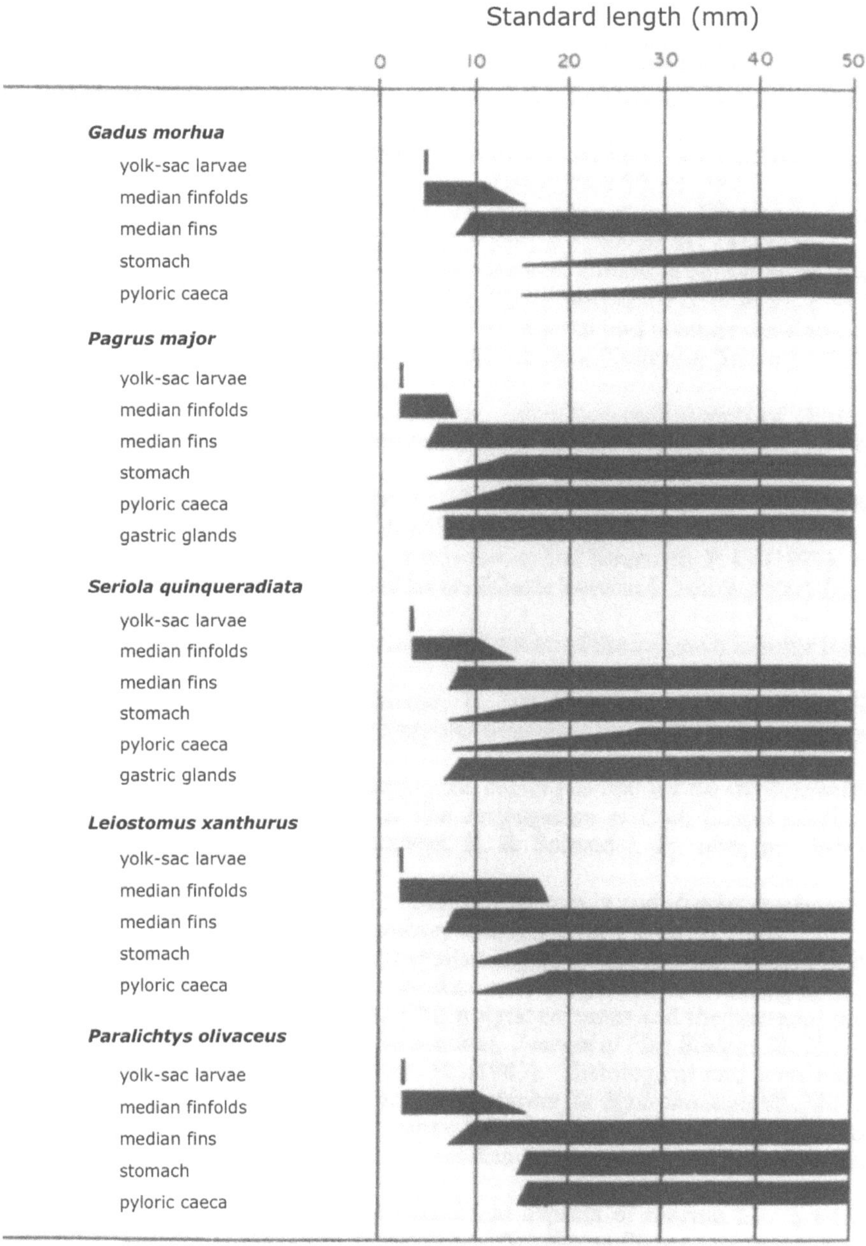

Fig. 3 Development of the pyloric caeca and stomach related to the development of external morphology in some teleosts. After Pedersen & Falk-Petersen (1992). Reprinted with permission from Blackwell Publishing.

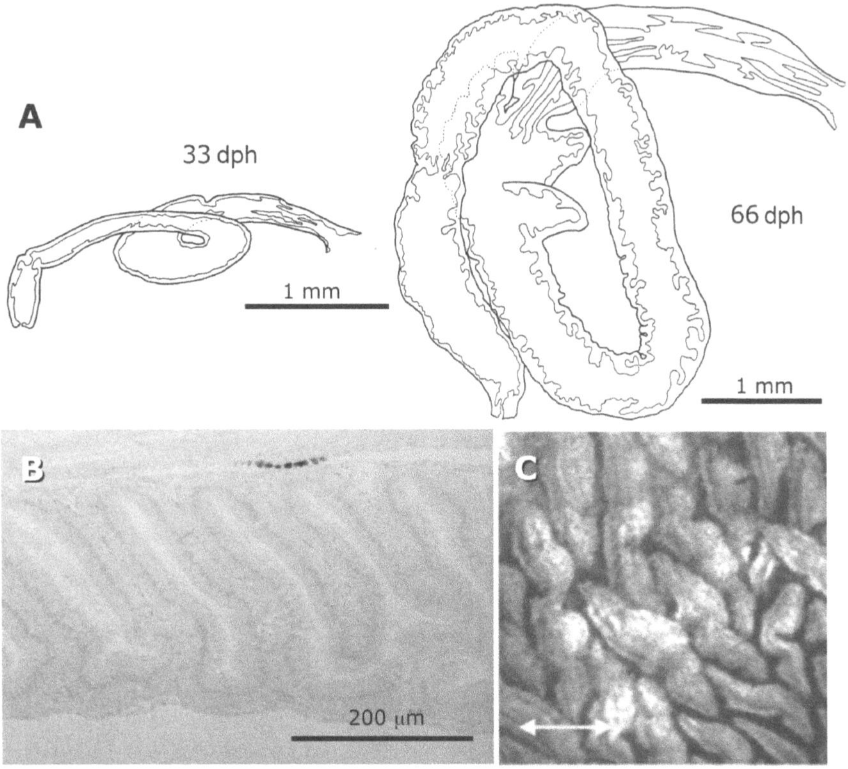

Fig. 4 Development of intestinal surface area and mucusal folding. (A) Atlantic halibut (*Hippoglossus hippoglossus*) at 33 or 66 days post hatching (dph) (After Kamisaka *et al.* (2001). Reprinted with permission from Elsevier). (B) Regular folding in the midgut of ayu (*Plecoglossus altivelis*) larvae (age: 30 dph; after Kamisaka *et al.* (2001). Reprinted with permission from Elsevier). (C) Folding in the anterior intestine of adult zebrafish showing irregular shaped folds. The double arrow defines anterior–posterior axis. After Wallace *et al.* (2005). Reprinted with permission from Elsevier.

all over the midgut (Fig. 2). On the other hand, in species with a rotated gut, CCK-producing cells were detected later in the larval stage and were distributed in the anterior part of the midgut (Kamisaka *et al.*, 2005).

The ultrastructural characteristics of the enterocytes of pre-feeding fera (*Coregonus fera*), turbot (*Scophthalmus maximus*), European seabass (*Dicentrarchus labrax*) and gilthead seabream larvae were described as being essentially the same as in adult fish (Loewe & Eckman, 1988; Segner *et al.*, 1994; Diaz *et al.*, 1997; Calzada *et al.*, 1998; Elbal *et al.*, 2004) and the existence of functional lipid absorption structures, such as well developed endoplasmatic reticulum (ER) and Golgi apparatus at the time of first-feeding has been noted (Deplano *et al.*, 1991; Segner *et al.*, 1994; Diaz *et al.*, 1997). Therefore, Segner *et al.* (1994) considered that the enterocytes are not only cytologically differentiated but also physiologically functional at first-feeding. Nevertheless, Deplano *et al.* (1991) noted a lower development of the ER and Golgi system, compared to later stages, and considered that the enterocytes are still not completely developed in European seabass at the start of exogenous feeding.

Histological studies indicate regional differences along the gut epithelial cells which also include the shape of the enterocyte brush border microvilli (Fig. 5, D-G). In the foregut of Atlantic cod (*Gadus morhua*), the villi are separate thin and elongated finger-like structures, while in the hindgut the villi appear to have an irregular pattern with extensive basal webbing (Kjørsvik *et al.*, 1991a). The villi significantly increase the surface area available for absorption of digested nutrients into the enterocytes.

Stomach

Altricial species develop a functional stomach during metamorphosis. In pleuronectiform species, metamorphosis is an easily recognisable event by the transformation of the external morphology (see Power *et al.*, this volume). However, in other species metamorphosis is less clear. For instance, in Atlantic cod the development of the stomach, gastric glands and also of pyloric caeca occurs gradually and is not complete until the fish is >4 cm, which is long after the median finfold is resorbed and the median fins have differentiated at 1.2 cm (Kjørsvik *et al.*, 1991a; Fig. 3). One of the primary roles of the fully functional stomach is the breakdown of large food particles, which is achieved by mechanical grinding in combination with acid denaturation and enzyme degradation. Together, these actions allow the regulated release of partly digested food (chyme) into the midgut, in order to maintain an optimal ratio of food to digestive enzymes in the midgut (Jobling & Hjelmeland, 1992).

The stomach develops from a transitional region between the oesophagus and the intestine (Fig. 1). In larval stages that have a rotated gut it can first be recognised as a pouch-like bulb in the foregut with functional oesophageal and pyloric valves before the onset of gastric secretion (Fig. 6; Rønnestad *et al.*, 2000a). This suggests that the presumptive stomach may serve as a short-term food store.

The ontogeny of gastric HCl production and secretion has been established by histological observations and pH measurements, and by reverse transcription-polymerase chain reaction (RT-PCR) (Tanaka, 1973; Murray *et al.*, 2006). All histological data demonstrates that marine fish larvae do not possess HCl-producing cells (gastric glandular acini) at the first-feeding stage. Measurements of the pH in the presumptive stomach have been performed by a variety of methods, including microelectrodes and indicator strips using dissected gut sections (Mahr *et al.*, 1983; Buddington, 1985; Walford & Lam, 1993; Hoehne-Reitan *et al.*, 2001a), and immersion or tube-feeding of pH indicators *in vivo* (Verreth *et al.*, 1992; Bengtson *et al.*, 1993; Rust *et al.*, 1993; Rust, 1995; Rønnestad *et al.*, 2000a). None of these studies could demonstrate an acid pH at first-feeding. This is also supported by an RT-PCR study on Winter flounder (*Pseudopleuronectes americanus*), which demonstrated that the expression of proton pumping genes occurred at 20 days post hatch (dph) and coincided with the appearance

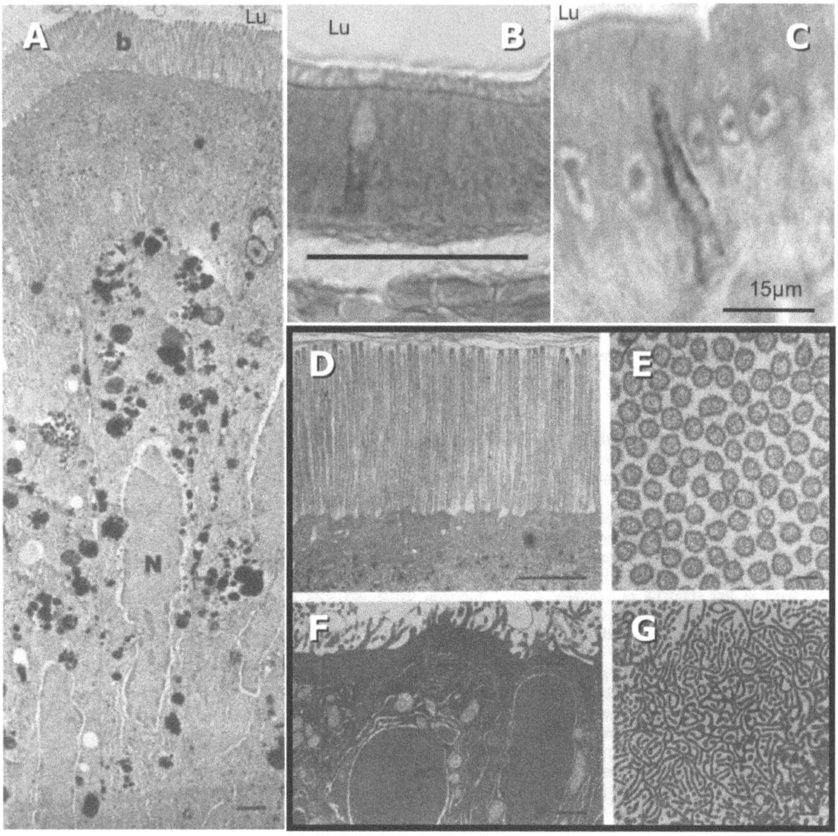

Fig. 5 (A-C) Cell types in intestinal epithelium. (A) enterocyte in 17 day-old Atlantic cod (*Gadus morhua*) larvae; note large electron dense vesicles; b: brush border, N: nucleous (from Kjørsvik *et al.,* 1991a). (B) goblet cell from Atlantic herring (*Clupea harengus*) larvae and (C) enteroendocrine cell, in this case a CCK-producing cell from Atlantic halibut (*Hippoglossus hippoglossus*) larvae. After Kamisaka (2005). Reprinted with permission from the author. Lu: gut lumen. (D-G) Details of microvilli from Atlantic cod (D,E) foregut villi, 11 day-old, longitudinal (D) and transversal (E) sections, and (F,G) hindgut villi with irregular pattern and basal webbing, longitudinal (F) and transversal (G) sections, 14 day-old. After Kjørsvik *et al.* (1991a). Reprinted with permission from Blackwell Publishing. Scale bars = 15 µm.

of gastric glands (Douglas *et al.*, 1999). A gradual fall in luminal pH, starting at the onset of metamorphosis (Fig. 6C), suggests a low but increasing capacity for HCl production and secretion into the stomach lumen. At the end of metamorphosis, the pH in the stomach of Asian seabass (*Lates calcarifer*), Japanese flounder (Fig. 6) and turbot, is lower than 4 (Walford & Lam, 1993; Rønnestad *et al.*, 2000a; Hoehne-Reitan *et al.*, 2001a). Studies on the ontogeny of the gastric glandular acini (Fig. 6D) support this notion.

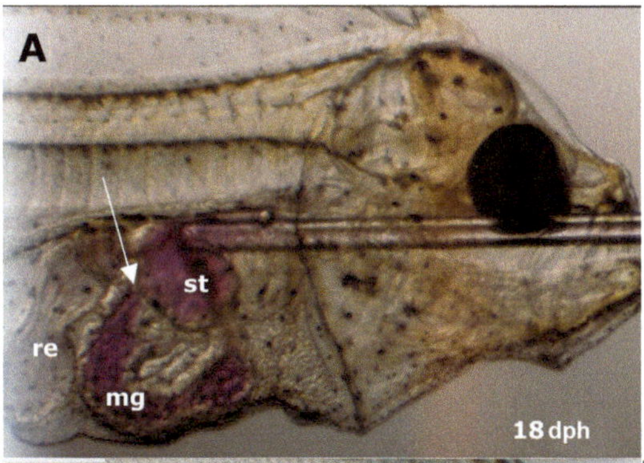

C

dph	Stomach	Midgut
18	8.0 - 8.5	8.0 - 8.5
24	6.5 - 8.0	8.0 - 8.5
26	6.5 - 7.0	8.0 - 8.5
29	6.0 - 7.0	8.0 - 8.5
31	< 4.0	8.0 - 8.5
33	< 4.0	8.0 - 8.5

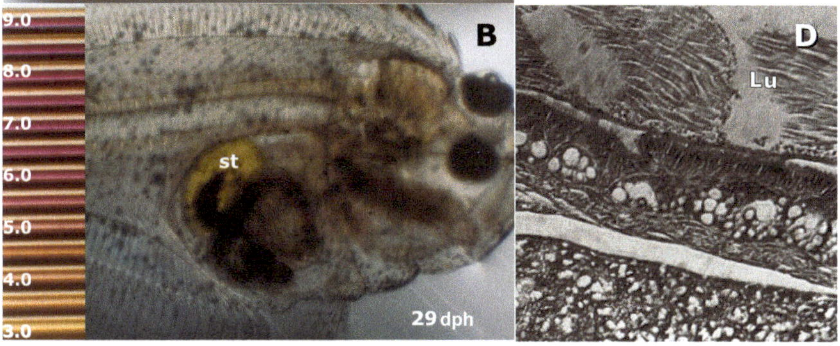

Fig. 6 (A-B): Ontogeny of a functional stomach (st) in Japanese flounder (*Paralichthys olivaceus*), at 18 (A) and 29 (B) days post hatch (dph). (A) Note that the functional sphincter (white arrow) in this species develops faster than the acid secretion. After Rønnestad *et al.* (2000a). (C) pH in the gut of developing Japanese flounder presented as a range based on visual observations of colour changes of two tube-fed pH mixtures (colour corresponding to each pH, from 3.0 to 9.0, is shown in the left corner of B). (D) Typical appearance of a functional gastric gland in Atlantic cod (*Gadus morhua*) larvae. After Pedersen & Falk-Petersen (1992). Reprinted with permission from Blackwell Publishing). Lu: lumen; mg: midgut; re: rectum; st: stomach.

Gastric pepsin-like activity seems to be very low at hatching (Cahu & Zambonino Infante, 2001). In Winter flounder, sequential expressions of two forms of pepsinogens have been shown (Douglas *et al.*, 1999). Pepsinogen IIa expression occurred one week before the onset of metamorphosis, while the expression of pepsinogen IIb coincided with expression of the proton pump, presence of gastric glands and the initiation of pepsin activity (Douglas *et al.*, 1999). *In situ* hybridisation demonstrated that gastric gland cells contain mRNA transcripts of pepsinogen IIa and IIb as well as proton pump genes (Gawlicka *et al.*, 2001). This suggests synchronised activation of the mechanisms responsible for the production and secretion of pepsin and HCl in gland cells, which would ensure HCl-mediated pepsin activation from its zymogen form, as well as establishing the optimum pH range for pepsin activity. However, the physiological role of an earlier expression of pepsinogen IIa (Douglas *et al.*, 1999) remains to be clarified.

It is also interesting to note that Gawlicka *et al.* (2001) were able to demonstrate the presence of mRNA for alpha-proton pump in mucus cells, suggesting that cells other than gastric glands are capable of acid production.

Pyloric caeca

Pyloric caeca develop during the transition from larvae to juveniles and originate in the anterior part of the intestine. The pyloric caeca are blind-ended appendages that serve to increase intestinal absorptive surface area without increasing the length or thickness of the digestive tract itself (Buddington *et al.*, 1997) (Fig. 7). The number of caeca varies from a few to >100 and they initially grow proportionally faster in length than the rest of the intestine, until they reach a constant ratio to the length of the intestine in the juvenile stage (Pedersen & Falk-Petersen, 1992). They often develop a folding of epithelial lining which contributes significantly to increase the surface area. The pyloric caeca are filled and emptied through the same opening. A study on Atlantic halibut juveniles (Rønnestad *et al.*, 2000b) demonstrated retrograde contractions in the anterior midgut that enabled chyme to flow into the caeca (Fig. 7C). Functionally, these retrograde contractions also appear to be involved in mixing the chyme with digestive secretions in the region of the pyloric caeca. Secretions in this region of the gut include enzymes from the exocrine pancreas, HCO_3^-, and bile. It has been suggested (Rønnestad *et al.*, 2000b) that peristaltic movements mix the chyme with enzymes before it enters the pyloric caeca. This supports the findings that the enzyme activities in the caeca are similar to that of the midgut (Morrison, 1987).

Based on the observations in Atlantic halibut juveniles, emptying of the pyloric caeca appears to involve a powerful contraction of the circular muscle along the whole length of a caecum, starting at the distal end. Furthermore, the length of the caecum remains more or less constant, which indicates that the longitudinal muscles do not play a significant role during evacuation (Rønnestad *et al.*, 2000b).

Midgut

The midgut plays a key role in the digestion and absorption of ingested feeds in larval fish. In most species, the main degradation and absorption takes place in the midgut segment and it is often not possible to recognise the feed when the chyme is passed into the hindgut/rectum.

The midgut appears to be alkaline throughout development (Walford & Lam, 1993; Rønnestad *et al.*, 2000a). In pre-gastric marine fish larvae, this alkaline condition has two likely sources: it may be due to drinking seawater (pH ~8.2), which occurs from hatching (Mangor Jensen & Adoff, 1987; Tytler *et al.*, 1993), to pancreatic or bile secretions of alkaline fluid (rich in HCO_3^-) into the gut lumen, or to a combination of both processes. At present, no firm conclusions can be drawn regarding the stage at which the alkaline secretion starts, since the analysed pH values in the presumptive stomach, midgut, and seawater are all within similar alkaline range from first-feeding (Rønnestad *et al.*, 2000a; Fig. 6A). Immunohistochemical analysis for trypsin-like

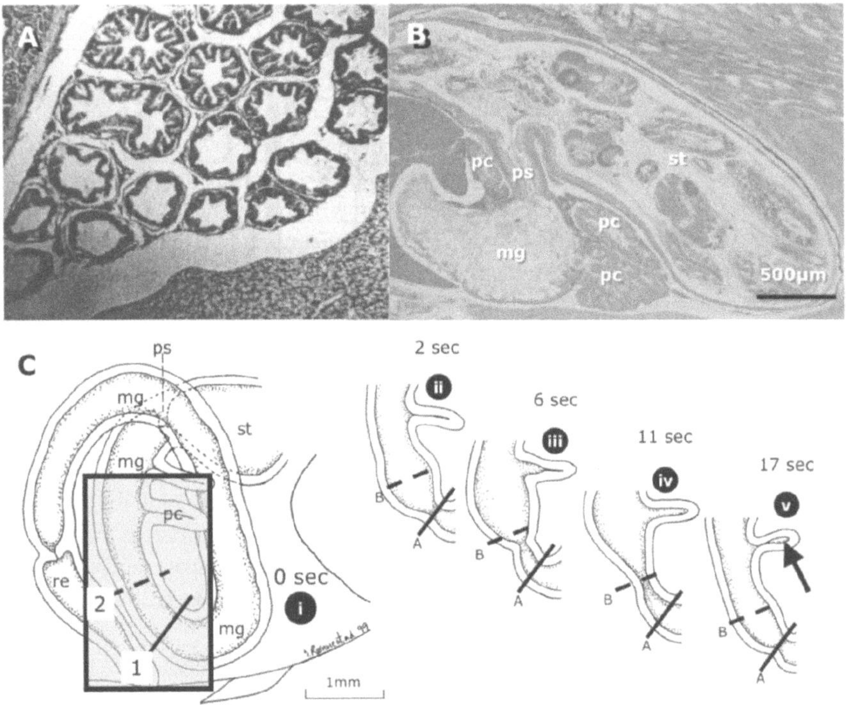

Fig. 7 Pyloric caeca in post larval fish. (A) Atlantic cod (*Gadus morhua*), transverse section. Note the extensive folding (After Pedersen & Falk-Petersen (1992). Reprinted with permission from Blackwell Publishing). (B) Bluefin tuna *(Thunnus thynnus)*. Pyloric caeca (pc) are an extension of the anterior midgut volume and area (After Kamisaka *et al.* (2002). Reprinted with permission from Taylor & Francis AS). (C) Mechanism for filling the pyloric caeca in post-larval Atlantic halibut (*Hippoglossus hippoglossus*). The retrograde peristaltic wave starts caudal to the pyloric caeca (line "1") and proceeds cranially to become extinct in the region of the pyloric caeca (line "2"). The propulsion of one single wave is shown in i-v. The shaded rectangle shows the area which is redrawn in ii-v (After Rønnestad *et al.* (2000b). Reprinted with permission from Blackwell Publishing). mg: midgut; pc: pyloric caeca; ps: pyloric sphincter; re: rectum; st: stomach.

enzymes has demonstrated secretion into the gut lumen from first-feeding (Kurokawa & Suzuki, 1996) and it is likely that the production and secretion of pancreatic HCO_3^- occurs concurrently. It is clear that in post-gastric marine fish larvae the high pH of the midgut is closely regulated. As an indication of this, Rønnestad *et al.* (2000a) demonstrated an immediate rise in pH as indicator solution passed from the stomach to the midgut of Japanese flounder (Fig. 6B).

Hindgut/rectum

The hindgut is a separate compartment from the onset of first-feeding (Figs. 1 and 2). In adult vertebrates the hindgut is mainly associated with final adjustments in the water and ion composition of the faeces (Guyton & Hall, 1996). In larval stages of fish very little is known about these processes. Histological studies of the epithelial lining show frequent pinocytotic invaginations and intracellular vesicles (Kurokawa *et al.*, 1996). *In vivo* studies have also confirmed endocytotic absorption of macromolecules such as horseradish protein from the gut lumen and also an active intracellular degradation of these proteins, albeit at a slow rate (Watanabe, 1984). Based on such observations, this route of absorption of dietary proteins was originally proposed to have an important nutritional role in larvae before the differentiation of a functional gastric stomach (Watanabe, 1984), although Wallace *et al.* (2005) discussed this phenomenon in an immunological context, similar to that of mammals and adult fish (McLean *et al.*, 1999).

Accessory Digestive Organs

Pancreas

The pancreas serves important exocrine as well as endocrine functions in the digestion and post-prandial metabolism of ingested feeds. The exocrine pancreas is the main source for the release of a variety of digestive enzymes (proteases, lipases and amylases) into the intestinal lumen while its endocrine role is the release of key metabolic hormones, such as insulin and glucagon, into the plasma.

The exocrine pancreas is an acinar gland and based on its morphology it appears to be functional prior to the onset of exogenous feeding (Govoni *et al.*, 1986; Boulhic & Gabaudan, 1992; Segner *et al.*, 1994; Bisbal & Bengtson, 1995; Kurokawa & Suzuki, 1996; Ribeiro *et al.*, 1999a). The secretory cells produce characteristic zymogen granulae that contain a battery of proenzymes. In addition to this, the excretory duct cells produce and release an alkaline fluid that participates in the adjustments of luminal pH. In contrast to adults, the exocrine pancreas in teleost larvae is a compact and distinct organ – *pancreas compactum* (Fig. 1). The adult *pancreas diffusum* is acquired during the late larval and early juvenile phases (Fig. 1).

There appears to be many varieties of excretory ducts delivering the pancreatic juice; from one major duct leading directly into the intestine, to a large number of smaller ducts and also a duct that joins with the bile duct before entering the intestine (Hoehne-Reitan & Kjørsvik, 2004).

Liver

In terms of digestion, the main role of the liver is the production of bile. Bile is composed of water, biliary lipids, bile salts, ions, and pigments, and serves an important role in the digestion (mostly in the emulsification) of lipids. Studies have shown that the hepatocytes produce biliary lipids, possibly from plasma lipoproteins, before the onset of exogenous feeding. These lipids, which appear as very low density lipoprotein (VLDL)-like particles in the ER and Golgi, are excreted into the bile canaliculi by exocytosis (Hoehne-Reitan & Kjørsvik, 2004).

Together with the pancreas, the liver and gall bladder (see below) are differentiated at hatching (Fig. 1) and functional before first-feeding in many fish species (Govoni *et al.*, 1986; Boulhic & Gabaudan, 1992; Segner *et al.*, 1994; Bisbal & Bengtson, 1995; Kurokawa & Suzuki, 1996; Ribeiro *et al.*, 1999a).

Gall bladder

The gall bladder (Fig. 1) serves as a storage site for bile produced by the liver and is responsible for the release of bile into the midgut. Bile is gradually concentrated by resorption of water in periods of fasting. The colour of the gall bladder can therefore give an indication of time since the last meal. In vertebrates, including fish, the emptying and excretion of its contents via the bile duct into the gut lumen is controlled by neuronal and humoral factors (Rønnestad, 2002; Hoehne-Reitan & Kjørsvik, 2004; Kamisaka, 2005).

INGESTION AND GUT TRANSIT RATES

Food intake is intimately associated with gut transit rate and, consequently, with the digestive and absorptive efficiency of dietary nutrients, therefore controlling how much of the dietary constituents will actually be assimilated. Conversely, the biochemical composition of the diet may also indirectly influence ingestion and gut transit rates through effects in the digestive and absorption efficiency and through post-absorptive regulatory mechanisms. These interactions should be considered when examining diet ingestion, digestion and absorption and for this reason they will be briefly approached here. Nonetheless, knowledge of the factors that control food intake in larvae and their relationship with larval digestive physiology is still largely incomplete. Studies are hampered by the small size and fragility of both the larvae and food particles, as well as by the difficulty in controlling the nutritional composition of the larval diets and of accurately quantifying their ingestion. The biochemical composition of live preys is relatively rigid and can only be slightly altered (mostly in terms of their FA composition), and, more importantly, it is normally difficult to predict or anticipate exactly how the dietary composition of the prey will be affected by a given enrichment formulation or protocol. On the other hand, in terms of inert formulated diets, there are still problems of acceptability by certain species (possibly due to lack of visual and/or chemical stimulus) and also technical problems of leaching of soluble dietary components.

Prey Density, Food Intake, Gut Transit Time and Digestive Efficiency

The interrelationship between prey density, ingestion rate, gut transit and absorption efficiency has long been of interest to researchers. For instance, Sorokin & Panov (1966) conducted a study to determine food intake and absorption efficiency at different live prey concentrations, in order to establish the optimum concentration for maximising food utilisation. A few studies based on visual observations of live prey inside the larval gut appeared to suggest that the degree of digestion was inversely related to the rate of gut transit, which in turn seemed linearly related to prey density, with food passing through the gut more rapidly at high prey densities (Werner & Blaxter, 1980; Pedersen, 1984). Through the use of radiotracers, allowing quantitative determinations, Boehlert & Yoklavich (1984) were able to confirm these studies. These authors showed that the percent of carbon retained from ^{14}C-algae-labelled rotifers and *Artemia* decreased significantly with increasing food concentration, even if the increased ingestion by the larvae was enough to compensate for the decreased carbon assimilation and therefore resulted in a higher total absorption, albeit at a lower efficiency. More recently, even though the gut transit time was not measured, it was seen that ^{14}C- labelled *Artemia* enriched on a higher dose of fish oil emulsion (HF) was ingested more than that enriched on a lower level of fish oil emulsion (LF) by gilthead seabream larvae. However, a lower amount of the more highly ingested diet was absorbed into the gut and body tissues (Fig. 8). A more rapid passage through the gut of the HF *Artemia*, allowing less time for it to be fully digested and absorbed, was suggested to explain these results (Morais *et al.*, 2006a).

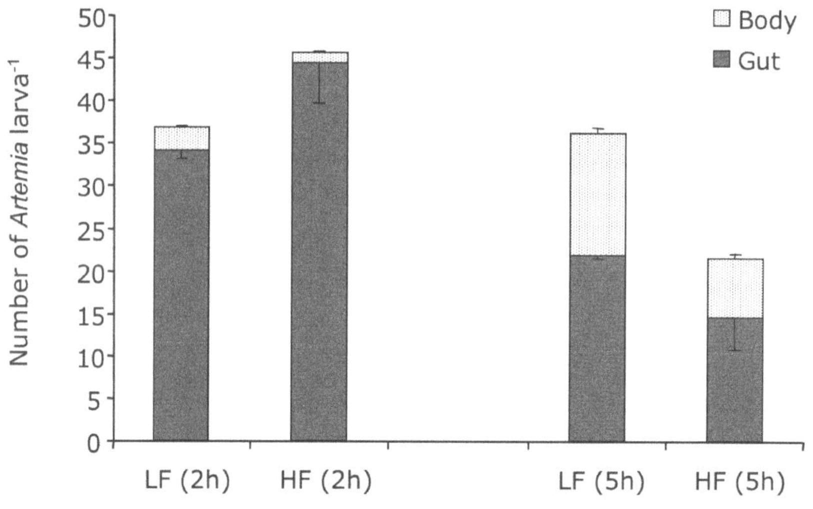

Fig. 8 Food intake and absorption (mean ± SD) in 33 days post hatch gilthead seabream (*Sparus aurata*) larvae, 2 h and 5 h after feeding on ^{14}C-radiolabelled *Artemia*, enriched on either a low (LF) or high dose (HF) of fish oil emulsion. Using a cold chase approach, larvae were allowed to feed on the labelled *Artemia* during 2 h, after which a group of larvae (25-30 larvae, in triplicate) were sampled. The remaining radioactive diet and much of the water was siphoned and fresh seawater with non-labelled treatment *Artemia* was added. Results at 2 h indicate total *Artemia* intake (mostly in the gut but already some slight absorption into the body), while at 5 h after start feeding (or 3 h after start of cold chase) all radiolabelled diet has been evacuated and results represent the fraction of the diet that has been absorbed into the body and gut tissues. After Morais *et al.* (2006a).

Studies conducted mostly with clupeoid larvae e.g. Atlantic herring and Pacific herring (*Clupea pallasi*) support the idea that these larvae fit the characterisation of number maximisers suggested by Griffiths (1975). In other words they are adapted to utilise high food concentrations associated with plankton patches in the pelagic environment by continually increasing their ingestion rate (Boehlert & Yoklavich, 1984). Checkley (1984) however, described an asymptotic relation between fish larval growth rates and ingestion and determined that growth efficiency (i.e. the fraction of ingested food incorporated into tissue) is maximal at intermediate ingestion rates. Consequently, this means that fish larvae in the sea maximise their ingestion rate and thus grow rapidly, but with a reduced efficiency – i.e. fish larvae have evolved so that the rate of growth rather than the growth efficiency is maximal (Checkley, 1984). For example, larval Pacific herring decrease carbon assimilation from individual food particles at high prey densities but the increased ingestion compensates for the decreased carbon assimilation and thus larvae gain greater total energy under conditions of high food concentration (Boehlert & Yoklavich, 1984). Conversely, during periods of low food availability, slowing evacuation rate with decreasing gut content may allow for a higher retention time of the diet in the digestive tract, with more time for it to be completely digested, thus leading to increased assimilation efficiency. Hence, it seems that fish maximise food intake during periods of high prey availability and maximise absorption efficiency during periods of low prey availability (Ryer & Boehlert, 1983).

Effect of Prey Type

Specific factors found in live prey and not in inert microdiets elicit a physiological response by the larvae, in terms of ingestion, transit rate and digestive efficiency. Furthermore, important differences may exist between different live prey types. Therefore, in order to have a complete overview of the digestive physiology of marine fish larvae, studies are necessary to identify and isolate these physiological stimulatory effects, as well as to understand their interdependence.

Besides prey density and feeding rate, other factors, such as the type and dimension of the prey organisms and their intrinsic proteolytic activity, have been suggested to be important in regulating gut transit time in Atlantic herring larvae (Pedersen, 1984). Considerable differences were found by Werner & Blaxter (1980) and by Pedersen (1984) in terms of the evacuation rates of similar-age Atlantic herring larvae fed on different densities of *Artemia* nauplii or copepods, respectively. Pedersen (1984) found a much more rapid gut evacuation of copepods and explained these differences by the fact that a higher ingestion rate was measured in larvae feeding on copepods than on *Artemia* nauplii, possibly due to that fact that copepods rather than *Artemia* are the natural prey of herring larvae. Additionally, the author suggested that *Artemia* is less well digested than copepods, either as a result of a more indigestible exoskeleton (i.e. less permeable to hydrolysing enzymes) or for supplying less exogenous proteolytic enzymes, i.e. for having a lower autolytic capacity, as the products of live prey autolysis may act in stimulating the larval digestive mechanisms. Indeed, Munilla-Moran *et al.*

(1990) measured very high enzymatic levels in adult copepods, compared to *Artemia*, with the lowest enzyme contents being measured in rotifers. In fact, several authors have suggested that fish larvae have a low endogenous proteolytic and lipolytic capacity and that the enzymatic contents of their prey (exogenous source) contributes to the larval digestive process (Dabrowski & Glogowski, 1977; Lauff & Hofer, 1984; Walford & Lam, 1993). However, considerable controversy exists around this notion as several other studies have quantified the contribution of live prey enzymes to the overall digestive process and have concluded that this contribution is negligible (Pedersen *et al.*, 1987; Cahu *et al.*, 1995; Cahu & Zambonino-Infante, 1997; Kurokawa *et al.*, 1998; Gawlicka *et al.*, 2000; Applebaum & Holt, 2003; Perez-Casanova *et al.*, 2006). Discrepancies between different studies may result from the fact that the potential enzymatic contribution from the prey may vary according to its species and even within a species, depending on its origin, age and nutritional state (Munilla-Moran *et al.*, 1990). Additionally, this relative contribution of exogenous enzymes will also depend on the digestive capacity of the fish species, at the developmental stage being analysed, and varies considerably depending on the specific digestive enzyme – e.g. zooplankton may supply over 50% or even 100% of the amylase content in the digestive tract of carnivorous larvae (Gawlicka *et al.*, 2000; Perez-Casanova *et al.*, 2006). Nonetheless, the commonly accepted idea at present is that larvae have an endogenous enzymatic capacity that is efficient to deal at least with their natural diet from first-feeding and that dietary enzymatic contribution is insignificant.

It is also possible that ingested live prey may regulate gut transit rate and digestive efficiency in larval fish through effects on the direct stimulation of pancreatic secretion or activation of zymogens within the larval gut or still by the stimulation of secretion of gastrointestinal hormones (e.g. bombesin and CCK) that are involved in the regulation of peristalsis as well as on the secretion of bile and pancreatic enzymes (Pedersen, 1984; Hjelmeland *et al.*, 1988; Person-Le Ruyet *et al.*, 1993; Kolkovski & Tandler, 1995; Koven *et al.*, 2001). For instance, it was observed that when *Artemia* was co-fed with an inert diet there was an increase in the peristaltic movements of the larval digestive tract, and this was correlated with an increased activity of the digestive hormone bombesin, either from an endogenous or an exogenous source (Kolkovski & Tandler, 1995). Additionally, in an experiment performed by Person-Le Ruyet *et al.* (1993), the evacuation rates were much faster in European seabass larvae fed on *Artemia*, compared to a group fed a biologically inactive formulated diet. In terms of live prey type, one of the factors that may eventually have a direct consequence on the release of these hormones is the composition of the exoskeleton which represents a chemical stimulus (Pedersen, 1984). On the other hand, as the stimulation of the release of gastrointestinal hormones is also induced by the presence of digestion products in the gut, differences in prey digestibility or in their autolytic proteolysis, as discussed above, would also exert an effect. Finally, the size and shape (e.g. spherical or oblong) of the prey may also have an important effect (see below).

A phenomenon that is frequently observed just after the onset of exogenous feeding of some species is that live food organisms may not be killed at the time of ingestion

(T. Harboe, IMR, Austevoll, Norway, pers. comm.). The mechanism of killing probably involves larval endogenous factors such as alkaline lysis by larval enzymes combined with hypoxia. The efficiency of these factors will depend upon the gut transit time and prey type. Killing of live prey is aided by pharyngeal teeth that probably rupture the exoskeleton as the plankton is swallowed. Thus, the strength of the exoskeleton could also determine the vulnerability of the ingested planktonic prey. In some fish species it is frequently reported that *Artemia* nauplii may survive ingestion and transit through the digestive tract or at least can be found intact or partially intact in the intestinal lumen (Luizi *et al.*, 1999). This is frequently reported for *Artemia* fed to Atlantic halibut during the first few weeks after the onset of first-feeding, particularly when prey concentrations are high and the transit rates are fast (T. Harboe, IMR, Austevoll, Norway, pers. comm.). It is evident that an intact and a ruptured feed organism may stimulate the digestive process very differently.

Stimulation of enzyme release is both under hormonal and neural control. The cephalic phase in mammals involves neural stimulation of pancreatic secretion and is mediated by the vagus nerve (Hjelmeland *et al.*, 1988; Pedersen & Andersen, 1992; Holmberg *et al.*, this volume). As a result, there is an enhanced enzyme secretion which may be induced by sight, smell or taste of food. However, in this respect, Pedersen & Andersen (1992) did not observe a stimulation of trypsinogen secretion during swallowing of copepodites in Atlantic herring larvae and this led the authors to suggest an absence of a cephalic phase in fish larvae, as it is known in mammals. Stomach distension has also been shown to induce a neural reflex in mammals. Hjelmeland *et al.* (1988) studied whether this might also occur in fish larvae and noted that non degradable polystyrene spheres, with no nutritional value, induced trypsinogen secretion in Atlantic herring larvae, although the response was lower than that observed with live prey (possibly due to differences in the number of particles ingested). The same was also later described by Pedersen & Andersen (1992). In fish larvae, neural stimulation of the pancreas could theoretically result from the distension of both the oesophagus and the intestine (as there is no functional stomach). Also in Atlantic herring larvae, Pedersen *et al.* (1987) and Pedersen & Andersen (1992), noted that the amount of pancreatic trypsin and trypsinogen secreted into the gut increased as the number of ingested preys increased, until a plateau was reached at high levels of food intake. Prey size has also been hypothesised to affect intestinal peristaltic movements through an effect on the distension of the gut, with larger preys (such as adult copepods) inducing a more pronounced mechanical stimulation (Pedersen, 1984). On the other hand, the visual observation of the digestion processes in the transparent gut of Atlantic herring larvae tube fed pure lipid suggested that the gut is highly sensitive to the physical pressure exerted by the food bolus, with gut contractions concentrating always in the larger lipid droplets or where there was a higher concentration of lipid droplets (Morais *et al.*, 2005a, Fig. 9C,D). This effect is mediated by the autonomic nervous system of the fish, which appears to be similar to that of mammals (Pedersen, 1984; Holmgren *et al.*, this volume).

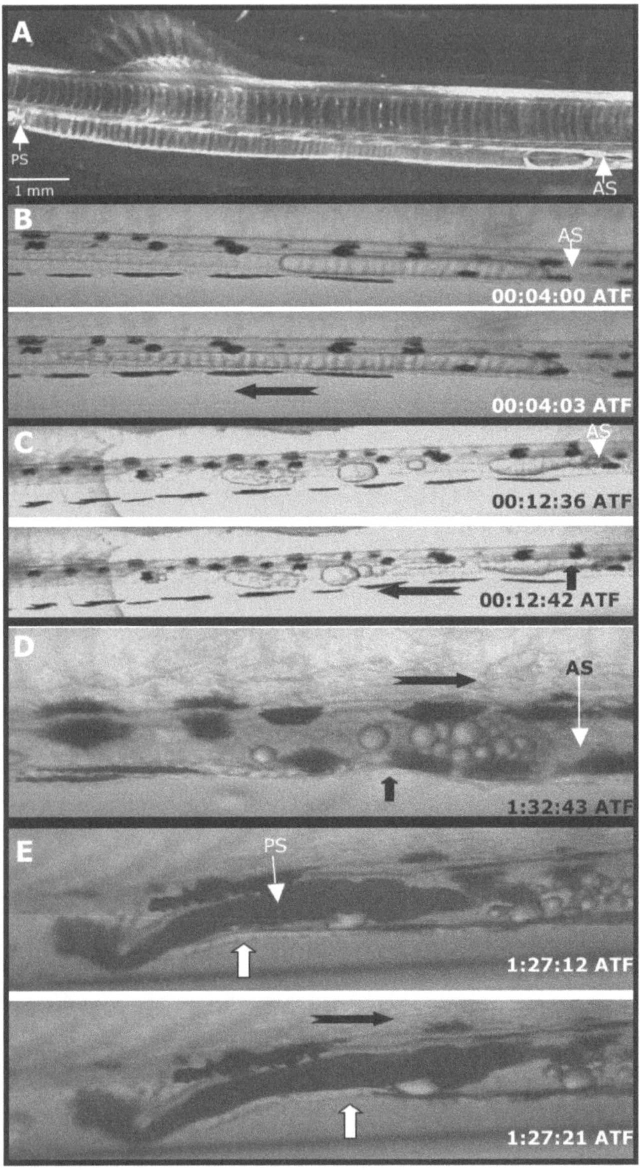

Fig. 9 Lipid digestion and gut motility in one Atlantic herring (*Clupea harengus*) larva, immediately after tube feeding 50.6 nl of triolein (A) and over time (B-E). The anterior and posterior midgut sphincters are indicated (AS and PS, respectively), vertical block arrows indicate the approximate location of origin of the peristaltic contraction and horizontal arrows show the direction of the movement. After Morais *et al.* (2005a). Reprinted with permission from Blackwell Publishing.

In conclusion, diets for fish larvae should necessarily meet several requirements: to be easily ingested (i.e. need for chemical and visual stimuli and have the correct size), cover the nutritional needs of the larva, contain ingredients that may be easily digested and absorbed by the larva's digestive tract and promote a neural and an endocrine response (release of gastrointestinal hormones) that will stimulate gut motility, enzyme secretion or activate zymogens in the larval gut (see Kolkovski, 2001 and Koven *et al.*, 2001 for a more complete review).

Effect of Dietary Nutrient Composition

The difficulty in controlling the nutritional composition of larval diets has seriously hampered studies on the effect of dietary nutrient composition (both in quantitative and qualitative terms) on larval food intake and their digestive physiology. These interactions have hardly been examined in larval fish, while a few studies have been conducted with juvenile and adult fish, particularly with respect to dietary lipid intake. However, caution should be used to extrapolate the results obtained from larger fish as fish larvae have much higher requirements for energy and structural components, while dietary lipid (particularly essential FA-EFA) is also extremely important for early organ development and for physiologically demanding processes such as metamorphosis.

The effect of gut transit time on the ability of fish larvae to utilise the ingested proteins and hydrolysates was observed in a study where the objective was to analyse dietary protein utilisation in early stages of fish, using a model ^{14}C-salmon serum protein (SSP) administered to Atlantic halibut by tube-feeding, followed by a hot chase (Tonheim *et al.*, 2005; Fig. 10). This study supported previous observations of an inverse relationship between the utilisation efficiency and evacuation rate of labelled protein. By regular sampling of larvae from 0.5 to 20 h after tube-feeding, it was seen that although the mean apparent digestibility of intact SSP in larvae was low, some individuals showed efficiencies that were almost comparable to juvenile Atlantic halibut or Atlantic halibut larvae that were administered a pre-hydrolysed protein, in which the absorption efficiency was considerably higher (see section on the efficiency of protein absorption, below). These results clearly suggested increased utilisation efficiency by those larvae that, for some reason, had a more prolonged gut transit time. In contrast to larvae administered the intact protein, the individual variation in absorption efficiency and evacuation in larvae administered the same pre-hydrolysed protein was low. Any evacuation occurring before absorption was complete would most likely have been at the expense of absorption efficiency. The lower variation and higher absorption of pre-hydrolysed protein in larvae could thus be explained by its higher absorption rate (Fig. 10; see discussion below). Based on this, Tonheim *et al.* (2005) suggested that intestinal transit rate is critical for the efficient utilisation of slowly digested and absorbed feed components in fish larvae, while it will have a lower effect on the utilisation efficiency of rapidly absorbed feed constituents such as free amino acids (FAA), peptides and pre-hydrolysed protein.

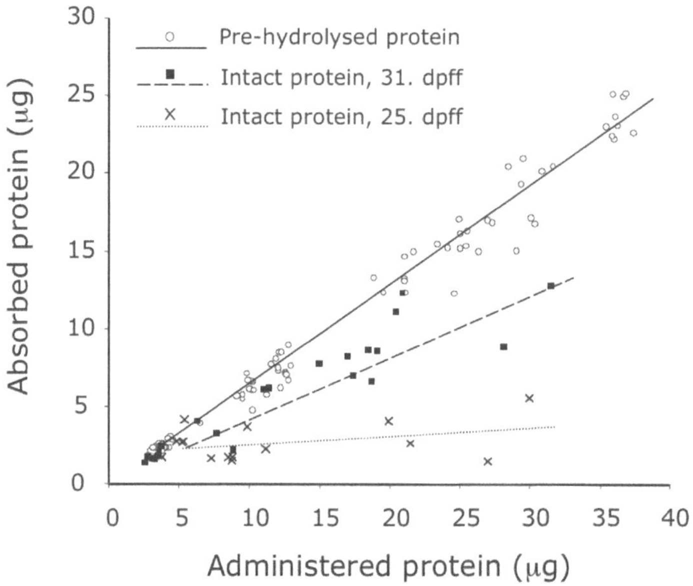

Fig. 10 Effects of increasing intestinal loading on the absorption of ^{14}C-labelled model salmon serum protein tube-fed to Atlantic halibut (*Hippoglossus hippoglossus*) larvae, 20 h after its administration. For intact protein there were significant differences caused by differences in the evacuation rates at the experimental dates (25 and 31 days post first-feeding, dpff). Larvae fed the pre-hydrolysed protein preparations did not differ significantly between days or diets and were therefore pooled into a common regression model. After Tonheim *et al.* (2005). Reprinted with permission from Elsevier.

In terms of dietary lipids, and in spite of the lack of studies at the larval stage, the effect of dietary total lipid level on food intake has been one of the most frequently addressed questions in juvenile and adult fish nutrition studies, where it is normally aimed to partly replace protein, which is the most expensive component of fish diets, without inducing negative effects on growth and flesh quality (i.e. adiposity). The concept that fish are able to adjust their energy intake is well established (Lee & Putnam, 1973; Boujard & Médale, 1994; Ogata & Shearer, 2000; Yamamoto *et al.*, 2000; Gélineau *et al.*, 2001; Sæther & Jobling, 2001; Boujard *et al.*, 2004; Skalli *et al.*, 2004). Dietary digestible energy content is thus a well documented factor that has been shown to affect food intake in juvenile and adult fish. Nonetheless, even this may be complex to establish as, in extreme cases, when fish are fed diets of very high or low energy density, stomach fullness and the control of gastric emptying may be responsible for the regulation of food intake (Lee & Putnam, 1973; Ogata & Shearer, 2000; Gélineau *et al.*, 2001).

The regulation of food intake according to the dietary neutral lipid level and FA composition was recently examined for the first time in larvae (Morais *et al.*, 2006a). In this study it was found that, contrary to what is commonly observed in juvenile and adult fish, in seabream larvae there is not a stringent control of food intake according to dietary energy level or a lipostatic regulation of food intake (Fig. 8). In addition, it was suggested that dietary FA composition may have a more important role in controlling ingestion, through indirect effects on gut transit time caused by differences in FA digestibility and absorption efficiency and/or by differential stimulation of secretion of gastrointestinal hormones (Morais *et al.*, 2006a). In a previous work conducted with Atlantic herring (Morais *et al.*, 2005a), it was suggested that an increment of lipid level above the digestive and absorptive capacity of the larvae could potentially lead to a faster evacuation of the diet. However, in this study, an artificial situation of feeding pure lipid was induced (Fig. 9) and it is uncertain whether the same would occur in larvae feeding a complete diet in which emulsifying agents such as phospholipids (PL), proteins and carbohydrates, are also present. In fact, a likely possibility is that, when feeding a complete diet, the digestive system of the larvae will function in a way that it will digest a meal as completely as possible before evacuation.

DIGESTION

Most of the knowledge gathered concerning the processes of digestion and absorption in fish originates from studies conducted with juveniles and adults. However, it has been established that the quantitative and qualitative dietary requirements of marine fish larvae are greatly different from those of adult fish (Person Le Ruyet *et al.*, 1993). Furthermore, the particular morphological and physiological characteristics of the larval digestive system, as well as the changes occurring during larval ontogeny and in response to dietary regime must be considered. Specific knowledge on larval digestive physiology is essential, as it will clearly affect dietary nutrient utilisation and limit the type of diet that should be offered.

Studies during larval stages are rare, due to methodological difficulties related to the extremely small and fragile nature of fish larvae. For this reason, many of the larval studies use whole body homogenates to quantify enzyme activity. This procedure is disadvantageous, as not only it implies a large "dilution" of the enzyme being analysed by other proteins, but mostly it does not distinguish between activities of different origins (intracellular enzymes or those present in tissues other than in the digestive tract and exogenous prey or microbial enzymes). In recent years, the development and more widespread use of sensitive molecular tools has enabled the identification, localisation and quantification of gene transcripts involved in larval digestive and absorptive physiology, in conjunction with biochemical determinations (Murray *et al.*, 2003, 2006; Perez-Casanova *et al.* 2004, 2006; Marza *et al.*, 2005).

Digestion of Proteins, Peptides and Amino Acids

Processes and mechanisms

Digestion of proteins is initiated in the stomach, when present, by acid denaturation and enzymatic cleavage by pepsin. Pepsin is produced and secreted in its inactive form (pepsinogen) by oxynticopeptic cells in the mucosa (Morrison, 1987; Sire & Vernier, 1992) and in fish these cells also secrete HCl, which makes the acid environment necessary to activate pepsinogen. Pepsin is an endopeptidase and hydrolyses proteins at bonds between aromatic and acidic AA (Jobling & Hjelmeland, 1992). Pepsin digestion yields polypeptides of very diverse sizes, but few FAA (Lied & Solbakken, 1984; Jobling & Hjelmeland, 1992; Espeland, 2006). The combined effects of HCl and pepsin make the stomach a highly efficient organ for solubilisation of collagenous connective tissues (Gildberg, 2004). In addition, the stomach possesses strong peristaltic and contractile activity that contributes to a mechanical degradation of the ingested feed and to an enhanced mixing of the luminal contents, allowing more of the proteins to be exposed to the acid proteolytic conditions. Peristalsis is also responsible for passing the chyme on to the anterior intestine for further hydrolysis.

Once the chyme with the partially degraded proteins and peptides enters the midgut, it is mixed with pancreatic secretions and bile. Qualitatively, the digestive enzymes in fish are similar to those of higher vertebrates (Ash, 1985; Sire & Vernier, 1992). Trypsin and chymotrypsin are the major endopeptidases secreted from the pancreas (Dabrowski, 1983; Gildberg, 2004) but other enzymes are also found (Fig. 11). Trypsin is a serine protease that hydrolyses peptide bonds on the carboxyl side of the basic AA, while chymotrypsin primarily hydrolyses peptide bonds next to the aromatic AA (Jobling & Hjelmeland, 1992). These enzymes degrade the peptides to small sizes. The pancreatic enzymes are secreted as proenzymes and are only activated in the gut lumen. Trypsin is a key factor in the activation of all pancreatic enzymes, and trypsin is itself activated when it contacts brush border-bound enterokinase. Endopeptidase activity yields protein fragments and small peptides in the luminal contents, while further hydrolysis is performed by exopeptidases, secreted by both pancreas and the apical membrane of the enterocytes (Buddington *et al.*, 1997).

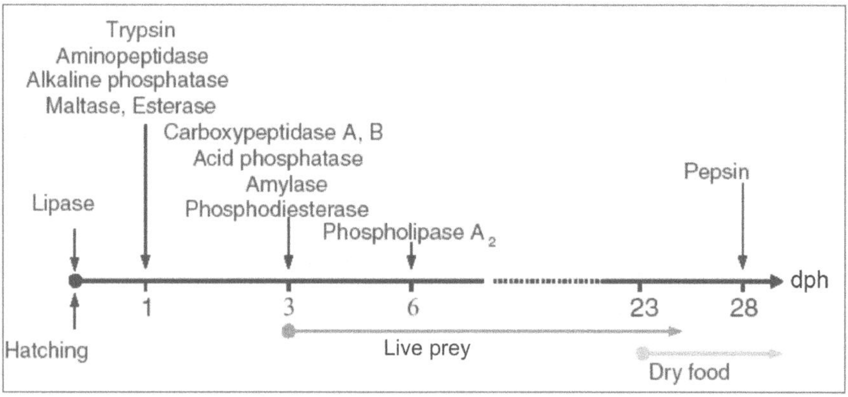

Fig. 11 Digestive enzyme activity in turbot (*Scophthalmus maximus*), showing when enzyme activity could first be detected. After Hoehne-Reitan & Kjørsvik (2004). Reprinted with permission from American Fisheries Society Symposium.

Peptidases in the brush border membranes also contribute to the end result of digestion, which are small peptides and FAA. Evidence suggests that digestion is not complete prior to absorption, and recent studies from a variety of vertebrate species suggest that entry of peptides over the apical membrane of the enterocyte is an important route of absorption of digested proteins (Daniel, 2004). Peptides that are transported over the brush border membrane are normally hydrolysed by cytoplasmatic peptidases, while FAA are mainly transported out of the basal membrane of the mucosa cells and into the portal circulation. However, small peptides that are resistant to hydrolysis may enter the blood, and some peptides may enter the systemic circulation. Peptide transport may therefore be an important route for absorption of both dietary proteins and peptides in teleost fishes.

Due to the absence of HCl^- and pepsin-secreting cells in altricial fish larvae at the onset of exogenous feeding, there is no preparatory acid denaturation of ingested proteins. Cahu & Zambonino-Infante (2001) suggested that the lack of a stomach does not hamper enzymatic protein digestion in fish larvae, since this is ensured by the pancreatic and intestinal enzymes. However, native proteins present a smaller surface area than denatured proteins for alkaline enzymatic attack, meaning that pancreatic and intestinal hydrolysis may be less efficient. In the stomachless gibel carp (*Carassius auratus gibelio*) hydrolysis of intact protein was about 12% of that of denatured protein (Jany, 1976). In addition, studies on Atlantic halibut (Tonheim *et al.*, 2004, 2005) as well as on zebrafish, striped bass (*Morone saxatilis*) and walleye (*Sander vitreus*) (Rust, 1995), suggest lower rates of digestibility and assimilation of protein in pregastric larvae than in postmetamorphic fish (see below).

Changes in proteolytic activity during larval development

The general pattern of proteolytic activity during ontogeny (Fig. 11) has been the focus of many studies (Pedersen *et al.*, 1990; Ribeiro *et al.*, 1999b; Gawlicka *et al.*, 2000; Cahu & Zambonino-Infante, 2001). Recent studies using PCR techniques have been able to detect transcripts of many proteolytic enzymes from hatch (Perez-Casanova *et al.*, 2006). The activity of proteolytic enzymes in the midgut is generally found to be low when marine fish larvae commence exogenous feeding, but it rises as metamorphosis approaches (Hjelmeland, 1995; Gawlicka *et al.*, 2000; Cahu & Zambonino Infante, 2001). Besides the general trend of developmental changes of proteolytic enzymes, there appear to be species-specific as well as dietary and temperature induced differences. Presently, the mechanisms underlying the physiological function of some of these observed differences are not known. For instance, Perez-Casanova *et al.* (2006) reported that the ontogenetic development of trypsin-like activity in whole body homogenates of haddock (*Melanogrammus aeglefinus*) and Atlantic cod reared on the same protocol was very different. In haddock the activity was high at mouth opening and then it declined and remained at a low level until 530 day degrees. In Atlantic cod, trypsin-like activity was low at mouth opening, but it increased after a switch in prey from non-enriched to enriched rotifers, and declined again when *Artemia* was gradually introduced (Perez-Casanova *et al.*, 2006).

Alkaline phosphatase (AP) is an enzyme that is routinely analysed in many studies of digestive function in larval fish. AP reportedly reflects the development of brush border membranes of the enterocytes (Vu, 1983). However, the biologic function of intestinal AP is poorly understood. Based on studies on humans, Poelstra *et al.* (1997) suggested that AP in general protects against endotoxins produced by bacteria that may cause lethal complications after an infection.

The activity of cytosolic peptidases like leucine-alanine peptidase are high around first-feeding but tend to decrease as larvae develop, concurrent with rising levels of AP (Cahu & Zambonino-Infante, 2001; Kolkovski, 2001). The ontogenetic pattern of trypsin, AP and cytosolic peptidases activity during larval development reflects changes in the mode of larval digestion, which becomes increasingly dependent on luminal digestion and less so on intracellular digestion (Cahu & Zambonino-Infante, 2001).

Larval proteolytic capacity

Several studies have tried to quantify the capacity of fish larvae to digest different proteins, as well as changes in digestive capacity during larval development. However, comparison of the results obtained in different studies on larval protein digestive capacity is made difficult by the fact that different types of proteins (with inherent differences in digestibility), species, stages of development and experimental conditions have been used. Technical limitations in the available techniques for labelling proteins have seriously restricted the experimental designs and thereby the advances in this field, and much effort has been done in order to produce a suitable labelled model protein (Rojas-García & Rønnestad, 2003a, b; Rønnestad *et al.*, 2003; Tonheim *et al.*, 2004, 2005). The commercially available ^{14}C-labelled proteins are almost exclusively labelled by ^{14}C-methylation, which is a process known to alter protein digestibility, as the electrical charge, molecular weight and bulkiness of the labelled lysine and arginine residues are affected (Rojas-García & Rønnestad, 2003a, b; Tonheim, 2005). In addition, methylation has been shown to negatively affect the digestion of proteins, hindering the action of proteases that attack and hydrolyse peptide bonds adjacent to lysine residues (Murakami & Etlinger, 1987). In order to overcome the lack of suitable commercially available radiolabelled proteins, Tonheim *et al.* (2004) developed a method of *in vivo* incorporation of uniformly labelled AA into SSP. Other studies attempted a more technically easy solution, such as to label algae protein by addition of $H^{14}CO_3$ to algal cultures but this method has been hampered by a low protein yield and purity, caused by contamination by labelled carbohydrates (Tonheim *et al.*, 2004).

Results from experiments where larval Atlantic halibut were administered an equal intestinal loading of ^{14}C-SSP preparations with a different degree of hydrolysis demonstrated a higher digestibility when the model protein was pre-hydrolysed prior to feeding (64% versus 36% for the intact protein) (Fig. 12A) (Tonheim *et al.*, 2005). In addition, the digestibility of intact protein, unlike pre-hydrolysed protein, was quite variable and dependent on gut transit time and the absorption efficiency of the intact/native ^{14}C-SSP protein decreased dramatically when the administered dose size increased (Fig. 10). These results suggest that the larval digestive tract capacity to process dietary protein is limited by proteolytic rather than by its absorptive capacity, and agrees with previous findings that indicated a rapid and efficient absorption of FAA, peptides and hydrolysed proteins in larvae (Rust *et al.*, 1993; Rust, 1995; Rønnestad *et al.*, 1999; 2000c; 2001; Rojas-Garcia & Rønnestad, 2003a; Applebaum & Rønnestad, 2004).

Nonetheless, it is generally believed that the digestive capacity of fish larvae and their ability to digest intact protein may increase gradually throughout development and different studies have tried to describe these changes. For instance, Rust (1995), using tube feeding to deliver labelled protein (^{35}S-labelled *Escherichia coli* protein), studied a variety of species and stages of development, finding average values of protein absorption of 29-62% in striped bass (19-47 dpff), 42-58% in walleye (10-40 dpff) and 64-94% in zebrafish (5-13 dpff). An increase in protein absorption during development was noted by Rust (1995) in 19-47 dpff striped bass, which was later also supported by studies of Tonheim *et al.* (2004), when comparing pre-metamorphic and juvenile Atlantic halibut tube fed ^{14}C-SSP (25% vs. 59%, respectively; Fig. 12B). Nonetheless, Govoni *et al.* (1982) did not find a significant relationship in carbon utilisation and absorption with development, in 7-47 days after hatching (dph) larval spot (*Leiostomus xanthurus*).

Morais *et al.* (2004a) have also examined ontogenetic changes in the capacity of Senegalese sole (*Solea senegalensis*) larvae to digest *Artemia* protein and AA using *Artemia* nauplii and metanauplii radiolabelled by adding a [U-^{14}C] protein hydrolysate to the hatching or enrichment media. This study established that Senegalese sole larvae, even young stages (12 dph), have a high capacity for digesting live *Artemia* (at least when given a single meal). The digestibility of *Artemia* was high in all tested ages, with label absorption varying between 77 and 83% at 24 h after feeding. A rapid digestion and absorption of *Artemia* AA was noted, with most of the absorption into the body occurring during the first 3 h after feeding. In general, very high *Artemia* protein digestibility values have been obtained through the use of [U-^{14}C] protein hydrolysate-labelled *Artemia*, even though species-specific differences have been found using exactly the same methodology – e.g. 77-83% in 12-35 dph (Morais *et al.*, 2004a) or 75-76% in 27 and 33 dph Senegalese sole (Morais *et al.*, 2005b), compared to 66% in 31-33 dpff Atlantic herring (Morais *et al.*, 2004b). This might eventually reflect species-specific differences in the morphology of the gut (straight in herring vs. coiled in sole). In general, the use of radiolabelled *Artemia* results in higher values of protein digestibility compared to other labelled protein sources, such as purified ^{14}C-labelled algal protein in Atlantic halibut post-larvae (42%) (Rønnestad *et al.*, 2001), radiolabelled SSP in pre-metamorphic and juvenile Atlantic halibut (25% and 59%; Tonheim *et al.*, 2004) and ^{35}S-labelled *E. coli* protein in the above mentioned fish species studied by Rust (1995). It may be hypothesised that the relatively higher absorption of AA label in studies using radiolabelled *Artemia* as the source of protein instead of tube feeding a liquid tracer could be partly related to the physical nature of the diet. It is likely that *Artemia* is retained in the larval gut for a longer period of time than a liquid test solution, thereby allowing more time for protein digestion and absorption. Similarly,

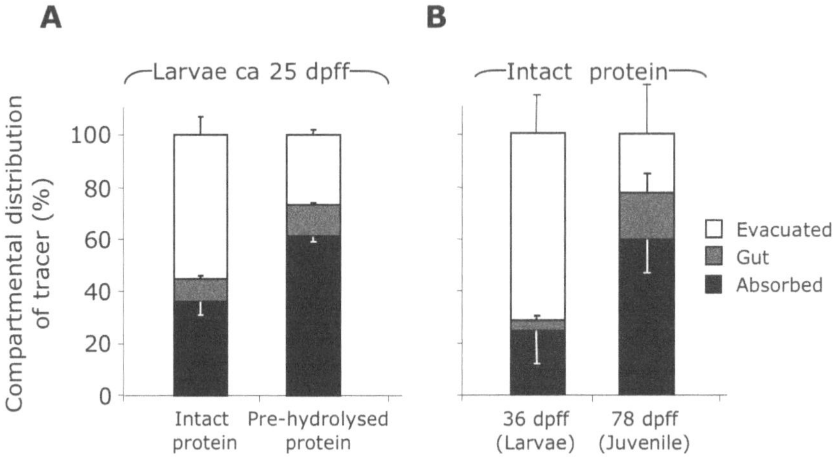

Fig. 12 Distribution of tracer 20 h after administration of ^{14}C-labelled model salmon serum protein in larvae and juvenile Atlantic halibut (*Hippoglossus hippoglossus*) by tube feeding. (A) distribution of ^{14}C in 25 days post first-feeding (dpff) larvae, 20 h after being administered intact or pre-hydrolysed protein (B) Distribution in larvae and juveniles administered the protein in intact form in equal doses relative to body mass wet weight. After Tonheim (2005) and Tonheim *et al.* (2005). Reprinted with permission from the author and Elsevier.

very high ^{14}C absorption (67-99%) has been recorded in larval spot using radiolabelled rotifers (*Brachionus plicatilis*) containing 66% of the ^{14}C-label in the protein fraction (Govoni *et al.*, 1982). It should also be noted that tube feeding followed by a hot chase may overestimate protein digestibility, compared to field-data and under culture conditions, as a result of larvae being given a single meal and subsequently deprived of food.

Digestion of Lipids

Processes and mechanisms

Lipids are an important source of metabolic energy and of EFA, structural components of biological membranes and precursors of essential metabolites (e.g. eicosanoids) (Sargent *et al.*, 1989). These functions are particularly important during the fast growing and developing larval stages, which have high demands for energy and structural components. However, in spite of their central role in larval nutrition, aspects such as digestion, absorption and transport of dietary FA have been insufficiently studied in fish in general and have only started to be investigated in fish larvae (Izquierdo *et al.*, 2000). In particular, very little is known concerning how dietary lipid level and composition may affect the digestive and absorptive capacity in fish larvae. However, parallels have been found between the digestive and absorptive processes of fish and mammals and it is presumed that the mechanisms are generally similar (Sire *et al.*, 1981; Honkanen *et al.*, 1985).

The process of lipid digestion can be divided in two distinct parts: (1) the hydrolytic cleavage of dietary FA esters in the gastrointestinal tract, and (2) the absorption of the resulting FA into the enterocytes (Olsen *et al.*, 1998). Within the first phase, three further steps can be distinguished: 1) the dispersion of bulk lipid droplets into finely divided emulsion particles, 2) enzymatic hydrolysis of FA at the emulsion-water interface, and 3) desorption and dispersion of insoluble lipid products into an absorbable form (Carey *et al.*, 1983). There is a large volume of literature dealing with all aspects of mammalian and human lipid digestion and absorption (Borgström, 1977; Carey *et al.*, 1983; Tso & Fujimoto, 1991; Nordskog *et al.*, 2001). This section, as well as that on lipid absorption, will briefly describe the main mechanisms based on the mammalian model, comparing it with what is known for fish in general, as very little specific data exists for larval fish.

In humans and most vertebrates the digestion of dietary lipids begins in the stomach, through the chemical action of gastric (acid) lipase, which is secreted by the gastric mucosa. Gastric lipase digests triacylglycerols (TAG) but not PL or cholesteryl esters. Its main hydrolytic products are diacylglycerols (DAG) and FA. Given that dietary lipids may be retained in the stomach for some time, up to 30% of the total dietary TAG may be digested (Borgström, 1977; Carey *et al.*, 1983; Nordskog *et al.*, 2001). The stomach also plays a very important role in the mechanical emulsification of dietary fat, which is an essential prerequisite for the efficient hydrolysis by pancreatic lipase in the intestinal lumen. The mechanical emulsification of lipids is aided by the DAG and FA resulting from acid lipolysis, as well as by dietary PL, peptides from protein digestion, complex polysaccharides and membrane-derived PL, which function as emulsifiers in the acid milieu of the stomach (Borgström,1977; Carey *et al.*, 1983; Nordskog *et al.*, 2001). A lipase resembling the gastric, lingual and lysosomal acid lipase of vertebrates has been partially sequenced in Winter flounder by Murray *et al.* (2003) but its expression in several adult tissues, including the stomach, determined by RT-PCR, was found to be low. In addition, no transcripts of this enzyme could be localised in any adult digestive organs by *in situ* hybridisation. On the other hand, earlier studies looking at the *in vivo* lipid digestion in Atlantic cod and turbot have shown that gastric digestion appears to be insignificant in these species (Lie *et al.*, 1987; Koven *et al.*, 1994a, b; Izquierdo & Henderson, 1998). Therefore, in terms of gastric lipolysis, substantial differences may exist between fish and humans, as acid lipase may not contribute significantly to lipid digestion in adult and juvenile fish. While this may be even more so for fish larvae, which lack a functional stomach (Govoni *et al.*, 1986), acid lipase expression has been detected by RT-PCR in Winter flounder larvae at low levels in all larval developmental stages, from hatch to after metamorphosis (Murray *et al.*, 2003).

The bulk of TAG digestion in vertebrates occurs in the duodenum (Tso & Fujimoto, 1991). The lipid is in the form of crudely emulsified droplets when entering the small intestine, which are then mixed with fluids arising from the pancreas, gall bladder and small intestine. As a result of the abrupt rise in pH, the FA in the emulsion become partially ionised and migrate to the interface of the emulsion (Carey *et al.*, 1983; Nordskog *et al.*, 2001). At the same time, biliary lipids released from the gall bladder in the form of lipid micelles come into contact with the emulsion particles and are transferred to the surface of the emulsions. These two events markedly increase the surface area of the emulsion and, together with the mechanical emulsification, create smaller stabilised particles. As the enzyme rich pancreatic secretions are released, pancreatic lipase-colipase, phospholipase A_2 (PLA_2) and non-specific esterases hydrolyse both core and surface components. The digestion products create an increased pressure in the emulsion surface, which causes the "budding off" of bilayered fragments, probably as unilamellar liposomes (Carey *et al.*, 1983).

Work conducted on Atlantic herring larvae demonstrated that gut motility can be an efficient emulsifier of pure lipid (triolein) that had been tube fed to the larvae (Morais *et al.*, 2005a; Fig. 9). Contractions of the midgut wall caused the mechanical breakdown of large lipid droplets and active dispersion and emulsification throughout the midgut, and they most probably ensured considerable mixing between the lipid droplets and the bile salts and pancreatic secretions that are released close to the anterior midgut sphincter (Pedersen, 1984) (Fig. 9B, C, D). As the mass of the lipid droplets moved to the posterior midgut, strong retrograde contractions of the midgut and hindgut likely increased the retention time in the gut and also allowed a better emulsification and mixing with bile and pancreatic secretions (Fig. 9E). Additionally, the gut wall seemed very sensitive to physical pressure, as the gut contractions were mainly concentrated on the largest lipid droplets. On the other hand, signs of chemical action during lipid digestion were also noted (Morais *et al.*, 2005a). For instance, it was observed that the smaller lipid droplets did not fuse back together, indicating the presence of bile salts in the gut lumen. In addition, the gradual change in the appearance of the lipid droplets, from a translucent and clear to an opaque and whitish look, was most probably a result of the chemical mechanisms occurring during digestion, such as the diffusion of protonated long-chain FA from the bulk of the micelles to the surface of the TAG droplets (where they become partially ionised) or the transfer of biliary lipids to the surface of the emulsions.

Honkanen *et al.*, (1985), working with killifish (*Fundulus heteroclitus*), a stomachless saltwater fish species, pointed out several similarities in the digestive processes of fish and mammals, from the location of the major site of lipid digestion and absorption (proximal one third of the intestine), to the concentration of bile salts in the gallbladder, as well as the concentrations of bile salts and lipolytic products in the intestinal lumen during lipid digestion. In some fish species, such as Atlantic cod and Arctic charr (*Salvelinus alpinus*), lipolysis and lipid absorption have been seen to occur mainly in the pyloric caeca and anterior ileum regions, where pancreatic enzymes are secreted and gastrointestinal contents are temporarily retained (Lied & Lambertsen, 1982; Lie *et al.*, 1987; Olsen *et al.*, 1998, 1999). The organisation of the pyloric caeca may enhance lipid digestion by promoting a prolonged exposure of lipids to digestive enzymes (Gjellesvik *et al.*, 1992). However, in other species, such as turbot, lipid digestion and absorption appears to take place mainly in the hindgut and rectal regions (Koven *et al.*, 1994a, b; Izquierdo & Henderson, 1998). The continuation of lipid digestion in the posterior gastrointestinal tract may be an advantage for certain fish species, particularly in strict carnivores with short digestive tracts (see Koven *et al.*, 1994a).

Two different neutral lipase activities have been identified in the pancreatic juice of higher animals – the classic pancreatic lipase or the pancreatic lipase-colipase system (bile salt inhibited and substrate specific) and the bile salt-dependent lipase (BSDL) or carboxyl ester lipase (non-specific and bile salt stimulated) (Borgström, 1977). In mammals, digestion of TAG is mostly due to pancreatic lipase, which is more active at the *sn*-1 and *sn*-3 positions of the TAG molecule, releasing 1,2-DAG, 2-monoacylglycerols (2-MAG) and free FA (FFA). This requires the presence of a cofactor, called colipase, that binds at the lipid-aqueous interface and provides a high affinity anchor site for the pancreatic lipase molecule which would otherwise be inhibited by the bile salts that induce its desorption from the lipid-water interface (Borgström, 1977; Carey *et al.*, 1983; Tso & Fujimoto, 1991; Nordskog *et al.*, 2001). Although 1-MAG may be formed through isomerisation of 2-MAG in an aqueous medium, this transformation is slower than the uptake of MAG by the small intestine and therefore the predominant form in which MAG is absorbed is as 2-MAG (Nordskog *et al.*, 2001). Pancreatic BSDL is also found, but in minor amounts in the mammalian pancreatic juice. It is a glycoprotein with broad substrate specificity that requires the presence of bile salts to hydrolyse insoluble FA esters and presents a relatively high affinity for DAG and particularly for esters of long-chain polyunsaturated FA (PUFA), which are more resistant to hydrolysis by pancreatic lipase (Borgström, 1977; Carey *et al.*, 1983).

The exact nature of the neutral lipases found in fish is still not completely established as different studies of digestive lipolysis have yielded results that differ as to which types of lipases can be found. In addition, the existence of a diffuse pancreas in most fish species and the higher difficulty in obtaining purified enzyme extracts renders the investigation of digestive lipases in fish more challenging than in mammals (Tocher & Sargent, 1984). Brockerhoff (1966), when analysing the isomeric forms of the glycerides found in the intestinal tract of Atlantic cod during lipid digestion, initially assumed that fish neutral lipases were similar to the mammalian pancreatic lipase, with a specificity for 1,3 ester bonds. Léger *et al.* (1977) also found evidence of a mammalian type pancreatic lipase in rainbow trout (*Oncorhynchus mykiss*), which appeared to be activated by a cofactor, but its optimum activity required bile salts. Léger & Bauchart (1972) had earlier suggested that rainbow trout lipase has a mechanism of action that differs from that of mammals, combining a 1,3 specificity with a specificity for the nature of the FA, which may superimpose each other. On the other hand, Patton *et al.* (1975), found significant complete hydrolysis (including 2-MAG) as well as a faster cleavage of C20 esters (which are normally resistant to pancreatic lipase) than of C18 esters in the intestinal extracts of several fish species. Tocher & Sargent (1984) reported that, during lipid digestion in rainbow trout, TAG was rapidly hydrolysed to FFA and 2-MAG, which may be further hydrolysed but only at a very slow rate, indicating a classical mammalian specificity for primary ester bonds. Even so, the enzymatic activity was dependent on the presence of bile salts and the presence of a cofactor similar to the mammalian colipase could not be shown. Other studies have also found the activity of neutral lipase to be bile salt-stimulated (Lie & Lambertsen, 1985; Ozkizilcik *et al.*, 1996). In addition, the only lipolytic enzyme found in the pancreas and pyloric caeca of Atlantic cod was a pancreatic BSDL or non-specific lipase that has been found to be homologous to the mammalian one (Gjellesvik, 1991; Gjellesvik *et al.*, 1992). Therefore, several authors have described lipolytic enzymes in fish with characteristics differing from that of mammalian pancreatic lipase. This might be expected, given that a considerable proportion of the energy intake of marine fish is in the form of lipids rich in PUFA which are frequently located in the position 2 of the TAG molecule and that are not readily hydrolysed by mammalian 1,3 specific pancreatic lipase (Gjellesvik *et al.*, 1992; Iijima *et al.*, 1998). Therefore, even if the presence of a 1,3 specific pancreatic lipase-colipase or even other types of lipase cannot be excluded, it is now commonly accepted that the major digestive lipase in teleosts, including larva, appears to be bile salt-dependent and non-specific, producing mainly FFA and glycerol (Lied & Lambertsen, 1982; Tocher & Sargent, 1984; Lie & Lambertsen, 1985; Lie *et al.*, 1987; Gjellesvik, 1991; Gjellesvik *et al.*, 1992; Koven *et al.*, 1994 a, b; Iijima *et al.*, 1998; Olsen *et al.*, 1998; Hoehne-Reitan *et al.*, 2001b, c; Murray *et al.*, 2003; Perez-Casanova *et al.*, 2004, 2006).

The FA specificity of lipolytic enzymes has been substantially studied and the effects of FA on the activity of pancreatic lipase are related to both the acyl chain length and degree of saturation (Brannon, 1990; Linscheer & Vergroesen, 1994). Gjellesvik (1991) has performed a detailed comparative study between human and Atlantic cod BSDL and determined that although the enzymes are homologous, there are minor but significant differences in FA specificity – while the human enzyme was found to be more effective in hydrolysing shorter chain saturated FA esters (even though its major physiological substrates are long chain FA), in Atlantic cod it was more active towards esters of long chain PUFA (up to 22 carbon atoms). Since then, many other authors have reported FA specificity with a preference for PUFA as substrate (regardless of position), followed by monounsaturated FA (MUFA) and finally saturated FA (SFA). Within MUFA and SFA, digestibility appears to decrease with increasing chain lengths (Austreng *et al.*, 1979; Lie & Lambertsen, 1985, 1991; Gjellesvik, 1991; Koven *et al.*, 1994b; Lie *et al.*, 1987; Iijima *et al.*, 1998; Olsen *et al.*, 1998; Johnsen *et al.*, 2000).

The digestion of dietary PL occurs entirely in the small intestine by pancreatic PLA_2, which acts at the *sn*-2 position to yield a lysophospholipid and a FA. PLA_2 is secreted as an anionic zymogen and is activated by tryptic cleavage in the presence of calcium ions. Its activity is also dependent on the presence of bile salts, at a 2:1 bile salt to phosphatidylcholine (PC) molar ratio for optimal activity (Carey *et al.*, 1983; Nordskog *et al.*, 2001). PLA_2 activity has also been found in several fish species, with similar activities and dependence characteristics as the mammalian form (Iijima *et al.*, 1990; 1997; Ozkizilcik *et al.*, 1996; Izquierdo & Henderson, 1998).

Lipolytic capacity during larval development

Several studies have shown significant levels of pancreatic and intestinal digestive enzymes at the onset of exogenous feeding (Fig. 11) (Cousin *et al.*, 1987; Izquierdo *et al.*, 2000; Hoehne-Reitan *et al.*, 2001c) and have found nutrient absorption capacities even before complete absorption of yolk reserves (Pedersen *et al.*, 1987; Loewe & Eckman, 1988; Kjørsvik *et al.*, 1991a; Segner *et al.*, 1994; Diaz *et al.*, 1997; Calzada *et al.*, 1998; Ribeiro *et al.*, 1999a, b). Morphological observations of the digestive tract just after the start of exogenous feeding revealed the presence of lipid vacuoles in the larval intestine of several species, e.g. Dover sole (*Solea solea*), summer flounder (*Paralichthyus dentatus*), gilthead seabream, and Senegalese sole, showing evidence of efficient absorption of dietary FA (Boulhic & Gabaudan, 1992; Bisbal & Bengtson, 1995; Sarasquete *et al.*, 1995; Ribeiro *et al.*, 1999a).

Lipase activity was detected as early as 2 dph – corresponding to first-feeding – in the brush border and cytoplasm of enterocytes of Senegalese sole (Ribeiro *et al.*, 1999a). In red drum (*Sciaenops ocellatus*), lipase specific activity was detectable at hatching, peaked on day 3, just prior to first-feeding, and subsequently decreased (Lazo *et al.*, 2000). Similarly, Bolasina *et al.* (2006) found a high lipase activity at the time of first-feeding (2 dph) in Japanese flounder, decreasing significantly from 4 to 6 dph and then gradually rising until the period of settlement, when a second decrease was measured. Additionally, Hoehne-Reitan *et al.* (2001c) noted that BSDL is present in turbot immediately after hatching (Fig. 11) and increased at the time of first-feeding, while in haddock and Atlantic cod BSDL transcripts and enzymatic activity were noted at hatching and transcript level increased during the first 4 days in haddock (Perez-Casanova *et al.*, 2004). In gilthead seabream (Izquierdo *et al.*, 2000) and winter flounder (Murray *et al.*, 2003) larvae, BSDL activity was detected at first-feeding and increased significantly during the first days of seabream larval development, just as in Atlantic halibut, where BSDL expression was detectable in the larval pancreatic tissue by RT-PCR and *in situ* hybridisation just prior to first-feeding (Murray *et al.*, 2006). In the case of California halibut (*Paralichthys californicus*), an increase in lipase specific activity was measured from hatching to 15 dph, followed by a slight decrease until 18 dph and a new progressive increase afterwards, until 30 dph (Alvarez-González *et al.*, 2006). Changes in lipolytic activity during the ontogeny of striped bass were studied by Ozkizilcik *et al.* (1996), who found low activities of TAG hydrolase, PLA_2 and wax ester (WE) hydrolase until 5-6 dph. At the time of hatching, PLA_2 was the dominant lipolytic enzyme and with the onset of feeding all enzymes increased significantly, although less drastically for PLA_2. TAG hydrolase became the lipolytic enzyme with the highest activity after feeding was initiated. In addition, feeding stimulated the enzymatic activity of TAG hydrolase and PLA_2, whose activity was significantly higher in fed, compared to unfed larvae. The authors have estimated that first-feeding striped bass larvae are able to digest 47% of their daily lipid ingestion – more specifically, 17, 71 and 94% of dietary TAG, WE and PL, respectively.

There are thus strong indications that fish larvae are capable of digesting and absorbing lipids from the start of exogenous feeding, which is not surprising having in mind that they feed mostly on zooplankton that may contain up to 20% of their dry weight as lipids (Sargent *et al.*, 1989). However, there is very little information concerning the maximal digestive and absorptive capacity for lipids and FA in marine fish larvae and how it may be affected by compositional differences of lipids in the diet.

Lipolytic efficiency

In the larvae of teleost fishes studied so far, the exocrine pancreas is functional since pancreatic and intestinal enzyme activity is detected before first-feeding, suggesting that enzymatic capacity is not induced by the diet (Ribeiro *et al.*, 1999a, b; Hoehne-Reitan *et al.*, 2001c; Zambonino-Infante & Cahu, 2001). Therefore, as in mammals (Henning, 1987), the temporal pattern of development is genetically determined, at least in the early larval stages, although this may change later in development. In turbot larvae, for instance, BSDL activity appears to follow an ontogenetically programmed pattern of development during the initial larval period, up to 5 dph, while a dietary effect started to be noticeable at 7 dph, when the BSDL content increased exponentially in fed larvae and decreased in starved larvae (Hoehne-Reitan *et al.*, 2001c). In Pacific threadfin (*Polydactylus sexfilis*) larvae a similar stimulation of lipase activity by feeding and a decrease in unfed larvae was also observed at the time of first-feeding (Kim *et al.*, 2001). Lipase activity in Japanese flounder larvae was also found to be very sensitive to nutritional condition, decreasing when larvae stop feeding and increasing rapidly when feeding restarted (Bolasina *et al.*, 2006). However, Alvarez-González *et al.* (2006) did not find significant differences in lipase specific or individual activity between fed and fasted California halibut larvae when the first three days after first-feeding were compared.

Diet quality appears to have a direct effect on the onset of the maturation processes of the digestive tract (Cahu *et al.*, 1995; and see also reviews by Cahu & Zambonino-Infante, 2001; and Zambonino-Infante & Cahu, 2001). In particular, dietary lipids have been observed to induce an earlier maturation of enterocytes in European seabass (Zambonino-Infante & Cahu, 1999) and red drum larvae (Buchet *et al.*, 2000). In addition, total dietary lipid content has been reported to stimulate lipolytic enzymes both in mammals (Wicker & Puigserver, 1989; Brannon, 1990; Spannagel *et al.*, 1996) and fish (Borlongan, 1990; Zambonino-Infante & Cahu, 1999). Studies looking at the quantitative lipid supply in marine fish larvae have described a stimulation of the lipolytic activities and mRNA levels of pancreatic lipase and PLA_2 by an increase in their substrate level in the diet, until a plateau is attained (Zambonino-Infante & Cahu, 1999; Buchet *et al.*, 2000) (Fig. 13). The existence of these plateaus of lipolytic enzyme expression thus suggests that there is a maximal capacity for enzyme synthesis (Zambonino-Infante & Cahu, 1999, 2001; Buchet *et al.*, 2000). In Atlantic herring, tube feeding increasing amounts of radiolabelled triolein resulted in a steep rise in evacuation, while only marginally enhancing label absorption (Morais *et al.*, 2005a;

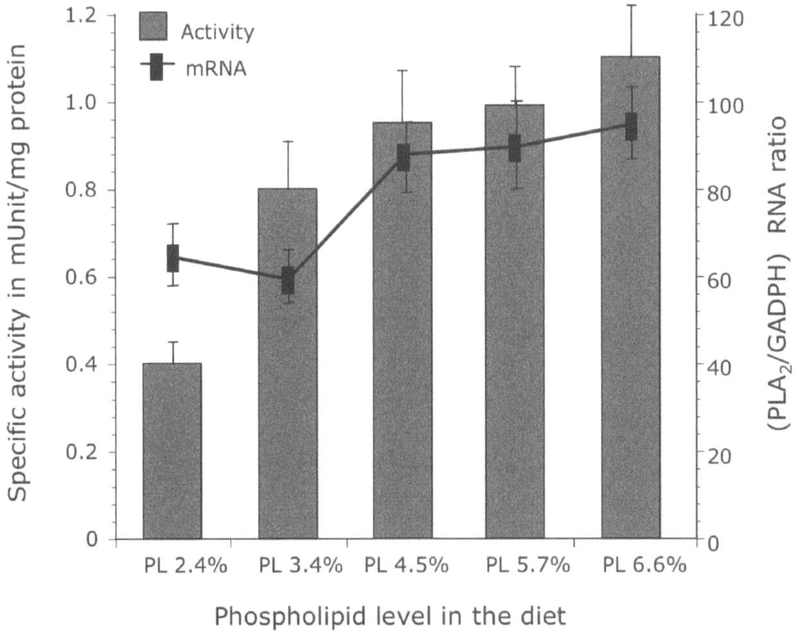

Fig. 13 Activities and mRNA levels of phospholipase A2 (PLA_2) in European seabass (*Dicentrarchus labrax*) larvae fed different dietary phospholipid levels. After Zambonino-Infante & Cahu (2001). Reprinted with permission from the author and Elsevier. mRNA level of PLA_2 was normalised by the mRNA level of the housekeeping glyceraldehyde-3-phosphate dehydrogenase (GADPH) gene. PLA_2 enzymatic activity is expressed as specific activity in mUnit/mg protein, being U = μmol substrate hydrolysed min^{-1}.

Fig. 14). Even though an inadequate digestive capacity could not fully explain the results, that suggested an important effect at the level of lipid absorption, the massive and rapid evacuation of the tube fed lipid at higher doses might have been caused by it being well beyond the plateau of enzymatic capacity of the Atlantic herring larvae (Morais *et al.*, 2005a).

In white sturgeon (*Acipenser transmontanus*) an increased activity of the brush border lipase was shown histochemically in larvae fed higher lipid diets (Gawlicka *et al.*, 2002). However, in larvae of turbot (Hoehne-Reitan *et al.*, 2001b), European seabass (Morais *et al.*, 2004c) and Senegalese sole (Morais *et al.*, 2006b), no stimulation of lipase synthesis was obtained by an increase in dietary lipid level. It has been suggested that the intestinal receptors for intraluminal stimulants require only a very low concentration for activation, meaning that the length of the intestine exposed to these products is the main determinant of pancreatic secretory response (Singer, 1987). If this is also the case in marine fish larvae, the ingestion rate of the diet may play a major role in determining lipase activity. The results of Hoehne-Reitan *et al.* (2001b), showing that the BSDL content of turbot larvae was not significantly affected by the lipid level of

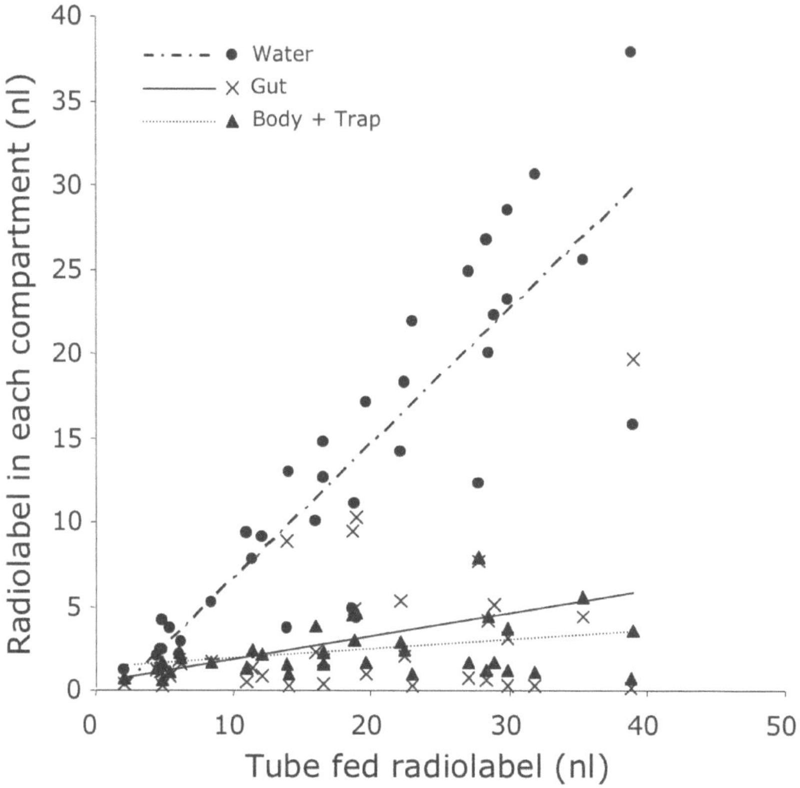

Fig. 14 Volume of radiolabel (nL) in each compartment (evacuated into the water; absorbed into gut; or absorbed into body and catabolised, i.e., found in body + metabolic trap), 24 h after tube feeding Atlantic herring (*Clupea harengus*) larvae a range of ^{14}C-triolein volumes. Lines correspond to linear regression lines fitted to the data points. After Morais *et al.* (2005a). Reprinted with permission from Blackwell Publishing. The Ho: $\beta=0$ was rejected by ANOVA, for the three analysed compartments. ANCOVA: $p<0.001$, with the regression slope of "water" being significantly different from "gut" and "body + trap".

the prey, but appeared to be a function of the ingestion rate, are in agreement with this hypothesis. In addition, Perez-Casanova *et al.* (2004) also suggested that BSDL synthesis in haddock larvae might be stimulated by the amount of ingested prey.

In spite of the scarcity of studies, evidence so far indicates that lipase activity may not be limiting for larval growth in normal hatchery conditions, at least when feeding on live prey. In larval red drum fed a microparticulate diet, live prey or both, digestive enzyme activity was not the limiting factor for growth (Lazo *et al.*, 2000). Furthermore, in European seabass larvae, despite lipase specific activity being shown to be affected by dietary lipid (in relation to FA composition but not lipid level), these effects were not sufficient to affect the larval growth (Morais *et al.*, 2004c). On the other hand, a high enzyme-substrate ratio has also been measured for proteolytic enzymes in larval

Atlantic herring, and the possibility for a very fast proteolysis may explain the fast evacuation rates observed in these larvae (Pedersen, 1984; Pedersen *et al.*, 1987). Lipase may thus be produced in sufficient amounts for the digestion of all ingested prey, as suggested by Hoehne-Reitan *et al.* (2001b).

Radiotracer studies which analysed the digestion and absorption of tube fed TAG (^{14}C-triolein), compared to FFA (^{14}C-oleic acid), in larvae of Atlantic herring (Morais *et al.*, 2005a) and Senegalese sole (Morais *et al.*, 2005c, 2006b), revealed that the FFA were absorbed at higher rates. This indicates that the digestive capacity was not able to fully deal with the tube fed neutral lipid. On the other hand, when oleic acid (OA) esterified in PC (as L-3-phosphatidylcholine 1,2, di-[1-^{14}C]oleoyl) was tube fed, less was evacuated and more absorbed compared to OA esterified in TAG. This suggests that, in Senegalese sole larvae, neutral lipase activity may be less efficient than that of PLA_2 or that triolein may have had a more deficient emulsification (Morais *et al.*, 2005c; Fig. 15). In addition, PC and free OA were absorbed and metabolised by the larvae in a comparable way, which further indicates a very high efficiency of PLA_2 digestion and absorption of the products of PL digestion (Morais *et al.*, 2005c).

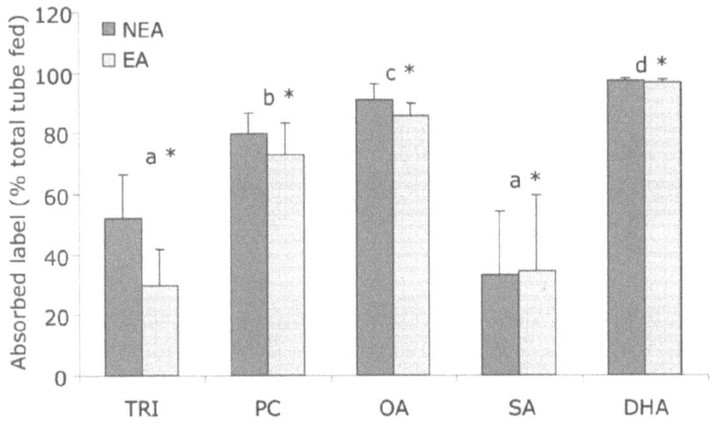

Fig. 15 Percentage of absorbed label (total tube fed minus evacuated), 24 h after tube feeding Senegalese sole (*Solea senegalensis*) with one of several ^{14}C-labelled lipids (triolein, TRI or L-3-phosphatidylcholine 1,2, di-oleoyl, PC) or free fatty acids (oleic acid, OA; stearic acid, SA or docosahexaenoic acid, DHA). Larvae were reared on non-enriched *Artemia* (NEA) or *Artemia* enriched on a soybean oil emulsion (EA) from 16 to 32 days after hatching (DAH) and prior to tube feeding, at 32 DAH, larvae were fed a meal of the corresponding treatment – NEA or EA. A two-way ANOVA revealed significant differences between the NEA and EA treatments (indicated by an asterisk) and between the different labels, independently of the dietary treatment (indicated by different letters). Although the analysis performed with all tube fed labels indicated significant differences between the NEA and EA treatments, a one-way ANOVA revealed significant differences only for the TRI treatment and a non-significant trend was noted for the PC and OA treatments, while no differences between dietary treatments occurred for SA and DHA. After Morais *et al.* (2005c). Reprinted with permission from The Nutrition Society.

The quality of the dietary lipid source is also an important parameter that may affect gastrointestinal tract development and digestive enzyme activity. TAG is quantitatively the most important lipid class in fish diets, but the constituent FA composition is known to affect digestion and absorption (Austreng *et al.*, 1979; Koven *et al.*, 1994b; Linscheer & Vergroesen, 1994; Olsen *et al.*, 1998; Johnsen *et al.*, 2000; Morais *et al.*, 2005c). Studies of the FA specificity of lipolytic enzymes revealed that the effects of FA on the activity of pancreatic lipase are related to both the acyl chain length and degree of saturation. The FA content of the diet might act directly on the regulation of enzymatic activity, either through increased enzyme-substrate stability and/or enhanced pancreatic secretion (Singer, 1987; Brannon, 1990). Therefore, when studying the effect of dietary lipid on lipase activity, its quantity cannot be dissociated from its source (i.e. FA composition). However, very few studies have focused on the effects of lipid quality on lipase activity, in fish in general and particularly in fish larvae. Izquierdo *et al.* (2000) reported that neutral lipase activity is influenced by the FA composition of dietary lipids, as gilthead seabream larvae showed an increased lipase activity when they were fed rotifers containing TAG rich in 20:5n-3 instead of MUFA. Similarly, in seabass larvae, at 24 dph, Morais *et al.* (2004c) observed a higher lipase activity in larvae that were fed a PUFA-rich fish oil diet, compared to a MUFA-rich triolein diet. However, this effect was age dependent, as at 52 dph the diet containing fish oil caused a significant reduction in lipase specific activity. A third diet formulated with coconut oil was also tested by Morais *et al.* (2004c) where the effect on lipase activity was more constant, with a high lipase activity being induced at 24 and 52 dph. The age-dependence of the digestive response to the fish oil-based diets could not be fully explained and should be the focus of further studies.

The results obtained with European seabass larvae suggest the existence of a regulatory mechanism of neutral lipolytic enzyme secretion and activity according to the dietary FA composition. Studies of this nature are particularly difficult given the multiplicity of factors (some of which are still not well studied in fish) that may intervene and interact – the effect of FA structure on lipid digestibility, stimulation of neuroendocrine pathways and other post-absorptive mechanisms. For instance, in mammals, unsaturated lipids induce lipase activity more than saturated lipids, but only when dietary lipid intake is moderate and the lipase content is not maximal (Brannon, 1990).

ABSORPTION

Absorption of Proteins, Peptides and Amino Acids

Processes and mechanisms

Fish larvae have been seen to absorb large molecules, including proteins, by endocytosis followed by intracellular digestion in the mucosal cells of the hindgut (Iwai & Tanaka, 1968; Watanabe, 1984; Kurokawa *et al.*, 1996; Luizi *et al.*, 1999). On the basis of these observations it has been suggested that low levels of intestinal proteolytic activity may be compensated for by hindgut protein endocytosis. However, Watanabe (1984) found that horseradish peroxidise, taken up by hindgut pinocytosis, took 10–24 hr to be degraded intracellularly in larvae of pond smelt (*Hypomesus transpacificus nipponensis*). Therefore, it has been argued (Rønnestad *et al.*, 2001) that such processing rates of dietary proteins are too low to satisfy the high metabolic and anabolic demands of rapidly growing fish larvae. McLean *et al.* (1999) suggested that a primary role of macromolecule absorption in the fish hindgut may be antigen sampling, as found in mammals. A similar conclusion was drawn by Wallace *et al.* (2005), who administered horseradish peroxidise and termed the cells that had been absorbing it as antigen sampling cells. Experimentally, it has not been possible to quantify this route of absorption and its dietary importance remains obscure.

In quantitative terms the main absorption of dietary proteins, peptides and AA in larval fish has been proposed to occur in the midgut (Rønnestad & Conceição, 2005), as in other vertebrates (Stevens & Hume, 1996). However, almost nothing is known regarding the mechanisms and ontogenetic changes of protein, peptide and FAA absorption in fish. In general, AA absorption involves many processes, some of them overlapping. The brush border membrane of the adult enterocytes expresses numerous AA transporters (AATs) (Ray *et al.*, 2002). The transport rates for FAA observed *in vivo* (Rønnestad *et al.*, 2000c; Applebaum & Rønnestad, 2004) are a function of several factors including the luminal concentration of AA, the transport affinity and capacity for each AAT, and the amount of each AAT present in the epithelium. In mammals, several of the genes coding for AATs have been recently cloned. These transporters are a diverse group of integral membrane proteins sharing several attributes (Ray *et al.*, 2002; Wipf *et al.*, 2002). They typically exhibit broad substrate specificity (several AA may be transported by one type of transport system) allowing them to recognise, bind and transfer AA across cell membranes. Existing data suggest that there are many genetic and splice variants within a group of what were once thought to be distinct AATs. Recent work has started to characterise *in vivo* FAA absorption rates in fish larvae (Rønnestad *et al.*, 2000c; Applebaum & Rønnestad, 2004). However, current data have not demonstrated a saturable component for FAA absorption (Applebaum & Rønnestad, 2004), and more studies using higher FAA concentrations need to be done in order to understand the absorption kinetics.

Vertebrate enterocytes express a variety of transporters responsible for AA transport, but so far only one intestinal transporter has been described for peptides; the oligopeptide transporter 1 (PepT1; Nielsen & Brodin, 2003; Daniel, 2004). The molecular characterisation of peptide transport has just begun in fish. The complete mRNA coding sequence for PepT1 has recently been reported in two teleosts, the zebrafish (Verri *et al.*, 2003) and Atlantic cod (Rønnestad *et al.*, 2006, 2007). For both transporters hydropathy analysis predicted at least 12 potential membrane spanning domains with a large extracellular loop between transmembrane domains IX and XI. Based on *in vitro* studies combined with computer based simulation experiments, there is a detailed knowledge concerning the minimum requirement and rigidity of the chemical design of the molecules that will cross the human PepT1 transporter (Nielsen & Brodin, 2003;

Vig *et al.*, 2006). Comparable studies of PepT1 from teleosts will give new insight into the constraints and possibilities for delivering peptides to fish, including during their early stages.

In teleosts, the presently available data suggests that the PepT1 transporter is expressed at hatching, both in zebrafish (Verri *et al.*, 2003) and Atlantic cod (Rønnestad *et al.*, 2006). The study with zebrafish demonstrated that PepT1 is highly expressed in the proximal intestine. *In situ* hybridisation data from Atlantic cod confirmed that the PepT1 transporter is expressed in the enterocytes (Rønnestad *et al.*, 2007). Functional analysis by *in vitro* expression systems categorised the zebrafish PepT1 as a classical low affinity/high capacity system (Verri *et al.*, 2003). The zebrafish PepT1 is a H^+-dependent peptide co-transporter, but in contrast to higher vertebrates, where the maximal transport activity is independent of extracellular pH, the maximal transport rates increase at alkaline extracellular pH (Verri *et al.*, 2003).

Studies from mammals have shown that proteins may not be fully degraded prior to absorption, and the entry of peptides through the apical membrane of the enterocyte is an important route of absorption of dietary protein (Daniel, 2004). This has not been investigated in larval fish, but recent studies have demonstrated that rainbow trout may be reared from the onset of exogenous feeding on diets using synthetic dipeptides as the source of AA (Dabrowski *et al.*, 2003, 2005). Another recent study has also demonstrated that the supply of peptides to balance the dietary AA profile increases AA retention in fish (Aragão *et al.*, 2004). Several studies suggest that the dietary inclusion of hydrolysed protein may eventually compensate for any deficiency in the larval capacity for processing dietary intact proteins (Dabrowski, 1984; Kaushik, 1995; Carvalho *et al.*, 1997; Zambonino-Infante & Cahu, 2001). An obvious step for future studies would be to functionally characterise and quantify the differences in absorption kinetics of peptides and FAA by the digestive system, and equally important, how the larval metabolism reacts to differences in the rates of AA influx in terms of catabolism and retention (Rønnestad & Conceição, 2005).

Effect of the degree of protein hydrolysis on absorption rates

The methodological problems associated with *in vivo* analysis of absorption rates of AA in larval fish equals those that assess their digestibility. In a recent approach, *in vivo* tube feeding followed by hot chase, has been used to acquire data on absorption rates and efficiency of various FAA, hydrolysates and whole proteins (Rust *et al.*, 1993; Rønnestad *et al.*, 2001). The hot chase approach consists of a mass balance method that follows the compartmental distribution of a ^{14}C-labelled nutrient in solution or as part of a diet that has been administered directly into the digestive tract. In combination with staged sampling, dissection of the digestive tract, quantification of $^{14}CO_2$ excretion and scintillation counting, the transfer kinetics into the body can be assessed (Tonheim *et al.*, 2004, 2005; Fig. 16). In these studies, the ^{14}C-SSP was administered to Atlantic halibut by tube-feeding. Tonheim *et al.* (2004) digested ^{14}C-SSP (which was a mixture of peptides from 15 to > 250 kDa) at two levels; A) with pepsin (pH 4.0), and B) with

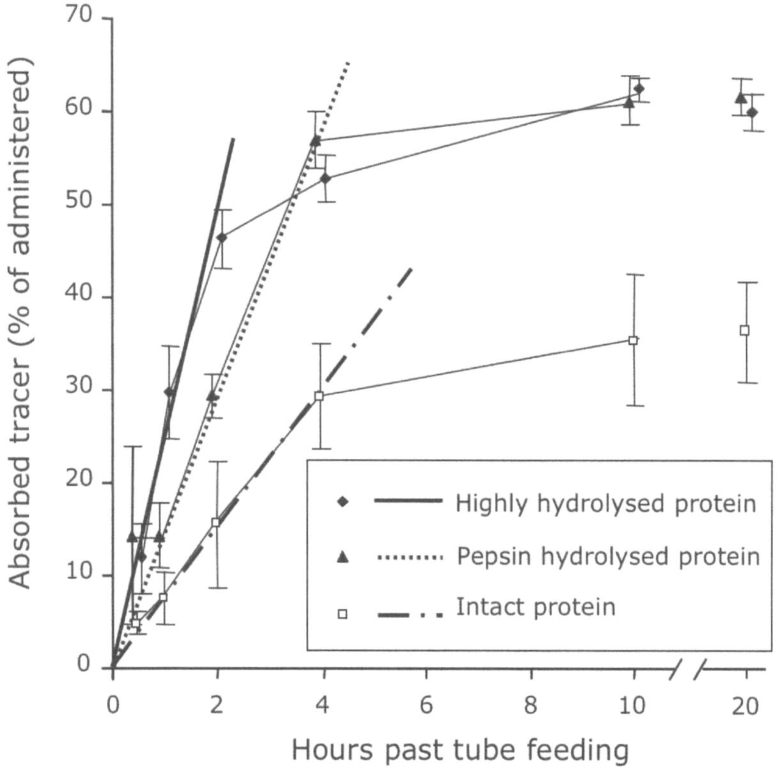

Fig. 16 Absorption rates of a ^{14}C-labelled salmon serum protein, SSP, into body tissues of Atlantic halibut (*Hippoglossus hippoglossus*) larvae. After Tonheim *et al.* (2005). Reprinted with permission from Elsevier. The SSP was administered either intact or hydrolysed to two different levels (Tonheim *et al.*, 2004). The absorption process was analysed using tube feeding followed by a hot chase approach involving time sampling, dissection, and scintillation counting (Rønnestad *et al.*, 2001).

pepsin, followed by additional hydrolysis with trypsin and other proteases (pH 8.0). SDS-PAGE analysis showed that pepsin hydrolysis resulted in a variety of peptides with a broad size range, while further treatment with trypsin and additional proteases resulted in a product with almost no peptide sizes above ~14 AA. This treatment produced preparations that differed in the molecular complexity of AA but with identical total AA profile and AA specific activity. A time-course study using the ^{14}C-SSP administered by tube-feeding to Atlantic halibut, revealed that the absorption rates of the pepsin hydrolysed protein and the more extensively hydrolysed protein were 2.2 and 3 times faster than the intact protein, respectively (Tonheim *et al.* 2005; Fig. 16).

Absorption of Lipids

Processes and Mechanisms

Until recently it was thought that the uptake of lipid digestion products by the enterocytes was through a passive diffusion process. The dispersed lipids, being bilayers themselves, are very soluble in membranes. Given their high intraluminal concentration, it is possible that lipids can move passively down their concentration gradient from the gut lumen via apical membranes of the enterocytes to the cytosol (Carey *et al.*, 1983; Nordskog *et al.*, 2001). However, findings by several investigators (reviewed by Nordskog *et al.*, 2001) have raised the possibility that some lipids may be taken up via energy-dependent carrier mediated processes and have indicated the existence of a FA binding protein (FABP) associated with the brush border membrane which plays a role in the absorption of FA (particularly of long chain FA) by the enterocytes. Sire & Vernier (1981) suggested the presence of a transport protein with characteristics similar to the mammalian FABP, which was used to explain the early absorption peak when rainbow trout were fed on a PUFA rich diet. Later on, André *et al.* (2000) isolated an intestinal FABP cDNA in zebrafish larvae and described its expression during gut morphogenesis. These authors showed that, contrary to mammals, where intestinal FABP mRNA and protein levels decrease gradually along the antero-posterior axis of the gut epithelium, in zebrafish larvae there is a strong regional expression of this gene. The expression is established before complete yolk resorption and first-feeding, and sharply demarcates the anterior section of the intestine (André *et al.*, 2000; Fig. 17). This confirms that the major site of lipid absorption and intracellular storage of lipid droplets in the enterocytes is in the proximal intestine in teleost larvae. Sharma *et al.* (2004) subsequently characterised the sequence of the gene encoding an intestinal FABP in zebrafish and detected its expression as early as 12 hours post-fertilisation and in several organs, such as in the primordial intestine, liver and pancreas. As for the transport of lyso-PC, it is still controversial in mammals and questions exist as to whether it is passive or carrier-mediated (Tso & Fujimoto, 1991).

Once absorbed by the enterocyte, the products of lipid digestion migrate from the site of absorption to the smooth ER, where complex lipids are resynthesised, being deposited in large lipid droplets of the mucosal epithelial cells. In the enterocyte, 2-MAG and FA enter the monoacylglycerol pathway, in which 2-MAG are reacylated into TAG through the consecutive action of MAG acyl transferase and DAG acyl transferase. The enzymes involved in this pathway are known as the TAG synthetase complex and are located on the cytoplasmic surface of the smooth ER (Tso & Fujimoto, 1991; Nordskog *et al.*, 2001). There is a second pathway in the intestinal mucosa that forms TAG, which is the α-glycerophosphate pathway. In this case, there is a stepwise acylation of glycerol-3-phosphate to form phosphatidic acid that is hydrolysed to form DAG and then converted to TAG. PL, as well as cholesterol esters, may also be synthesised through this route. This pathway may occur in both the rough and smooth ER. The relative importance of the MAG pathway and the α-glycerophosphate pathway depends on the supply of 2-MAG and FA but the MAG pathway is normally the most

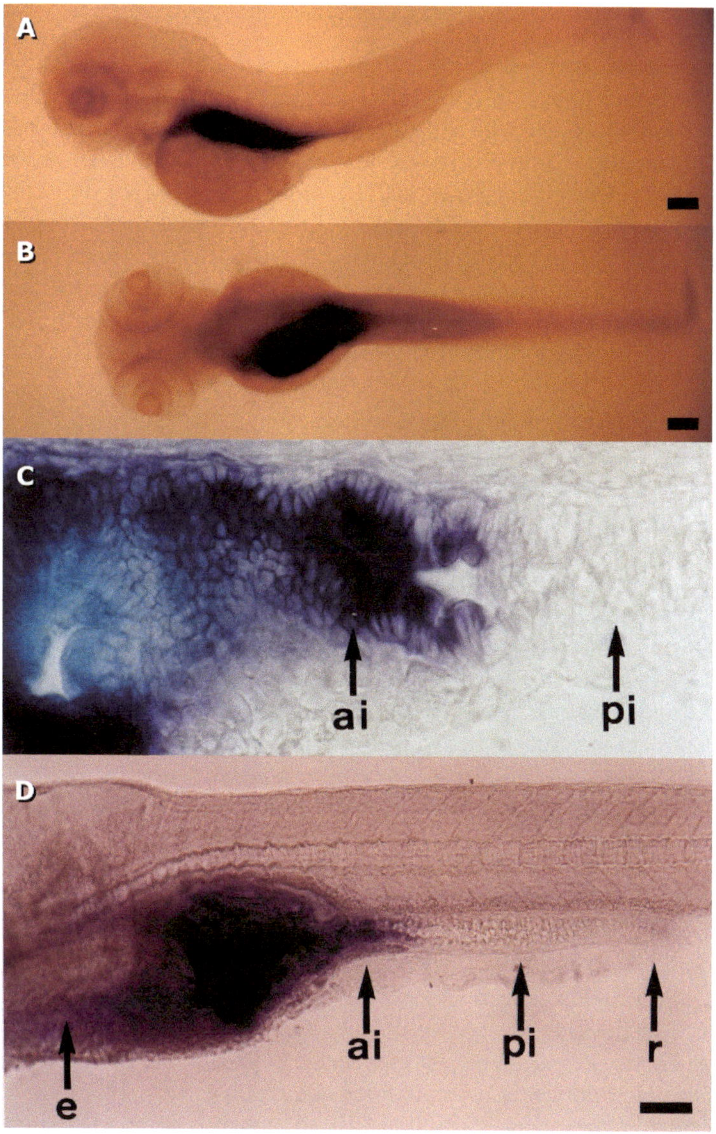

Fig. 17 Whole-mount *in situ* hybridisation of intestinal fatty acid binding protein (*i-fabp*) expression during zebrafish (*Danio rerio*) larval development. (A,B) Four-day larva, lateral (A) and dorsal (B) view. (C) Parasagittal section of a 7-day larva, showing the transition of the anterior intestine (ai) to the posterior intestine (pi). (D) Fifteen-day larva, lateral view. The hybridisation signal is coloured dark brown to blue. A strong *i-fabp* expression was observed in the anterior region of the intestinal tube, while only low levels of transcript were found in the posterior intestine (pi) region. e: oesophagus; r: rectum. Scale bars = 100 μm. After André *et al.* (2000). Reprinted with permission from UBC Press.

important for the resynthesis of TAG. During normal lipid absorption, when 2-MAG is present in sufficient amounts, the MAG pathway is preponderant and inhibits the alternative pathway. Only when the supply of 2-MAG is lacking or is insufficient, the α-glycerophosphate pathway becomes the major route for the formation of TAG (Tso & Fujimoto, 1991; Nordskog *et al.*, 2001). As for the absorbed lyso-PC, a fraction is reacylated into PC which may rapidly appear in lipoprotein membranes on the enterocyte and be used for the transport of the lipids out of the cell as chylomicrons. The remaining is further hydrolysed to glycero-3-phosphorylcholine (which is then transported via the portal blood for use in the liver), glycerophosphate (partly re-utilised to form TAG), glycerol and phosphate (Borgström, 1977; Tso & Fujimoto, 1991; Nordskog *et al.*, 2001).

After the products of lipid digestion have been reconverted to TAG, they are packed together with PL, dietary cholesterol and specific proteins (apolipoproteins) into large lipoprotein complexes, mainly chylomicron particles and VLDL, which consist of a hydrophobic core containing neutral lipids (TAG and cholesterol ester) and a hydrophilic coat with the more polar components (PL and protein). These are then collected in the lymphatic system, enter the blood vascular system through the thoracic duct (in mammals) and are transported to the liver and other tissues, for storage or energy metabolism (Rogie & Skinner, 1985).

Studies performed on several fish species have shown that the digested FA are absorbed and re-esterified into TAG in the intestinal cells, resulting in the production of a lipoprotein aggregate (chylomicron or VLDL-type) (Sire *et al.*, 1981; Honkanen *et al.*, 1985; Iijima *et al.*, 1990; Lie & Lambertsen, 1991). Nevertheless, in fish, as a result of the non-specificity of pancreatic lipase, the hydrolysis of dietary TAG is mostly complete, resulting in the production of FFA and glycerol. For this reason, in teleosts, the α-glycerophosphate (or glycerol-3-phosphate) pathway appears to be the major pathway responsible for the biosynthesis of both TAG and PL. In common carp (*Cyprinus carpio*) (Noaillac-Depeyre & Gas, 1974) and rainbow trout (Bauermeister *et al.*, 1979; Rogie & Skinner, 1981, 1985), chylomicrons are produced in the intestine in a manner that is similar to that of mammals, with TAG being synthesised in the smooth ER, processed in the Golgi system and finally discharged into the lateral and basal intercellular spaces and drained into the *lamina propria*. However, the precise role of the Golgi system during the absorption of dietary lipids in fish is still controversial, as movement into the intercellular spaces has also been reported to occur directly from the ER. Thus, the traffic of lipoproteins may follow two alternative pathways: (1) direct exocytosis from the ER or (2) via the Golgi complex, as occurs in mammals (Noaillac-Depeyre & Gas, 1974; Sire *et al.*, 1981; Vernier & Sire, 1986).

The presence of a gene coding for a microsomal triglyceride transfer protein (MTP) large subunit similar to the human form (54% identity, with the most conserved regions corresponding to important functional domains) has been recently described in zebrafish adults and larvae (Marza *et al.*, 2005). In larvae, MTP expression was identified in the yolk syncytial layer during early embryogenesis and in the anterior intestine and liver from 48 h post-fertilisation onwards. Additionally, a significant pre-translational up-regulation of MTP expression was shown in the larval anterior intestine after feeding (Marza *et al.*, 2005). This protein catalyses the lipid transport between membranes and is essential for the assembly and secretion of apolipoprotein B-containing lipoproteins. Its main roles are the transfer of neutral lipids from the ER membrane to nascent apoB in the lumen and to physically interact with apoB, acting as a chaperone, preventing apoB degradation (Jamil *et al.*, 1995; Marza *et al.*, 2005). Although a few differences were noted and discussed, the study conducted by Marza *et al.* (2005) indicated that human and fish MTP have similar properties for membrane binding and lipid transfer activity, thus reinforcing the suggestion of the existence of important parallels between the digestive and absorptive processes of fish and mammals.

Fish are characterised by producing smaller lipoproteins with more polar components at the surface than their human counterparts. Due to the predominance of the α-glycerophosphate pathway, PL synthesis in fish is higher than in mammals, for the same dietary intake of lipids. On the other hand, fish diets contain high concentrations of protein and the anterior intestine is extensively capable of absorbing the AA resulting from protein digestion. Taken together, these observations could explain why fish epithelial cells mainly synthesise particles closer in size to VLDL compared to mammalian chylomicrons (Sire & Vernier, 1981; Sire *et al.*, 1981; Babin & Vernier, 1989). In addition, there is evidence that the size of the lipoprotein varies according to the nature of the ingested meal. Just as in mammals, the degree of saturation of the micellar FA affects the nature of the lipoprotein particles and increasing the unsaturation of dietary lipids increases the lipoprotein size. Thus, after feeding SFA, the particle size is that of VLDL, while feeding unsaturated FA results in the production of chylomicron-type particles (Sire *et al.*, 1981; Olsen *et al.*, 1999). Furthermore, the intracellular pathways of TAG and PL reacylation may be affected by the nature of dietary lipids (Sire & Vernier, 1981; Izquierdo *et al.*, 2000; Olsen *et al.*, 2000a) which, in turn, will determine the amount of lipoprotein surface material being produced (Sire & Vernier, 1981). For instance, van Greevenbroek *et al.* (1995), working with Caco-2 cells derived from a human colorectal carcinoma, noted that palmitic acid (16:0) increased PL synthesis, while the presence of linoleic acid (18:2n-6) in the culture media led to higher amounts of TAG being synthesised. Olsen *et al.* (2000a) found that replacing a part of dietary linseed oil by 16:0 reduced significantly lipid droplet accumulation in the enterocytes of Arctic charr, while Pérez *et al.* (1999) noted that different radiolabelled FA incubated *in vitro* with trout enterocytes showed a different post-absorptive re-esterification fate, particularly concerning PL synthesis. Nevertheless, in a recent experiment conducted by Oxley *et al.* (2005), the replacement of fish oil by vegetable oil in the diet of Atlantic salmon (*Salmo salar*) did not significantly affect the lipid class composition of the gut mucosa or the activity of the enzymes MAG acyltransferase, DAG acyltransferase or DAG cholinephosphotransferase, which are involved in the intestinal reacylation of digested lipid into TAG or PL.

Efficiency

The intensive larval rearing of the majority of marine fish species continues to be based on the provision of cultured live food during the early larval stages, which commonly have a deficient nutritional composition, particularly in terms of EFA. To overcome this nutritional deficiency of *Artemia* and rotifers, enrichment products and protocols have been developed over the last decades to supply these zooplankton with sufficient levels of EFA (Rainuzzo *et al.*, 1994, 1997; Rodríguez *et al.*, 1996; Han *et al.*, 2000). Therefore, in order to meet the EFA requirements of the larvae and thus achieve good growth, survival and larval quality, the general approach is to simulate the biochemical composition of the larva's natural diet (marine plankton), which is particularly rich in (n-3) PUFA. However, the natural marine phytoplankton and zooplankton is composed mostly of (n-3) PUFA-rich PL, while many of the enrichment methodologies use lipid sources rich in TAG, which have a lower (n-3) PUFA content compared to polar lipids (Sargent *et al.*, 1989). Consequently, in order to supply high levels of EFA, live prey fed to larvae may in some cases have excessive total lipid content, with a high neutral lipid composition. In fact, a few studies have reported poor larval growth, even when the EFA composition of the diets appeared to meet larval requirements (Pousão-Ferreira *et al.*, 1999). In order to explain these results, the authors have hypothesised that quantitative and qualitative imbalances in the macronutrient composition of the diet, particularly neutral lipid excess, may adversely affect larval digestion and absorption ability (Kjørsvik *et al.*, 1991b; Pousão-Ferreira *et al.*, 1999; Izquierdo *et al.*, 2000; Olsen *et al.*, 2000b; Hoehne-Reitan *et al.*, 2001b; Gisbert *et al.*, 2005; Morais *et al.*, 2005a; b; c; 2006b) resulting in poor growth and larval performance.

On the other hand, larvae fed TAG rich diets commonly show an accumulation of lipid vacuoles in the basal zone of the enterocytes, which indicates a good digestion and absorption of dietary TAG but a reduced transport capacity (Diaz *et al.*, 1997; Fontagné *et al.*, 1998, 2000; Salhi *et al.*, 1999; Izquierdo *et al.*, 2000; Morais *et al.*, 2005b, c, 2006b). Current data indicate that, when rearing larvae on high lipid diets, lipid absorption into the enterocyte and subsequent transport into the body may be a larger problem than lipid digestion *per se*, although the two factors are probably involved (Izquierdo *et al.*, 2000; Morais *et al.*, 2005a, c; 2006b).

When Senegalese sole larvae where fed *Artemia* enriched with a higher and lower level of neutral lipid, an accumulation of lipid droplets within the enterocytes was induced in the larvae fed on the higher neutral lipid diet (Morais *et al.*, 2005b; c; 2006b), and resulted in a tendency for lower growth (Morais *et al.*, 2005b; c). This enabled studying to what extent these lipid accumulations may influence the capacity of sole larvae to digest and absorb dietary FA and how this might affect larval growth. When tube feeding lipid mixtures containing ^{14}C-OA either in the free form or esterified in TAG ([1-^{14}C]triolein) or in PL (L-3-phosphatidylcholine 1,2, di-[1-^{14}C]oleoyl; PC), a significant effect of the *Artemia* diet was noted on the absorption of the labels, particularly of triolein (Morais *et al.*, 2005c; Fig. 15). The larvae which had been fed a higher neutral lipid diet showed a significantly increased evacuation of the labelled FA, suggesting that the long term feeding of high lipid diets can affect the capacity of larvae to efficiently absorb dietary FA. Hence, the accumulation of lipid droplets observed within the gastrointestinal mucosa may present a physical barrier to efficient lipid absorption, which in turn may explain the lower growth observed in larvae fed the higher neutral lipid diet (Morais *et al.*, 2005b, c). In early studies performed with rats, it was thought that the rate limiting step for the esterification in the intestinal mucosa was the FA absorption from the lumen, but it was later discovered that an increase in the esterification rate resulted in an increased OA flux from the lumen (Borgström, 1977). Therefore, the intracellular FA pool may regulate the uptake into the enterocytes. This may also explain why, in fish larvae, a higher accumulation of lipid in the enterocytes might eventually lead to an impaired FA absorption and increased evacuation. However, the results presented by Morais *et al.* (2005c) also suggest that this effect depends also on the FA nature, as different FA may show a specific pattern of digestion, absorption and utilisation (Fig. 15).

Loewe & Eckman (1988) reported a lipid accumulation in the epithelium of the intestine of coregonid larvae, which was particularly intense between 7 and 13 dpff, but diminished after two weeks. The authors claimed that this was due to a lower basal transport rate across the membrane of the enterocyte during the earlier period of larval development, which becomes more efficient later on. It was suggested that even if the morphological appearance of the intestinal cells remains unchanged, the "biochemical machinery" or the intracellular compartments responsible for the assembly of the lipoproteins might be less developed in the early larval stages (Loewe & Eckman, 1988). Deplano *et al.* (1991) and Sarasquete *et al.* (1995) also observed that in first-feeding European seabass and gilthead seabream there was an accumulation of lipid droplets in the enterocytes that decreased in a later larval stage and explained this by a low capacity for lipoprotein synthesis in early stages as well. An increased capacity for lipoprotein synthesis in more advanced stages of larval development might be explained by a higher availability of surface material for lipoprotein synthesis as a result of an increase in PL synthesis capacity and/or an improved protein digestion in later stages of development (Elbal *et al.*, 2004). In addition, Deplano *et al.* (1991) hypothesised that in the more advanced stages of larval development the capacity for cellular intestinal absorption may be even greater than in the adult, which would explain their rapid growth in spite of the apparently lower differentiation of their digestive tract.

Similar lipid accumulations have been reported in adults of many fish species and they are probably a natural occurrence resulting from slow lipid processing, where there is a temporal separation between FA absorption and secretion by the intestinal tissue (Noaillac-Depeyre & Gas, 1974; Sire & Vernier, 1981). In ectothermic fish there are differences between the rate of luminal lipid digestion and diffusion of FA into the gut epithelia, which are not affected by water temperature, and the rate of reacylation into TAG and lipoprotein synthesis, which are slowed down by low water temperatures (Austreng *et al.*, 1979; Sire & Vernier, 1981; Ruyter *et al.*, 2006). Noaillac-Depeyre & Gas (1974) observed two types of lipid inclusions in the enterocytes of the common carp: small lipid particles or lipoproteins (diameter <100 nm) that normally accumulate in the intercellular spaces and were implicated in the direct transport of absorbed FA into the blood circulation, and lipid droplets (that may reach several microns in diameter), which seem to act as temporary sites of lipid storage. However, unbalanced diets may lead to even greater lipid accumulation in the form of lipid droplets by enhancing the discrepancy between the rates of absorption and lipoprotein synthesis. Thus, the causes and consequences of these accumulations and the mechanisms involved in the changes in enterocyte absorptive efficiency are still not completely understood and further work is necessary to fully understand the mechanisms of FA absorption in the larval enterocytes and how these may be affected by dietary lipid level and composition.

In terms of qualitative lipid imbalances in diets for fish larvae, the effect of PL deficiency has been quite extensively studied (Coutteau *et al.*, 1997), while changes in lipid assimilation and transport (more specifically an abnormal lipid accumulation in the gut enterocytes) due to imbalances in the EFA in the diet, and in particular to an insufficient level of DHA, has only recently been reported in larval striped trumpeter (*Latris lineata*) (Bransden *et al.*, 2005). A beneficial effect of dietary PL supplementation has been observed on larval growth and survival of several fish species (Kanazawa *et al.*, 1983; Kanazawa, 1993; Geurden *et al.*, 1995). A role of PL in enhancing lipid emulsification, in the absence of sufficient levels of larval bile salts, had been suggested by Koven *et al.* (1993) but could not be substantiated in subsequent studies (Geurden *et al.*, 1995; Fontagné *et al.*, 1998; Hadas *et al.*, 2003). The more important effect of dietary PL appears to be related to its role in lipid transport from the enterocytes into the body tissues (Fontagné *et al.*, 1998; Olsen *et al.*, 1999; Salhi *et al.*, 1999; Hadas *et al.*, 2003). Numerous studies have shown that larval diets deficient in PL lead to the accumulation of large amounts of lipid vacuoles in the enterocytes, probably due to insufficient lipoprotein synthesis (Diaz *et al.*, 1997; Fontagné *et al.*, 1998, 2000; Olsen *et al.*, 1999; Salhi *et al.*, 1999; Izquierdo *et al.*, 2000; Morais *et al.*, 2005b, c, 2006b). Lipoprotein production by the enterocyte involves a complex sequence of biosynthetic events that bring together specific apoproteins, PL (mostly PC), cholesterol, cholesterol esters, carbohydrate and reesterified TAG (Vernier & Sire, 1986). Iritani *et al.* (1984) reported that the α-glycerophosphate acyltransferase activity of fish is extremely low compared to other animals and it is now well established that marine fish, particularly in the larval stages, have a limited capacity for endogenous *de novo* PL biosynthesis, which may be insufficient to maintain an optimal rate of lipoprotein synthesis. Therefore, a specific dietary requirement for PL for the transport of lipids through the intestinal membranes (by enhancing lipoprotein synthesis) has been suggested (Geurden *et al.*, 1995; Fontagné *et al.*, 1998; Olsen *et al.*, 1999; Salhi *et al.*, 1999; Hadas *et al.*, 2003). In addition, a higher dietary requirement for PL in the earlier larval stages would be necessary to sustain the fast growth and organogenesis which likely require a high rate of membrane synthesis and turnover. However, in what concerns its role in lipoprotein assembly and lipid export, there is probably not an absolute dietary requirement for PL, as it should depend on the total dietary lipid level (including neutral lipid), as well as on the dietary FA composition (if different FA are used preferentially for TAG or PL biosynthesis). For instance, in common carp larvae, Fontagné *et al.* (2000) observed that 2% PL supplementation in a high lipid diet (12% total lipid) was not sufficient for the optimal absorption of neutral lipid, while the same PL level was efficient in the lower lipid diets (6-8%) tested by Geurden *et al.* (1995). So far, no studies have examined the balance between total NL and PL levels, in order to establish optimal ratios for FA digestion and absorption, enabling an optimised energetic and structural utilisation of the dietary total lipid fraction.

The chemical form in which a FA is supplied in the diet affects its utilisation (digestion, absorption and metabolism) and therefore its bioavailability (Favé *et al.*, 2004). This question is of considerable nutritional interest and should be considered when formulating diets, particularly for marine fish larvae, which are presumed to have immature digestive and absorptive systems. Due to their different chemical structures, TAG and PL have different metabolic pathways, mainly in enzymatic hydrolysis, absorption and incorporation into lipoproteins (Sala-Vila *et al.*, 2004). The advantages of using PL as n-3 highly unsaturated FA (HUFA) carriers instead of TAG in the nutrition of marine fish larvae has been analysed (Geurden *et al.*, 1998; Salhi *et al.*, 1999; Izquierdo *et al.*, 2000, 2001; Gisbert *et al.*, 2005; Morais *et al.*, 2005c). Potential beneficial effects of using PL as the FA vector may be a more efficient digestion by PLA_2, compared to neutral lipases, a lower risk of oxidation of the FA esterified to PL (particularly important for the essential long-chain PUFA) and a more efficient transport of the product of lipolysis to the enterocyte caused by higher levels of lysophospholipids (Favé *et al.*, 2004). Gisbert *et al.* (2005) showed that the best growth and survival performance in European seabass larvae was achieved with a microdiet supplying moderate levels of DHA and eicosapentaenoic acid (EPA) in the PL fraction rather than in TAG, and this was explained by the ability of young larvae to better modulate PLA_2 expression than that of neutral lipase. Indeed, the results obtained by Morais *et al.* (2005c) suggest that PLA_2 activity may be more efficient than neutral lipase in Senegalese sole. On the other hand, Geurden *et al.* (1998) observed a higher incorporation of HUFA in postlarval turbot fed a diet containing HUFA-PL than when providing these FA in the diet as ethyl esters supplemented with PC. This was thought to result from the fact that a part of the PL is not hydrolysed, so that the HUFA are absorbed esterified to the PC molecule, which might be more efficient than the uptake of HUFA in the FFA form. Given the importance of these nutrients for larval growth and proper development, this subject deserves further attention.

Another aspect which needs to be carefully considered is the FA composition of the dietary lipid, as it has been hypothesised that the dietary FA composition may influence not only the composition, morphology and fluidity of the intestinal cells but also the physiological mechanisms involved in intestinal lipid metabolism and transport (Sire & Vernier, 1981; Caballero *et al.*, 2002, 2003). For instance, in carnivorous juvenile and adult fish, the inclusion of plant-derived oils in the diet has been shown to potentially result in a reduction of the transport rate across the gut epithelia and thus in the accumulation of lipid droplets in the enterocytes, possibly leading to tissue damage and compromised gut integrity (Olsen *et al.*, 1999, 2000a, 2003; Caballero *et al.*, 2002;

2003). One of the potential metabolic effects of different sources of dietary lipids might be to induce changes in the intracellular pathways of TAG and PL reacylation (Sire & Vernier, 1981; Pérez *et al.*, 1999; Izquierdo *et al.*, 2000; Olsen *et al.*, 2000a), even though Oxley *et al.* (2005) did not find evidence of this, following replacement of fish oil by vegetable oil in the diet of Atlantic salmon. Additionally, an increase in the dietary lipid level causing a change in the FA composition of the brush border membrane (more specifically, a decrease in the level of PUFA and a concomitant increase in MUFA) of European seabass juveniles has been reported by Cahu *et al.* (2000) and was linked to a significant reduction in the brush border enzyme activities, possibly as a result of changes in membrane fluidity. Work carried out with Senegalese sole larvae by tube feeding FFA differing on chain length and degree of saturation (Morais *et al.*, 2005c) showed quite clearly that different FA are absorbed to differing degrees. The results confirmed that, in fish larvae, FA absorption increases with unsaturation, with the order of decreasing absorption having been DHA (22:6n-3) > OA (18:1n-9) > stearic acid (18:0) (Fig. 15). Therefore, just as noted for the dietary effect of lipid on lipase activity, when analysing the effect of neutral lipid on absorption efficiency, its level cannot be dissociated from its biochemical composition.

FUTURE PERSPECTIVES

A substantial amount of empirical knowledge has been gathered in the last decades and in some cases contradictory results have been obtained either due to species-specific differences or to differences in experimental design. There is still a very poor knowledge of basic physiological mechanisms and responses to dietary variations in fish larvae. Such studies are of great importance in order to be able to predict and anticipate dietary effects. While the larval digestive capacity and the timing of development of digestive organs have been well studied, very little is known concerning the absorptive process, which will ultimately determine how much of the dietary nutrients will actually be assimilated from larval diets. Other areas of importance in the future, which have been little examined so far or studied in a dispersed way, are the interactions between different mechanisms and phases of larval digestive physiology, such as interactions between ingestion/satiety, gut transit rate, digestion and absorption, as well as their regulation by post-absorptive mechanisms (e.g. stimulation of neuroendocrine pathways). Furthermore, a better understanding of the basic physiological and metabolic processes in fish larvae in response to different diets is considered essential for the development of a sustainable aquaculture with growing needs, namely in terms of the development of formulated diets which can satisfy larval requirements from first-feeding. The need for more integrated and holistic approaches that include classical zootechnical and biochemical studies together with newer approaches based on molecular biology and functional genomics will certainly lead to important progress in the field. It will be important to understand the ontogeny of basic physiological mechanisms and pathways of nutrient digestion and intestinal absorption in relation to the type and composition of diets.

Acknowledgements

The authors acknowledge support from the Research Council of Norway (141990/120, 165203/S40, 174229/S40, 175021/D15), and "Fundação para a Ciência e a Tecnologia", Portugal (SFRH/BD/4902/2001). The authors are very grateful to the editorial offices of the different journal publishers and to the respective publication authors for authorising the reproduction or adaptation of the figures used to illustrate this chapter.

References

Alvarez-González CA, Cervantes-Trujano M, Tovar-Ramírez D, Conklin DE, Nolasco H, Gisbert E, Piedrahita R (2006) Development of digestive enzymes in California halibut (*Paralichthys californicus*) larvae. Fish Physiol Biochem 31: 83-93

André M, Ando S, Ballagny C, Durliat M, Poupard G, Briançon C, Babin PJ (2000) Intestinal fatty acid binding protein gene expression reveals the cephalocaudal patterning during zebrafish gut morphogenesis. Int J Dev Biol 44: 249-252

Applebaum SL, Holt GJ (2003) The digestive protease, chymotrypsin, as an indicator of nutritional condition in larval red drum (*Sciaenops ocellatus*). Mar Biol 142: 1159-1167

Applebaum SL, Rønnestad I (2004) Absorption, assimilation and catabolism of individual free amino acids by larval Atlantic halibut (*Hippoglossus hippoglossus*). Aquaculture 230: 313-322

Aragão C, Conceição LEC, Martins D, Rønnestad I, Gomes E, Dinis MT (2004) A balanced dietary amino acid profile improves amino acid retention in post-larval Senegalese sole (*Solea senegalensis*). Aquaculture 233: 293-304

Ash R (1985) Protein digestion and absorption. In: Cowey CB, Mackie A, Bell JB (eds) Nutrition and Feeding in Fish. Academic Press, New York, p 69-91

Austreng E, Skrede A, Eldegard Å (1979) Effect of dietary fat source on the digestibility of fat and fatty acids in rainbow trout and mink. Acta Agric Scand 29: 119-126

Babin PJ, Vernier J-M (1989) Plasma lipoproteins in fish. J Lipid Res 30: 467-489

Balon EK (1975) Terminology of intervals in fish development. J Fish Res Bd Can 32: 1663-1670

Bauermeister AEM, Pirie BJS, Sargent JR (1979) An electron microscopic study of lipid absorption in the pyloric caeca of rainbow trout (*Salmo gairdneri*) fed wax ester-rich zooplankton. Cell Tissue Res 200: 475-486

Bengtson DA, Borrus DN, Leibovitz HE, Simpson KL (1993) Studies on structure and function of the digestive system of *Medidia beryllina* (Pisces, Atherinidae). In: Walther BT, Fyhn HJ (eds) Physiology and Biochemistry of Fish Larval Development. University of Bergen, Norway, p 199-208

Bisbal GA, Bengtson DA (1995) Development of the digestive tract in larval summer flounder. J Fish Biol 47: 277-291

Boehlert GW, Yoklavich MM (1984) Carbon assimilation as a function of ingestion rate in larval pacific herring, *Clupea harengus pallasi* Valenciennes. J Exp Mar Biol Ecol 79: 251-262

Bolasina S, Pérez A, Yamashita Y (2006) Digestive enzymes activity during ontogenetic development and effect of starvation in Japanese flounder, *Paralichthys olivaceus*. Aquaculture 252: 503-515

Borgström B (1977) Digestion and absorption of lipids. In: Crane RK (ed) International Review of Physiology, Gastrointestinal Physiology II, Volume 12. University Park Press, Baltimore, USA, p 305-323

Borlongan IG (1990) Studies on the digestive lipases of milkfish, *Chanos chanos*. Aquaculture 89: 315-325

Boujard T, Médale F (1994) Regulation of voluntary feed intake in juvenile rainbow trout fed by hand or by self feeders with diets containing two different protein/energy ratios. Aquat Living Resour 7: 211-215

Boujard T, Gélineau A, Covès D, Corraze G, Dutto G, Gasset E, Kaushik S (2004) Regulation of feed intake, growth, nutrient and energy utilization in European sea bass (*Dicentrarchus labrax*) fed high fat diets. Aquaculture 231: 529-545

Boulhic M, Gabaudan J (1992) Histological study of the organogenesis of the digestive system and swim bladder of the Dover sole, *Solea solea* (Linnaeus 1758). Aquaculture 102: 373-396

Brannon PM (1990) Adaptation of the exocrine pancreas to diet. Annu Rev Nutr 10: 85-105

Bransden MP, Cobcroft JM, Battaglene SC, Morehead DT, Dunstan GA, Nichols PD, Kolkovski S (2005) Dietary 22:6n-3 alters gut and liver structure and behaviour in larval striped trumpeter (*Latris lineata*). Aquaculture 248: 275-285

Brockerhoff H (1966) Digestion of fat by cod. J Fish Res Bd Can 23: 1835-1839

Buchet V, Zambonino-Infante JL, Cahu CL (2000) Effect of lipid level in a compound diet on the development of red drum (*Sciaenops ocellatus*) larvae. Aquaculture 184: 339-347

Buddington RK (1985) Digestive secretions of lake sturgeon, *Acipenser fulvescens*, during early development. J Fish Biol 26: 715-723

Buddington RK, Krogdahl Å, Bakke-McKellep AM (1997) The intestines of carnivorous fish: structure and functions and the relations with diet. Acta Physiol Scand 161: 67-80

Caballero MJ, Obach A, Rosenlund G, Montero D, Gisvold M, Izquierdo MS (2002) Impact of different dietary lipid sources on growth, lipid digestibility, tissue fatty acid composition and histology of rainbow trout, *Oncorhynchus mykiss*. Aquaculture 214: 253-271

Caballero MJ, Izquierdo MS, Kjørsvik E, Montero D, Socorro J, Fernández AJ, Rosenlund G (2003) Morphological aspects of intestinal cells from gilthead seabream (*Sparus aurata*) fed diets containing different lipid sources. Aquaculture 225: 325-340

Cahu CL, Zambonino-Infante JL (1997) Is the digestive capacity of marine fish larvae sufficient for compound diet feeding? Aquac Int 5: 151-160

Cahu C, Zambonino-Infante J (2001) Substitution of live food by formulated diets in marine fish larvae. Aquaculture 200: 161-180

Cahu CL, Zambonino-Infante JL, Le Gall MM, Quazuguel P (1995) Early weaning of seabass: are digestive enzymes limiting? In: Lavens P, Jaspers E, Roelants I (eds) LARVI '95 – Fish & Shellfish Larviculture Symposium. European Aquaculture Society, Special Publication No 24, Gent, Belgium, p 268-271

Cahu CL, Zambonino-Infante JL, Corraze G, Coves D (2000) Dietary lipid level affects fatty acid composition and hydrolase activities of intestinal brush border membrane in seabass. Fish Physiol Biochem 23: 165-172

Calzada A, Medina A, González de Canales ML (1998) Fine structure of the intestine development in cultured sea bream larvae. J Fish Biol 53: 340-365

Carvalho AP, Escaffre AM, Teles AO, Bergot P (1997) First feeding of common carp larvae on diets with high levels of protein hydrolysates. Aquac Int 5: 361-367

Carey MC, Small DM, Bliss CM (1983) Lipid digestion and absorption. Annu Rev Physiol 45: 651-677

Checkley DM (1984) Relation of growth to ingestion for larvae of Atlantic herring *Clupea harengus* and other fish. Mar Ecol Prog Ser 18: 215-224

Cousin JCB, Baudin-Laurencin F, Gabaudan J (1987) Ontogeny of enzymatic activities in fed and fasting turbot, *Scophthalmus maximus* L. J Fish Biol 30: 15-33

Coutteau P, Geurden I, Camara MR, Bergot P, Sorgeloos P (1997) Review on the dietary effects of phospholipids in fish and crustacean larviculture. Aquaculture 155: 149-164

Dabrowski K (1983) Comparative aspects of protein digestion and amino acid absorption in fish and other animals. Comp Biochem Physiol 74: 417-425

Dabrowski K (1984) The feeding of fish larvae: present 'state of the art' and perspectives. Reprod Nutr Dev 23: 807-833

Dabrowski K, Glogowski J (1977) Studies on the role of exogenous proteolytic enzymes in digestion processes in fish. Hydrobiologia 54: 129-134

Dabrowski K, Lee KJ, Rinchard J (2003) The smallest vertebrate, teleost fish, can utilize synthetic dipeptide-based diets. J Nutr 133: 4225-4229

Dabrowski K, Terjesen BF, Zhang YF, Phang JM, Lee KJ (2005) A concept of dietary dipeptides: a step to resolve the problem of amino acid availability in the early life of vertebrates. J Exp Biol 208: 2885-2894

Daniel H (2004) Molecular and integrative physiology of intestinal peptide transport. Annu Rev Physiol 66: 361-384

Deplano M, Diaz JP, Connes R, Kentouri-Divanach M, Cavalier F (1991) Appearance of lipid-absorption capacities in larvae of the sea bass *Dicentrarchus labrax* during transition to the exotrophic phase. Mar Biol 108: 361-371

Diaz JP, Guyot E, Vigier S, Connes R (1997) First events in lipid absorption during post-embryonic development of the anterior intestine in gilt-head sea bream. J Fish Biol 51: 180-192

Douglas SE, Gawlicka A, Mandla S, Gallant JW (1999) Ontogeny of the stomach in winter flounder: characterization and expression of the pepsinogen and proton pump genes and determination of pepsin activity. J Fish Biol 55: 897-915

Elbal MT, García-Hernández MP, Lozano MT, Agulleiro B (2004) Development of the digestive tract of gilthead sea bream (*Sparus aurata* L.). Light and electron microscopic studies. Aquaculture 234: 215-238

Espeland S (2006) Postprandial changes in intestinal, plasma, urine, and bile free amino acids of Atlantic cod (*Gadus morhua*) fed a pelleted commercial feed. MSc Thesis, University of Bergen, Bergen, Norway

Falk-Petersen IB (2005) Comparative organ differentiation during early life stages of marine fish. Fish Shellfish Immunol 19: 397-412

Favé G, Coste TC, Armand M (2004) Physicochemical properties of lipids: new strategies to manage fatty acid bioavailability. Cell Mol Biol 50: 815-831

Fontagné S, Geurden I, Escaffre A-M, Bergot P (1998) Histological changes induced by dietary phospholipids in intestine and liver of common carp (*Cyprinus carpio* L.) larvae. Aquaculture 161: 213-223

Fontagné S, Burtaire L, Corraze G, Bergot P (2000) Effects of dietary medium-chain triacylglycerols (tricaprylin and tricaproin) and phospholipids supply on survival, growth and lipid metabolism in common carp (*Cyprinus carpio* L.) larvae. Aquaculture 190: 289-303

Gawlicka A, Parent B, Horn MH, Ross N, Opstad I, Torrissen OJ (2000) Activity of digestive enzymes in yolk-sac larvae of Atlantic halibut (*Hippoglossus hippoglossus*): indication of readiness for first feeding. Aquaculture 184: 303-314

Gawlicka A, Leggiadro CT, Gallant JW, Douglas SE (2001) Cellular expression of the pepsinogen and gastric proton pump genes in the stomach of winter flounder as determined by *in situ* hybridization. J Fish Biol 58: 529-536

Gawlicka A, Herold MA, Barrows FT, de la Noüe J, Hung SSO (2002) Effects of dietary lipids on growth, fatty acid composition, intestinal absorption and hepatic storage in white sturgeon (*Acipenser transmontanus* R.) larvae. J Appl Ichthyol 18: 673-681

Gélineau A, Corraze G, Boujard T, Larroquet L, Kaushik S (2001) Relation between dietary lipid level and voluntary feed intake, growth, nutrient gain, lipid deposition and hepatic lipogenesis in rainbow trout. Reprod Nutr Dev 41: 487-503

Geurden I, Radünz-Neto J, Gergot P (1995) Essentiality of dietary phospholipids for carp (*Cyprinus carpio* L.) larvae. Aquaculture 131: 303-314

Geurden I, Reyes OS, Bergot P, Coutteau P, Sorgeloos P (1998) Incorporation of fatty acids from dietary neutral lipid in eye, brain and muscle of postlarval turbot fed diets with different types of phosphatidylcholine. Fish Physiol Biochem 19: 365-375

Gildberg A (2004) Digestive enzyme activities in starved pre-slaughter farmed and wild-captured, Atlantic cod (*Gadus morhua*). Aquaculture 238: 343-353

Gisbert E, Villeneuve L, Zambonino-Infante JL, Quazuguel P, Cahu CL (2005) Dietary phospholipids are more efficient than neutral lipids for long-chain polyunsaturated fatty acid supply in European sea bass *Dicentrarchus labrax* larval development. Lipids 40: 609-618

Gjellesvik DR (1991) Fatty acid specificity of bile salt-dependent lipase: enzyme recognition and super-substrate effects. Biochim Biophys Acta 1086: 167-172

Gjellesvik DR, Lombardo D, Walther BT (1992) Pancreatic bile salt dependent lipase from cod (*Gadus morhua*): purification and properties. Biochim Biophys Acta 1124: 123-134

Govoni JJ, Peters DS, Merriner JV (1982) Carbon assimilation during larval development of the marine teleost *Leiostomus xanthurus* Lacépède. J Exp Mar Biol Ecol 64: 287-299

Govoni JJ, Boehlert GW, Watanabe Y (1986) The physiology of digestion in fish larvae. Environ Biol Fishes 16: 59-77

Griffiths D (1975) Prey availability and food of predators. Ecology 56: 1209-1214

Guyton AC, Hall JE (1996) Textbook of Medical Physiology. 9th Edition, WB Saunders Company, Philadelphia, USA

Hadas E, Koven W, Sklan D, Tandler A (2003) The effect of dietary phosphatidylcholine on the assimilation and distribution of ingested free oleic acid (18:1n-9) in gilthead seabream (*Sparus aurata*) larvae. Aquaculture 217: 577-588

Han K, Geurden I, Sorgeloos P (2000) Enrichment strategies for *Artemia* using emulsions providing different levels of n-3 highly unsaturated fatty acids. Aquaculture 183: 335-347

Henning SJ (1987) Functional development of the gastrointestinal tract. In: Johnson LR (ed) Physiology of the Gastrointestinal Tract. Raven Press, New York, p 285-300

Hjelmeland K (1995) Trypsin in fish – Studies of the enzyme and its inhibitors in the digestive system and epidermis of fish. PhD Thesis. The Norwegian College of Fishery Science, Tromsø, Norway

Hjelmeland K, Pedersen BH, Nilssen EM (1988) Trypsin content in intestines of herring larvae, *Clupea harengus*, ingesting inert polystyrene spheres or live crustacea prey. Mar Biol 98: 331-335

Hoehne-Reitan K, Kjørsvik E (2004) Functional development of the liver and exocrine pancreas in teleost fish. Am Fish Soc Symp 40: 9-36

Hoehne-Reitan K, Kjørsvik E, Reitan KI (2001a) Development of the pH in the intestinal tract of larval turbot. Mar Biol 139: 1159-1164

Hoehne-Reitan K, Kjørsvik E, Reitan KI (2001b) Bile salt-dependent lipase in larval turbot, as influenced by density and lipid content of fed prey. J Fish Biol 58: 746-754

Hoehne-Reitan K, Kjørsvik E, Gjellesvik DR (2001c) Development of bile salt-dependent lipase in larval turbot. J Fish Biol 58: 737-745

Honkanen RE, Rigler MW, Patton JS (1985) Dietary fat assimilation and bile salt absorption in the killifish intestine. Am J Physiol Gastr Liver Physiol 249: G399-G407

Iijima N, Aida S, Mankura M, Kayama M (1990) Intestinal absorption and plasma transport of dietary triglyceride and phosphatidylcholine in the carp (*Cyprinus carpio*). Comp Biochem Physiol 96A: 45-55

Iijima N, Chosa S, Uematsu K, Goto T, Hoshita T, Kayama M (1997) Purification and characterisation of phospholipase A_2 from the pyloric caeca of red sea bream, *Pagrus major*. Fish Physiol Biochem 16: 487-498

Iijima N, Tanaka S, Ota Y (1998) Purification and characterization of bile salt-activated lipase from the hepatopancreas of red sea bream, *Pagrus major*. Fish Physiol Biochem 18: 59-69

Iritani N, Ikeda Y, Fukuda H, Katsurada A (1984) Comparative study of lipogenic enzymes in several vertebrates. Lipids 19: 825-835

Iwai T, Tanaka M (1968) The comparative study of digestive tract of teleost larvae-III. Epithelial cells in the posterior gut of halfbeak larvae. Bull Jap Soc Sci Fish 34: 44-48

Izquierdo MS, Henderson RJ (1998) The determination of lipase and phospholipase activities in gut contents of turbot (*Scophthalmus maximus*) by fluorescence-based assays. Fish Physiol Biochem 19: 153-162

Izquierdo MS, Socorro J, Arantzamendi L, Hernández-Cruz, CM (2000) Recent advances in lipid nutrition in fish larvae. Fish Physiol Biochem 22: 97-107

Izquierdo MS, Tandler A, Salhi M, Kolkovski S (2001) Influence of dietary polar lipids' quantity and quality on ingestion and assimilation of labelled fatty acids by larval gilthead seabream. Aquac Nutr 7: 153-160

Jany KD (1976) Studies on the digestive enzymes of the stomachless bonefish *Carassius auratus gibelo* (Bloch). Endopeptidases. Comp Biochem Physiol 53B: 31-38

Jamil H, Dickson JK, Chu C-H, Lago MW, Rinehart JK, Biller SA, Gregg RE, Wetterau JR (1995) Microsomal triglyceride transfer protein. Specificity of lipid binding and transport. J Biol Chem 270: 6549-6554

Jobling M, Hjelmeland K (1992) Ernaering og metabolisme. In: Døving K, Reimers, E (eds) Fiskens Fysiologi. John Grieg Forlag, Bergen, Norway, p 234-276

Johnsen RI, Grahl-Nielsen O, Roem A (2000) Relative absorption of fatty acids by Atlantic salmon *Salmo salar* from different diets, as evaluated by multivariate statistics. Aquac Nutr 6: 255-261

Kamisaka Y (2005) Gastrointestinal hormone cholecystokinin (CCK) in teleosts: the spatial and temporal distribution in the digestive tract of larvae and juveniles. PhD Thesis. Kyoto University, Kyoto, Japan

Kamisaka Y, Totland GK, Tagawa M, Kurokawa T, Suzuki T, Tanaka M, Rønnestad I (2001) Ontogeny of cholecystokinin-immunoreactive cells in the digestive tract of Atlantic halibut, *Hippoglossus hippoglossus*, larvae. Gen Comp Endocrinol 123: 31-37

Kamisaka Y, Kaji T, Masuma S, Tezuka N, Kurokawa T, Suzuki T, Totland GK, Rønnestad I, Tagawa M, Tanaka M (2002) Ontogeny of cholecystokinin-immunoreactive cells in the digestive tract of bluefin tuna, *Thunnus thynnus*, larvae. Sarsia 87: 258-262

Kamisaka Y, Fujii Y, Yamamoto S, Kurokawa T, Rønnestad I, Totland GK, Tagawa M, Tanaka M (2003) Distribution of cholecystokinin-immunoreactive cells in the digestive tract of the larval teleost, Ayu, *Plecoglossus altivelis*. Gen Comp Endocrinol 134: 116-121

Kamisaka Y, Drivenes O, Kurokawa T, Tagawa M, Rønnestad I, Tanaka M, Helvik JV (2005) Cholecystokinin mRNA in Atlantic herring, *Clupea harengus* – molecular cloning, characterization, and distribution in the digestive tract during the early life stages. Peptides 26: 385-393

Kanazawa A (1993) Essential phospholipids of fish and crustaceans. In: Kaushik SJ, Luquet P (eds) Fish Nutrition in Practice, IV International Symposium on Fish Nutrition and Feeding. INRA, France, p 519-530

Kanazawa A, Teshima S, Inamori S, Matsubara H (1983) Effects of dietary phospholipids on growth of the larval red sea bream and knife jaw. Mem Fac Fish Kagoshima Univ 32: 109-114

Kaushik SJ (1995) Nutrient requirements, supply and utilisation in the context of carp culture. Aquaculture 129: 225-241

Kim BG, Divakaran S, Brown CL, Ostrowski AC (2001) Comparative digestive enzyme ontogeny in two marine larval fishes: Pacific threadfin (*Polydactylus sexfilis*) and bluefin trevally (*Caranx melampygus*). Fish Physiol Biochem 24: 225-241

Kjørsvik E, Van der Meeren T, Kryvi H, Arnfinnson J, Kvenseth PG (1991a) Early development of the digestive tract of cod larvae, *Gadus morhua* L., during start-feeding and starvation. J Fish Biol 38: 1-15

Kjørsvik E, Olsen Y, Rosenlund G, Vadstein O (1991b) Effect of various lipid enrichments in rotifers and the development of early stages in turbot. In: Lavens P, Sorgeloos P, Jaspers E, Ollevier F (eds) Larvi '91 – Fish and Crustacean Larviculture Symposium. European Aquaculture Society, Special Publication, Vol 15, Gent, Belgium, p 20-22

Kolkovski S (2001) Digestive enzymes in fish larvae and juveniles – implications and applications to formulated diets. Aquaculture 200: 181-201

Kolkovski S, Tandler A (1995) Why microdiets are still inadequate as a viable alternative to live zooplankters for developing marine fish larvae. In: Lavens P, Jaspers E, Roelants I (eds) LARVI '95 – Fish & Shellfish Larviculture Symposium. European Aquaculture Society, Special Publication No 24, Gent, Belgium, p 265-266

Koven WM, Kolkovski S, Tandler A, Kissil GWm, Sklan D (1993) The effect of dietary lecithin and lipase, as a function of age, on n-9 fatty acid incorporation in the tissue lipids of *Sparus aurata* larvae. Fish Physiol Biochem 10: 357-364

Koven WM, Henderson RJ, Sargent JR (1994a) Lipid digestion in turbot (*Scophthalmus maximus*) II: Lipolysis *in vitro* of ^{14}C-labelled triacylglycerol, cholesterol ester and phosphatidylcholine by digesta from different segments of the digestive tract. Fish Physiol Biochem 13: 275-283

Koven WM, Henderson RJ, Sargent JR (1994b) Lipid digestion in turbot (*Scophthalmus maximus*). I: Lipid class and fatty acid composition of digesta from different segments of the digestive tract. Fish Physiol Biochem 13: 69-79

Koven WM, Kolkovski S, Hadas E, Gamsiz K, Tandler A (2001) Advances in the development of microdiets for gilthead seabream, *Sparus aurata*: a review. Aquaculture 194: 107-121

Kurokawa T, Suzuki T (1996) Formation of the diffuse pancreas and the development of digestive enzyme synthesis in larvae of the Japanese flounder *Paralichthys olivaceus*. Aquaculture 141: 267-276

Kurokawa T, Tanaka H, Kagawa H, Otha H (1996) Absorption of protein molecules by the rectal cells in eel larvae *Anguilla japonica*. Fish Sci 62: 832-833

Kurokawa T, Shiraishi M, Suzuki T (1998) Quantification of exogenous protease derived from zooplankton in the intestine of Japanese sardine (*Sardinops melanotictus*) larvae. Aquaculture 161: 491-499

Lauff M, Hofer R (1984) Proteolytic enzymes in fish development and the importance of dietary enzymes. Aquaculture 37: 335-346

Lazo JP, Holt GJ, Arnold CR (2000) Ontogeny of pancreatic enzymes in larval red drum *Sciaenops ocellatus*. Aquac Nutr 6: 183-192

Lee DJ, Putnam GB (1973) The response of rainbow trout to varying protein/energy ratios in a test diet. J Nutr 103: 916-922

Léger C, Bauchart D (1972) Hydrolyse de triglycérides par le système lipasique du pancréas de truite (*Salmo gairdneri* Rich.). Mise en évidence d'un nouveau type de spécificité d'action. C R Acad Sci, Ser D Sci Nat 275: 2419-2422

Léger C, Bauchart D, Flanzy J (1977) Some properties of pancreatic lipase in *Salmo gairdneri* Rich.: K_m, effects of bile salts and Ca^{2+}, gel filtrations. Comp Biochem Physiol 57B: 359-363

Lie Ø, Lambertsen G (1985) Digestive lipolytic enzymes in cod (*Gadus morhua*): fatty acid specificity. Comp Biochem Physiol 80B: 447-450

Lie Ø, Lambertsen G (1991) Lipid digestion and absorption in cod (*Gadus morhua*), triacylglycerols, wax esters and diacylalkylglycerols. Comp Biochem Physiol 98A: 159-163

Lie Ø, Lied E, Lambertsen G (1987) Lipid digestion in cod (*Gadus morhua*). Comp Biochem Physiol 88B: 697-700

Lied E, Lambertsen G (1982) Apparent availability of fat and individual fatty acids in Atlantic cod (*Gadus morhua*). Fisk Dir Skr Ernaering 2: 63-75

Lied E, Solbakken R (1984) The course of protein digestion in Atlantic cod (*Gadus morhua*). Comp Biochem Physiol 77A: 503-506

Linscheer WG, Vergroesen AJ (1994) Lipids. In: Shils ME, Olson JA, Shike M (eds) Modern Nutrition in Health and Disease. Williams & Wilkins, Baltimore, USA, p 47-88

Loewe H, Eckmann R (1988) The ontogeny of the alimentary tract of coregonid larvae: normal development. J Fish Biol 33: 841-850

Luizi FS, Gara B, Shields RJ, Bromage NR (1999) Further description of the development of the digestive organs in Atlantic halibut (*Hippoglossus hippoglossus*) larvae, with notes on differential absorption of copepod and *Artemia* prey. Aquaculture 176: 101-116

Mahr K, Graubner R, Hofer R, Moser H (1983) Histological and physiological development of the stomach in *Coregonus* sp. Pol Arch Hydrobiol 98: 344-356

Mangor-Jensen A, Adoff GR (1987) Drinking activity of the newly hatched larvae of cod *Gadus morhua* L. Fish Physiol Biochem 3: 99-103

Marza E, Barthe C, André M, Villeneuve L, Hélou C, Babin PJ (2005) Developmental expression and nutritional regulation of a zebrafish gene homologous to mammalian microsomal triglyceride transfer protein large subunit. Dev Dyn 232: 506-518

McLean E, Rønsholdt B, Sten C, Najamuddin JF (1999) Gastrointestinal delivery of peptide and protein drugs to aquacultured teleosts. Aquaculture 177: 231-247

Morais S, Lacuisse M, Conceição LEC, Dinis MT, Rønnestad I (2004a) Ontogeny of the digestive capacity of Senegalese sole (*Solea senegalensis* Kaup 1858), with respect to digestion, absorption and metabolism of amino acids from *Artemia*. Mar Biol 145: 243-250

Morais S, Conceição LEC, Dinis MT, Rønnestad I (2004b) A method for radiolabeling *Artemia* with applications in studies of food intake, digestibility, protein and amino acid metabolism in larval fish. Aquaculture 231: 469-487

Morais S, Cahu C, Zambonino-Infante JL, Robin J, Rønnestad I, Dinis MT, Conceição LEC (2004c) Dietary TAG source and level affect performance and lipase expression in larval seabass (*Dicentrarchus labrax*). Lipids 39: 449-458

Morais S, Rojas-Garcia CR, Conceição LEC, Rønnestad I (2005a) Digestion and absorption of a pure triacylglycerol and a free fatty acid by *Clupea harengus* L. larvae. J Fish Biol 67: 223-238

Morais S, Koven W, Rønnestad I, Dinis MT, Conceição LEC (2005b) Dietary protein/lipid ratio affects growth and amino acid and fatty acid absorption and metabolism in Senegalese sole (*Solea senegalensis* Kaup 1858) larvae. Aquaculture 246: 347-357

Morais S, Koven W, Rønnestad I, Dinis MT, Conceição LEC (2005c) Dietary protein:lipid ratio and lipid nature affects fatty acid absorption and metabolism in a teleost larva. Br J Nutr 93: 813-820

Morais S, Torten M, Nixon O, Lutzky S, Conceição LEC, Dinis MT, Tandler A, Koven W (2006a) Food intake and absorption are affected by dietary lipid level and lipid source in seabream (*Sparus aurata* L.) larvae. J Exp Mar Biol Ecol 331: 51-63

Morais S, Caballero MJ, Conceição LEC, Izquierdo MS, Dinis MT (2006b) Dietary neutral lipid level and source in Senegalese sole (*Solea senegalensis*) larvae: effect on growth, lipid metabolism and digestive capacity. Comp Biochem Physiol 144B: 57-69

Morrison CM (1987) Histology of the Atlantic cod, *Gadus morhua*: An atlas. Part One. Digestive tract and associated organs. Can Spec Publ Fish Aquat Sci 98: 219 p

Munilla-Moran R, Stark JR, Barbour A (1990) The role of exogenous enzymes in digestion in cultured turbot larvae (*Scophthalmus maximus* L.). Aquaculture 88: 337-350

Murakami K, Etlinger JD (1987) Degradation of proteins with blocked amino-groups by cytosolic proteases. Biochem Biophys Res Commun 146: 1249-1255

Murray HM, Gallant JW, Perez-Casanova JC, Johnson SC, Douglas S (2003) Ontogeny of lipase expression in winter flounder. J Fish Biol 62: 816-833

Murray HM, Gallant JW, Johnson SC, Douglas SE (2006) Cloning and expression analysis of three digestive enzymes from Atlantic halibut (*Hippoglossus hippoglossus*) during early development: predicting gastrointestinal functionality. Aquaculture 252: 394-408

Nielsen CU, Brodin B (2003) Di/tri-peptide transporters as drug delivery targets: regulation of transport under physiological and patho-physiological conditions. Curr Drug Targets 4: 373-88

Noaillac-Depeyre J, Gas N (1974) Fat absorption by the enterocytes of the carp (*Cyprinus carpio* L.). Cell Tissue Res 155: 353-365

Nordskog BK, Phan CT, Nutting DF, Tso P (2001) An examination of the factors affecting intestinal lymphatic transport of dietary lipids. Advanced Drug Delivery Reviews 50: 21-44

Ogata HY, Shearer KD (2000) Influence of dietary fat and adiposity on feed intake of juvenile red sea bream *Pagrus major*. Aquaculture 189: 237-249

Olsen AI, Attramadal Y, Reitan KI, Olsen Y (2000b) Food selection and digestion characteristics of Atlantic halibut (*Hippoglossus hippoglossus*) larvae fed cultivated prey organisms. Aquaculture 181: 293-310

Olsen RE, Henderson RJ, Ringø E (1998) The digestion and selective absorption of dietary fatty acids in Arctic Charr, *Salvelinus alpinus*. Aquac Nutr 4: 13-21

Olsen RE, Myklebust R, Kaino T, Ringø E (1999) Lipid digestibility and ultrastructural changes in the enterocytes of Artic char (*Salvelinus alpinus* L.) fed linseed oil and soybean lecithin. Fish Physiol Biochem 21: 35-44

Olsen RE, Myklebust R, Ringø E, Mayhew TM (2000a) The influences of dietary linseed oil and saturated fatty acids on caecal enterocytes in Artic char (*Salvelinus alpinus* L.): a quantitative ultrastructural study. Fish Physiol Biochem 22: 207-216

Olsen RE, Dragnes BT, Myklebust R, Ringø E (2003) Effect of soybean oil and soybean lecithin on intestinal lipid composition and lipid droplet accumulation of rainbow trout, *Oncorhynchus mykiss* Walbaum. Fish Physiol Biochem 29: 181-192

Oxley A, Torstensen BE, Rustan AC, Olsen RE (2005) Enzyme activities of intestinal triacylglycerol and phosphatidylcholine biosynthesis in Atlantic salmon (*Salmo salar* L.). Comp Biochem Physiol 141B: 77-87

Ozkizilcik S, Chu F-LE, Place AR (1996) Ontogenetic changes of lipolytic enzymes in striped bass (*Morone saxatilis*). Comp Biochem Physiol 113B: 631-637

Patton JS, Nevenzel JC, Benson AA (1975) Specificity of digestive lipases in hydrolysis of wax esters and triglycerides studied in anchovy and other selected fish. Lipids 10: 575-583

Pedersen BH (1984) The intestinal evacuation rates of larval herring (*Clupea harengus* L.) predating on wild plankton. Dana 3: 21-30

Pedersen BH, Andersen KP (1992) Induction of trypsinogen secretion in herring larvae (*Clupea harengus*). Mar Biol 112: 559-565

Pedersen BH, Nilssen EM, Hjelmeland K (1987) Variations in the content of trypsin and trypsinogen in larval herring (*Clupea harengus*) digesting copepod nauplii. Mar Biol 94: 171-181

Pedersen BH, Ugelstad I, Hjelmeland K (1990) Effects of a transitory, low food supply in the early life of larval herring (*Clupea harengus*) on mortality, growth and digestive capacity. Mar Biol 107: 61-66

Pedersen T, Falk-Petersen I-B (1992) Morphological changes during metamorphosis in cod (*Gadus morhua* L.), with particular reference to the development of the stomach and pyloric caeca. J Fish Biol 41: 449-461

Pérez JA, Rodríguez C, Henderson RJ (1999) The uptake and esterification of radiolabelled fatty acids by enterocytes isolated from rainbow trout (*Oncorhynchus mykiss*). Fish Physiol Biochem 20: 125-134

Perez-Casanova JC, Murray HM, Gallant JW, Ross NW, Douglas SE, Johnson SC (2004) Bile salt-activated lipase expression during larval development in the haddock (*Melanogrammus aeglefinus*). Aquaculture 235: 601-617

Perez-Casanova JC, Murray HM, Gallant JW, Ross NW, Douglas SE, Johnson SC (2006) Development of the digestive capacity in larvae of haddock (*Melanogrammus aeglefinus*) and Atlantic cod (*Gadus morhua*). Aquaculture 251: 377-401

Person Le Ruyet J, Alexandre JC, Thébaud L, Mugnier C (1993) Marine fish larvae feeding: formulated diets or live prey? J World Aquac Soc 24: 211-224

Poelstra K, Bakker WW, Klok PA, Hardonk MJ, Meijer DKF (1997) A Physiologic function for alkaline phosphatase: endotoxin detoxification. Lab Invest 76: 319-327

Poirier H, Mathieu Y, Besnard P, Bernard A (1997) Intestinal lipid esterification and aging in mice and rats. Comp Biochem Physiol 116: 253-260

Pousão-Ferreira P, Morais S, Dores E, Narciso L (1999) Eggs of gilthead seabream *Sparus aurata* L. as a potential enrichment product of *Brachionus* sp. in the larval rearing of gilthead seabream *Sparus aurata* L. Aquac Res 30: 751-758

Rainuzzo JR, Reitan KI, Olsen Y (1994) Effect of short- and long-term lipid enrichment on total lipids, lipid class and fatty acid composition in rotifers. Aquac Int 2: 19-32

Rainuzzo JR, Reitan KI, Olsen Y (1997) The significance of lipids at early stages of marine fish: a review. Aquaculture 155: 103-115

Ray EC, Avissar NE, Sax HC (2002) Growth factor regulation of enterocyte nutrient transport during intestinal adaptation. Am J Surgery 183: 361-371

Ribeiro L, Sarasquete C, Dinis MT (1999a) Histological and histochemical development of the digestive system of *Solea senegalensis* (Kaup, 1858) larvae. Aquaculture 171: 293-308

Ribeiro L, Zambonino-Infante JL, Cahu C, Dinis MT (1999b) Development of digestive enzymes in larvae of *Solea senegalensis*, Kaup 1858. Aquaculture 179: 465-473

Rodríguez C, Pérez JA, Izquierdo MS, Cejas JR, Bolaños A, Lorenzo A (1996) Improvement of the nutritional value of rotifers by varying the type and concentration of oil and the enrichment period. Aquaculture 147: 93-105

Rogie A, Skinner ER (1981) Some aspects of lipid transport in the rainbow trout. Biochem Soc Trans 9: 59-60

Rogie A, Skinner ER (1985) The roles of the intestine and liver in the biosynthesis of plasma lipoproteins in the Rainbow trout, *Salmo gairdneri* Richardson. Comp Biochem Physiol 81B: 285-289

Rojas-García CR, Rønnestad I (2003a) Assimilation of dietary free amino acids, peptides and protein in post-larval Atlantic Halibut (*Hippoglossus hippoglossus*). Mar Biol 142: 801-808

Rojas-García CR, Rønnestad I (2003b) Comment on the article "Assimilation of dietary free amino acids, peptides and protein in post-larval Atlantic halibut (*Hippoglossus hippoglossus*)". Mar Biol 142: 799

Rønnestad I (2002) Control and efficiency of digestive function of marine fish larvae. In: Cruz-Suárez LE, Ricque-Marie D, Tapia-Salazar M, Gaxiola-Cortés MG, Simões N (eds.) Avances en Nutrición Acuícola VI. Memorias del VI Simposium Internacional de Nutrición Acuícola. 3 al 6 de Septiembre del 2002. Cancún, Quintana Roo, México. ISBN 970-694090-1. p 152-165

Rønnestad I, Conceição LEC (2005) Aspects of protein and amino acid digestion and utilization by marine fish larvae. In: Starck JM, Wang T (eds) Physiological and Ecological Adaptations to Feeding in Vertebrates. Science Publishers, Enfield, NH, New Hampshire, USA, p 389-416

Rønnestad I, Rojas-García CR. (2006) Absorption kinetics of free amino acids in larval Atlantic herring (*Clupea harengus*) In: Abstracts from AQUA 2006 Annual Meeting of the World Aquaculture Society. Fortezza da Basso Convention Centre, Firenze (Florence), Italy May 9 - 13

Rønnestad I, Thorsen A, Finn RN (1999) Fish larval nutrition: A review of recent advances in the roles of amino acids. Aquaculture 177: 201-216

Rønnestad I, Peréz Dominguez R, Tanaka M (2000a) Ontogeny of digestive tract functionality in Japanese flounder (*Paralichthys olivaceus*) studied by *in vivo* microinjection: pH and assimilation of free amino acids. Fish Physiol Biochem 22: 225-235

Rønnestad I, Rojas-Garcia CR, Skadal J (2000b) Retrograde peristalsis, a possible mechanism for filling the pyloric cecae? J Fish Biol 56: 216-218

Rønnestad I, Conceição LEC, Aragão C, Dinis MT (2000c) Free amino acids are absorbed faster and assimilated more efficiently than protein in postlarval Senegal sole (*Solea senegalensis*). J Nutr 130: 2809-2812

Rønnestad I, Rojas-García CR, Tonheim SK, Conceição LEC (2001) *In vivo* studies of digestion and nutrient assimilation in marine fish larvae. Aquaculture 201: 161-175

Rønnestad I, Tonheim SK, Fyhn HJ, Rojas- García CR, Kamisaka Y, Koven W, Finn RN, Terjesen BF, Barr Y, Conceição LEC (2003) The supply of amino acids during early feeding stages of marine fish larvae: a review of recent findings. Aquaculture 227: 147-164

Rønnestad I, Gavaia P, Viegas C, Cancela LM (2006) PepT1 in a coldwater marine teleost larvae- Atlantic cod: cloning and preliminary studies of expression and phylogeny. Acta Bio Medica 77: Suppl 3: 82

Rønnestad I, Gavaia PJ, Viegas CSB, Verri T, Romano A, Nilsen TO, Jordal A-EO, Kamisaka Y, Cancela ML (2007) Oligopeptide transporter PepT1 in Atlantic cod (*Gadus morhua* L.): cloning, tissue expression and comparative aspects. J Exp Biol 210: 3883-3896

Rust MB (1995) Quantitative aspects of nutrient assimilation in six species of fish larvae. PhD Thesis. School of Fisheries: University of Washington, Seattle

Rust MB, Hardy RW, Stickney RR (1993) A new method for force-feeding larval fish. Aquaculture 116: 341-352

Ruyter B, Moya-Falcón C, Rosenlund G, Vegusdal A (2006) Fat content and morphology of liver and intestine of Atlantic salmon (*Salmo salar*): effects of temperature and dietary soybean oil. Aquaculture 252: 441-452

Ryer CH, Boehlert GW (1983) Feeding chronology, daily ration, and the effects of temperature upon gastric evacuation in the pipefish, *Syngnathus fuscus*. Environ Biol Fishes 9: 301-306

Sæther B-S, Jobling M (2001) Fat content in turbot feed: influence on feed intake, growth and body composition. Aquac Res 32: 451-458

Sala-Vila A, Castellote AI, Campoy C, Rivero M, Rodriguez-Palmero M, López-Sabater MC (2004) The source of long-chain PUFA in formula supplements does not affect the fatty acid composition of plasma lipids in full-term infants. J Nutr 134: 868-873

Salhi M, Hernández-Cruz CM, Bessonart M, Izquierdo MS, Fernández-Palacios H (1999) Effect of different dietary polar lipid levels and different n-3 HUFA content in polar lipids on gut and liver histological structure of gilthead seabream (*Sparus aurata*) larvae. Aquaculture 179: 253-263

Sarasquete MC, Polo A, Yúfera M (1995) Histology and histochemistry of the development of the digestive system of larval gilthead seabream, *Sparus aurata* L. Aquaculture 130: 79-92

Sargent J, Henderson RJ, Tocher DR (1989) The lipids. In: Halver JE (ed) Fish Nutrition. Academic Press, London, p 154-218

Segner H, Räsch R, Schmidt H, von Poeppinghausen KJ (1989) Digestive enzymes in larval *Coregonus lavaretus* L. J Fish Biol 35: 249-263

Segner H, Storch V, Reinecke M, Kloas W, Hanke W (1994) The development of functional digestive and metabolic organs in turbot, *Scophthalmus maximus*. Mar Biol 119: 471-486

Sharma MK, Denovan-Wright EM, Degrave A, Thisse C, Thisse B, Wright JM (2004) Sequence, linkage mapping and early developmental expression of the intestinal-type fatty acid-binding protein gene (*fabp2*) from zebrafish (*Danio rerio*). Comp Biochem Physiol 138B: 391-398

Singer MV (1987) Pancreatic secretory response to intestinal stimulants: a review. Scand J Gastroenterol 22 : Suppl 139: 1-13

Sire MF, Vernier JM (1981) Étude ultrastructurale de la synthèse de chylomicrons au cours de l'absorption intestinale des lipides chez la Truite. Influence de la nature des acides gras ingérés. Biol Cell 40: 47-62

Sire MF, Vernier JM (1992) Intestinal absorption of protein in teleost fish. Comp Biochem Physiol 103A: 771-781

Sire MF, Lutton C, Vernier JM (1981) New views on intestinal absorption of lipids in teleostean fishes: an ultrastructural and biochemical study in the rainbow trout. J Lipid Res 22: 81-94

Skalli A, Hidalgo MC, Abellán E, Arizcun M, Cardenete G (2004) Effects of the dietary protein/lipid ratio on growth and nutrient utilization in common dentex (*Dentex dentex* L.) at different growth stages. Aquaculture 235: 1-11

Sorokin JI, Panov DA (1966) The use of ^{14}C for the quantitative study of the nutrition of fish larvae. Int Rev Gesamt Hydrobiol 51: 743-756

Spannagel AW, Nakano I, Tawil T, Chey WY, Liddle RA, Green GM (1996) Adaptation to fat markedly increases pancreatic secretory response to intraduodenal fat in rats. Am J Physiol Gastr Liver Physiol 33: G128-G135

Stevens CE, Hume ID (1996) Comparative Physiology of the Vertebrate Digestive System. Cambridge University Press, Cambridge

Tanaka M (1973) Studies on the structure and function of the digestive system of teleost larvae. PhD Thesis, Kyoto University, Kyoto, Japan

Tocher DR, Sargent JR (1984) Studies on triacylglycerol, wax ester and sterol ester hydrolases in intestinal caeca of rainbow trout (*Salmo gairdneri*) fed diets rich in triacylglycerols and wax esters. Comp Biochem Physiol 77B: 561-571

Tonheim SK (2005) Protein digestion and postprandial amino acid utilisation in stomach-less marine fish larvae. PhD Thesis, University of Bergen, Norway

Tonheim SK, Espe M, Raae AJ, Darias MJ, Rønnestad I (2004) *In vivo* incorporation of [U]C-^{14}-amino acids: an alternative protein labelling procedure for use in examining larval digestive physiology. Aquaculture 235: 553-567

Tonheim SK, Espe M, Hamre K, Rønnestad I (2005) Pre-hydrolysis improves utilisation of dietary protein in the larval teleost Atlantic halibut (*Hippoglossus hippoglossus* L.). J Exp Mar Biol Ecol 321: 19-34

Tso P, Fujimoto K (1991) The absorption and transport of lipids by the small intestine. Brain Res Bull 27: 477-482

Tytler P, Bell MV, Robinson J (1993) The ontogeny of osmoregulation in marine fish, with particular reference to the effects of changes in salinity and temperature. In: Walther BT, Fyhn HJ (eds) Physiology and Biochemistry of Fish Larval Development. University of Bergen, Norway, p 249-258

van Greevenbroek MMJ, Voorhout WF, Erkelens DW, van Meer G, de Bruin TWA (1995) Palmitic acid and linoleic acid metabolism in Caco-2 cells: different triglyceride synthesis and lipoprotein secretion. J Lipid Res 36: 13-24

Verreth JAJ, Torreele E, Spazier E, van der Sluiszen A, Rombout JHWM, Booms R (1992) The development of a functional digestive system in the African catfish *Clarias gariepinus* (Burchell). J World Aquac Soc 23: 286-298

Vernier JM, Sire MF (1986) Is the Golgi apparatus the obligatory final step for lipoprotein secretion by intestinal cells? Tissue Cell 18: 447-460

Verri T, Kottra G, Romano A, Tiso N, Peric M, Maffia M, Boll M, Argenton F, Daniel H, Storelli C (2003) Molecular and functional characterisation of the zebrafish (*Danio rerio*) Pept1-type peptide transporter. FEBS Letters 549: 115-122

Vig BS, Stouch TR, Timoszyk JK, Quan Y, Wall DA, Smith RL, Faria TN (2006) Human PEPT1 pharmacophore distinguishes between dipeptide transport and binding. J Med Chem 49: 3636-3644

Vu TT (1983) Etude histoenzymologique des activités protéasiques dans le tube digestif des larves et des adultes de bar, *Dicentrarchus labrax*. Aquaculture 109: 57-69

Watanabe, Y., 1984. Morphological and functional changes in rectal epithelium cells of pond smelt during post embryonic development. Bull Jap Soc Sci Fish 50: 805-814

Walford J, Lam TJ (1993) Development of digestive tract and proteolytic enzyme activity in seabass (*Lates calcarifer*) larvae and juveniles. Aquaculture 109: 187-205

Wallace KN, Pack M (2003) Unique and conserved aspects of gut development in zebrafish. Dev Biol 255: 12-29

Wallace KN, Akhter S, Smith EM, Lorent K, Pack M (2005) Intestinal growth and differentiation in zebrafish. Mech of Dev 122: 157-173

Werner RG, Blaxter JHS (1980) Growth and survival of larval herring (*Clupea harengus*) in relation to prey density. Can J Fish Aquat Sci 37: 1063-1069

Wicker C, Puigserver A (1989) Changes in mRNA levels of rat pancreatic lipase in the early days of consumption of a high-lipid diet. Eur J Biochem 180: 563-567

Wipf D, Ludewig U, Tegeder M, Rentsch D, Koch W, Frommer WB (2002) Conservation of amino acid transporters in fungi, plants and animals. Trends Biochem Sci 27: 139-147

Yamamoto T, Shima T, Unuma T, Shiraishi M, Akiyama T, Tabata M (2000) Voluntary intake of diets with varying digestible energy contents and energy sources, by juvenile rainbow trout *Oncorhynchus mykiss*, using self-feeders. Fish Sci 66: 528-534

Zambonino-Infante JL, Cahu CL (1999) High dietary lipid levels enhance digestive tract maturation and improve *Dicentrarchus labrax* larval development. J Nutr 129: 1195-1200

Zambonino-Infante JL, Cahu CL (2001) Ontogeny of the gastrointestinal tract of marine fish larvae. Comp Biochem Physiol 130C: 477-487

CHAPTER

9

Nitrogen Excretion

Bendik F Terjesen

INTRODUCTION

Generation of Nitrogenous End-products

Several of the macromolecules necessary for life, in particular nucleic acids and proteins, contain nitrogen. These macromolecules are continuously turned over through degradation to nucleotides and amino acids, and then re-synthesised. However, some nucleotides and amino acids escape re-synthesis and are further degraded to smaller nitrogen containing end-products. The majority of the nitrogenous end-products are initially in the form of ammonia (Mommsen & Walsh, 1991) which is highly toxic, and necessitates efficient excretion or detoxification. Furthermore, nutrients such as proteins and nucleic acids are taken in with the food or are present in endogenous yolk-stores to provide for maintenance, activity and growth. If the amino acids or nucleotides are present in excess of requirements however, these compounds must also be degraded since they cannot be stored as such. Thus, two processes can generate nitrogenous end-products, macromolecular turnover and catabolism of nutrients that are in excess of requirements.

Author's address: AKVAFORSK, Institute of Aquaculture Research, N-6600, Sunndalsøra, Norway.
E-mail: bendik.terjesen@akvaforsk.no

The research field that covers synthesis, conversion, and excretion of nitrogenous end-products is collectively termed excretory nitrogen metabolism. Several excellent reviews have been made with focus on juvenile and adult fish (e.g. Goldstein & Forster, 1970; van Waarde, 1988; Mommsen & Walsh, 1992; Walsh & Wright, 1995; Wright, 1995; Walsh, 1997; Anderson, 2001; Ip *et al.*, 2001; Wright & Anderson, 2001; Wilkie, 2002), while fewer have been devoted to the life stages from fertilisation to metamorphosis, with the notable exception of Wright & Fyhn (2001). In the present review, embryos and larvae are the main focus, however, since this field is still in its infancy, supplementary information from experiments on juvenile and adult fish is provided wherever necessary.

Constraints on Nitrogen Excretion

The toxicity of ammonia, especially the gas NH_3, dictates that this compound must be kept at low concentrations *in vivo*. In most terrestrial animals, considerable quantities of water would have to be used in voiding ammonia if this compound was the only vehicle for nitrogen excretion available to the animal. Thus, excretory nitrogen metabolism interacts closely with water homeostasis. During the evolution of terrestrial animals, water balance has been a major driving force and has therefore resulted in adaptations of nitrogen metabolism to lessen the consumption of water for excretion, by for instance urea synthesis and a concentrated urine. In contrast, the aquatic habitat of most juvenile and adult fishes facilitates the voiding of ammonia directly to the external water via the gills, and hence saves valuable energy. Even so, this excretion must be conducted in a way that does not adversely affect ionic and osmotic regulation. In many species of fish, a major route of ammonia excretion is NH_3 diffusion down a partial pressure gradient across the gills. This is possible due to the higher total ammonia level in blood, and because CO_2 excretion creates a lower pH in the unstirred boundary water layer close to the gills, compared to blood pH. Because NH_3 is converted to NH_4^+ at the lower pH in the water next to the gills, an effective ammonia sink is created (Randall & Wright, 1989; Wilkie & Wood, 1996).

There are however, particular habitats and physiological situations in which fish encounter serious challenges related to nitrogen excretion. This is especially apparent when studying the African lungfishes (Protopteridae), and other species of fish living in alkaline lakes and in mangroves. These are habitats which occasionally or continuously hamper the use of ammonia for nitrogen excretion (Smith, 1930; Randall *et al.*, 1989; Frick & Wright, 2002). For example the Kenyan Lake Magadi tilapia (*Alcolapia grahami*) live and grow in highly buffered water at pH 10. Under such conditions, maintenance of a NH_3 partial pressure gradient downward from blood to gill water is not possible, since the high water pH promotes accumulation of NH_3 in the gill water. As will be discussed later, Lake Magadi tilapia have evolved a novel non-ammonotelic strategy.

During the early life stages of fishes, developmental constraints, due to the apparent absence of specialised excretory structures such as gills, can impair ammonia excretion and promote evolution of other means of nitrogen excretion or ammonia detoxification mechanisms. Teleost embryos and yolk-sac larvae constitute a semi-closed system with reliance on endogenous yolk reserves to cover energy demands for activity, growth, and maintenance of cellular function. The production of nitrogenous end-products can thus only originate from the catabolism of a fixed amount of substrate acquired during oogenesis. The possible fates of ammonia generated by embryonic teleosts can therefore be described by a simple model: a) it can be excreted from the embryo through the perivitelline space (PVS), across the chorion, and then to the ambient water, b) it can be stored in an acidic yolk-sac, or c) ammonia can be converted to other compounds in the embryonic tissues, and subsequently stored or excreted. Fish, in general, use amino acids as an important energy substrate (Morris, 1991), and during their early life the use of amino acids is even more important both to sustain a high rate of protein synthesis, but also as an energy source (Finn *et al.*, 2002; Rønnestad *et al.*, 2003). This is especially apparent in some marine teleost embryos and yolk-sac larvae in which free amino acids (FAA) during certain periods constitute the major substrate for energy dissipation (Fyhn, 1989). Thus, fishes must have evolved efficient mechanisms for handling the substantial amounts of nitrogenous end-products that must be formed during their early development.

CLASSES OF EXCRETED NITROGENOUS COMPOUNDS

Properties and Toxicity of Ammonia

Total ammonia exists as the sum of the gas NH_3 and the cation NH_4^+. The NH_3 fraction in total ammonia increases with pH, and the equilibrium $NH_3 + H_3O^+ \Leftrightarrow NH_4^+ + H_2O$, has a pK_a of about 9.5. Further, the NH_3 fraction increases with temperature while it declines with ionic strength. At physiological pH however, most ammonia is present as NH_4^+. The NH_3 gas moves freely across most biological membranes, while NH_4^+ is usually less permeable. Ammonia is distributed according to pH if NH_3 is the only membrane permeant species, but according to the transmembrane potential if only NH_4^+ is permeant (Wood, 1993).

The toxic effects of ammonia on living tissues are manifold, including interference with nerve conduction, membrane transport, neurotransmitter metabolism, cellular metabolism and pH regulation. High internal concentrations of NH_4^+ are thought to activate N-methyl-D-aspartate (NMDA) glutamate receptors in the central nervous system, resulting in a large Ca^{2+} and Na^+ influx that eventually can lead to ATP depletion, oxygen radical formation and cell death (Felipo & Butterworth, 2002; Finn, 2007). Ammonia interferes with normal cell metabolism by activation of glycolysis, while it affects the tri-carboxylic acid cycle through inhibition of enzymes, and to some extent depletion of the intermediate α-ketoglutarate. Ammonia can cause alkalinisation of cells when NH_3 enters, since it consumes H^+ ions to form NH_4^+. The resulting change in H^+ concentration can affect important pH-dependent cellular functions, such as enzymes or signal cascades. Since NH_4^+ has a similar ionic radius to K^+, it can substitute for K^+ in ionic channels. Increases in extracellular NH_4^+ concentration can therefore cause membrane depolarisation. It has been suggested that the oriental weatherloach (*Misgurnus anguillicaudatus*), which shows a high tolerance towards external ammonia exposure, has evolved specific K^+ channels that are relatively impermeable to NH_4^+ (Tsui *et al.*, 2004).

Symptoms of ammonia exposure in juvenile and adult fish include reduced feed intake, growth and swimming ability, as well as hyperventilation, coma, convulsions, and eventual death (Ip *et al.*, 2001). However, teleost fishes appear to be more tolerant than mammals (Campbell, 1991), especially during the embryonic stage (Rice & Stokes, 1975; Calamari *et al.*, 1981; Holt & Arnold, 1983; Daniels *et al.*, 1987; Hetland, 1992; Terjesen *et al.*, 1998). While several excretory and detoxification mechanisms for handling ammonia exist in teleosts, it should be noted that some recent experiments suggest that ambient ammonia may not always be a harmful factor. In juvenile rainbow trout (*Oncorhynchus mykiss*), low levels of aqueous ammonia can increase growth rates (Wood, 2004). In the ureogenic goby (*Mugilogobius abei*), 2 mM NH_4Cl exposure led to elevated growth rates and ^{15}N ammonia was incorporated in tissue proteins (Iwata *et al.*, 2005). Further, Barimo & Walsh (2005) recently reported that Gulf toadfish (*Opsanus beta*) chronically exposed to 300 μM total ammonia from the embryonic stage showed higher juvenile wet mass than toadfish reared in control seawater.

Properties of Urea

Compared to ammonia, urea is a relatively inert molecule with low solubility in lipid bilayers. It is uncharged, has a high aqueous solubility, and contains two amino groups. The deleterious hydrophobic interaction effects of urea are only prevalent at relatively high concentrations and tend to be less pronounced than the toxic effects of ammonia (Walsh, 1998; Timasheff & Xie, 2003). However, the effects of urea on enzyme activity can be detected at low concentrations. Several enzymes show tolerance towards urea, but for other proteins these effects are counteracted by compounds such as trimethylamines (Ballantyne, 1997; Seibel & Walsh, 2002).

Partitioning between Nitrogenous End-products

Most teleosts are ammonotelic, excreting predominately ammonia as their nitrogenous end-product. For an excellent overview of actual rates of urea and ammonia excretion in several species, the reader is referred to Wood (1993, 2001) for juveniles and adults and to Wright & Fyhn (2001) for earlier life stages. Urea-N (moles of nitrogen contained in urea) usually accounts for less than 20% of the total nitrogen excretion in adult fishes when calculated from the sum of ammonia and urea-N excretion. There are, however, important exceptions to this rule, as shown by Lake Magadi tilapia in which nitrogen excretion is entirely in the form of urea, synthesised through the ornithine urea cycle (OUC) (Randall *et al.*, 1989). It has not been possible to induce significant ammonia excretion in this species, either by feeding, reduced temperature, exercise or exposure to neutral pH (Wood *et al.*, 1989). Urea excretion is an adaptation to the alkaline waters of Lake Magadi, which severely hamper ammonia excretion (Randall *et al.*, 1989).

Several other nitrogenous compounds have been detected during excretion measurements of fishes, including amino acids, peptides, uric acid, creatine, creatinine, and trimethylamines. Mass balance studies, especially in adult fish, do not always show agreement between the total amount of nitrogen excreted and the sum of ammonia and urea-N. In a study on juvenile rainbow trout (Kajimura *et al.*, 2004), several compounds were detected in respirometry samples, including free amino acids, but there were still unidentified compound(s) excreted that accounted for up to 20% of total nitrogen excretion. Amino acids and peptides have been reported to be excreted from fish embryos (Yarzhombek & Maslennikova, 1971). In Atlantic halibut (*Hippoglossus hippoglossus*) yolk-sac larvae however, no free amino acid excretion was detected, despite the very high internal yolk concentration (Rønnestad, 1993). Furthermore, the amount of nitrogen calculated from the catabolism of amino acids and nitrogen measured in the end-products balanced to within 6% in yolk-sac and fasted larvae of the African catfish (*Clarias gariepinus*) (Terjesen *et al.*, 1997), and other researchers (Rønnestad *et al.*, 1992a; Finn *et al.*, 1995a) have obtained comparable results in other species. Thus, it may be that unidentified excreted end-products are quantitatively more important in adult fishes, but a detailed study on this problem in early life stages of several teleost species is still warranted.

With respect to embryonic and larval teleosts, much less is known concerning the proportion of end-products other than ammonia in nitrogen excretion. In a pioneering study, Smith (1947) found that rainbow trout embryos excreted ammonia as the major nitrogenous end-product and although he detected urea in excretory media, he (Smith, 1947) ascribed this predominately to injury of the embryos during incubation. Later it was shown that teleost embryos excrete both ammonia and urea (Yarzhombek & Maslennikova, 1971; Rice & Stokes, 1974; Kaushik *et al.*, 1982; Oliva-Teles & Kaushik, 1987, 1990; Wright *et al.*, 1995; Pilley & Wright, 2000), but uric acid excretion has been shown to be of minor importance (Yarzhombek & Maslennikova, 1971; Kaushik *et al.*, 1982), or was not detected (Rice & Stokes, 1974; Terjesen *et al.*, 2002). Korsgaard (1994) demonstrated that urea production can be substantial during ovarian development of the viviparous eelpout (*Zoarces viviparous*). In some studies on early life stages of oviparous marine teleosts, excreted nitrogenous end-products other than ammonia have been included in the analyses (Buckley & Dillmann, 1982; Cetta & Capuzzo, 1982; Torres *et al.*, 1996; Chadwick & Wright, 1999; Terjesen *et al.*, 2002), but in other experiments only ammonia was measured (Davenport, 1983; Finn *et al.*, 1991, 1995a; Rønnestad *et al.*, 1992a, b).

METABOLIC ORIGINS OF EXCRETED NITROGEN

Amino acids, a major source of ammonia following deamination, occupy a pivotal position in the nitrogen metabolism of animals. Not only do amino acids constitute the building blocks for proteins, they are also precursors in the synthesis of nucleotides, hormones, neurotransmitters, certain lipids, the heme molecule, and glucose through gluconeogenesis. Amino acids provide energy when their carbon skeletons are funnelled into the TCA cycle. Ingested purine nucleotides can also represent a significant source of nitrogen in end-products. However, if *de novo* purine synthesis occurs, then nitrogen in end-products originates from amino acids (Campbell, 1991).

Mass balance and respiratory studies have contributed towards the current understanding of which substrates contribute to the energy dissipation during teleost development. In particular, marine species that spawn pelagic eggs rely heavily on endogenous amino acid sources in the yolk for energy dissipation during embryonic and yolk-sac larvae development (reviewed by Wright & Fyhn, 2001). When calculated for the entire yolk-sac larval stage, the Atlantic halibut cover as much as 74% of the energy dissipation by amino acid catabolism and around hatching amino acids apparently constitute the sole energy fuel in this species (Finn *et al.*, 1991, 1995a).

Pathways for Ammonia Synthesis

Transfer and deamination of amino groups

Ammonia is usually the immediate end-product of nitrogen metabolism in aquatic vertebrates (Mommsen & Walsh, 1991). As reviewed by Walsh (1998), NH_4^+ is produced in teleosts primarily during the catabolism of amino acids by the sequential action of cytosolic amino transferases and mitochondrial glutamate dehydrogenases (GDH), a process traditionally termed trans-deamination (pathway 1, Fig. 1). Amino acids such as serine, glutamine, and asparagine can also be directly deaminated. Some studies have been conducted on pathways for ammonia production in teleost embryos and larvae (e.g. Hinterleitner *et al.*, 1987; Segner & Verreth, 1995; Seoka *et al.*, 1997). Shahsavarani *et al.* (2001) measured GDH distribution in rainbow trout and found it to be predominately expressed in the embryonic tissues, and not in yolk. Regarding later life stages, Segner & Verreth (1995) measured glutamate-pyruvate and glutamate-oxaloacetate amino transferase activities in African catfish first-feeding and starved larvae, and observed a high capacity for amino acid transamination up to metamorphosis with declining activity thereafter. In contrast, during development of the sea lamprey (*Petromyzon marinus*), the amino acid catabolic capacity, as inferred from GDH and amino transferase enzyme activities and protein levels, increases several-fold when the ammocoete larvae metamorphose to a parasitic lamprey that ingests large protein-rich blood meals from its host (Wilkie *et al.*, 2006). In marine teleosts, Rønnestad *et al.* (2001) inferred fom CO_2 measurements that post-larval Senegal sole (*Solea senegalensis*) conserve indispensable amino acids such as lysine and arginine, while the dispensable amino acids glutamate and alanine were mostly catabolised. Such catabolism likely resulted in elevated ammonia production, but was not measured in the study.

Glutamine synthetase (GSase) catalyses the formation of glutamine from ammonia and glutamate (Fig. 1), and is a first-line defense against ammonia toxicity in mammals,

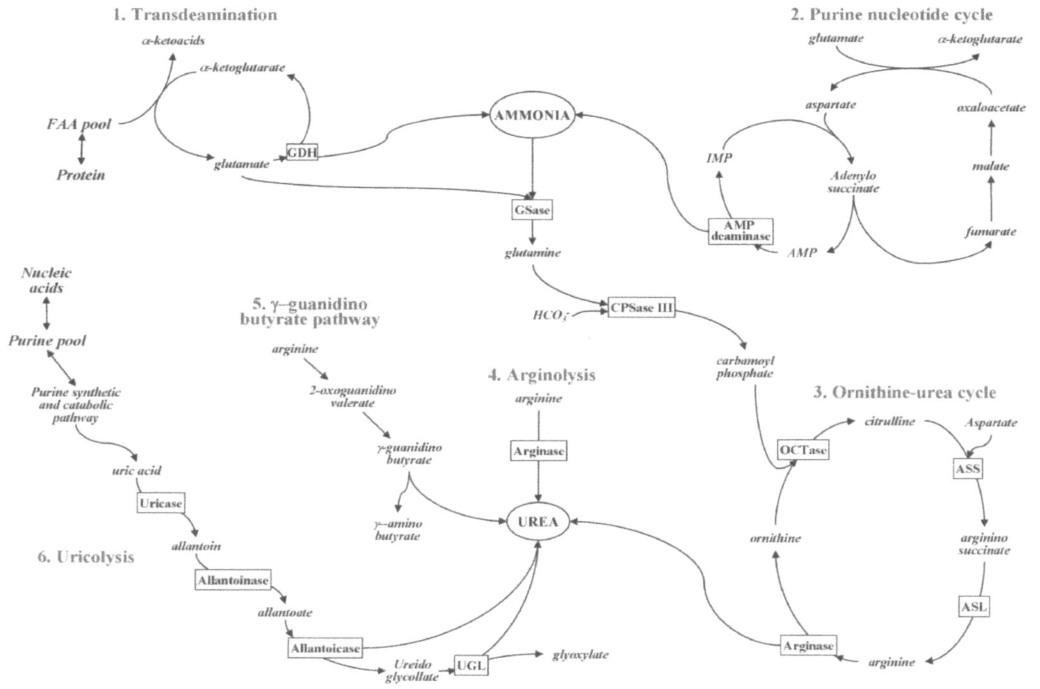

Fig. 1 Simplified view of the major metabolic pathways generating nitrogenous end-products in fish. After Wood (1993), Walsh (1998) and Terjesen (2001). Refer to the numbers next to each pathway for discussion in the text. Several enzymatic steps, substrates, products, and co-factors have been omitted for clarity. Furthermore, tissue and sub-cellular localisations are not indicated. Enzymes (in boxes): GDH, glutamate dehydrogenase; CPSase III, carbamoyl phosphate synthetase III, OCTase, ornithine carbamoyltransferase; GSase, glutamine synthetase; UGL, ureidoglycollate lyase; ASS, argininosuccinate synthetase; ASL, argininosuccinate lyase. Compounds: AMP, adenosine monophosphate; IMP, inosine monophosphate; FAA, free amino acids.

especially in nervous tissues (Cooper, 2001). This appears also to be the case in fish, since GSase activity increases in brain upon ammonia exposure. Indeed, treatment of Gulf toadfish with the GSase inhibitor methionine sulfoximine (MSO) prior to ammonia exposure, prevented an increase in brain glutamine and caused complete mortality of the experimental fish, compared to ammonia exposure without MSO (Veauvy *et al.*, 2005). In contrast to mammals, where a single gene codes for GSase (Kuo & Darnell, 1989), multiple genes are known in teleosts (Murray *et al.*, 2003; Walsh *et al.*, 2003). In embryos and alevins of rainbow trout, the mRNA expression profile of the four GSase isoforms differed to those of adult trout (Fig. 2). Essex-Fraser *et al.* (2005) proposed that the early Onmy-GS01 and Onmy-GS03 expression result in elevated capacity for glutamine formation from ammonia in the embryo, during a period when increased internal ammonia concentrations are observed in rainbow trout and other teleost embryos.

Nucleotide metabolism

The purine nucleotide cycle produces NH_3 by deamination of adenosine monophosphate (pathway 2, Fig. 1), and is important in ammonia production by adult teleost muscle following exercise (Walsh, 1998). However, except for observations in

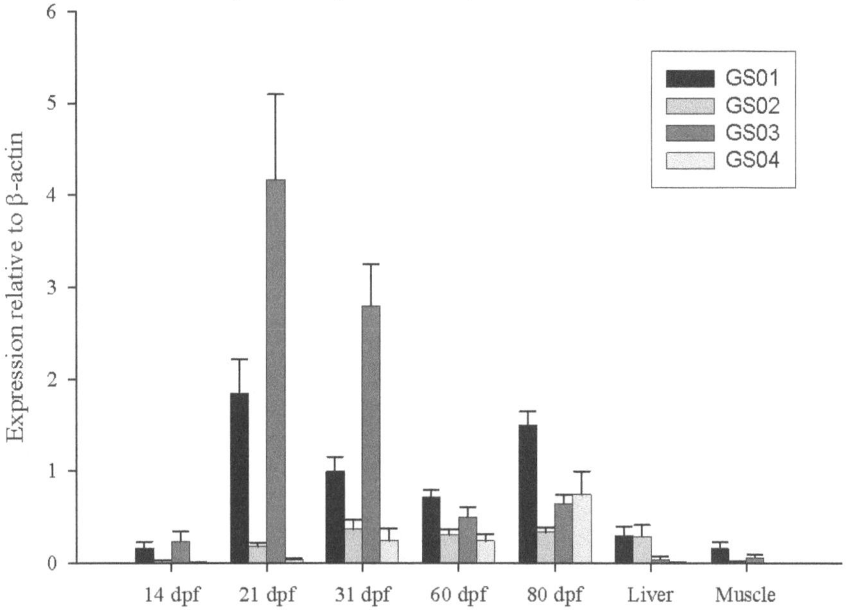

Fig. 2 Changes in mRNA levels of four glutamine synthetase genes (*Onmy-GS01* to *Onmy-GS04*) in early life stages of rainbow trout (*Oncorhynchus mykiss*). Hatching occurred between 25 and 31 dpf (days post-fertilisation). Values are means ± SEM. The experiments indicated differential expression of the genes compared to adult liver and muscle (Essex-Fraser *et al.*, 2005). Reprinted with permission from The American Society for Biochemistry and Molecular Biology.

chum salmon (*Oncorhynchus keta*) where adenylate degradation may be a source of ammonia following fertilisation (Yamamoto, 1984), the contribution by the purine nucleotide cycle during early teleost development is not known. Similarly, pyrimidine catabolism also generates ammonia in animals (Weihrauch *et al.*, 2004), but the importance in developing fish remains to be studied.

Pathways for Urea Synthesis

Nitrogen in urea can originate from arginine, purine nucleotides, ammonia and aspartate via the OUC. Four biochemical pathways can produce urea in teleosts (Fig. 1): the OUC (pathway 3), hydrolysis of arginine (arginolysis, pathway 4), the γ-guanidinobutyrate pathway from arginine (pathway 5), and purine catabolism and uricolysis (pathway 6). However, the expression of these pathways, especially enzymes of the OUC, differs greatly between species and life stages.

In the following the pathways are divided into those that can detoxifiy ammonia in fish and those that do not. To be classified as an ammonia detoxification pathway, free ammonia must be converted to a less toxic form, and then stored or excreted. Mechanisms such as partial and decreased amino acid catabolism, employed by several adult teleosts (reviewed by Ip *et al.*, 2001), are not ammonia detoxification mechanisms *per se*, but reduce the chance of ammonia toxicity by lowering production.

Urea synthetic pathways that can detoxify ammonia

Overview of the OUC The OUC was the first metabolic cycle to be discovered, when it was reported by Sir Hans Krebs and Kurt Henseleit in 1932 (Krebs & Henseleit, 1932; Krebs, 1982). In mammals, the OUC is intimately involved in maintaining safe levels of tissue ammonia, and there is a large reserve capacity for urea synthesis available at any time. Only GSase, the OUC and uricolysis, if coupled to *de novo* purine synthesis, can serve to detoxify ammonia in animals (Campbell, 1991). However, urea production through the OUC is energetically costly since 2.5 mol ATP are consumed for each mol N in urea. In addition, certain of the OUC enzymes, with the exception of arginase, are involved in a multi-tissue mechanism that produces net arginine *de novo* in mammals (Meijer, 1995).

Expression of the OUC in adult or juvenile fishes Current evidence indicates that, of the urea producing pathways in fishes, only the OUC can detoxify ammonia. When expressed, the OUC can form urea in teleosts or elasmobranchs through the utilisation of glutamine, HCO_3^-, ATP, and aspartate (Anderson, 1995a). Thus, to function as an ammonia detoxification mechanism, mitochondrial GSase is utilised as a feeder enzyme to the OUC in teleost and elasmobranchs (Walsh, 1997). The genes encoding the OUC enzymes were for a long time thought to have been lost during teleost evolution (Brown & Cohen, 1960; Huggins *et al.*, 1969; Wilson, 1973). Read (1971) however, found significant activities of the OUC enzymes in adult oyster toadfish (*Opsanus tau*). Subsequently, a number of fish species have been found to have OUC enzyme activities. To date, significant activities in adults have been found mostly in lungfishes, elasmobranchs or in some teleost species living in environments that continuously or occasionally hamper ammonia excretion (Saha & Ratha, 1987; 1989; Mommsen & Walsh, 1989; Randall *et al.*, 1989). In the case of the Indian walking catfish (*Clarias batrachus*), there appears to be intraspecific variations in OUC expression. This species may occasionally experience adverse environmental conditions (Saha & Ratha, 1998), and significant activities of key OUC enzymes have been reported by one research group analysing Indian fish (Saha *et al.*, 1999, 2002; Kharbuli *et al.*, 2006). However, significant key OUC enzyme activities were not detected by others analysing fish raised in Singapore, even after specimens from this stock of walking catfish were subjected to ammonia exposure (Ip *et al.*, 2004).

In most ammonotelic teleosts the activities of key OUC enzymes are low or below detection limits in the adult stage. In those species that express significant OUC activities, evidence for a role for this pathway in ammonia detoxification appears the most convincing (Randall *et al.*, 1989; Walsh, 1997; Saha & Ratha, 1998; Iwata *et al.*, 2000), although other hypotheses related to osmoregulation (Dépêche & Schoffeniels, 1975; Dépêche *et al.*, 1979), and arginine synthesis (Chiu *et al.*, 1986; Buentello & Gatlin, 2000) have been proposed.

Expression of the OUC in fish embryos and larvae In a seminal review, Griffith (1991) reasoned that urea synthesis is an essential ammonia detoxification mechanism during early development of teleosts. Activities of the OUC during early teleost development was first investigated by Rice & Stokes (1974) in rainbow trout, but these authors did not succeed in detecting carbamoyl phosphate synthetase (CPSase), argininosuccinate synthetase and argininosuccinate lyase (ASL), only ornithine carbamoyl transferase (OCT) and arginase. When Rice & Stokes (1974) conducted tracer studies *in vivo*, no incorporation of label was found in urea. In contrast, Dépêche et al. (1979) detected significant *in vivo* incorporation of label into urea in rainbow trout embryos. These findings constitute evidence that the OUC was active *in vivo* in early life stages of this teleost species. Later, Wright *et al.* (1995) employed sensitive assay methods and detected GSase, OCT, and arginase, and most importantly also CPSase III from hatching and during a considerable period of rainbow trout development. The latter authors (Wright *et al.*, 1995) suggested that all teleosts express the OUC enzymes during early development, but that expression is suppressed in the adult state except for a few species inhabiting adverse environments. More recently, a number of teleost species have been shown to express key OUC enzymes during early life stages, while in the adult state, only low levels or no activity can be detected (Table 1). Thus, the original hypotheses by Griffith (1991) and Wright *et al.* (1995) regarding OUC enzyme expression seems to hold for those species studied to date. Furthermore, if net *de novo* arginine synthesis is the primary reason for expression of CPSase III and certain of the other OUC enzymes, it could be expected that arginine would not be essential for growth in these life stages. However, when feeding larvae of rainbow trout with diets virtually devoid of arginine, growth and survival was negatively affected (Terjesen

Table 1 Presence of ornithine-urea cycle enzyme activity, mRNA or label recovered *in vivo* as ^{14}C-urea in fish early life stages (embryos and larvae). Additionally, tissue localisation of carbamoyl phosphate synthetase III is indicated in cases where this has been investigated. Data for juveniles or adults are provided only for those species that have been studied also during early life stages. Thus, this table is not exhaustive in terms of studies on adults and juveniles.

	Embryos or larvae					**Juveniles or adults**				
	Enzyme activity or mRNA expression			*CPS III localisation*	*In vivo: ^{14}C label in urea*	*Enzyme activity or mRNA expression*[A]			*CPS III localisation*	*References*
Species	GS	CPS III	OCT			GS	CPS III	OTC		
Freshwater teleosts										
Rainbow trout (*Oncorhynchus mykiss*)	-	No	Yes	-	-	-	-	-	-	Rice & Stokes (1974)
Rainbow trout (*O. mykiss*)	Yes	Yes	Yes	-	-	Yes	No[B]	No	-	Wright *et al.* (1995)
Rainbow trout (*O. mykiss*)	Yes	Yes	Yes	EH	-	-	-	-	-	Steele *et al.* (2001)
Rainbow trout (*O. mykiss*)	-	-	-	-	Yes	-	-	-	-	Dépêche *et al.* (1979)
Rainbow trout (*O. mykiss*)	-	-	-	-	-	Yes	Yes	Yes	M	Korte *et al.* (1997)
Guppy (*Poecilia reticulata*)	-	-	-	-	Yes	-	-	-	-	Dépêche *et al.* (1979)
African catfish (*Clarias gariepinus*)	Yes	Yes	Yes	EH	-	Yes	Trace	Yes	-	Terjesen *et al.* (2001)
African catfish (*C. gariepinus*)	-	-	-	-	-	Yes	No	Yes	-	Ip *et al.* (2004)
Walking catfish (*Clarias batrachus*)	Yes	Yes	Yes	-	-	Yes	Yes	Yes	L, K, M	Kharbuli *et al.* (2006)

Table 1 Contd. ...

Table 1 Contd. ...

Walking catfish (*C. batrachus*)	-	-	-	-	-	Yes	No	Yes	-	Ip *et al.* (2004)
Brazilian catfish (*Pseudoplatystoma coruscans*)	-	Yes	Yes	-	-	-	-	-	-	Moraes *et al.* (2002)
Pacu (*Piaractus mesopotamicus*)	Yes	-	Yes	-	-	-	-	-	-	Monzani *et al.* (1998)
					-					
<u>Marine teleosts</u>					-					
Atlantic cod (*Gadus morhua*)	Yes	Yes	Yes	-	-	Yes	No	Yes	-	Chadwick & Wright (1999)
Atlantic halibut (*Hippoglossus hippoglossus*)	-	Yes	-	-	-	-	Yes	-	M	Terjesen *et al.* (2000)
Atlantic halibut (*H. hippoglossus*)	Yes	Yes	Yes	-	-	Yes	-	Yes	-	Terjesen *et al.* (2002)
Gulf toadfish (*Opsanus beta*)	Yes	Yes	Yes	-	-	-	-	-	-	Barimo *et al.* (2004)
Gulf toadfish (*O. beta*)	-	-	-	-	-	Yes	Yes	Yes	L, M	Julsrud *et al.* (1998)
<u>Elasmobranchs</u>										
Big skate (*Raja binoculata*)	-	-	Yes	-	-	-	-	-	-	Read (1968)
Pacific spiny dogfish (*Squalus suckleyi*)	-	-	Yes	-	-	-	-	-	-	Read (1968)

Table 1 Contd. ...

Table 1 Contd. ...

Little skate (*Raja erinacea*)	Yes	Yes	Yes	M[C]	-	-	-	-	-	Steele *et al.* (2004)
Little skate (*R. erinacea*)	-	-	-	-	-	Yes	Yes	Yes	L, M	Steele *et al.* (2005)
<u>Agnatha</u>										
Sea lamprey (*Petromyzon marinus*)	Yes	No	Yes	-	-	-	-	-	-	Wilkie *et al.* (1999)
Sea lamprey (*P. marinus*)	-	-	-	-	-	Yes	Yes	Yes	L	Wilkie *et al.* (2004)

[A] Indicates if activity or mRNA have been detected in any of the investigated tissues.
[B] Wright *et al.* (1995) assayed adult liver for CPS III activity, not muscle as in Korte *et al.* (1997).
[C] CPS III activity in muscle *vs*. the whole embryo (8-month-old) indicated muscle localisation and possibly also other tissues.
"-" indicates not studied; EH, extra-hepatic localisation in the larval body; M, muscle; K, kidney; L, liver.
GS, glutamine synthetase; CPS III, carbamoyl phosphate synthetase III; OCT, ornithine carbamoyl transferase.

et al., 2006), despite expression of the OUC enzymes during this life period (Wright *et al.*, 1995). Finally, experiments on larvae of the Indian walking catfish show upregulation of OUC enzyme activities following ammonia exposure (Kharbuli *et al.*, 2006). Taken together, these findings support the hypotheses that OUC enzymes are expressed in teleosts during early life stages primarily for ammonia detoxification purposes (Griffith, 1991; Wright *et al.*, 1995).

Regarding elasmobranchs, the OUC is expressed in embryos for urea synthesis to maintain near-isosmotic pressure with seawater (Read, 1968; Kormanik, 1995; Steele *et al.*, 2004). The activities of CPSase III in the little skate (*Raja erinacea*) embryos are sufficient to account for the observed urea excretion (Steele *et al.*, 2004). In this species, the necessary mechanisms for using urea and other compounds as osmolytes are established by 4 months post-fertilisation (pf), when the egg capsule is opened and the embryo is directly exposed to the external medium (Steele *et al.*, 2004). Viviparous teleosts have high ovarian urea concentrations (Korsgaard, 1994), and for embryos of guppy (*Poecilia reticulata*) *in vivo* tracer experiments have demonstrated a functional OUC (Dépêche *et al.*, 1979). Barimo *et al.* (2004) demonstrated that in the facultative ureotelic Gulf toadfish, the key OUC enzymes were present already at the embryonic stage, and a considerable increase in activity was noted during early development.

Carbamoyl phosphate synthetase CPSase III and GSase are key enzymes of the OUC in teleosts, and together with OCT, their detection is a criteria for classifying whether species have a functional OUC. However, *in vitro* OUC enzyme activities only demonstrate the potential for urea synthesis via this pathway (Mommsen & Walsh, 1991). Thus, *in vivo* tracer studies are necessary for conclusive evidence. CPSase III was first discovered in invertebrates (Trammel & Campbell, 1971). This enzyme deviates from mammalian CPSase I mostly by the utilisation of glutamine instead of NH_3 as the primary substrate, and the synergistic relationship between the binding of glutamine and the allosteric effector N-acetyl-L-glutamate (AGA) to the enzyme (Anderson, 1976, 1995a; Casey & Anderson, 1983). In fish, Anderson (1976) characterised CPSase III for the first time in largemouth bass (*Micropterus salmoides*). CPSase II, an enzyme involved in pyrimidine synthesis and otherwise unrelated to the OUC, is also expressed in fish, and complicates assays of low levels of CPSase III in crude extracts (Anderson, 1995b). This was especially evident when searching for CPSase III in Atlantic halibut larvae, due to low AGA-induced activity, and high levels of CPSase II (Terjesen *et al.*, 2000). In this latter case, gel filtration was necessary to conclusively establish the presence of CPSase III. Detection of CPSase III in the African catfish larvae (Terjesen *et al.*, 2001), was less complicated due to the enormous AGA activation factor, approaching that of mammalian CPSase I (Rubio *et al.*, 1983). In the Gulf toadfish, CPSase III is also very active already at the embryonic stage, and the relative proportion of CPSase III in total CPSase activity increases as development progresses (Barimo *et al.*, 2004).

Despite the special properties of the Atlantic halibut CPSase III, significant AGA activation is still likely to occur during early development in this species assuming that cellular ATP and glutamine is present at levels equal to those that can be calculated from literature (Finn *et al.*, 1995a). Thus, as in adult Gulf toadfish (Julsrud *et al.*, 1998), short-term regulation of carbamoyl phosphate formation in exogenously feeding fish larvae may be influenced by variations in AGA concentrations due to changes in amino acid absorption. Molecular tools made feasible through the studies on fish CPSase III (Hong *et al.*, 1994; Korte *et al.*, 1997) and AGA synthetase genes (Caldovic *et al.*, 2002) will likely advance the understanding of how the OUC is regulated during early development in fish.

Tissue distribution of the OUC in fishes There is accumulating evidence that CPSase III is expressed exclusively in muscle or in both muscle and liver in a number of adult teleost species (Korte *et al.*, 1997; Felskie *et al.*, 1998; Julsrud *et al.*, 1998; Kong *et al.*, 1998; Lindley *et al.*, 1999; Terjesen *et al.*, 2000). Also in adult elasmobranchs, OUC enzymes have been detected in muscle (Steele *et al.*, 2005). The muscle location of key OUC enzymes in fishes is unique among the vertebrates. Regarding early life stages, Chadwick (1998) and Steele *et al.* (2001) found that in newly hatched rainbow trout, CPSase III is expressed in non-hepatic tissues of the embryonic body and not to any significant extent in yolk. In elasmobranchs, OUC enzyme activities were not found in yolk of Pacific spiny dogfish (*Squalus suckleyi*), big skate (*Raja binoculata*), or little skate, but were present in the embryonic tissues (Read, 1968; Steele *et al.*, 2004). Similar findings to that of rainbow trout have been reported for exogenously feeding larvae of the African catfish (Terjesen *et al.*, 2001), suggesting that muscle CPSase III expression is initiated early in development. Various hypotheses have been proposed to explain the muscle location of OUC enzymes in fishes. Chadwick (1998) proposed that enzyme capacities in larval liver are too low to scavenge all potentially toxic ammonia before it reaches sensitive tissues such as the brain. Muscle GSase and CPSase III expression may thus be advantageous. Regarding adult Lake Magadi tilapia, Lindley *et al.* (1999) reasoned that due to the low turnover rate of CPSase III and high rate of urea synthesis in this species, it would be impossible to package all the CPSase III needed into the small liver, and muscle was therefore utilised for expression. In Atlantic halibut larvae, a distinct liver is not present until about 24 dpf (Pittman *et al.*, 1990), which is after CPSase III is detectable (Terjesen *et al.*, 2000, 2002). Even at hatching there are only a few poorly developed hepatocytes (Kjørsvik & Reiersen, 1992). Adult Atlantic halibut express CPSase III in muscle, although at very low levels. In the African catfish at hatching, only an undifferentiated liver anlage is present. Thus, anatomical constraints of developing fish larvae may dictate that muscle is primarily utilised for CPSase III expression, in contrast to liver.

Pathways for urea synthesis that cannot detoxify ammonia

Hydrolysis of arginine Arginine is classified as an indispensable amino acid in teleosts for growth and survival (Wilson, 2002). This is in contrast to mammals, where it is a conditional indispensable amino acid (Flynn *et al.*, 2002). In mammals, net arginine is synthesised by certain of the OUC enzymes combined with enzymes for ornithine formation from glutamate, by the intestine-renal axis (Wu & Morris, 1998). The hydrolysis of arginine, catalysed by arginase, results in the formation of urea (pathway 4, arginolysis, Fig. 1). Urea can also be formed from arginine through the γ-guanidinobutyrate pathway (pathway 5, Fig. 1). However, this pathway has only been reported for adult fish (reviewed by Vellas & Serfaty, 1974), and a study on the activities and properties of this pathway during early development would be useful. Arginase is ubiquitous, and can be expressed in animals that do not have an active OUC (Jenkinson *et al.*, 1996). Arginine can be utilised for agmatine, creatine and nitric oxide synthesis; pathways which do not result in urea synthesis, and polyamine, proline and glutamate formation, which are pathways that do result in urea formation. While these latter pathways have been rarely studied in fish (Buentello & Gatlin, 1999, 2000, 2001), recent studies on larvae of rainbow trout have shown that enzymes for proline formation from pyrroline-5-carboxylate precursor are expressed in liver and the intestine (Dabrowski *et al.*, 2005).

In teleosts, urea production through arginolysis has not been regarded as major pathway due to the indispensability of arginine (Walton & Cowey, 1982). However, when arginine in excess of requirement is present in the diet, adult teleosts show elevated plasma urea levels (Berge *et al.*, 1997; Alam *et al.*, 2002; Gouillou-Coustans *et al.*, 2002; Fournier *et al.*, 2003), suggesting that in these situations arginolysis is a major route for urea production. Better tools for investigating arginolysis have now become available since teleost arginases were recently cloned (Wright *et al.*, 2004). As in mammals and amphibians, two distinct arginases exist, I and II, in rainbow trout, torafugu (*Takifugu rubripes*), and probably also zebrafish (*Danio rerio*) (Wright *et al.*, 2004). However, both teleost arginases contain mitochondrial targeting signals, whereas in mammals, only arginase II is mitochondrial (Wright *et al.*, 2004). During fasting in juvenile trout, total arginase activity increased, but only arginase II mRNA was upregulated in liver, suggesting that the types of arginase have distinct metabolic functions (Wright *et al.*, 2004).

During early development in fishes, the role of arginolysis in urea production is not known in detail. In African catfish yolk-sac and fasted larvae, it was estimated that arginine depletion from the free and bound pool could account for a maximum of one third of urea-N excretion (Terjesen *et al.*, 1997, 2001). In Atlantic halibut larvae a maximum of 75% of the urea-N can originate from arginine (Terjesen *et al.*, 2002), implying that other pathways for urea production were functional in both species during early development.

Purine catabolism and uricolysis Uricolysis in fishes was first demonstrated by Brunel (1937). This pathway (no. 6, Fig. 1) is involved in the breakdown of purines from the diet and nucleic acid turnover and may be an important pathway for urea production in teleosts (Goldstein & Forster, 1965; Vellas & Serfaty, 1974), but as for arginolysis, its relative importance is likely related to dietary concentrations of purines. Uricolysis forms two urea molecules from the breakdown of uric acid by the sequential action of uricase, allantoinase, allantoicase, and ureidoglycollate lyase (UGL, Takada &

Noguchi, 1986). As studied in adult teleosts, uricase, allantoicase and UGL are localised in perioxsomes, while allantoinase is found either in the cytosol or perioxsomes (Goldenberg, 1977; Noguchi *et al.*, 1979; Takada & Noguchi, 1983, 1986). Vellas & Serfaty (1974), cited in Wood (1993), found that when common carp (*Cyprinus carpio*) liver was perfused with glutamine, urea and uric acid production increased, but when [^{14}C]HCO_3^- was injected, no label was recovered in urea, suggesting that in this case uricolysis was functional and that any OUC activity *in vivo* in this tissue was below the level of detection.

In early life stages of teleosts however, pathways for urea production other than the OUC have received little attention. Rice & Stokes (1974) hypothesised that since uric acid concentration did not change appreciably during early development, the uricolytic pathway did not contribute significantly to urea production in rainbow trout embryos and larvae. However, this observation can also mean that the uricolytic pathway in trout is effective and only allows minor changes in uric acid concentrations. Vigetti *et al.* (2000) reported that in African clawed frog (*Xenopus laevis*) embryos, allantoicase mRNA is present from 24 hours pf, although detectable activity was not found until 5 dpf. Furthermore, xanthine dehydrogenase, the enzyme preceeding uricolysis in purine catabolism, has been detected in zebrafish embryos (Ziegler *et al.*, 2000).

In adult teleosts, the uricolytic pathway is thought to mainly function in degradation of dietary purines (Wood, 1993). During early development of zebrafish however, uricase *in situ* hybridisation shows that expression occurs not only in the liver, but also in the brain and retina (Fig. 3A). Such extra-hepatic expression of uricase can be also observed in adult Atlantic salmon (*Salmo salar*) (Fig. 3B; Andersen *et al.*, 2006). Expression of uricase is initiated in the Atlantic salmon embryo at 150 day-degrees, corresponding to the onset of organogenesis (Fig. 3C). This is consistent with the detection of uricase enzyme activity along with other uricolytic enzymes at the embryonic stage in African catfish (Fig. 3D) and Atlantic halibut (Terjesen *et al.*, 2002).

These findings raise questions concerning the tissue-specific roles of the uricolytic enzymes during early development. Nucleotides are energetically expensive to synthesise, and their catabolism may seem wasteful in an organism that relies solely on endogenous yolk sources for energy and building blocks. Terjesen *et al.* (2001) suggested that a functional uricolytic pathway prior to first-feeding would be essential for handling the first dietary nucleotide absorption. Furthermore, urate is present at relatively high levels in rainbow trout embryos (150-500 μmol kg^{-1}) (Rice & Stokes, 1974) and in fish seminal plasma (122-736 μmol L^{-1}) (Ciereszko *et al.*, 1999). Expression of uricolytic enzymes early in development may also be linked to catabolism of urate that has been deposited in yolk during oocyte maturation as suggested by Andersen *et al.* (2006). During fasting or starvation periods early in development, purine nucleotide breakdown may be inevitable, and uricase activity was higher in starved compared to fed larvae of the African catfish (Fig. 3D). Uric acid accumulated in the Atlantic halibut (Terjesen *et al.*, 2002) when these larvae enter a negative energy balance (Finn *et al.*, 1995a), and RNA content per individual starts to decline (Pittman, 1991).

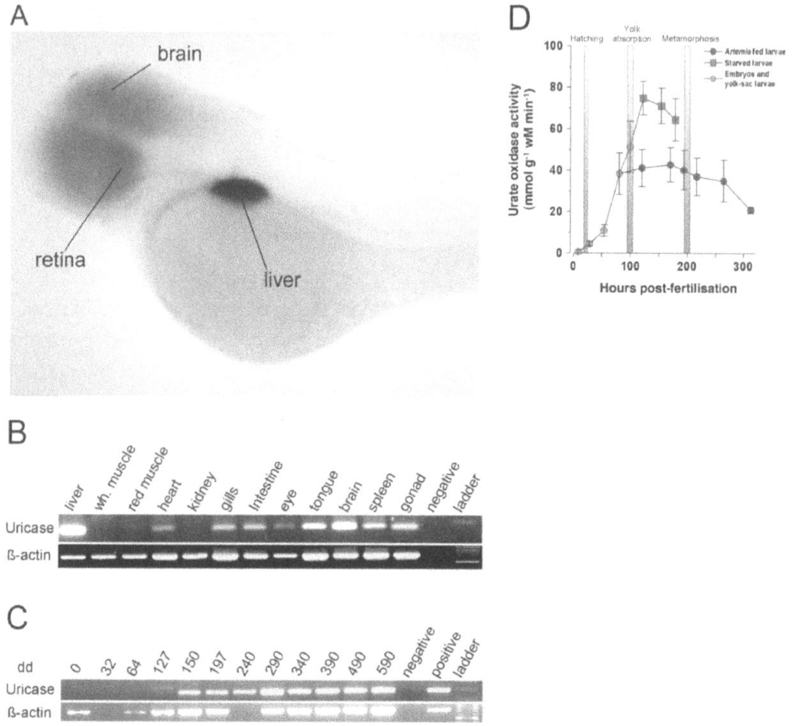

Fig. 3 Ontogeny of uricase expression and activity in fish, showing the presence of mRNA or enzyme activity in early life stages of three teleosts species. Panel A: Uricase mRNA distribution studied by *in situ* hybridisation in zebrafish (*Danio rerio*) embryos. After Thisse *et al.* (2001). Reprinted with permission from the author and ZFIN. Panel B: Uricase mRNA tissue distribution in juvenile Atlantic salmon (*Salmo salar*). After Andersen *et al.* (2006). Reprinted with permission from Blackwell Publishing. Panel C: Temporal expression of uricase mRNA during development of Atlantic salmon (dd = day-degrees). The lack of a ß-actin signal at 32 and 240 dd was due to an application error. After Andersen *et al.* (2006). Reprinted with permission from Blackwell Publishing. Panel D: Uricase activity during early development of the African catfish (*Clarias gariepinus*). Values are means ± SD. wM= wet mass. After Terjesen *et al.* (2001). Reprinted with permission from The Company of Biologists.

MECHANISMS AND PATTERNS OF NITROGEN EXCRETION

Ammonia

Ammonia excretion

Knowledge about ammonia excretion mechanisms is scarce in embryonic teleosts. In adult teleosts, ammonia excretion can be facilitated by several mechanisms, depending on environmental parameters such as pH and ionic strength. These mechanisms include NH_3 diffusion down a favourable concentration gradient and entrapment as NH_4^+ due to concurrent H^+ or CO_2 excretion. Furthermore, ammonia can be excreted through Na^+/NH_4^+ exchange, and NH_4^+ diffusion, the latter mechanism only occurring in marine fishes (Randall & Wright, 1987; Wilkie, 1997; Evans *et al.*, 2005). In some species that experience air exposure, ammonia excretion can also occur from cutaneous surfaces by volatilisation (Davenport & Sayer, 1986; Litwiller *et al.*, 2006).

In one of the few studies on early life stage excretion mechanisms in fishes, Rahaman-Noronha *et al.* (1996) showed that an acidic boundary layer external to the chorion was present in rainbow trout, suggesting a linkage between NH_3 and CO_2 excretion also in embryos as in adult trout. In support of this, the rate of excretion from rainbow trout embryos was partly dependent on the NH_3 partial pressure gradient between the PVS and the unstirred boundary water layer next to the embryo. Recently it was shown that proteins of the rhesus glycoprotein family are expressed in fish, and likely involved in ammonia excretion (Hung *et al.*, 2007; Nakada *et al.*, 2007a, b). Nakada *et al.* (2007b) demonstrated that Rhcg1 is expressed on the yolk sac, gill, and renal tubules of zebrafish larvae, and that expression is detectable on the yolk-sac already at 3 dpf. Interestingly, ammonia excretion correlated with expression of the Rhcg1 in zebrafish larvae (Nakada et al., 2007b). In early life stages of marine teleosts however, little is known regarding mechanisms for ammonia excretion. Excretion in Atlantic cod (*Gadus morhua*) and Atlantic halibut is elevated a short period following fertilisation (Finn *et al.*, 1991, 1995b; Terjesen *et al.*, 2002), which may be due to ammonia liberated from the cross-linking of the γ-carbonyl group of glutamine and the ε-amino group of lysine in the chorion during hardening (Oppen-Berntsen *et al.*, 1990). A relatively high ammonia excretion following fertilisation has also been observed in lumpsucker (*Cyclopterus lumpus*), but in this case it was suggested to originate from formation of a sticky outer chorion layer (Davenport *et al.*, 1983). A similar explanation was offered for Siberian sturgeon (*Acipenser baerii*), which also shows an elevated ammonia excretion shortly after fertilisation (Dabrowski & Kaushik, 1986). However, Yamamoto (1984) suggested that the ammonia excretion from chum salmon eggs may be due to deamination of adenylates. More detailed studies of ammonia formation in the period following fertilisation would be valuable, especially to be able to distinguish between metabolically formed ammonia (e.g. amino acid catabolism) and from structural origin (e.g. chorion processing).

During subsequent developmental stages, rates of ammonia excretion rise gradually in the marine teleosts studied, and increase semi-exponentially during the last period leading up to hatching (Davenport *et al.*, 1983; Finn *et al.*, 1991, 1995b; Terjesen *et al.*, 1998, 2002; Chadwick & Wright, 1999). In the rainbow trout however, rates of ammonia excretion increase linearly up to around 30 dpf, and remain at this level until after hatching, followed by a rapid increase in yolk-sac larvae (Wright *et al.*, 1995).

Ammonia accumulation

Several teleost species, especially marine, accumulate considerable quantities of ammonia during early development, concurrent to excretion. Griffith (1991) proposed that ammonia accumulates in teleost embryos, based on the general utilisation of amino acids as an energy substrate in fishes, their protracted embryonic development, and low membrane permeabilities. The vitelline membrane has been shown to be virtually devoid of intramembranous inclusions in the redchin panchax killifish (*Epiplatys dageti*) (Jørgensen & Schmalbruch, 1984), and has a very low permeability for water and solutes (Potts & Eddy, 1973; Riis-Vestergaard, 1982, 1987; Mangor-Jensen, 1987). In contrast, the chorion is freely permeable to low-molecular mass solutes (Potts & Rudy, 1969; Eddy, 1974; Riis-Vestergaard, 1982; Hølleland & Fyhn, 1986; Mangor-Jensen, 1987).

As reasoned by Griffith (1991) ammonia does indeed accumulate with age during early development of a number of species, including rainbow trout (Rice & Stokes, 1974; Wright *et al.*, 1995), African catfish (Terjesen *et al.*, 1997), Indian walking catfish (Kharbuli *et al.*, 2006), Atlantic halibut (Finn *et al.*, 1991; Terjesen *et al.*, 1998; Terjesen *et al.*, 2002), plaice (*Pleuronectes platessa*) (Skjærven *et al.*, 2003), Atlantic cod (Fyhn & Serigstad, 1987; Finn *et al.*, 1995b; Chadwick & Wright, 1999), turbot (*Scophthalmus maximus*) (Rønnestad *et al.*, 1992a; Finn *et al.*, 1996), lemon sole (*Microstomus kitt*) (Rønnestad *et al.*, 1992b), gilthead seabream (*Sparus aurata*) (Rønnestad *et al.*, 1994), and red sea bream (*Pagrus major*) (Seoka *et al.*, 1997). The only species so far reported which does not show ammonia accumulation during the embryonic stage is the Asian seabass (*Lates calcarifer*) (Sivaloganathan *et al.*, 1998). However, in the Asian seabass ammonia content trends after hatching were similar to other marine species. The eggs are small (0.8 mm diameter), the embryonic stage is short, and FAA is of lesser importance as an energy substrate (Sivaloganathan *et al.*, 1998). In many of the above species, the total ammonia content, calculated per individual, increases up to hatching, followed by a rapid decline. In contrast to other marine species, ammonia continues to accumulate in the yolk of Atlantic halibut larvae for several days after hatching.The reason for the observed ammonia accumulation in early life stages of teleosts is unclear. However, accumulation may occur if excretion is impaired due to the absence of gills in the earliest life stages and to a lack of sufficient water convection (Wright *et al.*, 1995). Furthermore, in marine species with pelagic eggs ammonia accumulation has been hypothesised to be connected with NH_4^+/K^+ exchange and buoyancy regulation (Fyhn, 1993; Terjesen *et al.*, 1998).

Compartments for ammonia accumulation

Where in the teleost embryo the accumulating ammonia is located and how gradients are maintained has not been extensively studied. However, Rahaman-Noronha *et al.* (1996), showed that ammonia was distributed according to transmembrane potential between yolk and the PVS, but according to pH between the PVS and the unstirred boundary water layer external to the embryo. Chadwick & Wright (1999) found that whole-animal ammonia concentrations in Atlantic cod peaked at about 20 mmol N kg^{-1} wet $mass^{-1}$ shortly before hatching. Since the chorion of Atlantic cod is permeable to low molecular mass solutes, including charged species (Mangor-Jensen, 1986), ammonia was probably not located to any quantitatively significant extent in the PVS.

Fyhn (1990) proposed that the yolk serves as a storage compartment during periods of ammonia accumulation in marine species with pelagic early life stages. In the Atlantic halibut rates of whole-animal ammonia accumulation are higher than in the larval body, suggesting that the yolk is the major storage compartment for ammonia (Terjesen *et al.*, 2002). Yolk ammonia concentration increases during embryonic and yolk-sac larval development in Atlantic halibut until about 35 days pf. Yolk pH of two marine species with pelagic embryos has been measured by *in vivo* NMR (Grasdalen & Jørgensen, 1985, 1987; Jørgensen & Grasdalen, 1990), indicating that the pH declined to 5.3 in plaice during gastrulation, and was 5.6 in Atlantic halibut yolk-sac larvae. In contrast, Atlantic halibut yolk-sac larval body tissues had an average pH of 7.4 (Jørgensen & Grasdalen, 1990). Although yolk storage cannot eliminate ammonia, it may serve to separate this toxic compound from the nervous tissues of the developing embryo during times when all ammonia produced is apparently not excreted, and to keep NH_3 levels low due to the acidic pH in yolk. In the freshwater African catfish yolk-sac larvae, whole animal ammonia accumulation is of lesser quantitative importance compared to the Atlantic halibut, since only 2% of the total nitrogenous end products formed accumulates in the larvae, while the remainder is likely excreted (Terjesen *et al.*, 1997).

Although in rainbow trout, yolk pH is less acidic (6.35) than in marine species, the yolk has the highest total ammonia level (Rahaman-Noronha *et al.*, 1996; Essex-Fraser *et al.*, 2005). During acute or chronic ammonia exposure trials with trout, it was found that yolk ammonia levels increased significantly, but the larval body concentration was unaffected, indicating that the rainbow trout has an efficient system for avoiding excessive ammonia levels in the respiring tissues by storage in yolk (Steele *et al.*, 2001). In the African catfish, the larval body maintains a lower ammonia concentration than in the yolk (Terjesen *et al.*, 1997). Interestingly, spiny dogfish shark (*Squalus acanthias*) pups may utilise a functionally similar system for ammonia accumulation. The uterine water has a pH down to 5.9, which would act to trap NH_3 as NH_4^+ (Kormanik & Evans, 1986). Total ammonia levels in the uterine fluid was found to be tenfold higher than in the pup plasma (Kormanik & Evans, 1986).

Urea

Urea excretion mechanisms

Urea has only limited permeability across lipid bilayers. Thus, in cases where urea concentrations must be regulated between compartments or a high transport rate maintained, for instance between blood and urine, the presence of specific carriers that can be modulated is advantageous (Walsh & Smith, 2001). Therefore, urea is transported in animal tissues also by other mechanisms than simple diffusion. In mammals, specific urea transporters (UT) have been described for medullary collecting ducts, erythrocytes, brain, liver, heart and testicular tissues. There are three main urea transporter gene families recognised, UT-A, UT-B, and UT-C; the latter recently described from the kidney of Japanese eel (Sands, 2003; Bagnasco, 2005; Mistry *et al.*, 2005).

In elasmobranchs, shark urea transporters (ShUT) were first isolated from spiny dogfish shark brain and kidneys. The ShUT is thought to be involved in reabsorption of urea, thus making it possible for the shark to maintain high blood and tissue urea concentrations for osmotic regulatory purposes (Smith & Wright, 1999). In the Lake Magadi tilapia, which only excretes urea and no ammonia, the calculated urea permeability of its gills far exceeds other studied fish species and what would have been possible if urea only diffused passively as in a lipid bilayer model (Walsh *et al.*, 2001b). In this tilapia, gill urea transporter proteins are expressed, as an adaptation that enables it to maintain the high rates of urea excretion necessary to survive in an environment where ammonia excretion is impossible (Walsh *et al.*, 2001b). Studies on the Gulf toadfish has given valuable insights into mechanisms of urea excretion in adult teleosts. It has been found that urea excretion occurs in 0.5-3 h long pulses in the toadfish, and that UT-A type urea transporter proteins are expressed, which greatly enhance gill urea permeability during the episodic pulse events (Wood *et al.*, 1995; Walsh, 1997, 2000; McDonald *et al.*, 2000). Recent results suggest that a signal conveyed by 5-hydroxytryptamine is responsible for the urea pulses in adult toadfish (McDonald & Walsh, 2004). The adaptive mechanism of the urea pulses may be related to predator avoidance, since it was recently found that a common predator of Gulf toadfish, the grey snapper (*Lutjanus griseus*), shows reduced behavioural response to a ammonia/urea mix compared to the ammonia attractant alone, indicating that urea functions as a masking agent (Barimo & Walsh, 2006). Urea transporter-like mRNAs have been detected in many ammonotelic marine species (Walsh *et al.*, 2001a), although the functional roles of these UT remain to be elucidated. However, in juvenile Japanese eel (*Anguilla japonica*), the UT may be linked to osmoregulation, since mRNA expression increase in gills (eUT) and kidney (eUT-C) upon transfer to seawater. In this environment urine output must be minimised, and it has been proposed that the eUTs assist in urea reabsorption and therefore water retention, from the glomerular filtrate to the blood, followed by urea excretion via the gill eUT (Mistry *et al.*, 2001, 2005).

Mechanisms for urea excretion have been little investigated in fish early life stages. However, in a series of experiments Pilley & Wright (2000) showed that urea excretion in rainbow trout embryos is inhibited by urea analogues and by inhibitors of the UT-A family of urea transporters. When rainbow trout embryos were exposed to external urea, the rate of influx increased up to about 5 mM urea and then plateaued, suggesting that a putative transporter became saturated. However, at very high non-physiological external urea levels (20 mM), the rate of influx increased again, suggesting a simple non-saturable diffusion process at this concentration. The authors concluded that urea is excreted in part through a facilitated transporter system in trout embryos, resembling that of mammalian collecting ducts and elasmobranch kidney (Pilley & Wright, 2000).

Patterns of urea excretion and accumulation

Significant urea excretion at the embryonic stage of fishes was demonstrated in common carp (Kaushik *et al.*, 1982) and whitefish (*Coregonus lavaretus*) (Dabrowski & Kaushik, 1984). In common carp, the relative urea excretion averaged 27% during the embryonic stage, and appeared to follow a diurnal rhythm, as observed for ammonia excretion (Kaushik *et al.*, 1982).

In winter flounder (*Pseudopleuronectes americanus*), Cetta and Capuzzo (1982) found that embryos not only excreted ammonia, but also primary amines. Similar observations were reported for summer flounder (*Paralichthys dentatus*) larvae after hatching (Buckley & Dillmann, 1982). In ship sturgeon (*Acipenser nudiventris*) it was concluded that the majority of the nitrogen excreted was in the form of urea (Yarzhombek & Maslennikova, 1971). Such observations have also been made on African catfish, where urea excretion accounted for as much as 62% of total nitrogen excretion in embryos (measured as urea-N + ammonia excretion) (Terjesen *et al.*, 1997). In rainbow trout however, urea excretion follows a quite different pattern, since it has been reported to be undetectable or low during the first days pf (Yarzhombek & Maslennikova, 1971; Oliva-Teles & Kaushik, 1990; Wright *et al.*, 1995; Essex-Fraser *et al.*, 2005). A very high relative level of urea excretion (50%-100%) has, however, been found in Atlantic cod during the first part of embryonic development (Chadwick & Wright, 1999), and in the Atlantic halibut (Terjesen *et al.*, 2002). It is unclear to what extent the urea excreted during the early part of the embryonic stage represents maternally transferred urea, which accumulated during ovarian development or metabolically produced urea by the embryo after fertilisation. However, in the Atlantic halibut there is a net production (taking into account accumulation and excretion) of urea during this period, such that urea is the dominating end-product measured up to 7 dpf (Terjesen *et al.*, 2002). Further studies measuring urea and ammonia excretion and individual content on a fine time-scale are however necessary to obtain conclusive results regarding the origin of urea during this early embryonic period.

In the period following hatching, rates of urea excretion and the relative importance of urea in total nitrogen excretion differs between species. The oyster toadfish and the Gulf toadfish, are primarily ammonotelic at the embryonic stage, but switch to ureotelism as yolk-sac larvae or early juveniles, respectively (Stephen & Griffith, 2001; Barimo & Walsh, 2005). The Gulf toadfish however, is more ureotelic than the oyster toadfish as juveniles (Stephen & Griffith, 2001; Barimo & Walsh, 2005). Within one week of complete yolk-sac absorption, Gulf toadfish excretes urea in pulses as in adults of this species (Barimo *et al.*, 2004). In the primarily ammonotelic rainbow trout however, the relative urea excretion increases up to around 22% at 60 dpf, which is shortly after the time of complete yolk absorption (CYA) (Essex-Fraser *et al.*, 2005). In the African catfish, the relative urea excretion fluctuates around 20%, but after metamorphosis urea accounts for as much as 44% of total nitrogen excreted (Terjesen *et al.*, 1997). Similar observations have been made for the Indian walking catfish, which excretes around 35-40% urea in exogenously feeding stages (Kharbuli *et al.*, 2006). Larval sea lamprey excrete about 15-20% of their total nitrogen excretion as urea, although significant activities of key OUC enzymes were not detected in the larval lamprey liver (Wilkie *et al.*, 1999). In both the Atlantic cod and Atlantic halibut, the relative importance of urea excretion appears to decline with age, and in the Atlantic halibut averages 7% during the yolk-sac stage (Chadwick & Wright, 1999; Terjesen *et al.*, 2002). Thus, urea is a major component of the total nitrogen excretion of several species that have been investigated, and measurements of this compound should be routinely included in studies investigating nitrogen balance during early development.

In contrast to ammonia, accumulation of urea in early fish life stages appears to be more evenly divided between the yolk and larval body compartments, possibly reflecting the low toxicity of urea compared to ammonia. This has been observed in the Atlantic halibut (Terjesen *et al.*, 2002), and is also the case when comparing 4 and 8 month old embryos of the elasmobranch little skate, where urea concentrations were not significantly different between body and yolk tissues (Steele *et al.*, 2004). In the rainbow trout, urea was found to be predominately located to the yolk in embryos (Steele *et al.*, 2001), but recent measurements around hatching showed that urea levels were significantly higher in the larval body than in the yolk (Essex-Fraser *et al.*, 2005). Measurements of tissue urea concentrations in larval African catfish showed that urea increased with age, especially around the time of CYA and metamorphosis (Terjesen *et al.*, 2001). In the Indian walking catfish an increase in urea concentration (whole-body) was observed until close to CYA (Kharbuli *et al.*, 2006). A considerable post-hatch increase in whole-body urea tissue levels has also been observed in the Gulf toadfish (Barimo & Walsh, 2005).

EFFECTS OF ENVIRONMENT AND NUTRITION ON NITROGEN EXCRETION

Environmental Ammonia

In rainbow trout embryos and yolk-sac larvae, external ammonia exposure (0.2 or 0.5 mM) inhibited ammonia excretion and led to ammonia accumulation that varied with developmental stage (Wright & Land, 1998). Both the embryos and yolk-sac larvae showed increased whole-body ammonia concentrations following exposure, while in yolk-sac larvae ammonia levels were not significantly changed (Wright & Land, 1998). The yolk-sac larvae were also the only stage that showed significantly higher urea excretion in absolute terms when exposed to elevated ambient ammonia, although all free-swimming stages showed significantly higher relative urea excretion after exposure (Wright & Land, 1998). In a later study Steele *et al.* (2001) showed that embryos responded to either acute or chronic ammonia exposure by increased urea tissue levels in the body, while ammonia concentration was maintained, despite the inward gradient. Interestingly, no significant changes were observed in OUC enzyme activities in embryos exposed to ammonia; however, the authors calculated that the measured CPSase III activity *in vitro* was still sufficient to cover approximately 50% of the elevated urea excretion (Steele *et al.*, 2001).

When continuously exposing Atlantic halibut embryos to 15-264 μM NH_3 for 6 days, no significant ammonia accumulation occurred during the first 2 days (Terjesen *et al.*, 1998), suggesting that Atlantic halibut embryos are not especially susceptible to excessive whole-animal ammonia accumulation during exposure, despite its large and acidic yolk (Jørgensen & Grasdalen, 1990). During this ammonia exposure, the Atlantic halibut embryos appeared able to maintain normal K^+, Cl^- and Na^+ levels (Terjesen *et al.*, 1998), whereas levels of K^+ in turbot and K^+, Cl^-, and Na^+ in rainbow trout yolk-sac larvae declined during NH_4Cl exposure (Hetland, 1992; Paley *et al.*, 1993). Thus, tolerance to ambient ammonia is substantial in Atlantic halibut embryos as observed in several other species during the embryonic stage compared to later developmental stages (Rice & Stokes, 1975; Calamari *et al.*, 1981; Holt & Arnold, 1983; Daniels *et al.*, 1987; Hetland, 1992). In fact, exposure to 211 μM NH_3 at 10°C (rainbow trout), or 514 μM NH_3 at 25°C (red drum, *Sciaenops ocellatus*) for 24 h did not affect embryo survival (Rice & Stokes, 1975; Holt & Arnold, 1983). Larval stages of gilthead seabream or Senegal sole however, are sensitive to ammonia exposure with 24 h LC_{50} values at 19.5°C as low as 14 μM and 78 μM NH_3, respectively (Parra & Yufera, 1999).

When exposing larval sea lamprey to elevated ambient ammonia, urea excretion was significantly elevated and ammonia accumulated, but no significant OUC activities were detected, which raises the question about the origin of the excreted urea (Wilkie *et al.*, 1999). However, in a later study using adult sea lamprey, CPSase III was detected in liver, although any activity in muscle or intestine was below the level of detection (Wilkie *et al.*, 2004).

Indian walking catfish larvae 15 dpf, when exposed to 10 mM NH_4Cl over 72h accumulate ammonia, but respond with increased urea production through excretion and internal accumulation (Kharbuli *et al.*, 2006). This increased urea production may originate from the OUC, since increased GSase, CPSase III, and ASL activities were observed in the catfish. In addition, levels of GSase and CPSase III increased after ammonia exposure (Kharbuli *et al.*, 2006).

The toadfishes are known for an exceptional tolerance towards environmental ammonia; adults of the oyster toadfish show an 96h LC_{50} as high as 691 μM NH_3 at 28°C (Wang & Walsh, 2000). These authors suggested that the low sensitivity of adult batrachoid fishes is due to their functional OUC, active GSase and glutamine accumulation, and inherent low ammonia sensitivity of the nervous system. Recently, the ammonia tolerance of Gulf toadfish embryos were investigated, and the 96 h LC_{50} was found to be 2077 μM NH_3 at 19-24°C, even more ammonia resistant than the adults (Barimo & Walsh, 2005). Toadfish larvae were less tolerant, although still showed low sensitivity (96 h LC_{50} of 200 μM NH_3) (Barimo & Walsh, 2005). Both embryos and larvae of the Gulf toadfish showed an increase in whole-body ammonia concentration with higher levels of ammonia exposure, but only the embryos showed elevated internal urea levels (Barimo & Walsh, 2005).

Water pH

Weakly acidic ambient water (pH 6.0) increases ammonia excretion two-fold from trout embryos, and lowers their internal ammonia concentration (Rahaman-Noronha *et al.*, 1996). However, when trout are exposed to elevated water pH (pH 9.0) a decrease in ammonia excretion rate is found in embryos and yolk-sac larvae, while urea excretion is significantly elevated in yolk-sac larvae exposed to pH 9.5 (Wright &

Land, 1998). Relative rates of urea excretion have been found to increase for all life stages exposed to elevated pH except trout embryos (Wright & Land, 1998).

In a different set of experiments on trout yolk-sac larvae, Wright and co-workers showed that pH 9.5 exposure raised urea excretion 1.7-2.4 fold (Wright & Fyhn, 2001). However, alkaline water also increased considerably the whole body ammonia levels while internal urea concentration and OUC enzyme activities remained unchanged (Wright & Fyhn, 2001). The authors suggested that the level of OUC enzyme activity present was either sufficient to account for the increased urea production, or that other pathways such as uricolysis contributed to the urea production (Wright & Fyhn, 2001). Regarding uricolysis, it is unclear if this urea production mechanism originated from already available purines or a proper ammonia detoxification mechanism, since in the latter case this would require *de novo* nucleotide biosynthesis by utilising the originally exposed ammonia as discussed by Anderson (2001).

Ionic Strength

The effects of ionic strength on nitrogen excretion and metabolism has been little investigated in fish early life stages. In guppy, transfer of females to hypertonic conditions, resulted in significantly elevated concentrations of urea in mature oocytes and embryos (Dépêche & Schoffeniels, 1975). In trout embryos, elevation of external NaCl concentration increased ammonia excretion two-fold, without affecting internal ammonia content (Rahaman-Noronha *et al.*, 1996). Expression of rhesus glycoprotein *rhcg1* mRNA, which may be involved in ammonia excretion, was recently shown to be altered by environmental ionic strength in zebrafish larvae (Nakada *et al.*, 2007b). In elasmobranchs, studies on embryos of the little skate show that exposure to dilute seawater (75% strength) leads to a five- to ten-fold increase in urea excretion, and a decrease in the tissue concentration of urea and other osmolytes compared to controls in full-strength seawater (Steele *et al.*, 2004). OUC enzyme activities did not change as a consequence of exposure to dilute seawater. These results indicate that embryos of the little skate are well equipped to handle environmental changes in osmotic pressure (Steele *et al.*, 2004).

Feed Intake and Nutrient Composition

Feed intake and composition is a major determinant of nitrogen excretion in fish (Wood, 2001). In adult fishes, feed intake leads to a higher ammonia excretion and plasma levels, while changes in feed composition such as elevated nucleic acid content significantly increase urea excretion and urea concentrations in plasma, muscle and liver (Fournier *et al.*, 2002; Wicks & Randall, 2002; Andersen *et al.*, 2006). There is limited information available regarding effects of feed intake and feed composition on nitrogen excretion in developing fish. This paucity of information is partly due to the difficulties associated with producing experimental small-sized feed particles suitable for fish early life stages. The use of live feeds are only of limited value since their composition is difficult to manipulate.

Some information regarding early life stages are, however, available. In common carp larvae fed zooplankton, the post-prandial peak of ammonia excretion declined with increasing wet mass of the larvae (Kaushik & Dabrowski, 1983). Postprandial urea and ammonia peaks have also been observed in common carp and Siberian sturgeon larvae during their first period of exogenous feeding, and is influenced by the type of feed (Kaushik & Dabrowski, 1983; Dabrowski *et al.*, 1987). In coregonid larvae (*Coregonus schinzi palea*), post-prandial peaks in ammonia and urea excretion occurred earlier when the larvae were fed brine shrimp (*Artemia* spp.) compared to when they were offered formulated diets. Unlike ammonia, significant differences in urea excretion between fed and fasted larvae persisted 14 h after the meal (Dabrowski & Kaushik, 1984). Grass carp larvae (*Ctenopharyngodon idella*) may excrete up to 64% of total nitrogen excretion as urea when fed a mix of zooplankton and oligochaetes while older larvae excrete only 1-5% as urea when fed oligochaetes and pellets (Urban, 1984). Ammonia and urea excretion patterns differed in rainbow trout larvae and juveniles fed diets that were either formulated using free amino acids, dipeptides or protein to supply the amino acid portion of the diets (Dabrowski *et al.*, 2003). These observations suggest that urea excretion in early exogenously feeding stages of fish not only represents low-level breakdown of purines from routine turnover of nucleic acids, but also represents catabolism of feed components such as purines, arginine and utilisation of precursors for the OUC.

In marine species, a distinctive post-prandial ammonia and primary amine excretion peak was found in summer flounder larvae fed natural zooplankton (Buckley & Dillmann, 1982). Observations in red drum larvae are different however, since Torres *et al.* (1996) found a higher urea excretion in 24 h starved larvae 6 and 10 days old, although not in 14-day-old larvae, compared to larvae fed rotifers to satiation. It would be useful to study the origin of the excreted urea during starvation in this species. In a detailed study on the Atlantic cod, it was found that ammonia excretion rates (but not oxygen uptake) of starved juveniles increased less with body mass than in fed individuals (given natural zooplankton), suggesting an increased ability to regulate ammonia excretion as development progressed (Finn *et al.*, 2002).

FUTURE PERSPECTIVES

Considerable progress has been made in the last decades on the understanding of nitrogen metabolism and excretion in early life stages of fishes. This is a captivating, but complex field of research. During experimentation, not only the treatments have effects that must be understood, but the experimental animal in itself changes rapidly. Explanations of mechanisms and processes may have validity for a certain developmental stage but be erroneous for another stage. These factors pose great challenges to the researcher.

Regarding nitrogen metabolism and excretion, several aspects should be further investigated. The use of the yolk compartment as an ammonia sink during development should be more fully described. What are the primary reasons for this strategy? Possibly, answers can be found in the priorities and developmental constraints of the developing embryo. Furthermore, how is the yolk/embryonic tissue ammonia gradient maintained at different developmental stages? Molecular and electrophysiological studies on the yolk membrane will likely be rewarding in this regard.

Studies on the OUC and its likely role in ammonia detoxification during fish early development have been one of the major findings in this field. The presence or absence of significant OUC enzyme mRNA levels and activities should be determined in more species. Furthermore, it would be useful to know quantitatively the relative amounts of urea originating from the OUC versus other urea generating pathways during ammonia exposure in fish early life stages. Part of this problem may be answered through radiotracer studies.

The rapid development of new tools and discovery of special life history strategies will surely bring many interesting findings on nitrogen metabolism and excretion in developing fish.

References

Alam MS, Teshima S-I, Koshio S, Ishikawa M (2002) Arginine requirement of juvenile Japanese flounder *Paralichthys olivaceus* estimated by growth and biochemical parameters. Aquaculture 205: 127-140

Andersen Ø, Aas T, Stugor S, Takle H, van Nes S, Grisdale-Helland B, Helland SJ, Terjesen BF (2006) Purine-induced expression of urate oxidase and enzyme activity in Atlantic salmon (*Salmo salar*). Cloning of urate oxidase liver cDNA from three teleost species and the African lungfish *Protopterus annectens*. FEBS J 273: 2839-2850

Anderson PM (1976) A glutamine- and N-acetyl-L-glutamate-dependent carbamyl phosphate synthetase activity in the teleost *Micropterus salmoides*. Comp Biochem Physiol 54B: 261-263

Anderson PM (1995a) Molecular aspects of carbamoyl phosphate synthesis. In: Walsh PJ, Wright PA (eds) Nitrogen Metabolism and Excretion, CRC Press, Boca Raton, p 33-49

Anderson PM (1995b) Urea cycle in fish: molecular and mitochondrial studies. In: Wood CM, Shuttleworth TJ (eds) Ionoregulation: Cellular and Molecular Approaches to Fish Ionic Regulation, Vol 14, Academic Press, New York, p 57-79

Anderson PM (2001) Urea and glutamine synthesis: Environmental influences on nitrogen excretion. In: Wright PA, Anderson PM (eds) Fish Physiology Vol 20, Academic Press, New York, p 239-277

Bagnasco SM (2005) Role and regulation of urea transporters. Pflueg Arch Eur J Physiol 450: 217-226

Ballantyne JS (1997) Jaws: The Inside Story. The Metabolism of Elasmobranch Fishes. Comp Biochem Physiol 118B: 703-742

Barimo JF, Walsh PJ (2005) The effects of acute and chronic ammonia exposure during early life stages of the gulf toadfish, *Opsanus beta*. Aquat Toxicol (Amst) 75: 225-237

Barimo JF, Walsh PJ (2006) Use of urea as a chemosensory cloaking molecule by a bony fish. J Exp Biol 209: 4254-4261

Barimo JF, Steele SL, Wright PA, Walsh PJ (2004) Dogmas and controversies in the handling of nitrogenous wastes: Ureotely and ammonia tolerance in early life stages of the gulf toadfish, *Opsanus beta*. J Exp Biol 207: 2011-2020

Berge GE, Lied E, Sveier H (1997) Nutrition of Atlantic salmon (*Salmo salar*): The requirement and metabolism of arginine. Comp Biochem Physiol 117A: 501-509

Brown GWJ, Cohen PP (1960) Comparative biochemistry of urea synthesis-III. Activities of urea cycle enzymes in various higher and lower vertebrates. Biochem J 75: 82-91

Brunel A (1937) Métabolism de l'azote d'origine purique chez les Poissons et chez les Batraciens. II - Catabolisme de l'azote d'origine purique chez les Téléostéens. Bull Soc Chim Biol 19: 1027-1036

Buckley LJ, Dillmann DW (1982) Nitrogen utilization by larval summer flounder, *Paralichthys dentatus* (Linnaeus). J Exp Mar Biol Ecol 59: 243-256

Buentello JA, Gatlin DM (1999) Nitric oxide production in activated macrophages from channel catfish (*Ictalurus punctatus*): Influence of dietary arginine and culture media. Aquaculture 179: 513-521

Buentello JA, Gatlin DM (2000) The dietary arginine requirement of channel catfish (*Ictalurus punctatus*) is influenced by endogenous synthesis of arginine from glutamic acid. Aquaculture 188: 311-321

Buentello JE, Gatlin DM (2001) Plasma citrulline and arginine kinetics in juvenile channel catfish, *Ictalurus punctatus*, given oral gabaculine. Fish Physiol Biochem 24: 105-112

Calamari D, Marchetti R, Vailati G (1981) Effects of long-term exposure to ammonia on the developmental stages of rainbow trout (*Salmo gairdneri* Richardson). Rapp P-V Reun Cons Int Explor Mer 178: 81-86

Caldovic L, Morizono H, Gracia Panglao M, Gallegos R, Yu X, Shi D, Malamy MH, Allewell NM, Tuchman M (2002) Cloning and expression of the human N-acetylglutamate synthase gene. Biochem Biophys Res Commun 299: 581-586

Campbell JW (1991) Excretory nitrogen metabolism. In: Prosser CL (ed) Environmental and Metabolic Animal Physiology, Vol 1, Wiley-Liss, New York, p 277-325

Casey CA, Anderson PM (1983) Glutamine-dependent carbamoyl phosphate synthetase and N-acetyl-L-glutamate-dependent carbamoyl phosphate synthetase from *Micropterus salmoides*: Purification, properties and inhibition by glutamine analogs. J Biol Chem 258: 8723-8732

Cetta CM, Capuzzo JM (1982) Physiological and biochemical aspects of embryonic and larval development of the winter flounder *Pseudopleuronectes americanus*. Mar Biol 71: 327-337

Chadwick TD (1998) Nitrogen excretion and expression of urea cycle enzymes during early life stages of Atlantic cod and rainbow trout. MSc Thesis, University of Guelph, Canada

Chadwick TD, Wright PA (1999) Nitrogen excretion and expression of urea cycle enzymes in the Atlantic cod (*Gadus morhua* L.): a comparison of early life stages with adults. J Exp Biol 202: 2653-2662

Chiu YN, Austic RE, Rumsey GL (1986) Urea cycle activity and arginine formation in rainbow trout (*Salmo gairdneri*). J Nutr 116: 1640-1650

Ciereszko A, Dabrowski K, Kucharczyk D, Dobosz S, Goryczko K, Glogowski J (1999) The presence of uric acid, an antioxidative substance, in fish seminal plasma. Fish Physiol Biochem 21: 313-315

Cooper A (2001) Role of glutamine in cerebral nitrogen metabolism and ammonia neurotoxicity. Ment Retard Dev Disabil Res Rev 7: 280-286

Dabrowski K, Kaushik SJ (1984) Rearing of coregonid (*Coregonus schinzi palea* Cuv. et Val.) larvae using dry and live food. II. Oxygen consumption and nitrogen excretion. Aquaculture 41: 333-344

Dabrowski K, Kaushik SJ (1986) Utilization of endogenous reserves during embryonic development and larval life of sturgeon, *Acipenser baeri* Brandt (Chondrostei). Zool Anz 216: 367-380

Dabrowski K, Kaushik SJ, Fauconneau B (1987) Rearing of sturgeon (*Acipenser baeri* Brandt) larvae. III. Nitrogen and energy metabolism and amino acid absorption. Aquaculture 65: 31-41

Dabrowski K, Lee K-J, Rinchard J (2003) The smallest vertebrate, teleost fish, can utilize synthetic dipeptide-based diets. J Nutr 133: 4225-4229.

Dabrowski K, Terjesen BF, Zhang Y, Phang J, Lee K-J (2005) A concept of dietary dipeptides: a step to resolve the problem of amino acid availability in early life of vertebrates. J Exp Biol 208: 2885-2894

Daniels HV, Boyd CE, Minton RV (1987) Acute toxicity of ammonia and nitrite to spotted seatrout. Prog Fish-Cult 49: 260-263

Davenport J (1983) Oxygen and the developing eggs and larvae of the lumpfish, *Cyclopterus lumpus*. J Mar Biol Assoc UK 63: 633-640

Davenport J, Sayer MD (1986) Ammonia and urea excretion in the amphibious teleost *Blennius pholis* (L.) in sea-water and air. Comp Biochem Physiol 84: 189-194

Davenport J, Lønning S, Kjørsvik E (1983) Ammonia output by eggs and larvae of the lumpsucker, *Cyclopterus lumpus*, the cod, *Gadus morhua* and the plaice, *Pleuronectes platessa*. J Mar Biol Assoc UK 63: 713-723

Dépêche J, Schoffeniels E (1975) Changes in electrolytes, urea and free amino acids of *Poecilia reticulata* embryos following high salinity adaptation of the viviparous female. Biochem Syst Ecol 3: 111-119

Dépêche J, Gilles R, Daufresne S, Chapello H (1979) Urea content and urea production via the ornithine-urea cycle pathway during the ontogenic development of two teleost fishes. Comp Biochem Physiol 63A: 51-56

Eddy FB (1974) Osmotic properties of the perivitelline fluid and some properties of the chorion in Atlantic salmon eggs (*Salmo salar*). J Zool Lond 174: 237-243

Essex-Fraser PA, Steele SL, Bernier NJ, Murray BW, Stevens ED, Wright PA (2005) Expression of four glutamine synthetase genes in the early stages of development of rainbow trout (*Oncorhynchus mykiss*) in relationship to nitrogen excretion. J Biol Chem 280: 20268-20273

Evans D, Permarini P, Choe K (2005) The multifunctional fish gill: Dominant site of gas exchange, osmoregulation, acid-base regulation, and excretion of nitrogenous waste. Physiol Rev 85: 97-177

Felipo V, Butterworth R (2002) Neurobiology of ammonia. Prog Neurobiol 67: 259-279

Felskie AK, Anderson PM, Wright PA (1998) Expression and activity of carbamoyl phosphate synthetase III and ornithine urea cycle enzymes in various tissues of four fish species. Comp Biochem Physiol 119B: 355-364

Finn RN (2007) The physiology and toxicology of salmonid eggs and larvae in relation to water quality criteria. Aquat Toxicol 81: 337-354

Finn RN, Fyhn HJ, Evjen MS (1991) Respiration and nitrogen metabolism of Atlantic halibut eggs (*Hippoglossus hippoglossus* L.). Mar Biol 108: 11-19

Finn RN, Rønnestad I, Fyhn HJ (1995a) Respiration, nitrogen and energy metabolism of developing yolk-sac larvae of Atlantic halibut (*Hippoglossus hippoglossus* L.). Comp Biochem Physiol 111A: 647-671

Finn RN, Fyhn HJ, Henderson RJ, Evjen MS (1995b) Physiological energetics of developing embryos and yolk-sac larvae of Atlantic cod (*Gadus morhua*). I. Respiration and nitrogen metabolism. Mar Biol 124: 355-369

Finn RN, Fyhn HJ, Henderson RJ, Evjen MS (1996) The sequence of catabolic substrate oxidation an enthalpy balance of developing embryos and yolk-sac larvae of turbot (*Scophthalmus maximus* L.). Comp Biochem Physiol 115A: 133-151

Finn RN, Rønnestad I, van der Meeren T, Fyhn HJ (2002) Fuel and metabolic scaling during the early life stages of Atlantic cod *Gadus morhua*. Mar Ecol Prog Ser 243: 217-234

Flynn NE, Meininger CJ, Haynes TE, Wu G (2002) The metabolic basis of arginine nutrition and pharmacotherapy. Biomed Pharmacother 56: 427-438

Fournier V, Gouillou-Coustans M, Metailler R, Vachot C, Moriceau J, Le Delliou H, Huelvan C, Desbruyeres E, Kaushik S (2002) Nitrogen utilisation and ureogenesis as affected by dietary nucleic acid in rainbow trout (*Oncorhynchus mykiss*) and turbot (*Psetta maxima*). Fish Physiol Biochem 26: 177-188

Fournier V, Gouillou-Coustans MF, Metailler R, Vachot C, Moriceau J, Delliou HL, Huelvan C, Desbruyeres E, Kaushik SJ (2003) Excess dietary arginine affects urea excretion but does not improve N utilization in rainbow trout *Oncorhynchus mykiss* and turbot *Psetta maxima*. Aquaculture 217: 559-576

Frick NT, Wright PA (2002) Nitrogen metabolism and excretion in the mangrove killifish *Rivulus marmoratus* I. The influence of environmental salinity and external ammonia. J Exp Biol 205: 79-89

Fyhn HJ (1989) First feeding of marine fish larvae: Are free amino acids the source of energy? Aquaculture 80: 111-120

Fyhn HJ (1990) Energy production in marine fish larvae with emphasis on free amino acids as a potential fuel. In: Mellinger J (ed) Animal Nutrition and Transport Processes, Vol 5, Comparative Physiology, Karger, Basel, p 176-192

Fyhn HJ (1993) Multiple functions of free amino acids during embryogenesis in marine fishes. In: Walther BT, Fyhn HJ (eds) Physiological and Biochemical Aspects of Fish Development, University of Bergen, Norway, p 299-308

Fyhn HJ, Serigstad B (1987) Free amino acids as energy substrate in developing eggs and larvae of the cod *Gadus morhua*. Mar Biol 96: 335-341

Goldenberg H (1977) Organization of purine degradation in the liver of a teleost (carp; *Cyprinus carpio* L.). A study of its subcellular distribution. Mol Cell Biochem 16: 17-21

Goldstein L, Forster RP (1965) The role of uricolysis in the production of urea by fishes and other aquatic vertebrates. Comp Biochem Physiol 14: 567-576

Goldstein L, Forster RP (1970) Nitrogen metabolism in fishes. In: Campbell JW (ed) Comparative Biochemistry of Nitrogen Metabolism, Vol 2, Academic Press, New York, p 495-515

Gouillou-Coustans MF, Fournier V, Metallier R, Vachot C, Desbruyeres E, Huelvan C, Moriceau J, Le Delliou H, Kaushik SJ (2002) Dietary arginine degradation is a major pathway in ureagenesis in juvenile turbot (*Psetta maxima*). Comp Biochem Physiol 132A: 305-319

Grasdalen H, Jørgensen L (1985) ^{31}P-NMR studies on developing eggs and larvae of plaice. Comp Biochem Physiol 81B: 291-294

Grasdalen H, Jørgensen L (1987) *In vivo* NMR-studies of fish eggs – Monitoring metabolite levels, intracellular pH, and the freezing and permeability of water in developing eggs of plaice, *Pleuronectes platessa* L. Sarsia 72: 359-362

Griffith RW (1991) Guppies, toadfish, lungfish, coelacanths and frogs: A scenario for the evolution of urea retention in fishes. Environ Biol Fishes 32: 199-218

Hetland RM (1992) Effects of ammonia on survival, oxygen uptake, ammonia accumulation and potassium content in eggs and larvae of turbot (*Scophthalmus maximus*). MSc Thesis, University of Bergen, Norway

Hinterleitner S, Platzer U, Wieser W (1987) Development of the activities of oxidative, glycolytic and muscle enzymes during early larval life in three families of freshwater fish. J Fish Biol 30: 315-326

Holt GJ, Arnold CR (1983) Effects of ammonia and nitrite on growth and survival of red drum eggs and larvae. Trans Am Fish Soc 112: 314-318

Hong J, Salo WL, Lusty CJ, Anderson PM (1994) Carbamyl phosphate synthetase III, an evolutionary intermediate in the transition between glutamine-dependent and ammonia-dependent carbamyl phosphate synthetases. J Mol Biol 243: 131-140

Huggins AK, Skutsch G, Baldwin E (1969) Ornithine-urea cycle enzymes in telostean fish. Comp Biochem Physiol 28: 587-602

Hung CYC, Tsui KNT, Wilson JM, Nawata CM, Wood CM, Wright PA (2007) Rhesus glycoprotein gene expression in the mangrove killifish *Kryptolebias marmoratus* exposed to elevated environmental ammonia levels and air. J Exp Biol 210: 2419-2429

Hølleland T, Fyhn HJ (1986) Osmotic properties of eggs of the herring *Clupea harrengus*. Mar Biol 91: 377-383

Ip YK, Chew SF, Randall DJ (2001) Ammonia toxicity, tolerance, and excretion. In: Wright PA, Anderson PM (eds) Fish Physiology, Vol 20, Academic Press, New York, p 109-148

Ip YK, Zubaidah RM, Liew PC, Loong AM, Hiong KC, Wong WP, Chew SF (2004) African sharptooth catfish *Clarias gariepinus* does not detoxify ammonia to urea or amino acids but actively excretes ammonia during exposure to environmental Ammonia. Physiol Biochem Zool 77: 242-254

Iwata K, Kajimura M, Sakamoto T (2000) Functional ureogenesis in the gobiid fish, *Mugilogobius abei*. J Exp Biol 203: 3703-3715

Iwata K, Sakamoto T, Iwata I, Nishiguchi E, Kajimura M (2005) High ambient ammonia promotes growth in a ureogenic goby, *Mugilogobius abei*. J Comp Physiol 175B: 395-404

Jenkinson C, Grody W, Cederbaum S (1996) Comparative properties of arginases. Comp Biochem Physiol 114B: 107-132

Julsrud EA, Walsh PJ, Anderson PM (1998) N-Acetyl-L-glutamate and the urea cycle in gulf toadfish *(Opsanus beta)* and other fish. Arch Biochem Biophys 350: 55-60

Jørgensen L, Grasdalen H (1990) Phosphate metabolites in larvae of halibut (*Hippoglossus hippoglossus*) studied by *in vivo* NMR. Program and abstracts of "Development and Aquaculture of Marine Larvae". University of Bergen, Bergen, August 12-15

Jørgensen NC, Schmalbruch H (1984) The eggs of the freshwater fish *Epiplatys dageti* have tight plasma membranes without intramembranous particles. Cell Tissue Res 235: 643-646

Kajimura M, Croke SJ, Glover CN, Wood CM (2004) Dogmas and controversies in the handling of nitrogenous wastes: The effect of feeding and fasting on the excretion of ammonia, urea and other nitrogenous waste products in rainbow trout. J Exp Biol 207: 1993-2002

Kaushik SJ, Dabrowski K (1983) Postprandial metabolic changes in larval and juvenile carp (*Cyprinus carpio*). Reprod Nutr Develop 23: 223-234

Kaushik SJ, Dabrowski K, Luquet P (1982) Patterns of nitrogen excretion and oxygen consumption during ontogenesis of common carp (*Cyprinus carpio*). Can J Fish Aquat Sci 39: 1095-1105

Kharbuli ZY, Datta S, Biswas K, Sarma D, Saha N (2006) Expression of ornithine-urea cycle enzymes in early life stages of air-breathing walking catfish *Clarias batrachus* and induction of ureogenesis under hyper-ammonia stress. Comp Biochem Physiol 143B: 44-53

Kjørsvik E, Reiersen AL (1992) Histomorphology of the early yolk-sac larvae of the Atlantic halibut (*Hippoglossus hippoglossus* L.) – An indication of the timing of functionality. J Fish Biol 41: 1-19

Kong H, Edberg D, Korte J, Salo W, Wright P, Anderson P (1998) Nitrogen excretion and expression of carbamoyl-phosphate synthetase III activity and mRNA in extrahepatic tissues of largemouth bass (*Micropterus salmoides*). Arch Biochem Biophys 350: 157-168

Kormanik GA (1995) Maternal-fetal transfer of nitrogen in chondrichthyans. In: Walsh PJ, Wright PA (eds) Nitrogen Metabolism and Excretion, CRC Press, Boca Raton, p 243-258

Kormanik GA, Evans DH (1986) The acid-base status of prenatal pups of the dogfish, *Squalus acanthias*, in the uterine environment. J Exp Biol 125: 173-179

Korsgaard B (1994) Nitrogen distribution and excretion during embryonic post-yolk sac development in *Zoarces viviparus*. J Comp Physiol 164B: 42-46

Korsgaard B, Mommsen TP, Wright PA (1995) Nitrogen excretion in teleostean fish: adaptive relationships to environment, ontogenesis, and viviparity. In: Walsh PJ, Wright PA (eds) Nitrogen Metabolism and Excretion, CRC Press, Boca Raton, p 259-284

Korte JJ, Salo WL, Cabrera VM, Wright PA, Felskie AK, Anderson PM (1997) Expression of carbamoyl-phosphate synthetase III mRNA during the early stages of development and in muscle of adult rainbow trout (*Oncorhynchus mykiss*). J Biol Chem 272: 6270-6277

Krebs H, Henseleit K (1932) Untersuchungen über die Harnstoffbildung im Tierkörper. Hoppe Seylers Z Physiol Chem 210: 33-66

Krebs HA (1982) 50 years ago: Discovery of the ornithine cycle of urea synthesis. Trends Biochem Sci 7: 76-78

Kuo CF, Darnell JJE (1989) Mouse glutamine synthetase is encoded by a single gene that can be expressed in a localized fashion. J Mol Biol 208: 45-56

Lindley TE, Scheiderer CL, Walsh PJ, Wood CM, Bergman HL, Bergman AL, Laurent P, Wilson P, Anderson PM (1999) Muscle as a primary site of urea cycle enzyme activity in an alkaline lake-adapted tilapia, *Oreochromis alcalicus grahami*. J Biol Chem 274: 29858-29861

Litwiller SL, O'Donnell MJ, Wright PA (2006) Rapid increase in the partial pressure of NH_3 on the cutaneous surface of air-exposed mangrove killifish, *Rivulus marmoratus*. J Exp Biol 209: 1737-1745

Mangor-Jensen A (1986) Osmoregulation in eggs and larvae of the cod (*Gadus morhua* L.). Basic studies and effects of oil exposure. In: Fyhn HJ (ed) Fish Larval Physiology and Anatomy Basic Research and Effects of Oil, University of Bergen, Norway, p 117-166

Mangor-Jensen A (1987) Water balance in developing eggs of the cod *Gadus morhua* L. Fish Physiol Biochem 3: 17-24

McDonald MD, Walsh PJ (2004) Dogmas and controversies in the handling of nitrogenous wastes: 5-HT2-like receptors are involved in triggering pulsatile urea excretion in the gulf toadfish, *Opsanus beta*. J Exp Biol 207: 2003-2010

McDonald MD, Wood CM, Wang Y, Walsh PJ (2000) Differential branchial and renal handling of urea, acetamide and thiourea in the gulf toadfish *Opsanus beta*: evidence for two transporters. J Exp Biol 203: 1027-1037

Meijer AJ (1995) Urea synthesis in mammals. In: Walsh PJ, Wright P (eds) Nitrogen Metabolism and Excretion, CRC Press, Boca Raton, p 193-205

Mistry AC, Honda S, Hirata T, Kato A, Hirose S (2001) Eel urea transporter is localized to chloride cells and is salinity dependent. Am J Physiol 281: R1594-1604

Mistry AC, Chen G, Kato A, Nag K, Sands JM, Hirose S (2005) A novel type of urea transporter, UT-C, highly expressed in proximal tubule of seawater eel kidney. Am J Physiol Renal Physiol 288: F455-F465

Mommsen TP, Walsh PJ (1989) Evolution of urea synthesis in vertebrates: The piscine connection. Science 243: 72-75

Mommsen TP, Walsh PJ (1991) Urea synthesis in fishes: evolutionary and biochemical perspectives. In: Hochachka PW, Mommsen TP (eds) Biochemistry and Molecular Biology of Fishes, Vol 1, Elsevier, Amsterdam, p 137-163

Mommsen TP, Walsh PJ (1992) Biochemical and environmental perspectives on nitrogen metabolism in fishes. Experientia 48: 583-592

Monzani P, Ramos S, Ramos R, Moraes G (1998) Expression of glutamine synthetase and ornithine carbamoyltransferase in the course of *Piaractus mesopotamicus* development. In: Randall DJ, MacKinlay D (eds) International Congress on the Biology of Fishes. American Fisheries Society Physiology Section, Towson University, Baltimore, USA

Moraes G, Altran A, Avilez I (2002) Ornithine-urea cycle in the Brazilian catfish, *Pseudoplatystoma corruscans* during early life stages. In: Wright PA, MacKinlay D (eds) International Conference on Biology of Fish. American Fisheries Society Physiology Section, Vancouver, Canada, p 45-51

Morris JG (1991) Nutrition. In: Prosser CL (ed) Environmental and Metabolic Animal Physiology, Vol 1, Wiley-Liss, New York, p 231-276

Murray BW, Busby ER, Mommsen TP, Wright PA (2003) Evolution of glutamine synthetase in vertebrates: multiple glutamine synthetase genes expressed in rainbow trout (*Oncorhynchus mykiss*). J Exp Biol 206: 1511-1521

Nakada T, Westhoff CM, Kato A, Hirose S (2007a) Ammonia secretion from fish gill depends on a set of Rh glycoproteins. FASEB J 21: 1067-1074

Nakada T, Hoshijima K, Esaki M, Nagayoshi S, Kawakami K, Hirose S (2007b) Localization of ammonia transporter Rhcg1 in mitochondrion-rich cells of yolk sac, gill, and kidney of zebrafish and its ionic strength-dependent expression. Am J Physiol 293: 1743-1753

Noguchi T, Takada Y, Fujiwara S (1979) Degradation of uric acid to urea and glyoxylate in peroxisomes. J Biol Chem 254: 5272-5275

Oliva-Teles A, Kaushik SJ (1987) Nitrogen and energy metabolism during the early ontogeny of diploid and triploid rainbow trout (*Salmo gairdneri* R.). Comp Biochem Physiol 87A: 157-160

Oliva-Teles A, Kaushik SJ (1990) Effect of temperature on utilization of endogenous energy reserves during embryonic development of diploid and triploid rainbow trout (*Salmo gairdneri* R.). Aquaculture 84: 373-382

Oppen-Berntsen DO, Helvik JV, Walther BT (1990) The major structural proteins of cod (*Gadus morhua*) eggshells and protein crosslinking during teleost egg hardening. Dev Biol 137: 258-265

Paley RK, Twitchen ID, Eddy FB (1993) Ammonia, Na^+, K^+ and Cl^- levels in rainbow trout yolk-sac fry in response to external ammonia. J Exp Biol 180: 273-284

Parra G, Yufera M (1999) Tolerance response to ammonia and nitrite exposure in larvae of two marine fish species (gilthead seabream *Sparus aurata* L. and Senegal sole *Solea senegalensis* Kaup). Aquac Res 30: 857-863

Pilley CM, Wright PA (2000) The mechanisms of urea transport by early life stages of rainbow trout (*Oncorhynchus mykiss*). J Exp Biol 203: 3199-3207

Pittman K, Skiftesvik AB, Berg L (1990) Morphological and behavioural development of halibut, *Hippoglossus hippoglossus* (L.) larvae. J Fish Biol 37: 455-472

Pittman KA (1991) Aspects of the early life history of the Atlantic halibut (*Hippoglossus hippoglossus* L.): Embryonic and larval development and the effects of temperature. PhD Thesis, University of Bergen, Norway

Potts WTW, Rudy JPP (1969) Water balance in the eggs of the Atlantic salmon *Salmo salar*. J Exp Biol 50: 223-237

Potts WTW, Eddy FB (1973) The permeability to water of the eggs of certain marine teleosts. J Comp Physiol 82: 305-315

Rahaman-Noronha E, O'Donnell MJ, Pilley CM, Wright PA (1996) Excretion and distribution of ammonia and the influence of boundary layer acidification in embryonic rainbow trout (*Oncorhynchus mykiss*). J Exp Biol 199: 2713-2723

Randall DJ, Wright PA (1987) Ammonia distribution and excretion in fish. Fish Physiol Biochem 3: 107-120

Randall DJ, Wright PA (1989) The interaction between carbon dioxide and ammonia excretion and water pH in fish. Can J Zool 67: 2936-2942

Randall DJ, Wood CM, Perry SF, Bergmann H, Maloiy GMO, Mommsen TP, Wright PA (1989) Urea excretion as a strategy for survival in a fish living in a very alkaline environment. Nature 337: 165-166

Read LJ (1968) Ornithine-urea cycle enzymes in early embryos of the dogfish *Squalus suckleyi* and the skate *Raja binoculata*. Comp Biochem Physiol 24: 669-674

Read LJ (1971) The presence of high ornithine-urea cycle enzyme activity in the teleost *Opsanus tau*. Comp Biochem Physiol 39B: 409-413

Rice SD, Stokes RM (1974) Metabolism of nitrogenous wastes in the eggs and alevins of rainbow trout, *Salmo gairdneri* Richardson. In: Blaxter JHS (ed) The Early Life History of Fish, Springer-Verlag, Berlin, p 325-337

Rice SD, Stokes RM (1975) Acute toxicity of ammonia to several developmental stages of rainbow trout, *Salmo gairdneri*. Fish Bull 73: 207-211

Riis-Vestergaard J (1982) Water and salt balance of halibut eggs and larvae (*Hippoglossus hippoglossus* L.). Mar Biol 70: 125-139

Riis-Vestergaard J (1987) Physiology of teleost embryos related to environmental challenges. Sarsia 72: 351-358

Rubio V, Britton HG, Grisolia S (1983) Mitochondrial carbamoyl phosphate synthetase activity in the absence of *N*-acetyl-L-glutamate. Mechanism of activation by this cofactor. Eur J Biochem 134: 337-343

Rønnestad I (1993) No efflux of free amino acids from yolk-sac larvae of Atlantic halibut (*Hippoglossus hippoglossus* L.). J Exp Mar Biol Ecol 167: 39-45

Rønnestad I, Fyhn HJ, Gravningen K (1992a) The importance of free amino acids to the energy metabolism of eggs and larvae of turbot (*Scophthalmus maximus*). Mar Biol 114: 517-525

Rønnestad I, Finn RN, Groot EP, Fyhn HJ (1992b) Utilization of free amino acids related to the energy metabolism of developing eggs and larvae of lemon sole *Microstomus kitt* reared in the laboratory. Mar Ecol Progr Ser 88: 195-205

Rønnestad I, Conceição LEC, Aragao C, Dinis MT (2001) Assimilation and catabolism of dispensable and indispensable free amino acids in post-larval Senegal sole (*Solea senegalensis*). Comp Biochem Physiol 130C: 461-466

Rønnestad I, Koven WM, Tandler A, Harel M, Fyhn HJ (1994) Energy metabolism during development of eggs and larvae of gilthead sea bream (*Sparus aurata*). Mar Biol 120: 187-196

Rønnestad I, Tonheim SK, Fyhn HJ, Rojas-García CR, Kamisaka Y, Koven W, Finn RN, Terjesen BF, Barr Y, Conceição LEC (2003) The supply of amino acids during early feeding stages of marine fish larvae: A review of recent findings. Aquaculture 227: 147-164

Saha N, Ratha BK (1987) Active ureogenesis in a freshwater air-breathing teleost, *Heteropneustes fossilis*. J Exp Zool 241: 137-141

Saha N, Ratha BK (1989) Comparative study of ureogenesis in freshwater, air-breathing teleosts. J Exp Zool 252: 1-8

Saha N, Ratha BK (1998) Ureogenesis in Indian air-breathing teleosts: adaptation to environmental constraints. Comp Biochem Physiol 120A: 195-208

Saha N, Das L, Dutta S (1999) Types of carbamoyl phosphate synthases and subcellular localization of urea cycle and related enzymes in air-breathing walking catfish, *Clarias batrachus*. J Exp Zool 283: 121-130

Saha N, Kharbuli ZY, Bhattacharjee A, Goswami C, Haussinger D (2002) Effect of alkalinity (pH 10) on ureogenesis in the air-breathing walking catfish, *Clarias batrachus*. Comp Biochem Physiol 132A: 353-364

Sands JM (2003) Molecular mechanisms of urea transport. J Membr Biol 191: 149-163

Segner H, Verreth J (1995) Metabolic enzyme activities in larvae of the African catfish, *Clarias gariepinus*: response to developmental and dietary influences. Fish Physiol Biochem 14: 385-398

Seibel BA, Walsh PJ (2002) Trimethylamine oxide accumulation in marine animals: relationship to acylglycerol storage. J Exp Biol 205: 297-306

Seoka M, Takii K, Takaoka O, Nakamura M, Kumai H (1997) Biochemical phases in embryonic red sea bream development. Fish Sci 63: 122-127

Shahsavarani A, Thomas ZC, Ballantyne JS, Wright PA (2001) A novel technique for the separation of yolk from the developing embryonic tissue in a teleost fish, *Oncorhynchus mykiss*. Fish Physiol Biochem 24: 321-326

Sivaloganathan B, Walford J, Ip YK, Lam TJ (1998) Free amino acids and energy metabolism in eggs and larvae of seabass, *Lates calcarifer*. Mar Biol 131: 695-702

Skjærven K, Finn R, Kryvi H, Fyhn H (2003) Yolk resorption in developing plaice (*Pleuronectes platessa*). In: Browman H, Skiftesvik A (eds) The Big Fish Bang, 26th Annual Larval Fish Conference. Institute of Marine Research, Bergen, p 192-209

Smith CP, Wright PA (1999) Molecular characterization of an elasmobranch urea transporter. Am J Physiol 276: R622-R626

Smith HW (1930) Metabolism of the lungfish, *Protopterus aethipicus*. J Biol Chem 88: 97-130

Smith S (1947) Studies on the development of the rainbow trout (*Salmo irideus*). I. The heat production and nitrogenous excretion. J Exp Biol 23: 357-377

Steele SL, Chadwick TD, Wright PA (2001) Ammonia detoxification and localization of urea cycle enzyme activity in embryos of the rainbow trout (*Oncorhynchus mykiss*) in relation to early tolerance to high environmental ammonia levels. J Exp Biol 204: 2145-2154

Steele SL, Yancey PH, Wright PA (2004) Dogmas and controversies in the handling of nitrogenous wastes: Osmoregulation during early embryonic development in the marine little skate *Raja erinacea*; response to changes in external salinity. J Exp Biol 207: 2021-2031

Steele SL, Yancey PH, Wright PA (2005) The little skate *Raja erinacea* exhibits an extrahepatic ornithine urea cycle in the muscle and modulates nitrogen metabolism during low-salinity challenge. Physiol Biochem Zool 78: 216-226

Stephen BJ, Griffith RW (2001) Excretion of ammonia and urea during development in the oyster toadfish, *Opsanus tau*. Mar Freshwat Behav Physiol 34: 257-264

Takada Y, Noguchi T (1983) The degradation of urate in liver perioxsomes. Association of allantoinase with allantoicase in amphibian liver but not in fish and invertebrate liver. J Biol Chem 258: 4762-4764

Takada Y, Noguchi T (1986) Ureidoglycollate lyase, a new metalloenzyme of peroxisomal urate degradation in marine fish liver. Biochem J 235: 391-397

Terjesen BF (2001) Aspects of nitrogen metabolism and excretion during ontogeny of the Atlantic halibut (*Hippoglossus hippoglossus*) and the African catfish (*Clarias gariepinus*). PhD Thesis, University of Bergen, Norway

Terjesen BF, Verreth J, Fyhn HJ (1997) Urea and ammonia excretion by embryos and larvae of the African catfish *Clarias gariepinus* (Burchell 1822). Fish Physiol Biochem 16: 311-321

Terjesen BF, Mangor-Jensen A, Fyhn HJ (1998) Ammonia dynamics in relation to hatching in Atlantic halibut (*Hippoglossus hippoglossus* L.). Fish Physiol Biochem 18: 189-201

Terjesen BF, Rønnestad I, Norberg B, Anderson PM (2000) Detection and basic properties of carbamoyl phosphate synthetase III during teleost ontogeny: a case study in the Atlantic halibut (*Hippoglossus hippoglossus* L.). Comp Biochem Physiol 126B: 521-535

Terjesen BF, Chadwick TD, Verreth JAJ, Rønnestad I, Wright PA (2001) Pathways for urea production during early life of an air-breathing teleost, the African catfish *Clarias gariepinus* Burchell. J Exp Biol 204: 2155-2165

Terjesen BF, Finn RN, Norberg B, Rønnestad I (2002) Kinetics and fates of ammonia, urea, and uric acid during oocyte maturation and ontogeny of the Atlantic halibut (*Hippoglossus hippoglossus* L.). Comp Biochem Physiol 131A: 443-455

Terjesen BF, Lee K-J, Zhang Y, Failla M, Dabrowski K (2006) Optimization of dipeptide-protein mixtures in experimental diet formulations for rainbow trout (*Oncorhynchus mykiss*) alevins. Aquaculture 254: 517-525

Thisse B, Pflumio S, Fürthauer M, Loppin B, Heyer V, Degrave A, Woehl R, Lux A, Steffan T, Charbonnier X, Thisse C (2001) Expression of the zebrafish genome during embryogenesis (NIH R01 RR15402). ZFIN Direct Data Submission http://zfin.org/cgi-bin/webdriver?MIval=aa-fxallfigures.apg&OID=ZDB-PUB-010810-1&fxallfig_probe_zdb_id=ZDB-EST-030829-56:

Timasheff SN, Xie G (2003) Preferential interactions of urea with lysozyme and their linkage to protein denaturation. Biophys Chem 105: 421-448

Torres JJ, Brightman RI, Donnely J, Harvey J (1996) Energetics of larval red drum *Sciaenops ocellatus*. Part 1: Oxygen consumption, specific dynamic action, and nitrogen excretion. Fish Bull (Seattle) 94: 756-765

Trammel PR, Campbell JW (1971) Carbamoyl phosphate synthesis in invertebrates. Comp Biochem Physiol 40B: 395-406

Tsui TKN, Randall DJ, Hanson L, Farrell AP, Chew SF, Ip YK (2004) Dogmas and controversies in the handling of nitrogenous wastes: Ammonia tolerance in the oriental weatherloach *Misgurnus anguillicaudatus*. J Exp Biol 207: 1977-1983

Urban E (1984) Energy balance and nitrogen changes of grass carp (*Ctenopharyngodon idella* Val.) larvae and fry under different food conditions. II. Assimilation and balance of nitrogen compounds. Ekol Pol 32: 341-382

van Waarde A (1988) Biochemistry of non-protein nitrogenous compounds in fish including the use of amino acids for anaerobic energy production. Comp Biochem Physiol 91B: 207-228

Veauvy CM, McDonald MD, Audekerke JV, Vanhoutte G, Camp NV, Linden AVd, Walsh PJ (2005) Ammonia affects brain nitrogen metabolism but not hydration status in the Gulf toadfish (*Opsanus beta*). Aquat Toxicol 74: 32-46

Vellas F, Serfaty A (1974) L'ammonique et I'urée chez un Téléostéen déau douce: la carpe (*Cyprinus carpio* L.). J Physiol 68: 591-614

Vigetti D, Monetti C, Pollgioni L, Taramelli R, Bernardini G (2000) *Xenopus* allantoicase: molecular cloning, enzymatic activity and developmental expression. Arch Biochem Biophys 379: 90-96

Walsh PJ, Wright PA (1995) Nitrogen Metabolism and Excretion. CRC Press, Boca Raton

Walsh PJ (1997) Evolution and regulation of urea synthesis and ureotely in (Batrachoid) fishes. Annu Rev Physiol 59: 299-323

Walsh PJ (1998) Nitrogen excretion and metabolism. In: Evans DH (ed) The Physiology of Fishes, 2nd Edition, CRC Press, Boca Raton, p 199-214

Walsh PJ, Smith CP (2001) Urea transport. In: Wright PA, Anderson PM (eds) Fish Physiol, Vol 20, Academic Press, New York, p 279-307

Walsh PJ, Heitz MJ, Campbell CE, Cooper GJ, Medina M, Wang YS, Goss GG, Vincek V, Wood CM, Smith CP (2000) Molecular characterization of a urea transporter in the gill of the gulf toadfish (*Opsanus beta*). J Exp Biol 203: 2357-2364

Walsh PJ, Wang Y, Campbell C, De Boeck G, Wood CM (2001a) Patterns of nitrogenous waste excretion and gill urea transporter mRNA expression in several species of marine fish. Mar Biol 139: 839-844

Walsh PJ, Grosell M, Goss GG, Bergman HL, Bergman AN, Wilson P, Laurent P, Alper SL, Smith CP, Kamunde C, Wood CM (2001b) Physiological and molecular characterization of urea transport by the gills of the Lake Magadi tilapia (*Alcolapia grahami*). J Exp Biol 204: 509-520

Walsh PJ, Mayer GD, Medina M, Bernstein ML, Barimo JF, Mommsen TP (2003) A second glutamine synthetase gene with expression in the gills of the gulf toadfish (*Opsanus beta*). J Exp Biol 206: 1523-1533

Walton MJ, Cowey CB (1982) Aspects of intermediary metabolism in salmonid fish. Comp Biochem Physiol 73B: 59-79

Wang YX, Walsh PJ (2000) High ammonia tolerance in fishes of the family Batrachoididae (Toadfish and Midshipmen). Aquat Toxicol 3: 205-219

Weihrauch D, Morris S, Towle DW (2004) Ammonia excretion in aquatic and terrestrial crabs. J Exp Biol 207: 4491-4504

Wicks BJ, Randall DJ (2002) The effect of sub-lethal ammonia exposure on fed and unfed rainbow trout: the role of glutamine in regulation of ammonia. Comp Biochem Physiol 132A: 275-285

Wilkie MP (1997) Mechanisms of ammonia excretion across fish gills. Comp Biochem Physiol 118A: 39-50

Wilkie MP (2002) Ammonia excretion and urea handling by fish gills: Present understanding and future research challenges. J Exp Zool 293: 284-301

Wilkie MP, Wood CM (1996) The adaptations of fish to extremely alkaline environments. Comp Biochem Physiol 113B: 665-673

Wilkie MP, Wang Y, Walsh PJ, Youson JH (1999) Nitrogenous waste excretion by the larvae of a phylogenetically ancient vertebrate: the sea lamprey (*Petromyzon marinus*). Can J Zool 77: 707-715

Wilkie MP, Turnbull S, Bird J, Wang YS, Claude JF, Youson JH (2004) Lamprey parasitism of sharks and teleosts: high capacity urea excretion in an extant vertebrate relic. Comp Biochem Physiol 138A: 485-492

Wilkie MP, Claude JF, Cockshutt A, Holmes JA, Wang YS, Youson JH, Walsh PJ (2006) Shifting patterns of nitrogen excretion and amino acid catabolism capacity during the life cycle of the sea lamprey (*Petromyzon marinus*). Physiol Biochem Zool 79: 885-899

Wilson RP (1973) Nitrogen metabolism in channel catfish, *Ictalurus punctatus* II. Evidence for an apparent incomplete ornithine-urea cycle. Comp Biochem Physiol 46: 625-634

Wilson RP (2002) Amino acids and proteins. In: Halver JE, Hardy RW (eds) Fish Nutrition, Academic Press, San Diego, p 143-179

Wood CM (1993) Ammonia and urea metabolism and excretion. In: Evans DH (ed) The Physiology of Fishes, CRC Press, Boca Raton, p 379-425

Wood CM (2001) Influence of feeding, exercise, and temperature on nitrogen metabolism and excretion. In: Wright PA, Anderson PM (eds) Fish Physiology, Vol 20, Academic Press, New York, p 201-238

Wood CM (2004) Dogmas and controversies in the handling of nitrogenous wastes: Is exogenous ammonia a growth stimulant in fish? J Exp Biol 207: 2043-2054

Wood CM, Perry S, Wright PA, Bergman H, Randall DJ (1989) Ammonia and urea dynamics in the Lake Magadi tilapia, a ureotelic teleost fish adapted to an extremely alkaline environment. Respir Physiol 77: 1-20

Wood CM, Hopkins TE, Hogstrand C, Walsh PJ (1995) Pulsatile urea excretion in the ureagenic toadfish *Opsanus beta*: an analysis of rates and routes. J Exp Biol 198: 1729-1741

Wright PA (1995) Nitrogen excretion: three end products, many physiological roles. J Exp Biol 198: 273-281

Wright PA, Land MD (1998) Urea production and transport in teleost fishes. Comp Biochem Physiol 119A: 47-54

Wright PA, Fyhn HJ (2001) Ontogeny of nitrogen metabolism and excretion. In: Wright PA, Anderson PM (eds) Fish Physiology, Vol 20, Academic Press, New York, p 149-200

Wright PA, Anderson PM (2001) Nitrogen Excretion, Academic Press, San Diego

Wright PA, Felskie A, Anderson PM (1995) Induction of ornithine-urea cycle enzymes and nitrogen metabolism and excretion in rainbow trout (*Oncorhynchus mykiss*) during early life stages. J Exp Biol 198: 127-135

Wright PA, Campbell A, Morgan R, Rosenberger AG, Murray BW (2004) Dogmas and controversies in the handling of nitrogenous wastes: Expression of arginase type I and II genes in rainbow trout: influence of fasting on liver enzyme activity and mRNA levels in juveniles. J Exp Biol 207: 2033-2042

Wu G, Morris SM (1998) Arginine metabolism: nitric oxide and beyond. Biochem J 336: 1-17

Yamamoto TS (1984) Ammonia release by chum salmon eggs at the initiation of their embryonic development. Dev Growth Differ 26: 95-104

Yarzhombek AA, Maslennikova NV (1971) Nitrogenous metabolites of the eggs and larvae of various fishes. J Ichthyol (English transl) 11: 276-281

Ziegler I, McDonaldo T, Hesslinger C, Pelletier I, Boyle P (2000) Development of the pteridine pathway in the zebrafish, *Danio rerio*. J Biol Chem 275: 18926-18932

PART 4

Sensory Physiology

CHAPTER

10

Mechanoreception

Patricia M Pankhurst

INTRODUCTION

The lateral line organs and the organs of the inner ear, the semicircular canals and otolith organs, are mechanical displacement detectors that share a common hair cell receptor structure (Flock, 1971; Blaxter, 1987; Platt, 1988; Popper *et al.*, 1988). Receptor cells and adjacent support cells make up the sensory epithelia of the mechanosensory organs; the neuromasts, cristae and maculae of the mechanosensory lateral line, semicircular canals and otolith organs, respectively. In juvenile and adult fish mechanosensory organ physiology, morphology and whole animal ecological responses to mechanosensory stimuli, have received considerable attention. Mechanosensory organs, either in isolation, or in conjunction with other sensory organs, are reported to mediate orientation to and position holding of fish in water currents (rheotaxis), the localisation and orientation to conspecifics (schooling) and prey, auditory perception, maintenance of body posture and equilibrium (vestibular control), and in some blind fish, spatial mapping of the environment in order to avoid obstacles while swimming.

Our understanding of mechanosensory physiology and function of fish larvae is less well understood. This is partly due to the small size, early stage of development and delicate nature of fish larvae, making them challenging experimental candidates to work with. In addition, the rapid change in both mechanosensory organ morphology

Author's address: School of Tropical and Marine Biology, Faculty of Science, Engineering and Information Technology, James Cook University, Townsville, Queensland 4811, Australia.
E-mail: Tish.Pankhurst@jcu.edu.au

and functional responses during the larval period, requires investigation throughout larval development (Fuiman *et al.*, 2004). The capacity to do this has been greatly enhanced with recent advances in rearing protocols for fish larvae. Much early research focused on investigating the morphological development of mechanosensory organs and in some instances the concurrent development of larval behaviours. Subjective associations between mechanosensory organ structure and developing behaviours were frequently made but were often unsubstantiated. Recently, more rigorous correlations between developing morphological complexity of mechanosensory organs and response characteristics of fish larvae to a range of mechanosensory stimuli have been made. These include laboratory studies in which whole animal behavioural responses to prey and predators were assessed using pharmacological agents to selectively block the mechanosensory lateral line organs (their use in fish larvae reviewed by Fuiman *et al.*, 2004), and investigation of the physiological response characteristics to auditory stimuli using heart rate conditioning (Iwashita *et al.*, 1999) and the Auditory Brainstem Response (ABR) (Higgs, 2002; Higgs *et al.*, 2003; Wright *et al.*, 2005). In addition, the behavioural responses of pre-settlement marine reef fish larvae to auditory stimuli have been assessed in the field revealing that orientation to reef sounds may play an important role in recruitment of some reef fish larvae. In comparison to the mechanosensory lateral line and the otolith organs, the semicircular canals of fish larvae have received little attention.

This chapter will briefly review the form and function of the mechanosensory hair cell organs of adult fish, providing the necessary background for review of the development and function of mechanosensory organs in fish larvae. The 'special' electroreceptive organs of the lateral line system, the tuberous and ampullary organs (Blaxter, 1987; Gibbs, 2004) will not be considered here because they respond to very different stimuli (Coombs *et al.*, 1988) and are dealt with elsewhere (see Kirschbaum and Denizot, this volume).

FUNCTIONAL ANATOMY: ADULT FISH

A Common Mechanosensory Receptor

The mechanosensory receptor cell has an arrangement of displacement-sensitive 'hairs' that project from the apical surface to form an asymmetric assemblage of stereocilia that progressively increase in length along an axis approaching a longer, single, more rigid kinocilium (Flock, 1971; Lowenstein, 1971; Blaxter, 1987; Gibbs, 2004). The hair cells have dual innervation, afferent and efferent fibres, and have directional sensitivity with respect to the stereocilia-kinocilium axis (Flock, 1971; Lowenstein, 1971; Blaxter, 1987; Coombs *et al.*, 1988; Kalmijn, 1989; Fay & Edds-Walton, 2000; Montgomery *et al.*, 2001a). Displacement of the stereocilia towards the kinocilium opens stretch-sensitive ion receptors (Montgomery *et al.*, 2001a) that result in a depolarisation of the receptor cell potential and concurrent change in the tonic rate of firing of the afferent nerve. Displacement away from the kinocilium results in hyperpolarisation of the cell, whereas a displacement at an oblique angle produces a change in receptor cell potential that is a cosine function of the direction of bending relative to the stereocilia-kinocilium axis (Blaxter, 1987; Popper *et al.*, 1988; Myrberg & Fuiman, 2002). The directional sensitivity of the mechanosensory receptor cells is a fundamental requirement for detecting stimulus direction (Rogers & Cox, 1988; Fay & Edds-Walton, 2000; Montgomery *et al.*, 2001b; Tolimieri *et al.*, 2004). The sensory epithelia of all mechanosensory organs have characteristic arrays of receptor cells with like and dislike directional sensitivity. Afferent nerve bundles of both neuromast (Montgomery *et al.*, 2001a) and otolith organs (Fay & Edds-Walton, 2000) are reported to only contact sensory hair cells with like directional sensitivity, providing the potential for central processing of the stimulus direction by integrating information derived across all receptor cell populations in the receptor array (Fay & Edds-Walton, 2000; Montgomery *et al.*, 2001a).

The Sensory Cell Displacement Stimulus

The stereocilia-kinocilium complex of mechanosensory lateral line and semicircular canal organs project into an acellular gelatinous mass, the cupula, which is secreted by support cells within the sensory epithelium. In lateral line neuromasts this generally forms a cylindrical or hemispherical cap over the sensory epithelium, which is either directly or indirectly exposed to the hydrodynamic environment immediately adjacent to the fish. This will depend upon whether the neuromast is located on the skin (superficial), or is contained within a lateral line canal (Fig. 1 A-C). The cupula precisely transfers viscous drag forces to the cilia of the receptor cells in proportion to the velocity of the water currents in the immediate environment of the fish (Blaxter, 1987; Kalmijn, 1989; Gibbs, 2004). In the sensory cristae of the semicircular canals, the gelatinous secretion forms a vane-like cupula diaphragm (Platt, 1983, 1988). During head accelerations, inertia of the endolymph fluid contained within the canals result in deformation of the cupula diaphragm and bending of receptor cilia (Platt, 1983; Kalmijn, 1989; Myrberg & Fuiman, 2002). In the otolith organs the gelatinous cupula mass is absent and an otolith membrane separates and possibly attaches to both the sensory epithelium and the overlying calcified otolith mass (Platt, 1983; Blaxter, 1987). Some components of the sound stimulus and body accelerations, including movement relative to gravity, result in the relative displacement between the macula and the overlying calcareous mass, creating a shear force that is transferred via the otolith membrane to the cilia of the receptor cells (Platt, 1983; Rogers & Cox, 1988).

The Mechanosensory Lateral Line

The neuromasts of the mechanosensory lateral line are variously situated: superficially on the epidermis of the fish, within epidermal pits, or in the dermis enclosed in rigid fluid-filled canals situated on the head and trunk (Coombs *et al.*, 1988; Webb, 1989a, b; Gibbs, 2004). Canal neuromasts maintain continuity with the external hydrodynamic

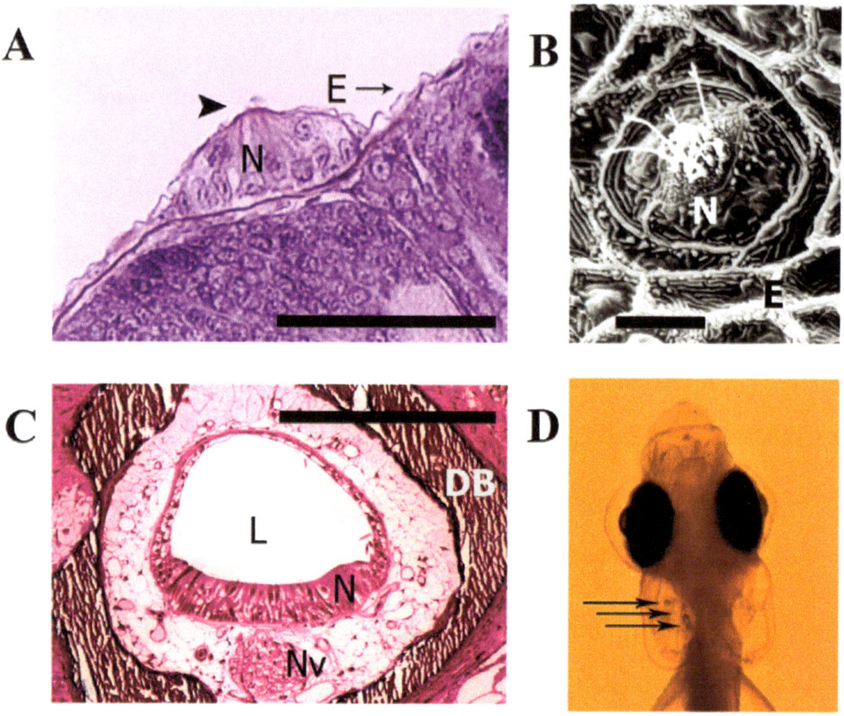

Fig. 1 A) Histological section of a superficial neuromast on the head of a squirefish (*Chrysophrys auratus*) larva 3 days post hatching. The neuromast comprises just a few sensory and support cells. Arrowhead indicates location of the stereocilia/kinocilia projections (not visible at this magnification). B) Scanning electron micrograph of a superficial neuromast situated on the head of a greenback flounder (*Rhombosolea tapirina*) larva 29 days post hatching in which long kinocilia and short stereocilia project from the apical surface. C) Histological section through a lateral line canal on the head of a juvenile rainbow trout (*Oncorhynchus mykiss*). The canal is supported by a sheath of dermal bone (DB), which has fractured during section preparation. The gelatinous cupula associated with the apical surface of the neuromast in A-C has been removed during tissue preparation. D) Dorsal view of a triplefin (*Forsterygion varium*) larva 15 days post hatching in which 3 otolith masses (arrows) are visible in highly prominent auditory/vestibular structures. E – epidermis; L – canal lumen; N – neuromast; Nv – nerve bundle. Scale bars are 10 μm (A & B) and 200 μm (C).

environment via regularly spaced pores (Webb, 1989a, b). The pattern of distribution of lateral line canals is highly conserved and present as variations of a basic 7 canal pattern, the supraorbital, infraorbital, mandibular, otic, temporal, supratemporal and trunk canals (Coombs *et al.*, 1988; Webb, 1989a, b; Gibbs, 2004), each with distinct innervations (Gibbs, 2004). A wide diversity is observed in canal diameter and in the relative prevalence of superficial or canal neuromasts. The variation is associated with differences in the ecology of the animal and variation in functional responses (Webb, 1989a; Gibbs, 2004).

Superficial neuromasts and canal neuromasts are both sensitive to changes in water velocity in the immediate vicinity of the fish, the sensitivity of which extends just a few body lengths at most (Kalmijn, 1989). However, they display different subsets of physiological responses (Engelmann *et al.* 2000; Coombs *et al.*, 2001; Montgomery *et al.*, 2002). Superficial neuromasts provide an important sensory role in rheotaxis (Montgomery *et al.*, 1997, 2002, 2003; Engelmann *et al.*, 2000), responding to the velocity of water flow over the skin in relation to the fish's own swimming movement, thus determining the net velocity between the swimming fish and the water current flow through which the fish is moving (Coombs *et al.*, 2001). On the other hand, canal neuromasts respond to fine-scale water velocity stimuli, those of most significance likely of biological origin (Coombs *et al.*, 2001), such as prey, predators or conspecifics. The inherent signal filtering properties of lateral line canals (Denton & Gray, 1988; Montgomery *et al.*, 2002) provide for very precise spatial localisation of the stimulus against the background noise of water current flow, and that of the fish's own swimming movements (Engelmann *et al.*, 2000). Thus canal neuromasts play an important role in the localisation of nearby prey (Bleckman, 1986; Montgomery, 1989; Jones & Janssen, 1992; Janssen *et al.*, 1995; New *et al.*, 2001; Montgomery *et al.*, 2002; Fuiman *et al.*, 2004) and the maintenance of highly coordinated orientation to conspecifics in a school (Partridge & Pitcher, 1980). Both systems have been implicated in the tracking of piscine prey using the tell-tale hydrodynamic signature or 'wake' (the spatial and temporal properties of which have been described by Hanke *et al.*, 2000) left behind when a fish swims through water (Pohlmann *et al.*, 2001; Montgomery *et al.*, 2002). Both systems are involved in the ability of blind cave fish to develop a three dimensional spatial map of the environment, a function that may be a ubiquitous feature among fishes, but has yet to be demonstrated (Montgomery *et al.*, 2001a). Finally, the dichotomy of the two sub-modalities of the mechanosensory lateral line organs is further demonstrated by the differential frequency response ranges of each to man-made vibrating stimuli (Engelmann *et al.*, 2000; Coombs *et al.*, 2001; Montgomery *et al.*, 2002) and accumulating evidence that information from superficial and canal neuromasts is processed separately in the central nervous system (Engelmann *et al.*, 2000; Montgomery *et al.*, 2002).

The Inner Ear

The paired organs of the inner ear, the semicircular canals and otolith organs, are an interconnected system of endolymph-filled ducts and chambers located on each side of the cranium, sometimes referred to as the vestibular labyrinth (Fig. 1D). Dorsally there are three orthogonally arranged semicircular canals (lateral, anterior and posterior) on each side of the head, each with a swelling, an ampulla, housing an area of sensory epithelium, the crista (Lowenstein, 1971; Platt, 1983, 1988). In addition, most fish have three otolith organs located in the ventral portion of the inner ear, the utricule, saccule and lagena, their sensory maculae and otolith masses, the lapillus, sagitta and astericus, respectively (Lowenstein, 1971; Platt, 1983, 1988; Popper *et al.*, 1988). Some fish have an additional "otolith" organ, the macula neglecta, that lacks an overlying otolith mass. The inner ear mediates sound perception and vestibular control. While the semicircular canals and otolith organs of all fish have the potential to detect body accelerations coincident with postural change and swimming, and the particle displacement component of the sound stimulus (Lowenstein, 1971; Platt, 1988), the primary role of the semicircular canals is considered to be vestibular, that of the saccule and lagena otolith organs primarily auditory, whereas that of the utricle otolith organ may be either vestibular or auditory, depending upon the species (Lowenstein, 1971; Platt, 1988; Popper & Fay, 1999). Montgomery *et al.* (2006) suggest that separation of vestibular and auditory roles between the otolith organs could be achieved on the basis of differences in the frequency and amplitude characteristics of these stimuli. It remains unclear however, whether these roles are in fact separated amongst the otolith end organs, or whether a single otolith organ performs a mixed vestibular and acoustic role. It may be that distinguishing vestibular and auditory functions of the inner ear is an entirely human artifice (Montgomery *et al.*, 2006). The end organs of the inner ear may in fact perceive 'vestibular' and 'auditory' stimuli as a continuum of self- and environmentally-induced body accelerations and acoustic stimuli.

The orthogonal arrangement of the semicircular canals provides for the detection of change in body orientation in three dimensions simultaneously (Platt, 1983). During accelerations of the head, the high viscosity of the endolymph fluid contained within the semicircular canals and frictional drag against the narrow rigid canal walls, results in endolymph fluid inertia relative to the head tilt movements. This results in bending the cupula diaphragm and ensuing mechanical displacement of the receptor cell cilia. Because the vane-like cupula diaphragm is thought to effectively stop fluid leakage around the diaphragm margins, cupula bending ceases only when the endolymph fluid reaches a steady state relative to the cupula; i.e. constant velocity. Upon deceleration, the cupula is bent in the opposite direction (Platt, 1983).

Accepting that otolith organs have the capacity to detect both linear accelerations associated with body movements and components of the acoustic stimulus, the following discussion addresses only the auditory response of otolith organs. Several reviews have provided detailed descriptions of the physical properties of underwater sound (Rogers & Cox, 1988; Montgomery *et al.*, 2006). In essence, underwater sound is propagated as a compression (pressure) wave with an accompanying particle motion (displacement) component (Popper *et al.*, 1988; Rogers & Cox, 1988). Particle motion dominates the sound stimulus close to its source and is referred to as the 'near field' sound stimulus, whereas the sound pressure component propagates further and is referred to as the 'farfield' sound stimulus. Because the density of a fish and water are nearly equal, a fish will vibrate in the sound field in unison with the particle displacement term of the sound stimulus. The otolith mass is however more dense than the rest of the fish so that the vibrational response of the otolith is out-of-phase relative to the sensory macula, resulting in receptor cell stimulation (Popper *et al.*, 1988; Rogers & Cox, 1988). The majority of fishes are hearing generalists, lacking special accessory hearing structures and may only detect the particle displacement sound term, a response referred to as the 'direct sound stimulus pathway' (Popper *et al.*, 1988; Rogers & Cox, 1988) or 'otolith hearing' (Montgomery *et al.*, 2001b) and which is most effective at low frequencies < 200 Hz (Popper *et al.*, 1988). As an aside, it should be noted that the particle displacement sound term is also an effective stimulus for canal neuromasts of the mechanosensory lateral line (Rogers & Cox, 1988; Kalmijn, 1989; Kenyon, 1996; Coombs *et al.*, 2001; Higgs *et al.*, 2003), a possibility that we must be aware of when investigating sound perception in fish. Hearing specialists, those characterised by the presence of accessory hearing structures, either the swim bladder (Coombs & Popper, 1979; Popper *et al.*, 1988; Rogers & Cox, 1988; Webb & Smith, 2000), or other gas-filled organ (Yan *et al.*, 2000; Wysocki & Ladich, 2001) located in close proximity, or with a direct physical link to the inner ear and/or lateral line, have increased hearing sensitivity and an expanded frequency response range compared to hearing generalists (Lowenstein, 1971; Rogers & Cox, 1988; Wysocki & Ladich, 2001; Higgs *et al.*, 2003). Accessory hearing organs are highly compliant gas-filled structures that resonate in direct response to the frequency of the sound pressure wave. Accordingly, they transduce the sound pressure stimulus to a particle motion stimulus that is effective for auditory receptor cell stimulation. This is referred to as the 'indirect sound stimulus pathway'. Sound-induced resonance of the swim bladder of some hearing specialists is directly transmitted to the inner ear via morphological connections between the two organs. For example, in the Ostariophysi (suborder Otophysi), a series of Weberian ossicles link the swim bladder and inner ear (Popper *et al.*, 1988; Rogers & Cox, 1988; Higgs *et al.*, 2003) and in the Clupeomorpha, resonance of paired otic bulla, gas-filled extensions of the swim bladder, is transduced to a particle motion stimulus in the fluid filled chambers of the inner ear and cephalic lateral line via an intermediary pressure-compliant membrane (Allen *et al.*, 1976; Blaxter *et al.*, 1981; Hoss & Blaxter, 1982). It has been proposed that even in the absence of accessory hearing structures directly linking, or in close association with the inner ear, hearing generalists may be able to detect the sound pressure stimulus if acoustic resonance of the swim bladder is effectively re-radiated through the intervening tissues to stimulate the inner ear (Popper *et al.*, 1988; Rogers & Cox, 1988; Montgomery *et al.*, 2001b, 2006; Myrberg & Fuiman, 2002). Yan *et al.* (2000), however, provide evidence that this may not be the case.

DEVELOPMENT AND FUNCTION: LARVAE

Embryonic Origins

The lateral line, inner ear and their innervations are ectodermal in origin (Kunz, 2004) and develop from a series of closely associated but discrete neurogenic placodes (Northcutt, 2003; Gibbs, 2004; Gibbs & Northcutt, 2004; Hall, this volume). Six placodes on each side of the head give rise to neuromasts of the mechanosensory lateral line, and in those fish that have them, the specialised electroreception organs. A seventh placode gives rise to the organs of the inner ear. As development progresses the placodes dissociate into sensory ridges, and neuroblasts of presumptive nerve cells and sensory organs begin to migrate. Nerve cell neuroblasts proceed centrally to contact the medulla, while those of the sensory end organs migrate to a predetermined peripheral location prior to differentiation (Northcutt, 2003; Gibbs, 2004; Gibbs & Northcutt, 2004). Development of mechanoreception organs is incomplete at hatching; post-embryonic growth characterised by continued sensory cell and neuron addition concomitant with ontogenetic development and organ growth into adult life (Popper & Hoxter, 1984; Platt, 1988; Kenyon, 1996; Higgs, 2002; Ledent, 2002).

The Mechanosensory Lateral Line

Post-embryonic development

Primary neuromasts, those arising directly from neurogenic placodes during embryonic development, are present at hatching, appearing first on the head and later on the trunk (Blaxter 1986, 1987; Fuiman *et al.*, 2004). They develop from neuromast primordia, comprising just a few support and receptor cells (Fig. 1A, B) that erupt through the overlying ectoderm in a process that involves both ectodermal retraction and upward projection of the neuromast (Gibbs, 2004). A cupula is then secreted by the neuromast support cells, and although cupulae are readily dislodged, regeneration is rapid, occurring within 24 hours in Atlantic herring (*Clupea harengus*) and willow shiner (*Gnathopogon elongatus caerulescens*) larvae (Blaxter, 1984; Mukai *et al.*, 1994). Neuromast eruption does not occur in an ordered sequence along a sensory ridge, but proceeds in successive waves, initially providing a dispersed neuromast array, only later 'filling in the gaps' (Ledent, 2002). Fuiman *et al.* (2004) suggest that such early development of a skeletal array of neuromasts may confer greater functional ability on larvae earlier. Typically, there are relatively few superficial neuromasts at hatching (Cobcroft & Pankhurst, 2003). Newly hatched striped trumpeter (*Latris lineata*) larvae (~ 3.5 mm TL) had just two neuromasts on each side of the head and one on the trunk. In comparison, soon after hatching, Atlantic herring larvae (8-10 mm TL) had ten primary neuromasts on each side of the trunk and 12-16 on each side of the head (Blaxter *et al.*, 1983). Primary neuromasts continue to proliferate over the surface of larvae for much of the larval period, the number and subsequent rate of addition relating to the rate of ontogenetic development of a species. As development proceeds, neuromasts increase in size (Iwai, 1963a; Blaxter *et al.*, 1983; Neave, 1986; Blaxter & Fuiman, 1989) by addition of hair cells (Blaxter & Fuiman, 1989; Mukai & Kobayashi, 1994) and cupula enlargement (Iwai, 1963a; Blaxter *et al.*, 1983; Mukai *et al.*, 1994). For further details about neuromast development and proliferation see: Iwai, 1963a, b; Cahn *et al.*, 1968; Blaxter *et al.*, 1983; Neave, 1986; Margulies, 1989; Kawamura *et al.*, 1990; Su & Wang, 1990; Jones & Janssen, 1992; Pankhurst & Butler, 1996; Ledent, 2002; Boglione *et al.*, 2003; Cobcroft & Pankhurst, 2003; Fuiman *et al.*, 2004.

Canal formation commences first on the head in late ontogenetic stage larvae (Fuiman *et al.*, 2004), the supraorbital and infraorbital canals forming first and trunk canals forming last (Blaxter, 1987; Jones & Janssen, 1992; Pankhurst & Butler, 1996; Cobcroft & Pankhurst, 2003; Fuiman *et al.*, 2004). The process starts with primary neuromasts sinking into dermal furrows. Ectodermal ridges then form parallel to the long axis of the presumptive canals, these growing to eventually enclose the neuromasts

(Fig. 1C). Myrberg and Fuiman (2002) suggest that canal formation is delayed until the late larval period because of the constraints imposed by non-ideal flow properties of very small canals. While most primary neuromasts will become canal neuromasts, some are destined to remain on the skin as 'accessory' canal neuromasts, or in the situation where canal formation is reduced, or canals are discontinuous, they have been referred to as replacement neuromasts (Webb, 1989a; Montgomery *et al.*, 2002). On the basis of their pharmacological and functional responses, all primary neuromasts are considered to be homologous with canal neuromasts and physiologically distinct from secondary neuromasts that develop later (Montgomery *et al.*, 2002).

The mechanosensory lateral line has a demonstrated role in the detection of prey and in predator escape behaviours in some fish larvae and has been implicated in rheotaxis and schooling behaviour.

Role in Feeding

While vision is considered to be the primary sensory modality used for prey detection and capture in the majority of fish larvae (Blaxter, 1986; Loew & Wahl, this volume) there is good evidence that some species use neuromasts to detect hydrodynamic disturbances generated by prey organisms, either in isolation, or in conjunction with other senses. Because vision is a light-dependent sensory modality, examination of the feeding ability of fish larvae in the light and dark enables separation of visual and non-visual (chemosensory and mechanosensory) senses, while testing feeding responses of larvae in the dark with live prey (providing chemosensory and mechanosensory stimuli) and dead prey (providing chemosensory stimuli only), separates mechanosensory and chemical cues. In this manner, Salgado and Hoyt (1996), showed that fathead minnow (*Pimephales promelas*) larvae fed in both the light and dark from the onset of first-feeding (3 days post hatching), and that feeding in the dark was the result of mechanosensory rather than chemosensory detection. In an effort to further tease apart the sensory basis of prey detection by fish larvae Jones and Janssen (1992) used the aminoglycoside antibiotic, streptomycin sulphate, to pharmacologically block both superficial and canal neuromasts of mottled sculpin (*Cottus bairdi*) larvae and clearly demonstrated a mechanosensory role in feeding by this species on *Artemia* prey in the dark. The authors determined that the reactive distance to *Artemia* prey in the dark was short, less than a body length. This is consistent with both the lateral line sensitivity range of no more than a few body lengths in adult fish (Kalmijn, 1989) and with the short reactive distance (~ 1 body length) reported for visually mediated feeding of striped trumpeter larvae (Cobcroft & Pankhurst, 2006). In mottled sculpin larvae there appeared to be a reduction in sensitivity of neuromasts to detect *Artemia* prey coincident with the onset of canal formation (Jones & Janssen, 1992). This appears to be transitory because use of the lateral line for selective feeding on zooplankton in the dark has been confirmed in adults of the species (Janssen *et al.*, 1995). The reasons for the loss in sensitivity remain unclear but may be associated with differences in the frequency response ranges of superficial and canal neuromasts resulting in lower detection by canal neuromasts of the low frequency beating of *Artemia* appendages. Alternatively, it may be a result of neural restructuring coincident with canal formation (Jones & Janssen, 1992). Other larval fish also have the capacity to feed in the dark. Blaxter (1969) reported that sole (*Solea solea*) larvae fed in the dark soon after yolk absorption was complete, whereas plaice (*Pleuronectes platessa*) larvae were initially strongly visual feeders and only developed the capacity to feed in the dark at metamorphosis. Only later, using pharmacological blocking of the lateral line of juvenile sole and plaice, was it determined that chemosensory input was entirely responsible for prey detection by plaice larvae feeding in the dark, and confirmed a secondary mechanosensory role in prey detection by sole larvae feeding in the dark (Batty & Hoyt, 1995). Larvae of the willow shiner, zooplantivores with particularly long neuromast cupulae (~ 100 μm), also feed in the dark (Mukai *et al.*, 1994). The feeding success of willow shiner larvae in the dark was reduced after chemical removal of the cupulae with unbuffered tricaine methansulfonate anaesthetic, steadily improving thereafter as the cupulae regenerated.

The growing body of evidence that fish larvae are able to detect the hydrodynamic disturbances created by prey organisms, prior to canal formation, appears at odds with the juvenile and adult situation in which prey detection has been attributed to canal neuromasts, their inherent filtering properties allowing for prey localisation in hydrodynamic environments complicated by the animals own swimming movements and that of water current flow (Bleckman, 1986; Montgomery, 1989; Jones & Janssen, 1992; Janssen *et al.*, 1995; New *et al.*, 2001; Montgomery *et al.*, 2002; Fuiman *et al.*, 2004). Montgomery *et al.* (2002) explain this apparent contradiction by suggesting that primary superficial neuromasts are functionally homologous with canal neuromasts; however, questions remain as to the physiological basis of prey detection by superficial neuromasts of fish larvae.

Role in predator evasion

Teleost larvae respond to a perceived predator threat with a stereotypic predator escape behaviour called a startle response (see Müller, this volume). This is a short latency, rapid, whole body, acceleration away from the stimulus that is mediated by large mauthner-like cells in the brain stem (Eaton & DiDomenico, 1986) receiving input from a range of sensory organs. Startle responses to tactile stimuli are apparent in many fish embryos even prior to hatching (Noakes & Godin, 1988). Thereafter, an expanding suite of sensory organs is known to play a part in eliciting startle responses to a perceived predation threat, including the mechanosensory lateral line, auditory and visual organs (Blaxter & Batty, 1985; Noakes & Godin, 1988; Blaxter & Fuiman, 1989, 1990; Fuiman *et al.*, 1999; Myrberg & Fuiman, 2002).

The ontogenetic improvement in responsiveness of fish larvae to attack by predatory fishes has been well established (Margulies, 1989; Blaxter & Fuiman, 1990; Fuiman, 1994) and yet relatively few studies have experimentally assessed the relative role of the mechanosensory lateral line and other sensory modalities in predator avoidance behaviours (reviewed by Fuiman *et al.*, 2004). Those that have employed pharmacological agents to temporarily block neuromast function in conjunction with sensory separation techniques. These approaches were similar to those used to investigate the role of the mechanosensory lateral line and other senses in the detection of prey by fish larvae. The roles of visual and neuromast organ input in eliciting a startle response to a simulated predator stimulus, the approach of a probe toward but not touching the larva, was identified by testing larvae in the light (both senses) and dark (mechanoreception only) and with and without use of pharmacological disruption of the mechanosensory lateral line. Removing both sensory inputs resulted in no, or low responsiveness (% of larvae responding) to a simulated predator stimulus in larvae of Sciaenidae: Atlantic croaker (*Micropogonias undulatus*), spotted sea trout (*Cynoscion nebulosus*), red drum (*Sciaenops ocellatus*), Clupeidae: Atlantic menhaden (*Brevoortia tyrannus*) and Engraulidae: bay anchovy (*Anchoa michilli*) (Higgs & Fuiman, 1996, 1998; Poling & Fuiman, 1997, 1999). However, the relative importance of visual and neuromast input in eliciting a startle response varied with species (Higgs & Fuiman, 1996; Poling & Fuiman, 1999) and in some species, changed during ontogeny (Poling & Fuiman, 1997). In intact larvae (no pharmacological treatment) of Atlantic menhaden and bay anchovy tested in the light, reactive distance to a simulated predator stimulus was short (less than one body length) and ranged between 1-3 mm in the smallest larvae (Higgs & Fuiman, 1996, 1998). Reactive distance increased with increasing body length in these larvae; this can be primarily attributed to neuromast proliferation and visual development in Atlantic menhaden and bay anchovy, respectively. Under similar test conditions, reactive distance of Atlantic croaker larvae did not change during ontogeny (mean of 5.51 $\pm$ 3.79 mm). This was despite improvement in visual acuity almost certainly allowing larger larvae to detect the predator stimulus at greater distances (Poling & Fuiman, 1997). It appears that in Atlantic croaker larvae neuromast input is involved in delaying the timing of the avoidance response until the approaching stimulus reaches a critical set point (distance from the approaching predator) at which time the startle response is initiated; a mechanism that may increase the likelihood of success of the evasive response (Poling & Fuiman, 1997). In larvae of three sciaenid and two clupeid species, there was no evidence of a rapid change in responsiveness, or reactive distance, to a simulated predator stimulus coincident with formation of the cephalic lateral line (Higgs & Fuiman, 1996, 1998; Poling & Fuiman, 1997, 1999), or with gas-filling of the otic bulla in the two clupeids (Higgs & Fuiman, 1996, 1998). This is despite the fact that clupeids are hearing specialists in which there is potential for increased sensitivity via transduction and amplification of the 'sound' pressure component of the predator stimulus to a particle displacement stimulus in the cephalic lateral line canal. In contrast, when Atlantic herring larvae were subjected to real fish predators, responsiveness improved markedly after closure of the cephalic lateral line and gas-filling of the otic bulla (Blaxter & Fuiman, 1990). This apparent disparity in responsiveness of larvae to a simulated and real predator threat may be explained by the very different stimulus characteristics of an advancing probe and a swimming predator (Blaxter & Fuiman, 1990; Higgs & Fuiman, 1996, 1998; Myrberg & Fuiman, 2002). A further suite of experiments confirmed that responsiveness of Atlantic herring larvae to real fish predators was low, both prior to, and after gas-filling of the otic bulla (between 9-15% of larvae), increasing markedly to 59% of larvae responding once the cephalic lateral line closure was complete (Blaxter & Fuiman, 1990). This suggests that enclosure of cephalic neuromasts within canals is a critical ontogenetic event for enhancement of the predator avoidance response in Atlantic herring larvae. Further to this, streptomycin sulphate treatment of Atlantic herring larvae after cephalic lateral line closure reduced but did not eliminate the response to a real fish predator threat (21% responsiveness in streptomycin-treated fish responding in light conditions). This confirmed the important contribution of both vision and the cephalic lateral line input in eliciting the predator escape response (Blaxter & Fuiman, 1990; Fuiman *et al.*, 2004). Nevertheless, the reactive distance of Atlantic herring larvae to a real fish predator under these laboratory conditions was short, just 2-4 cm.

Role in orientation behaviours

Rheotaxis of adult fishes is a complex behaviour facilitating position holding in water currents. Along with other sensory organs, the mechanosensory lateral line, specifically superficial neuromasts and not canal neuromasts (Montgomery *et al.*, 1997, 2002, 2003; Engelmann *et al.*, 2000) are reported to play a critical role in the rheotactic response. Superficial neuromasts provide information about the velocity and direction of water currents relative to the fish's own swimming movements (Coombs *et al.*, 2001). In contrast, little is known about rheotaxis of fish larvae. Masuda and Tsukamoto (1996) reported the onset of a rheotactic response in larvae of striped jack (*Pseudocaranx dentex*) to a water flow of 1.7 ± 0.5 cm s^{-1}, soon after hatching at 4.5 mm TL, when only primary superficial neuromasts were present, and long before canal formation commenced. While this provides supporting evidence of a role of superficial neuromasts in rheotaxis of fish larvae, further research is required to confirm this. The ability of striped jack larvae to hold position in water currents increased with age, likely a result of increasing sensory and swimming ability (Masuda & Tsukamoto, 1996).

Precise localisation of fine-scale water velocity stimuli against the background noise of water current flow and that of the fish's own swimming movement, has been attributed to canal neuromasts (Denton & Gray, 1988; Engelmann *et al.*, 2000; Montgomery *et al.*, 2002), implicating them as likely candidates for a role in the highly co-ordinated swimming behaviours observed in schooling fishes. While selective blocking of vision and mechanosensory lateral line input in adult fish demonstrated that both senses are involved in schooling behaviour (Partridge & Pitcher, 1980), no such physiological evidence has yet been obtained for fish larvae. Onset of schooling behaviours in sibling *Menidia* sp. larvae, as evident by the increasing tendency and duration of parallel body orientations (Cahn *et al.*, 1968), and in striped jack larvae by the reduction of inter-sibling distances and angles (Masuda & Tsukamoto, 1998), appears to be delayed until relatively late in larval life. This is consistent with the delayed enclosure of superficial neuromasts within canals. Fully synchronised schooling behaviour was only apparent in striped jack at 30 mm TL, once trunk canal closure was complete (Masuda &

Tsukamoto, 1988) and in Atlantic herring did not occur until 50-55 mm TL, well after metamorphosis (Gallego & Heath, 1994).

The Inner Ear

Post-embryonic development and physiology

Few studies have investigated the morphological development of the otolith organs and fewer still that of the semicircular canals of fish larvae (Higgs, 2002; Myrberg & Fuiman, 2002). In addition, until quite recently understanding of the auditory physiology during early ontogeny of fish was restricted to a single study in which heart rate conditioning was used to determine the auditory threshold of 0- to 2-year-old red seabream (*Pagrus major*) (Iwashita *et al.*, 1999) and investigations of whole animal behaviours (startle responses) to broadband auditory stimuli (Blaxter & Hoss, 1981; Blaxter *et al.*, 1981; Blaxter & Batty, 1985; Fuiman *et al.*, 1999). Recent adoption of auditory brainstem response (ABR) techniques to investigate fish auditory physiology (Kenyon *et al.* 1998) and subsequent application in larval fish research, has provided a relatively non-invasive method (Higgs, 2002; Higgs *et al.*, 2003) for investigation of the physiological auditory response to pure tone and ecologically relevant auditory stimuli from delicate early-stage larvae through to juvenile transformation and beyond (Wysocki & Ladich, 2001; Higgs *et al.*, 2003; Egner & Mann, 2005; Wright *et al.*, 2005). The investigation of auditory physiology of fish larvae is, however, still in its infancy.

Otolith organs are present in fish larvae at hatching (Blaxter, 1986; Sokolowski & Popper, 1988; Pankhurst & Butler, 1996; Cobcroft & Pankhurst, 2003) (Fig. 1D). In late-stage embryos of Gulf toadfish (*Opsanus tau*) the sensory epithelia of the saccule organs were reported to be differentiated with innervation and synaptic vesicles present, suggesting that they were functional prior to hatching (Sokolowski & Popper, 1988). In comparison, two pairs of otolith organs, the utricles and saccules, were present in newly hatched greenback flounder (*Rhombosolea tapirina*), while the third pair, the lagena organs, did not develop until the end of the larval period (Pankhurst & Butler, 1996). In this species, development of the semicircular canals was also a protracted process. Development of paired lateral semicircular canals and associated sensory epithelia, had commenced soon after hatching with development of the anterior semicircular canals commencing soon thereafter. The development of the posterior canals was not completed until well after settlement (Pankhurst & Butler, 1996). Behavioural responses of larvae support the commencement of vestibular function, the maintenance of body equilibrium, soon after hatching (Neave, 1985; Nunan & Noakes, 1985; Blaxter, 1986).

Addition of sensory cells to the sensory epithelium of otolith organs occurs during organ development in larvae and continues into adult life (Popper & Hoxter, 1984; Kenyon, 1996; Higgs *et al.*, 2001, 2003; Higgs, 2002). Growth of the saccule macula in larval zebrafish (*Danio rerio*) was isometric, occurring at the perimeter of the organ (Higgs *et al.*, 2003), and in cichlid oscar (*Astonotus ocellatus*), relative densities of auditory sensory cells in different regions of the saccule remain fairly constant across a wide size range of fish (Popper & Hoxter, 1984). In damselfishes (Pomacentridae), which are hearing generalists, a correlation has been suggested between the growth-related increase in auditory sensory cell number and the large increase in auditory sensitivity (threshold level for sound detection) reported across the entire frequency response range of juvenile bicolour damselfish (*Pomacentrus partitus* and *P. variables*) using behavioural conditioning (Kenyon, 1996). Other studies however, provide ambiguous or contradictory results. Zero- to 2-year-class red sea bream showed a small improvement in auditory sensitivity (using heart rate conditioning) across a narrower frequency response range (Iwashita *et al.*, 1999), whereas another damselfish, the Nagasaki damsel (*P. nagasakiensis*), showed a small improvement in auditory sensitivity from pre- to post-settlement stages (using ABR) in which auditory thresholds of pre-settlement fish were 8 dB (re 1 μPa) higher (i.e. lower sensitivity) than those of post-settlement fish at just two audible frequencies, 100 and 600 Hz (Wright *et al.*, 2005) (Fig. 2). ABR also revealed a small ontogenetic increase in auditory sensitivity within the most sensitive auditory frequency range of juvenile croaking gourami (*Trichopsis vittata*), members of the Anabantidae which are hearing specialists in which air-filled suprabranchial chambers located laterally to the ears were deemed to be functional (Wysocki & Ladich, 2001). In contrast, in zebrafish, otophysan hearing specialists, the ABR detected no difference in auditory sensitivity across the frequency response range (100-4000 Hz) and size range of fish (10-45 mm TL) tested (Higgs *et al.*, 2003) (Fig. 3). The absence of a change in auditory sensitivity with growth of the peripheral auditory organ in larval and post-larval zebrafish supports an alternative proposition that increase in auditory sensory cell number is required to maintain rather than increase hearing sensitivity during growth (Popper *et al.*, 1988; Rogers *et al.*, 1988; Higgs *et al.*, 2003). This is consistent with the significant increase in auditory sensory cell number with growth of adult zebrafish without change in auditory sensitivity (Higgs *et al.*, 2001). The factors underpinning ontogenetic improvement in auditory sensitivity remain unclear, but may simply reflect fish diversity, particularly in relation to the presence, absence and/or variation in form and function of auxiliary hearing structures (Higgs *et al.*, 2003).

Development of auxiliary hearing structures in Clupeidae (otic bullae) and Ostariophysi (Weberian ossicles) commences during larval development coincident with development of the inner ear (Allen *et al.*, 1976; Hoss & Blaxter, 1982; Higgs & Fuiman, 1996; Higgs *et al.*, 2003). Pressure compliance of communicating membranes in the otic bulla inner-ear cephalic lateral line system indicated that transduction of the sound pressure stimulus to a particle motion stimulus in the fluid filled compartments of the inner ear and cephalic lateral line, commenced at 18-20 mm TL in Atlantic menhaden (Hoss & Blaxter, 1982) and 30-35 mm TL in Atlantic herring (Blaxter & Hoss, 1979) larvae. The auditory sensitivity and frequency response range of clupeid larvae before and after development of the auxiliary hearing structures has yet to be described; however, startle responses in Atlantic herring larvae to acoustic stimuli did not commence until otic bullae were gas inflated (Blaxter & Hoss, 1981), indicative of

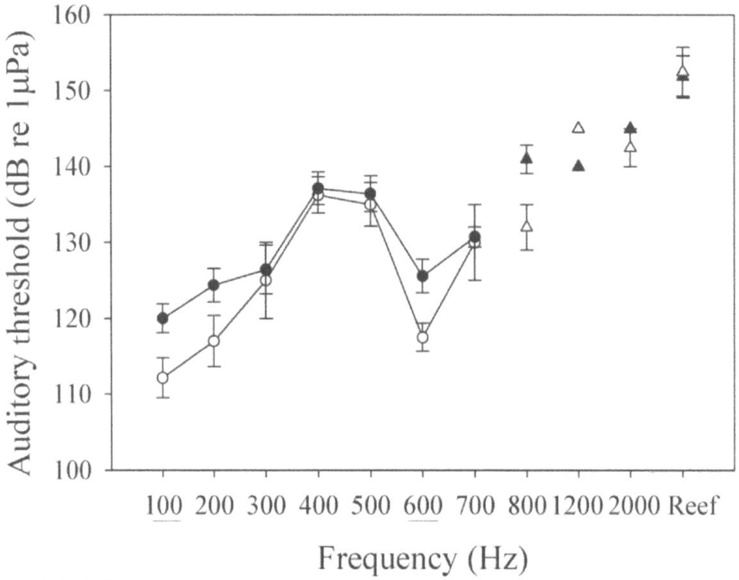

Fig. 2 Auditory thresholds for pre- (solid symbols) and post-settlement (open symbols) Nagasaki damselfish (*Pomacentrus nagasakiensis*). Values are means ± SEM. Responses for frequencies greater than 700 Hz are represented by triangles, indicating that these measurements are minimum estimates due to the equipment limitations. Underlined frequencies indicate a significant difference ($P<0.05$) between pre- and post-settlement responses. Please note that the values for 600 Hz should be ca 8 dB higher than indicated for both pre- and post-settlement larvae due to a software problem with the auditory brainstem response apparatus (KJ Wright, pers. comm.) From Marine Biology 147 (2005), page 1429, Auditory and olfactory abilities of pre-settlement larvae and post-settlement juveniles of a coral reef damselfish (Pisces: Pomacentridae), K.J. Wright, D.M. Higgs, A.J. Belanger and J.M. Leis, figure 3, © Springer-Verlag 2005: with kind permission of Springer Science and Business Media.

the important role of the auxiliary hearing structures in hearing of clupeid larvae. In zebrafish larvae, Weberian ossicle formation was first apparent in larvae 7 mm TL (Fig. 4); however, there remained large gaps between the ossicles. A contiguous arrangement of ossicles extending from the swim bladder to the inner ear was not apparent until larvae attained 19.5 mm TL (Higgs *et al.*, 2003). The ABR of zebrafish larvae to pure tone auditory stimuli confirmed, as predicted from previous studies of hearing specialists (see section 2.4), that development of accessory hearing structures coincided with the marked expansion of the sound frequency response range, the maximum detectable frequency range increasing from 200 Hz in 10 mm TL larvae, to 4,000 Hz in fish = 25-35 mm TL (Higgs *et al.*, 2003) (Fig. 4). Comparable electrophysiological data spanning larval to post-larval stages of hearing generalists is not available; however, an ABR was detected in pre- (12-15 mm SL) and post-settlement (15-17 mm SL) Nagasaki damsels across the entire frequency range tested (100 to 2000 Hz to pure tones and recordings of broad frequency night reef noise) (Fig. 2) (Wright *et al.*, 2005).

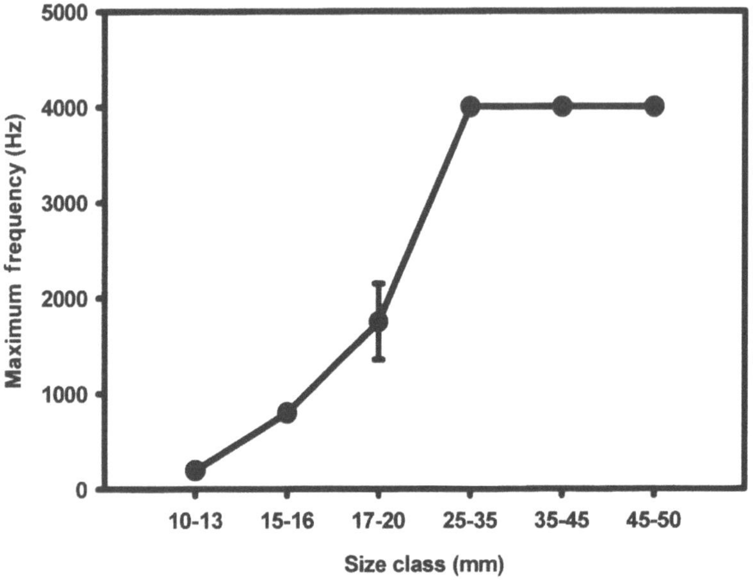

Fig. 3 The maximum frequency to which zebrafish (*Danio rerio)* showed an auditory brainstem response gradually increased from 200 Hz in 10-13 mm larvae up to 4000 Hz in larvae larger than 20 mm. The > 20 mm size class has been subdivided to visually demonstrate that maximum frequency of detection plateaus at 4000 Hz for zebrafish. Symbols represent mean ± SEM for 10-13 (n=4), 15-16 (n=3), 17-20 (n=8), 25-36, 35-45 and 45-50 mm TL size classes, respectively. The n value for all animals over 20 mm TL = 6 Reused with permission from Dennis M. Higgs, Audrey K. Rollo, Marcy J. Souza and Arthur N. Popper, The Journal of the Acoustical Society of America, 113, 1148 (2003). © 2003, Acoustical Society of America.

Hearing in the orientation to reefs by settlement-stage reef larvae

Pre-settlement stage larvae of reef-associated marine fishes are not passive particles but show active movement towards reefs (Stobutzki & Bellwood, 1997; Armsworth, 2000) with the result that there is capacity for significant recruitment onto natal reefs (Jones *et al.*, 1999; Swearer *et al.*, 1999). The role of hearing in the recruitment of marine fish larvae has been the focus of considerable debate in recent years. On the basis of information available at the time, Myrberg and Fuiman (2002) concluded that hearing in fish larvae is poor and that sound emanating from reefs was unlikely to provide a biologically relevant signal for reef localisation and subsequent settlement by late-stage pelagic fish larvae. There is now a more compelling body of evidence from field-based studies that some pre-settlement fish larvae use ambient sounds as a reef orientation cue (Tolimieri *et al.*, 2000, 2002, 2004; Leis *et al.*, 2003; Simpson *et al.*, 2004, 2005; Leis & Lockett, 2005), and in addition, physiological evidence from a laboratory based

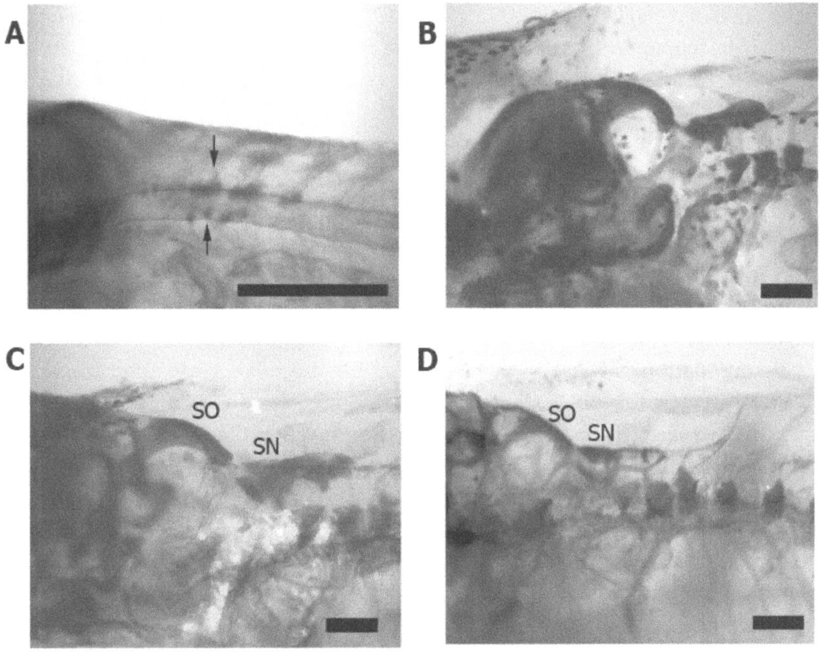

Fig. 4 Weberian ossicles are first evident at 7 mm total length (TL) in zebra fish (*Danio rerio)* larvae (arrows in (A)) but are very small and poorly connected. By 13 mm (B) the ossicles are larger but large gaps remain between individual elements. By 17 mm TL (C) the dorsal plate has expanded but there are still gaps between individual elements and the supraoccipital (SO) is not connected to the supraneural (SN) Weberian elements. By 19.5 mm TL (D) the supraoccipital bone is well attached to the supraneurals, forming an unbroken chain from the Weberian apparatus to the inner ear. Scale bars = 0.1 mm. Reused with permission from Dennis M. Higgs, Audrey K. Rollo, Marcy J. Souza and Arthur N. Popper, The Journal of the Acoustical Society of America, 113, 1151 (2003). © 2003, Acoustical Society of America.

study (using ABR) that pre-settlement larvae of the Nagasaki damsel are able to detect reef noise (Wright *et al.*, 2005). Reviews by Montgomery *et al.* (2001b, 2006) and Higgs (2005) provide comprehensive accounts of recent developments in this field of research.

Reef environments are noisy, much of which is of biological origin (Cato, 1978; McCauley & Cato, 2000). The combined chorus of reef organisms can reach 20-30 dB (re 1 μPa) above background noise levels and is detected for distances of up to 5-8 km (McCauley & Cato, 2000), consistent with the extraordinarily low attenuation of sound underwater (Rogers & Cox, 1988). This raises the question as to whether the auditory thresholds recorded for pre-settlement larvae (e.g. 110-150 dB re 1 μPa for Nagasaki damsel (Wright *et al.*, 2005) using ABR) (Fig. 2) are sufficiently sensitive to detect reef noises at ecologically relevant distances to facilitate orientation towards and location of reefs (Leis *et al.*, 2003; Higgs, 2005; Wright *et al.*, 2005). Assuming that physiological measurements of auditory thresholds underestimate the absolute auditory thresholds of fish (Kenyon *et al.*, 1998), Wright *et al.* (2005) suggest that reef noise may be detectable by the pre-settlement stage of the Nagasaki damsel at distances of many kilometers.

In order for fish larvae to use reef noise as an orientation cue they must be able to determine the direction (origin) of the sound stimulus. In most fish the saccule organs of left and right inner ears are primarily responsible for otolith hearing; detection of the particle motion (direct) sound stimulus. At the most fundamental level, determination of the direction of the sound stimulus is dependent upon the directional sensitivity of individual auditory receptor cells, as well as the receptor orientations within the entire saccule population. Typically, the sensory epithelium of the saccule comprises a patchwork of receptor cells, each with distinctive directional sensitivity. Montgomery *et al.* (2006) explain that in theory just two orthogonally oriented receptor populations within a single plane of the saccule epithelium are sufficient in order to determine the directional axis of the sound stimulus within that plane. Sound sources intermediate (oblique) to the primary directional sensitivities of these two receptor populations can be determined from the cosine responses of the receptors to oblique stimulus directions. If the sensory epithelia of saccule organs are curved rather than planar, and given that left and right saccule organs are obliquely oriented in the head relative to the long axis of fish, it is possible that localisation of the directional axis of the sound stimulus can be obtained in three dimensions by processing information from left and right auditory organs (Montgomery *et al.*, 2006). However, because fish tissue is acoustically 'transparent', the particle displacement sound stimulus will be received by left and right ears simultaneously, with no difference in inter-aural sampling time and stimulus intensity (Fay & Edds-Walton, 2000); this potentially giving rise to a 180 degree ambiguity in determining the direction of origin of the sound stimulus (Rogers & Cox, 1988; Fay & Edds-Walton, 2000; Montgomery *et al.*, 2001b; 2006; Myrberg & Fuiman, 2002). In other words, the fish will be able to determine the axis of travel of the sound stimulus (e.g. north–south) but not whether it originated from the north or south. In order to resolve this 180 degree ambiguity it has been proposed that fish must integrate both the direct and indirect (pressure) sound stimuli (Rogers & Cox, 1988; Montgomery *et al.*, 2001b, 2006). Montgomery *et al.* (2006) point out that binary choice chamber experiments conducted in the field (Tolimieri *et al.*, 2002, 2004) demonstrate that pre-settlement larvae can resolve the 180 degree ambiguity and determine the direction of origin of the sound stimulus. Larvae of hearing specialists would potentially develop directional sensitivity once pressure-compliant accessory hearing structures are functional; but what of larvae of hearing generalists? One possibility is that the swim bladder provides for detection of the indirect sound pressure stimulus by transmission of swim bladder resonance through the intervening tissues. This pathway has yet to be demonstrated and warrants further investigation given that there is some evidence, albeit for a limited number of species, that does not support this indirect sound stimulus pathway (Yan *et al.*, 2000). Further to this, not all fishes have swim bladders.

Alternatively, Montgomery *et al.* (2006) suggest that the air-water surface interface may provide for acoustic pressure transduction suitable for indirect sound stimulation of the inner ear, or that an as yet un-described pressure reference system may operate for hearing generalists in order to resolve the 180 degree sound source ambiguity.

FUTURE PERSPECTIVES

Pharmacological blocking of superficial and canal neuromasts of fish larvae used in conjunction with sensory separation techniques has confirmed the use of the mechanosensory lateral line, either in isolation or in combination with other sense organs, to detect prey. Given that the response characteristics of canal, and not superficial neuromasts, are attributed to prey localisation in adult fish, the physiological basis of prey detection by fish larvae prior to canal formation remains unclear. Treatment of fish larvae with pharmacological agents to block neuromast function has also been used to show that responsiveness of fish larvae to threat of a real fish predator improves during ontogeny, coincident with development of the mechanosensory lateral line and vision. Associations between growth and development of mechanosensory lateral line organs and the onset of rheotaxis and schooling behaviour have been proposed but have yet to be confirmed experimentally. Approaches described previously to investigate the role of mechanosensory organs and other senses in predator avoidance and prey detection by fish larvae could be employed to address this. Very few studies have investigated the development of the semicircular canals and inner ear, and fewer still the ontogeny and physiology of vestibular function in fish larvae. Addition of receptor cells in the otolith end organs coincident with development and growth of the inner ear has engendered debate as to whether growth-related addition of auditory receptor cells is required to increase or maintain auditory sensitivity; however, available evidence is either equivocal or conflicting. Investigation of auditory thresholds of a broad range of fish larvae, including both hearing specialists and generalists, and extending across the entire larval duration, may resolve this issue, or support an alternative view that auditory sensitivity may be more closely related to the development and function of auditory specialisations. ABR has provided a suitable technique for laboratory investigation of auditory physiology of delicate early stage larvae through to metamorphosis and beyond, paving the way for further investigation of auditory responses to ecologically relevant sound stimuli. Field-based binary choice chamber experiments have demonstrated that larvae are able to solve the proposed 180 degree ambiguity in determining the sound stimulus direction. As the majority of marine fish are considered to be hearing generalists, the question remains as to how larvae of hearing generalists detect the pressure component of the sound stimulus in order to solve the directional ambiguity of the sound stimulus.

Acknowledgements

PMP was supported by a James Cook University Aquaculture and Emerging Species Research Advancement Program grant.

References

Allen JM, Blaxter JHS, Denton EJ (1976) The functional anatomy and development of the swimbladder-inner ear-lateral line system in herring and sprat. J Mar Biol Assoc UK 56: 471-486

Armsworth PR (2000) Modelling the swimming response of late stage larval reef fish to different stimuli. Mar Ecol Prog Ser 195: 231-247

Batty RS, Hoyt RD (1995) The role of sense organs in the feeding behaviour of juvenile sole and plaice. J Fish Biol 47: 931-939

Blaxter JHS (1968) Light intensity, vision, and feeding in young plaice. J Exp Mar Biol Ecol 2: 293-307

Blaxter JHS (1969) Visual thresholds and spectral sensitivity of flatfish larvae. J Exp Biol 51: 221-230

Blaxter JHS (1984) Cupular growth in herring neuromasts. J Mar Biol Assoc UK 64: 935-938

Blaxter JHS (1986) Development of sense organs and behaviour of teleost larvae with special reference to feeding and predator avoidance. Trans Am Fish Soc 115: 98-114

Blaxter JHS (1987) Structure and development of the lateral line. Biol Rev 62: 471-514

Blaxter JHS, Batty RS (1985) The development of startle responses in herring larvae. J Mar Biol Assoc UK 65: 737-750

Blaxter JHS, Fuiman LA (1989) Function of the free neuromasts of marine teleost larvae. In: Coombs S, Görner P, Münz H (eds) The Mechanosensory Lateral Line: Neurobiology and Evolution. Springer-Verlag, New York, p 481-499

Blaxter JHS, Fuiman LA (1990) The role of sensory systems of herring larvae in evading predatory fishes. J Mar Biol Assoc UK 70: 413-427

Blaxter JHS, Hoss DE (1979) The effect of rapid changes of hydrostatic pressure on the Atlantic herring, *Clupea harengus* L. II. The response of the auditory bulla system in larvae and juveniles. J Exp Biol Ecol 41: 87-100

Blaxter JHS, Hoss DE (1981) Startle responses in herring: the effect of sound stimulus frequency, size of fish and selective interference with the *acoustico-lateralis* system. J Mar Biol Assoc UK 61: 871-879

Blaxter JHS, Denton EJ, Gray JAB (1981) *Acousticolateralis* system in clupeid fishes. In: Tavolga WN, Popper AN, Fay RR (eds) Hearing and Sound Communications in Fishes. Springer-Verlag, New York, p 39-59

Blaxter JHS, Gray JAB, Best ACG (1983) Structure and development of the free neuromasts and lateral line system of the herring. J Mar Biol Assoc UK 63: 247-260

Bleckmann H (1986) Role of the lateral line in fish behaviour. In: Pitcher TJ (ed) The Behaviour of Teleost Fishes. Croom Helm, London, p 177-202

Boglione C, Giganti M, Selmo C, Cataudella S (2003) Morphoecology in larval fin-fish: a new candidate species for aquaculture, *Diplodus puntazzo* (Sparidae). Aquacult Int 11: 17-41

Cahn PH, Shaw E, Atz EH (1968) Lateral-line histology as related to the development of schooling in the atherinid fish *Menidia*. Bull Mar Sci 18: 660-670

Cato DH (1978) Marine biological choruses observed in tropical waters near Australia. J Acoust Soc Am 64: 736-743

Cobcroft JM, Pankhurst PM (2003) Sensory organ development in cultured striped trumpeter larvae *Latris lineata*: implications for feeding behaviour. Mar Freshw Res 54: 669-682

Cobcroft JM, Pankhurst PM (2006). Visual field of cultured striped trumpeter *Latris lineata* (Teleostei) larvae feeding on rotifer prey. Mar Freshw Behav Physiol 39: 193-208

Coombs S, Popper AN (1979) Hearing differences among Hawaiian squirrel fishes (family Holocentridae) related to difference in peripheral auditory system. J Comp Physiol 132: 203-207

Coombs S, Janssen J, Webb, JF (1988) Diversity of lateral line systems: evolutionary and functional considerations. In: Atema J, Fay RR, Popper AN, Tavolga WN (eds) Sensory Biology of Aquatic Animals. Springer-Verlag, New York, p 553-593

Coombs S, Braun CB, Donovan B (2001) The orienting response of Lake Michigan mottled sculpin is mediated by canal neuromasts. J Exp Biol 204: 337-348

Denton EJ, Gray JAB (1988) Mechanical factors in the excitation of the lateral line of fishes. In: Atema J, Fay RR, Popper AN, Tavolga WN (eds) Sensory Biology of Aquatic Animals. Springer-Verlag, New York, p 595-617

Eaton RC, DiDomenico R (1986) Role of the teleost escape response during development. Trans Am Fish Soc 115: 128-142

Engelmann J, Hanke W, Mogdans J, Bleckmann H (2000) Hydrodynamic stimuli and fish lateral line. Nature 408: 51-52

Egner SA, Mann DA (2005) Auditory sensitivity of sergeant major damselfish *Abudefduf saxatilis* from post-settlement juvenile to adult. Mar Ecol Prog Ser 285: 213-222

Fay RR, Edds-Walton PL (2000) Directional encoding by fish auditory systems. Phil Trans R Soc Lond B 355: 1281-1284

Flock A (1971) The lateral line organ mechanoreceptors. In: Hoar, WS, Randall, DJ (eds) Fish Physiology, volume V, Sensory Systems and Electric Organs. Academic Press, New York, p 241-263

Fuiman LA (1994) The interplay of ontogeny and scaling in the interactions of fish larvae and their predators. J Fish Biol 45 (Suppl A): 55-79

Fuiman LA, Smith ME and Malley VN (1999) Ontogeny of routine swimming speed and startle responses in red drum, with comparison of responses to acoustic and visual stimuli. J Fish Biol 55 (Suppl A): 215-226

Fuiman LA, Higgs DM, Poling KR (2004) Changing structure and function of the ear and lateral line system of fishes during development. Am Fish Soc Symp 40: 117-144

Gallego A, Heath MR (1994) The development of schooling behaviour in Atlantic herring *Clupea harengus*. J Fish Biol 45: 569-588

Gibbs MA (2004) Lateral line receptors: where do they come from developmentally and where is our research going? Brain Behav Evol 64: 63-181

Gibbs MA, Northcutt RG (2004) Development of the lateral line system in the shovelnose sturgeon. Brain Behav Evol 64: 70-84

Hanke W, Brücker C, Bleckmann H (2000) The aging of the low-frequency water disturbances caused by swimming goldfish and its possible relevance to prey detection. J Exp Biol 203: 1193-1200

Higgs DM (2002) Development of the fish auditory system: How do changes in auditory structure affect function? Bioacoustics 12: 80-182

Higgs DM (2005) Auditory cues as ecological signals for marine fishes. In: Weissburg MJ, Browman HI (eds) Sensory Biology: Linking the Internal and External Ecologies of Marine Organisms. Mar Ecol Prog Ser 287: 263-307

Higgs DM, Fuiman LA (1996) Ontogeny of visual and mechanosensory structure and function in Atlantic menhaden *Brevoortia tyrannus*. J Exp Biol 199: 2619-2629

Higgs DM, Fuiman LA (1998) Associations between behavioural ontogeny and habitat change in clupeoid larvae. J Mar Biol Assoc UK 78: 1281-1294

Higgs DM, Souza MJ, Wilkins HR, Presson JC, Popper AN (2001) Age- and size-related changes in the inner ear and hearing ability of adult zebrafish (*Danio rerio*). JARO 3: 174-184

Higgs DM, Rollo AK, Souza MJ, Popper AN (2003) Development of form and function in peripheral auditory structures of the zebrafish (*Danio rerio*). J Acoust Soc Am 113: 1145-1154

Hoss DE, Blaxter JHS (1982) Development and function of the swimbladder-inner ear-lateral line system in the Atlantic menhaden *Brevoortia tyrannus* (Latrobe). J Fish Biol 20: 131-142

Iwai T (1963a) Development of lateral-line cupulae in the goboid fish, *Tridentiger trigonocephalus* (Gill). Bull Misaki Mar Biol Inst 4: 1-20

Iwai T (1963b) Sensory cupulae found in newly hatched larvae of *Blennius yatebei* Jordan et Snyder. Bull Jap Soc Sci Fish 29: 503-506

Iwashita A, Sakamoto M, Kojima T, Watanabe Y, Soeda H (1999) Growth effects on the auditory threshold of Red Sea bream. Nippon Suisan Gakkaishi 65: 833-838

Janssen J, Jones WR, Whang A, Oshel PE (1995) Use of the lateral line in particulate feeding in the dark by juvenile alewife (*Alosa pseudoharengus*). Can J Fish Aquat Sci 52: 358-363

Jones WR, Janssen J (1992) Lateral line development and feeding behavior in mottled sculpin, *Cottus bairdi* (Scorpaeniformes: Cottidae). Copeia 1992: 485-491

Jones GP, Millicich MJ, Emslie MJ, Lunow C (1999) Self-recruitment in a coral reef fish population. Nature 402: 802-804

Kalmijn AJ (1989) Functional evolution of the lateral line and inner ear sensory systems. In: Coombs S, Görner P, Münz H (eds) The Mechanosensory Lateral Line: Neurobiology and Evolution. Springer-Verlag, New York, p 187-215

Kawamura G, Takimoto M, Sobajima N (1990) Larval growth and age- and size-related variation in development of sense organs in the halfbeak, *Hemiramphus sajori*. In: Hirano R, Hanyu I (eds) The Second Asian Fisheries Forum, Asian Fisheries Society, Manila, p 407-410

Kenyon TN (1996) Ontogenetic changes in the auditory sensitivity of damselfishes (Pomacentridae). J Comp Physiol A 179: 553-561

Kenyon TN, Ladich F, Yan HY (1998) A comparative study of hearing ability in fishes: the auditory brainstem response approach. J Comp Physiol A182: 307-318

Kunz YW (2004) Developmental Biology of Teleost Fishes. Springer, Dordrecht

Ledent V (2002) Postembryonic development of the posterior lateral line in zebrafish. Development 129: 597-604

Leis JM, Lockett MM (2005) Localization of reef sounds by settlement-stage larvae of coral reef fishes (Pomacentridae). Bull Mar Sci 76: 715-724

Leis JM, Carson-Ewart BM, Hay AC, Cato DH (2003) Coral-reef sounds enable nocturnal navigation by some reef-fish larvae in some places and at some times. J Fish Biol 63: 724-737

Lowenstein O (1971) The Labyrinth. In: Hoar WS, Randall DJ (eds) Fish Physiology, Volume V, Sensory Systems and Electric Organs. Academic Press, New York, p 207-240

Margulies D (1989) Size-specific vulnerability to predation and sensory system development of white seabass *Atractoscion nobilis*, larvae. Fish Bull 87: 537-552

Masuda R, Tsukamoto K (1996) Morphological development in relation to phototaxis and rheotaxis in striped Jack *Pseudocaranx dentex*. Mar Freshw Behav Physiol 28: 75-90

Masuda R, Tsukamoto K (1998) The ontogeny of schooling behaviour in the striped jack. J Fish Biol 52: 483-493

McCauley RD, Cato DH (2000) Patterns of fish calling in a nearshore environment in the Great Barrier Reef. Phil Trans R Soc Lond B 355: 1289-1293

Montgomery JC (1989) Lateral line detection of planktonic prey. In: Coombs S, Görner P, Münz H (eds) The Mechanosensory Lateral Line: Neurobiology and Evolution. Springer-Verlag, New York, p 561-574

Montgomery JC, Baker CF, Carton AG (1997) The lateral line can mediate rheotaxis in fish. Nature 389: 960-963

Montgomery JC, Coombs S, Baker CF (2001a) The mechanosensory lateral line system of the hypogean form of *Astyanax fasciatus*. Environ Biol Fish 62: 87-96

Montgomery JC, Tolimieri N, Haine OS (2001b) Active habitat selection by pre-settlement reef fishes. Fish Fisheries 2: 261-277

Montgomery JC, Macdonald F, Baker CF, Carton AG (2002) Hydrodynamic contributions to multimodal guidance of prey capture behavior in fish. Brain Behav Evol 59: 190-198

Montgomery JC, McDonald F, Baker CF, Carton AG, Ling N (2003) Sensory integration in the hydrodynamic world of rainbow trout. Proc R Soc Lond B Suppl 270: S195-S197

Montgomery JC, Jeffs A, Simpson SD, Meekan M, Tindle C (2006) Sound as an orientation cue for the pelagic larvae of reef fish and decapod crustaceans. Adv Mar Biol 51: 143-196

Mukai Y, Kobayashi H (1994) Development of free neuromasts in larvae of cyprinid fish. Mem Fac Agric Kinki Uni 27: 1-14

Mukai Y, Yoshikawa H, Kobayashi H (1994) The relationship between the length of the cupulae of free neuromasts and feeding ability in larvae of the willow shiner *Gnathopogon elongatus caerulescens* (Telesostei, Cyprinidae). J Exp Biol 197: 399-403

Myrberg AA Jr, Fuiman LA (2002) The sensory world of coral reef fishes. In: Sale PF (ed) Coral Reef Fishes: Dynamics and Diversity in a Complex Ecosystem. Academic Press, Amsterdam, p 123-148

Neave DA (1985) The dorsal light reactions of larval and metamorphosing flatfish. J Fish Biol 26: 629-640

Neave DA (1986) The development of the lateral line system in plaice (*Pleuronectes platessa*) and turbot (*Scophthalmus maximus*). J Mar Biol Assoc UK 66: 683-693

New JG, Fewkes LA, Khan AN (2001) Strike feeding behavior in the muskellunge, *Esox masquinongy*: contributions of the lateral line and visual sensory systems. J Exp Biol 204: 1207-1221

Noakes DLG, Godin JJ (1988) Ontogeny of behavior and concurrent developmental changes in sensory systems in teleost fishes. In: Hoar WS, Randall DJ (eds) Fish Physiology, Vol X1B, Academic Press, New York, p 345-395

Northcutt RG (2003) Development of the lateral line system in channel catfish. In: Browman HI, Skiftesvik AB (eds) The Big Fish Bang. Proceedings of the 26th Annual Larval Fish Conference, Institute of Marine Research, Bergen, Norway, p 137-159

Nunan CP, Noakes DLG (1985) Response of rainbow trout (*Salmo gairdneri*) embryos to current flow in simulated substrates. Can J Zool 63: 1813-1815

Pankhurst PM, Butler P (1996) Development of the sensory organs in the greenback flounder, *Rhombosolea tapirina*. Mar Freshw Behav Physiol 28: 55-73

Partridge BL, and Pitcher TJ (1980) The sensory basis of fish schools: relative roles of lateral line and vision. J Comp Physiol 135: 315-325

Platt C (1983) The peripheral vestibular system of fishes. In: Northcutt RG, Davis RE (eds) Fish Neurobiology, Vol 1, Brainstem and Sense Organs. University of Michigan Press, Ann Arbor, p 89-123

Platt C (1988) Equilibrium in the vertebrates: signals, senses and steering underwater. In: Atema J, Fay RR, Popper AN, Tavolga WN (eds) Sensory Biology of Aquatic Animals. Springer-Verlag, New York, p 783-809

Poling KR, Fuiman LA (1997) Sensory development and concurrent behavioural changes in Atlantic croaker larvae. J Fish Biol 51: 402-421

Poling JR, Fuiman LA (1999) Behavioral specialization in developing sciaenids and its relationship to morphology and habitat. Environ Biol Fish 54: 119-133

Pohlmann K, Grasso FW, Breithaupt T (2001) Tracking wakes; the nocturnal predatory strategy of piscivorous catfish. Proc Natl Acad Sci USA 98: 7371-7374

Popper AN, Fay RR (1999) The auditory periphery in fishes. In: Fay RR, Popper AN (eds) Comparative Hearing: Fish and Amphibians. Springer-Verlag, New York, p 43-100

Popper AN, Hoxter B (1984) Growth of a fish ear: 1. Quantitative analysis of hair cell and ganglion cell proliferation. Hearing Res 15: 133-142

Popper AN, Rogers PH, Saidel WM, Cox M (1988) Role of fish ear in sound processing. In: Atema J, Fay RR, Popper AN, Tavolga WN (eds) Sensory Biology of Aquatic Animals. Springer-Verlag, New York, p 687-710

Rogers PH, Cox M (1988) Underwater sound as biological stimulus. In: Atema J, Fay RR, Popper AN, Tavolga WN (eds) Sensory Biology of Aquatic Animals. Springer-Verlag, New York, p 131-150

Rogers PH, Popper AN, Cox M, Saidel WM (1988) Processing of acoustic signals in the auditory system of bony fish. J Acoust Soc Am 83: 338-349

Salgado SD, Hoyt RD (1996) Early behaviour formation in fathead minnow larvae, *Pimephales promelas*: implications for sensory function. Mar Freshw Behav Physiol 28: 91-106

Simpson SD, Meekan MG, McCauley RD, Jeffs A (2004) Attraction of settlement-stage coral reef fishes to reef noise. Mar Ecol Prog Ser 276: 263-268

Simpson SD, Meekan M, Montgomery J, McCauley R, Jeffs A (2005) Homeward sound. Science 308: 221

Sokolowski BHA, Popper AN (1988) Transmission electron microscopy study of the saccule in the embryonic, larval, and adult toadfish *Opsanus tau*. J Morphol 198: 49-69

Stobutzki IC, Bellwood DR (1997) Sustained swimming abilities of the late pelagic stages of coral reef fishes. Mar Ecol Prog Ser 149: 35-41

Su J, Wang E (1990) The development of free neuromasts and lateral line system in the black sea bream, *Sparus macrocephalus* (Basilewsky). In: Hirano R, Hanyu I (eds) The Second Asian Fisheries Forum, Asian Fisheries Society, Manila, p 403-406

Swearer SE,Caselle JE, Lea DW, Warner RR (1999) Larval retention and recruitment in an island population of a coral-reef fish. Nature 402: 799-802

Tolimieri N, Jeffs A, Montgomery JC (2000) Ambient sound as a cue for navigation by pelagic larvae of reef fishes. Mar Ecol Prog Ser 207: 219-224

Tolimieri N, Haine O, Montgomery JC, Jeffs AG (2002) Ambient sound as a navigation cue for larval reef fish. Bioacoustics 12: 214-217

Tolimieri N, Haine O, Jeffs A, McCauley R, Montgomery J (2004) Directional orientation of pomacentrid larvae to ambient reef sound. Coral Reefs 23: 184-191

Webb JF (1989a) Gross morphology and evolution of the mechanoreceptive lateral-line system in teleost fishes. Brain Behav Evol 33: 34-53

Webb JF (1989b) Developmental constraints and evolution of the lateral line system in teleost fishes. In: Coombs S, Görner P, Münz H (eds) The Mechanosensory Lateral Line: Neurobiology and Evolution. Springer-Verlag, New York, p 79-97

Webb JF, Smith WL (2000) The laterophysic connection in chaetodonid butterfly fish: morphological variation and speculations on sensory function. Phil Trans R Soc Lond B 355: 1125-1129

Wysocki LE, Ladich F (2001) The ontogenetic development of auditory sensitivity, vocalization and acoustic communication in the labyrinth fish *Trichopsis vittata*. J Comp Physiol 187A: 177-187

Wright KJ, Higgs DM, Belanger AJ, Leis JM (2005) Auditory and olfactory abilities of pre-settlement larvae and post-settlement juveniles of a coral reef damselfish (Pisces: Pomacentridae). Mar Biol 147: 1425-1434

Yan HY, Fine ML, Horn NS and Colón WE (2000) Variability in the role of the gas bladder in fish audition. J Comp Physiol A186: 435-445

CHAPTER

11

Chemoreception

Kjell B Døving[1,*] and Alexander O Kasumyan[2]

INTRODUCTION

A common nominator for the chemosensory systems throughout the vertebrate phylum is the early appearance and functionality. Thus, the sense of smell and taste do function in new born infants and fish larvae alike. This point indicates the importance of these senses for early survival. Human babies <24 h, react to good and bad odours, and to sweet and sour tastes with readily interpreted facial expressions. These simple facts point out that chemical sense organs are functional, mediate information to the brain, and initiate the appropriate muscle activity. The grimaces are of course part of the communication between mother and child. Fishes do not display grimaces, but their chemosensory organs need to be developed at the early stages of life to be able to find food, to escape predators, and possibly to find the right place to shelter. In a multitude of species, the yolk-sac is small, the point of no return comes early, and the predators are many.

The chemical world in the aquatic environment is tremendously rich, and fishes have evolved sophisticated sensory systems to decipher this plethora of substances. The variation in form and function of chemosensory organs throughout the many thousands

Authors' addresses: [1]Department of Molecular Bioscience, University of Oslo, PO Box 1041 Blindern, N-0316 Oslo, Norway.
[2]Department of Ichthyology, Faculty of Biology, Moscow State University, R-119991, Russia.
**Corresponding author:* E-mail: K.b.doving@imbv.uio.no

of fish species is enormous and fascinating. It should be recalled that what we know is based on inspection of a small number of species, and that most species have not been studied. Thus, the present overview of the development is an attempt to cover some features of these systems, but many aspects and features are far from being known.

There are three chemosensory systems in fish, which can be distinguished due to form, distribution and function. The sense of smell has a localised peripheral organ on the snout. The cells that detect the chemical substances are primary sensory neurones, each with a tiny axon running in to a relay station. The sensory cells of the gustatory taste buds are secondary sensory receptor cells innervated by cranial nerves. The common chemical sense is the third chemosensory system. This system is presented by free nerve endings of the cranial and spinal nerves and by solitary chemosensory cells (SCC). These cells are also secondary sensory receptor cells innervated by cranial and spinal nerves. Their function remains mostly mysterious.

The most salient features of the **olfactory system** are that it can detect substances in minute quantities in nano- or picomolar concentrations; that activation of the sense of smell elicits distinct behaviours, and that these behaviours are related to essential life processes. Thus activation of the sense of smell can evoke reproductive behaviour, feeding and avoidance. The sense of smell has also been implied in the homing performance of many fish species, in particular salmonids (Stabell, 1984; Døving & Stabell, 2003).

The **gustatory system** detects chemicals at a somewhat higher concentration than the sense of smell. The substances activating this organ are all related to feeding. The substances can initiate food search, snapping of the jaws, swallowing, but also rejection and spitting out items wrongly accepted as food. The sensory structures of the gustatory system, the taste buds are heterogeneously spread in the oral cavity, but can also be found all over the body surface in many fish species.

The **SCC** are usually distributed over the entire body surface but can be confined in some species to fins and barbels. It has been shown that this system responds to particular substances such as mucus, but there are no certain ideas about its function.

In the following, the term "embryo" refers to the stage before hatching and the term "post fertilisation", pf and "post hatch", ph are frequently used. The term "prolarva" includes the stages between hatching and first-feeding, and "larva" from first-feeding until completion of metamorphosis. At this latter stage, the typical larval characteristics such as yolk-sac, fin-fold, etc. disappear, and fish adopt the shape and characteristics of the adult. TL is total length, BL body length and SL standard length.

OLFACTION

To achieve a high sensitivity to different types of odours, the olfactory organ must contain a large number of odorant receptors (OR) (Buck & Axel, 1991). To this end there must be a large number of sensory neurones expressing such receptors and that implies a large surface area. Sensory neurones are indeed many and varied, and often in the form of lamellae arranged in a rosette.

The other requirement for an efficient olfactory organ is that ambient water is made to pass over, or be transported through the organ. The "simplest" arrangemnent is found in garfish (*Belone belone*) where the sensory neurones are found on a fungiform structure situated in a shallow groove. Thus, swimming movements are sufficient to wash ambient water over the olfactory sensory neurones. In most fishes the mechanisms of water transport are more sophisticated, and fishes can be divided in two classes according to how water is transported through the peripheral olfactory organ. Water can either be pumped with the help of accessory sacs, or it can be transported in a steady stream with the aid of ciliary activity. The former are called the cyclosmates, and the latter isosmates (Døving *et al.*, 1977). In both types of fishes the construction of the olfactory cavities are masterpieces of miniature hydrodynamic creations. In the sea stickleback (*Spinachia spinachia*), a cyclosmate, there is only one opening to the olfactory cavity, and water passes in and out this opening via pumping mechanisms. Yet, the refined hydrodynamic construction of the olfactory cavity causes water to flush in over the epithelium only on the way into the cavity (Theisen, 1982). In other cyclosmates, with both anterior and posterior nares, there might be valves or flaps that permit water to pass only in one direction over the sensory organ. The anterior naris is seldom flush with the skin surface, but often extends exteriorly like a tube. This structure prevents boundary layer water immediately around the fish from entering the olfactory cavity (H. Westerberg. pers. comm.).

It is important to note that chemical stimuli do not have a vector. Thus, to be able to orient in relation to a chemical trail or towards a source, fishes must rely on other senses that can interpret water currents (Døving & Stabell, 2003). In this connection it should be remembered that the threshold for detection of infrasound, and thus linear acceleration is remarkably low in fishes, $\sim 1 \cdot 10^{-4}\ \mathrm{m \cdot s^{-2}}$ (Sand & Karlsen, 2000).

Recent developments on the functional anatomy and physiology of the fish olfactory system reveal three parallel pathways from the sensory epithelium via the olfactory bulb to the telencephalon. There are three morphological types of sensory neurones spread in a seemingly random arrangement in the olfactory epithelium. The axons of each type of sensory neurone converge to a specific region of the olfactory bulb and connect to a set of secondary or relay neurones. The axons of these relay neurones leave in three bundles to the telencephalon. Each bundle conveys specific information that elicits characteristic behaviour. One pathway is tuned to alarm substances and bile acids, another to sex pheromones, and the third to food odours (Hamdani & Døving, 2005). Evidence has been presented indicating that the ciliated sensory neurones are tuned to alarm and bile pheromones, the microvillous sensory neurones to food odours and crypt cells to sex pheromones (for review, see Døving *et al.*, 2005). As reproduction is a life process occurring at a late stage, it is probable that the sensory neurones involved in reproduction, i.e. the crypt cells, are not present or not mature at the early larval stages. But both ciliated and microvillous sensory neurones are present.

The first studies on the development of olfactory organs in fish were made about 100 years ago (Hoffmann, 1884; His, 1892; Keibel, 1893; Holm, 1894; Dieulaffé, 1905; Gawrilenko, 1910) and occurred after the first description of olfactory sensory neurones in fish (Schultze, 1856). Later on, many authors studied the development of the olfactory organ in a range of fish species (e.g. Watling & Hilleman, 1964; Zeiske *et al.*, 1976; Breucker *et al.*, 1979; Theisen *et al.*, 1980; Elston *et al.*, 1981; Derivot, 1984, and many others). During this period, studies concentrated on the gross morphology of fish olfactory organ (Kleerekoper, 1969). More detailed studies of the early stages of olfactory organ development have only become available during the last twenty years or so.

Embryogenesis

Olfactory placode

During development, different sense organs originate from different regions of the ectoderm. The process is illustrated in Fig. 1. During development of fish embryos (see Hall, this volme), part of the ectoderm called the neural plate forms a solid cylinder which secondarily forms a hollow tube. This tube gives rise to the central nervous system (CNS). The sensory organs or more specifically, the peripheral components of these systems in the vertebrate head arise from structures called placodes. The term placode refers to ectodermal thickenings, which originate during early embryogenesis. These thickenings are located in the head of embryos and give rise to various organs and components of the peripheral nervous system. The most rostral pair of placodes are the olfactory placodes, the "anlage" of the olfactory organ (Kupffer, 1894).

Placodes form from the ectoderm adjacent to the neural plate. In zebrafish (*Danio rerio*), the olfactory placode becomes microscopically visible at the stage when the developing eye evaginates from the diencephalon forming the optic vesicles (12-13 h pf) (Hansen & Zeiske, 1993). Placodal cells appear in the rostral and slightly ventral region where they fill the space between the eye and the forebrain. At this stage, a placode resembles a more or less continuous line of cells under the epidermis. This subepidermal layer of cells differs considerably from the very flat cells of the epidermis, and skirts the anterior and lateral sides of the head of the embryo. The cells do not interdigitate with the epidermal cells, but are usually widely separated from the epidermis and have large intercellular spaces. There is no basal lamina under the epidermis or the subepidermal cell layer. Two or three hours later (14-16 h pf), the placode appears as an aggregation of cells under the epidermis. Placodal cells show a higher electron density compared to epidermal cells, but are less electron-dense than brain or eye cells. The nuclei of the placodal cells are round and dark, whereas epidermal cell nuclei are long, oval and very light. Brain and eye nuclei are only slightly lighter than placodal nuclei, but very variable in shape. The placodal cells are loosely packed with large intercellular spaces surrounded by a dense extracellular matrix. The cells have numerous small pseudopodia-like extensions that make contact with other placodal cells. Occasionally there are rudimentary cilia, which are covered by a sheath, but do not seem to reach the adjacent cell surface (Hansen & Zeiske, 1993).

During the following hours of development, placodal cells increase in number and begin to fill the space between the developing forebrain and the eyes. By 24 h pf, the shape of the olfactory placode is oval in a rostrocaudal direction and semicircular in the dorsoventral direction, while the epidermis still consists of only one layer of cells with microridges beginning to form at the cell surface. The surface above the placode is completely closed, and there is no sign of cell lysis or degeneration in the epidermis. The junctional complex becomes now more distinct, but the epidermis and placodes do not yet show a basal lamina. The placodal cells become more closely packed, and nearly all cells facing the forebrain emit axonal processes. Some axons have already invaded the forebrain. Once the axons of the future sensory neurons have reached the brain, the placodal cells begin to elongate. About 26-27 h pf, almost all perikarya in the medial region of the placode are long and slender, reaching from the base of the placode to the epidermis. At the base of the placode the basal lamina begins to form. The space between the placode and the brain is now completely filled with long, slim cells between which bundles of axons grow towards the forebrain (Hansen & Zeiske, 1993).

The olfactory placodes in zebrafish develop in parallel with the olfactory bulbs (Whitlock & Westerfield, 2000). To understand how the placodes were formed, Whitlock and Westerfield (2000) injected single cells with lineage tracer dyes at the anterior end of the neural plate prior to the morphological appearance of the olfactory placode. The fates of these cells were followed until after the olfactory placode formed. A large field of cells was seen to be widely distributed along the edge of the anterior neural plate forming the olfactory placode (Fig. 1). This field of cells is regionalised such that cells in specific anterior-posterior locations give rise to specific parts of the developing olfactory organ.

Cells adjacent and medial to this olfactory placode field give rise to the telencephalon. This telencephalic field is also regionalised along the anterioposterior axis, such that the anterior olfactory bulb arises from cells adjacent to the region from which the first sensory neurones derive. Thus, in contrast to the hypothesis that the olfactory placodes arise from groups of cells that separate from the neural plate, Whitlock & Westerfield (2000) show that the placodes arise from large fields of cells along the anterior-lateral neural plate. They suggested that the placodes form by anterior convergence of these cellular fields and point out that as early as the 4-somite stage (12 h pf), olfactory fates are already regionalised within the developing neural plate.

Olfactory pit and nostrils

In the embryo at 30-32 h pf, epidermal cells, which cover the placode begin to separate, and immediately under the epidermis many long pseudopodia-like extensions of placodal cells appear. At 34-36 h pf, a small slit indicates the olfactory pit. The slit is surrounded by small spherical cells of the epidermis, and soon after the slit opens, the surface of the placode-bearing microvilli and kinocilia becomes visible. The basal lamina of the olfactory placode as well as of the forebrain is easily perceptible.

The developing olfactory pit continues to widen, and by 48-50 h pf, its shape is oval to round. Olfactory receptor cells with knobs bearing cilia, microvillous receptor cells,

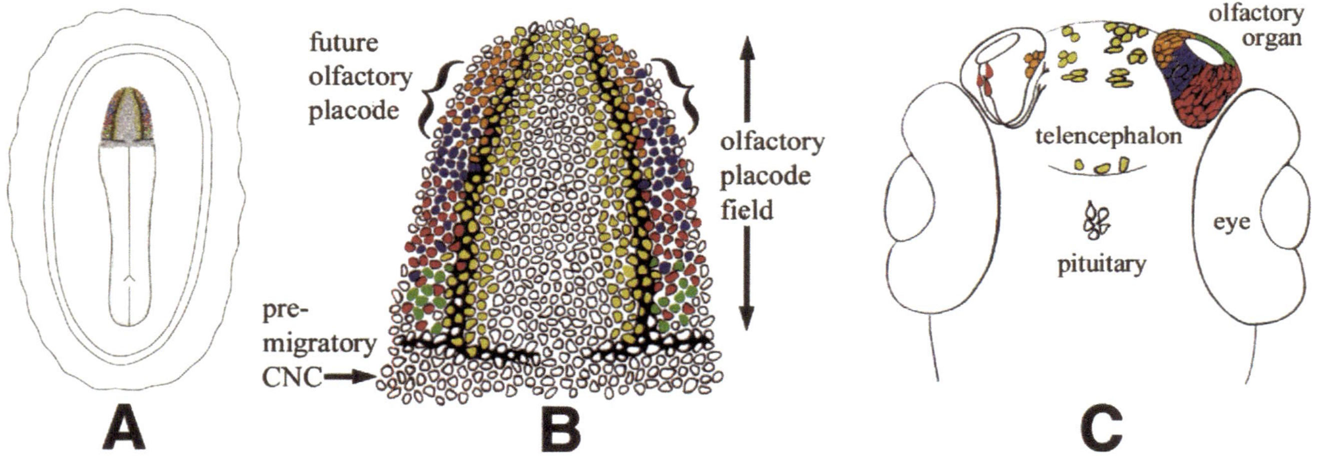

Fig. 1 Origin of the olfactory placode. A, dorsal view of a zebrafish (*Danio rerio*) embryo 12 h pf. The coloured area represents the olfactory placode field enlarged in B. B, the olfactory placode field 12 h pf. The blackened border extending along the anteroposterior axis represents cells giving rise to the telencephalon. The coloured cells represent the fields of cells giving rise to the olfactory organ. The transverse blackened border represents the anterior limits of premigratory cranial neural crest. The cells are colour-coded with respect to their future fates in the developing olfactory organ and bulb at 50 h pf. C, schematic view of a 50 h pf zebrafish head with anterior towards the top of the page. Right-hand olfactory organ: the olfactory organ has been subdivided into four coloured areas. Clones in these areas were colour-coded in diagram B. Left hand olfactory organ: sensory neurons with axons extending into the CNS at 50 h pf come predominantly from red cells, pioneer neurones come from orange cells but only in the region abutting the telencephalic field anteriorly. The pituitary clones come from the non-coloured cells most anterior in the neural plate between the two blackened borders. After Whitlock & Westerfield (2000), Fig. 3. Reprinted with permission of The Company of Biologists.

supporting cells, and nonsensory cells with long kinocilia can be distinguished in the epithelium. At the base of the placode, occasionally cells are seen to penetrate the newly formed basal lamina, thus reaching into the space between the placode and the forebrain. The axons of the receptor cells aggregate, forming the olfactory nerve.

At hatching, which occurs between 2.5-4 days pf in zebrafish, cilia and kinocilia develop and the surface of the olfactory epithelium is obscured. The ciliated nonsensory cells are densely arranged especially at the rim of the olfactory pit, and build a thick and dense fringe of kinocilia. Soon after hatching about 2/3 of the olfactory organ is still covered by the epidermis (Fig. 2) (Hansen & Zeiske, 1993).

The olfactory organs remain as invaginated olfactory pits after hatching (Fig. 2). Further development of the olfactory apparatus occurs during the larval stage. The development of the primary olfactory pit, which occurs by invagination, initially forms a shallow pit that progressively deepens, and is found in selachians (Berliner, 1902), sturgeons (Zeiske *et al.*, 2003), and cyprinids (Pashchenko & Kasumyan, 1983, 1986). It is generally considered a plesiomorphic character (Kleerekoper, 1969; Zeiske *et al.*, 1992), as opposed to the derived mode found in gars (*Lepisosteus* sp.) (Balfour & Parker, 1882), in viviparous tooth carps (Poeciliidae) (Reinke, 1937; Zeiske *et al.*, 1976), and in the Northern pike (*Esox lucius*) (Devitsina, 1998). In these distantly

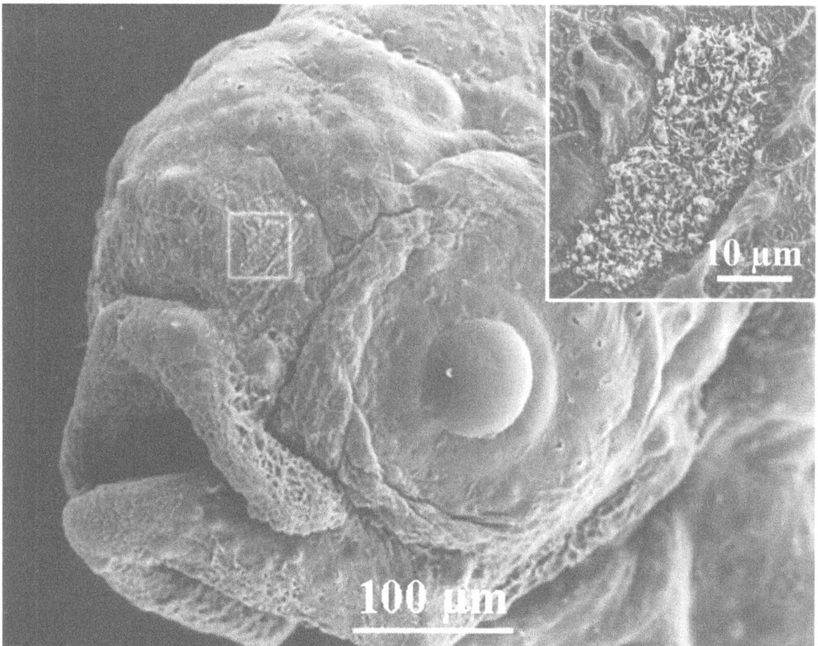

Fig. 2 SEM of a cod larva. The head of a day 4 ph Atlantic cod (*Gadus morhua*) larvae. The olfactory pit is seen inside the rectangle in front of the eye, and is enlarged in the inset. After Døving *et al.* (1994).

related actinopterygian fishes, an internal cavity forms first which later opens to the exterior.

Formation of the external nares starts with processes growing from medial and lateral margins of the olfactory groove. These processes grow towards each other and fuse to build an isthmus, or bridge above the olfactory groove elongating in the anterior-posterior direction. In Arctic grayling (*Thymallus arcticus*), this process starts around day 15 and ends at day 67 pf or day 45 ph (Watling & Hilleman, 1964). In the Arctic charr (*Salvelinus alpinus*) nostrils are formed between day 10 and day 50 ph when the larval BL increases from 16 to 20 mm (Olsén, 1993). In the Northern pike (Devitsina, 1998), the formation of the two nares is complete at the end of metamorphosis. Similar findings have been observed for zebrafish and other cyprinid species, e.g. European minnow (*Phoxinus phoxinus*), shemaya (*Chalcalburnus chalcoides*), grass carp (*Ctenopharyngodon idella*), silver carp (*Hypophthalmichthys molitrix*), common carp (*Cyprinus carpio*), bitterling (*Rhodeus sericeus amarus*) (Pashchenko & Kasumyan, 1983, 1986; Hansen & Zeiske, 1993; Kasumyan, 2004). In sturgeons, the formation of the anterior incurrent and posterior excurrent nostrils is completed before the onset of exogenous feeding (Fig. 3) (Zeiske *et al.*, 2003). Later, the incurrent naris become circular and the wall at the anterior end of each incurrent nostril and the nasal bridge expand to form an upright funnel.

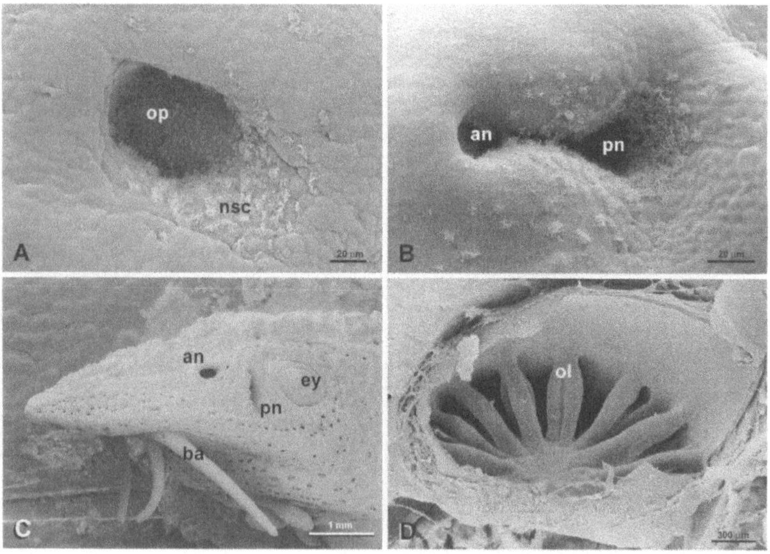

Fig. 3 Formation of the olfactory organ in the Siberian sturgeon (*Acipenser baerii*). A, lateral view of the olfactory pit at 132 h pf. B, formation of the nostrils at day 11 pf. C, head of a juvenile at day 33 pf (35 mm TL) showing the position of the anterior and posterior naris in front of the eye. D, the olfactory rosette with lamellae in a juvenile at day 52 pf. Abbreviations: an, anterior nostril; ba, barbel; ey, eye; nsc, non-sensory ciliated epithelium; ol, olfactory lamellae; op, olfactory pit; pn, posterior naris. After Zeiske *et al.* (2003), Fig. 4. Reprinted with permission of Springer Science and Business Media.

A lack of symmetry in nostril formation is often observed. In flatfishes, the olfactory organ at the ocular side of the body develops more rapidly than that on the abocular side (Harvey, 1996). In artificially propagated sturgeon larvae, incomplete olfactory bridge formation is found in 7 up to 52% of the individuals (Schmalhauzen, 1962; Goryunova *et al.*, 2000). The older juveniles or adult sturgeons do not seem to regain nostrils once these structures are impaired.

The formation of the nasal cavity and the anterior and posterior nostrils from the original single pit is considered to be phylogenetically primitive (plesiomorphic) among actinopterygians (Bertmar, 1969; Zeiske *et al.*, 1992). It closely resembles the formation of the naso-buccal grooves and ventilatory openings of the olfactory organ in Dipnoi (Bertmar, 1965; Bartsch, 1993), and the incomplete separation of incurrent and excurrent opening in Chondrichthyes (Berliner, 1902). This type of formation has also been described for sturgeons (Devitsina & Kazhlaev, 1993b), paddlefish (*Polyodon spathula*) (Ballard & Needham, 1964; Bemis & Grande, 1992; Devitsina & Kazhlaev, 1993a) and the bowfin (*Amia calva*) (Dean, 1897) as well as for many other teleosts (e.g. salmonids, cyprinids, atherinids, flatfishes) (Hoffmann, 1884; Holm, 1894; Gawrilenko, 1910; Reinke, 1937; Teichmann, 1964; Breucker *et al.*, 1979; Soin *et al.*, 1981; Appelbaum *et al.*, 1983; Pashchenko & Kasumyan, 1986; Hansen & Zeiske, 1993; Olsén, 1993; Appelbaum & Riehl, 1997). A second type of nostril formation, characterised by separate development of the two apertures, is found in two species of a viviparous cyprinidonts, sparkling panchax (*Aplocheilus lineatus*) and swordtail (*Xiphophorus helleri*) (Reinke, 1937; Zeiske *et al.*, 1976). In a third type, found in the garfish, the olfactory anlage gradually elongates into a groove but the olfactory cavity remains open (Sewertzoff, 1931; Theisen *et al.*, 1980).

Olfactory epithelium

While ultrastructural features and cellular compositions of the adult olfactory epithelium have been surveyed in many major teleost orders, only scattered information is available for its ontogeny in the Actinopterigii. The formation of the olfactory epithelium is very different in Teleostei, exemplified by the zebrafish, and the Chondrostei exemplified by two sturgeon species, sterlet sturgeon (*Acipenser ruthenus*) and Siberian sturgeon (*Acipenser baerii*). In zebrafish, sensory neurons and their supporting cells and ciliated non-sensory cells originate from a single cell layer, which is located between epidermis and the developing brain. During development, the subepidermal layer thickens to form the paired, typically single-layered olfactory placodes of this teleost (Hansen & Zeiske, 1993, 1995). In contrast, the olfactory placodes of sturgeons consists of two morphologically different cellular layers: a superficial non-neural layer, and a deeper neural layer. The latter layer is close to the neural tube and gives rise to the sensory neurons whereas supporting and ciliated non-sensory cells derive from epithelial surface cells (Zeiske *et al.*, 2003). This type of formation resembles that seen in the amphibian, the African clawed frog (*Xenopus laevis*) (Klein & Graziadei, 1983; Costanzo & Graziadei, 1987; Reiss & Burd, 1997). It was assumed that the type found in sturgeons and African clawed frog (the latter representing the sarcopterygian outgroup), is phylogenetically the plesiomorphic type, provided that similar processes did not evolve independently. However, detailed investigations of other actinopterygian and sarcopterygian fishes as well as amphibians are clearly necessary for a better understanding of the variations in early development of the olfactory epithelium in lower vertebrates.

Olfactory sensory neurones

At the ultrastructural level, the olfactory epithelium in fish contains three distinct olfactory sensory cell types: ciliated, microvillous, and crypt sensory neurones. In many teleost species studied, the ciliated sensory neurones ontogenetically precede the microvillous sensory neurones and often dominate in olfactory epithelium at larval stages (Breucker *et al.*, 1979; Evans *et al.*, 1982; Zielinski & Hara, 1988; Hansen & Zeiske, 1993; Zippel *et al.*, 1997; Devitsina, 1998). In rainbow trout (*Oncorhynchus mykiss*), the difference in appearance of ciliated and microvillous sensory neurones is about one week. The free epithelial surface of the two sensory cell types is formed in different manners. Ciliated sensory cell maturation includes olfactory knob expansion, centriole migration into the olfactory knob, centriole division, and a sequential growth of cilia. The free surface of the microvillous cell forms from a narrow dendritic apex that carries a small number of microvilli. As the number of microvilli increases, a low olfactory vesicle is formed. The newly appeared ciliated sensory neurones have relatively short cilia on the knob, and cilia grow sequentially as well as microvilli on the knob of microvillous sensory neurones. As the sensory cells of both types mature, the dendrite becomes shorter and the cytoplasm becomes less electron dense (Zielinski & Hara, 1988). It was suggested that olfactory sensory neurones, with different ratios of cilia and microvilli as described by Pyatkina (1976) and by Devitsina and Kazhlaev (1993b) for sturgeon species, merely represent different developmental stages of maturing ciliated sensory neurones, rather than distinct types of olfactory sensory neurones (Zeiske *et al.*, 2003).

Crypt sensory neurones are present in basic as well as in highly derived actinopterygians (Hansen & Finger, 2000). The crypt cells are found in the olfactory epithelium of juveniles, but not in the embryos or larvae (Zeiske *et al.*, 2003). However, this does not necessarily mean that they are absent at these early stages. Crypt sensory neurones are scarce and not visible in SEM, and they can easily escape detection in thin TEM sections.

In the olfactory epithelium of sea lamprey (*Petromyzon marinus*), the ciliated sensory neurones appear in larval ammocoetes and the density of sensory neurones in this stage is higher than in juveniles and upstream migrants (VanDenbossche *et al.*, 1995).

Lamellae and olfactory rosette

The first lamella appears as a fold in the posterior part of the floor of the olfactory groove. This fold becomes the first olfactory lamella in salmonid fish. In cyprinid larvae this first fold differentiates into the midline raphe; the posterior end of the first fold becomes the ordinary unpaired lamella with a short proximal part. In Northern pike, the median ridge starts growing in prolarvae at day 7-8 ph, just after first-feeding (Devitsina, 1998). The first lamella starts to differentiate 4 weeks after startfeeding (day 70 ph, 31-37 mm TL) in Arctic charr (Olsén, 1993) and 3 weeks (day 42 ph, 24 mm TL) in Arctic grayling (Watling & Hilleman, 1964).

New olfactory lamellae appear in the posterior part of the rosette, anterior to the first lamella. The sequence in which the second and following lamellae appear is different in different groups of fishes. In salmonids, subsequent lamellae appear on the opposite side of the first lamella and the time difference between the appearances of the following lamellae can take from 2 to 15 days in larvae, but and in fingerlings may take up to months. The sequence of lamellar appearance might be different in left and right organs as found in Arctic grayling (Watling & Hilleman, 1964). In cyprinid species new lamellae are simultaneously formed as symmetrical pairs on both sides of the midline raphe. Each new pair of lamellae is formed anterior and adjacent to the midline raphe (Pashchenko & Kasumyan, 1986).

The number of primary lamellae is related to age. It is noteworthy that in the elasmobranch, Oman shark (*Iago omanensis*), the first 12-14 olfactory lamellae appear simultaneously in embryos and the number of lamellae quickly increase with growth reaching up to 30-32 lamellae in adult specimens (Table 1) (Fishelson & Baranes, 1997). In salmonids the number of lamellae remains constant after the fish has reached a certain age, one year in rainbow trout (10-17 lamellae) (Halama, 1982), 18 months in Arctic charr (14-15 lamellae) (Olsén, 1993) and 3-4 years in Arctic grayling (14-15 lamellae) (Fig. 4) (Watling & Hilleman, 1964).

In cyprinids, the number of primary lamellae increases with age (Pashchenko & Kasumyan, 1986) (Fig. 4). Likewise, the number of lamellae increases with age in sunfishes (Centrarchidae, Elassomatidae). But in the banded pygmy sunfish (*Elassoma zonatum*), the folds stop to develop at about 23 mm BL. These features have been interpreted as a sign of neotony as the fish attains sexual maturity while many

Table 1 Age-related changes in gross morphology of olfactory organ in Oman shark (*Iago omanensis*) (Fishelson & Baranes, 1997).

Body length, cm	*Rosette length*, mm	*Distance between nostrils*, mm	*Number of lamellae*
2.4-2.6	0.8	1.8	12
3.4-3.8	2.0	2.0	15
4-5	2.3	2.2	16
6-7	2.6	4.2	18
8-10	2.6	5.0	20
12-14	3.2	6.6	23
16-17.5	4.0	7.5	25
40-50	8.0	16.0	29
60-70	10.0	18.5	30

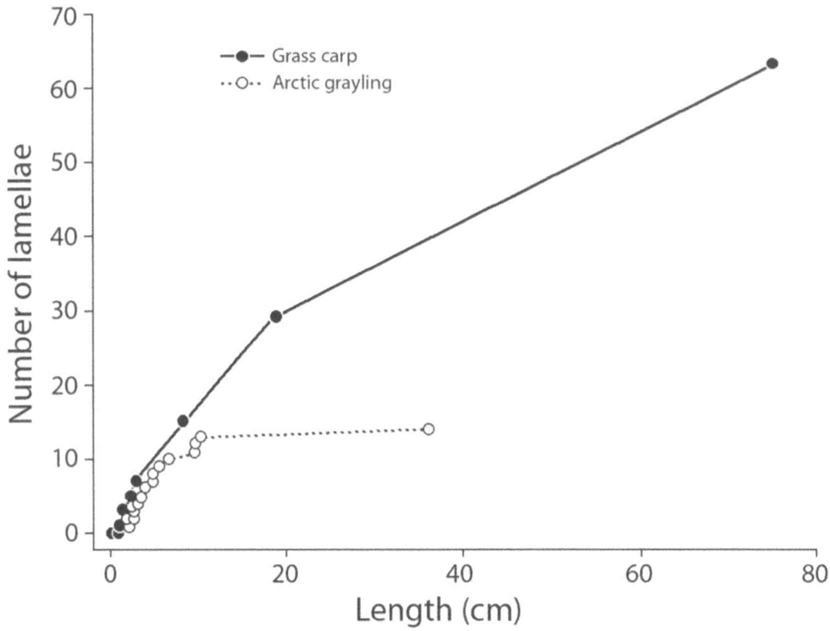

Fig. 4 Development of lamellae in the olfactory rosette. The graph shows the number of lamellae in the olfactory rosette as a function of the fish total length (TL) for grass carp (*Ctenopharyngodon idella*) (Pashchenko & Kasumyan, 1986) and Arctic grayling (*Thymallus arcticus*) (Watling & Hilleman, 1964). Note that in the arctic grayling the number of lamellae does not increase beyond a fish total length of about 12 cm.

morphological features still resemble those of larvae or juveniles (Eaton, 1956). Increasing the number and size of lamellae increases the size of the rosette with fish age (Table 2).

Secondary lamellae

In different fishes such as sharks and salmonids, the primary olfactory lamellae develop secondary lamellae. The function of these structures is obscure, but might be related to the passage of water through the rosette. In Northern pike, the first secondary lamellae can be seen at the end of metamorphosis at 30 mm BL (Devitsina, 1998). In Arctic grayling, secondary lamellae form in 4 month-old fish when the rosette has 10 lamellae, 4 months old fish. In Arctic charr and brown trout (*Salmo trutta*), secondary folding of the lamellae starts much later, when the fish is about 1.5 years old. The process starts in caudal lamellae, and in old fish caudal lamellae are the most strongly folded. In the 7 year-old Arctic charr, the 3 small anterior lamellae lack secondary foldings. Both ciliated and microvillar sensory neurons are located in the depressions between the secondary lamellae (Watling & Hilleman, 1964; Bertmar, 1972; Olsén, 1993). In salmonids the appearance of secondary lamellae coincides with fish age, when the

Table 2 The age-related changes in gross morphology of olfactory organ in grass carp (*Ctenopharyngodon idella*) (Pashchenko & Kasumyan, 1986).

Fish TL, cm	*Olfactory organ length*, mm	*Olfactory organ width*, mm	*Number of lamellae*	*Total surface of olfactory lamellae*, mm^2
0.46	0.17	0.14	0	0.005
0.75	0.25	0.21	0	0.017
0.85	0.29	0.25	0	0.052
1.02	0.33	0.27	0	0.068
1.36	0.43	0.32	1	0.11
1.76	0.60	0.53	3	0.26
2.4	0.80	0.70	5	0.6
3.08	1.0	0.83	7	1.5
8.5	2.5	2.4	15	14.1
19.0	5.4	5.3	29	112.8
75.0	10.0	8.5	63	773.7

development of new primary lamellae is completed (Fig. 4). In elasmobranchs, the secondary folds become visible at the time the primary lamella forms. The number of secondary folds on one lamella gradually increases and reaches 30 in adult Oman shark (Fishelson & Baranes, 1997).

Goblet cells

Goblet cells produce mucus which can help to remove particles entering the olfactory cavity or aid the water transport in the isosmates (Døving *et al.*, 1977). It has been hypothesised that mucus also plays an important role in chemoreception and that molecules of odorants are first introduced into the mucus and only then reach the receptor areas in the cellular membrane of cilia and microvilli. The mucus creates optimal conditions for the molecular processes involved into the interaction between the odorants and the receptor proteins (Sorensen & Caprio, 1998). Goblet cells appear in the olfactory epithelium much later than in the epidermis of zebrafish (Hansen & Zeiske, 1993) and prior to the formation of the lamellae, on day 19 pf in Arctic grayling (Watling & Hilleman, 1964). At this time, goblet cells are scattered within the epithelium, but a few extend to the surface of the epithelium. With development of lamellae, many goblet cells are confined to the free surface of the epithelium, but there are no patches in goblet cells distribution on the lamellar surface. When the lamellar secondary ridges arise, goblet cells continue to proliferate on the free margins of ridges (Watling & Hilleman, 1964).

Accessory sacs

The olfactory organ in some cyclosmates possess up to 4 accessory sacs extending from different parts of the olfactory cavity. They pump water through the olfactory organ.

Accessory sacs appear quite early in telesost ontogeny, earlier than the appearance of the first lamellae or formation of separate nostrils, as found in Arctic grayling and in pleuronectids (Reinke, 1937; Watling & Hilleman, 1964). All sacs do not form at the same time. In Arctic grayling larvae, the ventral accessory sac is formed on day 27 pf as of a blunt caudo-ventrally directed diverticulum from the posterior end of the nasal chamber. The first indication of the dorsal accessory sac appears one month later, as an out-pocketing which is directed from the posterior wall of the nasal chamber in the dorsal and posterior direction. The origin of this diverticulum is near the anterior limit of the eye orbit. The size of ventral and dorsal sacs increases with fish size and they reach 4-7 mm in length and 1-3 mm in diameter at their respective bases when the Arctic grayling is 11 cm BL (Watling & Hilleman, 1964).

Olfactory organ migration

Several earlier studies describe the movement of the olfactory groove from the ventral to the dorsal side of the head in salmonids (His, 1892; Keibel, 1893; Holm, 1894; Price, 1934; Jahn, 1972). It was suggested that the olfactory groove moves from a ventral to a dorsal position along the side of the snout (Gawrilenko, 1910) or forward over the terminal end of the snout (Watling & Hilleman, 1964). However, the movement of the olfactory groove is not a true migration because the olfactory groove seems not to change its position in relation to the brain and is situated in its ultimate location already when the placode is first seen. As the embryo's head gradually detaches from the yolk-sac and the jaws develop, the olfactory organ is pushed forward and upward together with the snout as a part of the head (Pashchenko & Kasumyan, 1983, 1986). In sharks the olfactory grooves are formed on the dorso-lateral part of the head as in teleosts and at a later stage they move to their final, ventral position (Fishelson & Baranes, 1997).

Pioneer neurones and formation of the olfactory nerve

Much of the work on the development of the fish olfactory system has focused on the sensory epithelium and olfactory organ but relatively few studies have examined the formation of the olfactory nerve, the bulb, and the olfactory tract and forebrain centres.

In Arctic charr, axons of the olfactory sensory neurones reach the olfactory bulbs (day 10 ph, 16 mm TL) (Olsén, 1993). The appearance of an identifiable olfactory nerve extending from the placode to the brain precedes the differentiation of the free surface of the sensory neurone dendrite, and it has been suggested that their synapses in the olfactory bulb are associated with the final differentiation of the sensory cells (Zielinski & Hara, 1988). As the fish grow the olfactory nerve increases in length. The addition of more fibres from the olfactory epithelium increases also the diameter of the olfactory nerve. In the white sucker (*Catostomus commersoni*), a cyprinid fish, the olfactory tract connecting the olfactory bulb with the telencephalon, is present as a thin fascicle at hatching. But two days later the olfactory tract seems 'solid' (Werner & Lannoo, 1994).

Olfactory nerve formation was studied in zebrafish by Whitlock & Westerfield (1998). It was found that the first axons projecting from the olfactory placodes and reaching the central nervous system (CNS) belong to pioneer neurones but not to the sensory neurones. The pioneer cells are the precocious class of neurones which precede the olfactory sensory neurones. These cells are antigenically, morphologically, spatially and developmentally distinct from olfactory sensory neurones. The pioneer neurones arise from a more anterior region of the lateral neural plate than do the first olfactory sensory neurones and originate from a distinct lineage that does not include olfactory sensory neurones. The pioneer cells do not to express any of the known OR genes. Rather the pioneer cells establish initial contact with the telencephalon, and then they undergo apoptotic death. Axons of pioneer cells first emerge from the olfactory placode at 24 h pf (Fig. 5). At this time and during the second day of development there is an average of 11 ± 2 pioneer neurones in each placode in their basal-medial region. These neurones had large round cell bodies and axons that extended along the lateral edge of the telencephalon. By 30 h pf, the axons of the pioneer neurones had extended into the telencephalon and had started to defasciculate and form axonal condensations. By 38 h pf, axons penetrated the surface of the telencephalon and extended into its most anterior region, the presumptive ganglia of what will become the olfactory bulb. The zns-2, a mouse monoclonal antibody, labelling pioneer neurone cell bodies in the olfactory epithelium diminishes greatly by 42-44 h and is no longer detectable beyond 48 h pf, although axonal labelling remained strong. The first sensory neurones extend axons into the developing olfactory bulb after the pioneer neurones had formed axonal condensation within the olfactory bulb. Some of the cells in the initial cluster of four to six sensory neurones had dendrites but no extended axons. As was shown in the same preparation, the sensory axons and the pioneer axons overlap with one another in the telencephalon. The axons of sensory neurones are proximal to the axons of the pioneer neurones at the exit of the olfactory nerve, and the axons of both cell types are co-localised at their terminations within the presumptive olfactory bulb. This indicates that, once leaving the olfactory organ, the growth cones of the later developing sensory neurones follow the pioneer axons (Whitlock & Westerfield, 1998).

The pioneer neurones form a transient population of cells. They undergo apoptosis as the axons of the first sensory neurones reach the differentiating olfactory bulb, around 48 h pf in zebrafish. The ablation of pioneer cells before axonogenesis (18-20 h pf) evokes severe mis-routing by the sensory axons that follow. The mis-routed sensory axons typically extend posteriorly toward the anterior or postoptic commissures, where they stray extensively in the dorsoventral plane, a trajectory never observed in normal embryos. If the pioneer neurones were ablated after they had establish axonal condensations in the telencephalon and the axons of sensory neurones had started growing into the CNS (38-48 h pf), there was no effect on axonal routing. This strongly suggests that pioneer neurones are necessary for sensory axon guidance from the olfactory placode to the region of the telencephalon in which the olfactory bulb will develop (Whitlock & Westerfield, 1998).

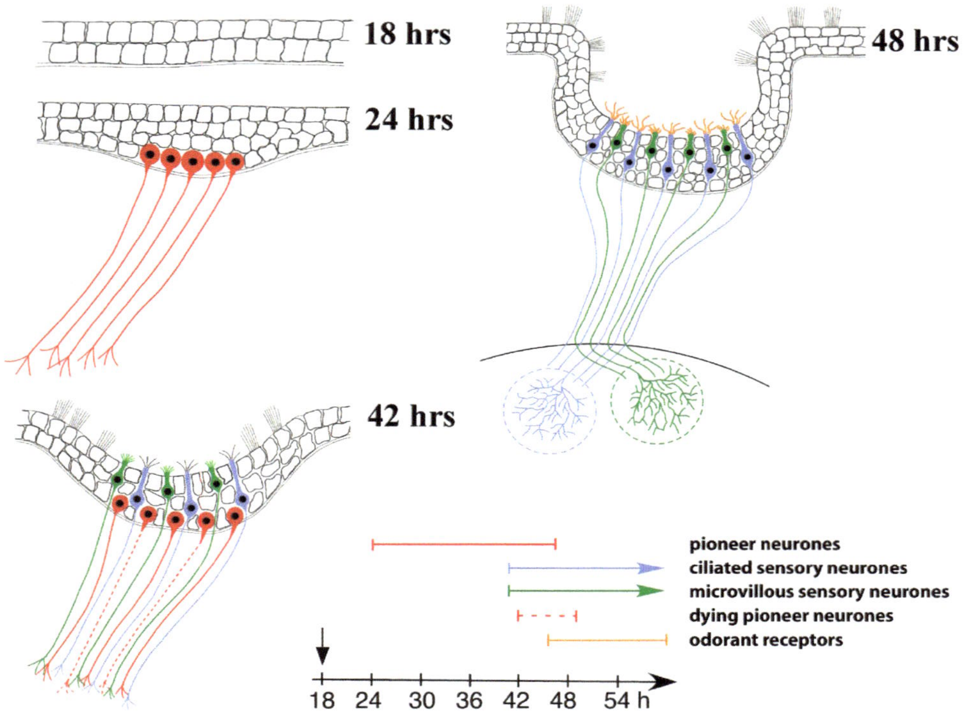

Fig. 5 Schematic illustration of the formation of the olfactory nerve. Illustration of how the axons of the sensory neurones connect to the olfactory bulb in zebrafish at 18, 24, 42, and 48 h pf. The arrow indicates the time when the olfactory placode becomes apparent with Nomariski optics in the live embryo. The pioneer cells (in red) establish the first contacts with the olfactory bulb, and then the ciliated and the microvillous sensory neurons differentiate and establish permanent contact with the bulb.

Olfactory bulb, tract, and secondary olfactory centres

The olfactory nerve layer of the olfactory bulb is relatively larger in juveniles than in adult fish (Poling & Brunjes, 2000). Neuronal differentiation in the developing olfactory bulb in zebrafish occurs during the second day of development judged by the mature neurochemical profile and sensitivity to an ionotropic NMDA glutamate receptor agonist (Sakata *et al.*, 2003).

In zebrafish, olfactory tract neurones can be traced to the dorsal telencephalon as early as one day pf (Wilson *et al.*, 1990). Telencephalic nuclei develop from three bilaterally symmetrical cell clusters, which are present by 16 h pf. At this time, axons interconnecting these three cell clusters are also present (Ross *et al.*, 1992). The number of axons of the olfactory nerve, the surface area of the bulb, and the number of mitral cells, as well as the average convergence of axons onto the cells of the bulbar relay (axons/mitral cells) increase exponentially late in the ontogeny of West African lungfish (*Protopterus annectens*), from 10 to 44 cm of fish BL (Table 3) (Derivot, 1984).

Table 3 Quantitative data concerning the olfactory system of West African lungfish (*Protopterus annectens*) (Derivot, 1984)

Fish TL, cm	*Number of lamellae*	*Surface area of medial lamella, mm^2*	*Axon density in olfactory nerve, μm^{-2}*	*Axons of olfactory nerve, ×1000*	*Mitral cells, ×100*	*Mitral cells density, μm^{-2}*	*Convergence*
10	11	0.41	11	190	54.5	7350	35/1
13	17	0.79	12	400	85	6200	47/1
16	24	0.89	9	950	91	5000	104/1
21	29	0.96	7	1200	106.5	4050	113/1
28	31	1.70	10	1700	130.5	4150	130/1
36	36	2.20	6	2200	146	3650	151/1
41	41	3.01	11	2750	149	2700	185/1

There are impressive increases in the volume of the olfactory bulb during growth from juvenile to adult in the Chinook salmon (*Oncorhynchus tshawytscha*) (Jarrard, 1997) (Fig. 6). The thickness or volume of the layers of the olfactory nerve and the glomeruli indicate the approximate number of axons entering the olfactory bulb. In chinook salmon, the volume of these layers increases from about 31 % in the juveniles, to 56 % at the end of the parr-smolt transformation to reach 74 % of the olfactory bulb in the adult (Jarrard, 1997). A similar change is also observed in the zebrafish as the volume of the olfactory nerve and the glomeruli are larger in the adult than in the juveniles (Poling & Brunjes, 2000). It would be of interest to observe what types of sensory neurones are responsible for these remarkable changes.

The morphogenesis of the "olfactory part" of the brain occurs concomitantly with the developments of the olfactory organ. The allometric growth of the peripheral and central part of the olfactory system is prominent in Oman shark. For example, in

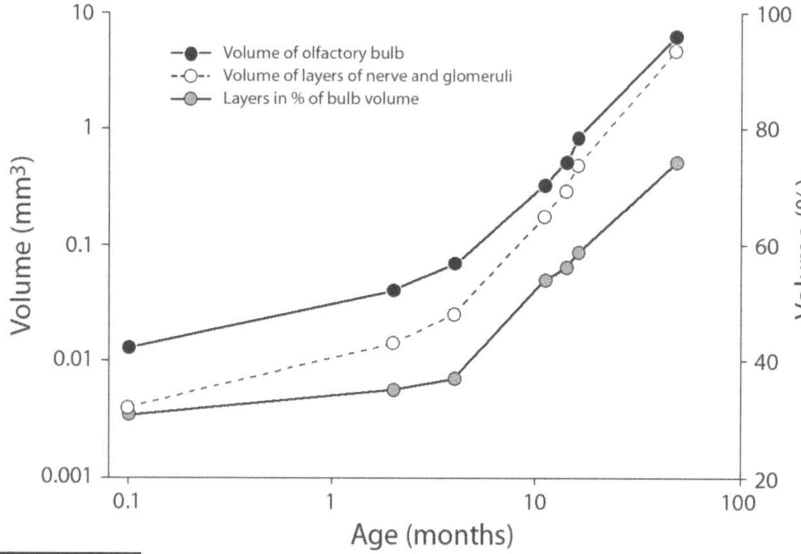

Fig. 6 The growth of the olfactory bulb in Chinook salmon (*Oncorhynchus tshawytscha*). The graph illustrates the volume of the olfactory bulb (solid line; closed circles), and of the layers of the nerve and glomerulus (dotted line; open circles) from the age of 0.1 (juvenile) to 48 months (adult). Note the jump in volumes at the parr-smolt transformation at 11-16 months. The lower curve (solid line; grey circles) gives the volume of the nerve and glomerulus layers in percent of the bulb volume. Data from Jarrad (1997).

embryos of 7-8 cm TL, the weight ratio between the olfactory organ and olfactory bulb is 0.8-0.9; in embryos of 12-15 cm TL, the ratio is 1.6-1.8; in adult the ratio is 2.4-2.6. Concomitant to this development of the olfactory rosette is an increase in mass of the olfactory bulbs and elongation of the olfactory tracts (Fishelson & Baranes, 1997).

Odorant receptors (ORs)

Fish have around 100 distinct genes encoding for ORs (Barth *et al.*, 1996). This is about one tenth of that found in mammals (Buck & Axel, 1991). In zebrafish, the first neurons expressing OR genes were detected at 24 h pf in 50% of the animals and by 30 h AF all preparations contained at least 2-3 cells per olfactory organ expressing OR genes. This time coincides with the arrival of the pioneer neuron axons in the telencephalon (Whitlock & Westerfield, 1998). The temporal patterns of neurones expressing these ORs did not undergo obvious changes, but the number of cells expressing the specific OR genes increased (Byrd *et al.*, 1996). These observations of neurones expressing detectable levels of OR occur 12 hours after axons from the olfactory placode enter the brain, but 12 hours before the olfactory neurone morphology displays a mature appearance (Hansen & Zeiske, 1993).

The appearance of ORs in the developing zebrafish was studied by Barth and collaborators (1996). They used six receptor probes and found that cells in the larval olfactory epithelium hybridised to individual OR probes in the adult zebrafish olfactory epithelium to varying degrees, from 176 cells per rosette for the OR6 probe and 1034 cells per rosette for the OR2 probe. None of the probes detected signals for these receptors at 24 h pf, and a small number for these receptors at 30 h pf. By 36-48 h of development, the number of signals increased dramatically, and sections of olfactory placodes contained multiple receptor-positive cells (Fig. 7). It is noteworthy that cells expressing RNA complementary to OR13 are not observed until 36-48 h of development; these cells increase in their frequency through 120 h pf. Probes for both OR6 and OR10 detect very few cells until 2-4 weeks of development. Neurones expressing specific ORs are randomly distributed in 48 h zebrafish embryos, and similar to that observed in the adult olfactory epithelium (Vogt *et al.*, 1997).

With the use of polymerase chain reaction techniques, a transient increase in expression of OR transcripts related to the parr-smolt transformation has been demonstrated (Dukes *et al.*, 2004). A receptor probe for an OR increases in the early part, while two vomeronasal receptor probes (VR) increase late in the parr-smolt transformation. It is probable that the ORs are expressed in ciliated sensory neurones with long dendrites. These neurones may be associated with the perception of pheromones, such as bile salts and alarm substances, and thus tuned to the "home"

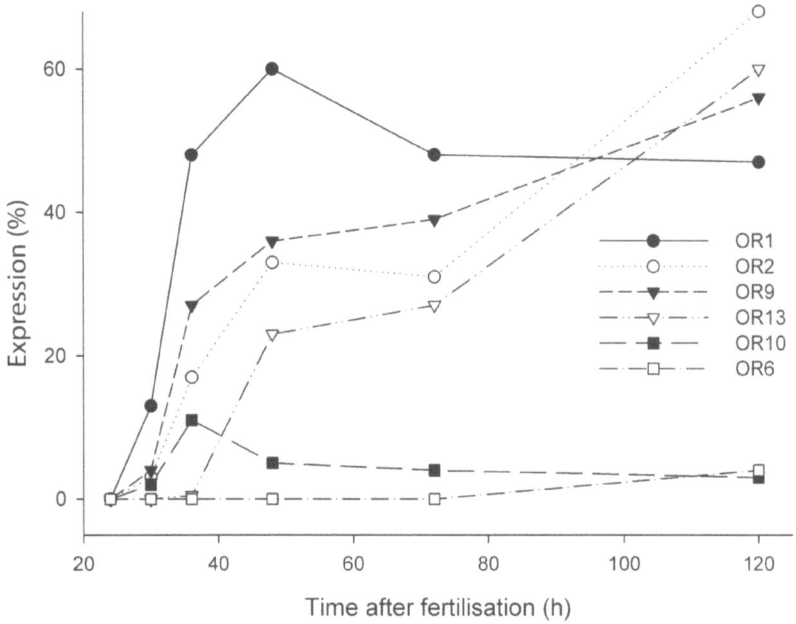

Fig. 7 Expression of odorant receptors (OR) in zebrafish (*Danio rerio*). The percentage of olfactory placode tissue sections demonstrating OR expression by RNA *in situ* hybridisation with the aid of probes for different ORs, OR. Data from Barth *et al.* (1996).

odours of the salmon. The VRs are expressed in the microvillar sensory neurones. In the species investigated, these neurones detect food odours and might be suitable for food search behaviour of salmon at sea. The authors take the expression of ORs as evidence of olfactory "learning" or imprinting; however, their results give no evidence for such a proposal.

A quantitative investigation of expressivity and penetrance of expression for ten zebrafish OR genes during the first month of development was performed by Argo *et al.* (2003). Their findings support the hypothesis that a stochastic choice from within deterministically defined gene groups shapes the expression dynamics during early developmental stages. Four of the analysed genes are expressed already at 24 h pf the earliest time-point investigated. The latest onset of expression was observed at 72 h pf. Some ZOR genes showed of a transient decrease in expressivity between day 5 and 12 pf. It was noted that olfactory sensory neurones select a single odorant receptor gene for expression out of a large gene family (Argo *et al.*, 2003).

Function

The development of olfactory function has been studied in several fish species. In short, the results of these studies show that the ability of a fish to respond to olfactory stimuli develops hours or days after the sensory neurones first appear in the olfactory organ. Recordings of the electro-olfactogram (EOG), and spike activity have been performed from the olfactory epithelium in rainbow trout embryos (Zielinski & Hara, 1988). The first sign of nervous activity was encountered when the sensory epithelium contained ciliated sensory neurones with 1-3 short cilia and an identifiable olfactory nerve was evident. The response threshold decreased from 10^{-5} to 10^{-9} M and a concentration-response relationship evolved as the embryos aged from day 18 to 29 pf (at 11° C). During this period, the density of ciliated sensory neurones increased, as well as the number of cilia per cell (from 3 to 12-13) and length of cilia (from 1 to 5 μm). During this period the microvillous sensory neurones also appeared at the surface of the epithelium.

Chemokinesis

It has been found that before transition from endogenous to exogenous feeding, prolarvae of Atlantic cod (*Gadus morhua*) and turbot (*Scophthalmus maximus*), show non-specialised behavioural responses to such chemical stimuli as food extracts and free amino acids. When exposed to chemical stimuli, the prolarvae reduce their swimming velocity, reduce their mean frequency of swimming trajectories, reduce their total distance moved per trajectory, and their distance between the start- and end-position in three-dimensional space (straightness index) is shortened (Fig. 8). This kind of behaviour was called positive chemokinesis (Døving *et al.*, 1994). It is hypothesised that such alterations of swimming activity are adaptations to the patchiness of food objects often found in the aquatic environment. The adaptive role of this behaviour includes an increase in the probability of food detection during the important and

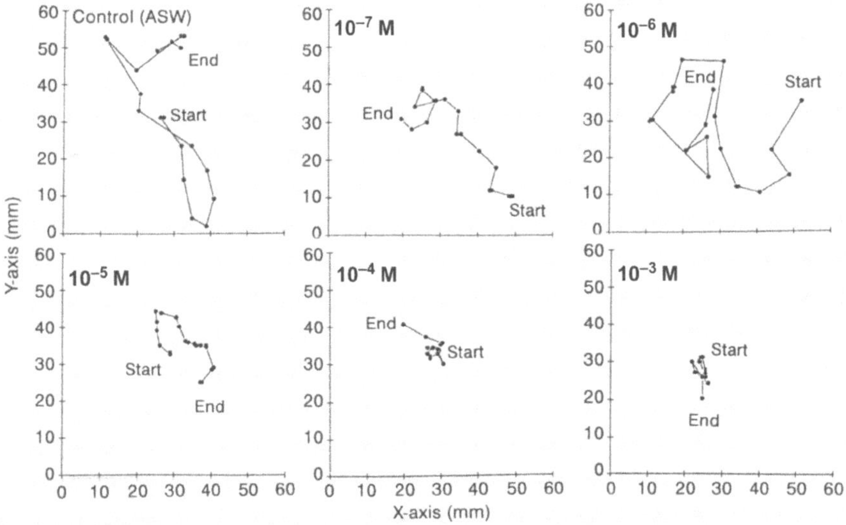

Fig. 8 Chemokinesis in Atlantic cod (*Gadus morhua*) larvae. Examples of Atlantic cod larval swimming trajectories in artificial sea water ASW (control) and different concentrations of arginine seen in the x-y plane. Each trajectory is one minute and the start and endpoint is marked. Note the distinct reduction in trajectory length as the concentration of arginine is 10^{-5} M or higher. After Døving *et al.* (1994).

critical period of transition to exogenous feeding (Døving *et al.*, 1994). It has been shown that marine fish larvae of the anchovy (*Engraulis mordax*) aggregated in patches of prey (Hunter & Thomas, 1974). Likewise, the planktonic newly hatched yolk-sac larvae of red seabream (*Pagrus major*) gathered in layers of rotifer (*Brachionus plicatilis*) extract (Tanaka *et al.*, 1991).

Certain substances induce negative chemokinesis and increase significantly the swimming activity of prolarvae, as do the amino acids glycine, histidine, and glutamine in turbot. The threshold concentration for amino acids detected by turbot prolarvae was 10^{-5} M. Such negative chemokinesis would lead the prolarvae away from areas where these substances are detected and could be an anti-predator response. Having chemical signals that allows for both a decrease and an increase in swimming parameters might be advantageous for prolarvae to optimise the expenditure of their limited energy reserve (Kasumyan *et al.*, 1998).

Alarm chemical signals

Many cyprinid fish species show the first signs of reacting to alarm pheromones at the beginning of the larval period. After 3-4 months, species of cyprinid fish attain the definitive level of sensitivity to alarm pheromones, and vigorous alarm responses are developed in 1 year-old fish (Fig. 9) (Pashchenko & Kasumyan, 1983, 1986; Kasumyan, 2004).

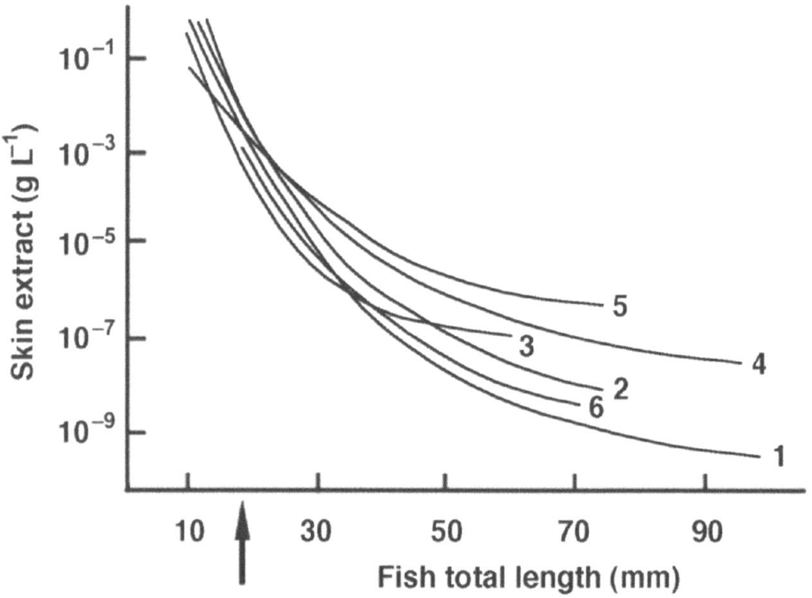

Fig. 9 Development of sensitivity to alarm pheromone in cyprinid fish. The graph illustrates changes in threshold concentration of alarm pheromone (water extract of skin from adult conspecifics) for six cyprinid fish. 1. grass carp (*Ctenopharyngodon idella*); 2. bitterling (*Rhodeus sericeus amarus*); 3. kutum (*Rutilis frisi kutum*); 4. asp (*Aspius aspius*); 5. Danube bleak (*Chalcalburnus chalcoides*); 6. silver carp (*Hypophthalmicthys molitrix*) during their first year of life. The arrow indicates the onset of feeding. After Kasumyan (2004).

The ability to distinguish the alarm pheromones of different species develops late in fish ontogeny. As was shown in experiments with bitterling (Cyprinidae); adult fish showed a more intense alarm reaction when exposed to skin extract from conspecifics, than when exposed to skin extract from spined loach (*Cobitis taenia*) (Cobitidae) (Fig. 10) (Kasumyan & Ponomarev, 1986). However, the intensity of fry response to two types of skin extract was the same. These experiments indicate that the olfactory system in adult fish discriminates alarm pheromones better than that of the fry (Døving *et al.*, 2005).

The developmental stage, rather than the age of the fish, correlates with the responsiveness to alarm pheromones. As was shown for zebrafish fry, smaller and morphologically stunted individuals do not react to alarm pheromones while the larger fry from the same clutch (siblings) show a definite response (Waldman, 1982). The difference in water temperature and other conditions, e.g. feeding, during fish rearing might be one of the reasons for the discrepancy between data obtained by different authors. As an example, Pfeiffer (1963) found that young zebrafish first reacted to alarm pheromone at day 32 ph, 20 days prior to the age found by Waldman (1982).

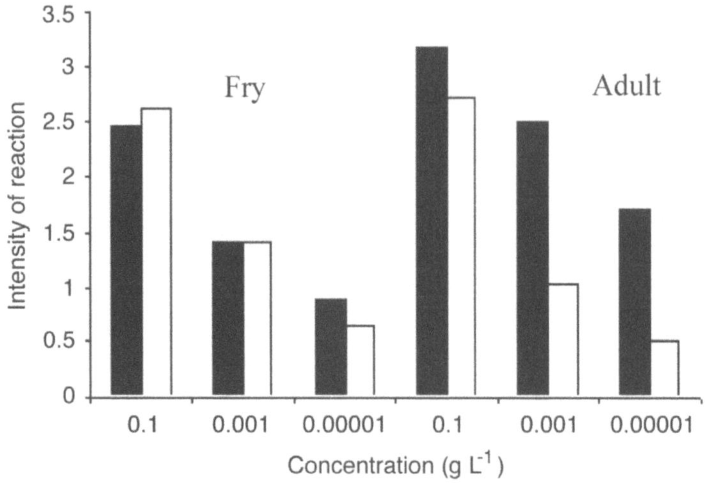

Fig. 10 Development of recognition. The bars indicate the intensity of the alarm reaction in fry and adult bitterlings to skin extracts of bitterling (*Rhodeus sericeus amarus*) (black bars) and of spined loach (*Cobitis taenia*) (white bars). Note the difference in efficiency of the lower concentrations of bitterling skin extract in adults, compared to extract from the spined loach. Data from Kasumyan & Ponomarev (1986).

The ability of cyprinid larvae to respond to predator odour develops at the same ontogenetic stage as they become sensitive to conspecific alarm pheromones (Pashchenko & Kasumyan, 1986). Fry of many other fish species are also sensitive to the chemical signals of predators, e.g. brook charr alevins (*Salvelinus fontinalis*) emerge from the nest earlier when exposed to chemical cues of slimy sculpin (*Cottus cognatus*), a common predator of brook charr eggs. By shifting the time of emergence alevins may escape predation in the nest but may pay a cost in being under-developed upon emergence into the open water (Mirza *et al.*, 2001). Larvae of the paradise fish (*Macropodus opercularis*) are able to react to the odour of sympatric and allopatric predators at the age of 15-25 days (Miklosi *et al.*, 1997).

Food chemical signals

Many fish start responding to food odours at the same age as they react to alarm pheromone in early larval stages. Newly hatched zebrafish prolarvae respond to a mixture of 12 amino acids (100 μM each) at the day ph increasing the number of trajectories and swimming speed and net-to-gross displacement i.e. the ratio of the shortest linear distance between start and end points of a path, and the total travel distance (Lindsay & Vogt, 2004). In Atlantic herring (*Clupea harengus*) prolarvae before first-feeding, the swimming activity increased in response to washings and extracts of *Balanus* nauplii and to the amino acids glycine and L-proline at a concentration of 100 μM in the aquarium. When the olfactory nerve was severed, however, this response no longer occurred (Dempsey, 1978). Certain chemical stimuli have a significant effect on the behaviour of turbot and of Dover sole (*Solea solea*) larvae at the start of exogenous feeding. Larvae of both species display behavioural patterns associated with feeding (swimming, attacking movements, snapping, yawning) more frequently when exposed to chemical stimuli (Knutsen, 1992). The 14 day-old Mozambique tilapia (*Oreochromis mossambicus*) fry are attracted by water from aquaria with live *Tubifex* sp. and 16 day-old red seabream larvae increased their activity when contacted with juice of clams (Iwai, 1980). Low concentrations of several free amino acids (5 μM) elicited feeding behaviour in the fry of Atlantic salmon (*Salmo salar*) and brown trout at the stage of first-feeding (Mearns, 1986). Certain chemical stimuli emanating from trochophores of the serpulid polychaete (*Hydroides dianthus*) can suppress feeding at day 11-13 ph larvae of weakfish (*Cynoscion regalis*) (Connaughton *et al.*, 1994).

Detailed studies of several acipenserid and cyprinid fish species revealed that development of the olfactory sensitivity to food stimuli occurs in close relationship with the feeding ecology of juveniles and their general sensory equipment: responses to food stimuli develop earlier and more quickly in fish to which olfaction has a major role in feeding behaviour. For example, in sturgeons, whose feeding behaviour is significantly based on olfaction, the response to food stimuli appear just after the transition of larvae to external feeding, ranging from day 8 to 17 ph for the Russian sturgeon (*Acipenser gueldenstaedtii*), stellate sturgeon, and Siberian sturgeon. Already when the juveniles are 1.5 months ph they display the same thresholds to food stimuli as found in older fish (Kasumyan & Kazhlaev, 1993b; Kasumyan, 1999).

Among cyprinid species studied, the ability to respond to food water extract appeared first in common carp (day 22-25 ph, 20°C) and later in zebrafish (day 32-38 ph, 24-25°C). The sensitivity to food extract reaches the maximum level in common carp juveniles of 50-60 days old and in zebrafish juveniles of 40-47 days old (Kasumyan & Ponomarev, 1990). At this age, zebrafish juveniles have a BL between 18-22 mm, thus the same length at which naïve (no previous exposure to alarm pheromone) zebrafish fry show a definite response to alarm pheromones (Waldman, 1982).

The remarkable metamorphosis of flatfishes is interesting particularly when considering the development of the olfactory organs, which in contrast to the eye remains approximately symmetrically situated. Flatfishes are cyclosmates, and both ciliated and microvillous sensory neurons are found in flounder larvae of Japanese flounder (*Paralichthys olivaceus*) (Kawamura & Ishida, 1985), sand sole (*Psettichthys melanostictus*), Rex sole (*Glyptocephalus zachirus*) (Stehr, 1983), and the Dover sole (Appelbaum *et al.*, 1983). During metamorphosis the olfactory sensory epithelium becomes larger on the ocular, than the abocular side (Appelbaum *et al.*, 1983). When the flatfish lies down, the anterior naris on the abocular side is close to the bottom, while the anterior naris on the ocular side is higher up. Another point is that there is a common accessory sac for the two olfactory organs. This means that water is sucked over the olfactory organs, making it possible to compare simultaneously the odour input from the ocular and abocular side.

To understand the advantage of such a system we need to consider the physical realities at the boundary layer. The odours from a prey animal hidden in the bottom diffuse slowly through the thin bottom boundary layer where vertical turbulence is weak. When these odours reach above the boundary layer they are transported with the current more rapidly. Thus downstream from the prey there will be a zone where there are prey odours present just above the boundary layer, but none in the boundary layer. Thus, flatfishes sampling odours in and above the boundary layer will have a possibility to determine the presence and distance to a prey. At closer range the position of the abocular naris makes it possible for the fish to follow the odour trail all the way to the source. These properties of the flatfish olfactory system have been pointed out as the possible evolutionary advantage explaining the asymmetry of flatfishes (Westerberg, 1990, and pers. comm. 2005).

Other types of chemical signals

The free-swimming fry of the Midas cichlid (*Cichlasoma citrinellum*) (day 1 to 10 ph, 5 mm BL) chose the water plume of their mother's odour. They were unable to distinguish between the odour of their own or another mother. It was suggested that predation pressure makes these responses adaptive (Barnett, 1977). It was also shown that fry of the jewelfish (*Hemichromis bimaculatus*) can recognise the odour of conspecific fry (Kühme, 1963).

Migrating European glass eel (*Anguilla anguilla*) show highly developed sensitivity to different substances in binary-choice experiments. European glass eels show attractive responses to bile salts (deoxycholate and taurochenodeoxycholate) solutions down to a concentration of 10^{-14} M. Fish respond to taurine down to 10^{-12} M and to many of free amino acids down to 10^{-9}-10^{-8} M. All earthy and green odour substances such as geosmin, pyrazines and thiazoles tested have clear attractiveness for European glass eels at extremely low concentrations, ranging between 10^{-9} -10^{-13} mg $\cdot$ L^{-1} (Sola & Tosi, 1993; Sola *et al.*, 1993; Tosi & Sola, 1993; Sola, 1995). American glass eels (*Anguilla rostrata*) are attracted by the odour of other glass eels and adult eels (Miles, 1968; Sorensen, 1986).

GUSTATION

In contrast to olfaction, the gustatory sensory system serves a role only in feeding behaviour. The gustation system guards the food intake and it determines the palatability of food. Based on information from the taste system, fish swallow appropriate food items and reject or neglect unpalatable ones. Thus, gustation provides the penultimate sensory evaluation in the feeding process and serves the consumatory phase of feeding behaviour. The final sensory evaluation occurs in the stomach or early intestine, where substances may cause a vomiting response. Compared with other groups of animals, the gustatory reception in fish seems to be highly developed. Electrophysiological recordings of gustatory responses to stimulation with free amino acids have demonstrated that in many fish species the threshold concentrations of these compounds are significantly lower than, for example in amphibians or mammals (Marui & Caprio, 1992). Unlike other vertebrates, the gustatory system in fish may be divided into two distinct subsystems, oral and extraoral, both of them mediating behavioural responses to food items brought in contact with the taste buds. When we talk about taste buds in the mouth and outside the mouth we shall use the terms oral and extraoral, meaning in the *oropharyngeal cavity* and outside the *oropharyngeal cavity*, respectively.

Due to numerous extraoral taste buds, the gustatory reception in fish generally plays a more prominent part in foraging behaviour. The role of extraoral taste reception in search for food by fish in the near space is indicated by the amount of data on the distribution of taste buds on the body surface. The highest density of taste buds (up to $300 \cdot mm^{-2}$) is found in those regions of the body that stand a high probability of being in contact with potential food items, namely on barbels, the lower surface of the head, opercula, pectoral and ventral fins (Gomahr *et al.*, 1992). Fish that have numerous extraoral taste buds on the body surface such as catfishes (*Ictalurus* sp.) may search for, and localise bait by means of the gustatory system alone (Bardach *et al.*, 1967). However, many fishes following experimental acute anosmia, lose their ability to react to food chemical signals and become unable to find odour sources even at high concentrations of test solutions (Kleerekoper, 1969; Kasumyan & Ponomarev, 1986; Kasumyan & Kazhlaev, 1993b). This means that the gustatory reception in many fish retains the properties of the contact sensory system.

The morphology of the gustatory system, taste bud structure and distribution has been studied extensively since the 19th century. But even today little is known about the development of taste buds in fish. In a thorough study, Landacre (1907) described the time course of appearance of taste buds in black bullhead (*Ameiurus melas*), but did not provide a detailed description of taste bud formation and its overall morphology. Studies about taste bud formation are few and have mostly appeared during the last 20 years (Pevzner, 1985; Hansen & Zeiske, 1993; Northcutt, 2005). In the same period, electro-physiological methods have been used to evaluate the functional properties of the gustatory system, but none of these methods have been applied the electrophysiological approach to study the functional development. The first studies on the functional development of the gustatory system in fish were made by behavioural methods (Kasumyan & Kazhlaev, 1993a; Kasumyan & Døving, 2003). Based on these behavioural approaches, the spectra of effective stimuli for extraoral (incitants and suppressants) and oral (stimulants and deterrents) gustatory systems were assessed for different age groups of fish larvae and juveniles.

Following the nomenclature used for classification of taste chemical stimuli, the **incitants** are substances that induce capture of the food item via the external gustatory system; **suppressants** are substances that decrease the rate of grasping of food items; **stimulants** are substances that promote food intake (swallowing); **deterrents** are substances that make the fish abandon food intake and evoke food rejection (Kasumyan & Døving, 2003).

With the term "orophayngeal cavity" we mean the region between the lips and the upper oesophagus. The oropharyngeal cavity is divided into the **oral** cavity, i.e. the region between the lips and the articulations of the lower jaw; the **buccal** cavity extends between the lower jaw articulations and the first branchial slits; the **branchial** cavity (anterior pharynx) is bordered by the gill arches, and the **pharyngeal** cavity (posterior pharynx) lies between the pharyngeal teeth and the chewing pad, caudally it converges into the narrow oesophagus.

Morphology of Gustatory System in Adult

The peripheral sensory organs that constitute the gustatory system are the taste buds. Taste buds are the structural basis of the gustatory system in all gnathostomes – fish, amphibians, reptiles, birds and mammals. Agnathans (hagfish, Myxinidae, and lamprey, Petromyzontidae) have end buds – organs that resemble the teleost taste buds (Reutter, 1986). Phylogenetically, the basic patterns of the taste bud are well conserved among fishes, although there are many fine-scale differences between groups (Kapoor *et al.*, 1975; Reutter, 1986; Jakubowski & Whitear, 1990).

In fish, taste buds are situated not only within the oral cavity, pharynx, oesophagus and gills, but may also occur on the lips, barbels, and fins and over the entire body surface in many species. These peculiarities distinguish fish from other groups of gnathostomes except the amphibians. In some developmental stages of amphibians, taste buds are found within the oral cavity and also on the skin of the head. The abundance of taste buds is another peculiarity of the fish gustatory system: taste buds are more numerous in fish than in any other animal. The taste bud density varies according to location and species and may reach up to 300-400 $\cdot$ mm^{-2} or even more in the palatal organ and in the gular region of some benthivorous cyprinids (Gomahr *et al.*, 1992; Osse *et al.*, 1997). In some of fish like rainbow trout, Northern pike, American eel, clupeids and elasmobranchs extraoral taste buds are not found (Kapoor *et al.*, 1975; Twongo & MacCrimmon, 1977; Jakubowski & Whitear, 1990; Devitsina, 1998).

A typical taste bud has an ovoid or pear-shaped form; its long axis is oriented vertically to the surface of the epithelium, which is normally not keratinised (Fig. 11). As a rule, the base of a taste bud is situated on top of a small ascending papilla of the corium (dermis) (Reutter *et al.*, 1974). Marginal cells form the border or margin between a taste bud and the neighbouring stratified squamous epithelium. Oral and extraoral taste buds consist of two types of gustatory sensory cells, the electron lucent light cells and the electron denser dark cells. The number of gustatory sensory cells in a taste bud greatly varies from a few to several dozens. Sensory cells of both types have an elongated shape and follow the long axis of the taste bud (Jakubowski & Whitear, 1990; Reutter, 1992). Each sensory cell bears apical microvilli which together form the taste bud pore exposed to the environment (Fig. 12: European eel, C and D). Between light cells it is possible to distinguish cells which end in one large apical microvillus, and cells which show different numbers of distinctly shaped microvilli. The most common dark cell subtype normally bears small and undivided microvilli. The micromorphology of

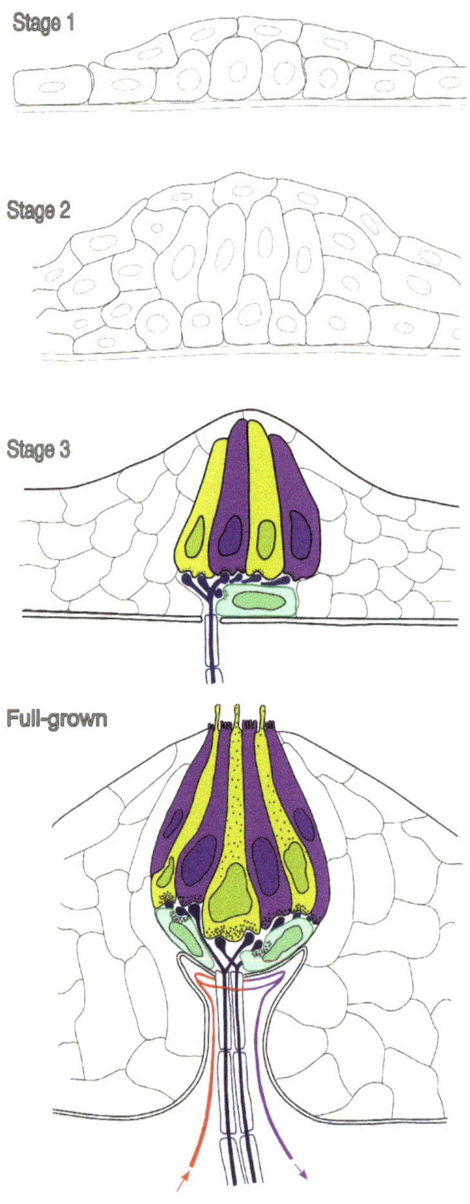

Fig. 11 Different stages in the development of taste buds. In stage 1, the first enlargement of the subepidermal cells grows in size and a shallow dome is formed. In stage 2, the primordial cells appear more prominent, and by stage 3 the sensory cells differentiate in light cells (yellow) and dark cells (blue). In the full-grown taste bud, a taste pore is formed and basal cells (green) are seen at the base. Note the blood supply of the full-grown taste bud.

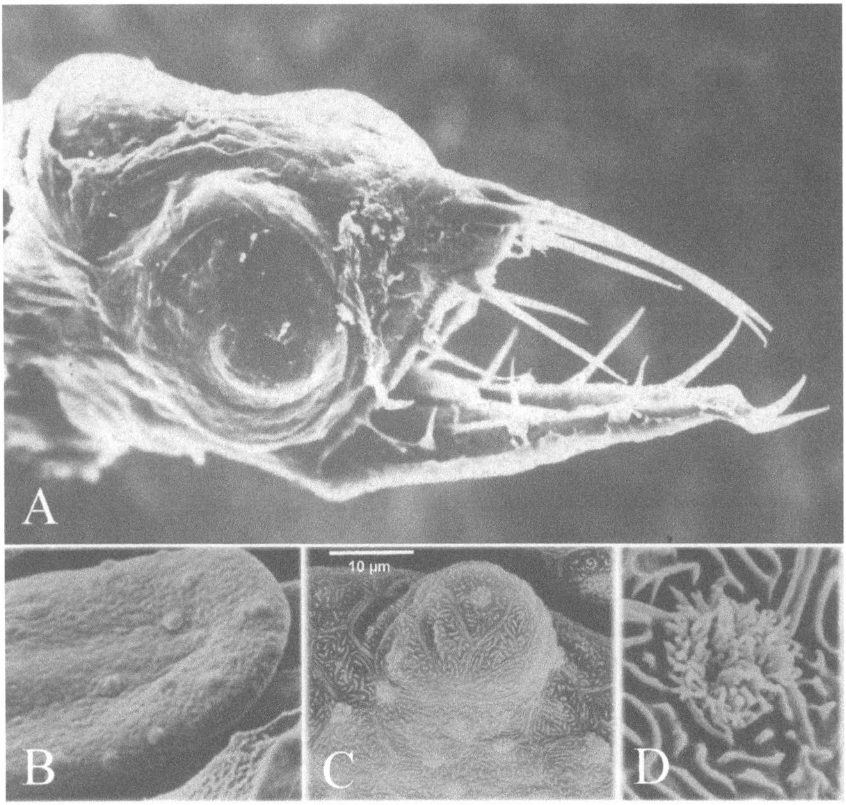

Fig. 12 SEM of European eel (*Anguilla anguilla*) larvae and taste buds on the tongue. A, head of a European eel larva. B, SEM of the tip of the eel larva tongue. C, the taste bud pore is seen on the top of the papilla dome. Other taste pores are seen in the tongue. D, enlarged picture of the taste pore. Photos by K.B. Døving, unpublished.

microvilli is species specific (Reutter & Hansen, 2005). Up to five basal cells are situated at the bottom of the taste bud. It has been suggested that both light and dark cells as well as basal cells have synaptic associations with nerves (Reutter, 1986).

Full-grown taste buds are not static structures. The cells in a taste bud undergo renewal from the epithelial cells that surround the base of the taste bud. These cells undergo division with some of their daughter cells migrating into the taste bud. The average lifespan of a taste cell is dependent on the ambient temperature: in channel catfish (*Ictalurus punctatus*) the taste cell turns over every 40 days at 14°C and every 12 days at 30°C (Raderman-Little, 1979).

Every taste bud has a network of nerves located between the basal cells and the basal part of the sensory cells which comprised unmyelinated and myelinated nerve fibres (Finger *et al.*, 1991). The network of nerves is the location of synaptic contacts within the taste bud. Synapses occur at the subnuclear part and at the basal processes of gustatory sensory cells, and also at basal cells. Nerve fibres innervating taste buds are derived from the facial (VII), glossopharyngeal (IX), and vagal (X) cranial nerves. The facial nerve supplies extraoral taste buds of the barbels and lips (*Nervus facialis maxillaris*), fins and body surface (*N. facialis recurrens*) and oral taste buds of rostral palate (*N. facialis palatinus*). The vagal nerve innervates most of the orobranchial taste buds, whereas the glossopharyngeal nerve plays a minor role in the gustatory supply to the oral cavity (Herrick, 1901; Atema, 1971; Finger, 1976). A single nerve fibre may innervate several taste buds and each sensory cell forms synapses with several nerve fibres (Jakubowski & Whitear, 1990).

The primary gustatory nuclei lie in the *medulla oblongata* and form the facial, vagal and glossopharyngeal lobes. Within the medulla, local gustato-motor networks and descending pathways form the neural substrate for reflexive movements such as food pick-up or biting and swallowing or reject (Kanwal & Finger, 1992). Based on the anatomical peculiarities, the gustatory system may be divided into two distinct, though interrelated subsystems: oral gustatory subsystem and extraoral gustatory subsystem (Finger & Morita, 1985).

The functional properties of the fish gustatory system have mainly been studied by electrophysiological means (Marui & Caprio, 1992). It was found that the fish gustatory system is highly responsive to many organic or inorganic chemicals as well as to natural food extracts. The taste specificities across the fish species studied is very different. The specificities and sensitivities of the taste buds innervated by the different branches of the facial nerve are similar in the same fish species (Caprio, 1975; Davenport & Caprio, 1982). The estimated thresholds for the most potent substances were less than 10^{-9} M. The oropharyngeal taste buds, which are innervated by both the glossopharyngeal and vagal nerves, are generally less sensitive to specific amino acids by 1.5 to 2 orders of magnitude than taste buds innervated by the facial nerve, although they have been shown to have similar amino acid specificity (Kanwal & Caprio, 1983).

Recently, results obtained by electrophysiological approach have been supported by data obtained from behavioural assays. Fish are able to recognise different taste substances. Fish taste preferences are highly species specific and the differences among fish species are apparent when comparing the width and composition of spectra for both stimulants and deterrents. What is evident is that there is strong similarity in taste preferences between geographically isolated fish populations of the same species and that taste preferences are similar in males and females although taste preferences at the individual level may vary dramatically among conspecifics. There is a correspondence between oral and extraoral taste preferences in the same fish species. Taste preferences in fish show low plasticity (in relation to the diet) and are determined genetically. An overview of the taste preferences in fish was recently published (Kasumyan & Døving, 2003).

Embryonic Origin of Taste Buds

There are two major views on the embryonic origin of taste buds. The first claims that taste buds are endodermal in origin and that, where they occur in parts of the ectoderm, it is the result of an outward growth of endodermal tissue. This view was put forward by Johnston (1908), who found that in *Petromyzon* sp., *Catostomus* sp., and *Coregonus* sp. taste buds developed only in the pharynx, appeared later on the lips and, finally, on the head and body. After studying the development of taste buds in *Squalus sp.*, and the common carp, Cook and Neal (1921) and Edwards (1930), respectively, supported the view of Johnston (1908). Goodrich (1958), while supporting the concept of endodermal origin, discusses a second view which postulates that taste buds may be ectodermal organs that develop originally in that part of the buccal cavity derived from the stomodeum. One of the arguments for this postulate is that no sensory cells are known to occur in the endodermal lining of the vertebrates except for taste buds; and that in general, all sensory and nervous cells are ultimately derived from either the ectoderm or neural crest. Landacre (1907) expressed the compromise view of dual origin, believing that oral taste buds were endodermal in origin while extraoral ones were ectodermal. This view was supported by Kapoor *et al.* (1975). Recently, experimental studies in amphibians demonstrate that oral taste buds are generated from endoderm, whereas the sensory neurons of cranial ganglia arise from neurogenic ectoderm (Barlow, 2000).

Embryogenesis

The first appearance of taste buds and their growth in number and densities has been studied in relation to fish systematic position and their feeding ecology. However, detailed studies of taste bud formation have been examined only in a few fish species, namely the Russian sturgeon (Pevzner, 1985), zebrafish (Hansen *et al.*, 2002), and channel catfish (Northcutt, 2005). In general, taste buds develop later than the olfactory organ and the solitary chemosensory cells.

Taste bud formation

The first few primordia of oral and extraoral taste buds in developing zebrafish appear simultaneously, 3 to 4 days pf. At 26°C this is the time of hatching. The early taste bud primordia are covered by epithelial cells and consist of only one type of cell, which are roundish and often interdigitate with the overlying epithelial cells (Fig. 11). No dermal papillae are evident beneath the early taste bud primordia. The primordial cell nuclei may have various shapes (roundish, oval, lobed, or angular), but they are never elongate as are the nuclei of the surrounding regular epithelial cells. As the future taste bud cells stretch vertically, the nuclei also stretch and become more ovoid in shape. Before the taste bud cells extend to the apical surface to build the open receptor area, dark and light cells can be distinguished. The morphological methods used by Hansen and co-authors (2002) are insufficient to determine whether there are different types of precursor cells or whether only one type of precursor cell gives rise to both dark and light cell. The development of light and dark receptor cells in the taste buds seems a rapid process in the zebrafish ontogeny. It only takes a few hours at a water temperature of 26° C (Hansen *et al.*, 2002). Taste bud primordia differentiate into young taste buds with a taste pore exposed to the environment one to two days after their detection by means of TEM.

In zebrafish, there are two ways the taste pore or receptor area of a taste bud can form. Either the taste pore appears at the junction of three, sometimes two, epithelial cells, and does not protrude above the surface of the epithelium. These taste buds are typically found on the outer region of the lips, on the head and on the gill rakers in zebrafish larvae. The cells of the taste bud primordium, whose cell bodies are covered apically by the epithelial cell, "push up" the epithelial cell adjacent to the exposed receptor area so that a little hillock is formed. These hillocky taste buds are mainly located at the inner side of the lips, in the mouth cavity, and on the tips of the early gill rakers (Hansen *et al.*, 2002). In sturgeons, the taste pore is formed by moving apart the epithelial cells. The first receptor cells to appear in the open taste pore are the light cells with a single apical microvillus. The apical microvilli begin to form in receptor cells after the young taste buds reach the epithelium surface and the receptor area is exposed to the environment (Pevzner, 1985).

In contrast, dark receptor cells with several small microvilli are the first cells to appear in the open taste pore in zebrafish (Hansen *et al.*, 2002). As taste pores with only these small microvilli of dark cells are seldom seen, the outgrowth of the large microvillus of light cells seems to follow immediately. The diameter of the early taste pore is 1-1.5 μm compared with approximately 5 μm in adult zebrafish. In addition to the cells of the dark and light type, a third cell type in taste buds in zebrafish was found, the brash-like cells. This cell type is lighter in electron density, than the dark cell and has a brush-like apical ending that consists of approximately ten thin microvilli. The microvilli of the brush-like cells appear in the open receptor area only after the microvilli of the other two cell types are already present (Fig. 13). In young taste buds, the cell with the brush-like apex is usually confined to the margins of taste buds, although later on it is also found in the centre of the receptor area. With respect to electron density, the difference between dark and light cells is not very pronounced, especially in young taste buds. The mature taste bud in a juvenile zebrafish contains approximately 30-40 cells. The ratio between light cells, dark cells and brush-like cells within a taste bud changes during development. In the earlier stages of taste bud maturation, only few more dark cells than light cells are present. In the mature taste bud, proportions are rather constant: the ratio of dark cells, light cells and brush-like cells being roughly 6:3:1 (Hansen *et al.*, 2002).

Basal cells are the last cell type to appear in taste bud differentiation (Pevzner, 1985). In zebrafish, very young taste buds have only one disk-like basal cell, which is horizontally oriented as in mature taste buds. Although basal cells lie directly above the basal lamina, they do not form hemidesmosomes as do sensory cells. Primordial taste bud cells develop without an underlying preformed dermal papilla. A dermal papilla is first seen when the primordium has differentiated into a pear-shaped young taste bud (Hansen *et al.*, 2002).

As the young taste bud grows, new cells are continuously added. Indistinct differentiating cells are seen in the basal and central region of the taste bud. Degenerating taste bud cells with very light cytoplasm and swollen mitochondria and membranous stacks of cell debris are present at any time. Also, intact-looking cells that

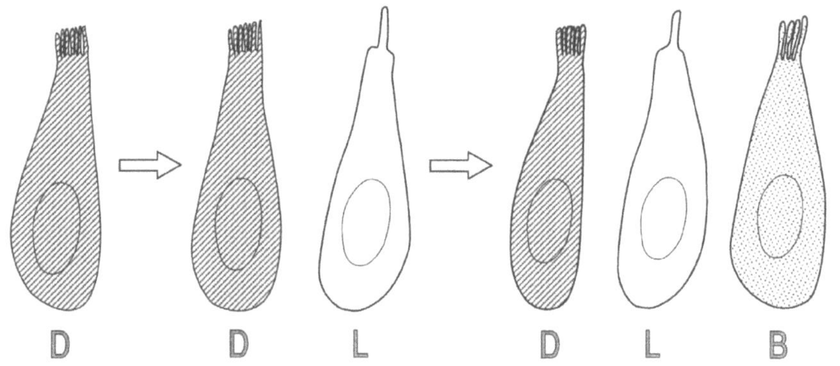

Fig. 13 Extension of cell types diversity during the taste bud formation in zebrafish (*Danio rerio*). D – dark cell; L – light cell; B – brush-like cell. After Hansen *et al.* (2002).

contain cell debris may occur at every stage of taste bud development, even in cells that have not yet reached the apical receptor area. Membranous stacks of cell debris seen in these cells indicate that the debris was phagocytised. Dying cells even in the very young taste buds indicate an early cell turnover in taste buds (Hansen *et al.*, 2002).

To observe the formation of taste buds one can use an antibody towards calretinin as a sensitive marker for at least one class of sensory cells. Visualisation of calretinin within the developing sensory receptors was used to construct the maps of the distribution of developing taste buds (Northcutt, 2005).The first indication of extraoral taste bud formation occurs on rudiments of the maxillary barbels day 6 pf in embryos of channel catfish (stage 39 according to the tables of normal development for brown catfish (*Ictalurus nebulosus*) (Armstrong & Child, 1962). In this stage, three to eight calretinin-positive cells occur singly along the rostral surface of barbels usually within a distinct taste bud primordium. During the next two days, the number of cells within taste bud primordium has increased substantially, and 3 to 4 calretinin-positive cells occur in each primordium. By day 9 pf (day 1 ph), taste bud primordia have formed a distinct spherical cluster of 10-20 calretinin-positive cells and at least one basal cell. Later, the taste bud primordia continue to elongate, and the calretinin-positive cells form long, tapering apical processes that characterise the sensory cells of mature taste buds. In spite of a high density of taste buds in juveniles and adult fish, taste bud primordia do not appear to arise from mature taste buds. In fact, taste bud primordia were never seen in contact with mature taste buds. Instead, these primordia appear to arise from a small number of cells, some of which are calretinin-positive, that are distributed within the ectoderm found between mature taste buds (Northcutt, 2005).

Based on TEM, four stages are recognised in the development of taste buds in zebrafish: stage 1, cells with nuclei of various shapes that have accumulated beneath a slightly elevated single layer of epithelium; stage 2, taste bud primordia whose cells have begun to elongate from the basal lamina of the epithelium; stage 3, taste bud primordia in which three cell types (basal, dark, and light cells) can be distinguished, as can basal nerve fibres; and stage 4, young taste buds with an open taste pore (Hansen *et al.*, 2002) (Fig. 11). The same number of morphological stages could be recognised in the development of taste buds in channel catfish based on the visualisation of at least one class of sensory cells, with antibody to calretinin, within taste bud primordia, and identification of developing taste buds, based on their morphology: stage 1, small cluster of three to five cells, including a single calretinin-positive cell; stage 2, primordia whose cells had begun to elongate from the basal lamina and that included two or three calretinin-positive cells; stage 3, primordia consisting of sensory and basal cells with large numbers of calretinin-positive cells; stage 4, mature taste buds with open taste pore (Northcutt, 2005). Although stages 3 and 4 in both channel catfish and zebrafish clearly present comparable stages of taste bud development, it is not clear whether stages 1 and 2 in channel catfish correspond to the same stages in zebrafish. Because the channel catfish stages were based primarily on the number of calretinin-positive cells, whereas the zebrafish stages were based on ultrastructural criteria, it is possible that the first zebrafish stage could not be recognised in channel catfish embryos, and those stages 1 and 2 in channel catfishes correspond to early and late stage 2 taste bud primordia in zebrafish.

The rate of taste bud development seems to be species specific, even between closely related fish. Among sturgeon species of the family Acipenseridae; taste bud formation was most rapid in beluga (*Huso huso*), followed by ship sturgeon (*Acipenser nudiventris*) and Russian sturgeon (Pevzner, 1985). The first calretinin-positive cells on the posterior palate of channel catfish appear at day 10 pf, and mature taste buds are seen two days later (Northcutt, 2005). In zebrafish, the duration of taste bud development is also about two days (Hansen *et al.*, 2002). In common carp, the period between the appearance of the first taste bud primordia and mature taste buds is about 20 h (Edwards, 1930). The ontogenetic processes of taste bud formation differ from the processes of taste bud formation observed during regeneration of barbels in adult fish (Reutter, 1978).

It is not obvious that the gustatory system becomes functional when the taste bud receptor area on the epithelium surface is formed. In sturgeon, the first synaptic contacts between the basal processes of the receptor cells and the nerve endings begin to form one day after formation of the taste bud receptor area, just 1-2 days before first-feeding (Pevzner, 1985).

Taste bud appearance in relation to first-feeding

As the gustatory system plays an important role in fish feeding, and provides the final sensory evaluation of a selected food items (Atema, 1980; Kasumyan & Døving, 2003), it is reasonable to compare the appearance of taste buds with the time of first-feeding; i.e. at the age when larvae start exogenous feeding and sensory evaluation of the food is required. Comparison of available data shows that in some fishes the first taste buds appear before the fish larvae start to take food, whereas in other species feeding begins before taste bud formation (Table 4). In some fish they differentiate

Table 4 The sequence of oral and extraoral taste bud appearance in fish ontogeny, in relation to the first-feeding (**FF**)

Species	*Sequence*	*Reference*
Adriatic sturgeon (*Acipenser naccarii*)	Extraoral ⇒ Oral ⇒ **FF**	Boglione *et al.* (1999)
Siberian sturgeon (*Acipenser baerii*)	Extraoral ⇒ Oral ⇒ **FF**	Devitsina & Kazhlaev (1993b)
Stellate sturgeon (*Acipenser stellatus*)	Extraoral ⇒ Oral ⇒ **FF**	Devitsina & Kazhlaev (1993b)
Ship sturgeon (*Acipenser nudiventris*)	Extraoral ⇒ Oral = **FF**	Devitsina & Gadzhieva (1996)
Channel catfish (*Ictalurus punctatus*)	Extraoral = Oral ⇒ **FF**	Northcutt (2005)
Nile tilapia (*Oreochromis niloticus*)	Oral ⇒ **FF** ⇒ Extraoral	Kawamura & Washiyama (1989)
Japanese halfbeak (*Hemiramphus sajori*)	Oral ⇒ **FF** ⇒ Extraoral	Kawamura *et al.* (1990)
Paddlefish (*Polyodon spathula*)	Oral = **FF** ⇒ Extraoral	Devitsina & Kazhlaev (1993a)
Black bullhead (*Ameiurus melas*)	Oral = **FF** = Extraoral	Landacre (1907)
Zebrafish (*Danio rerio*)	Oral = **FF** = Extraoral	Hansen *et al.* (2002)
Bonefish (*Albula* sp.), leptocephali	Oral ⇒ **FF**	Rasquin (1955)
Rainbow trout (*Oncorhynchus mykiss*)	Oral ⇒ **FF**	Twongo & MacCrimmon (1977)
Northern pike (*Esox lucius*)	Oral = **FF**	Devitsina (1998)
Largehead hairtail (*Trichiurus lepturus*)	**FF** ⇒ Oral	Kawamura & Munekiyo (1989)
Turbot (*Scophthalmus maximus*)	**FF** ⇒ Oral	Reutter *et al.* (1995)
Dover sole (*Solea solea*)	**FF** ⇒ Oral	Appelbaum *et al.* (1983)
Bluefin tuna (*Thunnus orientalis*)	**FF** ⇒ Oral	Kawamura *et al.* (2003)
Marble goby (*Oxyeleotris marmoratus*)	**FF** ⇒ Oral	Senoo *et al.* (1994)
Persian sturgeon (*Acipenser persicus*)	**FF** ⇒ Oral = Extraoral	Devitsina & Gadzhieva (1996)
Japanese flounder (*Paralichthys olivaceus*)	**FF** ⇒ Oral ⇒ Extraoral	Kawamura & Ishida (1985)
Largemouth bass (*Micropterus salmoides*)	**FF** ⇒ Oral ⇒ Extraoral	Kawamura & Washiyama (1989)
Common carp (*Cyprinus carpio*)	**FF** ⇒ Extraoral	Appelbaum & Riehl (1997)

first in the oral, pharyngeal or branchial cavities, and then on the body surface – lips, barbels, head, trunk and fins. In Nile tilapia (*Oreochromis nilotica*), the first taste buds appear in the epithelium of upper pharynx and gill arches, 36 h larvae; and in the epithelium of the oral cavity at 60 h. Taste bud densities increase in the front of the upper and lower oral valves, and on the lips at 120 h ph. (Many neoteleost fishes such as tilapias and perches have valves in dorsal and ventral sides of the oral cavity, one valve on each side.) Tilapia larvae begin first-feeding on day 4 pf (Kawamura & Washiyama, 1989). Even in rainbow trout, the taste buds appear before first-feeding, and mature looking taste buds are found on the tongue at day 8 ph, although the fish starts only to feed at day 27 ph. No taste buds are found in the epidermis of rainbow trout (Twongo & MacCrimmon, 1977; MacCrimmon & Twongo, 1980).

Leptocephalus larvae are characteristic of Elopomorpha (Fig. 12). This large group of marine teleosts contains more than 700 species within the orders Anguilliformes, Elopiformes, Notacanthiformes and Saccopharyngiformes. In general, premetamorphic leptocephali of Elopomorpha are much larger than most fish larvae, commonly reaching lengths of 50-100 mm and in some of species up to 1.8 m (Smith, 1979). In leptocephali of bonefish (*Albula* sp.), the mature taste buds do not appear until about halfway through metamorphosis but leptocephali do not begin exogenous feeding until near the end of metamorphosis, which is usually completed at about 25 mm TL (Rasquin, 1955; Pfeiler, 1989).

In Japanese halfbeak (*Hemiramphus sajori*), numerous taste buds appear before hatching, much earlier than in other species studied. The taste buds are present in the epithelium of the oropharyngeal cavity, specifically in the pharyngeal region and the anterior oesophagus. Halfbeak larvae are capable of exogenous feeding at hatching. After hatching, taste buds increased in density and also appeared in the epidermis of the upper and lower lips. Thereafter, the density of taste buds increased in the epithelium of the oral cavity, tongue, and gill arches but not in the pharyngeal region (Kawamura *et al.*, 1990).

In Northern pike and paddlefish, the first taste buds appear in the oral cavity just at the start of exogenous feeding, at 7-8 days ph pike larvae (13.0-13.5 mm) and at 13 dph paddlefish larvae (17-18 mm). In Northern pike larvae, the first taste buds are situated near the teeth on the jaws and palate and on the first gill arch (Devitsina, 1998). In paddlefish larvae, the first taste buds are present in small conical taste papillae on the vomer, along the edge of the mouth cavity and around the acrodont teeth. In 16 day-old paddlefish larvae (21 mm) the taste buds are found on the gill arches, along the midline of the palate and in the pharyngeal region of the palate and on the tongue. Extraoral taste buds appear much later than oral ones. Paddlefish larvae have the first extraoral taste buds on the barbels at day 23 ph. The taste buds in paddlefish are restricted to the basal part of the barbel and are absent on the tip of the barbel. Taste buds are found on the ribs of the barbels which are flattened in paddlefish. The total number of taste buds on the barbels is small (Devitsina & Kazhlaev, 1993a).

In several representatives of the Acipenseriformes, the extraoral taste buds appear first; then come the oral taste buds, and then the larvae start feeding. In Adriatic sturgeon (*Acipenser naccarii*), extraoral taste bud primordia begin to differentiate at day 2 ph. One day later, the taste buds on the inner and external upper lip and on the outside of both barbels rise above the surrounding epithelium. At day 4 ph, submerged gustatory receptor primordia are visible on the lateral mouth roof and the taste buds are seen in the pharynx. At day 12 ph, taste buds are organised in regular rows inside the lips and on the external and lateral sides of the barbels and are elevated with respect to the average level of the epidermis. At day 23 ph, when food is found in the stomach for the first time, and in the intestine along with the spiral valve, the taste buds were found on the oesophageal mucosa. Thus, the gustatory sensorial zones differentiate heterochronously during ontogenesis: the extraoral and pharyngeal taste buds are the first to appear, followed by the palatal and lingual ones (Boglione *et al.*, 1999). In ship sturgeon and stellate sturgeon, the palatal taste buds form after extraoral ones on the barbels and lips (Table 5) (Devitsina & Gadzhieva, 1993b, 1996).

Table 5 Fish age (days ph) for oral and extraoral taste bud appearance and first feeding in sturgeons and paddlefish ontogeny

Species	*Extraoral taste buds*	*Oral taste buds*	*First feeding*	*Reference*
Ship sturgeon (*Acipenser nudiventris*)	3-4	4-5	4-5	Devitsina & Gadzhieva (1996)
Persian sturgeon (*Acipenser persicus*)	8	8	7	Devitsina & Gadzhieva (1996)
Stellate sturgeon (*Acipenser stellatus*)	5	6	7-8	Devitsina & Kazhlaev (1993a)
Paddlefish (*Polyodon spathula*)	<23	<13	<13	Devitsina & Kazhlaev (1993b)

In several species the oral and extraoral taste buds appear at the same age of fish larvae. It was found that channel catfish possess both extraoral taste buds (maxillary barbel) and oral ones (rostral pharynx) at 8 dpf, i.e. two days before first-feeding (Northcutt, 2005). In Persian sturgeon (*Acipenser persicus*) extraoral and oral taste buds appear at 8 dpf, i.e. one day before first-feeding (Devitsina & Gadzhieva 1996). In zebrafish, oral (gill arches) and extraoral (lips) taste buds with open taste pores appear simultaneously, one or two dph. For zebrafish larvae this is the age when they start to feed (Hansen *et al.*, 2002). Coincidence between the appearance of mature taste buds and the onset of feeding was found also in black bullhead. Taste buds occur first on the lips and on the gill arches; taste buds on the maxillary barbel and in the oral cavity appear simultaneously (Landacre, 1907).

In some fishes, taste buds appear after first-feeding (Table 4). The Japanese flounder starts feeding day 3 ph, yet the taste buds differentiate first in the gill arches of larvae at day 12 ph. Extraoral taste buds on the lip surface are observed in larvae at day 15 ph (Kawamura & Ishida, 1985). It is interesting to note that food choice (food preference) in Japanese flounder larvae is only seen after taste bud differentiation. According to

Kuwahara and Suzuki (1982), larvae larger than 6-7 mm feed preferentially on the calanoid copepod (*Paracalanus parvus*) in spite of its low density in the sea. The size range of these larvae corresponds to the age when the first taste buds appear. It is possible that the lag of taste bud formation is connected with the well developed vision in Japanese flounder larvae which are visual feeders and detect and separate food organisms based on vision. The visual system is morphologically completed at 3.5 dph, at the age when Japanese flounder larvae start take a food (Kawamura & Ishida, 1985). The first taste buds appeared in the oral cavity in marble goby (*Oxyeleotris marmoratus*), a fresh water Eleotridae fish, 3 days after the onset of feeding on phytoplankton (Senoo *et al.*, 1994).

Taste bud appearance later than first-feeding has been observed in several marine teleosts, in Japanese sea bass (*Lateolabrax*), puffer (*Fugu*), stone flounder (*Kareius*) and sea bream (*Pagrus*) (Iwai, 1980). In bluefin tuna (*Thunnus orientalis*) taste buds appear in larvae day 8 ph (5.6 mm), 5 days after the onset of feeding (4.0 mm) (Kawamura *et al.*, 2003). They differentiate first in the upper pharynx and later in the epithelium of the oral cavity and gill arches day 10 ph, when canine-like teeth and pharyngeal teeth appeared. Taste buds are not present on the lips even in juveniles at day 33 ph (45.5 mm) (Kawamura *et al.*, 2003). In largehead hairtail (*Trichiurus lepturus*), the lag between the time of first-feeding and appearance of the first taste buds is much shorter: first-feeding takes place at 6.0 mm and first taste buds differentiate in the epithelium of the oral cavity of a 6.2 mm larva. Numerous taste buds appear in the pharyngeal region at 11.8 mm larva and on the gill arches at 16.5 mm larva (Kawamura & Munekiyo, 1989). However, stomach content analyses of 10 mm larvae do not show selective feeding in largehead hairtail (Munekiyo & Kuwahara, 1985). In common carp, differentiated taste buds can be seen on the lips at larvae age around 145 h ph (11 mm) long after the time when exogenous feeding has started, 45 h ph (7.5 mm) (Appelbaum & Riehl, 1997). In largemouth bass (*Micropterus salmoides*), the first taste buds appear in the epithelium of the oral cavity in 132 h larvae, hours after the first-feeding (120 h ph). At day 6, the taste buds are present in the epithelium of the branchial cavity, gill arches, and pharyngeal region. Thereafter, the taste buds increase in density and appear on the lips, on oral valves, tongue, and in the epidermis behind the lower jaw by day 22 ph (Kawamura & Washiyama, 1989). Based on the study by Edwards (1930), the first taste bud primordia become detectable in the oropharyngeal cavity of the common carp in larvae 22 h ph. Typical mature taste buds were observed in the pharynx of larvae 42 h ph.

The time of taste bud appearance shifts in fish maintained in different environmental conditions. In Japanese flounder, taste buds develop 2-3 days earlier in wild larvae compared to reared larvae (Kawamura *et al.*, 1989). Taste bud formation depends on external factors such as food availability. In goatfish (*Upeneus tragula*), the taste buds were large in pelagic stage larvae "starved every second day" and small in fully-fed larvae (McCormick, 1993). Starvation caused different changes in taste bud morphology and abundance depending on their localisation, fry age and duration of food deprivation. In the first-feeding larvae of Russian sturgeon, starvation caused an increase in the size of taste buds on the barbels and palate, whereas in non-starving larvae the number of taste buds remained the same. The taste buds of the tongue and gill arches remained the same size, but the number increased. Following a 10-day starvation period, taste buds on the gill arches disappeared completely and the number on the palate, tongue and on the lips decreased significantly (Devitsina & Gadzhieva, 1996). There are also individual variations in oral taste buds density in fish early larvae (Kawamura *et al.*, 1990).

Development of primary gustatory centres

The formation of the primary gustatory centres has been described in two acipenserids, ship sturgeon and Persian sturgeon (Devitsina & Gadzhieva, 1996). A bulk of densely packed undifferentiated neuroectodermal cells with similar oval form were found in the dorsomedial part of the *medulla oblongata* in 3 to 4 day-old larvae, several days before the appearance of the first mature taste buds. The cells had large nuclei, up to 80-90% of the cell volume, and two or more nucleoli. In the following days, the size of cells increased (by means of the perikarion volume mainly) and they gradually acquired a spindle-like form with two polar outgrowths and differentiated into migrating neuroblasts. During this period, part of the migrating neuroblasts moved in the lateral-caudal direction. The remaining migrating neuroblasts migrated in dorsomedial and ventrolateral directions and differentiated to become protoneurons. Relatively small protoneurons mainly occupied the dorsomedial part adjacent to the ventricular layer of ependyme glia and later on became sensory neurones. Relatively large protoneurons accumulated in the ventrolateral part of the *medulla oblongata* and later became motor neurones. At the end of this period, 5-7 dph, this facial region transformed into the facial lobe. The vagal lobes became outlined anatomically in 4 day-old larvae.

At the time when the yolk-sac disappeared, the facial lobe was anatomically differentiated from other brain regions, and the distance between neurones increased from 5-6 μm in 3-day-old larvae to 9-12 μm in 10-day-old larvae. Larger neurones predominated in the ventrolateral region and formed the motor zone. Smaller neurones, with round or spindle-like forms, were situated medially and dorsally making up a special zone of sensory projections (Devitsina & Gadzhieva, 1996).

Taste bud innervation

Before one can observe taste bud primordia in zebrafish larvae, there are nerve fibres present beneath the epidermis and between epidermal cells, but no obvious connection between nerve fibres and taste bud primordia. As soon as the primordial cells begin to differentiate, nerve fibres are observed in the underlying dermal connective tissue innervating the young taste bud (Hansen *et al.*, 2002). In Russian sturgeon, ship sturgeon and beluga, the first nerve fibres penetrate the basal membrane and seem to enter the taste bud only in taste buds with an open taste pore. One day later the first synaptic contacts between the basal processes of the sensory cells and the nerve endings begin to form (Pevzner, 1985). This is in contrast to taste bud development in turbot where even the very early taste bud primordia contain nerve fibre profiles (Reutter *et al.*, 1995). Similarly, during human taste bud development, all taste buds examined were found to be innervated (Witt & Reutter, 1996, 1998).

In the basal part of the taste bud of zebrafish, the basal processes of sensory cells intermingle with the ingrowing nerve fibres and form the taste bud's nerve fibre plexus. Nerve fibres from the dermal region adjacent to the taste bud primordium also grow to the upper portion of the young taste bud close to the taste pore. These fibres wrap around the apical portions of the elongated taste bud cells and are covered by the neighbouring cells, the apical marginal cells. All sensory cells seem to be in synaptic contact with nerve profiles (Hansen *et al.*, 2002). However, these contacts are inconspicuous and rare, a characteristic of taste buds in fish (Reutter & Witt, 1993).

In channel catfish larvae, nerve fibres appear to come into contact with taste-bud-forming epithelia prior to the development of these receptors. In the case of the maxillary barbel, facial fibres appear to enter the barbel day 5 pf, or approximately 24 hours prior to the appearance of the first calretinin-positive cells within early taste bud primordia. This is also true for the trunk, where fibres of the recurrent ramus of the facial nerve branch within the dorsal lobe of the caudal fin day 8 pf, or two days prior to the appearance of the first calretinin-positive cells and taste bud primordia (Northcutt, 2005).

The observations above seem consistent with the theory that nerve fibres induce taste buds. Moreover, it is possible that the occurrence of calretinin-positive cells does not indicate the earliest formation of taste bud primordia. There are also other markers, such as brain-derived neurotrophic factor, sonic hedgehog, and keratin 8, that label taste bud primordia before nerve fibres reach these taste buds (Knapp *et al.*, 1995; Jung *et al.*, 1999; Hall *et al.*, 2003; Mbiene & Roberts, 2003). There is substantial evidence from studies on amphibians (Barlow & Northcutt, 1998; Barlow, 2000) and mammals (Stone *et al.*, 1995; Fritzsch *et al.*, 1997; Mbiene *et al.*, 1997; Nosrat *et al.*, 1997; Mbiene & Roberts, 2003) that nerve fibres are not needed for the induction of taste buds. In a new model, which was termed the "early specification theory" (Barlow & Northcutt, 1998), it is proposed that oropharyngeal epithelium is specified to form taste buds during gastrulation and that taste bud primordia arise by cell to cell interactions within the specified epithelium. According to this theory, newly formed taste bud primordia signal to arriving nerve fibres and induce specific contacts with the primordia's developing sensory cells. These nerve fibres have a neutotrophic role in the maintenance of mature taste buds. The early specification theory is based on oropharyngeal epithelia and may or may not be applicable to the development of extraoral taste buds in fish.

How many taste buds are innervated by a single taste nerve? This is another aspect of gustatory system development. The recurrent facial nerve of channel catfish has been used as model for this question. This nerve innervates all of the taste buds along the body surface caudal to the operculum and is described as a pure gustatory nerve at its point of exit from the skull (Herrick, 1901). It contains neither motor nor general cutaneous sensory elements, and originates from a single size class of ganglion cells (Finger, 1976). The growth of the nerve fibres is extremely rapid, for example the first fibres of the recurrent ramus of the facial nerve that innervate all taste buds on the trunk of channel catfish, can be recognised at day 6 pf, and these fibres reach the level of the head-trunk juncture by day 7 pf, and have reached the tip of the caudal fin by day 8 pf, i.e. at the time of hatching (Northcutt, 2005).

Postlarval growth in the gustatory system was assessed by examination of the process of growth of the recurrent facial nerve and determination of changes of the degree of convergence of receptors onto primary sensory afferents during growth in channel catfish (Finger *et al.*, 1991). The results demonstrate that the gustatory system grows throughout life, and that far more taste buds than primary afferent fibres are added during postlarval growth. The diameter of the recurrent facial nerve of channel catfish increases linearly with increasing fish length. Approximately 2 500 new axon profiles (approximately 50:50, myelinated:unmyelinated) are added for each centimeter increase in fish BL, or roughly 70 new axon profiles per day. The number of taste buds in the recurrent nerve field was estimated by calculating an average density and multiplying this by the estimated surface area. These results show an increase from about 11 000 taste buds on a 5 cm channel catfish to over 600 000 taste buds on a 35 cm fish (Table 6) (Finger *et al.*, 1991).

Table 6 Comparison of fish body surface area and external taste bud density and number in small and large channel catfish (*Ictalurus punctatus*). After Finger *et al.* (1991).

	Surface area, mm^2		*Taste bud density, number* $\cdot$ mm^{-2}		*Number of taste buds*		
Fish size, cm	*Body*	*Fins*	*Body*	*Fins*	*Body*	*Fins*	*Total*
4.7-5.5	600	500	11.6	8.9	6 900	4 500	11 000
35-39.5	58 000	22 000	10.6	3.1	610 000	69 000	680 000

The results by Finger and co-authors demonstrate that the number of taste buds increases at a higher rate than the number of axon profiles. Thus, each nerve fibre innervates more taste buds in a large fish than a small one. This increase in ratio between sensory neurone and afferent fibre is similar to the situation in the developing lateral line (Zakon, 1984; Vischer, 1989a, b), auditory (Corwin, 1983), and visual systems (Johns & Easter, 1977). A possible consequence of these changes can be an increase in sensitivity or acuity of the gustatory system as well as increase in spectra of effective chemical stimuli and decrease the duration of food palatability assessment (see grayling Fig. 18).

Taste bud spreading

In many fish species the appearance of taste buds in the oropharyngeal cavity starts in the posterior part and proceeds in a rostral direction in a temporal sequence (Fig. 14). The first oral taste buds in channel catfish develop by day 6 pf, when 10-15 calretinin-positive cells occur on the rostral floor (basibranchials) of the pharynx. By day 7 pf, the

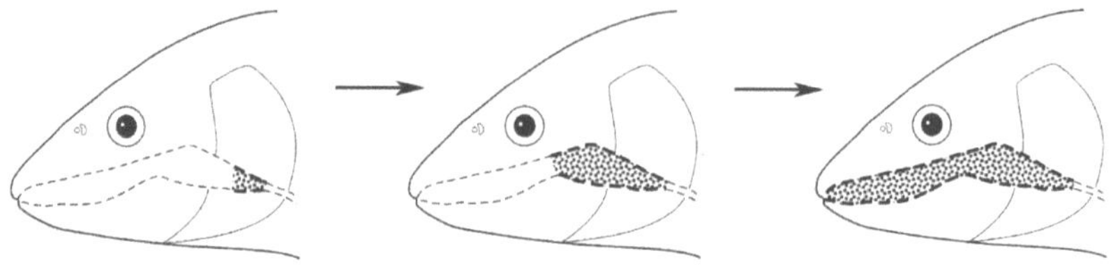

Fig. 14 Spread of oral taste buds. Schematic illustration of the appearance of taste buds close to the oesophagus and the spread towards the anterior part of the oropharyngeal cavity.

number of these cells doubles, and taste bud primordia can be recognised on the first three gill arches during ensuing several hours later. The first calretinin-positive cells on the rostral roof (anterior palatine) of the oral cavity are recognised at day 9 pf, and they appeared on the caudal roof (posterior palatine) of the oral cavity at day 10 pf. By day 12 pf, mature taste buds occur over all parts of the oropharynx (Northcutt, 2005).

In common carp, the taste bud primordia appear simultaneously in the oral and pharyngeal cavities in 22 h ph larvae, but soon after that, at 26 h ph, the pharyngeal taste buds are more numerous and better developed than the oral taste buds (Edwards, 1930). This observation is in agreement with that of Johnston (1910) for whitefish (*Coregonus*) and sucker (*Catostomus*), where taste buds first appear in the pharynx and oesophagus and then spread to the oral cavity. Taste buds in the pharyngeal region are more numerous than those in the oral cavity in alevins of rainbow trout and fry of mrigal (*Cirrhina mrigala*) (Moitra & Sinha, 1972; Twongo & MacCrimmon, 1977) and in several other marine and freshwater fish species like tilapia (*Tilapia*), goldfish (*Carassius*), smelt (*Hypomesus*), Japanese sea bass (*Lateolabrax*), puffer (*Fugu*), stone flounder (*Kareius*), sea bream (*Pagrus*), and big-scaled redfin (*Tribolodon hakonensis*) (Fig. 15) (Iwai, 1980; Komada, 1994).

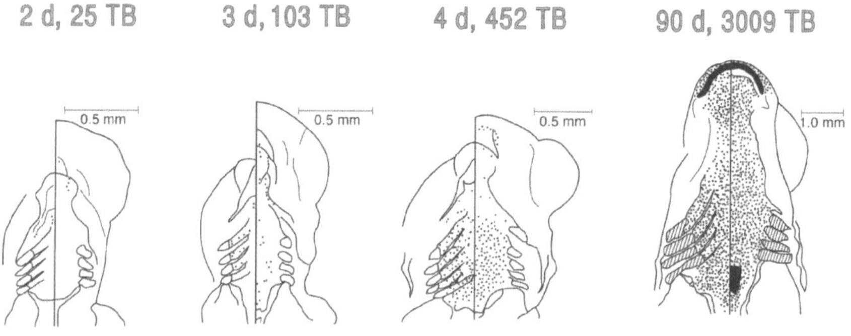

Fig. 15 Taste bud development in the mouth of big-scaled redfin (*Tribolodon hakonensis*). Distribution and total number of taste buds (black dots) in the oropharyngeal cavity at age 2, 3, 4, and 90 days post hatch. Left – ventral surface; right – dorsal surface; solid black indicates masticating plate. After Komada (1994).

Extraoral taste buds first appear on structures in the head, and development proceeds in a caudal direction (Landacre, 1907; Northcutt, 2005). Northcutt (2005) showed that the earliest indication of extraoral taste buds occurs as "anlages" in the maxillary barbels in channel catfish embryos during day 6 pf, and the first extraoral mature taste buds were recognised at day 10 pf. The first extraoral mature taste buds appear on the upper lip of the mouth. Taste bud maturation on the maxillary barbel and the other two barbels begins proximally and sweeps distally. At day 12 pf, mature taste buds occur over all parts of the head. The development of taste buds on both trunk and fins lags behind that of the head, with development of taste buds on the fins prior to the trunk. The first scattered calretinin-positive cells occurring on the lobes of the caudal fin at day 10 pf, on the dorsal fin at day 10-12 pf; on the anal, adipose and pectoral fins at day 12-13 pf. The pelvic fins are the last fins to develop calretinin-positive cells, which occur at first-feeding i.e. day 17 pf or day 8 ph (Fig. 16) (Northcutt, 2005).

On the trunk, taste bud primordia develop initially as separate rostral and caudal groups. Rostrally, they first appear at first-feeding on the belly and the dorsolateral surface of the trunk, adjacent to the dorsal fin. At this stage, taste bud primordia also occur along the caudal trunk; these taste bud primordia form dorsal and ventral lines, and appear to develop along a caudorostral gradient. Taste bud primordia continue to increase in number over most of the trunk, so that the rostral dorsolateral group is continuous with the dorsocaudal trunk line in 1 week-old larvae, although there is still a gap between the taste bud primordia on the belly and those of the ventrocaudal trunk line. Trunk taste buds begin to erupt in 1 week-old larvae, starting on the rostral dorsolateral surface of the trunk, adjacent to the dorsal fin, and on several of the fins. In 2 week-old larvae, taste buds have erupted along the entire length of the trunk, with the exception of the caudal lobe of the caudal fin, where eruption is just beginning (Northcutt, 2005).

Barbels

In fishes, the role of barbels in food search and preliminary sensory control of food quality seems obvious. The density of taste buds is usually highest on the barbels. The number of barbels in fish species varies from a single one in gadids, to several pairs in acipenserids, cyprinids, balitorids, silurids, cobitids, and many other groups of fish. Localisation and spreading of taste buds on the barbel surface during fish development differs between species.

In channel catfish, the first extraoral mature taste buds appear on the maxillary barbels. Taste bud maturation on all barbels begins proximally and sweeps distally (Northcutt, 2005). Sturgeons possess two pairs of each barbel; and the first taste buds appear before first-feeding on the rostral tip of the barbels and are seldom found in the middle or basal region of barbel (Devitsina & Kazhlaev, 1993b). The planktivorous paddlefish has only one pair of barbels; and taste buds are recognised only in the basal part of barbels and they appear after the larvae start feeding. In zebrafish, with one pair of barbels, mature taste buds on the barbel surface are not apparent before the juvenile stage. Mature taste buds with open taste pores are present early in the development at

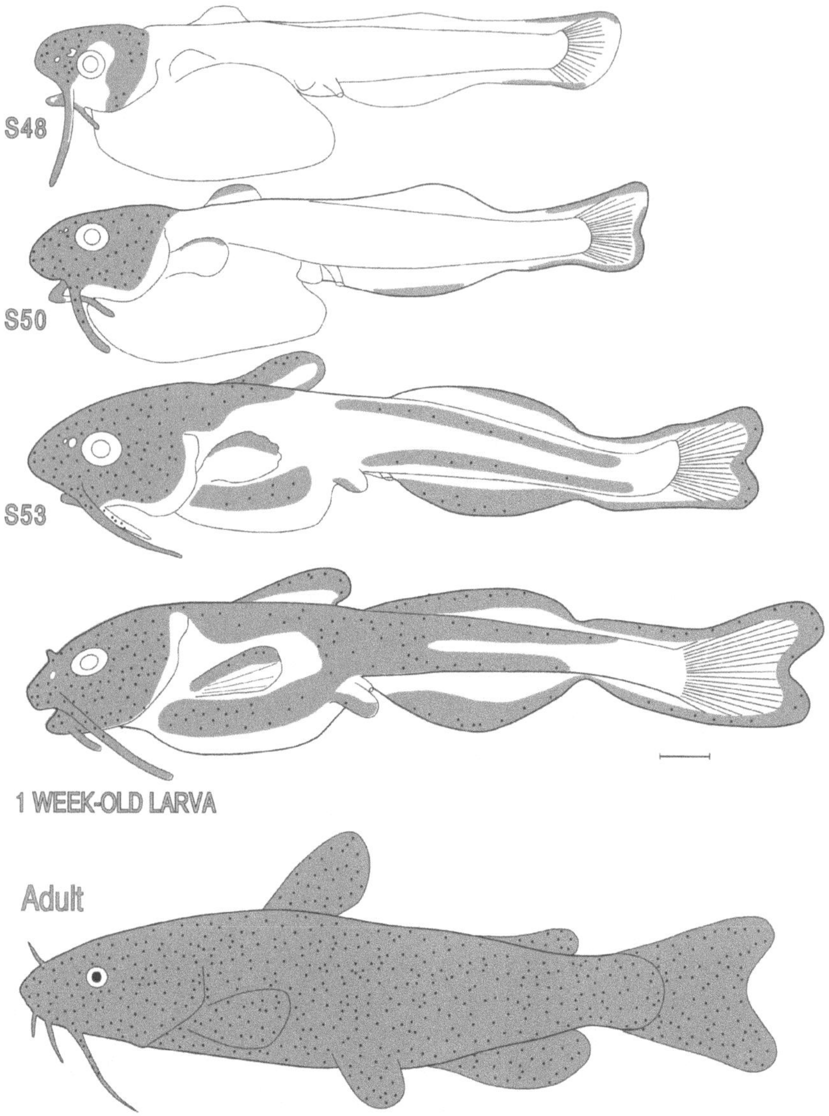

Fig. 16 Spread of extraoral taste buds in catfish. During development of the channel catfish (*Ictalurus punctatus*), the first taste buds appear on the head, and then they are found on the tail and fins, to be present over most of the skin at the age of one week. Larval stages are indicated according to the most of normal development for channel catfish, Armstrong & Child (1962). In adult yellow bullhead (*Ictalurus natalis*), the taste buds are present on the total body surface. Scale bar represents 1 mm for channel catfish, and 1 cm for the adult yellow bullhead. After Northcutt (2005), Fig. 7. Reprinted with permission of Wiley-Liss, Inc., a subsidiary of John Wiley & Sons, Inc., and Atema (1971), Fig. 1. Reprinted with permission of S. Karger AG, Basel.

the region where the barbels start growing, and these taste buds remain at the tip of the developing barbel. New taste buds are added to the barbel shaft between its base and tip (Hansen *et al.*, 2002). In Atlantic cod, the first taste buds can be recognised on the barbel in 20 mm larvae, much later than the appearance of the taste buds on the snout and lips (8 mm). In juvenile cod the highest density of taste buds is at the tip of the barbel and declined in the middle region (Harvey & Batty, 1998).

Growth and abundance of taste buds

The shape of taste buds in rainbow trout changes from globular to almost oval during the first weeks after first-feeding. This increase in taste bud height appears to be due to growth in the apical part of the taste bud (Twongo & MacCrimmon, 1977). The size of taste buds, the height and number of cells, as well as the diameter of the taste pore increase during fish development (Moitra & Sinha, 1972; Pevzner, 1985; McCormick, 1993; Harvey & Batty, 1998). In the reef goatfish (Mullidae), the size of taste buds on the barbels increases dramatically (up to 100%) during the 6-12 hours period of settlement. During this brief period, barbels show a rapid growth and differentiation which coincides with the transition of fish from the pelagic to benthic life (McCormick, 1993). In Atlantic cod larvae 20 mm BL, the mean taste pore diameter is 2.2 μm and increases to 7.2 μm in juveniles 90 mm BL. Beyond this size, mean taste pore diameters increase slowly, reaching a maximum of 10.4 μm in cod 225 mm TL (Harvey & Batty, 1998).

Taste bud size and densitiy increases dramatically during the life of fish. A 20 mm BL Atlantic cod larva, has two taste buds on a barbel 0.12 mm long, at 39 mm BL the number has increased to 90 taste buds and the barbel length to 1 mm. The total number of taste buds on the barbel increases from 3000 in an 80 mm fish to 6000 in a 420 mm long cod. The highest density of taste buds was found on the edge of the tube forming the anterior naris, reaching 850 · mm^{-2} in a cod 148 mm long (Harvey & Batty, 1998).

Changes associated with development have also been found in the oropharyngeal cavity. The mandibula, maxilla, tongue, palate, pharynx, hypopharyngeal and hyperpharyngeal pads had significantly higher density in juvenile (39-42 mm) Japanese croaker (*Argyrosomus japonicus*) than in fingerlings (106-121 mm). The total number of taste buds increases with fish growth (Lin *et al.*, 1998). An increase in the total number of taste buds during fish development was noted for other species (Twongo & MacCrimmon, 1977; Finger *et al.*, 1991). In mammals, there is a high number of taste buds at the time of birth, which is followed by taste bud loss (Mistretta *et al.*, 1988).

Ecological adaptations

The time and occurrence of the first taste buds in the oropharyngeal cavity or on the body surface has been associated with the nature of the food and the feeding behaviour of fishes. Feeding in various fish larvae is based on well developed visual perception and their feeding is linked mainly to daytime. Larvae of these fish behave like predators searching and pursuing the prey. The appearance of the first oral taste buds in the pharyngeal region of the oropharyngeal cavity of early larvae resemble the distribution of taste buds seen in adult predators or plankton feeding fish, that feed by sight only. These fish have relatively small taste buds seen in the pharynx and esophagus mucosa (Al-Hussaini, 1947; Pasha, 1964; Western, 1969). It was suggested that taste buds which cover the caudal part of oropharynx are sensitive to deterrent stimuli (Kasumyan & Morsi, 1996). The ability of the early larvae to respond to deterrents before they react to stimulants might be an advantage.

In sturgeons, extraoral taste buds develop prior to oral ones and the density of taste buds increases rapidly on the barbel surface. The advanced development of the extraoral gustatory system both morphologically and functionally, gives an obvious advantage for these fish having very poor vision. Touching a food item with barbels is the first step in the feeding sequence of sturgeons (Kasumyan & Kazhlaev, 1993a; Pavlov & Kasumyan, 1998). Extraoral taste is an important cue in searching and localisation of prey also in yellow bullhead (*Ictalurus natalis*) (Atema, 1971). In this species, taste buds on the barbel surface erupt several days before first-feeding. Taste buds on the trunk begin to form later in channel catfish (Northcutt, 2005). It is not clear why there is such a marked lag in the development of these receptors on the trunk, but it is possible that a later increase in the number of extraoral taste buds only enhances the possibilities for localisation of prey.

The changes in taste bud abundance are correlated with fish feeding behaviour at different life stages. In the mrigal, a cyprinid, the number of taste buds on the lips and in the oral cavity is low in the fry and increases in the juveniles and adults. This development is correlated with changes in feeding habits, from sight capture of zooplankton by fry, to the gustatory selection of benthos by older fish (Moitra & Sinha, 1972). However, in Nile tilapia, oral taste buds are abundant in larvae of 14-20 mm SL, and decrease as the fish length increases. This variation appears to be correlated with the change in diet from particulate feeding to filter feeding, and the consequent change in function of the pharyngeal pads from handling single prey items to mucus-bound phytoplankton (Northcott & Beveredge, 1988).

Function

Taste preferences

Many fish larva use vision when searching for prey. Yet, the evaluation of taste quality of food items and the decision to swallow or to reject prey is based on the information from the gustatory system. Many studies concerning ontogeny of the gustatory system in fish have focused on the morphology of the taste buds. According to these studies, the taste buds in many fish species appear several days or even hours before fish larvae start to catch food items, which is at the beginning of exogenous feeding, the time when the estimation of taste quality of food becomes essential and necessary.

Experiments performed on Siberian and stellate sturgeons have shown that there is good correspondence between morphology and function in the development of gustatory system. The larvae of these species at the "start-feed" stage have both oral and extraoral taste buds, and they discriminate artificial food pellets containing different taste stimuli. Taste behavioural responses mediated by oral or extraoral taste buds were the same to sucrose, sodium chloride and calcium chloride in larvae of "start-feed" age and in juveniles. Citric acid was an indifferent substance for sturgeon larvae, but evoked conspicuous oral and extraoral taste responses in sturgeon juveniles (Kasumyan, 1992; Kasumyan & Kazhlaev, 1993a).

Studies on the taste preferences of larvae and juveniles have demonstrated that larvae have a more restricted spectrum of effective taste stimuli than the juveniles. Comparison of oral and extraoral taste responses in Russian sturgeon shows that a smaller number of free amino acids acted as incitants and stimulants in larvae (12-15 days ph) than in juveniles (30-35 days ph). For larvae, 11 amino acids were incitants and only one amino acid a stimulant, while for juveniles the numbers were 16 and 6, respectively.

One can ask if the oral and extraoral taste preferences are similar within an individual, and if they changes with age. In the sturgeon larvae, the extraoral and oral taste preferences for amino acids were not correlated (Spearman rank correlation coefficient 0.36). In contrast, the juveniles showed a positive correlation between oral and extraoral preferences of the same amino acids (Spearman rank correlation coefficient 0.79). Interestingly, the oral taste preferences change dramatically (Spearman rank correlation coefficient 0.04), but the extraoral preferences were similar in larvae and juveniles (Spearman rank correlation coefficient 0.69) (Fig. 17) (Kasumyan, 1992; Kasumyan & Døving, 2003).

It has been demonstrated that older fish have a wider spectrum both of stimulants and deterrents. For 4 month-old juveniles (2.7-3.0 cm, TL) of the lake trout (*Salvelinus namaycush*), only L-cysteine was highly palatable, while 20 other free amino acids were ineffective. For the 11-12 months old juveniles (7-9 cm, TL) L-cysteine, L-glutamic acid, L-aspartic acid and L-phenylalanine were stimulants and three amino acids (L-arginine, glycine and L-lysine) were deterrents (Kasumyan & Sidorov, 2001).

There are other examples showing that fish larvae at the "start-feed" stage have the ability to discriminate among diets with different taste substances and select particular ones and ignore others. First-feeding chinook salmon fry (0.4 g body weight) significantly increase consumption of a basal fish diet upon adding glycine (0.01 M). The addition of trimethylamine (0.01 M) to the basal fish diet caused a drop in food consumption (Hughes, 1991). The effect of both substances, glycine and trimethylamine, was more prominent for the bigger chinook salmon (2.0 g in body weight) (Hughes, 1993). At the age of 1-3 weeks ph, larvae of common carp, rainbow trout, Northern pike, zander (*Sander lucioperca*), Nile tilapia and vendace (*Coregonus albula*) were offered live *Artemia* nauplii soaked in different taste substances. Larvae of all species rejected nauplii soaked in citric acid solution, except tilapia. All larvae preferred bitter taste (quinine), with the exception of pike and vendace. Salty taste was also preferable for all species. At the older stages, the taste preferences of some species may change slightly for some substances (Appelbaum, 1980).

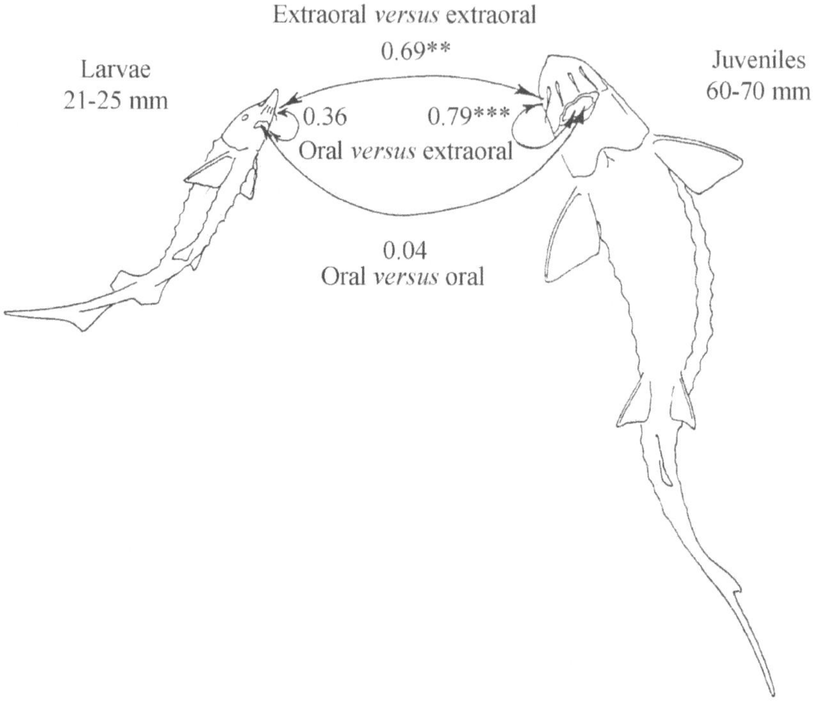

Fig. 17 Development of oral and extraoral taste preferences. Values of Spearman rank correlation coefficient between oral- and extraoral taste preferences for free amino acids in larvae and juveniles of Russian sturgeon (*Acipenser gueldenstaedtii*). After Kasumyan (1999). See text for details.

As the fish grows, there are other functional characters of the gustatory system that develop along with the taste preferences. Grayling (*Thymallus thymallus*) 2-3 month-old juveniles, retain pellets much longer than older graylings (5-6 months) before they swallow or reject the pellets. In older juveniles the retention time does not depend upon the palatability of pellets. In other words, older juveniles use a relatively short and constant period of time for evaluation of taste quality of the food items caught. The older juveniles swallow pellets or ultimately reject them at the first grasp, while younger juveniles need repeated tests, they usually spit out and catch the pellet again several times before they swallow or ultimately reject the pellet (Fig. 18) (A. Kasumyan, unpublished). The same changes in retention time and number of repeated grasps associated with fish development were found in behavioural taste responses of lake trout and brown trout (Kasumyan & Sidorov, 2001, 2005).

Thresholds

An important parameter of the fish gustatory system is the minimum concentration needed to give a positive or negative response of a fish towards taste stimuli. There is

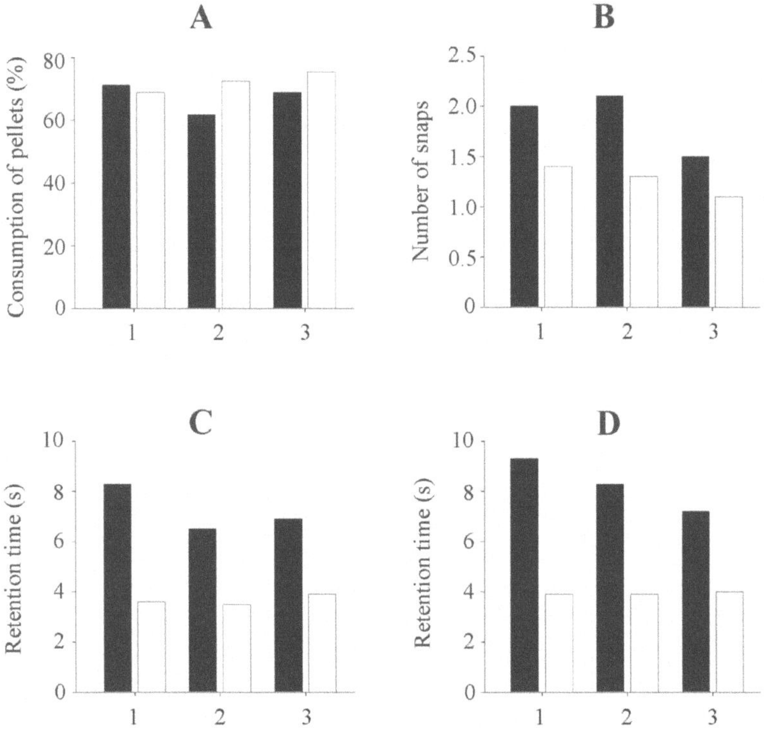

Fig. 18 Development of taste parameters in grayling (*Thymallus thymallus*). The bars show responses of two stages of juvenile graylings to agar pellets containing 0.1 M glutaric acid (1), 0.1 M α-keto-glutaric acid (2), and 0.1 M citric acid (3). The black bars are for fish with mean BL 55 mm and the white bars are for fish with mean BL 90 mm. A, consumption of pellets; B, number of snaps before the fish swallow or ultimately rejected the pellet; C, retention time after first capture; D, retention time for all captures.

only one published study where behavioural taste responses of fish in different ages were compared. It was found that the threshold concentration for effective substances tested is the same for fish in different ages. In Siberian sturgeon of three age groups (larvae at 12 days old, 17-19 mm; larvae at 17 days old, 22-24 mm; and juveniles at 38 days old, 49-52 mm), the threshold concentrations were 1.7 M for sodium chloride for oral taste and 9×10^{-2} M for calcium chloride for both oral and extraoral taste. In stellate sturgeon of two age groups (larvae at 12 days old, 22-23 mm and juveniles at 35 days old, 32-39 mm), the threshold concentrations for both oral and extraoral taste was 9×10^{-2} M for calcium chloride (Kasumyan & Kazhlaev, 1993b).

The combination of behavioural assays and morphological studies has shown that there is good correspondence between the development of the gustatory system and the ability to discriminate taste properties of food items when the larvae are at the start feeding stage. Larvae at this age can discriminate some taste substances only and release an adequate behavioural taste response. The sensitivity of the gustatory system does not change dramatically with fish development. Nevertheless, taste perception in fish larvae is less developed than in juveniles and adult fish and a relatively narrow spectrum of effective stimuli is found. Exceptions are several marine species in which the first taste buds appear long after start feeding. The limited taste perception may be one reason for mass mortality of fish larvae and early postlarvae among capelin (*Mallotus villosus*) and Atlantic herring which is caused by a dinoflagellate that is toxic to the fish (Gosselin *et al.*, 1989). The adult fish are highly sensitive towards the taste of toxins, which enable them to avoid poisonous food (Yamamori *et al.*, 1988).

SOLITARY CHEMOSENSORY CELLS (SCC)

The epidermis of lamprey as well as that of all groups of fish possesses SCC that are single, spindle-shaped, secondary epidermal sensory cells with single or several apical microvilli that protrude between the squamous cells of the superficial layer of epidermis. The SCCs are distributed relatively evenly over the body surface, but with higher densities on the head than on the trunk. These dispersed SCC create the generalised SCC system (Kotrschal, 1995). Densities of SCC on the body surface vary between the few fish species studied from several hundred up to several thousand per mm^2. The highest density of SCC ($100\,000 \cdot mm^{-2}$) was found in a peculiar, chemoreceptive dorsal fin of rockling (*Ciliata mustela*) (Gadidae) and was called the specialised SCC system (Kotrschal *et al.*, 1984; Kotrschal, 1995). The apical endings of the SCC vary in shape from brushlets of 2-7 small, equally sized microvilli, 0.5-1 μm in length, emerging independently from each other from the distal cell surface to single finger-like microvilli, which may exceed 3 μm in length (Fig. 19).

Fig. 19 SEM of the peripheral part of the solitary chemosensory cell. The apical part with microvilli of a single chemosensory cell in the oral disc papillae of an adult brook lamprey (*Lampetra planeri*). After Baatrup & Døving (1985).

There is only one study concerning the SCC development and that was made on the zebrafish (Kotrschal *et al.*, 1997). The first SCC apices were found by TEM at day 3 pf, just prior to hatching. They protrude through a superficial layer of epidermis simultaneously on the head and on the trunk. The density of SCC increases from hatching to day 25 pf from a few hundred up to $4{\times}10^3 \cdot mm^{-2}$. The density of SCC on the head is 2 to 5 times higher than on the body (Fig. 20). There are in total around $3.2{\times}10^5$ SCC on the head of a sexually mature individual at day 180 pf and more than 10^6 SCC on the entire body surface.

It seems that the shape of apical endings of solitary chemosensory cells changes from oligovillous to monovillous as the zebrafish develop from hatching to adult. It is not settled how SCCs reach the monovillous condition during fish growth; either via a change of individual receptor cells apices or via a succession of cells, or both.

It may be assumed that SCCs become functional upon the appearance of microvilli at the body surface at hatching (Kotrschal *et al.*, 1997). But data concerning functional development of a generalised SCC system as well as development of the specialised SCC system is lacking.

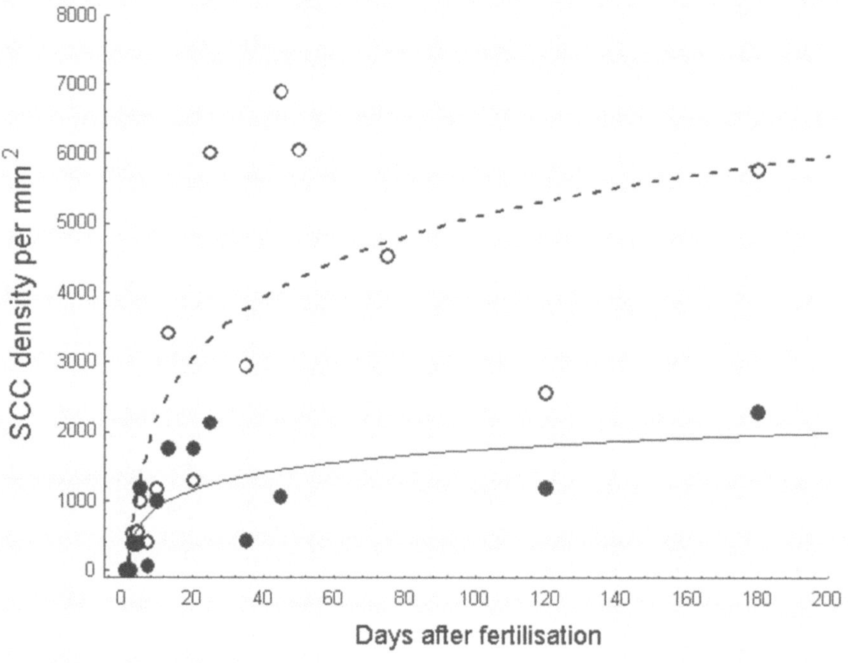

Fig. 20 Development of densities of solitary chemosensory cells (SCC) on the head and body surface of zebrafish (*Danio rerio*). The graph illustrates changes in density of SCC on the head (dashed line, open circles) and on the body (solid line, black circles) in zebrafish from hatching to sexual maturity. Data from Kotrschal *et al.* (1997).

FUTURE PERSPECTIVES

There has been a remarkable progress in our understanding of the anatomy and function of the chemosensory organs in fishes during the last decades. These advances are reflected in our understanding of the developmental processes of these organs. An outstanding feature in the development of the chemosensory systems in fishes is that the total number of olfactory sensory neurones, of taste buds, and SCC, all increase with age, or more accurately with BL. The same trend holds for other sensory systems, e.g. the number of hair cells in the macula and the number of superficial neuromasts in the lateral line, and the number of hair cells in both canal and superficial neuromasts (Fuiman *et al.*, 2004), all increase with size. This is in contrast to the situation in mammals, where the number of taste buds is not related to age (Arvidson, 1979; Mistretta & Baum, 1984; Bradley *et al.*, 1985).

Due to the use of specific cellular markers, considerable progress has been achieved in our understanding of olfactory placode formation (Whitlock & Westerfield, 2000). However, detailed investigations of various groups of actinopterygian and sarcopterygian fish are still needed to better understand the variations in the early development of the olfactory epithelium (Hansen & Zeiske, 1993; Zeiske *et al.*, 2003). We also need to relate such advances to evolution of chemosensory organs.

There has been more activity related to the description of the olfactory organ development than the other chemosensory systems. Thus, numerous articles describe different aspects of olfactory system formation, yet less is known concerning gustatory system development; and there is only one article on the development of the common chemical sense. The common chemical sense has been known as a separate chemosensory system in fish for nearly a century, but still our knowledge remains scant concerning the function of this chemical sense. It is also unclear if this sense is the same as the SCC system.

There are few studies on the functional aspects of development in comparison with those on the morphology as portrayed in this treatise. This difference is true for olfaction and taste. A possible reason for the lack of functional studies might be the difficulties in using electrophysiological methods on the larvae that are appropriate for studies of functional characteristics of sensory systems in adult fish, due to the small size and vulnerability. In this respect, the embryos of sharks, skates, and premetamorphic leptocephali of many elopomorph species which have an unusually large size might be suitable for electrophysiological studies of chemosensory development.

Of the many questions that are not posed, we may highlight the following: Why is it that the appearance of ciliated olfactory sensory neurones precedes the microvillous ones? How does development of ciliated and microvillous sensory neurones relate to the ability of the fish to respond to different types of odorants? Why do olfactory sensory neurones and other cell types of olfactory epithelium appear in such an early stages of fish ontogeny, even before hatching? Are there different types of precursor cells in the olfactory epithelium and taste buds, or does one type of precursor cell give rise to different sensory neurones? How do various environmental factors, such as water temperature alter the development of chemosensory systems? It can be stated that the formation and rate of development varies among the chemosensory systems: the olfactory system appears earlier than the gustatory system and common chemical sense. Is the difference between these systems due to the type of sensory cells? There are primary sensory neurones in the olfactory system, but secondary sensory cells in both the gustatory system and the common chemical sense. For the olfactory system, the interval between the appearance of olfactory sensory neurones and the functional olfactory system is longer than for the same processes in the gustatory system, where the period between taste bud appearance and the ability of fish larvae to assess the palatability of food items is relatively short. This difference also might be connected with the type of sensory cells (primary or secondary). Based on this suggestion, the assumption of Kotrschal *et al.* (1997) that the common chemical sense becomes functional upon the appearance of microvilli of SCC at the body surface at hatching looks quite reasonable.

The difference between olfactory and taste spectra in larvae and adult fish is poorly investigated and our notion about the rules of spectra formation is scarce (what types of stimuli are first and what stimuli are second or third; for instance, stimulants or deterrents? And between stimulants – are the most palatable substances for adult fish present as the first taste spectra of early larvae?). Basic knowledge about chemosensory system development might be of significant value when applied to fish cultivation. The majority of articles on development are focused on structural changes in the chemosensory end organs, i.e. olfactory organs, and taste buds. There are a few articles which describe the nerve and tract development, or primary and secondary centres for olfaction and taste. We still lack understanding of the adaptive role of the impressive increase in the number of olfactory sensory neurones, taste buds and common chemical sense with fish growth. Is the number of sensory neurones in the adult superfluous? This question is asked, because only 30-40% of the olfactory sensory neurone population is sufficient to maintain normal sensitivity to the alarm pheromone in cyprinids (Pashchenko & Kasumyan, 1984).

Acknowledgments

The present study was supported by the Russian Foundation for Basic Research (RFBR grant no. 07-04-00793) and the Research Council of Norway grant 159213/V40. We are grateful to Klaus Reutter, Finn-Arne Weltzien and Nigel Finn for comments on earlier versions of the manuscript. The artwork was made by John Reierstad, Faculty of Science, University of Oslo.

References

Al-Hussaini AH (1947) The anatomy and histology of the alimentary tract of the plankton feeder *Atherina forskali*. J Morphol 80: 251-286

Appelbaum S (1980) Versuche zur Geschmacksperzeption einiger Susswasserfische im larvalen und adulten Stadium. Archiv Fischereiwiss 31: 105-114

Appelbaum S, Adron JW, George SG, Mackie AM, Pirie BJS (1983) On the development of the olfactory and the gustatory organs of the dover sole, *Solea solea*, during metamorphosis. J Mar Biol Assoc UK 63: 97-108

Appelbaum S, Riehl R (1997) Scanning electron microscopic observation of the chemo- and mechanoreceptors of carp larvae (*Cyprinus carpio*) and their relationship to early behaviour. Aquat Living Resour 10: 1-12

Argo S, Weth F, Korsching SI (2003) Analysis of penetrance and expressivity during ontogenesis supports a stochastic choice of zebrafish odorant receptors from predetermined groups of receptor genes. Eur J Neurosci 17: 833-843

Armstrong PB, Child JS (1962) Stages in the Development of *Ictalurus nebulosus*. Syracuse University Press, Syracuse, p 23

Arvidson K (1979) Location and variation in number of taste buds in human fungiform papillae. Scand J Dent Res 87: 435-442

Atema J (1971) Structures and functions of the sense of taste in the catfish (*Ictalurus natalis*). Brain Behav Evol 4: 273-294

Atema J (1980) Chemical senses, chemical signals, and feeding behavior in fishes. In: Bardach JE, Magnuson JJ, Mary RC, Reinhart JM (eds) Fish Behavior and Its Use in the Capture and Culture of Fishes. Proceedings of the Conference on the Physiological and Behavioral Manipulation of Food Fish as Production and Management Tools, Bellagio, Italy, 3-8 November 1977, p 57-94

Baatrup E, Døving KB (1985) Physiological studies on solitary receptors of the oral disc papillae in the adult brook lamprey, *Lampetra planeri* (Bloch). Chem Senses 10: 559-566

Balfour FM, Parker WN (1882) On the structure and development of *Lepisosteus*. Phil Trans R Soc Lond 2: 359-442

Ballard WW, Needham RG (1964) Normal embryonic stages of *Polyodon spathula* (Walbaum). J Morphol 114: 465-478

Bardach JE, Todd JH, Crickmer RK (1967) Orientation by taste in fish of genus *Ictalurus*. Science 155: 1276-1278

Barlow LA (2000) Gustatory system development. In: Finger TE, Silver WL, Restrepo D (eds) The Neurobiology of Taste and Smell. John Wiley & Sons, New York, p 395-422

Barlow LA, Northcutt RG (1998) The role of innervation in the development of taste buds: insights from studies of amphibian embryos. Ann NY Acad Sci 855: 58-69

Barnett C (1977) Chemical recognition of the mother by the young of the cichlid fish, *Cichlasoma citrinellum*. J Chem Ecology 3: 461-466

Barth AL, Justice NJ, Ngai J (1996) Asynchronous onset of odorant receptor expression in the developing zebrafish olfactory system. Neuron 16: 23-34

Bartsch P (1993) Development of the snout of the Australian lungfish *Neoceratodus forsteri* (Krefft, 1870), with special reference to cranial nerves. Acta Zool 74: 15-29

Bemis WE, Grande L (1992) Early development of the actinopterygian head. I. External development and staging of the paddlefish *Polyodon spathula*. J Morphol 213: 47-83

Berliner K (1902) Die Entwicklung des Geruchsorgans der Selachier. Arch Mikrosk Anat Entw Mech 60: 386-406

Bertmar G (1965) The olfactory organ and upper lips in Dipnoi, an embryological study. Acta Zool 46: 1-40

Bertmar G (1969) The vertebrate nose, remarks on its structural and functional adaptation and evolution. Evolution 23: 131-152

Bertmar G (1972) Secondary folding of the olfactory organ in young and adult sea trout. Acta Zool 53: 113-120

Boglione C, Bronzi P, Cataldi E, Serra S, Gagliardi F, Cataudella S (1999) Aspects of early development in the Adriatic sturgeon *Acipenser naccarii*. J Appl Ichthyol 15: 207-213

Bradley RM, Stedman HM, Mistretta CM (1985) Age does not affect numbers of taste buds and papillae in adult rhesus monkeys. Anat Rec 212: 246-249

Breucker H, Zeiske E, Melinkat R (1979) Development of the olfactory organ in the rainbow fish *Nematocentris maccullochi* (Atheriniformes, Melanotaeniidae). Cell Tissue Res 200: 53-68

Buck L, Axel R (1991) A novel multigene family may encode odorant receptors: a molecular basis for odor recognition. Cell 65: 175-187

Byrd CA, Jones JT, Quattro JM, Rogers ME, Brunjes PC, Vogt RG (1996) Ontogeny of odorant receptor gene expression in zebrafish, *Danio rerio*. J Neurobiol 29: 445-458

Caprio J (1975) High sensitivity of catfish taste receptors to amino acids. Comp Biochem Physiol 52A: 247-251

Connaughton VP, Schuur A, Targett NM, Epifanio CE (1994) Chemical suppression of feeding in larvae weakfish (*Cynoscion regalis*) by trochophores of the serpulid polychaete *Hydroides dianthus*. J Chem Ecol 20: 1763-1771

Cook MH, Neal HV (1921) Are the taste-buds of elasmobranchs endodermal in origin? J Comp Neurol 33: 45-63

Corwin JT (1983) Postembryonic growth of the macula neglecta auditory detector in the ray, *Raja clavata*: continual increases in hair cell number, neural convergence, and physiological sensitivity. J Comp Neurol 217: 345-356

Costanzo RM, Graziadei PPC (1987) Development and plasticity of the olfactory system. In: Finger TE, Silver WL (eds) Neurobiology of Taste and Smell. John Wiley & Sons, New York, p 233-250

Davenport CJ, Caprio J (1982) Taste and tactile recording from the *ramus recurrens facialis* innervating flank taste buds in the catfish. J Comp Physiol A147: 217-229

Dean B (1897) On the larval development of *Amia calva*. Zool Jb, Abt Syst 9: 639-672

Dempsey CH (1978) Chemical stimuli as a factor in feeding and intraspecific behaviour of herring larvae. J Mar Biol Assoc UK 58: 739-747

Derivot JH (1984) Functional anatomy of the peripheral olfactory system of the African lungfish *Protopterus annectens* Owen: development of the primary olfactory pathway during postembryonic growth. Am J Anat 171: 15-23

Devitsina GV (1998) Development of chemoreception organ in the pike *Esox lucius*. J Ichthyol 38: 468-477

Devitsina GV, Gadzhieva AR (1996) Dynamics of morphological development of gustatory system during the early ontogenesis of two representatives of acipenserids *Acipenser nudiventris* and *A. persicus*. Voprosi Ikhtiologii 36: 642-653 (in Russian)

Devitsina GV, Kazhlaev AA (1993a) Chemoreceptor organs in early juvenile paddlefish, *Polyodon spathula*. J Ichthyol 33: 143-149

Devitsina GV, Kazhlaev AA (1993b) Development of chemosensory organs in Siberian sturgeon, *Acipenser baeri*, and stellate sturgeon, *A. stellatus*. J Ichthyol 33: 9-19

Dieulaffé L (1905) Les fosses nasals des vertébrés (morphologie et embryologie). J Anat Physiol (Paris) 41: 300-318, 478-560, 658-680

Dukes JP, Deaville R, Bruford MW, Youngson AF, Jordan WC (2004) Odorant receptor gene expression changes during the parr-smolt transformation in Atlantic salmon. Mol Ecol 13: 2851-2857

Døving KB, Dubois-Dauphin M, Holley A, Jourdan F (1977) Functional anatomy of the olfactory organ of fish and the ciliary mechanisms of water transport. Acta Zool 58: 245-255

Døving KB, Hamdani EH, Höglund E, Kasumyan AO, Tuvikene A (2005) A review on the chemical and physiological basis of alarm reactions in cyprinids. In: Reuter K, Kapoor BG (eds) Fish Chemosenses. Science Publishers Inc, Enfield, NH, USA, p 133-163

Døving KB, Mårstøl M, Andersen JR, Knutsen JA (1994) Experimental evidence of chemokinesis in newly hatched cod larvae (*Gadus morhua* L.). Mar Biol 120: 351-358

Døving KB, Stabell OB (2003) Trails in open waters: Sensory cues in salmon migration. In: Collin SP, Marshall NJ (eds) Sensory Processing in the Aquatic Environment. Springer-Verlag, New York, p 39-52

Eaton TH (1956) Notes on the olfactory organs in Centrarchidae. Copeia 1956: 196-199

Edwards LF (1930) Origin of taste buds in the oro-pharyngeal cavity of the carp (*Cyprinus carpio* Linnaeus). Ohio J Sci 30: 385-397

Elston R, Corazza L, Nickum JG (1981) Morphology and development of the olfactory organ in larval walleye, *Stizostedion vitreum*. Copeia 1981: 890-893

Evans RE, Zielinski B, Hara TJ (1982) Development and regeneration of the olfactory organ in rainbow trout. In: Hara TJ (ed) Chemoreception in Fishes. Elsevier, Amsterdam, p 15-37

Finger TE (1976) Gustatory pathways in the bullhead catfish. I. Connections of the anterior ganglion. J Comp Neurol 165: 513-526

Finger TE, Morita Y (1985) Two gustatory systems: facial and vagal gustatory nuclei have different brainstem connections. Science 227: 776-778

Finger TE, Drake SK, Kotrschal K, Womble M, Dockstader KC (1991) Postlarval growth of the peripheral gustatory system in the channel catfish, *Ictalurus punctatus*. J Comp Neurol 314: 55-66

Fishelson L, Baranes A (1997) Ontogenesis and cytomorphology of the nasal olfactory organs in the Oman shark, *Iago omanensis* (Triakidae), in the Gulf of Aqaba, Red Sea. Anat Rec 249: 409-421.

Fritzsch B, Sarai PA, Barbacid M, Silos-Santiago I (1997) Mice lacking the neurotrophin receptor trkB lose their specific afferent innervation but do develop taste buds. Int J Dev Neurosci 15: 563-576

Fuiman LA, Higgs DM, Poling KR (2004) Changing structure and function of the ear and lateral line system of fishes during development. Am Fish Soc Symp 40: 117-144

Gawrilenko A (1910) Die Entwicklungsweise des Geruchsorgans bei *Salmo salar* (Zur Stammesentwickelung des Jacobsonschen Organs). Anat Anz 36: 411-427

Gomahr A, Palzenberger M, Kotrschal K (1992) Density and distribution of external taste buds in cyprinids. Environ Biol Fishes 33: 125-134

Goodrich ES (1958) Studies on the Structure and Development of Vertebrates, Vol 2. Dover Publication, Inc, New York, p 486-837

Goryunova VB, Shagaeva VG, Shevchenko NF (2000) An analysis of developmental abnormalities in morphology of the sturgeon fishes larvae and juveniles in the Volga-Caspian basin under condition of artificial reproduction. J Ichthyol 40: 804-809

Gosselin S, Fortier L, Gagne JA (1989) Vulnerability of marine fish larvae to the toxic dinoflagellate *Protogonyaulax tamarensis*. Mar Ecol Prog Ser 57: 1-10

Halama L (1982) Morphometry of the olfactory organs of the rainbow trout (*Salmo gairdneri* Rich.). Acta Biol Cracow Ser Zool 24: 1-7

Hall JMH, Bell ML, Finger TE (2003) Disruption of sonic hedgehog signaling alters growth and patterning of lingual taste papillae. Dev Biol 255: 263-277

Hamdani EH, Døving KB (2005) Functional organization of the olfactory system in fish. In: Ladich F, Collin SP, Moller P, Kapoor BG (eds) Fish Communication in Fishes, Vol 1. Science Publishers, Enfield, NH, USA, p 223-257

Hansen A, Zeiske E (1993) Development of the olfactory organ in the zebrafish, *Brachydanio rerio*. J Comp Neurol 333: 289-300

Hansen A, Zeiske E (1995) Development of the olfactory organ in the zebrafish: an electron microscopic and immunocytochemical study of early differentiation and growth. Biophysics 40: 159-170

Hansen A, Finger TE (2000) Phyletic distribution of crypt-type olfactory receptor neurons in fishes. Brain, Behav Evol 55: 100-110

Hansen A, Reutter K, Zieske E (2002) Taste bud development in the zebrafish, *Danio rerio*. Dev Dynam 223: 483-496

Harvey R (1996) The olfactory epithelium in plaice (*Pleuronectes platessa*) and sole (*Solea solea*), two flatfishes with contrasting feeding behaviour. J Mar Biol Assoc UK 76: 127-139

Harvey R, Batty RS (1998) Cutatenous taste buds in cod. J Fish Biol 53: 138-149

Herrick CJ (1901) The cranial nerves and cutaneous sense organs of the Notrth American silurid fishes. J Comp Neurol Psychol 11: 177-249

His W (1892) Die Entwickelung der menschlichen und thierischen Physiognomien. Arch Anat Entwicklungsgeschichte: 384-424

Hoffmann CK (1884) Zur Ontogenie der Knochenfische. Arch Mikrosk Anat 23: 45-108

Holm JF (1894) The development of the olfactory organ in the Teleostei. Morphol Jahr 21: 620-624

Hughes SG (1991) Response of first-feeding spring chinook salmon to four potential chemical modifiers of feed intake. Prog Fish-Cult 53: 15-17

Hughes SG (1993) Single-feeding response of chinook salmon fry to potential feed intake modifiers. Prog Fish-Cult 55: 40-42

Hunter JR, Thomas GL (1974) Effect of prey distribution and density on the searching and feeding behaviour of larval anchovy *Engraulis mordax* Gigard. In: Blaxter JHS (ed) The Early Life History of Fish. Springer-Verlag, Berlin, p 559-574

Iwai T (1980) Sensory anatomy and feeding of fish larvae. In: Bardach JE, Magnuson JJ, May RC, Reinhart JM (eds) Fish Behaviour and its use in the Capture and Culture of Fishes, Manila, p 124-145

Jahn LA (1972) Development of the olfactory apparatus of the cutthroat trout. Trans Am Fish Soc 100: 282-289

Jakubowski M, Whitear M (1990) Comparative morphology and cytology of taste buds in teleosts. Z Mikrosk-Anat Fors 104: 529-560

Jarrard HE (1997) Postembryonic changes in the structure of the olfactory bulb of the chinook salmon (*Oncorhynchus tshawytscha*) across its life history. Brain Behav Evol 49: 249-260

Johns PR, Easter J (1977) Growth of the adult goldfish eye. II. Increase in retinal cell number. J Comp Neurol 176: 331-342

Johnston JB (1908) The Nervous System of Vertebrates. John Murray, London, p 370

Johnston JB (1910) The limit between ectoderm and endoderm in the mouth and the origin of taste buds. Am J Anat 10: 41-67

Jung H-S, Oropeza V, Thesleff I (1999) Shh, Bmp-2, Bmp-4 and fgf-8 are associated with initiation and patterning of mouse tongue papillae. Mech Dev 81: 179-182

Kanwal JS, Caprio J (1983) An electriphysiological investigation of the oro-pharyngeal (IX-X) taste system in the channel catfish, *Ictalurus punctatus*. J Comp Physiol A150: 345-357

Kanwal JS, Finger TE (1992) Central representation and projections of gustatory systems. In: Hara TJ (ed) Fish Chemoreception. Chapman and Hall, London, p 79-102

Kapoor BG, Evans HE, Pevzner RA (1975) The gustatory system in fish. Adv Mar Biol 13: 53-108

Kasumyan AO, Døving KB (2003) Taste preferences in fish. Fish and Fisheries 4: 1-59

Kasumyan AO (1992) Development of olfactory and taste responses to chemical signals in acipenserid fishes. In: Doty RL, Müller-Schwarze D (eds) Chemical Signals in Vertebrates. Plenum Press, New York, p 135-139

Kasumyan AO (1999) Olfaction and taste in sturgeon behaviour. J Appl Ichthyol 15: 228-232

Kasumyan AO (2004) The olfactory system in fish: structure, function, and role in behavior. J Ichthyol 44: 180-223

Kasumyan AO, Kazhlaev AA (1993a) Behavioral responses of early juveniles of Siberian sturgeon *Acipenser baeri* and stellate sturgeon *A. stellatus* (Acipenseridae) to gustatory stimulating substances. J Ichthyol 33: 85-97

Kasumyan AO, Kazhlaev AA (1993b) Formation of searching behavioral reaction and olfactory sensitivity to food chemical signals during ontogeny of sturgeons (Acipenseridae). J Ichthyol 33: 51-65

Kasumyan AO, Morsi AMK (1996) Taste sensitivity of common carp *Cyprinus carpio* to free amino acids and classical taste substances. J Ichthyol 36: 391-403

Kasumyan AO, Ponomarev VY (1986) About the species specificity of alarm pheromone in fish from the order of Cypriniformes. In: Sokolov VE (ed) Chemical Communication in Animals. Nauka, Moscow, p 202-207

Kasumyan AO, Ponomarev VY (1990) The ontogeny of feeding behavior in relation to natural chemical signals in cyprinid fishes. J Ichthyol 30: 89-100

Kasumyan AO, Sidorov SS (2001) The taste preferences in juveniles of lake trout, *Salvelinus namaycush* (Salmonidae). Problems of Fisheries Suppl 1: 121-125

Kasumyan AO, Sidorov SS (2005) Taste preferences of the brown trout *Salmo trutta* from three geographically isolated populations. J Ichthyol 45: 111-123

Kasumyan AO, Ryg M, Døving KB (1998) Effect of amino acids on the swimming activity of newly hatched turbot larvae (*Scophthalmus maximus*). Mar Biol 131: 189-194

Kawamura G, Ishida K (1985) Changes in sense organ morphology and behaviour with growth in the flounder *Paralichthys olivaceus*. Bull Jap Soc Sci Fish 52: 155-165

Kawamura G, Munekiyo M (1989) Development of the sense organs of ribbon fish *Trichiurus lepturus* larvae and juveniles. Nippon Suisan Gakkaishi 55: 2075-2078

Kawamura G, Washiyama N (1989) Ontogenetic changes in behavior and sense organ morphogenesis in largemouth bass and *Tilapia nilotica*. Trans Am Fish Soc 118: 203-213

Kawamura G, Mori H, Kuwahara A (1989) Comparison of sense organ development in wild and reared flounder *Paralichthys olivaceus* larvae. Nippon Suisan Gakkaishi 55: 2079-2083

Kawamura G, Takimoto M, Sobajima N (1990) Larval growth and age- and size-related variation in development of sense organs in the halfbeak, *Hemiraphus sajori*. In: Hirano R, Hanyu I (eds) Proceedings of Second Asian Fisheries Forum. Asian Fisheries Society, Manila, Philippines, p 407-410

Kawamura G, Masuma S, Tezuka N, Koiso M, Jinbo T, Namba K (2003) Morphogenesis of sense organs in the bluefin tuna *Thunnus orientalis*. In: Browman HI, Skiftesvik AB (eds) The Big Fish Bang. Proceedings of the 26th Annual Larval Fish Conference. Institute of Marine Research, Bergen, Norway, p 123-135

Keibel F (1893) Zur Entwicklungsgeschichte und vergleichenden Anatomie der Nase and des oberen Mundrandes (Oberlippe) beim Vertebraten. Anat Anz 8: 473-487

Kleerekoper H (1969) Olfaction in Fishes. Indiana University Press, Bloomington, IN, USA

Klein SL, Graziadei PPC (1983) The differentiation of the olfactory placode in *Xenopus laevis*: A light and electron microscope study. J Comp Neurol 217: 17-30

Knapp L, Lawton A, Oakley B, Wong L, Zhang C (1995) Keratins as markers of differentiated taste cells of the rat. Differentiation 58: 341-349

Knutsen JA (1992) Feeding behaviour of North Sea turbot (*Scophthalmus maximus*) and dover sole (*Solea solea*) larvae elicited by chemical stimuli. Mar Biol 113: 543-548

Komada N (1994) Distribution of taste buds in the oropharyngeal cavity of larval and juvenile stages of the cyprinid fish, *Tribolodon hakonensis*. Jap J Ichthyol 41: 307-311

Kotrschal K (1995) Ecomorphology of solitary chemosensory cell systems in fish: A review. Environ Biol Fishes 44: 143-155

Kotrschal K, Krautgartner WD, Hansen A (1997) Ontogeny of the solitary chemosensory cells in the zebrafish, *Danio rerio*. Chem Senses 22: 111-118

Kotrschal K, Whitear M, Adam H (1984) Morphology and histology of the anterior dorsal fin of *Gaidropsarus mediterraneus* (Pisces, Teleostei), a specialized sensory organ. Zoomorphol 104: 365-372

Kupffer Cvon (1894) Studien zur vergleichenden Entwicklungsgeschichte des Kopfes der Kranioten. 2. Heft. Die Entwickelung des Kopfes von Ammocoetes planeri. JF Lehmann, Munchen, Leipzig

Kuwahara A, Suzuki S (1982) Vertical distribution and feeding of a flounder *Paralichthys olivaceus* larva. Bull Jap Soc Sci Fish 48: 1375-1381

Kühme W (1963) Chemisch ausgeloste Brutpflege- und Schwarmreaktion bei *Hemichromis bimaculatus* (Pisces). Z Tierpsychol 20: 688-704

Landacre EL (1907) On the place of origin and method of distribution of taste buds in *Ameiurus melas*. J Comp Neurol Psychol 17: 1-66

Lin K-L, Huang B-Q, Chen C-T (1998) Ontogenetic development of taste buds in the oropharyngeal cavity of juvenile and young Japanese croaker, *Argyrosomus japonicus*. J Fish Soc Taiwan 25: 37-44

Lindsay SM, Vogt RG (2004) Behavioral responses of newly hatched zebrafish (*Danio rerio*) to amino acid chemostimulants. Chem Senses 29: 93-100

MacCrimmon HR, Twongo TK (1980) Ontogeny of feeding behaviour in hatchery-reared rainbow trout, *Salmo gairdneri* Richardson. Can J Zool 58: 20-26

Marui T, Caprio J (1992) Teleost gustation. In: Hara TJ (ed) Fish Chemoreception. Chapman and Hall, London, p 171-198

Mbiene JP, Roberts D (2003) Distribution of keratin 8-containing cell clusters in mouse embryonic tongue: evidence for a prepattern for taste bud development. J Comp Neurol 457: 111-122

Mbiene JP, MacCallum DK, Mistretta CM (1997) Organ cultures of embryonic rat tongue support tongue and gustatory palillae morphogenesis *in vitro* without intact sensory ganglia. J Comp Neurol 377: 324-340

McCormick MI (1993) Development and changes at settlement in the barbel structure of the reef fish, *Upeneus tragula* (Mullidae). Environ Biol Fishes 37: 269-282

Mearns KJ (1986) Sensitivity of brown trout (*Salmo trutta* L.) and Atlantic salmon (*Salmo salar* L.) fry to amino acids at the start of exogenous feeding. Aquaculture 55: 191-200

Miklosi A, Pongracz P, Csanyi V (1997) The ontogeny of antipredator behaviour in paradise fish larvae (*Macropodus opercularis*). II. The response to chemical stimuli of heterospecific fishes. Behaviour 134: 391-413

Miles SG (1968) Rheotaxis of elvers of the American eel (*Anguilla rostrata*) in the laboratory to water from different streams in Nova Scotia. J Fish Res Bd Can 25: 1591-1602

Mirza RS, Chivers DP, Godin J-G (2001) Brook charr alevins alter timing of nest emergence in response to chemical cues from fish predators. J Chem Ecol 27: 1775-1785

Mistretta CM, Baum BJ (1984) Quantitative study of taste buds in fungiform and circumvallate papillae of young and eged rats. J Anat 138: 323-332

Mistretta CM, Gurkan S, Bradley RM (1988) Morphology of chorda tympani fiber receptive fields and proposed neural rearrangements during development. J Neurosci 8: 73-78

Moitra SK, Sinha GM (1972) The structure, distribution and probable function of taste buds and mucous cells in the alimentary tract during the life history of *Cirrhina mrigala* (Hamilton). J Inland Fish Soc Ind 4: 103-114

Munekiyo M, Kuwahara A (1985) Food habits of ribbon fish in the western Wakasa Bay. Nippon Suisan Gakkaishi 51: 913-919

Northcott ME, Beveredge MCM (1988) The development and structure of pharyngeal apparatus associated with filter feeding in tilapias (*Oreochromis niloticus*). J Zool, Lond 215: 133-149

Northcutt RG (2005) Taste bud development in the channel catfish. J Comp Neurol 482: 1-16

Nosrat CA, Blomlof J, ElShamy WM, Ernfors P, Olson L (1997) Lingual deficits in BDNF and NT3 mutant mice leading to gustatory and somatosensory disturbances, respectively. Development 124: 1333-1342

Olsén HK (1993) Development of the olfactory organ of the Arctic charr, *Salvelinus alpinus* (L.) (Teleostei, Salmonidae). Can J Zool 71: 1973-1984

Osse JWM, Sibbing FA, Van Den Boogaart JGM (1997) Intra-oral food manipulation of carp and other cyprinids: adaptations and limitations. Acta Physiol Scand 161: 47-57

Pasha SMK (1964) The anatomy and histology of the alimentary canal of a carnivorous fish *Megalops cyprinoides*. Proc Indian Acad Sci 60: 107-115

Pashchenko NI, Kasumyan AO (1983) Some morpho-functional features of the olfactory organ development in European minnow *Phoxinus phoxinus* (Cypriniformes, Cyprinidae). Zool J (Zoologicheskiy Zhurnal) 26: 12-25 (in Russian)

Pashchenko NI, Kasumyan AO (1984) Degenerative and restorative processes in the olfactory lining of white amur, *Ctenopharyngodon idella* (Cyprinidae) after treatment with detergent triton X-100. J Ichthyol 24: 112-121

Pashchenko NI, Kasumyan AO (1986) Morpho-functional features of the olfactory organ in cyprinid fishes. I. Morphology and function of olfactory organ during ontogeny of the white amur, *Ctenopharyngodon idella*. J Ichthyol 26: 12-25

Pavlov DS, Kasumyan AO (1998) The structure of the feeding behavior of fishes. J Ichthyology 38: 116-128

Pevzner RA (1985) The fine structure of taste buds of the ganoid fishes. III. Fries during endogenous feeding. Cytology 27: 1240-1246 (in Russian, with English summary)

Pfeiffer W (1963) The fright reaction in North American fish. Can J Zool 41: 69-77

Pfeiler E (1989) Sensory systems and behavior of premetamorphic and metamorphic *Leptocephalus* larvae. Brain Behav Evol 34: 25-34

Poling KR, Brunjes PC (2000) Sensory deafferentation and olfactory bulb morphology in the zebrafish and related species. Brain Res 856: 135-141

Price JW (1934) The embryology of whitefish (*Coregonus clupeaformes* Mitchell). Ohio J Sci 34: 399-412

Pyatkina GA (1976) Receptor cells of various types and their proportional interrelation in the olfactory organ of larvae and adults of acipenserid fishes. Cytology 18: 1444-1449 (in Russian with English summary)

Raderman-Little R (1979) The effect of temperature on the turnover of taste bud cells in catfish. Cell Tissue Kinetics 12: 269-280

Rasquin P (1955) Observations on the metamorphosis bonefish, *Albula vulpes* (Linnaeus). J Morphol 97: 77-117

Reinke W (1937) Zur Ontogenie und Anatomie des Geruchsorgans der Knochenfische. Z Anat Entwicklungsgeschichte 100: 600-624

Reiss JO, Burd GD (1997) Cellular molecular interactions in the development of the *Xenopus* olfactory system. Semin Cell Dev Biol 8: 171-179

Reutter K (1978) Taste organ in the bullhead (Teleostei). Adv Anat Embryol Cell Biol 55: 1-98

Reutter K (1986) Chemoreceptors. In: Bereiter-Hahn J, Matoltsy AG, Richards KS (eds) Biology of the Integument, Vol 2. Vertebrates. Springer-Verlag, Berlin, p 586-604

Reutter K (1992) Sturcture of the peripheral gustatory organ, represented by the siluroid fish *Plotosus lineatus* (Thunberg). In: Hara TJ (ed) Fish Chemoreception. Chapman and Hall, London, p 60-78

Reutter K, Witt M (1993) Morphology of vertebrate taste organs and their nerve supply. In: Simon SA, Roper SD (eds) Mechanisms of Taste Transduction. CRC Press, Boca Raton, p 29-82

Reutter K, Hansen A (2005) Subtypes of light and dark elongated taste bud cells in fish. In: Reutter K, Kapoor BG (eds) Fish Chemosenses. Science Publishers, Inc, Enfield, NH, USA, p 211-230

Reutter K, Breipohl W, Bijvank GJ (1974) Taste bud types in fishes. II. Scanning electron microscopical investigations on *Xiphophorus helleri* Heckel (Poeciliidae, Cyprinodontiformes, Teleostei). Cell Tissue Res 153: 151-165

Reutter K, Witt M, Knutsen JA, Døving KB (1995) Taste bud development in turbot larvae (Teleostei). Chem Senses 20: 764-765

Ross LS, Parret T, Easter SS Jr (1992) Axogenesis and morphogenesis in the embryonic zebra fish brain. Neurosci 12: 467-482

Sakata Y, Olson JK, Michel WC (2003) Assessment of neuronal maturation and acquisition of functional competence in the developing zebrafish olfactory system. Methods Cell Sci 25: 39-48

Sand O, Karlsen HE (2000) Detection of infrasound and linear acceleration in fishes. Phil Trans R Soc Lond B 355: 1295-1298

Schmalhauzen OI (1962) Disturbance of olfactory organ development in the sturgeon under specific developmental conditions. Trudi Inct Morfol Zhivotn Akademii Nauk SSSR 40: 188-218 (in Russian)

Schultze M (1856) Uber die Endigungsweise des Geruchsnerven und die Epithelialgebilde der Nasenschleimhaut. Monatsber Konigl Preuss Akad Wiss, Berlin 21: 505-515

Senoo S, Ang KJ, Kawamura G (1994) Development of sense organs and mouth and feeding of reared marble goby *Oxyeleotris marmoratus*. Fish Sci 60: 361-368

Sewertzoff AN (1931) Morphologische Gesetzmassigkeiten der Evolution. Gustav Fisher-Verlag, Jena

Smith DG (1979) Guide to the leptocephali (Elopiformes, Anguilliformes and Notacanthiformes). NOAA tech Rep, Natl Mar Fish Serv Vircular 424: 1-39

Soin SG, Kasumyan AO, Pashchenko NI (1981) Ecological and morphological analysis of the development of the minnow *Phoxinus phoxinus* (Cyprinidae). J Ichthyol 21: 90-105

Sola C (1995) Chemoattraction of upstream migrating glass eels *Anguilla anguilla* to earthy and green odorants. Environ Biol Fishes 43: 179-185

Sola C, Spampanato A, Tosi L (1993) Behavioural responses of glass eels (*Anguilla anguilla*) towards amino acids. J Fish Biol 42: 683-691

Sola C, Tosi L (1993) Bile salts and taurine as chemical stimuli for glass eels *Anguilla anguilla*: a behavioural study. Environ Biol Fishes 37: 197-204

Sorensen PW (1986) Origins of the fresh water attractant(s) of migrating elvers of the American eel *Anguilla rostrata*. Environ Biol Fishes 17: 185-200

Sorensen PW, Caprio J (1998) Chemoreception. In: Evans DH (ed) The Physiology of Fishes. CRC Press, Boca Raton, p 375-405.

Stabell OB (1984) Homing and olfaction in salmonids: A critical review with special reference to the Atlantic salmon. Biol Rev 59: 333-388

Stehr CM (1983) Changes in the surface epithelium of the developing olfactory organ in the sand sole *(Psettichthys melanastictus)* and the rex sole *(Glyptocephalus zachirus)*. Scanning Electron Microscopy 4: 1947-1954

Stone LM, Finger TE, Tam PPL, Tan S-S (1995) Taste receptor cells arise from local epithelium, not neurogenic ectoderm. Proc Natl Acad Sci USA 92: 1916-1920

Tanaka Y, Mukai Y, Takii K, Kumai H (1991) Chemoreception and vertical movement in planktonic yolk-sac larvae of red sea bream *Pagrus major*. J Appl Ichthyol 7: 129-135

Teichmann H (1964) Experimente zur Nasenentwicklung der Regenbogenforelle (*Salmo irideus* W. Gibb.). Wilhelm Roux Arch EntwMech Org 155: 129-143

Theisen B (1982) Functional morphology of the olfactory organ in *Spinachia spinachia* (L.) (Teleostei, Gasterosteidae). Acta Zool 63: 247-254

Theisen B, Breucker H, Zeiske E, Melinkat R (1980) Structure and development of the olfactory organ in the garfish *Belone belone* (L.) (Teleostei, Atheriniformes). Acta Zool 61: 161-170

Tosi L, Sola C (1993) Role of geosmin, a typical inland water odour, in guiding glass eels *Anguilla anguilla* (L.) migration. Ethology 95: 177-185

Twongo TK, MacCrimmon HR (1977) Histogenesis of the oropharyngeal and oesophageal mucosa as related to early feeding in rainbow trout, *Salmo gairdneri* Richardson. Can J Zool 55: 116-128

VanDenbossche J, Seelye JG, Zielinski BS (1995) The morphology of the olfactory epithelium in larval, juvenile and upstream migrant stages of the sea lamprey, *Petromyzon marinus*. Brain Behav Evol 45: 19-24

Vischer HA (1989a) The development of lateral line receptors in *Eigenmannia* (Teleostei, Gymnotiformes). I. The mechanoreceptive lateral-line system. Brain Behav Evol 33: 205-222

Vischer HA (1989b) The development of lateral line receptors in *Eigenmannia* (Teleostei, Gymnotiformes). II. The electroreceptive lateral-line system. Brain Behav Evol 33: 223-236

Vogt RG, Lindsay SM, Byrd CA, Sun M (1997) Spatial patterns of olfactory neurons expressing specific odor receptor genes in 48-hour-old embryos of zebrafish *Danio rerio*. J Exp Biol 200: 433-443

Waldman B (1982) Quantitative and developmental analyses of the alarm reaction in the zebra danio, *Brachydanio rerio*. Copeia 1982: 1-9

Watling H, Hilleman HH (1964) The development of the olfactory apparatus of the grayling (*Thymallus arcticus*). J Fish Res Board Can 21: 373-396

Werner RG, Lannoo MJ (1994) Development of the olfactory system of the white sucker, *Catostomus commersoni*, in relation to imprinting and homing: a comparison to the salmonid model. Environ Biol Fishes 40: 125-140

Westerberg H (1990) Properties of aquatic odour trails. In: Døving KB (ed) Tenth International Symposium on Olfaction and Taste. GCS A/S, Oslo, p 45-54

Western JRH (1969) Studies on the diet, feeding mechanism and alimentary tract in two closely related teleosts, the freshwater *Cottus gobio* L. and the marine *Parenophrys bubalis* L. Acta Zool 50: 185-205

Whitlock KE, Westerfield M (1998) A transient population of neurons pioneers the olfactory pathway in the zebrafish. J Neurosci 18: 8919-8927

Whitlock KE, Westerfield M (2000) The olfactory placodes of the zebrafish form by convergence of cellular fields at the edge of the neural plate. Development 127: 3645-3653

Wilson SW, Ross LS, Parret T, Easter SS, Jr. (1990) The development of a simple scaffold of axon tracts in the brain of the embryonic zebrafish *Brachydanio rerio*. Development 108: 121-145

Witt M, Reutter K (1996) Embryonic and early fetal development of human taste buds: a transmission electron microscopical study. Anat Rec 246: 507-523

Witt M, Reutter K (1998) Innervation of developing human taste buds. An immunohistochemical study. Histochem Cell Biol 109: 281-291

Yamamori K, Nakamura M, Matsui T, Hara TJ (1988) Gustatory responses to tetrodoxin and saxitoxin in fish: a possible mechanism for avoiding marine toxins. Can J Fish Aquat Sci 45: 2182-2186

Zakon HH (1984) Postembryonic changes in the peripheral electrosensory system of a weakly electric fish: addition of receptor organs with age. J Comp Neurol 228: 557-570

Zeiske E, Kux J, Melinkat R (1976) Development of the olfactory organ of oviparous and viviparous cyprinodonts (teleostei). Zeitschrift fur Zoologische Systematik und Evolutionsfirschung 14: 34-40

Zeiske E, Theisen B, Breucker H (1992) Structure, development, and evolutionary aspects of the peripheral olfactory system. In: Hara TJ (ed) Fish Chemoreception. Chapman & Hall, London, p 13-39

Zeiske E, Kasumyan A, Bartsch P, Hansen A (2003) Early development of the olfactory organ in sturgeons of the genus *Acipenser*, a comparative and electron microscopic study. J Morphol Embryol 206: 357-372

Zielinski B, Hara TJ (1988) Morphological and physiological development of olfactory receptor cells in rainbow trout (*Salmo gairdneri*) embryos. J Comp Neurol 271: 300-311

Zippel HP, Hansen A, Caprio J (1997) Renewing olfactory receptor neurons in goldfish do not require contact with the olfactory bulb to develop normal chemical responsiveness. J Comp Physiol 181A: 425-437

CHAPTER

12

Photoreception

Ellis R Loew[1,*] and Christina M Wahl[2]

INTRODUCTION

The vast majority of teleost fishes pass through a planktivorous larval phase where they are obligate visual diurnal feeders. Failure to obtain adequate food to promote growth will obviously lead to high mortality (Hjort, 1926; Blaxter & Hempel, 1963; Lasker, 1978). It follows that the visual system of larvae, as in adults (Neumeyer, 2003), should be adapted as regards sensitivity (luminosity and chromaticity), structure (dioptric apparatus and retina) and analytical capability. An excellent review of larval fish vision and the fish retina in general has recently been published (Evans & Browman, 2004). Among the topics discussed first is exactly what events temporally define the larval period. For the visual ecologist the definition is simple – it starts when the larvae first show signs of 'visual' behaviour, which is almost always going to be detection and orientation towards prey. A general discussion of this topic can be found in Loew & McFarland (1990). The embryological development of the eye and neuronal pathways present before this event are of obvious importance, but it is what is present at the time of first-feeding and how this changes with further growth that matters here. The end of the larval period is associated with a metamorphic or lifestyle change almost always

Authors' addresses: [1]Cornell University, Department of Biomedical Sciences, Ithaca, NY 14853, USA.
[2]Wells College, Department of Biological and Chemical Sciences, Aurora, NY 13026, USA.
**Corresponding author:* E-mail: erl1@cornell.edu

correlated with different visual necessities than are required by young planktivores (reviewed in Evans & Fernald, 1990; Evans & Browman, 2004). Rather than simply reiterate Evans & Browman (2004), we will concentrate and expand on areas currently receiving increased attention as regards vision in this very important life-history stage. We will also highlight those areas that in our opinion should receive more attention.

OCULAR AND RETINAL ANATOMY

As befits their highly variable photic environment and visual tasks, there is a great deal of variation in the overall structure of the teleostean eye (reviewed in Nicol, 1989). However, there is a common pattern to the basic anatomy for all vertebrate eyes (Fig. 1). The globe is formed by a fibrous tunic, the sclera (A), which in fishes is largely or partly cartilaginous and may, in some cases, contain bony elements (scleral ossicles). Moving inward, there comes the pigmented and highly vascular choroid (B), followed by an epithelial layer, the retinal pigment epithelium (RPE, C). Overlaying the RPE is the retina consisting of the photoreceptor and neuronal layers (D). Axons from neurons of the innermost retinal layer, the ganglion cells, course over the retinal surface converging at the optic disk where they leave the eye as the optic nerve (Fig. 1,E).

Light enters through a transparent cornea (Fig. 1,F) and is brought to focus at the retinal plane by a refractile lens (1,G). Unlike mammals, accommodation in fish is

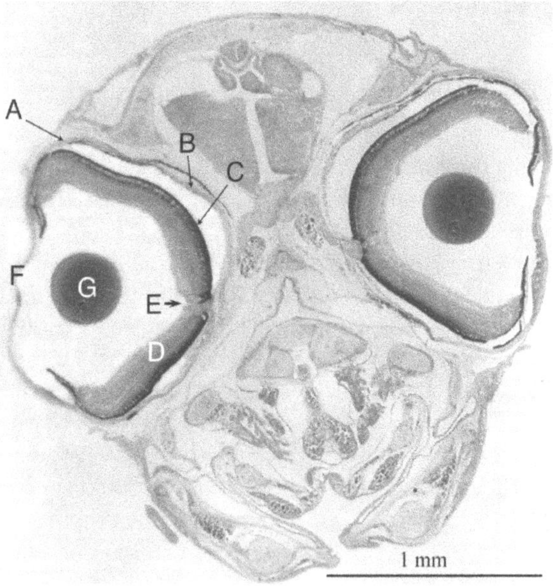

Fig. 1 Section through the head of a larval walleye (*Stizostedion vitreum*) at the level of the optic chiasm, showing the major ocular anatomy. A = sclera, B = choroid, C = pigment epithelium, D = neural retina, E = optic fissure (optic nerve head), F = cornea, and G = lens.

accomplished by movement of the lens along the optic axis mediated by the *retractor lentis* muscle.

Specific to fishes are specialisations associated with vision in an aqueous medium. The cornea is relatively flat and provides little in terms of refractive power. This is due to the fact that its refractive index is not all that different from water. Almost all of the refractive power derives from the spherical lens that can fill almost the entire posterior chamber.

Ocular Development

For those interested in 'pre-feeding' ocular development and fish development in depth, the excellent book by Yvette Kunz (2004) should be consulted. For our purposes it is sufficient to pick up with the eye immediately prior to pigmentation (Fig. 2). The basic anatomical features are evident although pigmentation of both the globe and iris is absent as is a functional retina (Fig. 2A, B). The lens can be clearly seen through the pupil. The embryonic fissure, a growth zone, may still be evident at this stage with fusion occurring before the start of pigmentation. Continued growth leads to dense pigmentation and retinal development sufficient to allow for feeding (note the green zooplankton in the gut of the first-feeder in Fig. 2C). The 'mature' larval eye is large relative to body length with a fully functional retina although changes in visual pigments and photoreceptor cell types may still occur (see below). The speed with which the eye develops from the non-functional unpigmented (Fig. 2A) to the functional (Fig. 2C) can be very rapid – less than 12 hours in the case of yellowfin tuna (*Thunnus albacares*) in culture. The eye continues to enlarge throughout the larval and juvenile stage with change, both ocular and retinal, often continuing throughout life and with each metamorphic event.

Spectral Sensitivity Changes during the Larval Period

The boundary conditions for vision are set by the absorption spectra of the visual pigments present in the visual photoreceptors (an exception to this statement might be in several deep sea species where photosensitisers such as chlorophyll apparently extend red sensitivity) (see Douglas *et al.*, 2003). Post-receptor processing underlies the perceptual experience of such constructs as colour, shape, texture, etc. with the retina performing a surprising amount of the analysis (for an interesting review see Werblin & Roska, 2007). But all processing must operate within the 'absorption boundaries' set by the photoreceptors. The number of visual pigments extant in the retina at any given time is determined by the expression pattern of the resident opsin genes (Bowmaker & Loew, 2008). The spectral location of a particular visual pigment is determined by the amino acid sequence of its opsin and the bound vitamin A-derived chromophore (Loew, 1995). In addition to a rod class of pigment (RH1), there are four spectrally distinct classes of cone pigments encoded by distinct opsin genes: a long- to

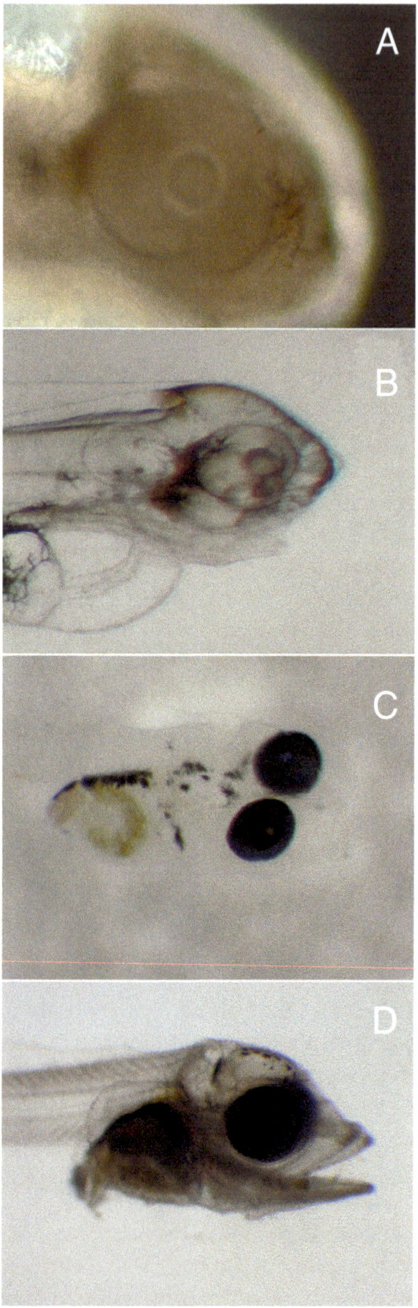

Fig. 2 Appearance of the developing fish eye, in this case yellowfin tuna (*Thunnus albacares*), from just before feeding to the early juvenile stage.

middle-wave class (LWS) maximally sensitive in the red-green spectral region from about 490-570 nm, a middle-wave class (RH2) sensitive in the green from about 480-535 nm, a short-wave class (SWS2) sensitive in the blue-violet from about 410-490 nm and a second short-wave class (SWS1) sensitive in the violet-ultraviolet from about 355-440 nm. Many sub-types within each class have been identified thereby multiplying the number of distinct opsins that may be present in the retina at any given time. All of these may be red-shifted to varying degrees through the use of vitamin A_2 versus vitamin A_1 as the visual pigment chromophore.

A rich literature has developed directed specifically at changes in visual pigment complement during the larval stage and at the larval-to-juvenile-to-adult transitions, from the points of view of A_1/A_2 chromophore shifting and opsin expression. Initial studies were limited to extraction studies on rods, showing chromophore shifts between rhodopsins (A_1-based) and porphyropsins (A_2-based), and therefore peak absorbance shifts associated with changing environmental parameters such as the spectrum of the background spacelight, day length, temperature, etc. (Bridges, 1972; Bowmaker, 1995; Loew, 1995). With the advent of microspectrophotometry (MSP) and, more recently, molecular capabilities that allow direct studies of opsins via expression and their respective visual pigments via reconstitution, information on cones is also available (Bowmaker, 1995; Bowmaker & Loew, 2008). It is this recent literature that will be addressed here.

Changes in opsin expression patterns seem to be quite common during the larval stage (Britt *et al.*, 2001; Loew *et al.*, 2002; Cheng *et al.*, 2006). An introduction to the types and complexity of such changes is illustrated by the following examples:

Winter flounder *(Pseudopleuronectes americanus)*

Early studies used histology and MSP to follow the retinal and visual pigment changes taking place during pre-metamorphic and post-metamorphic periods (Evans & Fernald, 1993; Evans *et al.*, 1993; Evans & Browman, 2004). To this has been added opsin expression studies using RT-PCR on pre- and post-metamorphic stages (Marder & Cameron, 2004). The first cells to appear during ontogeny in pre-metamorphic fish are single cones with λ_{max} at about 520 nm expressing a single RH2 opsin gene. With development, rods are added at 509 nm expressing an RH1 followed by single cones at 461 nm expressing an SWS2, and double cones with 529 nm and 547 nm pigments expressing and RH2 and LWS opsin, respectively. No evidence of a UV pigment was found at any stage, although the presence of an SWS1 gene cannot be excluded. Interestingly, the authors suggest that a single RH2 opsin produces both the 520-nm pigment in the pre-metamorphic fish and the 529-nm pigment in one member of the post-metamorphic double cone. They base this on their inability to identify more than one RH2 product using RT-PCR and cross-reactivity of *in situ* probes. However, they have the potential of possessing both an RH2A and an RH2B opsin gene and it is likely that a second RH2 gene is present, but has yet to be isolated.

Zebrafish (*Danio rerio*)

Nawrocki *et al.* (1985) used MSP to compare the larval and adult complement of visual pigments in zebrafish retina. The λ_{maxes} of the pigments they found in 1 to 2 year old adults were at 417 nm in a short single cone, 480 nm in long singles and the long members of double cones, 556 nm in the other member of the double cones and 501 nm in the rods. Except for the 556 nm pigment, which is known to be closer to 565 nm, their values are close to those more recently reported by Allison *et al.* (2004) and found by Loew (personal observation). Possible reasons for this discrepancy will be discussed below. Nawrocki *et al.* (1985) failed to identify the UV pigment (λ_{max} 360) known to be present in adult zebrafish (Robinson *et al.*, 1993; Allison *et al.*, 2004). The larval stages they examined were early (6–8 days post-fertilisation (dpf)) and late (11-17 dpf). Rearing temperature was kept constant at 28.5°C. Since feeding is known to commence at approximately 4 dpf (Kimmel *et al.*, 1995) these larvae were certainly visual. Starting with their early stage larvae, all four visual pigments were present, using their reporting method, λ_{max} values were like those for the adults. The early stage had only single cones and rods while their late stage had already formed the cone types found in the adult. However, re-examination of their Figure 5 and Table 1 shows average λ_{max} changes with development, particularly for the long wavelength pigment that shifts from 540 nm to 556 nm. Using partial bleaching they found no evidence suggesting that such a shift was due to chromophore mixing. A possible explanation for this is the temporal expression of opsin class subtypes. Note that many of us who have performed MSP in the past have assumed that much, if not all, of the variation we see in λ_{max} values is instrumental and due to 'noise' or the mixing of A_1- and A_2-based pigments. For the latter λ_{max} differences of less than 10 nm are difficult to ascribe using MSP and partial bleaching or template shape except for records with high density. Thus we simply identify a spectral class by λ_{max} grouping or association with cell type, calculate a mean and standard deviation and report the results. With the possibility of opsin subtypes and co-expression it now behooves "MSP'ers" to report range and perhaps mode along with scatter plots to make sure that any variation is not lost in the reporting!

More recent studies have identified a rich complement of opsin genes and the λ_{maxes} of their reconstituted visual pigments in the zebrafish – LWS-1 (λ_{max} 558 nm), LWS-2 (λ_{max} 548 nm), RH2-1 through RH2-4 (467 nm, 476 nm, 488 nm and 505 nm λ_{maxes} , respectively) and single copies of an SWS1 (λ_{max} 355 nm), SWS2 (λ_{max} 416 nm) and RH1 (rod with λ_{max} 501 nm), (Chinen *et al.* 2003). Within a class, the subtypes are known to produce visual pigments having different λ_{max}. However, these genes are not necessarily expressed all at once during the hours after fertilisation and throughout the larval stage nor are their expression patterns uniform over the retina at any stage (Takechi & Kawamura, 2005). Their staging has the larval period starting at 3 dpf (Kimmel *et al.*, 1995), which is well before the early larvae examined by Nawrocki *et al.* (1985) and before 4 dpf when feeding starts. However, examination of their data for the periods after feeding starts indicates temporal sequencing and regional variation of the LWS and RH2 family of opsin genes. At 7 dpf both RH2-1 and RH2-2 expression

Table 1 λ_{max} values for zebrafish (*Danio rerio*) visual pigments obtained using MSP and by reconstitution

Source	*Method*	*RH1*	*SWS1*	*SWS2*	*RH2*	*RH2-1*	*RH2-2*	*RH2-3*	*RH2-4*	*LWS*	*LWS-1*	*LWS-2*
Allison *et al*. (2004)	MSP	503	361	411	482					565		
Loew (adult) personal observation	MSP	502	363	415	478					565		
Loew (+5 dpf larvae) personal observation	MSP		358	413	480					564		
Nawrocki *et al*. (1985)	MSP	501	X	417	480	X	X	X	X	556	X	X
Chinon *et al*. (2003)	Reconstitution	501	355	416	-	467	476	488	505	-	558	548

was observed throughout the retina. At 16 dph RH2-1 expression disappeared in the marginal retinal areas while RH2-2 expression disappeared in the central retina. RH2-3 and RH2-4 expression was limited to the most marginal side of the nasal retina. However, with futher development the RH2-2 expression returned to the central and dorsal retina to become the dominant subtype expressed of all the RH2 opsins (see also Chinen *et al.*, 2003). The LWS-2 gene was expressed well before first-feeding while LWS-1 expression only became evident in a broad area of the marginal ventral retina at 3.5–5.5 dpf. At 7 dpf expression of LWS-1 was confined to the ventral retina while LWS-2 was present throughout the retina. In the adult retina the LWS-X expression remains regional although cells at the boundary appear to co-express the LWS genes.

The expression data suggest that during the visual larval and adult stages there are regional differences within the retina as regards spectral sensitivity. This should affect colour contrast thresholds for parts of the retina 'seeing' different parts of the visual scene. Unfortunately, no behavioural data are available suggesting the existence of such variation. More disturbing are the differences between the λ_{max} values for the visual pigments obtained by reconstitution versus those obtained using MSP and the failure of MSP to find the kind of λ_{max} variation suggested by the RT-PCR/reconstitution data of Chinen *et al.* (2003) and *in situ* hybridisation experiments of Takechi & Kawamura (2005) for the RH2 and LWS types. This is seen in Table 1. There is good agreement between the rod, SWS1 and SWS2 values. However, while the MSP data for the RH2 and LWS cones agree, it is hard to reconcile these to the reconstitution and *in vitro* data. With the exception of the data of Nawrocki *et al.* (1985), all the MSP values for the LWS pigment are shifted further into the red than the reconstituted LWS-1 at 558 nm. In addition, there is no evidence of LWS-2-based visual pigment in any of the MSP studies nor is there a suggestion of co-expression. Even with no region-specific MSP being done, one would have expected to see some evidence of LWS-2 cones or at least a much greater variation of LWS values than were noted. While the longer wavelength LWS λ_{max} using MSP could be explained by a vitamin A_1/A_2 mix (it is known that zebrafish are capable of synthesizing pigments using vitamin A_2 (see Allison *et al.*, 2004; Loew, personal observations)), this cannot explain the failure to find LWS-2 cones around 548 nm. The same kind of problem exists for the RH2 class. All the MSP data are essentially the same. However, none of them reasonably correspond to any of the RH2 subtypes. The closest match is to RH2-2 with a λ_{max} at 476 nm. This makes sense since this is the dominant subtype expressed in the adult retina. Again, the slightly higher values for the MSP data could be due to A_2 'contamination'. However, there is no evidence of any RH2-1 subtype cones and not enough variation to suggest failure of MSP to differentiate between RH2-1 and RH2-2 cones. Also, the 5+ dpf larval data of Loew (personal observation) showed no evidence for broad RH2 subtype representation or any more variation than was seen for the adults if the smaller size of the larval cones is taken into account. To our knowledge, there has been no evidence of co-expression within RH2-class cones. We are presently unable to explain these discrepancies. However, MSP and extraction data are often at odds with published reconstitution data and this is an area requiring further study.

Using MSP, Allison *et al.* (2004) identified five visual pigments in the adult zebrafish retina with λ_{max} at 361 ± 3 nm, 411 ± 5 nm, 482 ± 6 nm, 565 ± 10 nm and 503 ± 5 nm). Very similar values and variances have been found for larvae 5 dpf and older (Loew, personal observations).

Yellowfin Tuna (*Thunnus albacares*)

A recent study (Loew *et al.*, 2002) has found that as yellowfin tuna mature and their retinas change from all single-cones to a mixed population of rods, single- and twin-cones, the spectral sensitivity of their cones, as determined using MSP, can also change. This is mediated by variations in the expression of opsin genes including co-expression in individual cones. To summarise (see also Fig. 3):

1. There are four opsin genes expressed in yellowfin tuna: SWS2 (425 nm λ_{max} single cone), RH1 (485 nm λ_{max} rods), RH2 (485 nm λ_{max} single cones in larvae/juvenile, twin cones in juveniles/adults) and LWS (single and twin cones in larvae/juveniles) (Carleton & Loew, unpublished observations).
2. All cone opsins are expressed in the early larvae, which have only single cones and no rods. Only RH1, RH2 and SWS2 are expressed in the adult (within the sampling limits set by the MSP technique).
3. There is co-expression of the RH2 and LWS opsins in single and 'young' twin cones particularly during down-regulation of the LWS gene at the time of twinning.
4. The ratio of expression varies among cones from a given retina with some evidence of 'preferred' ratios based on limited MSP.
5. LWS expression decreases after twinning starts with none being found in adults (again within the limits inherent in MSP).
6. There is some evidence for either a second SWS2 gene or an SWS1 having a λ_{max} below 415 nm, although this has not yet been identified.

The changes seen in yellowfin tuna opsin expression seem to involve mostly the LWS gene(s). Expression changes at the long wavelength end of the spectrum are quite rare with the only other reported occurrence being in the goatfish (*Upeneus tragula*) (Shand, 1993). How this is adaptive for larval planktivory remains unknown, but it has been suggested that the presence of an LWS pigments may improve the contrast of chlorophyll-containing prey when viewed against the blue water background (McFarland & Loew, unpublished observations).

Salmonids

No group of fishes shows as much variation in spectral sensitivity with age as the salmonids. This is brought about by the loss and gain of UV-sensitive cones, chromophore switching and opsin expression changes involving SWS1 and SWS2 genes. While there is some controversy as to the specifics of the changes and differences that might be related to species, the following seems to hold:

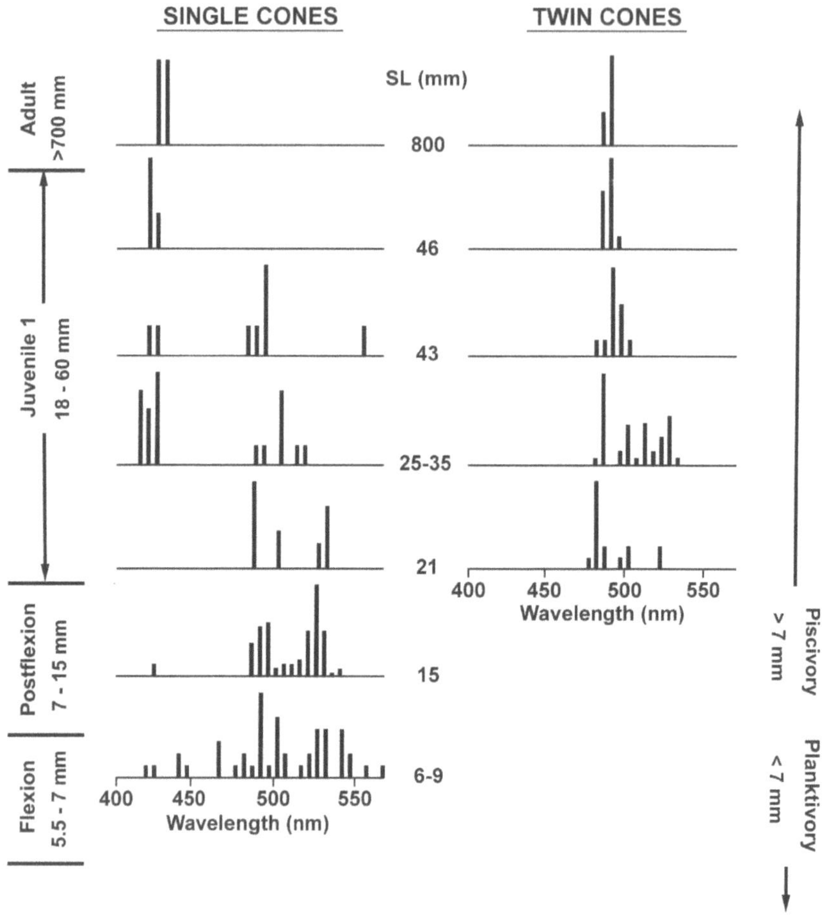

Fig. 3 Combined single and twin cone absorbance maximum frequency distribution for larval and juvenile yellowfin tuna (*Thunnus albacares*). The cone and λ_{max} data are correlated with stage/standard length (mm) on the left scale, and with feeding behaviour on the right. After Loew *et al.* (2002).

1. Larval and juvenile (parr) stages use mostly vitamin A_2 as the visual pigment chromophore with a switch to vitamin A_1 usage timed to downstream migration (see also Stefansson *et al.*, this volume). Upstream reproductive adults return to the A_2 condition with a return of corner cones, presumably UV (Bridges, 1972; Novales-Flamarique, 2000). It should be noted that a recent paper by Temple *et al.* (2006) reports for coho salmon (*Oncorhynchus kisutch*) the ratio of A_2- to A_1-based pigment varies seasonally with day length and temperature such that the change noted at the time of smoltification that normally takes place in the spring might be coincidental and not causally linked to smoltification itself.

2. Ultraviolet cones are present in larvae/juveniles although they may undergo changes in retinal distribution (Bowmaker & Kunz, 1987; Browman & Hawryshyn, 1994a, b; Novales-Flamarique, 2000; Allison *et al.*, 2003, 2006; Cheng *et al.*, 2006).
3. The timing of the reduction/redistribution of the UV cones is correlated with smoltification although not necessarily causally linked to it (Cheng *et al.*, 2006).
4. Thyroxine and retinoic acid appear to be involved in the timing of changes in UV cell presence and distribution (Browman & Hawryshyn, 1994a, b).
5. While there is some evidence that the loss of the UV cones occurs due to apoptosis, recent evidence suggests that the apparent loss is due to a switch from SWS1 to SWS2 expression in the short single cones (Cheng & Novales-Flamarique 2004; Cheng *et al.*, 2006). This may be species-specific.
6. The sequence of opsin expression is SWS1, LWS, RH1, RH2 and SWS2 (Cheng *et al.*, 2007). This is different from most other fishes where RH2 seems to be the first opsin expressed (see Bowmaker & Loew, 2008).

Cichlids

Typically, adult African cichlids possess three cone visual pigments (Fernald & Liebman, 1980; Van der Meer & Bowmaker, 1995; Carleton *et al.*, 2000, 2006 Carleton & Kocher, 2001; Parry *et al.*, 2005; Jordan *et al.*, 2006). MSP measurements of cones from a relatively small sample of species from Lake Malawi, suggest that cichlids inhabiting rock and sand habitats have significantly different cone visual sensitivities (Carleton & Kocher, 2001). The Mbuna (rock dwelling) species tend to have double cones with maximum sensitivities at about 535 and 488 nm and ultraviolet- or violet-sensitive single cones with λ_{max} at about 370 nm or 420 nm. In contrast, the more sand-dwelling non-Mbuna species possess double cones with maximum sensitivities at about 570 and 535 nm and blue-sensitive single cones with λ_{max} at about 450 nm (Levine & MacNichol, 1979; Carleton *et al.*, 2000; Parry *et al.*, 2005; Jordan *et al.*, 2006). However, in some species, notably *Pseudotropheus acei* (an Mbuna species), rare additional cones have been identified suggesting that at least seven spectrally distinct cone pigments may be expressed in the retina (Parry *et al.*, 2005).

Analysis of the opsin gene complement from two species, zebra mbuna (*Metriaclima zebra*) and the Nile tilapia (*Oreochromis niloticus*) (Parry *et al.*, 2005) demonstrate that of the seven genes, three are from the RH2 gene class, two from the SWS2 class, with single representatives of the LWS and SWS1 gene classes. All seven gene sequences have been amplified from retinal cDNA, expressed in mammalian cells, and regenerated with 11-*cis*-retinal in vitro (Parry *et al.*, 2005; Spady *et al.* 2006).

Variations in expression across the retina have not been described in these cichlids, but ontogenetic changes have been established, at least in tilapia. Using quantitative methods to determine levels of gene expression, Spady *et al.* (2006) have shown that larval tilapia primarily express four of the seven opsins: SWS1, RH2B, RH2Aα and

LWS at approximately equal levels. However during development there is a major switch in expression with a marked down-regulation of the SWS1 and RH2B genes and a significant up-regulation of the SWS2A and LWS gene. Adult tilapia express primarily only three opsins: SWS2A, RH2Aα and LWS, with the LWS noticeably dominant.

Thus it appears that larval tilapia are potentially tetrachromatic with sensitivity extending into the near ultraviolet, but that adults are limited to trichromacy and are primarily long-wave sensitive. This is supported by MSP data (Loew *et al.*, unpublished observations) where larval fish have been shown to possess at least four spectral cone classes, including UV-sensitive small single cones, whereas adults possess only three spectral cone classes, double cones that are either 'red/red' or 'red/green' and blue-sensitive single cones. The presence of 'red/red' and 'red/green' double cones reflects the predominant expression of the LWS gene in the adult fish.

Lingcod (*Ophiodon elongatus*)

This marine species also shows changes in SWS1 and SWS2 opsin expression during larval development. The larvae are near-surface diurnal planktivores, whereas adults can be found from the surface to depths in excess of 400 m (Allen & Smith, 1988). Using MSP it is found that at the time of hatching only UV cones are present, although these are few and disappear very quickly. At pre-flexion there are UV-, violet-, green- and red-sensitive single cones presumably expressing SWS1, SWS2(B), RH2 and LWS opsin genes respectively. Between pre-flexion and flexion there is a rapid loss of UV cones, presumably due to down-regulation of the SWS1 gene and most likely apoptosis of the UV cells. At the time of pre-juvenile to juvenile transformation, all of the violet-sensitive cones show a spectral shift towards the blue which can be interpreted as a change in opsin gene expression from the SWS2 opsin gene to either a second SWS2(A) opsin gene or an RH2 opsin. By the end of this transformation, which takes only about 48 hours, all of the violet-sensitive cones have become blue-sensitive. There is also an apparent small long-wave shift in the green-sensitive single cones from about 512 nm to 522 nm. This may well reflect a switch between two RH2 opsin genes. Adults express LWS, an RH2 and probably the SWS2A opsin genes. These MSP data suggest the ontological expression of at least six cone opsin genes, somewhat similar to that described above for cichlids. Lingcod, more closely related to the cottoids, is, like the cichlids, a percomorph where gene duplication of both the RH2 and SWS2 opsin genes has occurred. In salmonids, there is no evidence of transformation or co-expression in the 'green' RH2 or 'red' LWS cones (see above). These observations are summarised in Table 2.

Several developmental themes arise from the above examples. The first is that there is much plasticity at the short wavelength end of the spectrum affecting particularly the expression of the UV (SWS1) opsin. The simplest interpretation is that this spectral region is adaptive for planktivory as has been often proposed (Loew & McFarland, 1990). As this feeding stage ends there is a decrease in UV/violet opsin expression. This may be accompanied by upregulation of an SWS2 (violet/blue) gene or actual loss of

Table 2 Visual pigments (λ_{max} in nm ± SD) versus developmental stage for the lingcod (*Ophiodon elongatus*). After Britt (2006), with kind permission. Note that the rods do not appear until the flexion stage, which is similar to that reported for a number of other species e.g. tuna (see below) and salmonids (see above). The same is true for a number of other Pacific Northwest marine species. However, in several species rods were present at the pre-flexion stage along with single cones (Britt *et al.*, 2001)

Photoreceptor	*Preflexion*	*Felxion*	*Postflexion*	*Prejuvenile*	*Transformation*	*Juvenile*	*Adult*
Rod	—	500 ± 1	500 ± 1	499 ± 2	501 ± 3	500 ± 2	499 ± 2
Single Cones							
UV	364 ± 9	—	—	—	—	—	—
Violet	405 ± 4	402 ± 2	402 ± 1	407 ± 6	—	—	—
Violet/Blue	—	—	—	—	421 ± 19		
Blue	—	—	—	—	—	460 ± 5	461 ± 1
Green	514 ± 1	512 ± 1	512 ± 2	513 ± 2	517 ± 2	521 ± 1	523 ± 1
Yellow	564 ± 6	564 ± 2	565 ± 1	568 ± 3	565 ± 5		
Double Cones							
Green Member	—	—	—	—	519 ± 2	522 ± 1	522 ± 4
Yellow Member	—	—	—	—	567 ± 4	565 ± 4	563 ± 7
Twin Cones	—	—	—	—	519 ± 3	520 ± 2	521 ± 3

the UV cells. Of course, this begs the question as to why fish such as goldfish (*Carassius auratus*) and zebrafish retain their UV cells into adulthood. For freshwater species it appears that there is little change in overall spectral sensitivity at the long wavelength end. For pelagic species such as yellowfin tuna the major change appears to involve loss of LWS gene expression. This may or may not involve loss of SWS1 or second SWS2 gene expression. Spectral sensitivity changes in adults seem to involve changes in A_1/A_2-chromophore utilisation except for the salmonids where SWS1 opsin expression with a return of UV cones may accompany the return to freshwater. It is clear that what was once thought to be a relatively simple and straightforward trajectory of spectral sensitivity change through the larval period can be quite complex. The prevalence of co-expression either during transition from one expression pattern to another or permanently as in adult guppy (*Poecilia reticulata*) (Archer & Lythgoe, 1990) only adds to this complexity.

STRUCTURAL CHANGES DURING THE LARVAL PERIOD

It is to be expected that first-feeding larvae must have visual systems capable of detecting and recognising prey items if they are to survive. This means adequate luminosity and chromatic sensitivity over a spectral region that maximises prey contrast, an optical system with enough acuity to resolve prey features, and the necessary 'wiring' to extract relevant characteristics such as motion, shape and perhaps colour. As the larvae grow, changes in ocular size and retinal organisation must remain adaptive for the tasks of feeding and predator avoidance while preparing the fish for metamorphic events that can mean dramatic changes in environment and life style. The reader is referred to Evans & Fernald (1990) and Evans & Browman (2004) for an in-depth review of these aspects of larval vision.

The developmental trajectories for most fish can be divided into three broad types: 'indirect', 'intermediate' and 'direct' (Balon, 1985). In their 1990 paper, Evans & Fernald associate photoreceptor appearance with these developmental types as follows: 1). 'Indirect' in which cones are the only photoreceptors present throughout most of the larval period with rods appearing coincident with a major change in lifestyle and body type at metamorphosis (e.g. winter flounder, Atlantic halibut (*Hippoglossus hippoglossus*), yellow perch (*Perca flavescens*), walleye (*Stizostedion vitreum*); 2). 'Intermediate' in which only cones are present at hatching with rods appearing later in the larval period, but well before transformation (e.g. salmonids) and 3). 'Direct' in which both cones and rods are present before hatching (e.g. zebrafish, goldfish). Since their original assignations, many examples have appeared that smear the distinction between these three groups as regards the appearance of photoreceptors. While the designations are useful, it is more reasonable to define them based on the state of the cones at hatching and the delay, if any, until the appearance of rods. What is common among all of these is that cones are present at the time of first-feeding even if the species ultimately become nocturnal or crepuscular as late juveniles or adults (exceptions seem to be members of the Elopomorpha) all of which have leptocephalus larvae and pure rod retinas with cones only added well after first-feeding (Blaxter & Staines, 1970; Pankhurst, 1984; Taylor, 2005; Taylor & Grace, 2005) and deep-sea species that may never develop 'morphological' cones presumably due to the absence of bright light (Blaxter & Staines, 1970) although several spectral classes of 'morphological' rods may be present (Partridge *et al.*, 1988; Partridge, 1989; Denton, 1990) . Throughout the larval period there may also be dramatic changes in cone type and distribution either occurring before the appearance of rods or coincident with their appearance. As discussed above, these structural changes may also involve changes in visual pigment complement/distribution. For illustrative purposes, examples of each of these developmental types are given below.

Indirect Development

Well-studied examples of this developmental trajectory are the winter flounder (Evans & Fernald, 1993; Marder & Cameron, 2004; Hoke *et al.*, 2006), perch (Ahlbert, 1973; Guma'a, 1982; Wahl *et al.*, 1993), walleye (Wahl, 1990), and the Atlantic halibut (Helvik *et al.*, 2001 also lists more examples). Again, a common feature of all these species is that only morphologically identical single cones are present in the retina with no anatomical evidence of an organised mosaic until near the time of metamorphosis. However, there is considerable species variability in the visual pigment expression pattern even at the first-feeding stage (see above discussion). Winter flounder seem to express only a single green opsin at first-feeding (Evans *et al.*, 1993) while Atlantic halibut express all four of the adult cone opsins (Helvik *et al.*, 2001). The development of the adult square mosaic pattern including paired cones was a post-metamorphic event as was the appearance of rods. However, an 'expression mosaic' was noted in the morphologically similar single cones of the Atlantic halibut (Helvik *et al.*, 2001). For winter flounder the change in opsin expression pattern from all RH2 to the adult pattern proceeded separately from the development of the square mosaic pattern seen in the adult (Marder & Cameron, 2004; Hoke *et al.*, 2006).

Larval yellow perch and walleye are closely related pelagic diurnal planktivores and often school together in the upper meter of the water column (Forney, 1974; Neilson, 1980). As adults they assume different feeding habits - yellow perch are diurnal feeders while walleye are crepuscular/nocturnal feeders possessing a tapetum as befits their lifestyle. The changes in moving from similar all-cone retinas at first-feeding to the markedly different retinas of the adult do not occur abruptly, but almost continuously throughout the larval period. Among the changes noted are:

1. With growth there is a decrease in the density of cones although the actual number increases (Guma'a, 1982; Wahl, 1990). Initially the retinae of both yellow perch and walleye have uniform receptor distribution with cones concentrated around the optic nerve and central posterior retina. Density changes with age occur gradually with highest cone densities being found in the ventral and central posterior retina, and the lowest in the anterior dorsal

retina. Interestingly, walleye lack outer segments at hatching with obvious single cones appearing within 5 days post-hatch. For yellow perch, twinning can be seen as early as Day 6 post-hatch through what appears to be a fusion process (Wahl, 1990). At this time the central single cone is the same size as the twin members although it contains a different visual pigment, presumably and RH2, than both member of the LWS-expressing twins (Loew, personal observation). This suggests some order in the twinning process, as similar expressive cones must 'find' each other to fuse into twins. While it is possible that twins could arise due to fusion of cones expressing different opsins with one of them switching expression, no evidence for such a process has been seen in any species to date (see discussion in Loew *et al.*, 2002). Larval yellow perch are seen to develop a population of small single cones not seen in the walleye at any developmental stage. It is known that these are UV/violet sensitive and that they drop out of the corner position of the already developed square mosaic during transformation (Wahl, 1990; Loew & Wahl, 1991; Loew *et al.*, 1993).

2. The size and volume of the cones increases throughout the larval period for both species even after twinning has occurred (Fig. 4). The larger size of the walleye cones is consistent with their crepuscular/nocturnal habit.
3. Based on nuclear counts there is a sequential change in neural summation occurring throughout the larval period as is seen in Fig. 5 (from Wahl, 1990). As seen, at six weeks there is less convergence in the central retina of yellow perch as compared to walleye, which suggests better resolution although perhaps less sensitivity due to reduced pooling. Prior to 6 weeks there are no rod outer segments visible although rod nuclei may be present. The trend for there to be more convergence between photoreceptors and ganglion cells in the walleye than in the yellow perch continues through to the adult stage with the finding that there are many more rods in the walleye retina than in the yellow perch.
4. The single cones of both yellow perch and walleye are organised first into rows as is seen for both Atlantic halibut and winter flounder. With twinning at or about 6 dph, yellow perch start to develop the typical square mosaic seen in the adult with the exception of the small single cones occupying some of the corner positions as mentioned above. In walleye a square mosaic develops near the optic nerve, but there is a transition to a typical row mosaic as one moves towards the periphery.

The differences and similarities between the walleye and yellow perch are illustrative of the kinds of trajectories seen in other members of the 'indirect' development group. While all start out with single cones arrayed in rows, followed by twinning, the addition of rods and changes in cone types and opsin expression, the transition to the adult pattern seems to progress through a series of smooth gradations rather than abrupt steps.

The trajectory for scombrids is similar to that for yellow perch (Margulies, 1997; Loew *et al.*, 2002). Early preflexion retinas are all cone with no morphological indication of chromotype even given that different visual pigments are present (above and Loew

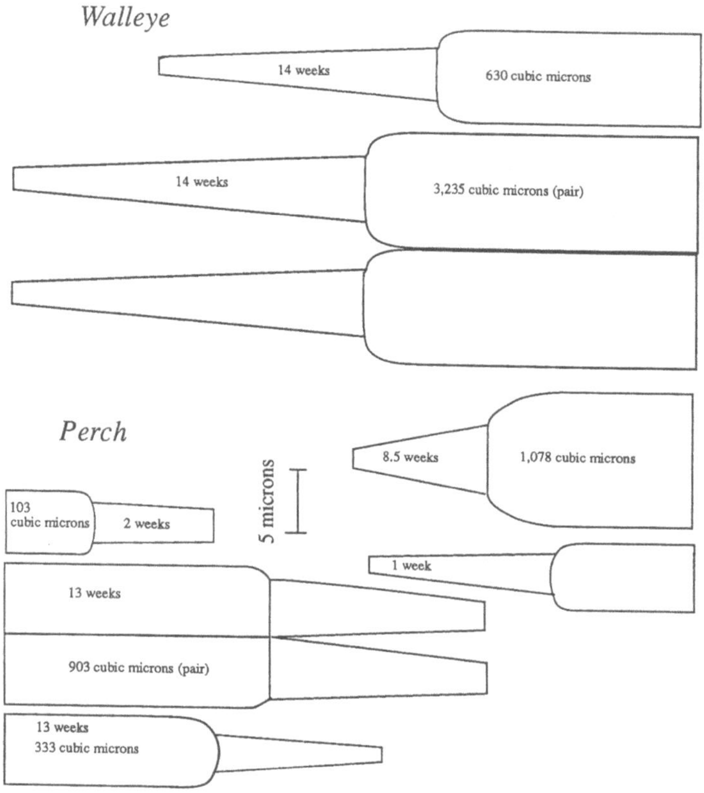

Fig. 4 Changes in cone size and volume for yellow perch (*Perca flavescens*) and walleye (*Stizostedion vitreum*) from the single cone, early larval stage to the transitional stage when both single- and twin-cones are present.

et al., 2002). There is a 1:1 ratio of cones to outer nuclei with overall count decreasing with growth. The highest density is in the temporal and ventro-temporal regions. Scattered double cones can be found in postflexion larvae. Rod outer segments appear simultaneously with twinning although the nuclei of rod precursors can be identified in the outer nuclear layer before this (Margulies, 1997).

Tectal data for scombrids are available and show that the tectum dominates the mesencephalon in all feeding larvae (Margulies, 1997). The tectum consists of two layers: an inner *stratum periventricular* (SPV) and an outer *stratum zonale* (SZ). The former is composed almost entirely of neurons while the latter is primarily fibrous tissue. With age the tectum thickens as neurons are recruited from the SPV to the SZ. The *torus longitudinalis* (TL), a sensory integration centre in the tectum, develops after feeding commences. This centre is associated with more complex visual processing tasks and the development of visual-motor and propioceptive co-ordination (Munro, 1984). In adults it is associated with saccadic eye movements that are also present in larvae (Northmore,

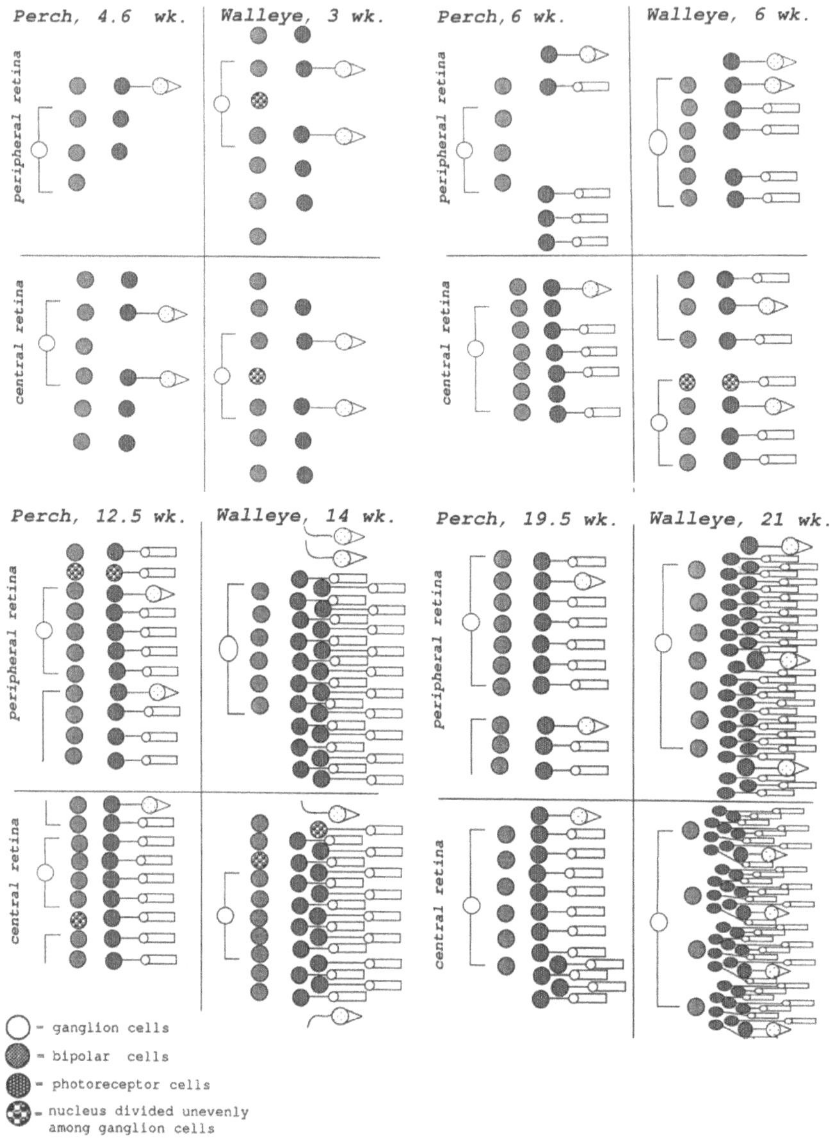

Fig. 5 Changes in convergence patterns with age for both yellow perch (*Perca flavescens*) and walleye (*Stizostedion vitreum*).

1984). Its development along with tasks such as piscivory and schooling is consistent with its supposed function. The relevance of this to the present discussion is that the early tasks of feeding do not require the higher order functions usually associated with the *torus*. It would appear that the processing localised to the retina itself as regards motion detection, edge and shape recognition and, perhaps, colour are enough for the larvae to detect, locate and capture prey.

Intermediate Development

The species developing along this trajectory have long 'larval' stages that can last a year or longer before metamorphosis to the adult form (Evans & Fernald, 1990). The Salmoninae of the genus *Oncorhynchus, Salmo* and *Salvelinus* are well-studied examples where the true larval stage is followed by an extended parr stage followed by smoltification to the adult form (see Stefansson *et al.*, this volume). A second metamorphic event associated with anadromy is often seen as well. As for 'indirect' species, first-feeders seem to have pure cone retinas with no evidence of rods at hatching (Ali, 1959; Blaxter, 1975; Novales-Flamarique & Hawryshyn, 1996). An exception to this is the finding of rods before hatching in sockeye salmon (*Oncorhynchus nerka*) (Novales-Flamarique & Hawryshyn, 1996). Also of interest here is the presence of a well-formed square mosaic before hatching. Thus, it would appear that first-feeders not only have mainly cone retinas, but can also have the mosaic organisation seen in the adult along with the attendant complement of visual pigments. Following hatching the retinal trajectory is like that seen for the 'indirect' group with the caveat that regional differences in density and chromatic cone populations may differ among species even at the same developmental stage.

Perhaps the most dramatic feature of premetamorphic development concerns the comings and goings of the small corner cones known to be either UV or blue depending on age (Bowmaker & Kunz, 1987; Cheng *et al.*, 2006). Both salmon and trout (genus *Oncorhynchus*) hatch with widely distributed UV corner cones. There follows a period of UV cone loss as well as a switch to blue opsin expression in those corner cones that remain (Cheng *et al.*, 2006). The timing of this event differs among members of the genus with a correlation between loss and the time of downstream migration (Cheng *et al.* 2006). The situation is different for the Atlantic salmon and the rainbow trout (*Oncorhynchus mykiss*) where corner cone loss occurs much later. Behavioural tests of spectral and polarisation sensitivity show changes correlating with the changes in corner cone presence and/or contained visual pigment(s) (Hawryshyn *et al.*, 1989, 1990; Beaudet *et al.*, 1993; Browman *et al.*, 1994; Novales-Flamarique, 2000; Novales-Flamarique & Browman, 2001; Deutschlander *et al.*, 2001). It is uncertain how these changes are adaptive for performing the visual tasks observed at different ages. Right now the data are correlative and not causative.

The structural development of the tectum in the brown trout (*Salmo fario*, a morph of *Salmo trutta*) and rainbow trout have been reported by Sharma (1975) and Pinganaud (1980), respectively. The adult tectum consists of three layers – superficial, intermediate and deep. Correlates of these exist in the embryonic and larval tectum. The intermediate is present at hatching while the 2 others appear progressively between 1 and 2 months post-hatch. Differentiation progresses latero-medially and from the anterior level to the posterior level. Again, this shows that first feeding larvae are able to successfully feed without the full repertoire of visual capabilities required for later juvenile and adult success.

Direct Development

This is perhaps the most thoroughly studied of the developmental types as it includes both the goldfish and the zebrafish. The latter is currently the species of choice for developmental studies of the retina and visual system (Detrich *et al.*, 2004; Fadool & Dowling, 2006).

As is definitive of the group, there is minimal delay between the appearance of cones and that of rods when compared to the other developmental groups (Branchek & Bremiller, 1984; Raymond *et al.*, 1995; Stenkamp *et al.*, 1996). While the histological data from zebrafish of Branchek & Bremiller (1984) suggest only single cones present at the time of first-feeding (4 dpf in their study), they hedge a bit by noting single synaptic ribbons consistent with rods at the time of hatching. A delayed appearance of rods has also been reported for goldfish (Johns & Hwang, 1981) although, again, this needs further study. Thus it is possible that they have both rods and cones at first-feeding by 4 dpf. Again, the appearance of obvious rods occurs simultaneously with that of double cones.

To summarise the above section, regardless of developmental type, almost all larvae, except members of the Elopomorpha and some deep-sea fishes, have almost exclusively single cones in their retinas at first-feeding. Neural convergence is minimal and only part of the tectum is developed. Whatever their ultimate role in vision it is clear that retinal mosaics and double cones as seen in later-stage larvae, juveniles and adults are not needed for successful feeding first-feeders.

VISUAL ACUITY

For larval fishes the reaction distance to planktonic prey increases with fish size and correlates with improved visual acuity, but at best only amounts to several body lengths (Confer *et al.*, 1978; Wanzenböck & Schiemer, 1989; Wahl *et al.*, 1993; Browman, *et al.*, 1994). In a typical visual predation cycle the detection event is followed by orientation, pursuit, strike and consumption events followed by a return to search mode (Holling, 1966; O'Brien, 1979, 1987; Evans & Browman, 2004). However, the fact that larval fish, such as yellow perch, can discriminate among detected prey (Durbin, 1979; Mills *et al.*, 1984, 1987; Wahl *et al.*, 1993; Wanzenböck *et al.*, 1996) suggests that a recognition step must be added to the predation cycle. Recognition requires the ability to extract spatial information about the target defined by edges and spatial contrast. Visual acuity, or resolution, will set the limit for this ability. The ultimate limiting factor is the 'effective' photoreceptor cell density (density modified by pooling). The minimum separable angle (MSA) is determined by the distance between adjacent photoreceptors or photoreceptor pools. In order to resolve two points, there must be a sufficient contrast edge between them so that differential stimulation above noise of adjacent photoreceptors can occur.

It is important to distinguish between 'detection' and 'resolution'. Imagine a circular array of equal radiance sources each separated by some distance **x** and focused by a perfect lens onto a detector array with equal element diameter and spacing **y**. The diameter of the source array is such that when placed at infinity its focused image falls on a single detector element. Resolution of the individual array sources is impossible, but detection of the array is not. All that would matter is that the signal produced by the illuminated detector element be discriminated from the noise in the contiguous elements (i.e. it must have sufficient contrast) (see Overington, 1976). The source array is moved closer to the lens until its image size is such that **x** = **y** in the image plane. That is, each source element is falling on a single, contiguous detector element. Again, resolution of the individual sources is impossible although probability of detection may be enhanced due to pooling of the detector output (for visual detection of small targets with diameter α and a 50% probability (denoted by ε) it is found that $\varepsilon\alpha^2 = \mathbf{constant}$ holds. This is known as Ricco's Law). The source array is again moved closer to the lens until **x** = 2**y** at the image plane. In this case unilluminated detector elements surround each illuminated one and, as long as each illuminated detector meets the detection criteria, the probability of resolving the individual sources is 100%. Between **x** = **y** and **x** = 2**y** the probability of resolving the sources will vary from 0 to 100%. The resolution distance can be equal to or less than the detection distance, but can obviously never be greater.

In the context of predation, the detection event stimulates an attentive response. The predator need extract no information other than target presence and location. This could theoretically be done at great distance, although the attentive range is probably limited to immediate surroundings. It is likely that this is the reaction distance reported in many studies. When considering the ability of a predator to recognise a prey, we need to include factors such as visual system optical characteristics, number of possible stimuli, complexity of form (pattern), previous experience, innate search images and association with a particular field of view, in addition to those factors influencing detection (see excellent review of this whole area in Overington, 1976). In considering eye design, we need to know the resolution limit of the optical system and the characteristics of the retina. Resolution in this context is the minimum distance between two features that allow the animal to sense that there are two, rather than one feature. Assuming 'perfect' optical quality, the variables of import are the entrance aperture of the system, the focal length and the detector 'grain', which can be approximated by measuring the photoreceptor spacing. For two bright sources, such as two scattering points inside a zooplankton or two bright spines on an appendage, the Rayleigh criterion sets the limit of resolution. This criterion is met when the peak of one of the object's point spread function lies on the first minimum of the other (see Loew, 1999). At this point the distance between the peaks (Δu) is given by

$$\Delta u = \frac{0.610\,\lambda}{n\,\sin(\alpha)}$$

where $n\,\sin(\alpha)$ is the numerical aperture (NA) of the lens. These relationships have two consequences as regards near-field recognition of zooplankton. The first is the fact that resolution improves as wavelength decreases. Larval fish having UV sensitive photoreceptors would not only be able to detect zooplankton at a greater distance due to the fact that they have greater contrast in the UV, but would also be able to resolve patterns at a greater distance (assuming that the UV system is capable of image encoding and there is not too much degradation due to scatter). The second concerns the relationship between numerical aperture and resolution. As the aperture (i.e. the lens diameter in fish) increases, the resolution increases. For yellow perch, Wahl (1990) found that the relationship between standard length (SL) and lens diameter (Θ) fit the equation

$$\Theta = 0.0638(SL)^{1.0671}$$

Thus as larvae grow, their resolving power increases. Δu can also be measured as the angular separation of two points in object space. This is the minimum separable angle (MSA) that, for yellow perch, is related to lens diameter (Θ) by the formula (Wahl, 1990)

$$MSA = 31.737(\Theta)^{-0.8064}$$

Combining these two formulas, it can be calculated that as SL increases from 10 to 25 mm, MSA decreases from 40.3 minutes of arc to 18.3 minutes of arc. If one assumes a pattern spacing of 0.2 mm, the resolution distances for the 10 mm and 25 mm SL fish would be 17.1 mm and 37.6 mm, respectively. Interestingly, the formulas also predict that if resolution distance scales directly with SL, as it appears to do for reaction distance (Wanzenböck & Schiemer, 1989), the separation distances at the limit of resolution are quite similar. For a strike distance of 2 × SL, the separation distance would be 0.23 mm and 0.27 mm, respectively. To see whether or not these calculations have any real significance for plankton recognition one would want to relate the resolution limit to the spatial frequencies present in plankton images.

A common feature of most planktivorous larvae is a large eye relative to body length. Generally, the larger the eye the better the resolution. By increasing the strength of the lens and enlarging the fundus, the number of visual cells across which an image falls increases. With this comes an increase in the MSA. This assumes adequate correction for aberrations and an entrance aperture providing sufficient image radiance entering the eye. Regardless of eye size, ocular proportions for fish remain remarkably constant. Thus the ratio of the distance of the lens centre to the lens radius is consistently 2.55 (Mattheissen's ratio) for a wide variety of juvenile and adult fishes (Mattheissen, 1880). However, at first-feeding this ratio can be substantially greater, dropping closer to its 'constant' value with larval growth (reviewed in Shand *et al.*, 1999). They hypothesise that the increased ratio is an adaptation for increased depth of field to make up for the fact that the *retractor lentis* muscle responsible for accommodation has not yet developed. The importance of receptor density has already been mentioned. The relationship between receptor density and MSA can be calculated using the formula of Tamura (1957):

$$\sin \alpha = 1/f\,(0.2/n^{-2})$$

where α equals the MSA in radians, f is the focal length in mm and n is the photoreceptor cell density (number per 0.1 mm^2). For first-feeders having only single cones the density is easily obtained from tangential sections or even with fresh preparations using Nomarski or other contrast improving microscope optics. This determination becomes problematic with the appearance of double cones since there is some uncertainty as to whether or not the members should be counted separately.

The change in MSA for a number of species during larval development has been obtained with the common finding that MSA decreases rapidly during early larval development (Wahl, 1990; Margulies, 1997). Examples of this change for yellow perch and walleye are seen in Fig. 6 (Wahl, 1990). First- feeders are much less discriminating in terms of prey capture than older larvae with the main selection factors being motion and gape-limited size (Evans & Browman, 2004).

RETINOMOTOR MOVEMENTS

There are two major types of retinomotor movement (RM) in fish: those of the retinal pigment epithelium (RPE) and those of the photoreceptor cells (reviewed by Ali, 1975; Ali & Wagner, 1975). The RPE of fishes contain various pigmented granules, usually of melanin but in some instances guanine or xanthine. During the day or under bright light conditions these granules move up processes of the RPE between the outer segments. This acts to reduce light exposure and reduce intra-ocular scatter between photoreceptors. Photoreceptor RM are not as universal as are pigment movements among teleosts, occurring chiefly among diurnal species having well-developed scotopic and photopic visual systems (i.e. having mixed retinas with both rods and cones) (see Table 1 in Ali & Wagner, 1975). When RM are initiated, the myoids of the photoreceptors expand or contract depending on cell class and direction of light intensity

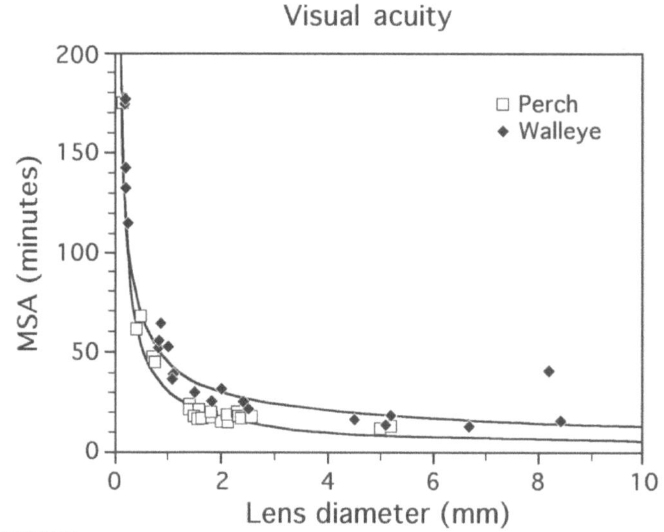

Fig. 6 Change in minimal separable angle (MSA) with increasing lens diameter for yellow perch (*Perca flavescens*) and walleye (*Stizostedion vitreum*). The lens diameter is directly proportional to standard length. After Wahl (1990).

change. At dawn or after exposure to bright light requiring light adaptation, the rod myoids extend moving the inner and outer segments away from the outer limiting membrane to line up sclerally along the basement membrane where they are shielded by pigment and out of the focal plane. At the same time, the cone myoids contract drawing the cones into position along the outer limiting membrane and the focal plane. The opposite happens at dusk or during dark adaptation. While being a mechanism for placing the appropriate photoreceptor type into focal position and adjusting potential quantum catch, this mechanism also allows for closer packing density at the focal plane by recovering the space taken up by the 'other' photoreceptor class.

The mechanisms underlying RM are complex involving both a direct effect of light and several circadian oscillators (Ali, 1975; Neave, 1984; Drenckenhahn & Wagner, 1885; Dearry & Barlow, 1987; Menger *et al.*, 2005). The contraction 'motor' involved is based on the interaction of actin and myosin for both rods and cones. Elongation of cones relies on a tubulin-associated mechanism while rod elongation is driven by actin polymerisation (see Drenckenhahn & Wagner, 1885). Circadian contraction of cones in fish seems to be regulated by dopamine via D2-receptors (Dearry & Burnside, 1986). Regulators of the other motors are unknown although a correlation has been found between cAMP levels and RM in brown trout (McCormack *et al.*, 1989). Interestingly, in constant darkness only cones in the green sunfish (*Lepomis cyanellus*) continued to show RM – they contracted during subjective day and elongated during subjective night (Dearry & Barlow, 1987). For light-induced RM it appears that it is the rods that provide the signal at least for the blue acara cichlid (*Aequidens pulcher*) (Kirsch *et al.*, 1989), and the red seabream (*Pagrus major*) (Kawamura *et al.*, 1997).

The development of RM in a number of species has been examined with the almost universal finding that no cone RM occurs until rods are present in the retina (Ali & Wagner, 1975; Neave, 1984; Wahl, 1990; Hodel *et al.*, 2006). This fits with the idea that rods control cone RM. However, pigment RM can occur before the appearance of rods when only single cones are present. For example, pigment granules are responsive to light and migrate as early as 5 dpf. At 20 dpf the pigment granules condense in the apical ends of the RPE mircovilli. Cone RM ensues at 28 dpf with the functional maturation of rods and the appearance of double cones.

The fact that the all single cone retina does not show photoreceptor RM is not all that surprising if one accepts the idea, mentioned above, that one of the main functions of photoreceptor RM is to increase the packing density of the appropriate photoreceptor at the focal plane. With no rods present it is unnecessary to move them out of the way to increase cone packing. The fact that pigment RM can be present handles the potential problems with cone light exposure.

FUTURE PERSPECTIVES

It is clear that the molecular and genomic techniques being applied to the zebrafish, medaka and perhaps salmon hold the promise of clearly elucidating many if not all of the processes underlying the development of the retina, the eye and higher centres. However, there is still the need for comparative studies applying what is learned from the 'model' species to others where adaptation has led to changes in both genotype and phenotype, including plasticity. Future surveys of visual pigments should include opsin identification and either MSP or expression/reconstitution. Two-dimensional opsin expression mapping is limited to a very few species and should be expanded to other species selected because of their visual tasks. This goes for both adults and earlier stages. Ultimately, the visual ecologist is interested in questions of fitness. The ability to use gene silencing or knock in/knock out techniques to affect retinal structure or spectral sensitivity could lead to fitness experiments that could go a long way towards answering why a particular developmental trajectory or visual system is adaptive in an evolutionary context.

References

Ahlbert IB (1973) Ontogeny of double cones in the retina of perch fry (*Perca fluviatilis*, Teleostei). Acta Zool 54: 241-254

Allen MJ, Smith GB (1988) Atlas and zoogeography of common fishes in the Bering Sea and northeastern Pacific. NOAA Tech Rept NMFS 66: 151

Ali MA (1959) The ocular structure, retinomotor and photobehavioural responses in juvenile Pacific salmon. Can J Zool 37: 965-996

Ali MA (1975) Retinomotor responses. In: Ali MA (ed) Vision in Fishes. Plenum Press, New York, p 313-356

Ali MA, Wagner HJ (1975) Distribution and development of retinomotor responses. In: Ali MA (ed) Vision in Fishes. Plenum Press, New York, p 369-396

Allison WT, Dann SG, Helvik JV, Bradley C, Moyer HD, Hawryshyn CW (2003) Ontogeny of ultraviolet-sensitive cones in the retina of rainbow trout (*Oncorhynchus mykiss*). J Comp Neurol 461: 294-306

Allison WT, Haimberger TJ, Hawryshyn CW, Temple SE (2004) Visual pigment composition in zebrafish: Evidence for a rhodopsin-porphyropsin interchange system. Visual Neurosci 21: 945-952

Allison WT, Dann SG, Veldhoen KM, Hawryshyn CW (2006) Degeneration and regeneration of ultraviolet cone photoreceptors during development in rainbow trout. J Comp Neurol 499: 702-715

Archer SN, Lythgoe JN (1990) The visual pigment basis for cone polymorphism in the guppy, *Poecilla reticulata*. Vision Res 30: 225-234

Balon EK (1985) Early Life Histories of Fishes: New Developmental, Ecological and Evolutionary Perspectives. Dr W Junk Publishers, The Hague, The Netherlands

Beaudet L, Browman HI, Hawryshyn CW (1993) Spectral sensitivity and retinal structure in rainbow trout of different sizes. Vision Res 33: 1739-1746

Blaxter JHS (1975) The eyes of larval fish. In: Ali MA (ed) Vision in Fishes. Plenum Press, New York, p 427-444

Blaxter JHS, Hempel G (1963) The influence of egg size on herring larvae. J Cons Perm Int Explor Mer 28: 211-240

Blaxter JHS, Staines M (1970) Pure-cone retinae and retinomotor responses in larval teleosts. J Mar Biol Assoc UK 50: 449-460

Bowmaker JK (1995) The visual pigments of fish. Prog Retin Eye Res 15: 1-31

Bowmaker JK, Loew ER (2008) Vision in fish. In: Hoy RR, Shepherd GM, Basbaum Al, Kaneko A, Westheimer G (eds) The Senses: A Comprehensive Reference. Elsevier, Oxford, UK. (in press)

Bowmaker JK, Kunz YW (1987) Ultraviolet receptors, tetrachromatic color vision and retinal mosaics in the brown trout (*Salmo trutta*): age-dependent changes. Vision Res 27: 2101-2108

Branchek T, Bremiller R (1984) The development of photoreceptors in the zebrafish, *Brachydanio rerio*. I. Structure. J Comp Neurol 224: 107-115

Bridges CDB (1972) The rhodopsin-porphyropsin visual system. In: Dartnall HJA (ed) The Handbook of Sensory Physiology VII/2. Springer-Verlag, Berlin, p 417-480

Britt LL, Loew ER, McFarland WN (2001) Visual pigments in the early life stages of Pacific northwest marine fishes. J Exp Biol 204: 2581-2587

Browman HI, Hawryshyn CW (1994a) Retinoic acid modulates retinal development in the juveniles of a teleost fish. J Exp Biol 193: 191-207

Browman HI, Hawryshyn CW (1994b) The developmental trajectory of ultraviolet photosensitivity in rainbow trout is altered by thyroxine. Vision Res 34: 1397-1406

Browman HI, Novales-Flamarique I, Hawryshyn CW (1994) Ultraviolet photoreception contributes to prey search behaviour in two species of zooplanktivorous fishes. J Exp Biol 186: 187-198

Carleton KL, Hárosi FI, Kocher TD (2000). Visual pigments of African cichlid fishes: evidence for ultraviolet vision from microspectrophotometry and DNA sequences. Vision Res 40: 879-890

Carleton KL, Kocher TD (2001) Cone opsin genes of African cichlid fishes: tuning spectral sensitivity by differential gene expression. Mol Biol Evol 18: 1540-1550

Carleton KL, Spady T, Kocher TD (2006) Visual communication in East African cichlid fishes: diversity in a phylogenetic context. In: Ladich F, Collin SP, Moller P, Kapoor BG (eds) Communication in Fishes, Vol 2. Science Publishers, Enfield, NH, USA, p 485-516

Cheng CL, Novales-Flamarique I (2004). New mechanism for modulating colour vision. Nature 428: 279

Cheng CL, Novales-Flamarique I, Harosi FI, Rickers-Haunerland J, Haunerland NH (2006) Photoreceptor layer of salmonid fishes: Transformation and loss of single cones in juvenile fish. J Comp Physiol 495: 213-235

Cheng CL, Gan KJ, Novales-Flamarique I (2007) The ultraviolet opsin is the first opsin expressed during retinal development of salmonid fishes. Invest Ophthalmol Vis Sci 48: 866-873

Chinen A, Hamaoka T, Yamada Y, Kawamura S (2003) Gene duplication and spectral diversification of cone visual pigments of zebrafish. Genetics 163: 663-675

Confer JL, Howick GL, Corzette MH, Kramer SL, Fitzgibbon S, Landesberg R (1978) Visual predation by planktivores. Oikos 31: 27-37

Dearry A, Barlow RB Jr (1987) Circadian rhythms in the green sunfish retina. J Gen Physiol 89: 745-770

Dearry A, Burnside B (1986) Dopaminergic regulation of cone retinomotor movement in isolated teleost retinas I. Induction of cone contraction is mediated by D2-receptors. J Neurochem 46: 1006-1021

Denton EJ (1990) Light and vision at depths greater than 200 metres. In: Herring PJ, Campbell AK, Whitfield M, Maddock L (eds) Light and Life in the Sea. Cambridge University Press, Cambridge, p 127-148

Detrich HW III, Westerfield M, Zon LI (eds) (2004) The Zebrafish: Cellular and Developmental Biology. Elsevier Academic Press, San Diego

Deutschlander ME, Greaves DK, Haimberger TJ, Hawryshyn CW (2001) Functional mapping of ultraviolet photosensitivity during metamorphic transitions in a salmonid fish, *Oncorhynchus mykiss*. J Exp Biol 204: 2401-2413

Douglas RH, Hunt DM, Bowmaker JK (2003) Spectral sensitivity tuning in the deep-sea. In: Collin SP, Marshall NJ (eds) Sensory Processing in Aquatic Environments. Springer, New York, p 323-342

Drenckhahn D, Wagner H-J (1985) Relation of retinomotor responses and contractile proteins in vertebrate retinas. Eur J Cell Biol 37: 156-168

Durbin AG (1979) Food selection by plankton feeding fishes. In: Clepper H (ed) Predator-Prey Systems in Fisheries Management. Sports Fishing Institute, Washington, p 203-218

Evans BI, Browman HI (2004) Variation in the development of the fish retina. Am Fish Soc Symp 40: 145-166

Evans BI, Fernald RD (1990) Metamorphosis and fish vision. J Neurobiol 21: 1037-1052

Evans BI, Fernald RD (1993). Retinal transformation at metamorphosis in the winter flounder (*Pseudopleuronectes americanus*). Visual Neurosci 10: 1055-1064

Evans BI, Harosi FI, Fernald RD (1993) Photoreceptor spectral absorbance in larval and adult winter flounder (*Pseudopleuronectes americanus*). Visual Neurosci 10: 1065-1071

Fadool JM, Dowling JE (2006) Zebrafish models of retinal development and disease. In: Sernagor E, Eglen S, Harris B, Wong R (eds) Retinal Development. Cambridge University Press, Cambridge, p 342-370

Fernald RD, Liebman PA (1980) Visual receptor pigments in the African cichlid fish, *Haplochromis burtoni*. Vision Res 20: 857-864

Forney JL (1974) Interactions between yellow perch abundance, walleye predation, and survival of alternate prey in Oneida Lake, New York. Trans Am Fish Soc 103: 15-24.

Guma'a R (1982) Retinal development and retino motor responses in perch (*Perca fluviatilis*). J Fish Biol 20: 611-618

Hawryshyn CW, Arnold MG, Chiasson D, Martin PC (1989) The ontogeny of ultraviolet photosensitivity in rainbow trout (*Salmo gardneri*). Visual Neurosci 2: 247-254

Hawryshyn CW, Arnold MG, Bowering E, Cole RI (1990) Spatial orientation of rainbow trout to plane-polarized light: the ontogeny of e-vector discrimination and spectral sensitivity characteristics. J Comp Physiol A 166: 565-574

Helvik JV, Drivenes O, Harboe T, Seo H-C (2001) Topography of different photoreceptor cell types in the larval retina of the Atlantic halibut (*Hippoglossus hippoglossus*). J Exp Biol 204: 2553-2559

Hjort J (1926) Fluctuations in the year classes of important fishes. J Cons Int Explor Mer 1: 5-38

Hodel C, Neuhauss SCF, Biehlmaier O (2006) Time course and development of light adaptation processes in the outer zebrafish retina. Anat Rec 288: 653-662

Hoke KL, Evans BI, Fernald RD (2006) Remodeling of cone photoreceptor mosaic during metamorphosis of flounder (*Pseudopleuronectes americanus*). Brain Behav Evol 68: 241-254

Holling CS (1966) The functional response of invertebrate predators to prey density. Mem Entomol Soc Ca 48: 1-86

Johns PR, Hwang Y (1981) Retinal development in larval goldfish: Delayed formation of rods. Invest Ophthalmol Vis Sci (Suppl) 20: 150

Jordan R, Kellogg K, Howe D, Juanes F, Stauffer J, Loew ER (2006) Photopigment spectral absorbance of Lake Malawi cichlids. J Fish Biol 68: 1291-1299

Kawamura G, Miyagi M, Anraku K (1997) Retinomotor movement of all spectral cone types of red seabream, *Pagrus major*, in response to monochromatic stimuli and UV sensitivity. Fish Sci 63: 233-235

Kimmel CB, Ballard WW, Kimmel SR, Ullmann B, Schilling TF (1995) Stages of embryonic development of the zebrafish. Dev Dyn 203: 253-310

Kirsch M, Wagner H-J, Douglas RH (1989) Rods trigger light adaptive retinomotor movements in all spectral cone types of teleost fish. Vision Res 29: 389-396

Kunz YW (2004) Developmental Biology of Teleost Fishes. Springer, Dordrecht, The Netherlands

Lasker R (1978) The relation between oceanographic conditions and larval anchovy food in the California current: Identification of factors contributing to recruitment failure. Rapp P-V Reun Cons Int Explor Mer. 173: 212-230

Levine JS, MacNichol EF Jr (1979) Visual pigments in teleost fishes: effects of habitat, microhabitat and behaviour on visual system evolution. Sensory Processes 3: 95-130

Loew ER (1999). Light and photoreception – physical principles. In: Archer SN, Djamgoz MBA, Loew ER, Partridge JC, Vallerga S (eds) Adaptive Mechanisms in the Ecology of Vision: Kluwer Academic Publishers, Dordrecht, The Netherlands, p 3-24

Loew ER, Wahl CM (1991) A short-wavelength sensitive mechanism in juvenile yellow perch, *Perca flavescens*. Vision Res 31: 353-360

Loew ER, McFarland WN (1990) The underwater visual environment. In: Douglas RH, Djamgoz MBA (eds) The Visual System of Fish. Chapman & Hall, London, p 1-44

Loew ER (1995) Determinants of visual pigment spectral location and photoreceptor cell spectral sensitivity. In: Djamgoz MBA, Archer SN, Vallerga S (eds) The Outer Retina. Chapman & Hall, London, p 57-78

Loew ER, McFarland WN, Margulies D (2002) Developmental changes in the visual pigments of the yellowfin tuna, *Thunnus albacares*. Mar Fresh Behav Physiol 35: 235-246

Loew ER, McFarland WN, Mills EL, Hunter D (1993) A chromatic action spectrum for planktonic predation by juvenile yellow perch, *Perca flavescens*. Can J Zool 71: 384-386

Marder MM, Cameron DA (2004) Photoreceptor differentiation during retinal development, growth and regeneration in a metamorphic vertebrate. J Neurosci 24: 11463-11472

Margulies D (1997) Development of the visual system and inferred performance capabilities of larval and early juvenile scombrids. Mar Fresh Behav Physiol 30: 75-98

Mattheissen L (1880) Untersuchungen über dem Aplanatismus und die Periscopie der Kristallinsen in den Augen der Fische. Pflug Arch 21: 287-307

McCormack CA, Hayden TJ, Kunz YW (1989) Ontogenesis of diunal rhythms of cyclic AMP concentration, outer segment disc shedding and retinomotor movements in the eye of the brown trout, *Salmo trutta*. Brain Behav Evol 34: 65-72

Menger GJ, Koke JR, Cahill GM (2005) Diurnal and circadian retinomotor movements in zebrafish. Visual Neurosci 22: 203-209

Mills EL, Confer JL, Ready RC (1984) Prey selection by young yellow perch: the influence of capture success, visual acuity and prey choice. Trans Am Fish Soc 113: 579-587

Mills EL, Widzowski DV, Jones SR (1987) Food conditioning and prey selection by young yellow perch (*Perca flavescens*). Can J Fish Aquat Sci 44: 549-555

Munro, AD (1984) The ontogeny of the retina and optic tectum in *Aequidens portalegrensis*. J Fish Biol 24: 377-394

Nawrocki L, BreMiller R, Streisinger G, Kaplan M (1985) Larval and adult visual pigments of the zebrafish, *Brachydanio rerio*. Vision Res 25: 1569-1576

Neave DA (1984) The development of retinomotor reactions in larval plaice, *Pleuronectes platessa*, and turbot, *Scophthalmus maximus*. J Exp Mar Biol Ecol 76: 167-176

Neilson LA (1980) Effect of walleye (*Stizostedion vitreum*) predation on juvenile mortality and recruitment of yellow perch (*Perca flavescens*) in Oneida Lake, New York. Can J Fish Aquat Sci 37: 11-19

Neumeyer C (2003) Color vision in fishes and its neural basis. In: Collin SP, Marshall NJ (eds) Sensory Processing in Aquatic Environments. Springer, New York, p 223-235

Nicol JAC (1989) The Eyes of Fishes. Clarendon Press, Oxford, UK

Northmore DPM (1984) Visual and saccadic activity in the goldfish *Torus longitudinalis*. J Comp Physiol A 155: 333-340

Novales-Flamarique I (2000) The ontogeny of ultraviolet sensitivity, cone disappearance and regeneration in the sockeye salmon, *Oncorhynchus nerka*. J Exp Biol 203: 1161-1172

Novales-Flamarique I, Hawryshyn CW (1996) Retinal development and visual sensitivity of young Pacific sockeye salmon (*Oncorhynchus nerka*). J Exp Biol 199: 869-882

Novales-Flamarique I, Browman HI (2001). Foraging and prey search behaviour of small juvenile rainbow trout (*Oncorhynchus mykiss*) under polarized light. J Exp Biol 204: 2415-2422

O'Brien WJ (1979) The pedator-prey interaction of planktivorous fish and zooplankton. Amer Sci 67: 572-581

O'Brien WJ (1987) Planktivory by freshwater fish: Thrust and parry in the pelagia. In: Kerfoot WC, Sih A (eds) Predation – Direct and Indirect Impacts on Aquatic Communities. University Press of New England, Hanover, p 3-16

Overington I (1976) Vision and Acquisition. Pentech Press, London

Pankhurst NW (1984) Retinal development in larval and juvenile European eel, *Anguilla anguilla* (L.). Can J Zool 62: 335-343

Parry JWL, Carleton KL, Spady TC, Carboo A, Hunt DM, Bowmaker JK (2005) Mix and match color vision: tuning spectral sensitivity by differential opsin gene expression in Lake Malawi cichlids. Curr Biol 15: 1734-1739

Partridge JC (1989) The visual pigments of deep-sea fishes: ecophysiology and molecular biology. Prog Underwater Sci 14: 17-31

Partridge JC, Archer SN, Lythgoe JN (1988) Visual pigments in the individual rods of deep-sea fishes. J Comp Physiol 162: 543-550

Pinganaud, G (1980) Development of the trout visual system (*Salmo irideus*). Arch d'Anat Micro Morph Experimentale 69: 215-231

Raymond PA, Barthel LK, Curran GA (1995) Developmental patterning of rod and cone photoreceptors in embryonic zebrafish. J Comp Neurol 359: 537-550

Robinson J, Schmitt EA, Harosi FI, Reese RJ, Dowling JE (1993) Zebrafish ultraviolet visual pigment: absorption spectrum, sequence and localization. Proc Natl Acad Sci USA 90: 6009-6012

Shand J (1993) Changes in the spectral absorption of cone visual pigments during the settlement of the goatfish *Upeneus tragula*: The loss of red sensitivity as a benthic existence begins. J Comp Physiol A 173: 115-121

Shand J, Døving KB, Collin SP (1999) Optics of the developing fish eye: comparisons of Matthiessen's ratio and the focal length of the lens in the black bream *Acanthopagrus butcheri* (Sparidae, Teleostei). Vision Res 39: 1071-1078

Sharma SC (1975) Development of the optic tectum in brown trout. In: Ali MA (ed) Vision in Fishes. Plenum Press, New York, p 411-418

Spady TC, Parry JWL, Robinson PR, Hunt DM, Bowmaker JK, Carleton KL (2006) Evolution of the cichlid visual palette through ontogenetic subfunctionalization of the opsin gene arrays. Mol Biol Evol 23: 1-10

Stenkamp DL, Hisatomi O, Barthel LK, Tokunaga F, Raymond PA (1996) Temporal expression of rod and cone opsins in embryonic goldfish retina predicts the spatial organization of the cone mosaic. Invest Ophthalmol Vis Sci 37: 363-376

Takechi M, Kawamura S (2005) Temporal and spatial changes in the expression pattern of multiple red and green subtype opsin genes during zebrafish development. J Exp Biol 208: 1337-1345

Tamura T (1957) A study of visual perception in fish, especially on resolving power and accommodation. Bull Jap Soc Sci Fish 22: 537-557

Taylor SM (2005). Ontogeny of Vision in Elopomorph Fishes: Evolution and Functional Morphology of a Continuously Changing Retina in a Diverse Teleost Clade. PhD Thesis, Florida Institute of Technology, Melbourne, Florida, USA

Taylor SM, Grace MS (2005) Development of retinal architecture in the elopomorph species *Megalops atlanticus*, *Elops saurus* and *Albula vulpes* (Elopomorpha:Teleostei). Contrib Mar Sci 37: 1-29

Temple SE, Plate EM, Ramsden S, Haimberger TJ, Roth W-M, Hawryshyn CW (2006) Seasonal cycle in vitamin A1/A2-based visual pigment composition during the life history of coho salmon (*Oncorhynchus kisutch*). J Comp Physiol A 192: 301-313

Van der Meer HJ, Bowmaker JK (1995) Interspecific variation of photoreceptors in four coexisting haplochromine cichlid fishes. Brain Behav Evol 45: 232-240

Wahl CM (1990) Development of the retinal mosaic in the walleye (*Stizostedion vitreum*) and the yellow perch (*Perca flavescens*), including daily changes in morphology. PhD Thesis, Cornell University, Ithaca, NewYork, USA

Wahl CM, Mills EL, McFarland WN, Degisi JS (1993) Ontogenetic changes in prey selection and visual acuity of the yellow perch, *Perca flavescens*. Can J Fish Aquat Sci 50: 743-749

Wanzenböck J, Schiemer F (1989) Prey detection in cyprinids during early development. Can J Fish Aquat Sci 46: 995-1001

Wanzenböck J, Zaunreiter M, Wahl CM, Noakes DLG (1996) Comparison of behavioural and morphological measures of visual resolution during ontogeny of roach (*Rutilus rutilus*) and perch (*Perca flavescens*). Can J Fish Aquat Sci 53: 1506-1512

Werblin F, Roska B (2007) The movies in our eyes. Sci Am 296: 72-80

CHAPTER

13

Electroreception

Frank Kirschbaum[1,2,*] and Jean-Pierre Denizot[3]

INTRODUCTION

Electroreceptors are special sensory organs of the lateral line system sensitive to weak electric fields. About one in six of the approximately 55,000 extant vertebrates are electroreceptive (Table 1) (Albert & Crampton, 2006). Of these electroreceptive species about 41% are amphibians, and another 41% are catfishes. The remaining electroreceptive species comprise lampreys, cartilagenous fishes, non-teleost and teleost species. The peripheral electroreceptor organs and the mechanosensory neuromasts are innervated in most species by one of six lateral line nerves and often develop from lateral line placodes (Northcutt, 2005).

The phylogenetic distribution of electroreception among extant vertebrates indicates a complex evolutionary history with several events (Fig. 1). Superficial electroreceptive organs originated in the common ancestor of lampreys and gnathostomes early in the Palaezoic and are found in extant lampreys. The electroreceptive organ of cartilaginous fishes, and basic non-teleost bony fishes (lungfishes, coelacanths, bichirs, reedfish, sturgeons and paddlefishes) is the ampullary organ, often named ampullae of Lorenzini.

Authors' addresses: [1]Leibniz-Institute of Freshwater Ecology and Inland Fisheries, Müggelseedamm 310, 12587 Berlin, Germany.
[2]Institute of Animal Sciences, Humboldt-University Berlin, Berlin, Germany.
[3]C.N.R.S., Unité de Neurosciences Intégratives et Computationnelles, 91198 Gif-sur-Yvette, Cedex, France.
**Corresponding author:* E-mail: fkirschb@igb-berlin.de

Table 1 Diversity estimates of 13 electroreceptive vertebrate taxa and salient features of electroreceptive systems. After Albert & Crampton (2006). Reprinted with permission of Taylor and Francis Group LLC.

Electroreceptive taxon	*Common name(s)*	*Spp.*	*Electroreceptive mode*	*Electroreceptor organs*	*Habitat*
Petromyzontiformes	Lampreys	38	Passive	End buds	Marine
Elasmobranchii	Sharks, skates, rays	1,000	Passive	Ampullary	Fresh water
Chimaeriformes	Chimeras	43	Passive	Ampullary	Marine
Coelacanthiformes	Coelacanths	2	Passive	Ampullary	Marine
Dipnoi	Lungfishes	9	Passive	Ampullary	Marine
Lisamphibia	Amphibians	3,850[1]	Passive	Ampullary	Fresh water
Monotremata	Platypus, echnidnas	3	Passive	Mucos glands	Fresh water
Polypteriformes	Bichirs	15	Passive	Ampullary	Fresh water
Chondrosteans	Sturgeon, paddlefishes	28	Passive	Ampullary	Fresh water
Xenomystinae	African knifefishes	2	Passive	Ampullary	Fresh water
Mormyriformes	African electric fishes	235	Passive, Active	Ampullary, tuberous	Fresh water
Siluriformes	Catfishes	3,571	Passive	Ampullary[2]	Fresh water[3]
Gymnotiformes	Neotropical electric fishes	173	Passive, Active	Ampullary, tuberous	Fresh water

Note: 41% of electroreceptive vertebrate species are amphibians and another 41% catfishes.

1. Excluding 1,500 spp. with direct development (Lisamphibia with 5,350 spp.).
2. Tuberous organs known in at least one cetopsid catfish (*Pseudocetopsis*).
3. Ariidae and Plotosidae are marine.

However, this primitive ampullary organ comprising a hair cell bearing an apical kinocilium (Fig. 1) appears to have been lost with the origin of the neopterygian bony fishes, the gars, bowfins and teleosts. Though there are sufficient anatomical data to support the presence of electroreception in the non-teleost bony fishes the behavioural correlates of the electric sense in lampreys, chimaeras, sturgeons, coelacanths and lungfish are not well investigated.

The phylogeny of teleost fishes is characterised by repeated evolution of electroreceptive and electrogenic (electric organ) systems (Fig. 2). The ampullary organs re-evolved three times among different groups of teleosts; independently in the African knifefishes (Xenomystinae) and elephantnoses (Mormyriformes), both members of the bony tongues (Osteoglossomorpha), and again in the the catfishes and the South American knifefishes, members of the Ostariophysi. In addition to these derived low-frequency sensitive ampullary organs (representing hair cells bearing apical microvilli), high-frequency sensitive (50 Hz to 2 KHz) tuberous electroreceptors developed in two teleost taxa (Elephantnoses and South American knifefishes) (Figs. 1 and 2) concomittantly with electric organs.

The ampullae of Lorenzini of cartilaginous fish and the ampullary organs of teleosts are sensitive to low frequency stimulations (0.1 to 50 Hz; e.g. Zakon, 1986). The sensory cells of these organs resemble the inner ear sensory cells. The ampullae of Lorenzini are excited by cathodal and inhibited by anodal stimuli (Bullock *et al.*, 1983) whereas the teleost ampullary organs are excited by anodal stimuli. These stimulations depolarise the apical membrane of the sensory cell. The depolarisation is propagated towards the basal membrane and causes the activation of the chemical synapse and finally leads to the stimulation of the afferent fibre (Bretschneider & Peters, 1992). The sensory cells and the afferent fibres in general are maintained in a state of partial activity or oscillation even in the absence of stimulation. Ampullary organs belong to the *octavolateralis* system, the ampullae of Lorenzini project to the dorsal *octavolateralis* nucleus and the teleost ampullary organs project to the electrosensory lateral line lobe (for details see Bodznick & Montgomery, 2005).

The electric systems comprising the electroreceptive and electrogenic components are involved in the following behavioural functions: passive electrolocation, active electrolocation, species recognition, and intraspecific communication. A detailed description of these different functions can be found in Zupanc & Bullock (2005).

Passive electrolocation is the ability to perceive due to very sensitive ampullary electroreceptors, weak, low-frequency electrical signals ($< 1\ \mu V\ cm^{-1}$, <50 Hz) emanating from physical or biological sources (Fig. 3). The passive electrosense is used for prey location and in some cases for detection of orientation cues.

Active electrolocation is performed by weakly electric fishes. They can detect with the tuberous electroreceptors objects that are different in conductivity than the surrounding water and cause distortions of their self-generated three-dimensional electric field. The electric image of an object depends on its electrical properties, size, and shape and on its distance from the fish. Active electrolocation in general is limited to a distance of several centimetres (Fig. 4).

Species recognition has been demonstrated in both wave and pulse-type species of mormyriform and gymnotiform weakly electric fishes. Invariant properties of the electric organ discharge (EOD) can be used for species identification, i.e. waveform, peak spectral frequency, and pulse repetition rate is used in pulse-type species, but fundamental frequency[1], waveform, and pulse repetition rate is used in wave-type species. Species recognition is enhanced by the increased sensory sensitivity to the cues of the EOD of the own species, while filtering out cues of foreign species.

Intraspecific communication. The electric system comprises the generation and perception of electric signals and therefore serves the function of communication with conspecifics (Moller & Serrier, 1986). This electrocommunication is used during social, agonistic, courtship and mating behaviour.

The terminology of primary developmental stages in fishes varies according to the authors. An overview can be found in Kendall *et al.* (1984). We apply here the terminology of Balon (1975) who defines the embryo stage from fertilisation up to the beginning of exogenous feeding. The larval stage starts with exogenous feeding and ends with the attainment of the juvenile features. However, not all fish have larvae, those with direct development pass from the embryo stage directly to the juvenile stage.

Which of the electroreceptive fishes possess larvae? Lampreys have a very pronounced larval stage; however, electroreception is poorly investigated in this taxon. Cartilaginous fishes produce large, yolky eggs and are characterised by direct development (Fiedler, 1991) as do the coelacanths (Balon, 2003). The remaining non-teleost bony fishes, the Dipnoi, bichirs and reedfish (Polypteriformes), and sturgeons and paddlefishes (Acipenseriformes) all pass through a larval stage during early ontogeny. The electroreceptive teleosts, the mormyrids (Kirschbaum & Schugardt, 2002) and the ostariophysans (Fuiman, 1984; Kirschbaum & Schugardt, 2006) all are characterised by larvae. The early ontogeny of species of the subfamily Xenomystinae is unknown.

[1]The sinuisodal waveform of wave-type species can be considered a harmonic; the fundamental frequency is the lowest frequency in a harmonic series.

Fig. 1 Evolution and distribution of electroreception in Agnatha and fishes (**A**) and anatomy and physiology of the sensory cell of the primitive (ampullae of Lorenzini, **B**) and advanced ampullary electroreceptor (**C**). **A**, in black taxa comprising electroreception and the corresponding processing brain structures. Hatched areas indicate the re-evolution of electroreception in Osteoglossomorpha (II) (Mormyriformes and Xenomystinae) and Ostariophysi (III) (Siluriformes (electrogenic in part) and Gymnotiformes (all electrogenic)) **B**, the primitive ampullary organ possesses a kinocilium (KC) and is characterised by anodal stimulation. **C,** derived ampullary organ bearing microvilli (MV) and stimulation is anodal. Note that the sensory cell of the ampullary organ of fishes of the subfamily Xenomystinae (see Fig. 10) differs from that of all other teleosts by the presence of both an apical cilium and micrivilli. **A**, M, medial octavolateral nucleus; D, dorsal nucleus; NALLv, ventral root of anterior lateral line nerve; NALLd, dorsal root of anterior lateral line nerve; CC, cerebellar corpus; L, electrosensitive lateral line lobe; NVIII, octaval nerve. **B**, **C**, N, nerve; RC, receptor cell. After Bullock & Heiligenberg (1986). Reprinted with permission from John Wiley and Sons, Inc.

Fig. 1 Contd. ...

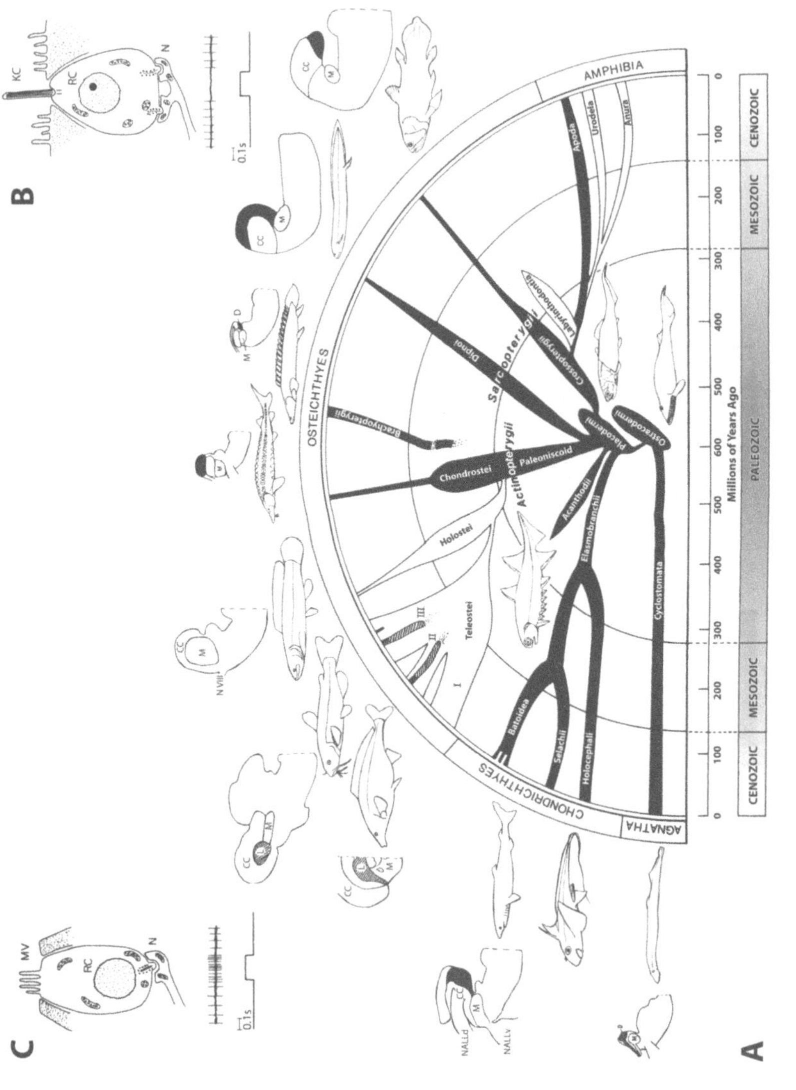
A
B
C
OSTEICHTHYES
CHONDRICHTHYES
AGNATHA
AMPHIBIA
Apoda
Urodela
Anura
Labyrinthodontia
Crossopterygii
Sarcopterygii
Dipnoi
Brachyopterygii
Chondrostei
Paleoniscoid
Actinopterygii
Holostei
Teleostei
Placodermi
Ostracodermi
Acanthodii
Elasmobranchii
Cyclostomata
Batoidea
Selachii
Holocephali
Millions of Years Ago
CENOZOIC
MESOZOIC
PALEOZOIC
0
100
200
300
400
500
600
KC
RC
N
MV
CC
M
D
NVIII
NALLd
NALLv
0.1s

Fig. 1 Contd. ...

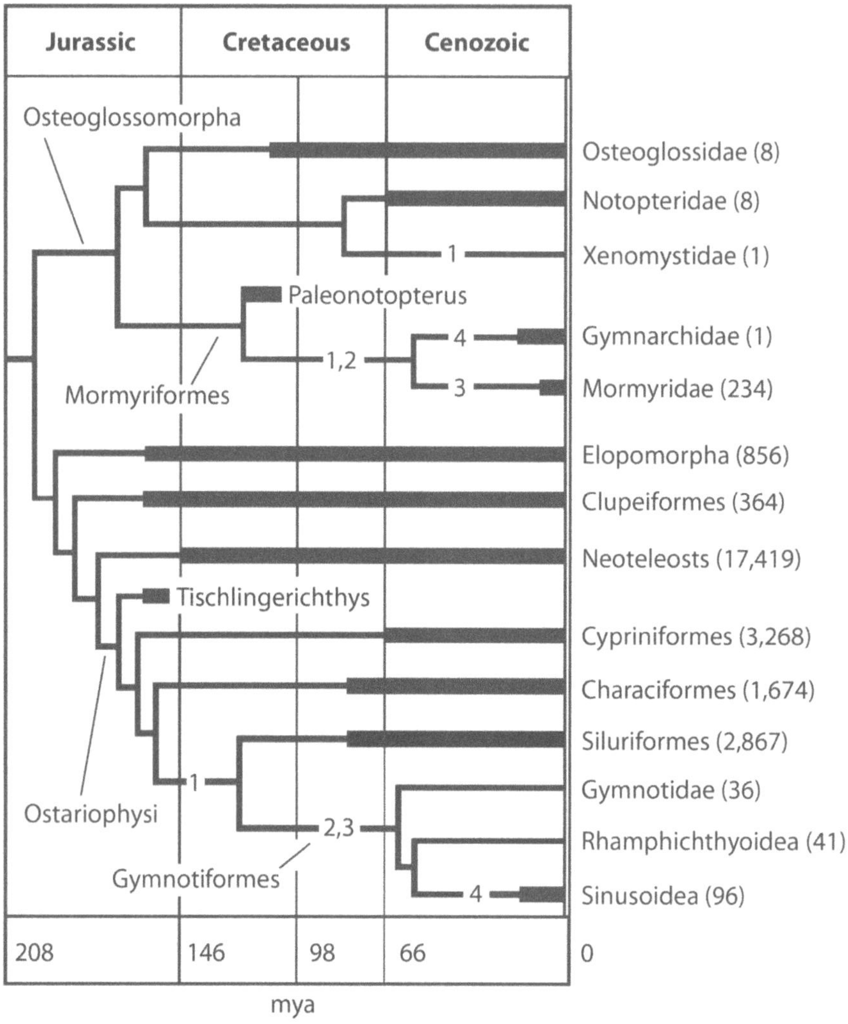

Fig. 2 Phylogeny of teleost fishes showing repeated evolution of electroreceptive and electrogenic systems. (1) Passive electroreception with low-frequency (DC to 50 Hz) sensitive ampullary organs, bearing hair cells with apical microvilli and anodal stimulation, (2) Active electroreception with high-frequency (50 Hz to 2 kHz) sensitive tuberous organs. (3) Pulse-type electric organ discharge (EOD). (4) Wave-type EOD. Some low diversity taxa omitted for clarity. Thick lines represent known minimum ages from fossils; thin lines are ghost lineages inferred from ages of sister taxa. Estimated numbers of extant species in parentheses. Rhamphichthyoidea = Rhamphichthyidae + Hypopomidae; Sinusoidae = Sternopygidae + Apteronoidae. mya: million years ago. After Albert & Crampton (2006). Reprinted with permission from Taylor and Francis Group LLC. Estimated numbers of species according to Nelson (2006) and Albert & Crampton (2006).

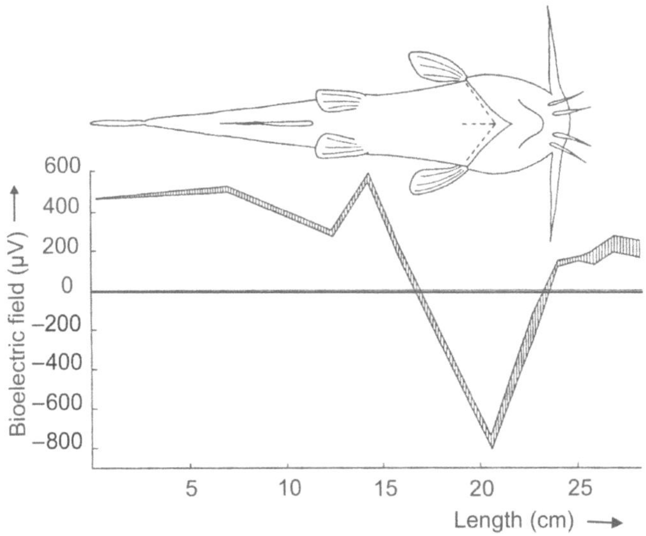

Fig. 3 Bioelectric field (DC field potentials) of the brown bullhead (*Ictalurus nebulosus*). The potentials were measured 5 mm from the ventral side of the animal with respect to a 1 mm distant reference electrode. The negative values illustrate the negative part of the dipol, the positive part, i.e. the brown bullhead represents electrically a double dipole with the negative pole located at the gills and the two positive poles at the head and the tale. The thickness of the line indicates the amplitude of the respiratory potential. After Peters & Bretschneider (1972). Reprinted with permission from Springer.

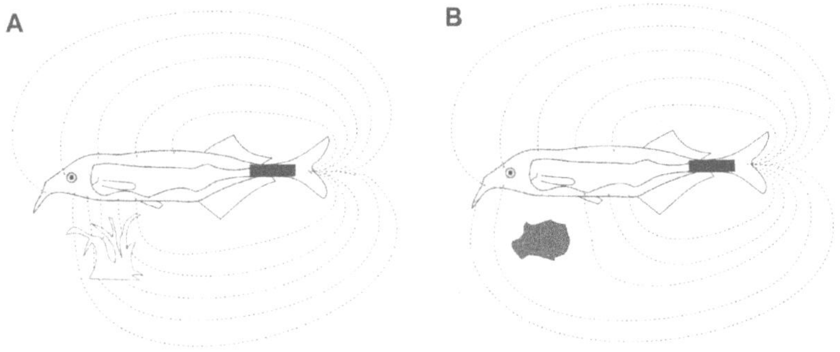

Fig. 4 Distortions of the electric field of the elephant nose fish (*Gnathonemus petersii*) caused by natural objects with impedances differing from that of the surrounding water. (**A**) Water plants are good conductors and 'attract' the electric field lines, causing more current to flow though the electroreceptor organs in that part of the skin located opposite to the object. (**B**) Stones as isolators repel the electric field lines and thus result in a reduced current flow in the corresponding electroreceptors compared to the current flow in the absence of the object. In addition to the field lines, the location of the electric organ in the caudal peduncle (black bar) and the electroreceptive body surface of the fish (gray shade) are shown. After von der Emde (1999). Reprinted with permission from Company of Biologists Ltd.

Though a number of fishes are electroreceptive (see above) and many possess larvae the information on electroreception in fish larvae is still very limited. The main reasons are the lack of knowledge concerning the developmental stages and the difficulty to work on the physiology and behaviour of these fish larvae which are often quite small and difficult to handle.

ELECTRORECEPTION IN FISH LARVAE

Lampreys (Agnatha)

The presence of a dorsal *octavolateralis* nucleus in the *medulla oblongata* of adult lampreys (Fig. 1) and direct recordings from single fibres of the epidermal sensory cells (Fig. 5) called end buds (Bodznick & Northcutt, 1981; Bodznick & Preston, 1983) are indications of an electroceptive sensory system in adult lampreys. However there are no behavioural data to support this assumption (Jørgensen, 2005).

Ammocoete larvae of lampreys seem to have a sense for weak electric fields (Ronan & Bodznick, 1986), though they do not possess end buds; instead, there are a number of multivillous sensory cells (Fig. 5) (Whitear & Lane, 1983) resembling sensory cells belonging to the *octavolateralis* system. These cells occur in small groups on the tail and in the gill vent papillae (Steven, 1951). These sensory cells probably are electroreceptors though definitive proof is still missing.

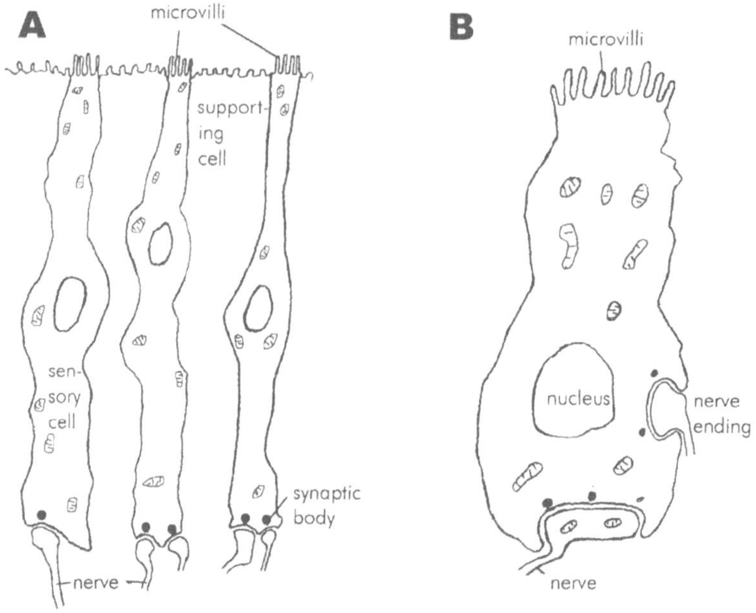

Fig. 5 Outlines of (**A**) sensory cells in the adult lamprey end bud and (**B**) multivilous cell from the lamprey ammocoete larva. After Jørgensen (2005). Reprinted with permission from Springer.

Lungfishes (Dipnoi)

The lungfishes are very old primary freshwater fishes comprising 6 extant species: The Australian lungfish (*Neoceratodus forsteri*) (Carvalho, 2003a), the South American lungfish (*Lepidosiren paradoxa*), and there are 4 species of African lungfishes of the genus *Protopterus* (Carvalho, 2003b). The ampullary organs of the lungfishes are embedded in a thick epidermis and they are not only found on the head, but also on the trunk and the caudal fin (Northcutt, 1986). In the South American lungfish electroreception has been demonstrated electrophysiologically. Pehrson (1949) has described the ontogeny of the lateral line placodes. However, data on electroreception in lungfish larvae are still lacking.

Bichirs and Reedfish, (Polypteriformes)

The bichirs (genus *Polypterus*) and the reedfish (genus *Calamoichthys*) are a small group of African fishes (Carvhalo & Schelly, 2003). Their ampullary organs comprise simple tubes in the skin. The sensory epithelium consists of electroreceptor cells separated by supporting cells. On the apex of the sensory cell a single cilium is present surrounded by microvilli. Fine structural features were described by Roth & Tscharntke (1976) and Jørgensen (1982). These ampullary organs are found on the head with concentrations on the snout region in front of the eyes (Northcutt, 1986). Neither the development of the electroreceptors nor the behaviour of larvae related to electroreception has been performed so far (Northcutt, 2005).

Sturgeons and Paddlefishes (Chondrostei)

There are 25 species of sturgeons of the family Acipenseridae (Birstein and Bemis, 1997) and two species of paddlefishes (Family Polyodontidae), the North American Mississippi paddlefish (*Polyodon spathula*) and the Chinese swordfish (*Psephurus gladius*).

Acipenseridae

The ampullary organs occur on the head, on the rostrum and in some species on the opercula (Fig. 6). Northcutt (1986) mapped their distribution in the shovelnose sturgeon (*Scaphirhynchus platorynchus*). The ampullary organs are comprised of numerous single organs (Fig. 6). The pear-shaped sensory cells bear an apical cilium. Boglione *et al.* (1999) described the ontogeny of the ampullary organs in Adriatic sturgeon (*Acipenser naccarii*). Four days post hatching the electrosensory organ anlagen are arranged in parallel rows on the ventral and lateral sides of the rostrum and on the opercular fold. 12 days post hatching the ampullary organs are arranged in many regular lines in the rostral ventral region (Fig. 7A). They have more than doubled in diameter compared to the 4 day-old larva and display typical calyces (Fig. 7B). A histological section of the ampullary organ of a juvenile is shown in Fig. 7C. These anatomical and histological data indicate (as in the Mississippi paddlefish, see below) an early function of the electrosensory system in sturgeons. However, experimental or behavioural evidence is still missing.

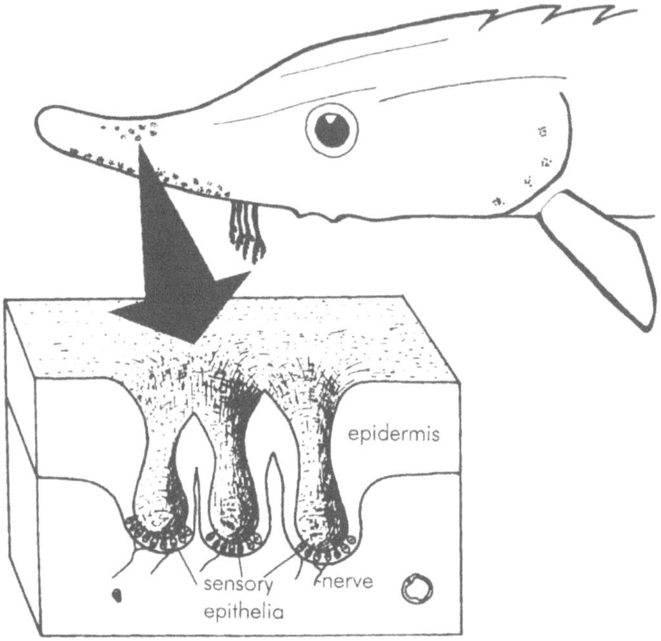

Fig. 6 The head of a sterlet (*Acipenser ruthenus*) with ampullary organs on the snout and gill cover. After Jørgensen (2005). Reprinted with permission from Springer.

Polyodontidae

Data on electroreception in the two species of paddlefishes are only available on the Mississippi paddlefish, a species quite common in the Mississippi and Missouri drainage system in North America. The paddlefish is a planktonic feeder for its entire life and is distinctive for its flat, elongated rostrum (Fig. 8). This organ is covered with numerous ampullary electroreceptors, called ampullae of Lorenzini (Jørgensen *et al.*, 1972). There is electrophysiological evidence (New & Bodznick 1985) that these sense organs are indeed electroreceptors. The function of the rostrum has long been unknown. However, there is recent evidence (Wilkens *et al.*, 1994, 1997, 2001, 2002; Wojtenek *et al.*, 2001a) that this organ is used for passive electroreception during plankton feeding in larvae and juveniles (Fig. 8). The fish are able to detect and catch prey at distances up to 8 cm (Fig. 9). In addition to the indirect evidence of passive electroreception deduced from the feeding experiments there is direct evidence for electrosensory feeding. Sinusoidal currents simulating the oscillating dipole field of the plankton (Wilkens *et al.*, 1997; Wojtenek *et al.*, 2001b) elicited feeding strikes in the dark preferentially for frequencies between 5 and 15 Hz; this is similar to the low-frequency signals recorded from the water flea (*Daphnia*) the preferred plankton food of Mississippi paddlefish. Extraneous electrical signals in the near-field environment at low noise levels (0.1 $\mu V\ cm^{-1}$) had little

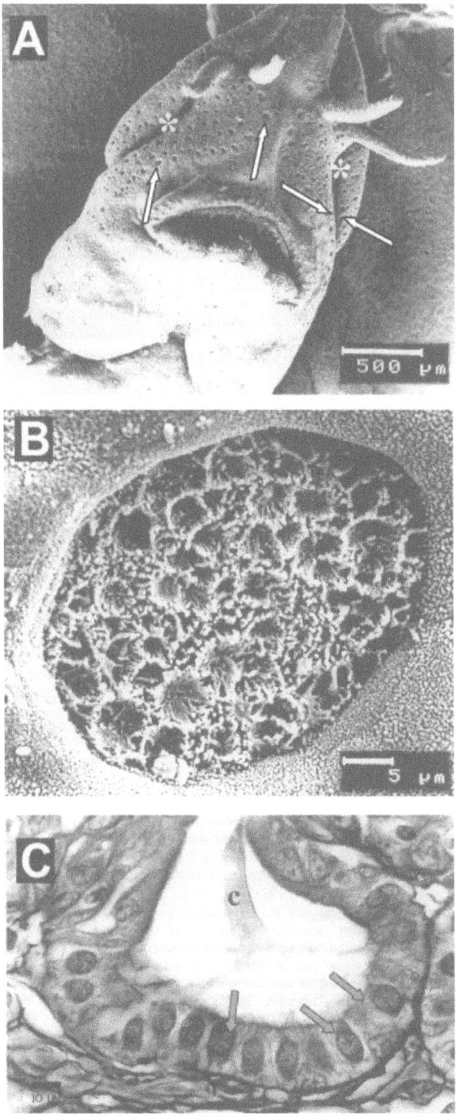

Fig. 7 Ontogeny of electroreceptors in Adriatic sturgeon (*Acipenser naccarii*). **A,** Ventral view of the rostrum of a 12 day-old larva. The external openings of the numerous electrosensory organs are arranged in regular rows. In the central region, from the apex of the rostrum to the mouth, no electroreceptors are observed. The free neuromasts of the cephalic-ventral lateral line are located in two lateral grooves (*, note this symbol is white in the Fig.). **B,** the ampullary organs consist of epidermal circular depressions layered with ciliate calyces. **C**, section of an ampullary receptor of a 180 day-old larva. The arrows indicate the darker sensorial cells. c, jelly-like substance that fills the lumen of the ampulla. After Boglione *et al.* (1999). Reprinted with permission from Blackwell Wissenschaftsverlag.

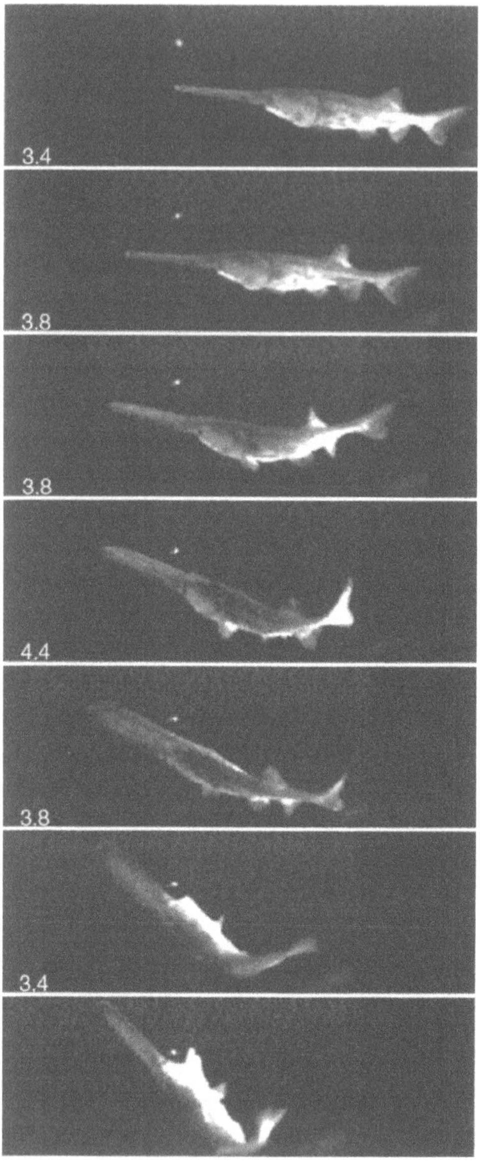

Fig. 8 Lateral view of the capture sequence of a juvenile Mississippi paddlefish (*Polyodon spathula*) equivalent to feeding in the open environment. Panels have been adjusted to align the water flea (*Daphnia*) (white dot) in the vertical sequence. Thus, the fish is viewed as moving forward normally to engulf the plankton. Swimming velocity throughout the sequence was relatively constant, as indicated numerically ($mm \cdot s^{-1}$) between panels. Velocity was calculated with reference to the horizontal position of the eye and relative time interval between frames. After Wilkens *et al.* (2002). Reprinted with permission from Elsevier Ltd.

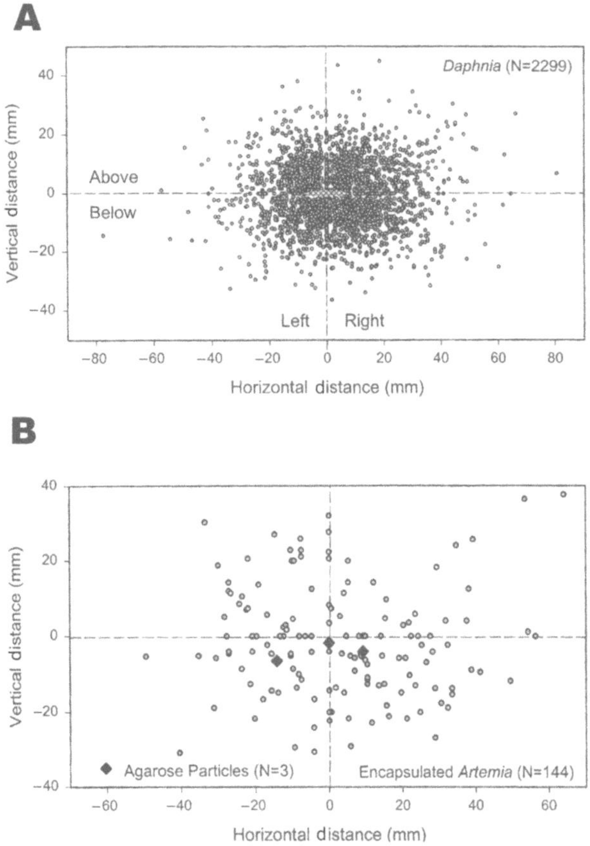

Fig. 9 Scatterplots of plankton distributions prior to capture in juvenile Mississippi paddlefish (*Polyodon spathula*) in feeding experiments. All results were obtained in the dark under infer-red illumination. (**A**) Pooled results for free-swimming water fleas (*Daphnia*). The proximate cross-sectional size of the rostrum is indicated by the hatched ellipsoid at the intersection of the horizontal and vertical axes. (**B**) Pooled results of experiments in which equal numbers of agarose-encapsulated plankton (brine shrimp (*Artemia),* circles), and control agarose particles of equal size (diamonds) were presented. Encapsulated plankton were taken at a rate nearly 50 times greater than agarose particles. After Wilkens *et al.* (2002). Reprinted with permission from Elsevier Ltd.

effect on plankton capture, higher intensities (= 1 μV cm^{-1}) limited prey capture to short distances from the rostrum, and high intensities (50 μV cm^{-1}) nearly abolished feeding (Russell *et al.*, 1999). In contrast, mechano- and chemosensory background noise did not affect feeding. There is good evidence that in early ontogeny the passive electrosense is the only sensory modality for the detection and capture of zooplanktonic prey. Interestingly, the paddlefish possess about 75,000 receptors (Nachtrieb, 1910); this is by far the largest number of electroreceptors of any passive or active electrosensory animal.

Teleostei

Osteoglossomorpha

The Osteoglossomorpha are an unusual group of teleost fishes (Kirschbaum, 2003) comprising about 250 species of freshwater fishes, most of which are in one family, the Mormyridae. The monospecific family Gymnarchidae comprises the large growing piscivorous aba (*Gymnarchus niloticus*). All species of both families are weakly electric fishes possessing an electric system. They are monophyletic and are grouped into the superfamily Mormyroidea; sometimes considered an order, Mormyriformes. Related are members of the family Notopteridae, of which fishes of the genera *Xenomystus* and *Papyrocranus* (subfamily Xenomystinae) are electroreceptive. Members of the remaining four families are all non-electroreceptive.

Xenomystinae

Electrophysiology and comparative neuroanatomy provided the initial evidence for electroreception in the Xenomystinae. In species of both genera the presence of two medullary structures, the medial octavolateral nucleus and the electrosensory lateral lobe could be confirmed (Braford, 1982a, b); both brain areas are characteristic of electroreceptive teleosts. Bullock & Northcutt (1982), using electrophysiological methods, could record evoked potentials from the electrosensory lateral lobe and the midbrain using stimulus intensities between 6 mV $\cdot$ cm^{-1} and 10 μV $\cdot$ cm^{-1}. The ampullary electroreceptor of the Xenomystinae (Fig. 10) differs from that of all other teleosts (see Fig. 1) by the presence of both an apical cilium and microvilli on its electroreceptive cell. However, there are no behavioural data concerning electroreception in the Xenomystinae (Wilkens & Hofmann, 2005).

There are also no data on the development of the species of this subfamily. However, recent unpublished data on the ontogeny of the related species the bronze featherback (*Notopterus notopterus*) (Weitkamp, pers. comm.) shows that this species passes through a larval stage. This indicates that the electrosensory species of the subfamily Xenomystinae are also characterised by larvae, which possess electroreceptors; because in all species studied so far ampullary electroreceptors develop very early during ontogeny (Denizot *et al.*, 1998; Northcutt, 2005).

Aba (Gymnarchus niloticus, Gymnarchidae)

The weakly electric aba possesses rather simple ampullary (Fig. 11A) and tuberous organs (Fig. 11C) compared to those of the closely related mormyrids. Comparison with the larval electroreceptors of mormyrids (see below) indicate that these electroreceptors represent an early stage in the evolution of these organs. Lissmann (1958) and Lissmann & Machin (1958) in their pioneering work demonstrated for the first time that this fish is able to detect objects via electroreception. Electrocommunication is not well studied in Aba. This fish breeds during the rainy season in swamps and builds a large nest from plant material. Several hundreds of large, yolky eggs are spawned, the

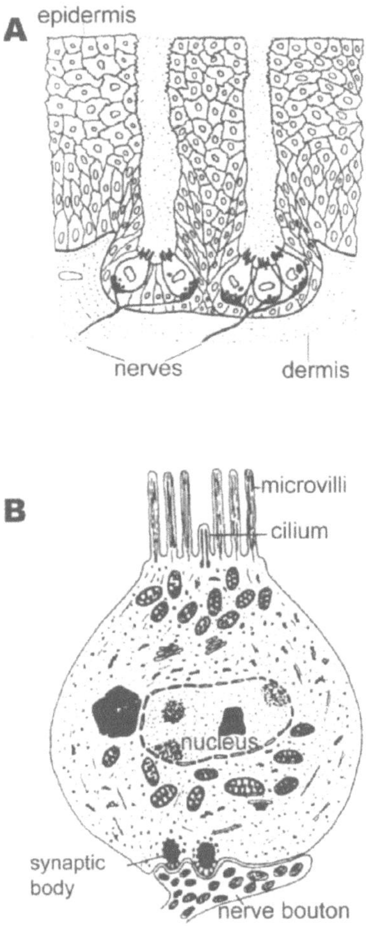

Fig. 10 (**A**) Drawing of two ampullary organs from the African knifefish (*Xenomystus nigri*). (**B**) An isolated sensory cell. Note that these sensory cells possess in contrast to other teleosts, both an apical cilium and micrivilli (compare to Fig.1). After Jørgensen & Bullock (1987). Reprinted with permission from Springer.

free embryos are guarded by the male. It is not known whether the early ontogenetic stages pass through a larval stage. The development of electroreceptors has not been described up to now.

Mormyridae

Mormyrids possess ampullary organs comprising several sensory cells and two types of tuberous organs, called knollenorgans and mormyromasts (Szabo, 1974) (Fig. 11C, D). The knollenorgan serves for electrocommunication, the mormyromast for active electrolocation. There is a vast literature on the behaviour of adult mormyrids

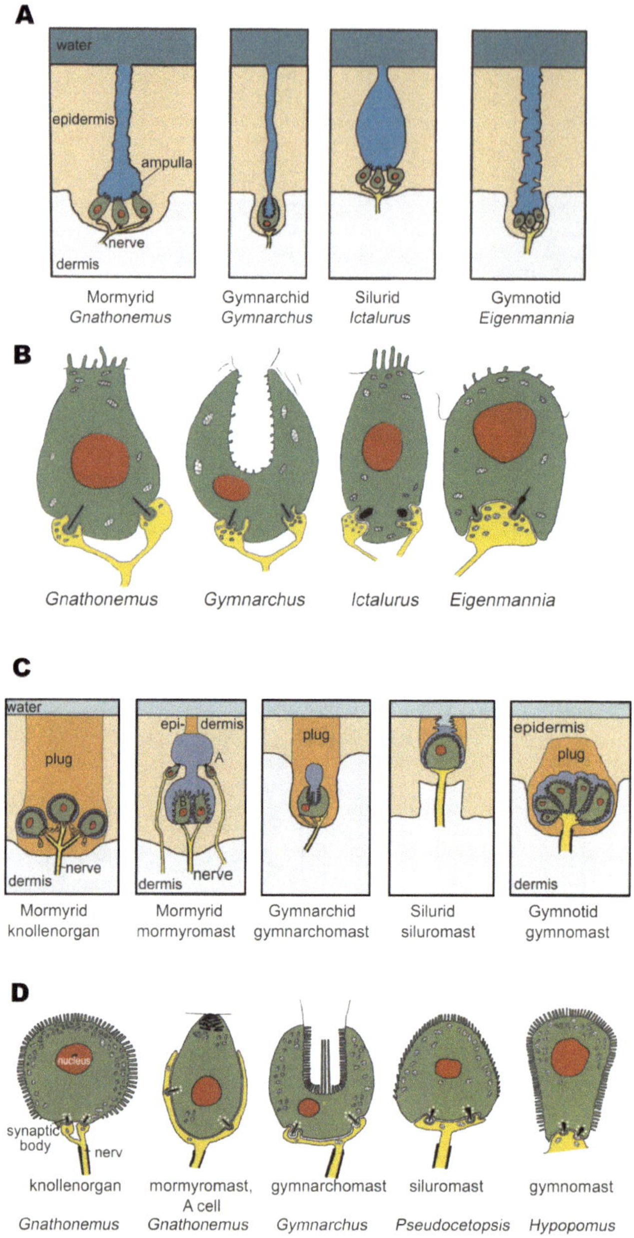

Fig. 11 (**A**) Teleost ampullary organs and their sensory cells (**B**) from representatives of four different families. (**C**) Teleost tuberous organs and their sensory cells (**D**) from representatives of four different families. After Jørgensen (2005). Reprinted with permission from Springer.

concerning these two behavioural functions (see reviews in Bullock & Heiligenberg, 1986; Kramer 1990, 1995; Moller, 1995; Bullock *et al.*, 2005; Albert & Crampton, 2006).

Successful breeding of several mormyrid species by experimentally imitating high water conditions (Kirschbaum, 1975; Kirschbaum & Schugardt, 2002; 2006) has allowed sequential studies on the ontogeny of the electric system of these fishes. Interestingly, mormyrid larvae are characterised by a complete larval electric system (Denizot *et al.*, 1998). The larval organ develops early in ontogeny and becomes functional at around day 8 post fertilisation (pf) in all species studied so far (Kirschbaum & Westby, 1975; Kirschbaum, 1977; Westby & Kirschbaum, 1977, 1978; Kramer & Postner, 1997; Baier *et al.*, 2006) and degenerates after the disappearance of larval characters (Kirschbaum, 1981).

The electroreceptors start to differentiate at around day 6 pf (Kirschbaum & Denizot, 1975) and are functional about day 8 pf. However, these electroreceptors comprise two larval tuberous electroreceptors, Types A and B (Bensouilah *et al.*, 2000; 2002) (Fig. 12A, B) and a larval mormyromast (Denizot *et al.*, 2006) (Fig. 13A). The larval electroreceptors are predominantly found in the head region (Fig. 12 C-E, Fig. 13B). In addition to these three types of larval electroreceptors, knollenorgans and ampullary organs, typical for adults, develop as well. The two kinds of larval tuberous organs (A and B) will degenerate at the onset of the juvenile stage, whereas the larval mormyromast differentiates at the same time into the more complicated mormyromast typical for adults (Denizot *et al.*, 2006). The larval mormyromast very much resembles the tuberous organ (gymnarchomast) of Aba (Fig. 11C, D).

Behavioural correlates for electroreception in mormyrid larvae only exist for *Pollimyrus adspersus*. In many original papers on reproduction and behaviour, the mormyrid species investigated was considered to be *Pollimyrus isidori* (e.g. Westby & Kirschbaum, 1977; 1978; Kirschbaum, 1987; Kramer, 1978). However, later examination of the specimens used indicated that the species investigated was always the elephant fish *P. adspersus* (see also Kramer & Postner, 1997).

During intraspecific communication mormyrids respond to the discharge of a conspecific at a fixed latency with a high probability. This electric behaviour is called preferred latency response (Kramer, 1978; Lücker & Kramer, 1981) and can be visualised in interval histograms (Fig. 14). Kramer & Postner (1997) investigating the ontogenetic development of this behaviour in the elephant fish showed that first signs of the preferred latency response occurred as early as in 11 dpf larvae (Fig. 14, Table 2). This behaviour was elicited when exposing the larvae to stimulus pulses at 5 Hz and at an amplitude of $120\,\mu V_{p\text{-}p} \cdot cm^{-1}$ ($V_{p\text{-}p,}$ amplitude measured between the maximum of the head negative and the head positive phases). Postner & Kramer (1995) determined electrosensory thresholds and tuning in larvae of the elephant fish using the stop response of the electric organ discharge to weak electric stimuli: 10 to 15 dpf larvae were sensitive to weak electric stimuli down to $10\,\mu V_{p\text{-}p} \cdot cm^{-1}$. The threshold curves for monopolar sine wave pulses indicated that the larvae were well adapted for sensing larval (monophasic) rather than adult electric discharges (triphasic). The 54- to 60 day-old larvae (pf) were

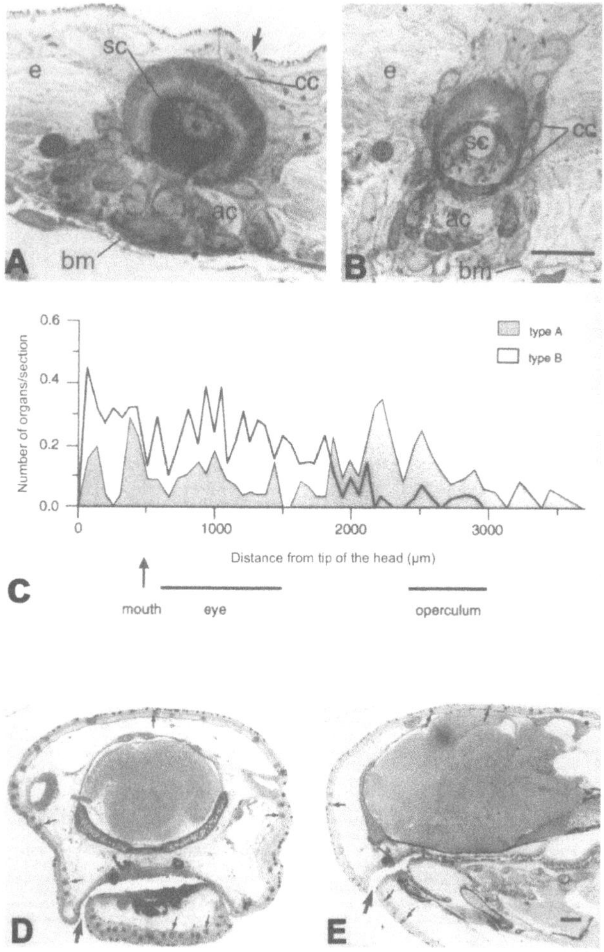

Fig. 12 Tuberous larval electroreceptors of types A (**A**) and B (**B**) in the epidermis (e) of the head of a 12 day post fertilisation (dpf) larva of the mormyrid *Mormyrus rume proboscirostris*. Each type of organ consists of a single sensory cell (sc) surrounded by microvilli of the surrounding covering cells (cc); the sensory cell sits on a platform of accessory cells (ac); the outer surface of the epidermis is indicated by a thick arrow. bm, basal membrane. Scale bar = 10 µm. **(C)** Diagram showing the distribution of the two types (A and B) of larval tuberous organs in the epidermis of a 11 dpf larva of the mormyrid elephantfish (*Campylomormyrus cassaicus*) from the tip of the head to the end of the operculum, comprising a total length of 3,700 µm, based on analysis of 2-µm-thick sections in the transverse plane. Note the difference in the distribution of the two types of organs. **D, E.** Transverse **(D)** and sagittal **(E)** section through the head of a 12 dpf larva of *M. rume proboscirostris* showing the distribution of the larval electroreceptors (small arrows) in the epidermis. Opening of the buccal cavity indicated by the large arrow. Scale bar = 10 µm. After Bensouilah *et al.* (2002). Reprinted with permission from Wiley-Liss Inc.

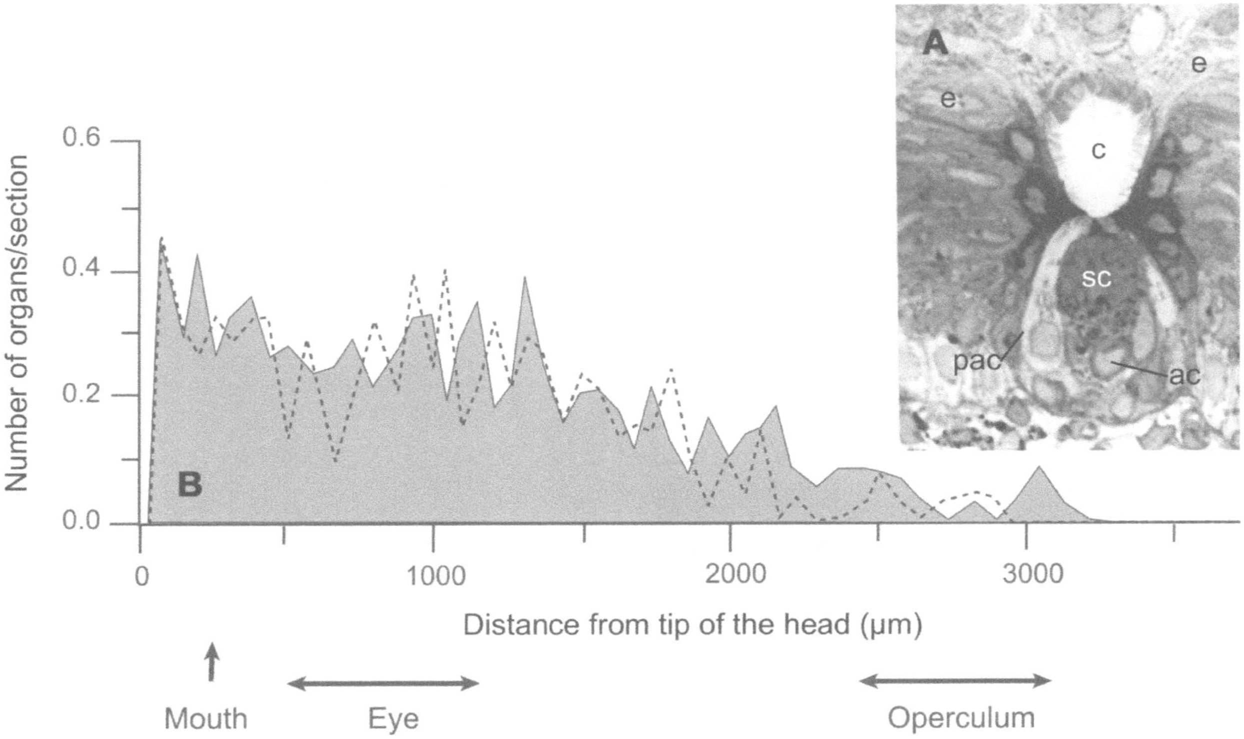

Fig. 13 Promormyromast anatomy (**A**) and distribution (**B**) on the head of a 12 day post fertilisation larva of the mormyrid *Mormyrus rume proboscirostris*. Hatched lines indicate distribution of larval electroreceptors of type B for comparison (see Fig. 12). ac, accessory cell; c, intra-epidermal cavity; e, epidermis; pac, piriform accessory cell; sc, sensory cell.

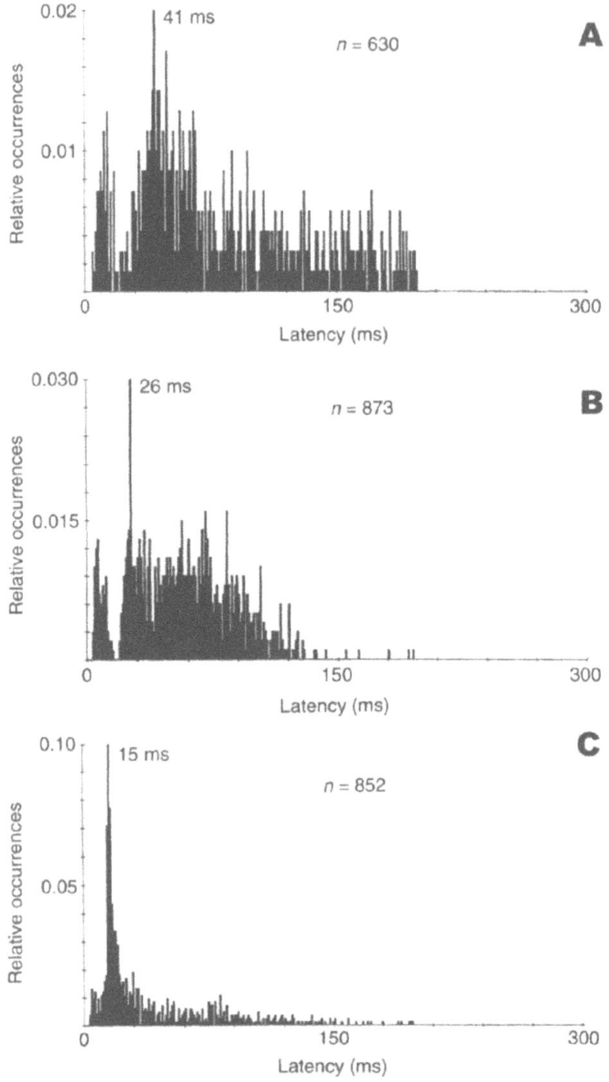

Fig. 14 During intraspecific communication mormyrids respond to the discharge of a conspecific at a fixed latency with a high probability. This electric behaviour is called preferred latency response and can be visualised in interval histograms. The ontogeny of the preferred latency response of larvae of the elephant fish (*Pollimyrus adspersus)* after artificial stimulation is shown. The latencies of the first EOD (decreasing from 41 ms (**A**) to 15 ms (**C**)) to a stimulus pulse only are illustrated; latencies shorter than 3 ms not shown for technical reasons (*n,* number of latencies for a histogram). Stimuli were biphasic, single-cycle sinusoids of 1 ms duration repeated at 5 Hz, and 120 $\mu V_{p\text{-}p}$/cm amplitude. (**A**) 11 day post fertilisation (dpf) larva (**B**) 31 dpf larva (**C**) 64 dpf larva. After Kramer & Postner (1997). Reprinted with permission from Blackwell Publishing Ltd.

Table 2 Development of preferred latency responses (shown as minima or maxima in the latency histogram, see Fig. 14) of larval electric organ discharges (EOD) of the elephant fish (*Pollimyrus adspersus*) to trains of stimulus pulses (5 Hz). *n*, number of histograms with a minimum or maximum the mean values of which are given in ms (±S.D.). Stimulus amplitude, 120 $\mu V_{p\text{-}p}$/cm. After Kramer & Postner (1997). Reprinted with permission from Blackwell Publishing Ltd.

Age (dpf)	*Minimum (ms)*	*Maximum (ms)*	*n larvae*
11	19±2 $n = 4$	36±5 $n = 11$	23
12	20 $n = 2$	32 $n = 1$	10
31	12±2 $n = 5$	19±5 $n = 8$	10
58	11±1 $n = 12$	16±1 $n = 12$	12
64	12±3 $n = 2$	18±5 $n = 10$	10

more sensitive (stimuli of 2.4 $\mu V_{p\text{-}p} \cdot cm^{-1}$) than the younger larvae, and larvae 18 mm long already displayed adult sensitivity (1 $\mu V_{p\text{-}p} \cdot cm^{-1}$).

Larvae of this length possess two functional electric organs, a so-called larval and an adult electric organ (Kirschbaum, 1977). Westby & Kirschbaum (1978) have measured the electric fields of both electric organs in a 17.5 mm larva (Fig. 15). The field strength of these electric fields are much higher than the behavioural electrosensory threshholds determined experimentally (see above). These data therefore indicate the possibility of intraspecific communication in larvae of the elephant fish.

There are no data concerning behavioural correlates for passive electroreception or active electrolocation in mormyrid larvae though the corresponding electroreceptors—ampullary and tuberous receptors—are well developed and apparently functional very early during ontogeny (Bensouilah *et al.*, 2002; Denizot *et al.*, 2006).

Siluriformes

Based on behavioural experiments catfish were quite early considered to be electroreceptive (Parker & van Heusen, 1917; Uzuka, 1934). The electroreceptors involved in electroreception are the small pit organs (ampullary organs) (Fig. 16) found over most of the head, but also on the trunk and in the epidermis of the fins. There is experimental evidence in several catfish species, which clearly shows these teleosts use their passive electrosense in prey detection, navigation, orientation, and social interactions. Recent overviews on the subject are found in Wilkens & Hofmann (2005) and Albert & Crampton (2006).

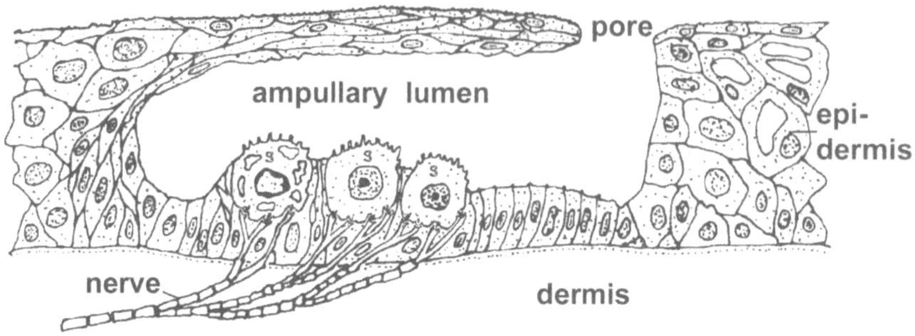

Fig. 15 An ampullary organ from the transparent catfish (*Kryptopterus bicirrhis*). After Jørgensen (2005). Reprinted with permission from Springer.

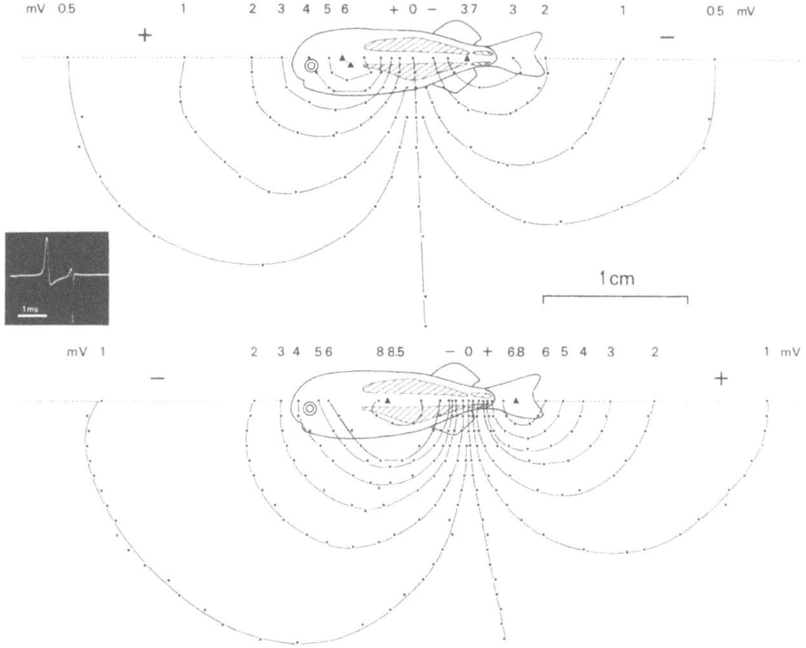

Fig. 16 Electric field of a 17.5 mm juvenile elephant fish (*Pollimyrus adspersus)* plotted separately for the larval electric organ discharge (EOD) (above) and the adult EOD (below). Equipotential lines were plotted for one half of the field only with the fish restrained on its side. Reference electrode was electrically infinitely distant. Inset shows approximately equivalent recorded amplitude of the two discharges (the larval discharge is represented by the head positive discharge) as seen between head and tail at time of measurement. Potentials are peak to peak maxima of EODs. Note differences in polarity and shapes of the fields and the different inversion points and maxima in the midline. Limits of the rostral (larval) electric organ and the caudal (adult) electric organ are shown as cross-hatched areas. After Westby & Kirschbaum (1978). Reprinted with permission from Springer.

The development of the lateral line system of the channel catfish (*Ictalurus nebulosus*) was described by Northcutt (2003). He (Northcutt, 2003) noted that first ampullary organ primordia could be recognised at hatching. On the head ampullary organ primordia begin to open at stage 49, and mature ampullary organs occur by stage 51 (Fig. 17). Behavioural correlates of passive electroreception in catfish larvae are completely lacking.

Gymnotiformes

Gymnotiforms possess ampullary organs comprising several sensory cells and tuberous organs (Fig. 11). Functionally there are two types of tuberous organs: Type T in wave species and type M in pulse species are used during electrocommunication, type P in wave species and types B and D in pulse species serve for active electrolocation (Kawasaki, 2005). Electrocommunication and active electrolocation has been intensively studied in adult gymnotiform species (see reviews in Heiligenberg, 1977; Bullock & Heiligenberg, 1986; Kramer, 1990; 1995; Moller, 1995; Bullock *et al.*, 2005; Albert & Crampton, 2006).

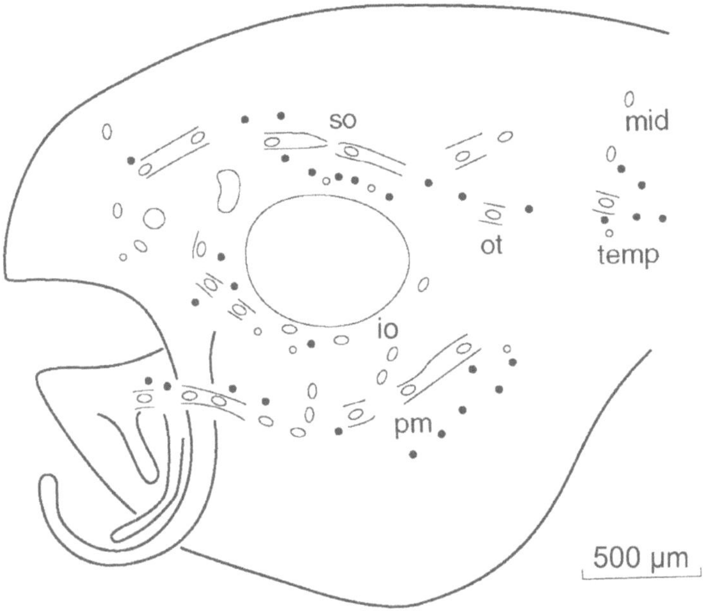

Fig. 17 A map of the flattened head ectoderm from a stage 51 channel catfish (*Ictalurus punctatus*) yolk-sac larva showing the positions of erupted neuromasts illustrated by open ovals, which are beginning to be enclosed in ectodermal grooves illustrated by the parallel lines. Mature ampullary organs illustrated by solid circles and ampullary organ primordial indicated by open circles occur immediately adjacent to the developing lines of neuromasts. io, infraorbital; mid, middle; ot, otic; pm, preoperculomandibular; so, supraorbital, temp, temporal lines. After Northcutt (2003). Reprinted with permission from Institute of Marine Research, Bergen, Norway.

As in mormyrids, experimental imitation of high water conditions allowed successful breeding of several gymnotiform species (Kirschbaum, 1975, 1979; Kirschbaum & Schugardt, 2002, 2006). Sequential studies on the ontogeny of the electric system have been performed mainly in glass knife fish (*Eigenmannia virescens* (*lineata*)) and longnosed black ghost (*Apteronotus leptorhynchus*).

Interestingly, as in mormyrids, larvae of two wave-type species of knife fish, the glass knife fish and longnosed black ghost are characterised by larval electric organs which are functional at the onset of exogenous feeding, degenerate at the beginning of the juvenile stage and are finally replaced by the adult organ (Kirschbaum & Westby, 1975; Kirschbaum 1977, 1983, 1995).

Glass knife fish (*Eigenmannia lineata*)

A complete description of the ontogeny of the lateral line receptors and the electroreceptors in the glass knife fish has been presented by Vischer and collegues (Vischer, 1989a, b, 1995; Vischer *et al.*, 1989). In 4 dpf embryos most of the lines of the neuromasts have already formed and the first electroreceptor primordia develop on the lateral edges of the neuromast lines. Differentiated tuberous organs occur for the first time on the head on day 7 pf, and ampullary organs on day 8 pf. These electroreceptor organs are found on the trunk at day 17 pf. The first electroreceptor primordia are found in the *stratum germinativum* of the epidermis and are probably induced by lateral line fibres. Therefore, lateral line receptors and both types of electroreceptors are functional at the onset of exogenous feeding in the glass knife fish.

The glass knife fish shifts its EOD frequency away from a similar frequency (frequency difference, Df) of a neighboring fish interfering with its own signal (Watanabe & Takeda, 1963); this electric behaviour is called Jamming Avoidance Response (JAR). The mixing of the fish's EOD with that of the neighbouring fish causes amplitude and phase modulations of the mixed signal and disturbs the electrolocation ability of the fish (Heiligenberg, 1977). The result of the JAR is the restoration of a high frequency difference towards the foreign signal and the re-establishment of the electrolocation ability. In case of a lower interfering frequency (negative Df) the fish increases its frequency, a higher foreign signal (positive Df) leads to a frequency decrease. The Df is retained as long as the interfering frequency is present (see Fig. 20). The JAR, expressed in Hz, comprises the Df between the fish's original frequency and the final frequency after the jamming stimulus. The JAR reaction can also be elicited by artificial stimuli. In such experimental design a fixed Df (frequency clamp) can be established: the jamming frequency follows the JAR behaviour. A maximal JAR is obtained at positive or negative Df values around 4 Hz. Adult glass knife fish produce maximal JARs up to 25 Hz. In rare cases, in particular during the ontogeny of the JAR behaviour, the fish shifts its frequency in the 'wrong' direction, i.e. towards the interfering frequency. In this case the reaction is termed "incorrect JAR" (see Fig. 18D) in contrast to the normal reaction ("correct JAR").

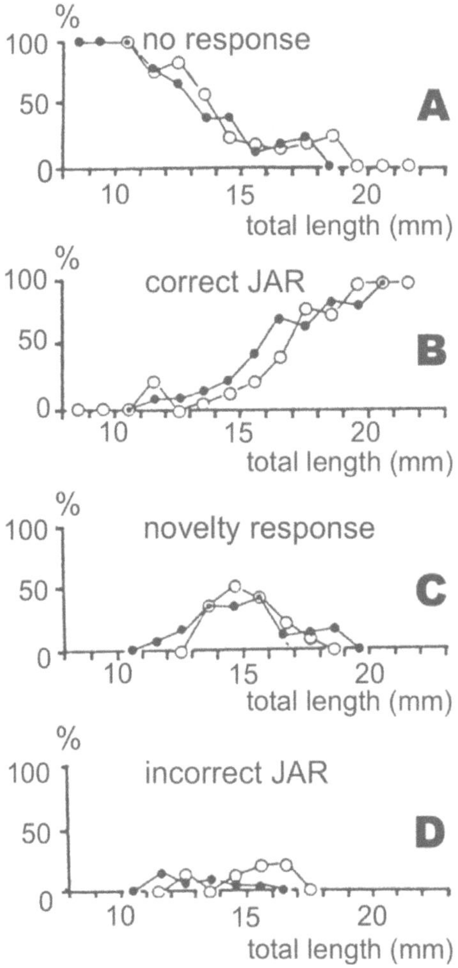

Fig. 18 Different forms of behavioural responses to jamming in larval glass knife fish (*Eigenmannia lineata)*. Behaviours depicted in **A-D** occur at particular total lengths (abscissa). Ordinates give the proportion of fish (in %) within different classes of total body length that showed the particular form of response. Open circles represent socially raised fish, filled circles represent fish raised in complete isolation. For both groups of fish, all 4 curves, one from each diagram (**A-D**), sum up to one for any point on the abscissa. Up to a size of 10 mm, no fish responded to the jamming regimen (**A**). As the number of non-responders declined with increasing body length, correct Jamming Avoidance Responses (JARs) eventually emerged in all fish (**B**). In some individuals, novelty responses (**C**) and incorrect JARs (**D**) occurred prior to the emergence of a correct JAR, at body lengths around 15 mm (**C**). Note that there is no difference between the developmental patterns of fish raised socially (total of 60) and those raised in complete isolation (total of 64). **C** incorporates only novelty responses that occurred without a concomitant JAR. After Viete & Heiligenberg (1991). Reprinted with permission from Springer.

Hagedorn *et al.* (1988) have investigated the ontogeny of the JAR in the glass knife fish and they have shown that a correct JAR emerges at a body length of 1.2-1.5 cm. The development of the JAR does not depend on social interactions, since 3 fish raised in complete isolation were able to perform a correct JAR upon their first exposure to the jamming stimulus. Viete & Heiligenberg (1991) conducted a more detailed study on this subject by testing in total 164 fish. They applied the frequency clamp situation and a Df of 4 Hz (positive or negative). 64 fish were raised in complete isolation and 100 fish together with their siblings. The individual variation of the ontogeny of the JAR was quite large. The first fish started to react to the artificial jamming stimulus at 11 mm length (Fig. 18 A), the last fish reacted at 19 mm length. Correct JARs occurred in fish varying in length between 11 and 20 mm (Fig. 18 B). At around 15 mm length, prior to the occurrence of a correct JAR, novelty responses (short frequency rises after the switch from a positive to a negative Df) occurred in a few fish (Fig. 18 C). In 6 of 164 fish, the emergence of a correct JAR followed an interim appearance of an incorrect JAR (Fig. 18 D). The time courses of the shift from the incorrect to the correct JAR in five of these six fish shows a large individual variation (Fig. 19). The incorrect JAR may last from two (Fig. 19 A and B) up to six days (Fig. 19 E). The isolated fish showed the same JAR behaviour than the socially raised fish. The authors therefore concluded that the JAR and its developmental schedule are innate.

Longnosed black ghost (Apteronotus leptorhynchus)

All species of the family Apteronotidae possess neurogenic electric organs as adult fishes (Bennett, 1971). However, in larvae, first a myogenic organ and subsequently the neurogenic organ develops. The larval myogenic organ degenerates at the onset of the juvenile stage (Kirschbaum, 1983).

Apteronotids produce JARs (Fig. 20) as in the glass knife fish, however the frequency shift is always a positive reaction, i.e. even if the jamming stimulus is higher that the fish's own frequency. In addition, the apteronotids communicate by producing short frequency rises, called chirps (Hagedorn & Heiligenberg, 1985; Dye & Heiligenberg, 1987), long lasting frequency rises called yodels (Dye, 1987) and active phase coupling (APC) (Gottschalk & Scheich, 1979). Chirps occur predominantly during a JAR reaction and yodels during the refractory period, i.e. during the establishment of the original frequency. Stams & Kirschbaum (1989) and Kirschbaum & Stams (1997) have described the ontogeny of these three electric behaviours in longnosed black ghost. They found first indications of JARs in 14 dpf larvae, i.e. short rises of frequency of 3 to 4 Hz at the beginning of the stimulus (clamped frequency, Df +4 Hz) of 3 minutes duration. Some fish in the beginning showed negative starting reactions, as decribed for larvae of the glass knife fish. Starting reactions occurred in 16 dpf longnosed black ghost consisting of frequency rises of about 4 Hz lasting for about 60 seconds (Fig. 21). The first clear JAR reactions were found in 29 dpf fish comprising frequency rises of about 10 Hz for the whole stimulus period of 3 minutes (Fig. 22). The maximum JAR values increased with age and reached values up to 36 Hz in adult fish (Fig. 23) and

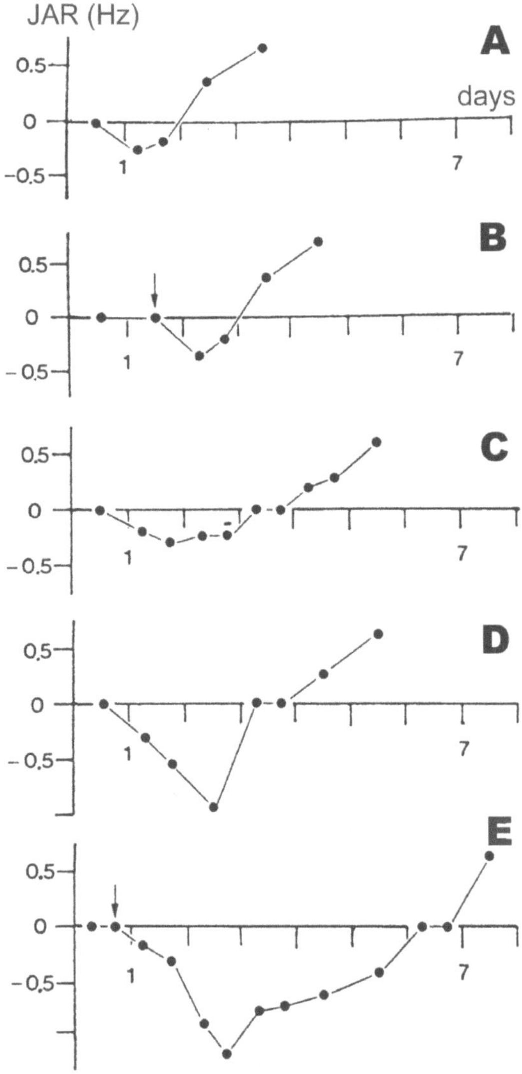

Fig. 19 Time courses of Jamming Avoidance Responses (JARs) development in 5 individuals of larval glass knife fish (*Eigenmannia lineata*) (**A**-**E**) that all showed incorrect JARs (negative values on ordinates) (see Fig. 18**D**) before acquiring a correct (positive) JAR. Fish were tested at least once a day. Whereas the fish in **A** showed an incorrect JAR for only one day, the fish in **E** maintained an incorrect JAR for 5 days. Days indicates the time after the start of the JAR response. Note that the magnitude of an incorrect JAR is gradually reduced as the fish nears the point of achieving a correct JAR, sometimes yielding no consistent response, i.e. a value of zero, at the point of transition. Arrows indicate instances of novelty responses. After Viete & Heiligenberg (1991). Reprinted with permission from Springer.

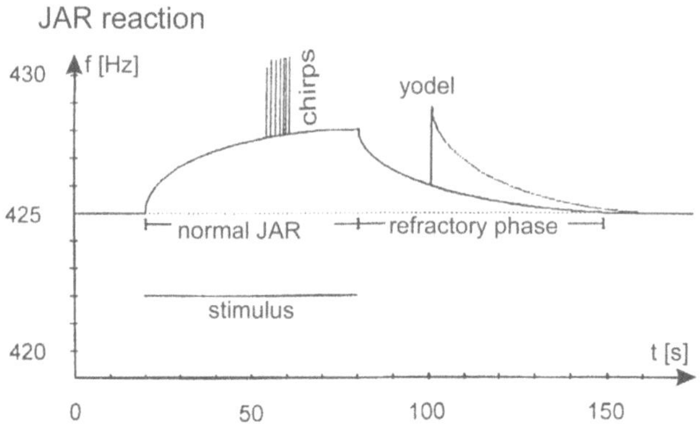

Fig. 20 Schematic drawing of the Jamming Avoidance Response (JAR) and associated electric behaviours in longnosed black ghost (*Apteronotus leptorhynchus*); chirps and yodels do not always occur. The frequency increase is retained as long as the stimulus (60 s in this case) is presented. The frequency difference (Df) between the stimulus and the fish's initial frequency was set to a Df of -3 Hz. The JAR expressed in Hz at the end of the experiment amounted to 3 Hz. After Kirschbaum & Stams (1997). Reprinted with permission from VNW Verlag Natur und Wissenschaft.

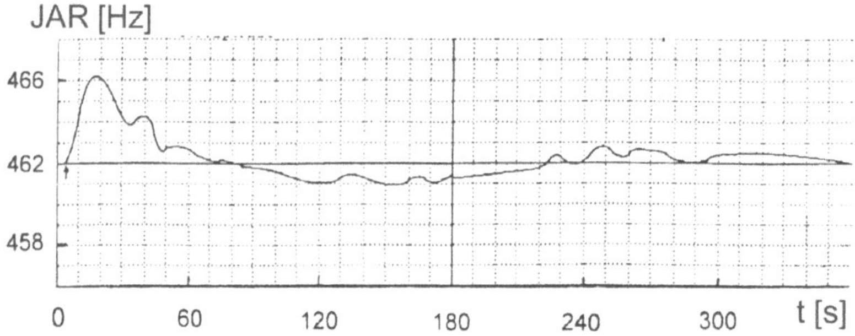

Fig. 21 Starting reaction (increase of frequency of about 3 Hz for about 60 s) of Jamming Avoidance Response (JAR) of a 16 day post fertilisation larva (12 mm long) of longnosed black ghost *(Apteronotus leptorhynchus)*. Frequency difference between the stimulus and the fish's frequency was +4 Hz, stimulus (clamped frequency, stimulus frequency in the beginning 466 Hz) duration 3 min; stimulus intensity 40 mV_{norm}. The arrow indicates the beginning of the stimulus. The zero line (462 Hz) represents the fish's original frequency. After Kirschbaum & Stams (1997). Reprinted with permission from VNW Verlag Natur und Wissenschaft.

depended on the stimulus intensity. Yodels occurred for the first time in 27 dpf larvae; the characteristics of this frequency rise were identical between fish of different age (Fig. 24) and even identical to that of adult fish. Chirps could be found for the first time in 18 dpf fish, i.e. before the establishment of a stable JAR reaction. The time course

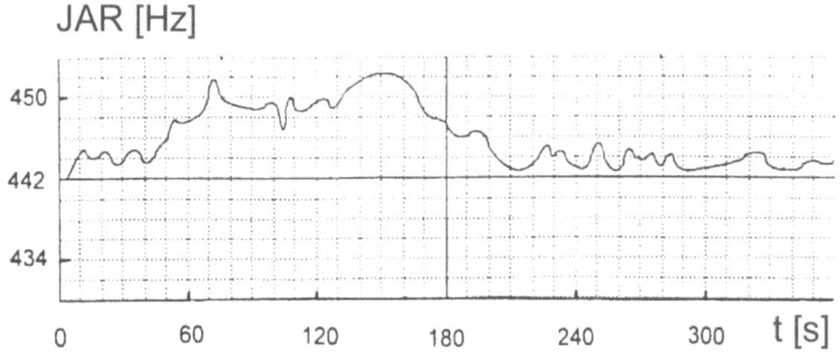

Fig. 22 First Jamming Avoidance Response (JAR) comprising a frequency increase for the whole duration of the stimulus (446 Hz in the beginning, +4Hz above the fish's original frequency, clamped frequency) of 3 min of a 29 day post fertilisation larva (18 mm long) of longnosed black ghost *(Apteronotus leptorhynchus)*; JAR maximum 10 Hz. After Kirschbaum & Stams (1997). Reprinted with permission from VNW Verlag Natur und Wissenschaft.

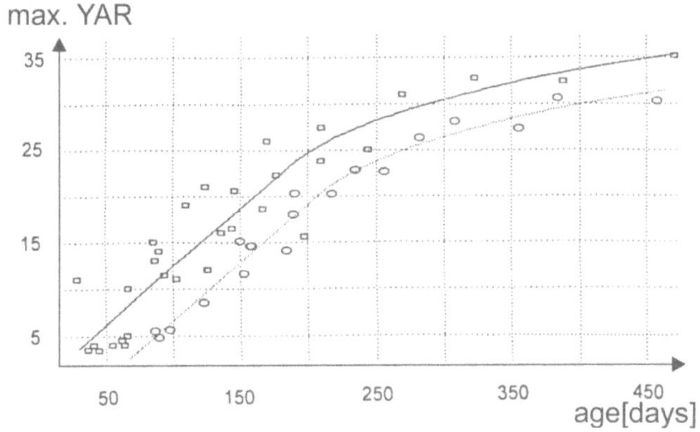

Fig. 23 Jamming Avoidance Response (JAR) maximum as a function of age and stimulus strength in larvae and juveniles of longnosed black ghost *(Apteronotus leptorhynchus)*. Clamped frequency difference +4 Hz above the fish's frequency. Solid line: stimulus strength 200 mV/cm standardised (□, 31 animals). Hatched line: stimulus strength 120 mV/cm standardised, (○, 19 animals). After Kirschbaum & Stams (1997). Reprinted with permission from VNW Verlag Natur und Wissenschaft.

and the amplitude of chirps did not change during ontogeny: a 14 mm larva and 70 mm juvenile produce nearly identical chirps (Fig. 25). Interestingly, the highest number of chirps was always produced by the largest fish when a group of fish were kept together. These larvae were physically isolated but not electrically. The first active phase coupling behaviour occurred in 36 day AF fish. This detailed study showed that the various elements of electrocommunication were already present in young larvae. In longnosed black ghost however, no data have been published concerning electroreceptor ontogeny.

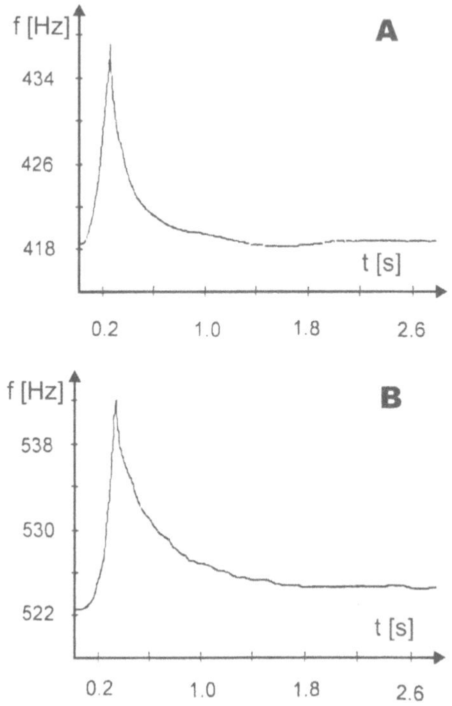

Fig. 24 Time course of the long lasting frequency increase (yodel) of two longnosed black ghost *(Apteronotus leptorhynchus)* of different age comprising a 27 day post fertilisation larva (13 mm long) (**A**) and of a 90 day-old juvenile (50 mm long) (**B**), respectively. In both cases the frequency increase above the fish's original frequency amounts to16 Hz and the decrease down to the original frequency level are nearly identical. Therefore, there is no change in the time course of the yodels during ontogeny and this kind of electrical behaviour is present very early during development. After Kirschbaum & Stams (1997). Reprinted with permission from VNW Verlag Natur und Wissenschaft.

FUTURE PERSPECTIVES

As shown above the knowledge about electroreception in fish larvae is still very limited. In particular the behavioural correlates are poorly investigated up to now. However, the available histological ontogenetic data show that electroreception is a very important sensory system in fish larvae.

Acknowledgements

We thank the editor and an unknown referee for helpful comments on the manuscript and Leonore Fischer and Alexandra Sakowsky for help concerning the technical improvement of the manuscript.

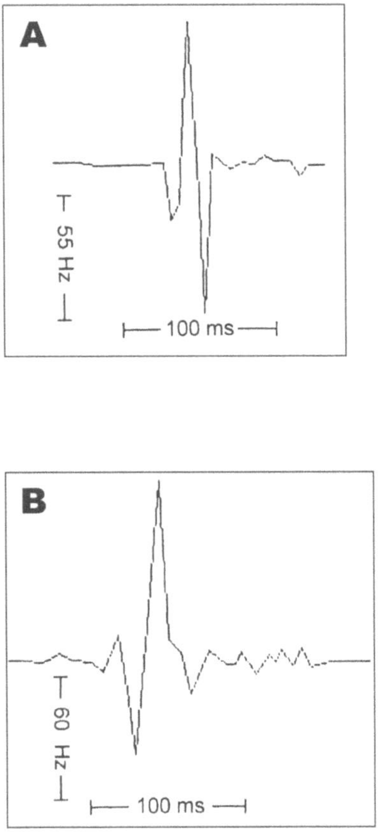

Fig. 25 Time course of two short frequency rises and drops (chirp) of two longnosed black ghost *(Apteronotus leptorhynchus)* of different length comprising a 13 mm larva (**A**) and a 30 mm juvenile (**B**), respectively. The maximum deviation from the fish's original frequency (baseline) is about 50 to 60 Hz in both cases. Therefore, the chirps produced at different stages of ontogeny are very similar. Clamped frequency, +4 Hz above the fish's frequency; stimulus strength 150 mV_{norm}. After Kirschbaum & Stams (1997). Reprinted with permission from VNW Verlag Natur und Wissenschaft.

References

Albert JA, Crampton WGR (2006) Electroreception and electrogenesis. In: Evans DH, Claiborne JB (eds) The Physiology of Fishes, 3rd edition, CRC Press, Taylor and Francis, New York, p 431-472

Balon E (1975) Terminology of intervals in fish development. J Fish Res Bd Can 32: 1663-1670

Balon E (2003) Coelacanthiformes. In: Thoney DA, Loiselle PV, Schlager N (eds) Grzimeks Animal Life Enzyclopedia, Vol 4, Fishes I. 2nd Edition, Farmington Hills: Gale Group, p 189-196

Baier B, Lamml M, Kramer B (2006) Ontogeny of the electric organ discharge in two parapatric species of the dwarf stonebasher, *Pollimyrus castelnaui* and *P. marianne* (Mormyridae, Teleostei). Acta Zool 87: 209-214

Bennett MVL (1971) Electric organs. In: Hoar WS, Randall DJ (eds) Fish Physiology, Vol 5, Academic Press, New York, p 347-491

Bensouilah M, Roesler R, Kirschbaum F, Schugardt C, Denizot J-P (2000) Électrorécepteurs larvaires dans l'épiderme des alevins de *Campylomormyrus cassaicus* et de *Pollimyrus isidori* (Momyridae, Osteoglossomorpha). Cybium 24: 41-47

Bensouilah M, Roesler R, Kirschbaum F, Schugardt C, Denizot J-P (2002) Larval electroreceptors in the epidermis of mormyrid fish. I. Tuberous organs of types A and B. J Comp Neurol 4: 309-322.

Birstein VJ, Bemis WE (1997) How many species are there within the genus *Acipenser*? Environ Biol Fishes 48: 157-163

Bodznick D, Northcutt RG (1981) Electroreception in lampreys; evidence that the earliest vertebrates were electroreceptive. Science 212: 465-467

Bodznick D, Montgomery JC (2005) The physiology of low-frequency electrosensory system. In: Bullock TH, Hopkins CD, Popper AN, Fay RR (eds) Electroreception. Springer Science and Business Media Inc, New York, p 132-153

Bodznick D, Preston DG (1983) Physiological characterization of electroreceptors in the lampreys *Ichthyomyzon unicuspis* and *Petromyzon marinus*. J Comp Physiol 152: 209-217

Boglione C, Bronzi P, Cataldi E, Serra S, Gagliardi F, Cataudella S (1999) Aspects of early development in the Adriatic sturgeon *Acipenser naccarii*, Proceedings of the 3rd International Symposium on Sturgeon. J Appl Ichthyol 15: 207-213

Braford MR (1982a) African, but not Asian, notopterid fishes are electroreceptive: Evidence from brain characters. Neurosci Lett 32: 35-39

Braford MR (1982b) Electroreceptive and mechanoreceptive afferents of the *torus semicircularis* in the notopterid fish, *Xenomystus nigri*. Soc Neurosci Abstr 8: 764

Bretschneider F, Peters RC (1992) Transduction and transmission in ampullary electroreceptors of catfish. J Comp Physiol A103: 245-252

Bullock TH, Northcutt G (1982) A new electroreceptive teleost: *Xenomystus nigri* (Osteoglossiformes: Notopteridae). J Comp Physiol A 148: 345-352

Bullock TH, Heiligenberg W (eds) (1986) Electroreception. John Wiley & Sons, New York

Bullock TH, Bodznick DA, Northcutt RG (1983) The phylogenic distribution of electroreception: evidence for convergent evolution of a primitive vertebrate sense modality. Brain Res 287: 25-46

Bullock TH, Hopkins CD, Popper AN, Fay RR (eds) (2005) Electroreception. Springer Science and Business Media, Inc, New York

Carvalho M (2003a) Ceratodontiformes. In: Thoney DA, Loiselle PV, Schlager N (eds) Grzimeks Animal Life Enzyclopedia, Vol 4, Fishes I. 2nd Edition, Farmington Hills: Gale Group, p 197-200

Carvalho M (2003b) Lepidosireniformes. In: Thoney DA, Loiselle PV, Schlager N (eds) Grzimeks Animal Life Enzyclopedia, Vol 4, Fishes I. 2nd Edition, Farmington Hills: Gale Group, p 201-207

Carvalho M, Schelly R (2003) Polypteriformes. In: Thoney DA, Loiselle PV, Schlager N (eds) Grzimeks Animal Life Enzyclopedia, Vol 4, Fishes I. 2nd Edition, Farmington Hills: Gale Group, p 209-212

Denizot JP, Kirschbaum F, Schugardt C, Bensouilah M (1998) Larval electroreceptors indicate a larval electric system in mormyrids. Neurosci Lett 241: 103-106

Denizot JP, Bensouilah M, Roesler R, Schugardt C, Kirschbaum F (2006) Larval Electroreceptors in the Epidermis of Mormyrid Fish: II. The Promormyromast. J Comp Neurol 501: 810-823

Dye J (1987) Dynamics and stimulus-dependence of pacemaker control during behavioural modulations in the weakly electric fish, *Apteronotus*. J Comp Physiol 161: 175-185

Dye J, Heiligenberg W (1987) Intercellular recording in the medullary pacemaker nucleus of the weakly electric fish, *Apteronotus*, during modulatory behaviours. J Comp Physiol 161: 187-200

Fiedler K (1991) Fische. Kaestner A (ed) In: Lehrbuch der Speziellen Zoologie, Vol II, Wirbeltiere, Starck D (ed), Gustav Fischer Verlag, Jena, 2 Teil

Fuiman LA (1984) Ostariophysi: Development and relationships. In: Moser HG, Richards WJ, Cohen DM, Fahay MP, Kendall Jr AW, Richardson SL (eds) Ontogeny and Systematics of Fishes. Allen Press Inc, Lawrence, USA, p 116-137

Gottschalk B, Scheich H (1979) Phase sensitivity and phase coupling: Common mechanisms for communication behaviours in gymnotid wave and pulse species. Behav Ecol Sociobiol 4: 395-408

Hagedorn M, Heiligenberg W (1985) Court and spark: Electric signals in the courtship and mating of gymnotoid fish. Anim Behav 33: 254-265

Hagedorn M, Heiligenberg W, Carr CE (1988) The development of the Jamming Avoidance Response in the weakly electric fish *Eigenmannia*. Brain Behav Evol 31: 161-169

Heiligenberg W (1977) Principles of Electrolocation and Jamming Avoidance. Studies of Brain Function. Vol 1, Springer-Verlag, Berlin

Jørgensen JM (1982) Fine structure of the ampullary organ of the bichir *Polypterus senegalus* Cuvier, 1929 (Pisces: Brachyopterygii) with some notes on the phylogenetic development of electroreceptors. Acta Zool 63: 211-217

Jørgensen JM (2005) Morphology of electroreceptive sensory organs. In: Bullock TH, Hopkins CD, Popper AN, Fay RR (eds) Electroreception. Springer Science and Business Media, Inc, New York, p 47-67

Jørgensen JM, Bullock TH (1987) Organization of the ampullary organs of the African knife fish *Xenomystus nigri* (Teleostei: Notopteridae) J Neurocyt 16: 311-315

Jørgensen JM, Flock A, Wersäll J (1972) The Lorenzinian ampullae of *Polyodon spathula*. Zeitschr Zellforsch 130: 362-377

Kawasaki M (2005) Physiology of tuberous electrosensory systems. In: Bullock TH, Hopkins CD, Popper AN, Fay RR (eds) Electroreception. Springer Science and Business Media Inc, New York, p 154-194

Kendall Jr AW, Ahlström EH, Moser HG (1984) Early life history stages of fishes and their characteristics. In: Moser HG, Richards WJ, Cohen DM, Fahay MP, Kendall Jr AW, Richardson SL (eds) Ontogeny and Systematics of Fishes. Allen Press Inc, Lawrence, USA, p 11-22

Kirschbaum F (1975) Environmental factors control the periodical reproduction of tropical electric fish. Experientia 31: 1159-1160

Kirschbaum F (1977) Electric organ ontogeny: Distinct larval organ precedes the adult organ in weakly electric fish. Naturwiss 64: 387-388

Kirschbaum F (1979) Reproduction of the weakly electric fish *Eigenmannia virescens* (Rhamphichthyidae, Teleostei) in captivity. I. Control of gonadal recrudescence and regression by environmental factors. Behav Ecol Sociobiol 4: 331-355

Kirschbaum F (1981) Ontogeny of both larval electric organ and electromotoneurones in *Pollimyrus isidori* (Mormyridae, Teleostei). In: Szabo T, Czech G (eds) Advances in Physiological Sciences, Vol. 31, Sensory Physiology of Aquatic Lower Vertebrates, Pergamon Press, Budapest, Hungary, p 129-157

Kirschbaum F (1983) Myogenic electric organ precedes the neurogenic organ in apteronotid fish. Naturwissenschaften 70: 305-307

Kirschbaum F (1987) Reproduction and development of the weakly electric fish *Pollimyrus isidori* (Mormyridae, Teleostei) in captivity. Environ Biol Fishes 20: 11-31

Kirschbaum F (1995) Reproduction and development in mormyriform and gymnotiform fishes. In: Moller P (ed) Electric Fishes - History and Behavior. Fish and Fisheries Series, Chapman and Hall, London, p 267-301

Kirschbaum F (2003) Osteoglossiformes. In: Thoney DA, Loiselle PV, Schlager N (eds) Grzimeks Animal Life Enzyclopedia, Vol 4, Fishes I. 2nd Edition, Farmington Hills: Gale Group, p 231-241

Kirschbaum F, Denizot JP (1975) Sur la différenciation des électrorécepteurs chez *Marcusenius* sp. (Mormyridés) et *Eigenmannia virescens* (Gymnotidés) poissons électriques à faible décharge. C R Acad Sc Paris 281, D, 419-422

Kirschbaum F, Schugardt C (2002) Reproductive strategies and developmental aspects in mormyrid and Gymnotiform fishes. J Physiol 96: 557-566

Kirschbaum F, Schugardt C (2006) Fortpflanzungsstrategien und entwicklungsbiologische Aspekte bei südamerikanischen Messerfischen (Gymnotiformes) und afrikanischen Nilhechten (Mormyridae) - Vergleichende Betrachtungen, Reproductive strategies and developmental features in South-American knifefishes (Gymnotiformes) and African elephantnoses (Mormyridae) - Comparative aspects. In: Greven H, Riehl R (eds) Biologie der Aquarienfische, Tetra Verlag GmbH, Germany, p 81-116.

Kirschbaum F, Stams H (1997) Jamming Avoidance Response (JAR) und assoziierte elektrische Verhaltensweisen bei dem schwach elektrischen Messerfisch *Apteronotus leptorhynchus*: Adultzustand und Ontogenese. Z Fischkunde 4: 163-200

Kirschbaum F, Westby GWM (1975) Development of the electric discharge in Mormyrid and Gymnotid fish (*Marcusenius* sp. and *Eigenmannia virescens*). Experientia 31: 1290-1293

Kramer B (1978) Spontaneous discharge rhythms and social signalling in the weakly electric fish *Pollimyrus isidori* (Cuvier et Valenciennes) (Mormyridae, Teleostei). Behav Ecol Sociobiol 4: 61-74

Kramer B (1990) Electrocommunication in Teleost Fishes: Behaviour and Experiments. Springer Verlag, Berlin

Kramer B (1995) Electroreception and Communication in Fishes. Georg Fischer Verlag, Stuttgart

Kramer B, Postner M (1997) Development of electrical signalling in larvae of the African fish, *Pollimyrus adspersus* (Mormyridae, Teleostei): the patterns of interdischarge intervals. J Zool 243: 329-340

Lissmann HW (1958) On the function and evolution of electric organs in fish. J Exp Biol 35: 156-191

Lissmann HW, Machin KE (1958) The mechanism of object location in *Gymnarchus niloticus* and similar fish. J Exp Biol 35: 451-486

Lücker H, Kramer B (1981) Development of a sex difference in the preferred latency response in the weakly electric fish *Pollimyrus isidori* (Cuvier et Valenciennes) (Mormyridae, Teleostei). Behav Ecol Sociobiol 9: 103-109

Moller P (1995) Electric Fishes: History and Behaviour. Chapman and Hall, London

Moller P, Serrier J (1986) Communication in weakly electric fish *Gnatonemus petersii* (Mormyridae). Interaction of electric organ discharge activities of two fish. Anim Behav 21: 501-512.

Nachtrieb HF (1910) The primitive pores of *Polyodon spathula* (Walbaum). J Exp Biol 9: 455-468

Nelson JS (2006) Fishes of the World. 4th Edition. John Wiley & Sons, Inc, Hoboken, New Jersey

New JG, Bodznick D (1985) Segregation of electroreceptive and mechanoreceptive lateral line afferents in the hindbrain of chondrostean fishes. Brain Res 336: 89-98

Northcutt RG (1986) Electroreception in nonteleost bony fishes. In: Bullock TH, Heiligenberg W (eds) Electroreception. John Whiley and Sons, New York, p 257-285

Northcutt RG (2003) Development of the lateral line system in the channel catfish. In: Browman HI, Skiftesvik AB (eds) The Big Fish Bang. Proceedings of the 26th Annual Larval Fish Conference, Institute of Marine Research, Bergen, Norway, p 137-159

Northcutt RG (2005) Ontogeny of electroreceptors and their neural circuitry. In: Bullock TH, Hopkins CD, Popper AN, Fay RR (eds) Electroreception. Springer Science and Business Media Inc, New York, p 112-131

Parker GH, van Heusen AP (1917) The response of the catfish, *Amiurus nebulosus*, to metallic and non-metallic rods. Am J Physiol 44: 405-420

Pehrson T (1949) The ontogeny of the lateral line system in the head of dipnoans. Acta Zool 30: 153-182

Peters RC, Bretschneider F (1972) Electric phenomena in the habitat of the catfish *Ictalurus nebulosus* LeS. J Comp Pathol 81: 345-362

Postner M, Kramer B (1995) Electrosensory thresholds in larvae of the weakly electric fish *Pollimyrus adspersus* (Mormyridae, Teleostei) during ontogeny. J Exp Biol 198: 783-791

Ronan MC, Bodznick D (1986) End buds: non-ampullary electroreceptors in adult lampreys. J Comp Physiol A 158: 9-15

Roth A, Tscharntke H (1976) Ultrastructure of the ampullary electroreceptors in lungfish and brachyopterygii. Cell Tissue Res 173: 95-108

Russell DF, Wilkens LA, Moss F (1999) Use of behavioural stochastic resonance by paddlefish for feeding. Nature 402: 291-294

Stams H, Kirschbaum F (1989) Adult and larvae of *Apteronotus leptorhynchus* show highly sophisticated phase coupling behaviour. In: Erler J, Pflüger HF, Todt D (eds) Neural Mechanisms of Behaviour, George Thieme, Stuttgart, p 97

Steven DM (1951) Sensory cells and pigment distribution in the tail of the ammocoete. Quart J Microscop Sci 92: 233-247

Szabo T (1974) Anatomy of the specialised lateral line organs of electroreception. In: Fessard A (ed). Handbook of Sensory Physiology III/3, Springer-Verlag, Berlin, p 13-58

Uzuka K (1934) Some notes on the behaviour of the catfish, *Parasilurus asotus*, as seen through the responses to weak electric current. Sci Rep Tohoku Imp Univ Ser 4 8: 369-381

Viete S, Heiligenberg W (1991) The development of the Jamming Avoidance Response (JAR) in *Eigenmannia*: An innate behaviour indeed. J Comp Physiol A 169: 15-23

Vischer HA (1989a) The development of lateral-line receptors in *Eigenmannia* (Teleostei, Gymnotiformes). I. The mechanoreceptive lateral-line system. Brain Behav Evol 33: 205-222

Vischer HA (1989b) The development of lateral-line receptors in *Eigenmannia* (Teleostei, Gymnotiformes). II. The electroreceptive lateral-line system. Brain Behav Evol 33: 223-236

Vischer HA (1995) Electroreceptor development in the electric fish *Eigenmannia*: a histological and ultrastructural study. J Comp Neurol 360: 81-100

Vischer HA, Lannoo MJ, Heiligenberg W (1989) Development of the electrosensory nervous in *Eigenmannia* (Gymnotiformes). I. The peripheral nervous system. J Comp Neurol 290: 16-40

von der Emde G (1999) Active electrolocation of objects in weakly electric fish, *Gnathonemus petersii*. J Exp Biol 202: 1205-1215

Watanabe A, Takeda K (1963) The change of discharge frequency by A.C. stimulus in a weak electric fish. J Exp Biol 40: 57-66

Westby GWM, Kirschbaum F (1977) Emergence and development of the electric organ discharge in the mormyrid fish *Pollimyrus isidori* I. The larval discharge. J Comp Physiol 122: 251-271

Westby GWM, Kirschbaum F (1978) Emergence and development of the electric organ discharge in the mormyrid fish *Pollimyrus isidori*. II. Replacement of the larval by the adult discharge. J Comp Physiol 127: 45-59

Whitear M, Lane EB (1983) Bar synapses in the end buds of lamprey skin. Cell Tissue Res 216: 445-448

Wilkens LA, Hofman MH (2005) Behavior of animals with passive, low-frequency electrosensory systems. In: Bullock TH, Hopkins CD, Popper AN, Fay RR (eds) Electroreception. Springer Science and Business Media Inc, New York, p 229-263

Wilkens LA, Cox M, Russell DF (1994) Evidence for electrosensitivity in planktivorous feeding in the paddlefish *Polyodon spathula*. Am Zool 34: 36A

Wilkens LA, Russell DF, Pei C, Gurgens C (1997) The paddlefish rostrum functions as an electrosensory antenna in plankton feeding. Proc R Soc Lond B264: 1723-1729

Wilkens LA, Wettring BA, Wagner E, Wojtenek W, Russell DF (2001) Prey detection in selective plankton feeding by the paddlefish: is the electric sense sufficient. J Exp Biol 204: 1381-1389

Wilkens LA, Hofmann MH, Wojtenek W (2002) The electric sense of the paddlefish: a passive system for the detection and capture of zooplankton prey. J Physiol 96: 363-377

Wojtenek W, Hofmann MH, Wilkens LA (2001a) Primary afferent electrosensory neurons represent paddlefish natural prey. Neurocomputing 38: 451-458

Wojtenek W, Pei C, Wilkens LA (2001b) Paddlefish strike at artificial dipoles simulating the weak electric fields of plankton prey. J Exp Biol 204: 1391-1399

Zakon HH (1986) The electroreceptive periphery. In: Bullock TH, Heiligenberg W (eds) Electroreception. John Whiley and Sons, New York, p 103-156

Zupanc KHG, Bullock TH (2005) From electrogenesis to electroreception: An overview. In: Bullock TH, Hopkins CD, Popper AN, Fay RR (eds) Electroreception. Springer Science and Business Media Inc, New York, p 5-46

CHAPTER

14

Magnetoreception

Krzysztof Formicki

INTRODUCTION

In recent decades, in view of difficulties encountered in attempts to explain a number of biological phenomena, e.g. orientation of birds and fishes and their migrations, more and more attention has been paid to effects of magnetic fields on living organisms. Magnetic fields may affect behavioural responses and a number of processes taking place in the body of an organism. Over millions of years of their evolution, living organisms have been having to develop adaptations to the changing environment. From today's perspective, it would be difficult to propose that the magnetic field, an important component of the environment, would be ignored in adaptive processes, and that living organisms would not attempt to take advantage of it to meet their needs. This is inherently related to the reception of magnetic field. Magnetoreception is an ability of an organism to perceive, and respond to, effects of the external magnetic field, changes in its direction, intensity, and gradient, the ability being manifested as specific responses. To understand the nature and mechanism of the process, it is necessary to possess knowledge on the principles of magnetic stimulus reception, on location of the appropriate sensory organs, on their structure and links with the nervous system.

Author's address: University of Agriculture in Szczecin, Department of Fish Anatomy and Embryology 4 K. Królewicza St. 71-550 Szczecin, Poland.
E-mail: kformicki@fish.ar.szczecin.pl

Many organisms are receptive and respond to magnetic stimuli, including bacteria, molluscs, arthropods, and vertebrate taxa such as fish, birds, turtles, and whales (Frankel *et al.*, 1979; Tenforde, 1979; Kirschvink *et al.*, 1985, 2001; Blank, 1993; Wiltschko & Wiltschko, 1995, 2005; Kirschvink, 1997).

Research on magnetoreception in fish has been focused mainly on those species that undertake long-distance migrations, as it has been assumed that they must possess particularly well-developed sensory organs geared at perceiving magnetic field and at orientation.

Early studies on the behaviour of eels (Anguillidae) in magnetic fields involved experiments on directional orientation of yellow eels, a stationary freshwater form of the European eel (*Anguilla anguilla*) (Branover *et al.*, 1971; Tesch & Lelek, 1973 a, b). Subsequent experiments (Tesch, 1974 a; Tesch *et al.*, 1992) on the silver (migratory) form of European eel confirmed the directional preferences in the species. Reversal of the horizontal component of the geomagnetic field resulted in a change of the direction of the eel's movement (Karlsson, 1985). Souza *et al.* (1988) who studied the American eel (*Anguilla rostrata*) both in the natural magnetic field and in a field of doubled strength showed most of the fish to prefer moving north. Reversing the direction of the magnetic field resulted in the eels choosing to move south. Magnetosensitivity was also detected in the Japanese eel (*Anguilla japonica*) and was manifested as bradycardia (Nishi *et al.*, 2004). The eels that were made anosmic lost their magnetic sense, which suggests that the magnetoreceptors in the Japanese eel are located in or around the nose (Nishi *et al.*, 2005). Observations on migration of the European eel in the open ocean are, further indicative of the fish being capable of orientation in the magnetic field (Tesch, 1974 b, 1978, 1980).

Salmonidae are another family of fishes showing directional responses and orientation ability. A magnetic field-dependent choice of direction of movement (magnetic compass) was revealed in the juvenile sockeye salmon (*Oncorhynchus nerka*) (Quinn, 1980). Effects of magnetic field on fish behaviour was reported from smolts of the sockeye salmon (Quinn and Brannon, 1982), chum salmon (*Oncorhynchus keta*) (Quinn & Groot, 1983), and chinook salmon (*Oncorhynchus tschawytscha*) (Taylor, 1986, 1987). The rainbow trout (*Oncorhynchus mykiss*) show orientation preferences as well, their compass orientation being directed towards the magnetic north. Shielding the laboratory resulted in the disappearance of direction preferences (Chew & Brown, 1989). A change in the magnetic field value, relative to the geomagnetic field, around entrances to chambers in the experimental device constructed to test the behaviour of larvae of the sea trout (*Salmo trutta*), and to entrances of fish traps deployed in a natural water body induced the fish to preferentially enter the chambers and traps with elevated magnetic field around the entrance (Formicki *et al.*, 2004 a, b).

Much attention was paid to the Chondrichthyes (sharks and rays), which are also capable of using magnetic field as an aid to their selection of the direction of movement and orientation. The leopard sharks (*Triakis semifasciata*) were found to prefer moving north. The round stingrays (*Urolophus halleri*) can be trained to depend on the magnetic field direction, changed by the experimenter, in their search for food in an enclosure

(Kalmijn, 1978). Studies on the *in situ* behaviour of the hammerhead sharks (*Sphyrna lewini*) showed the fish to move along certain routes overlapping with local magnetic field gradients (Klimley, 1993).

Some fish species can be trained in the presence of magnetic field. The aba (*Gymnarchus niloticus*) can be trained to feed, depending on the presence of a magnet on the aquarium wall (Lissmann, 1958). The round stingray described above can be trained (Kalmijn, 1978, 1981), as can be the yellowfin tuna (*Thunnus albacares*) (Walker, 1984; Walker *et al.*, 1985), to discriminate between magnetic stimuli. The longnosed black ghost (*Apteronotus leptorhynchus*) learned to respond to a change in magnetic north in response to food availability (Sherman, 1992). The juvenile rainbow trout can be trained to discriminate between the presence and absence, or a change in intensity, of an artificially created magnetic field superimposed on the background Earth's magnetic field (Walker *et al.*, 1997). Recent research on sharks showed them to be capable of learning to choose certain sites, depending on the prevalent magnetic field (Meyer *et al.*, 2005).

This chapter will review observations and data on changes in embryonic development, embryonic and larval locomotion, physiological responses, and directional responses of fish embryos and larvae to artificial and natural magnetic fields. Magnetoreception in fish embryos and larvae, frequently associated with organismic orientation, will be presented in a broader context, since an organism's orientation is a net result of, and can be inferred from, specific behavioural responses which can be observed and recorded.

So far, a few mechanisms have been proposed to explain how living organisms perceive magnetic fields. At present, three such mechanisms are regarded to sufficiently explaining the process, depending on the environment. They will be discussed in the final section on mechanisms of magnetoreception.

The research referred to involved highly diverse materials: embryos and larvae of a number of fish species, which required a diversity of methodological approaches. Therefore, individual problems will be presented in separate sub-sections. Each will describe the materials used, the methods applied, and the results obtained.

EFFECTS OF MAGNETIC FIELD ON EMBRYONIC DEVELOPMENT OF FISH

Compared to other fish families, embryonic development of Salmonidae is, by its very nature, relatively long-lasting, for which reason individual developmental stages can be observed in minute details. Therefore, research on magnetic field effects on embryonic development of the fish were focused primarily on eggs of the rainbow trout and sea trout incubated from fertilisation to hatch under constant magnetic field (superimposed on the Earth's field) and in the geomagnetic field (control) under identical conditions. The constant temperatures applied (8 ± 1 and 12 ± 1°C for the sea trout and rainbow trout, respectively) made it possible to determine, with a high degree of precision, the duration of individual developmental stages in thermal units (degree-days, °D). Magnetic fields were generated with magnets.

Magnetic fields of 1 to 5.5 mT were found to slow down the embryonic development and thus to extend incubation of sea trout and rainbow trout eggs. The newly hatched rainbow trout larvae, exposed to magnetic field, were, on average, longer (up to 2.48 mm) and heavier (up to 17 mg) than the control ($p < 0.05$; mean length 12.77 mm; mean weight 61.3 mg). Similarly, sea trout with resorbed yolk sacs were, on average, longer (even by 3.79 mm) and heavier (by 31.3 mg) than the control ($p < 0.05$; mean length 25.57 mm; mean weight 118.4 mg). The magnetic fields used produced a favourable effect in the embryonic and larval development of the fish studied. The effect was manifested as reduced egg mortality during incubation and a shorter duration of hatching. On the other hand, exposure of incubated rainbow trout eggs to a higher magnetic field (13 mT) resulted in a reduction of length and weight of newly hatched larvae (Formicki & Winnicki, 1998).

EFFECTS OF MAGNETIC FIELD ON FISH EMBRYO LOCOMOTION

Locomotor manifestations in fish embryos are diverse, from slight actomyosin contractions in the ectoplasm to the rotational movement of the entire oocyte within the egg membranes, reflected by a circular movement of the germ plate (particularly well-visible in the pike) to movements of structures serving vegetative functions (heart, circulation system, pectoral fins, etc.). The movements observed during embryogenesis are specific responses of an embryo to external stimuli.

Oocyte Movements

Studies of magnetic field effects on early embryonic movements in fish were conducted to determine the basis of directional responses of the embryos, and to follow their influence on initial cytoplasmic movements.

The studies were carried out on developing eggs of the Northern pike (*Esox lucius*) in a temperature-controlled laboratory. Constant (4 mT) and alternating magnetic fields tested were generated by a computer-controlled, water-cooled magnetic field source (15 mT, 50 Hz, sinusoid). After a 10-min acclimation period (treated as a control), the embryos were exposed to magnetic field switched on for 30 min. The locomotor behaviour was observed in individual embryos placed in an observation chamber, using a system consisting of a stereomicroscope (with non-magnetic base) placed between the coils, a CCD camera, an SVHS video-recorder, and a computer equipped with image-analysis software (Fig. 1).

The study tackled quasi-peristalsis, a form of movement occurring in the egg cell ectoplasm (external plasma layer) early in the embryonic development of some fish species, e.g. Northern pike, three-spined stickleback (*Gasterosteus aculeatus*), and common whitefish (*Coregonus lavaretus*). Quasi-peristalsis is visible as waves of ectoplasm moving laterally and longitudinally over the yolk surface, which results in spinning and circular movements of the germ plate and is translated to the movement of the entire yolk sphere

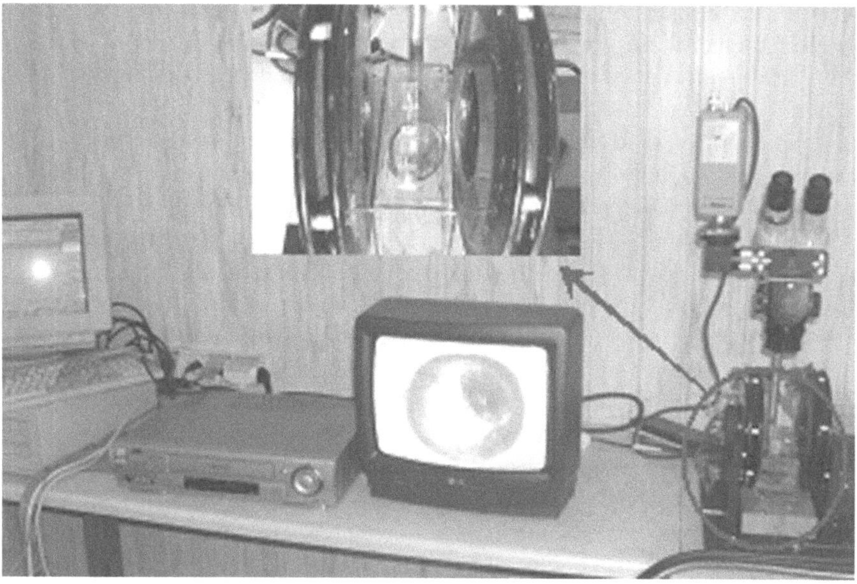

Fig. 1 A system used to record embryonic movements in magnetic fields (see text for description).

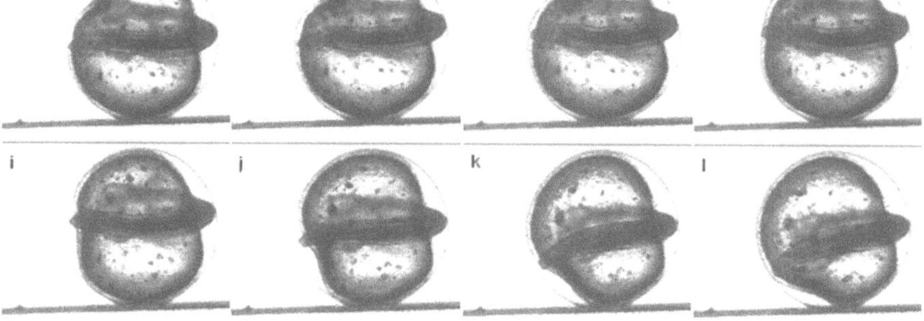

Fig. 2 A quasi-peristaltic wave moving over ectoplasm (protoplasm) surface in egg cell of Northern pike (*Esox lucius*) (frames frozen at 1 min intervals; lateral view). After Tanski *et al.* (2000).

(Fig. 2) (Korzelecka, 1999; Tanski *et al.*, 2000). Duration of one full circle of the quasi-peristaltic wave was measured before (control) and after the magnetic field was switched on. The values obtained were converted to angular velocity and compared with the control.

The experiments involved Northern pike embryos at the stages of the blastula (3 °D after fertilisation), gastrula (1/2 epiboly, 20 °D of development), and embryo performing its first somatic movements (50 °D of development) (Winnicki *et al.*, 2004).

The experiments showed the magnetic field (constant and alternating) induced changes in the quasi-peristalsis to be highly regular (Fig. 3). In the alternating magnetic field, the angular velocity of the quasi-peristaltic wave in the blastula increased as the movement progressed. During the first 8 cycles, the angular velocity increased slowly, until the statistically significant difference was detected (Fig. 3a). Characteristically, the ectoplasm wave height was found to increase. A similar tendency was also observed in the gastrula (Fig. 3b). In the fully formed embryo, the angular velocity increased at a faster rate, the statistically significant difference being detected in the fourth cycle of the ectoplasm wave (Fig. 3c).

Constant magnetic field also increased the quasi-peristaltic wave velocity, the statistically significant differences occurring in the eighth cycle in the blastula (Fig. 4a) and in the tenth cycle in the gastrula (Fig. 4b). On termination of gastrulation, the constant magnetic field produced no significant acceleration of ectoplasm wave in Northern pike embryos, although an increasing trend could be detected (Fig. 4c).

As shown by the experiments described, the egg quasi-peristalsis – a form of embryonic locomotor activity – responds to changes in constant and alternating magnetic field, the responses being visible as acceleration of ectoplasmic contractions. Most probably, the magnetic field affects ionic transport through membranes, thereby producing an effect in ectoplasm actomyosin complexes. A higher ectoplasm wave appearing after the magnetic field has been switched on is a result of stronger, or full, contractions. Sets of contractile filaments may play an important role in directional positioning of an embryo in the egg relatively to the magnetic field lines. Contraction and movement of the external plasma layer in the egg play an important part in ensuring the best respiratory conditions as they enhance intensive mixing of the perivitelline fluid surrounding the egg cell; moreover, the germ plate performs the spinning and circular movements described above.

Movements of the Circulation System in Fish Embryos and Larvae

As soon as the circulation system emerges in developing teleost embryos, various parts of the system take on the functions of embryonic respiration organs (see Pelster, this volume; Burggren & Bagatto, this volume). The heart as well as the major blood vessels become clearly visible through the egg membranes. Those structures are also visible before pigmentation appears in the newly hatched larvae and facilitate observation of magnetic field-induced effects on the activity of cardiac muscle of fish embryos and

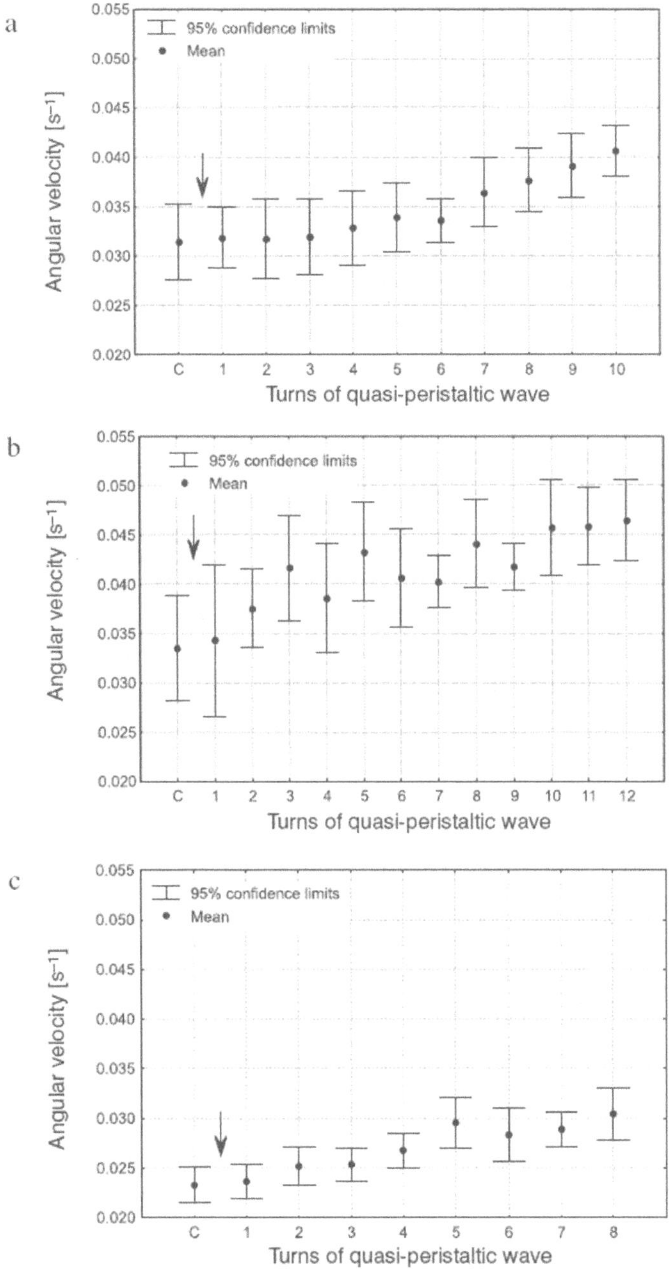

Fig. 3 Changes in angular velocity of quasi-peristaltic wave in Northern pike (*Esox lucius)* embryos at different stages, exposed to alternating magnetic field; n=11; a, blastula; b, gastrula; c, embryo; C, control; ↓, magnetic field switched on. After Winnicki *et al.* (2004).

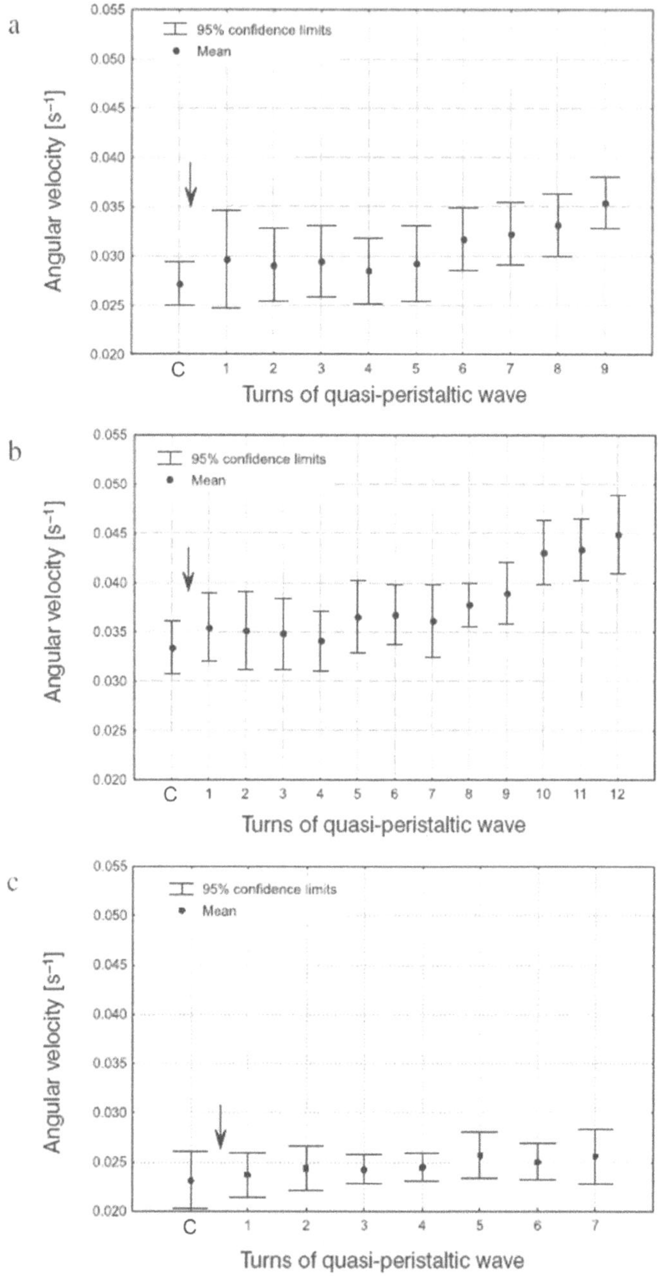

Fig. 4 Changes in angular velocity of quasi- peristaltic wave in Northern pike (*Esox lucius*) embryos at different stages, exposed to constant magnetic field; n=11; a, blastula; b, gastrula; c, embryo; C, control; ↓, magnetic field switched on. After Winnicki *et al.* (2004).

larvae. In addition, respiratory movements performed by the larvae with their pectoral fins were observed.

The observations were carried out on embryos and larvae of the Northern pike, sea trout, and common carp (*Cyprinus carpio*). The experiments involved embryos with emerging and visible central part of their circulation system, i.e. the heart (the rhythmically pulsating part of the system), and the newly hatched larvae. The embryos or larvae were placed, one each, in glass chambers kept at a constant temperature, on the compound microscope stage. The observation system was described above. Constant magnetic field was generated by magnets or a set of coils, the latter being used to generate alternating field. Constant magnetic field (4 and 50 mT) was used in experiments involving embryos and larvae of the Northern pike, sea trout, and common carp, while alternating magnetic field (15 mT, 50 Hz, sinusoid) effects were tested on embryos and larvae of the Northern pike (Winnicki & Formicki, 1990; Winnicki *et al.*, 1993, 2004; Formicki & Winnicki, 1996). After a period of adaptation (10–30 min) when the heart beat stabilised at a constant level (control), the embryos or larvae were exposed to magnetic field, the effects being observed and recorded for the subsequent 30 min. From the beginning of the experiment, the heart beat rate was recorded continuously. Between-individuals differences in the number of cardiac contractions at rest are natural in vertebrates; for this reason, the number of contractions during adaptation (control) was assumed to be 100 and was regarded as a reference level for the subsequent changes (Figs. 5–8).

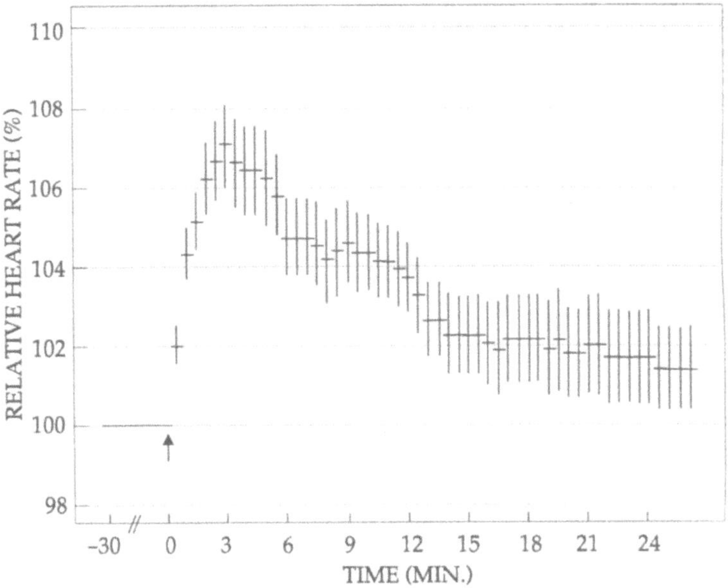

Fig. 5 The relative number of cardiac contractions in Northern pike (*Esox lucius)* embryos in the control and under exposure to magnetic field (50 mT; F=4.599; d.f.=58; p<0.001; mean ±SE). The number of contractions in the control was assumed to be 100%. ↑: magnetic field switched on. After Winnicki *et al.* (1993).

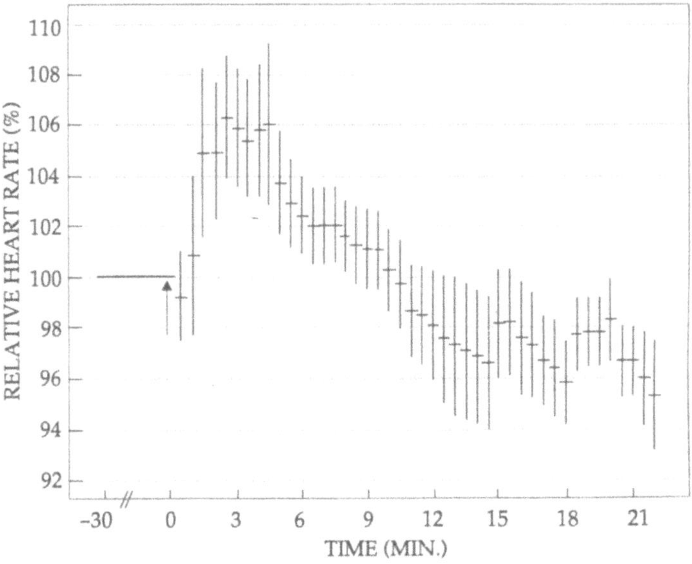

Fig. 6 The relative number of cardiac contractions in Northern pike *(Esox lucius)* larvae in the control and under exposure to magnetic field (50 mT; $F=2.179$; d.f.=50; $p<0.001$; mean ±SE). The number of contractions in the control was assumed to be 100%. ↑: magnetic field switched on. After Winnicki *et al.* (1993).

After the test magnetic field was switched on, changes in the heart rate were observed to occur with a high regularity. They were manifested as an increase in cardiac contractions, both in the embryos and in the larvae. The increase was regular in terms of both the degree of acceleration and the curve describing the history of changes. The number of contractions was observed to increase during the initial 3–4 min following the beginning of exposure to magnetic field. There was a noticeable difference between the responses of embryos and larvae of Northern pike and common carp in constant magnetic field during the first minute of exposure. In the embryos, the number of contractions increased monotonously to the maximum value, while the heart rate of the larvae dropped below the control level (Figs. 5 and 6). In the larvae, the heart rate was, for a short time, slowed down by constant magnetic field. Results obtained in the common carp were very similar to those produced by the Northern pike. It seems that the external stimulus provided by magnetic field directly affects embryonic membranes and the cardiac muscle. In contrast, the larvae, already in the possession of a developed central nervous system receive and process stimuli via the brain, which results in a short period of slow-down. The embryonic cardiac muscle responds differently to alternating magnetic field (Fig. 7b). The response is not rapid and the number of contractions increases gradually to reach its maximum after 25 min, the increase – relative to the control – being much higher.

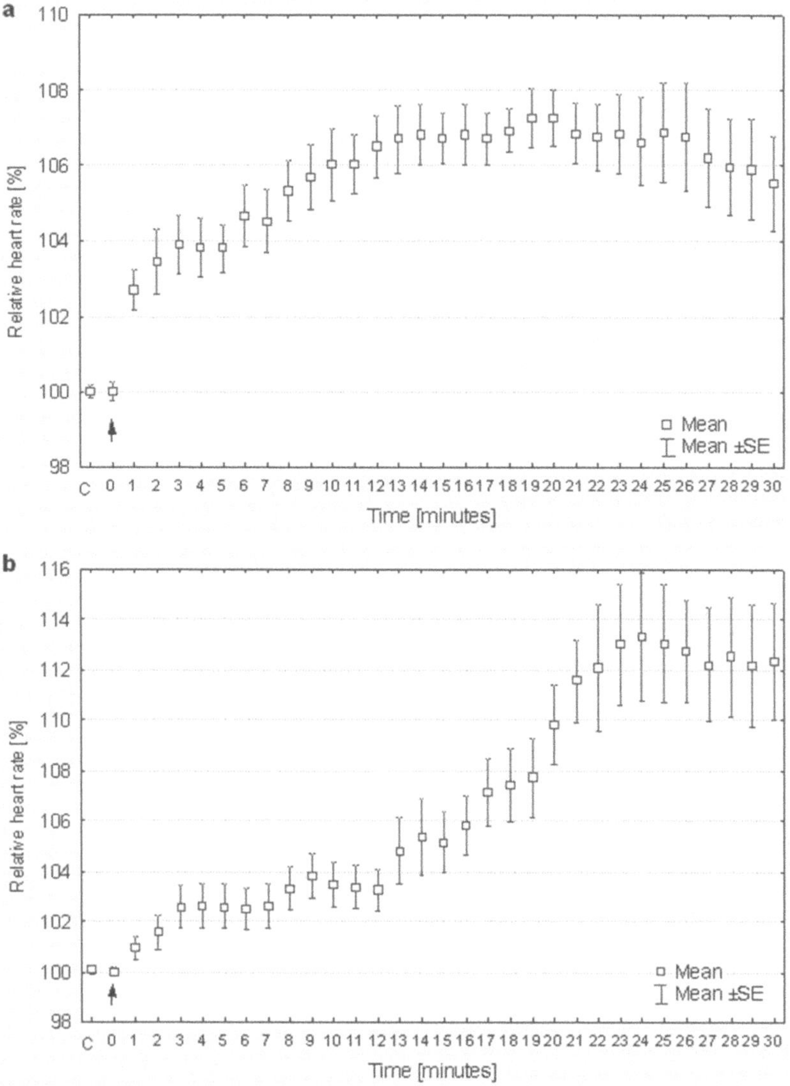

Fig. 7 Change in the heart rate of Northern pike (*Esox lucius*) embryos exposed to magnetic field; a, constant (4 mT); b, alternating (15 mT, 50 Hz, sinusoid); C, control; ↑, magnetic field switched on. After Winnicki *et al.* (2004).

Pectoral fins of the sea trout larvae move constantly or periodically, which is associated with the production of water movement around the body and enhancement of breathing. Magnetic field-induced changes in the cardiac muscle activity in the sea trout larvae (Fig. 8) are accompanied by changes in the pectoral fin movement rate (Fig. 9).

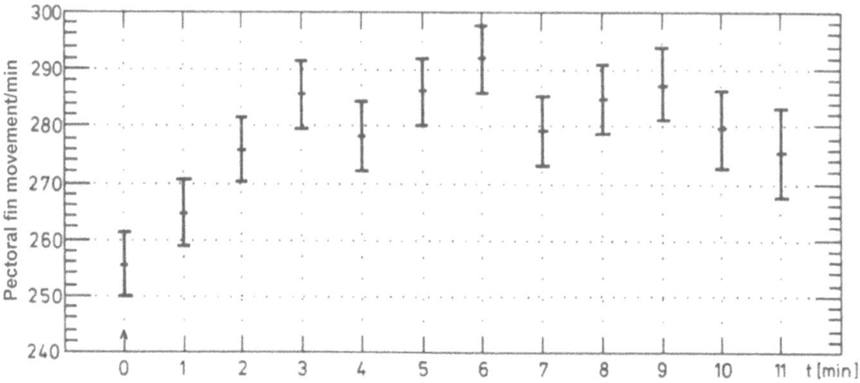

Fig. 8 Change in the heart rate of sea trout *(Salmo trutta)* larvae exposed to magnetic field (50 mT, n=11; p<0.001); C, control; ↑, magnetic field switched on. After Winnicki & Formicki (1990).

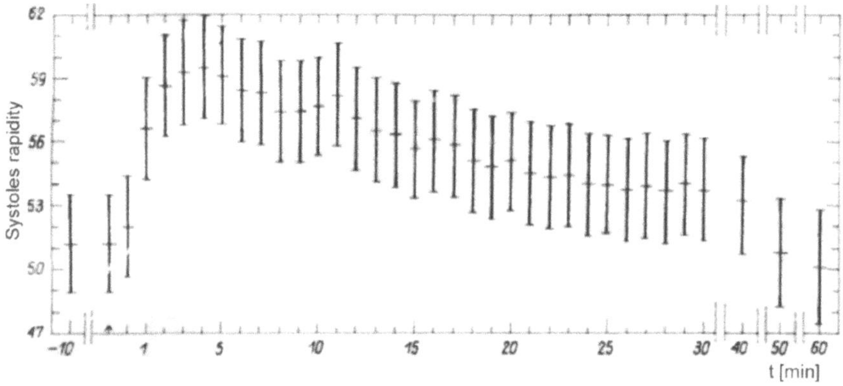

Fig. 9 Pectoral fin movement rate in sea trout *(Salmo trutta)* larvae; ↑, magnetic field (50 mT; n=11; p=0.001) switched on. After Formicki (1992).

Experiments performed in a system identical to that used for observing the heart rate showed the frequency of pectoral fin movement rate in the sea trout larvae to increase significantly under exposure to magnetic field (Formicki, 1992). A comparison between the heart rate and the frequency of pectoral fin movement of the sea trout larva exposed to magnetic field showed a certain time lag. The highest number of cardiac contractions occurred between minute 2 and 7 of exposure, the highest number of pectoral fin movements being recorded between minute 3 and 9. It may be assumed that the body produces a compensatory effect, the cardiac muscle being the first link in the compensation chain. The cardiac activity in overload triggers the second link of the compensation chain in the form of intensified movements of the pectoral fins, which increases the flow of oxygenated water to the larval breathing apparatus.

DIRECTIONAL RESPONSES IN MAGNETIC FIELD

As already mentioned in the introduction, adult forms of numerous fish species, e.g. rays, sharks, salmonids, eels, and tunas, possess a specific magnetic sense used in behavioural responses and for orientation in the environment (Kalmijn, 1978; Quinn, 1980; Karlsson, 1985; Tesch *et al.*, 1992; Walker *et al.*, 1997).

Directional Responses of Fish Embryos

Earlier observations on developing embryos which positioned (oriented) themselves in a certain way in the egg (within egg membranes) provided a basis for further detailed investigations (Formicki *et al.*, 1997; Formicki & Winnicki, 1998). They were aimed at determining magnetic field effects on directional responses (directional positioning, i.e., alignment of embryonic body axis relative to magnetic field lines); and finding out at which moment during embryogenesis the embryonic body axis becomes orientated, how the process proceeds, and whether it may be dependent on the value of magnetic field to which the embryo is exposed (Formicki *et al.*, 1997; Formicki & Winnicki, 1998; Tanski *et al.*, 2005).

The studies involved eggs of various fish species, which provided an additional opportunity of addressing a question if embryo positioning (arrangement of the body axis relative to magnetic field lines) occurs both in the species undertaking long-distance migrations (salmonids) and in the non-migratory ones. Observations were made on developing eggs of the Atlantic salmon (*Salmo salar*), sea trout, rainbow trout, vendace (*Coregonus albula*), Northern pike, and rudd (*Scardinius erythropthalmus*).

For incubation, the fertilised eggs were placed on a substratum consisting of plastic netting inserted in circular mini-aquaria. The mesh size of the netting, stretched between supports above the bottom of the mini-aquarium, was in each experiment adjusted to the size of the eggs being incubated, therefore the eggs remained immobile. The mini-aquarium water was changed (1/3 of the volume each time) using a specially constructed plumbing system which made it possible to change the water at a very slow rate (0.5 dm^3 h^{-1}), so that the developing eggs remained in an unchanged position (Fig. 10). The mini-aquaria, i.e. orientation arenas, were placed in artificially generated and natural magnetic fields, the latter serving as a control.

Constant magnetic fields of 0.5, 1.0, 2.0 or 4.0 mT, depending on the experimental treatment, were generated by magnets with expanders. The magnets made it possible to generate a constant, uniform magnetic field throughout the period of incubation. The experimental design involved increasing the field value, over the geomagnetic field, in the direction identical with the latter, as it was expected that an elevated magnetic field value may positively influence the orientation of the embryos. The treatments involving the geomagnetic field only (serving also as a control) utilised dummies of magnets and expanders. Orientation of the embryos was recorded also when no magnet dummy was used; the directions of embryonic body positioning were identical in dummy and no-dummy treatments. The experiments were carried out in

Fig. 10 A mini-aquarium, i.e. an experimental arena for egg incubation, with the netting supporting the developing eggs.

totally blacked-out laboratories to prevent the embryos' responding to light. Positioning of the embryos in all the experiments was recorded following the closure of the blastopore, as recording the orientation later in the embryonic development, when the first movements appear, could have distorted the results. The embryonic orientation was determined by matching the anterior part of the embryo's body with an appropriate position on a 16 compass point scale (wind rose).

Constant magnetic field and the geomagnetic field were found to affect the direction of embryonic body (or rather its axis of symmetry) alignment. Salmon embryos tested in an 0.5 mT magnetic field were found to orientate themselves primarily north-south. A considerable proportion of individuals showed also NNE-SSW and NNW-SSE orientation ($p = 0.001$). Similar results were obtained in a sample exposed to a 1.0 mT field ($p = 0.001$). The preferred directions at 2.0 and 4.0 mT were NNE-SSW and north-south ($p = 0.001$), respectively. Those embryos kept in the geomagnetic field and serving as a control were observed to position themselves along a north-south axis ($p = 0.01$). When exposed to an 0.5 mT field, sea trout embryos were found to prefer NNE-SSW orientation; a preference towards NNE-SSW was visible also at 1.0 and 4.0 mT, a north-south orientation being observed under exposure to 2.0 mT ($p = 0.001$). The developing sea trout embryos placed in the geomagnetic field preferred NE-SW orientation ($p = 0.006$). Similarly, rainbow trout embryos which were exposed, throughout incubation, to magnetic fields higher than the geomagnetic field were found to prefer north-south orientation, a preference observed also in the geomagnetic field ($p = 0.001$). The direction preferred by vendace embryos and dominant in their arrangement under exposure to an 0.5 mT magnetic field was NNE-SSW, the preferences observed in stronger fields and in the geomagnetic field being similar ($p = 0.01$. In the geomagnetic field, the pike embryos tested were observed to position their body axis of symmetry mainly NE-SW, but the difference was insignificant ($p = 0.12$). The artificially generated fields induced a change to mainly NW-SE (0.5 and 1.0 mT) and north-south (4.0 mT) ($p < 0.01$), no change being observed in the 2.0 mT field (NE-SW arrangement) ($p = 0.05$). Spatial arrangements of the rudd embryos tested proved similar.

It is thus evident that developing embryos of both migratory (Atlantic salmon, sea trout, rainbow trout) and non-migratory fish positioned their body axis of symmetry mainly in directions close to north-south.

Effects of magnetic fields of directions differing from that of the geomagnetic field on spatial arrangement of fish embryos

Another set of experiments tested spatial arrangements of fish embryos in a generated magnetic field superimposed (clockwise) on and rotated 90° relative to the geomagnetic field. The experimental system was identical to that described above, except that the generated magnetic field (0.5 mT) was rotated with respect to the geomagnetic field direction. The control consisted of eggs developing in the mini-aquaria exposed to the geomagnetic field only. Rotating the generated magnetic field (0.5 mT) poles resulted in embryos arranging their body axes of symmetry along the lines of that field. Those embryos remaining in the geomagnetic field did not change their orientation. The developing embryos of the Atlantic salmon, sea trout, and rainbow trout as well as those of the vendace, Northern pike, and rudd were observed to prefer mainly a north-south arrangement as well as closely related directions (NNW-SSE, NNE-SSW) of the generated magnetic field ($p < 0.05$) (Tanski *et al.*, 2005).

Timing of orientation emergence in the embryos and development of the process

An attempt was also made to pinpoint the stage in the embryonic development when orientation, i.e. alignment of embryonic body axis of symmetry along the lines of magnetic field (artificial or natural) appears, and whether the process depends on the value of the magnetic field to which an embryo is exposed. Developing eggs of the Atlantic salmon, sea trout, vendace, and Northern pike were transferred from an incubation apparatus insulated from magnetic field exposure to orientation arenas (circular mini-aquaria). The eggs of individual species differed in their stage of embryonic development, as the development is species-specific. The Atlantic salmon and sea trout eggs were transferred after 30, 60, and 90 °D of their embryonic development; the eggs of the vendace were transferred after 15, 35, and 50 °D, while the Northern pike eggs were transferred after 12, 25, and 37 °D. Each transfer resulted in a changed position of an egg, and hence in a changed spatial arrangement of embryo's axis of symmetry in the egg. Prior to transfer to the orientation arenas, eggs of the Atlantic salmon and trout had been kept in a hatching apparatus, eggs of the vendace and pike being kept in a Weiss apparatus. Following the transfer, the eggs were incubated in 0.5 and 2.0 mT magnetic fields superimposed on the geomagnetic field, and in the natural field. Lines of the artificially generated field were positioned identically to those of the geomagnetic field. As in the previous set of experiments, the spatial arrangement of the developing embryos was recorded after the closure of the blastopore (Formicki *et al.*, 2001; Tanski, 2002).

Only Atlantic salmon embryos older than 30 °D, i.e. after the closure of the blastopore, were found to show a statistically significant ($p < 0.05$) spatial arrangement in the 0.5 and 2.0 mT fields. A distinct trend was visible in the geomagnetic field, but the significance level was higher than 0.05. The developing embryos after 60 °D showed also a significant arrangement in the 2.0 mT magnetic field ($p = 0.01$). Embryos subjected to changes in orientation later in their development (after 90 °D) showed no tendency to any distinct spatial orientation.

Embryos of the sea trout, vendace, and Northern pike the spatial position of which was changed at 30 °D, 15 °D, and 12 °D, respectively of the embryonic development were found to orientate themselves in the geomagnetic field in a statistically significant manner ($p < 0.05$). A change in the position of sea trout eggs at 60 °D, vendance at 35 °D, and Northern pike at 25 °D of the development caused the embryos to show significant orientation in the 0.5 and 2.0 mT magnetic fields ($p < 0.05$), but not in the geomagnetic field ($p > 0.05$) (Tanski, 2002).

A change in orientation of embryos at a later stage resulted in the lack of directional preferences. It has to be inferred that orientation of the body of those embryos relative to the magnetic field lines depends on that time point in embryogenesis when the eggs cannot change their position, and on the magnetic field value: an increase in the value reduces the time preceding the appearance of orientation.

Directional Responses of Fish Larvae

The study on directional responses in fish larvae involved the larvae and, for comparison, early fry of the sea trout a few days after the yolk-sac had been completely resorbed.

Two age groups of the larvae were tested: one consisted of the newly hatched larvae before 2/3 of their yolk-sac were resorbed, the other consisting of the larvae whose yolk-sac resorption progressed from 2/3 to the total disappearance. For comparative purposes, the early fry which had just resorbed their yolk-sac were tested (Formicki *et al.*, 2004a).

The larvae belonging to the first group, which should be more appropriately termed the eleutheroembryos or prolarvae, overly burdened by the yolk-sac contents, remained immobile on the bottom and performed only slight movements in the water column. Those larvae belonging the second group were vigorously motile, which is related to their commencing active feeding; they are then aided by the exteroreception system which becomes indispensable at that stage. It was then assumed that the period of time between the resorption of 2/3 of the yolk-sac and its total disappearance should provide most information on morpho-physiological mechanisms enabling the larvae to position themselves suitably in magnetic field.

The larval behaviour in constant magnetic field was observed in a specially designed experimental set-up consisting of 4 chambers separated by a corridor. Each chamber could be entered by an opening provided with curtains that formed a slit, which made it difficult for the larva to back up once in the chamber (Fig. 11). The set was placed in the background geomagnetic field in a darkened laboratory room (Formicki *et al.*, 2004a).

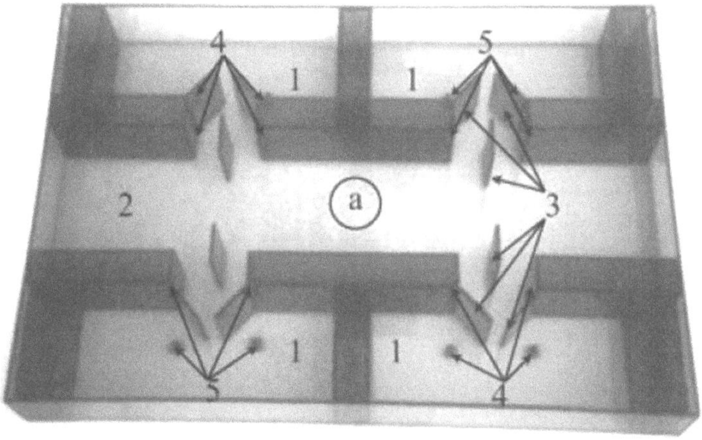

Fig. 11 Experimental set-up for the study of larval behaviour in the presence of magnetic field; 1, chambers; 2, corridor; 3, curtains; 4, magnets; 5, magnet dummies; a, the site of larvae release. After Formicki *et al.* (2004a).

The set design and the rectangular shape used emerged from an earlier study on directional responses of adult fish, observed *in situ* in a natural water body. That study involved fish traps rigged with magnets that generated a magnetic field around the entrance to the trap. The rigged traps were alternated with control ones (with magnet dummies placed where true magnets were attached in the treatment traps). The traps were deployed along the shore of the water body so that the trap axis was at right angles to the shoreline (Formicki *et al.*, 2004b).

The entire experimental set-up, before and during the experiment, was covered by a non-transparent cylindrical lid made of thin Plexiglas to exclude effects of physical stimuli other than magnetic field (e.g. light or sound) on the selection of direction by the larvae. The laboratory was also blacked out. The bottom of the set was illuminated by uniform, shade-free light of low intensity, which mimicked daytime conditions in the natural environment. Behaviour of the larvae was videotaped (Fig. 12). Magnets (two pairs each) were mounted at the entrances to and inside of two chambers located on both ends of the set's diagonal; the remaining two chambers were rigged with two magnet dummies of shape and colour identical with that of the true magnets

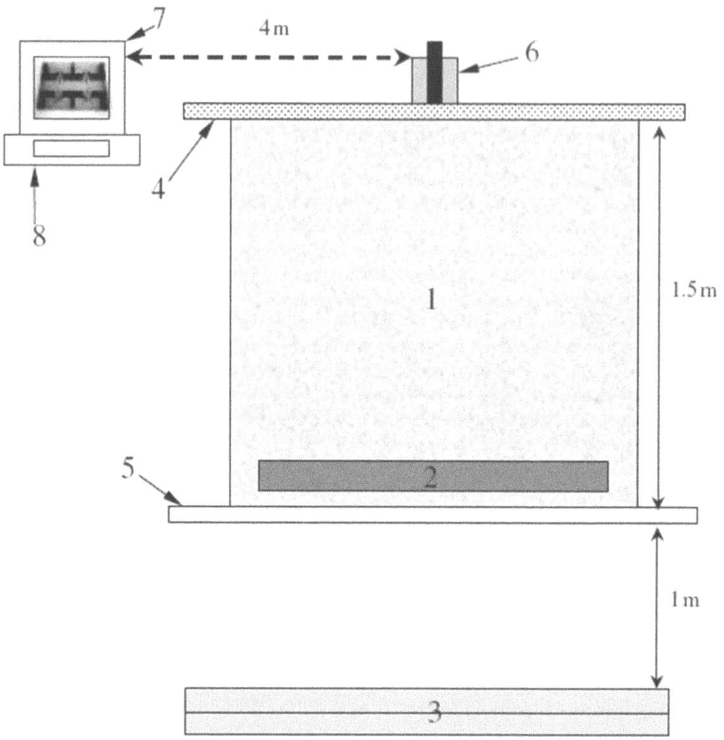

Fig. 12 A schematic representation of the arrangement of cylindrical sheath (1), experimental set (2), lighting (3), top and bottom covers (4, 5), and recording devices: camera (6), computer screen (7), and VCR (8). After Formicki *et al.* (2004a).

(Figs. 11 and 13). Figure 14 shows the distribution of magnetic field values between the pairs of magnets at the entrance to a chamber and inside it.

The larvae were transferred to the experimental set-up and acclimated to ambient conditions there (Fig. 12) for 30 min. Subsequently (the actual start of the experiment) the fish were released. During the next 150 minutes, the number of larvae in each chamber was counted. Those found in the magnet-rigged chambers were denoted "magnetic field"; the larvae which entered the dummy-rigged chamber were termed "control"; and those found in the corridor were classified as "corridor".

Sea trout larvae prior to resorption of 2/3 of their yolk-sac content

Of a total of 1380 larvae tested, 365 (26.4%) selected the magnet-rigged chambers, the control chambers being chosen by 232 larvae (16.8%). The remaining 783 (56.8%) larvae stayed in the corridor ($\chi^2 = 26.03$; $p < 0.0001$). Restricting the analysis to those larvae that selected a chamber, the results are as follows: 365 larvae (61.1%) selected

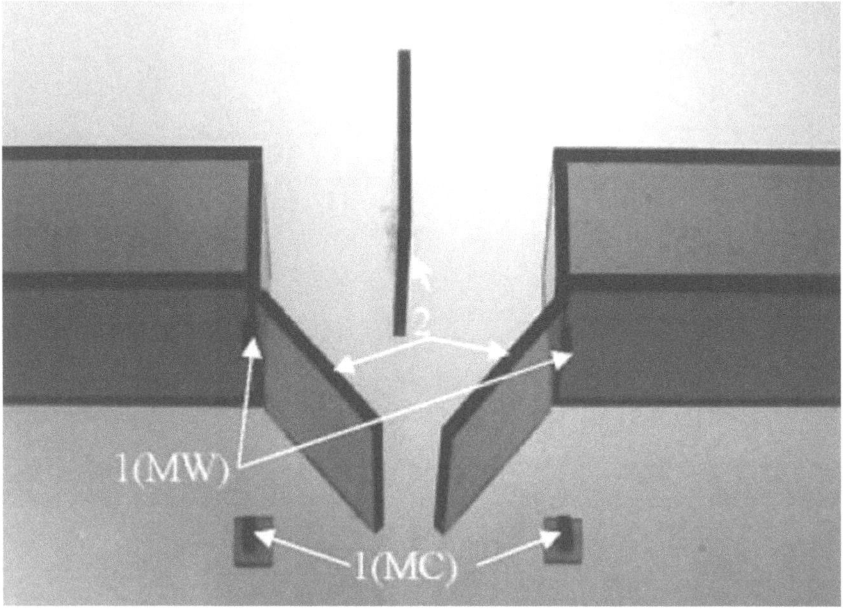

Fig. 13 Placement of magnets and magnet dummies in chambers and the curtains preventing larval escape: (1) magnets and magnet dummies in chamber walls (MW) and inside the chambers (MC); (2) curtains. After Formicki *et al.* (2004a).

the magnet-rigged chambers and 232 (38.9%) chose the control ones ($\chi^2 = 4.95$; $p = 0.026$).

Sea trout larvae with 2/3 of their yolk-sac resorbed

Of the 960 individuals tested, 382 (39.8%) selected the magnet-rigged chambers, 282 (29.4%) chose the chambers rigged with dummies. The remaining 296 larvae (30.8%) stayed in the corridor ($\chi^2 = 1.89$; $p = 0.387$). When the statistical test was restricted to the larvae that selected a chamber, 54.5% of the 664 individuals chose the magnet-rigged chambers, 42.5% moving to the control ones ($\chi^2 = 2.26$; $p = 0.131$).

To compare directional responses of sea trout larvae to stronger magnetic fields with behaviours displayed by the slightly older larvae, an identical experiment was performed with the early sea trout fry the yolk-sacs of which had been resorbed a few days earlier. That experiment involved 56 individuals 36 of which (64.4%) selected entering the magnet-rigged chambers, while 16 (28.5%) chose the chambers rigged with magnet dummies. The percentage of fish remaining in the corridor was very low (7.1%, i.e. 4 individuals) ($\chi^2 = 49.87$; $p < 0.0001$). Restricting the statistical testing to those individuals which entered a chamber resulted in finding out that 69.3% of them selected the magnet-rigged chambers and 30.7% moved to the control ones ($\chi^2 = 14.74$; $p = 0.0001$).

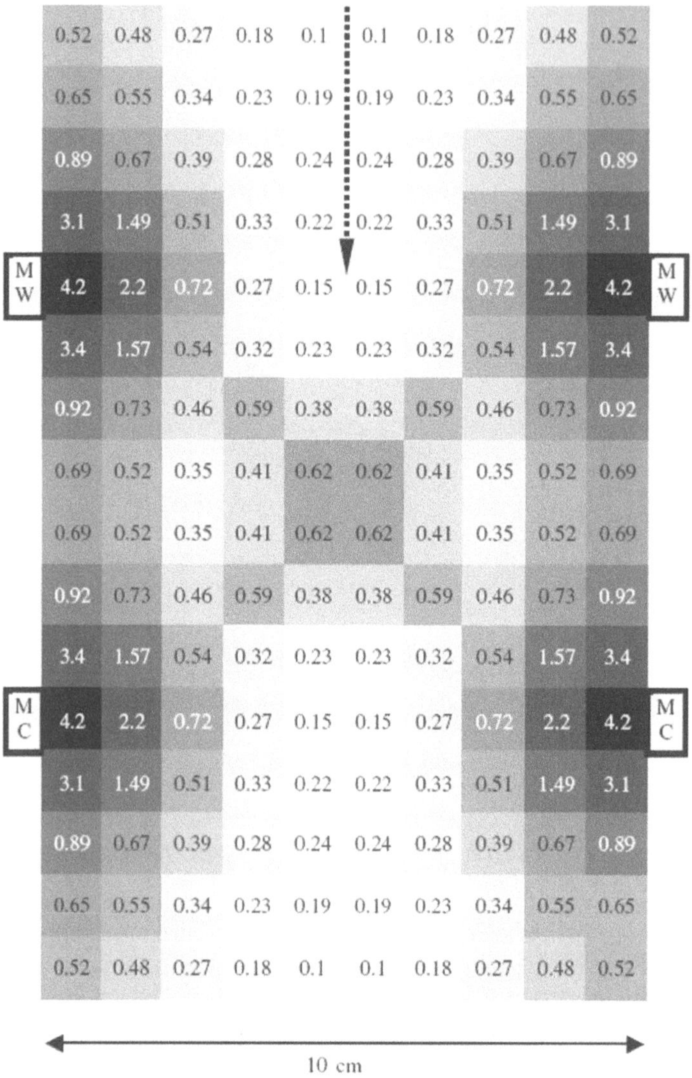

Fig. 14 Magnetic field values (mT) at the entrance to the chamber; the side of each small square is 1 cm; MW, magnets in chamber walls; MC, magnets inside chambers; ↓, direction of larval movement into the chambers. After Formicki *et al.* (2004a).

These data indicate that during the larval phase of life, when the fish begin to move freely and feed, a changed value (gradient) of magnetic field induces certain directional responses. It seems that the perception of a change in the intensity (gradient) of the generated magnetic field elicits an investigative response in the larvae and fry (Formicki *et al.*, 2004a).

The directional responses observed may be related to magnetic field affecting the magnetic material particles present in the fish body. This may in consequence result in a change in the permeability of the cell membrane ion transport channels.

MECHANISMS OF MAGNETORECEPTION

Scientists involved in magnetoreception research have for years pondered the question of how living organisms, including fish, perceive and receive external magnetic field and how they decipher information necessary for making appropriate decisions on how to respond to, and position themselves in their immediate environment. The search for a receptor or receptors capable of intercepting diverse information from magnetic field has been encountering numerous hurdles. Living organisms may take advantage of the direction and intensity of magnetic field. Moreover, magnetic field passes freely into live biological tissues whereby appropriate receptor cells (magnetoreceptors) or sensitive organs, as opposed to organs of other senses (e.g. vision, olfaction), do not have to be located on the outer surface of the body and remain in direct contact with the external environment. They may instead be located anywhere in the animal body. Moreover, they do not include additional structures, for instance those channeling the stimulus on the receptor itself. Furthermore, magnetoreceptors may be tiny and scattered widely within a large volume of a tissue or tissues. As a consequence, no primary receptors responsible for detecting magnetic field have been so far identified with any certainty (Kirschvink, 1982; Wiltschko & Wiltschko, 1995, 2005; Walker *et al.*, 1997; Lohmann & Johnsen, 2000; Kirschvink *et al.*, 2001; Johnsen & Lohmann, 2005).

Another extremely important issue is the process, or mechanism, of transduction of magnetic information to the organism's nervous system. It seems that the transduction may proceed via chemical reactions, thus eliminating any need for a specialised organ (Ritz *et al.*, 2000).

Living organisms are capable of extracting at least two types of information from the ambient magnetic field: the compass information and the field intensity. The intensity varies, so the information may be important for the organism to determine its location in space ("magnetic mapping") (Johnsen & Lohmann, 2005; Wiltschko & Wiltschko, 2005). It is worthwhile to mention that humans have the ability to perceive magnetic field (Baker, 1997), although the opinions differ. Constant magnetic field may induce epileptic activity in patients ready to undergo brain surgery (Fuller *et al.*, 1995). Polynesians seem to be capable of determining compass direction even in the absence of any evident cues.

Of the several mechanisms proposed so far, three seem to have withstood the trial of time, and two involve fish as the basis of their validity.

Electromagnetic induction-based magnetoreception

This theory, developed by Kalmijn (1974, 1978, 1981, 1982, 1984) who was inspired by Faraday's lectures on electricity, assumes that ambient magnetic information is received by electroreceptors – ampullae of Lorenzini, present in the Chondrichthyes and used in prey location and fish orientation (see Kirschbaum & Denizot, this volume). When swimming, sharks, skates and rays cut across magnetic field lines, i.e. the horizontal component of the geomagnetic field and induce an electric field proportional to the fish movement rate and direction. The ampullae of Lorenzini detect slight changes in the electric field. In sharks, the ampullary canals terminate with openings located on the head, while in rays the openings cover almost the entire anterior part of the body. The ampullary canal walls, as well as the ampullae proper, are very good "isolators" in that they have a very high mural resistance, the ampullae proper being lined by a single layer of electrosensitive epithelium. The jelly in the ampullary canals is almost as conductive as the sea water (Kalmijn, 1984). The Chondrichthyes sensitivity threshold is extremely low (5 nV/cm). By actively moving in salt water, the fish induce local electric field and receive induced voltage gradients, the gradient value depending on the direction in which that fish moves. The movement speed should amount to as little as a few cm per second. The calculated voltage gradient produced then is much higher than the experimentally determined sensitivity threshold of the fish and fully sufficient for orientation in magnetic field (Kalmijn, 1978, 1982). A change in the value of the electric field induced in this way carries information on the direction of fish movement. The Chondrichthyes themselves do not have to move (change position in the geomagnetic field) actively to produce the effect, it is sufficient that they be passively carried by the open-sea currents moving in the geomagnetic field. Ocean currents moving in the geomagnetic field induce electric fields in which the Chondrichthyes may orient themselves in a familiar territory or compensate for the passive drift (Kalmijn, 1981).

The mechanism described is based on the presence of sensitive electroreceptors in the form of ampullae of Lorenzini in the Chondrichthyes living in the sea water, a highly conductive medium. However, it is theoretically possible that other organisms may use an internal current loop. It would require a change in the position of the loop relative to the external magnetic field, and the presence of a specialised transduction organ (Purcell, 1985; Rosenblum *et al.*, 1985).

Magnetoreception based on magnetic field-dependent chemical reactions

This mechanism is explained by assuming that chemical reactions are modulated by external fields of values close to that of the geomagnetic field. Weak magnetic fields may affect biological molecules and chemical reactions. Of a number of descriptions of the process (including photoreception-involving magnetoreception; Leask, 1977), it was only the radical pair model that has gained recognition and is discussed (Ritz *et al.*, 2000; Johnsen & Lohmann, 2005; Wiltschko & Wiltschko, 2005).

The radical pair model as proposed by Schulten & Windemuth (1986) was further developed by Ritz *et al.* (2000). The model assumes the presence of a chemical compass based on interactions of radical pairs with the external magnetic field, depending on the direction. Without entering into details (as the mechanism may be applied rather to explain responses of birds, amphibians, and marine turtles), suffice it to say that magnetic field affects spin states of paired radical ions and alters dynamics of the transition between spin states.

Magnetoreception in birds is most probably restricted to the right eye. Migratory birds with patched right eye failed to orient, while those with patched left eye as well as those whose eyes had not been patched showed good orientation (Wiltschko *et al.*, 2002, 2003). In the salamander, the receptor seems to be located in the pineal gland, the third eye in amphibians sensitive to light. Changes occurred even when the pineal gland skull area was shaded with a colour filter (Deutschlander *et al.*, 1999; Phillips *et al.*, 2001).

The proposed theory of radical pair mechanism is supported by the fact that radical pair reactions can be disturbed by radio waves of the energy identical to that of interaction between spin states and the geomagnetic field (Henbest *et al.*, 2004; Thalau *et al.*, 2005).

Magnetoreception based on single-domain ferromagnetic particles (biogenic magnetite)

The physical basis of responses produced by living organisms, including fish, is a response of tiny crystals of single domain magnetite (Fe_3O_4) which occur in numerous organisms, from bacteria to higher vertebrates (Frankel *et al.*, 1979; Kirschvink & Gould, 1981; Kirschvink & Walker, 1985; Walker *et al.*, 1997; Diebel *et al.*, 2000). The ferromagnetic particles behave like constant miniature magnets. In the chain-like arrangement of those single-domain magnetite particles, individual magnetic moments of the particles point in the same direction.

The first reports on the presence of magnetite of biogenic origin in living organism date back to the early 1960s when magnetite was discovered in radula teeth of chitons (small molluscs) (Lowenstam, 1962). However, at that time the presence of magnetite was not associated with reception of magnetic field which may be used by organisms to produce certain behaviours and orientation. It was only after magnetotactic bacteria were discovered (Blakemore, 1975) that the question came into focus.

Magnetotactic bacteria have chains of single domain magnetite or greigite (Fe_3S_4) in their cells (Kirschvink & Walker, 1985; Schuler & Frankel, 1999). The external magnetic field (including the geomagnetic field) produces a magnetic moment that is capable of positioning, and even turning around, the single bacterial cells along the field lines, both in horizontal and vertical magnetic fields (Kalmijn & Blakemore, 1978). The bacteria move along the magnetic field lines. It is also possible – using the magnetic pulse – to reverse the magnetic poles of the bacteria, similarly to what can be done with small ferromagnets; the pole reversal is manifested as a change in the direction the bacteria swim (Kalmijn & Blakemore, 1978). In the northern hemisphere, magnetotactic bacteria swim north, while those in the southern hemisphere move south, along the geomagnetic field lines. Hence the bacteria are termed north- and south-seeking, respectively (Blakemore *et al.*, 1980; Kirschvink, 1980; Blakemore,

1982). Those simple organisms align with the geomagnetic field lines even when dead. Magnetotactic bacteria used to be a specific model in studies on magnetic field effects on biological systems.

Research reported in the literature revealed the presence of a magnetic material in numerous animal taxa: in crustaceans, insects, birds, and in mammals (Walcott, 1978; Zoeger *et al.*, 1981; Jones & McFadden, 1982; Lohmann, 1984; Bauer *et al.*, 1985; Edwards *et al.*, 1992).

Most migratory fish species were found to contain magnetite. Experiments revealed the presence of a magnetic material in numerous fish species: bluefin tuna (*Thunnus thynnus*), albacore (*Thunnus alalunga*), bigeye tuna (*Thunnus obesus*), Atlantic bonito (*Sarda sarda*), Californian anchovy (*Engraulis mordax*) (Walker *et al.*, 1984, 1985), Atlantic herring (*Clupea harengus*), mackerel (*Scomber scombrus*) (Hanson & Westerberg, 1987), in salmonids: chinook salmon (Kirschvink *et al.*, 1985), sockeye salmon, (Mann *et al.*, 1988; Sakaki *et al.*, 1990), chum salmon (Ogura *et al.*, 1992), rainbow trout (Walker *et al.*, 1997), Atlantic salmon (Moore *et al.*, 1980; Walker *et al.*, 1985), and in European eel (Hanson & Westerberg, 1987). The presence of a magnetic material was also found in non-migratory fish, e.g. common carp and European perch (*Perca fluviatilis*) (Hanson & Westerberg, 1987). A magnetic material synthesised biologically throughout an individual's life span is accumulated in, e.g. ethmoid and parietal areas of the skull and in the *lamina propria* layer within olfactory lamellae (Kirschvink *et al.*, 1985; Chew & Brown, 1989; Walker *et al.*, 1997). Recent studies on rainbow trout show that a candidate receptor cell in fish was identified (Walker *et al.*, 1997; Diebel *et al.*, 2000).

It is important to note that transmission electron microscope-aided identification is necessary to ensure that iron oxide minerals found in many living organisms are indeed biogenic magnetite (Kirschvink, 1997). Another question is the sensitivity threshold of a receptor – or receptors – the activity of which involves single-domain magnetite crystals. It is estimated that, for a receptor to reach the sensitivity threshold necessary to intercept, for instance small fluctuations of magnetic field it is necessary to integrate signals emitted by numerous (a few millions) receptor cells (Kirschvink & Walker, 1985; Kirschvink *et al.*, 2001). New understanding on the structure and function of magnetic sense in vertebrates was cast by research on fish, namely on the rainbow trout. The research involved studying behavioural responses: three young rainbow trout were individually trained to swim and strike a target to be fed. In addition, electrophysiological responses to magnetic field were recorded and an area in the rainbow trout nose that most probably houses magnetoreceptor cells was located (Walker *et al.*, 1997).

A confocal microscope-aided study on rainbow trout made it possible to develop a three-dimensional model of a magnetite-based sensory cell (Diebel *et al.*, 2000). The magnetoreceptor cells occur in the nose, in the basal lamina of the olfactory epithelium. Discussion is in progress on how information from the external magnetic field can be received by magnetite crystal chains in cells. It is assumed that the location of the crystals in each cell is indicative of possible mechanical connections between the crystals and ion channels in the membranes.

According to a detailed conceptual structure of a single receptor capable of receiving and determining the direction and intensity of magnetic field, a chain of single-domain magnetite particles such as those found in the rainbow trout (Walker *et al.*, 1997; Diebel *et al.*, 2000) is connected by microtubule-like strands with a number of ion channels in the receptor cell membrane. The channels may be thus mechanically opened or closed by the chain, whereby magnetic field-induced changes of its location are transferred. Opening or closure of the transmembrane ion channels alters the cell's membrane potential (Kirschvink *et al.*, 2001; Walker *et al.*, 2002). It may be also assumed that the biophysical mechanism of stimulus transmission in the nervous system that has been produced is shared by the bacteria and the higher animal taxa (Kirschvink *et al.*, 2001).

Discussion on changes in cell membrane permeability is tightly connected with a quest for explanation of mechanisms of the responses observed in the fish embryos and larvae exposed to magnetic fields. Changes in membrane permeability may modify a number of metabolic processes of a cell (an oocyte in this case) during embryonic development. Ionic transfer is associated with changes in contractile structures, which may translate into responses observed in embryonic movements in the egg cell. Directional responses of embryos developing in the egg seems to be coupled with a slow movement of the ectoplasm, because the early embryos – as opposed to those at the final stage of embryogenesis – perform no independent movement themselves. The locomotor activity of the circulation system of fish embryos and larvae, the latter with the nervous and exteroreception system more advanced than in the first, shows differences in circulation system's responses to magnetic field and seems to be related to magnetic field-induced changes in ion transfer through cell membranes. It may be justified to consider effects of magnetic field on chemical reactions, particularly that some of them – those that involve radical pair mechanism – are affected by a magnetic field of 1 mT (Taraban *et al.*, 1997). The magnetic field generated in studies on early developmental stages was frequently superimposed on the geomagnetic field, thereby increasing the overall field value. This seems to have affected metabolic processes and responses of embryos and larvae. Directional responses of fish larvae placed in magnetic field suggest the involvement of a magnetic material, magnetite chains, in ion flow through cell membranes, which is translated into behavioural responses.

FUTURE PERSPECTIVES

The magnetic sense, similar to other known senses, has been most probably evolving and has been subjected to the laws identical to those governing every other genetically controlled biological process (Kirschvink *et al.*, 2001). To continue this thread of thought, one may assume that magnetoreception could have evolved as the initial sensory system. This assumption is supported by the presence of magnetosomes and magnetosome chain structures in carbonate blebs of ALH 84001 Martian meteorite (Thomas-Keprta *et al.*, 2000, Friedmann *et al.*, 2001; Kirschvink *et al.*, 2001) which date back to 4 billion years ago.

The results obtained in various studies unequivocally demonstrate the specific ubiquity of magnetic field effects on metabolic processes and directional responses in early ontogenetic stages of fish, at numerous levels of organisation and morphological integration of internal structures.

As shown by the survey of relevant scientific achievements, many manifestations of magnetic field effects on fish have already been understood. However, the great complexity of a living organism and its life dynamics do not favour specific and precise conclusions as to the mechanism of the effects, despite the very good knowledge on the physical and behavioural aspects of the problem. There are still numerous questions that need to be addressed and many more experiments that have to be performed before the laws governing magnetic field effects on fish larvae can be unravelled and accurately described.

References

Baker RR (1987) Human navigation and magnetoreception: the Manchester experiments do replicate. Anim Behav 35: 691-704

Bauer GB, Fuller M, Perry A, Dunn JR, Zoeger J (1985) Magnetoreception and biomineralization of magnetite in cetaceans. In: Kirschvink JL, Jones DS, McFadden BJ (eds) Magnetite Biomineralization and Magnetoreception in Organisms. Plenum Press, New York, p 489-507

Blakemore RP (1975) Magnetotactic bacteria. Science 19: 377-379

Blakemore RP (1982) Magnetotactic bacteria. Ann Rev Microbiol 36: 217-238

Blakemore RP, Frankel RB, Kalmijn AJ (1980) South-seeking magnetotactic bacteria in the Southern hemisphere. Nature 286: 384-385

Blank M (ed) (1993) Electricity and Magnetism in Biology and Medicine. San Francisco Press, San Francisco.

Branover GG, Vasil'yev AS, Gleizer SI, Tsinober AB (1971) A study of the behavior of the eel in natural and artificial magnetic field and an analysis of its reception mechanisms. J Ichthyol 11: 608-614

Chew GL, Brown GE (1989) Orientation of rainbow trout (*Salmo gairdneri*) in normal and null magnetic fields. Can J Zool 67: 641-643

Deutschlander ME, Phillips JB, Borland SC (1999) The case for light-dependent magnetic orientation in animals. J Exp Biol 202: 891-908

Diebel CE, Proksch R, Green CR, Nielson P, Walker MM (2000) Magnetite defines a magnetoreceptor. Nature 406: 299-302

Edwards HH, Schnell GD, DuBois RL, Hutchison VH (1992) Natural and induced remanent magnetism in birds. Auk 109: 43-56

Formicki K (1992) Respiratory movements of trout (*Salmo trutta* L.) larvae during exposure to magnetic field. Acta Ichthyol Piscat 22, 2: 149-154

Formicki K, Winnicki A (1996) Effects of constant magnetic field on cardiac muscle activity in fish embryos. Publ Espec Inst Espanol Oceanogr 21: 287-292

Formicki K, Winnicki A (1998) Reaction of fish embryos and larvae to constant magnetic fields. Ital J Zool 65 Suppl: 479-482

Formicki K, Bonislawska M, Jasinski M (1997) Spatial orientation of trout (*Salmo trutta* L.) and rainbow trout (*Oncorhynchus mykiss* Walb.) embryos in natural and artificial magnetic field. Acta Ichthyol Piscat 27: 29-40

Formicki K, Tanski A, Winnicki A (2001) Axis of symmetry orientation in trout (*Salmo trutta* L.) embryos exposed to constant magnetic field: emergence of the process. 2001 Asia-Pacific Radio Science Conference, AP–RASC'01, Chuo University, Tokyo, Japan, Aug 1–4, 2001, Conference Digest: 407

Formicki K, Sadowski M, Tanski A, Korzelecka-Orkisz A, Winnicki A (2004a) Behaviour of trout (*Salmo trutta* L.) larvae and fry in a constant magnetic field. J Appl Ichthyol 20: 290-294

Formicki K, Tanski A, Sadowski M, Winnicki A (2004b) Effects of magnetic fields on fyke net performance. J Appl Ichthyol 20: 402-406

Frankel RB, Blakemore RP, Wolfe RS (1979) Magnetite in freshwater magnetotactic bacteria. Science 203: 1355-1356

Friedmannn IE, Wierzchos J, Ascaso C, Winklhofer M (2001) Chains of magnetite crystals in the meteorite ALH84001: Evidence of biological origin. Proc Natl Acad Sci USA 98: 2176-2181

Fuller M, Dobson J, Wieser HG, Moser S (1995) On the sensitivity of the human brain to magnetic fields: evocation of epileptiform activity. Brain Res Bull 36: 155-159

Hanson M, Westerberg H (1987) Occurrence of magnetic material in teleosts. Comp Biochem Physiol 86A: 169-172

Henbest KB, Kukura P, Rodgers CT, Hore PJ, Timmel CR (2004) Radio frequency magnetic field effects on a radical recombination reaction: a diagnostic test for the radical pair mechanism. J Am Chem Soc 126: 8102-8103

Johnsen S, Lohmann KJ (2005) The physics and neurobiology of magnetoreception. Nature Rev Neurosci 6: 703-712

Jones DS, McFadden BJ (1982) Induced magnetization in the monarch butterfly, *Danaus plexippus* (Insecta, Lepidoptera). J Exp Biol 96: 1-9

Kalmijn AJ (1974) The detection of electric fields from inanimate and animate sources other than electric organs. In: Fessard A (ed) Handbook of Sensory Physiology III, 3: Electroreceptors and Other Specialized Receptors in Lower Vertebrates. Springer-Verlag, Berlin, p 147-200

Kalmijn AJ (1978) Experimental evidence of geomagnetic orientation in elasmobranch fishes. In: Schmidt-Koenig K, Keeton WT (eds) Animal Migration, Navigation, and Homing. Springer-Verlag, Berlin, p 347-353

Kalmijn AJ (1981) Biophysics of geomagnetic field detection. IEEE Trans Magn 1: 1113-1124

Kalmijn AJ (1982) Electric and magnetic field detection in elasmobranch fishes. Science 218: 916-918

Kalmijn AJ (1984) Theory of electromagnetic orientation: a further analysis. In: Bolis L, Keynes RC, Maddrell SH (eds) Comparative Physiology of Sensory Systems. Cambridge University Press, Cambridge, p 525-560

Kalmijn AJ, Blakemore RP (1978) The magnetic behavior of mud bacteria. In: Schmidt-Koenig K, Keeton WT (eds) Animal Migration, Navigation, and Homing. Springer-Verlag, Berlin, p 354-355

Karlsson L (1985) Behavioural responses of European silver eel (*Anguilla anguilla*) to the geomagnetic field. Helgol Wiss Meeresunters 39: 71-81

Kirshvink JL (1980) South-seeking magnetic bacteria. J Exp Biol 86: 345-347

Kirshvink JL (1982) Birds, bees and magnetism: a new look at the old problem of magnetoreception. Trends Neurosci 5: 160-167

Kirshvink J (1997) Homing in on vertebrates. Nature 390: 339-340

Kirshvink JL, Gould JL (1981) Biogenic magnetite as a basis for magnetic field detection in animals. Biosystems 13: 181-201

Kirshvink JL, Walker MM (1985) Particle-size consideration for magnetite based magnetoreceptors. In: Kirschvink JL, Jones DS, McFadden BJ (eds) Magnetite Biomineralization and Magnetoreception in Organisms. Plenum Press, New York, p 243-256

Kirshvink JL, Jones DS, McFadden BJ (eds) (1985) Magnetite Biomineralization and Magnetoreception in Organisms. A New Biomagnetism. Plenum Press, New York

Kirshvink JL, Walker MM, Diebel CE (2001) Magnetite-based magnetoreception. Curr Opin Neurobiol 11: 462-467

Klimley AP (1993) Highly directional swimming by scalloped hammerhead sharks (*Sphyrna lewini*), and subsurface irradiance, temperature, bathymetry, and geomagnetic field. Mar Biol 117: 1-22

Korzelecka A (1999) Embryonic locomotor activity in teleost fishes. PhD Thesis, University of Agriculture in Szczecin [in Polish]

Leask MJM (1977) A physiochemical mechanism for magnetic field detection by migratory birds and homing pigeon. Nature 267: 144

Lissmann HW (1958) On the function and evolution of electric organs in fish. J Exp Biol 35: 156-191

Lohmann KJ (1984) Magnetic remanence in the Western Atlantic spiny lobster, *Panulirus argus*. J Exp Biol 113: 29-41

Lohmann KJ, Johnsen S (2000) The neurobiology of magnetoreception in vertebrate animals. Trends Neurosci 23: 153-159

Lowenstam HA (1962) Magnetite in denticle capping in recent chitons (Polypacophora). Geol Soc Am Bull 73: 435-438

Mann S, Sparks NHC, Walker MM, Kirschvink JL (1988) Ultrastructure, morphology and organization in biogenic magnetite from sockeye salmon, *Oncorhynchus nerka*: implications for magnetoreception. J Exp Biol 140: 35-49

Meyer CG, Holland KN, Papastamatiou YP (2005) Sharks can detect changes in the geomagnetic field. J R Soc Interface 2: 129-130

Moore A, Freake SM, Thomas IM (1980) Magnetic particles in the lateral line of the Atlantic salmon (*Salmo salar* L.). Phil Trans R Soc B 329: 11-15

Nishi T, Kawamura G, Matsumoto K (2004) Magnetic sense in the Japanese eel, *Anguilla japonica*, as determined by conditioning and electrocardiography. J Exp Biol 207: 2965-2970

Nishi T, Kawamura G, Sannomiya S (2005) Anosmic Japanese eel *Anguilla japonica* can no longer detect magnetic field. Fish Sci 71: 101-106

Ogura M, Kato M, Arai N, Sasada T, Sakaki Y (1992) Magnetic particles in chum salmon (*Oncorhynchus keta*): Extraction and transmission electron microscopy. Can J Zool 70, 5: 874-877

Phillips JB, Deutschlander ME, Freake MJ, Borland SC (2001) The role of extraocular photoreceptors in newt magnetic compass orientation: parallels between light-dependent magnetoreception and polarized light detection in vertebrates. J Exp Biol 204: 2543-2552

Purcell EM (1985) Electricity and Magnetism. McGraw Hill, New York

Quinn TP (1980) Evidence of celestial and magnetic compass orientation in lake migrating sockeye salmon fry. J Comp Physiol A137: 243-248

Quinn TP, Brannon EL (1982) The use of celestial and magnetic cues by orienting sockeye salmon smolts. J Comp Physiol A147: 547-552

Quinn TP, Groot C (1983) Orientation of chum salmon (*Oncorhynchus keta*) after internal and external magnetic field alteration. Can J Fish Aquat Sci 40: 1598-1606

Ritz T, Adem S, Schulten K (2000) A model for vision-based magnetoreception in birds. Biophys J 78: 707-718

Rosenblum B, Jungerman RL, Longfellow L (1985) Limits to induction-based magnetoreception. In: Kirschvink JL, Jones DS, McFadden BJ (eds) Magnetite Biomineralization and Magnetoreception in Organisms. A new biomagnetism. Plenum Press, New York, p 223-232

Sakaki Y, Motomiya T, Kato M, Ogura M (1990) Possible mechanism of biomagnetic sense organ extracted from sockeye salmon. IEEE Trans Magn 26: 1554-1556

Schuler D, Frankel RB (1999) Bacterial magnetosomes; microbiology, biomineralization and biotechnological applications. Appl Microbiol Biotechnol 52: 464-473

Shulten K, Windemuth A (1986) Model for a physiological magnetic compass. In: Maret G, Kiepnhauer J, Boccara N (eds) Biophysical Effects of Steady Magnetic Fields. Springer-Verlag, Berlin, p 99-106

Sherman K (1992) Classical conditioning to magnetic fields of *Apteronotus leptorhynchus*. MSc Thesis, State University of New York, Geneseo, USA

Souza JJ, Poluhowich JJ, Guerra RJ (1988) Orientation responses of American eels, *Anguilla rostrata*, to varying magnetic fields. Comp Biochem Physiol 90A: 57-61

Tanski A (2002) Orientation of fish embryos developing in magnetic field. PhD thesis, Agricultural University of Szczecin [in Polish]

Tanski A, Formicki K, Korzelecka-Orkisz A, Winnicki A (2005) Spatial orientation of fish embryos in magnetic field. Electr J Ichthyol 1: 21-34

Tanski A, Korzelecka A, Bonisławska M, Winnicki A, Formicki K (2000) New data on morphomechanical changes during embryogenesis of pike (*Esox lucius* L.). Folia Univ Agric Stetin 214, Piscaria 27: 207-213

Taraban MB, Leshina TV, Anderson MA, Grissom CB (1997) Magnetic field dependence of electron transfer and the role of electron spin in heme enzymes: horseradish peroxidase. J Am Chem Soc 119: 5768-5769

Taylor PB (1986) Experimental evidence for geomagnetic orientation in juvenile salmon, *Oncorhynchus tschawytscha*. J Fish Biol 28: 607-624

Taylor PB (1987) Experimental evidence for juvenile chinook salmon, *Oncorhynchus tschawytscha* Walbaum: orientation at night and in sunlight after a 7° change in latitude. J Fish Biol 31: 89-111

Tenforde TS (1979) Magnetic Field Effect in Biological Systems. Plenum Press, New York

Tesch FW (1974a) Influence of geomagnetism and salinity on the directional choices of eels. Helgol Wiss Meeresunters 26: 382-395

Tesch FW (1974b) Speed and direction of silver and yellow eels, *Anguilla anguilla*, released and tracked in the open North Sea. Ber Dtsch Wiss Komm Meerforsch 23: 181-197

Tesch FW (1978) Horizontal and vertical swimming of eels during the spawning migration at the edge of the continental shelf. In: Schmidt-Koenig K, Keeton WT (eds) Animal Migration, Navigation, and Homing. Springer-Verlag, Berlin, p 378-391

Tesch FW (1980) Migratory performance and environmental evidence of orientation In: Ali MA (ed) Environmental Physiology of Fishes. NATO Adv Study Inst Ser A Life Sci 35: 589-612

Tesch FW, Lelek A (1973a) An evaluation of the directional choice in the eel, in captivity. Arch Fischereiwiss 24: 237-251

Tesch FW, Lelek A (1973b) Directional behaviour of transplanted stationary and migratory forms of the eel, *Anguilla anguilla*, in a circular tank. Neth J Sea Res 7: 46-52

Tesch FW, Wendt T, Karlsson L (1992) Influence of geomagnetism and salinity on orientation of the eel *Anguilla anguilla* L., as evident from laboratory experiments. Ecol Freshw Fish 1: 52-60

Thalau P, Ritz T, Stapput K, Wiltschko R, Wiltschko W (2005) Magnetic compass orientation of migratory birds in presence of a 1.315 MHz oscillating field. Naturwissenschaften 92: 86-90

Thomas-Keprta KL, Bazylinski DA, Kirschvink JL, Clemett SJ, McKay DS, Wentworth SJ, Vali H, Gibson EK Jr, Romanek CS (2000) Elongated prismatic magnetite crystals in ALH84001 carbonate globules : potential martian magnetofossils. Geochim Cosmochim Acta 64: 4049-4081

Walcott C (1978) Anomalies in earth's magnetic field increase the scatter of pigeons' vanishing bearings. In: Schmid-Koening K, Keeton W (eds) Animal Migration, Navigation and Homing. Springer-Verlag, Berlin, p 143-157

Walker MM (1984) Learned magnetic field discrimination in yellowfin tuna, *Thunnus albacares*. J Comp Physiol A155: 673-679

Walker MM, Kirschvink JL, Chang S-BR, Dizon AE (1984) A candidate magnetic sense organ in the yellowfin tuna, *Thunnus albacares*. Science 224: 751-753

Walker MM, Kirschvink JL, Dizon AE (1985) Magnetoreception and biomineralization of magnetite in fish. In: Kirschvink JL, Jones WJ, McFadden BJ (eds) Magnetite Biomineralization and Magnetoreception in Organisms. Plenum Press, New York, p 417-497

Walker MM, Diebel CE, Haugh CV, Pankhurst PM, Montgomery JC, Green CR (1997) Structure and function of the vertebrate magnetic sense. Nature 390: 371-376

Walker MM, Dennis TE, Kirschvink JL (2002) The magnetic sense and its use in long-distance navigation by animals. Curr Opin Neurobiol 12: 735-744

Wiltschko R, Wiltschko W (1995) Magnetic Orientation in Animals. Springer-Verlag, Berlin

Wiltschko R, Wiltschko W (2005) Magnetic orientation and magnetoreception in birds and other animals. J Comp Physiol A191: 675-693

Wiltschko W, Traudt J, Güntürkün O, Prior H, Wiltschko R (2002) Lateralization of magnetic compass orientation in a migratory bird. Nature 419: 467-470

Wiltschko W, Munro U, Ford H, Wiltschko R (2003) Lateralisation of magnetic compass orientation in silvereyes, *Zosterops lataeralis*. Aust J Zool 51: 1-6

Winnicki A, Formicki K (1990) Effect of constant magnetic field on myocardium activity in larvae of trout (*Salmo trutta* L.). B Pol Acad Sci-Biol Sci 38: 57-60

Winnicki A, Formicki K, Korzelecka A, Sobocinski A (1993) Cardiac responses of pike (*Esox lucius* L.) embryos and larvae to constant magnetic field. Arch Pol Fish 1: 87-93

Winnicki A, Korzelecka-Orkisz A, Sobocinski A, Tanski A, Formicki K (2004) Effects of the magnetic field on different forms of embryonic locomotor activity of northern pike, *Esox lucius* L. Acta Ichthyol Piscat 34, 2: 193-203

Zoeger J, Dunn JR, Fuller M (1981) Magnetic material in head of the common Pacific dolphin. Science 213: 892-894

PART 5

Movement

CHAPTER

15

Buoyancy

John J Govoni[1,*] and Richard B Forward Jr[2]

INTRODUCTION

Buoyancy and the physiology of buoyancy compensation of adult fishes is well studied. Numerous reviews have detailed the specific gravity of adult fishes (Alexander, 1966) and the mechanisms employed in buoyancy compensation (Harden Jones & Marshall, 1953; Fänge, 1966; Steen, 1970; Neighbors & Nafpaktitis, 1982; Pelster, 1997). A synthesis of these reviews indicates that the principal function of the swimbladder is buoyancy compensation and the achievement of hydrostatic equilibrium or neutral buoyancy.

Studies of the specific gravity of fish eggs and larvae, and of the development of the mechanisms and structures that effect buoyancy compensation are spread over a wide array of literature, as well as the phylogenetic myriad of fishes. A synthesis for eggs (Craik & Harvey, 1987) and a review of swimbladder development over the entire early life of fishes (Pelster, 2004) have emerged, but there is no overall composite synthesis for eggs, yolk-sac larvae and larvae after yolk absorption.

Authors' addresses: [1] US Department of Commerce, National Oceanic and Atmospheric Administration, National Ocean Service, National Centers for Coastal Ocean Science, Center for Coastal Fisheries and Habitat Research, 101 Pivers Island Road, Beaufort, NC 28516-9727, USA.
[2] Duke University, Nicholas School of the Environment and Earth Sciences, Marine Laboratory, 135 Duke Marine Lab Road, Beaufort, NC 28516-9721, USA.
**Corresponding author:* E-mail: Jeff.Govoni@noaa.gov

Developing fishes are composed of structural materials that are denser than water: principally carbohydrates (chondromucin of cartilage), proteins (muscle), and in developmentally advanced larvae, metallic apatite (bone and calcified cartilage). Unless immersed in hyper-saline water, eggs (Sundby, 1991), yolk-sac larvae, and larvae (Blaxter & Ehrlich, 1973) sink. Other compositional and structural components of eggs and larvae that are less dense than water provide lift to counteract negative buoyancy (although often used, an oxymoron). These buoyant components are water (in seawater) and lipids of some eggs and yolk-sac larvae (Craik & Harvey, 1987; Hyslop & Bell, 1987; Riis-Vestergaard, 2002) and a gas-filled swimbladder (Pelster, 2004).

The composition of eggs, their embryos, and larvae change with development and so does their buoyancy. The water content of eggs and larvae declines with development in marine fishes. Lipids within the yolk, and oil globule, are catabolised during yolk and oil globule absorption. After yolk and oil globule absorption, the chief mechanism of buoyancy compensation is the swimbladder, a unique osteichthyan adaptation (Berenbrink *et al.*, 2005), yet the swimbladder develops at different times and at different rates, and functions differently, among fishes. Buoyancy of eggs and yolk-sac larvae is achieved through passive physiological mechanisms (constituent compounds); buoyancy of larvae is achieved through active mechanisms (swimbladder inflation and deflation) and vertical swimming.

Teleostean fishes are a highly diverse class of vertebrates and the development of fishes, including the structures and attendant physiological mechanisms that effect buoyancy compensation, develop differently, and with different timing (Blaxter & Ehrlich, 1973; Pelster, 2004) among and within phylogenetic groupings. This diversity of development translates into differences in the specific gravity of eggs and larvae, as well as to differences in the achievement of buoyancy compensation. Here, the physiology of buoyancy of developing osteichthyan fishes are considered, excluding fishes of the Subclass Sarcopterygii and the Orders Polypteriformes, Acipensiformes, Lepidosteiformes, Amiiformes, and Osteoglossiformes. The eggs of these fishes are demersal and not buoyant. Insofar as is known, the swimbladder of the larvae of air breathing fishes among this subclass and these orders, functions in buoyancy compensation, rather than in oxygen acquisition, though developmental progress of many, but not all, organ systems toward its adult state, and the function of an embryonic and larval organ is a continuum until metamorphosis (Govoni, 2004a, b). Among those teleostean fishes that aestivate (e.g. Zheng & Lui, 1988), mammalian-like alveoli do not appear in the swimbladder until late in development (e.g. Lui,1993).

BUOYANCY AND THE SPECIFIC GRAVITY OF EGGS AND LARVAE

Physical Principles of Buoyancy

According to Archimedes' principle, an egg or a larva that is partially or completely submerged in water will be buoyed by a force equal to the weight or mass of the water displaced by the egg or larva. For eggs and larvae, Archimedes principle is formulated as:

$$F = W = \left\{ \frac{\rho_{HOH}}{\rho_{component}} - 1 \right\}$$

where F is the buoyant force in dynes; W is the mass of a component of buoyancy (g); ρ_{HOH} is the density of ambient water ($g \cdot cm^{-3}$); and $\rho_{component}$ is the density of the component (Craik & Harvey, 1987). Elaboration of this formula for multiple components is available (e.g. Thorsen *et al.*, 1996; Riis-Vestergaard, 2002).

Since the upward or buoyant force on an object, is equal to the weight of the fluid displaced, the buoyant force depends upon the density of the object which is defined as the weight to volume ratio ($g \cdot cm^{-3}$: $kg \cdot m^{-3}$). Densities of different materials in water are related by calculating specific gravity, which is defined as the ratio of the weight of a body to the weight of an equal volume in pure water at 4°C with a specific gravity of 1 $g \cdot cm^{-3}$. When using these units, density is numerically equivalent to specific gravity. Since specific gravity of seawater increases as the salinity of ocean water increases (e.g. specific gravity at a salinity of 35 ppt and a temperature of 4°C is 1.0280 $g \cdot cm^{-3}$), the buoyant force will also increase as salinity increases, although as a ratio, specific gravity is dimensionless. At a fixed salinity, density decreases as temperature increases. Temperature effects upon specific gravity and density depend upon the salinity of the water. In pure water maximum density occurs at 4°C. As the salinity increases, the temperature of maximum density and specific gravity declines. Thus, in the ocean, colder water will reside under warmer water, and in an area with a decrease in temperature with depth, the buoyancy of the water will increase with depth.

Passive vertical movements owing to buoyancy change can be avoided if an egg or larva is neutrally buoyant and has a specific gravity equal to the surrounding water. Alternatively, passive upward movement or sinking implies that the overall specific gravity of an egg or larva is greater or lesser than the surrounding water. There are several methods used in determining the density or specific gravity of animals in water based on these principles. Coombs (1981) developed a method of measuring directly the specific gravity of eggs by immersion in a density-gradient column of seawater and determined specific gravity from the position of neutral buoyancy. The more sophisticated method of Power *et al.* (1991) weighed larvae in waters with identical osmolality, but different densities. Mass and volume were then determined by applying Archimedes' principle. Using these values, body density and weight in water were calculated. Power & Walsh (1992) successfully used this method to determine the relationship between growth of Atlantic menhaden (*Brevoortia tyrannus*) larvae and buoyancy.

The specific gravity, of eggs, yolk-sac larvae, and larvae has been determined in only a few species (Table 1), largely because of the difficulty and inherent errors of measurement (Power, 1996). The general trend is for specific gravity to increase with age.

Table 1 Measured, whole organism estimates of specific gravity of some teleostean fishes.

Stage	*Spawning habitat*	*Egg/Larva type*	*Species common name*	*Specific gravity* Initial[a]	Final[b]	Point estimate[c]	*Source*
Egg	Marine	Pelagic	blue whiting	1.0260	1.0306		Coombs *et al.* (1981)
	Brackish	Pelagic	Atlantic cod	1.011	1.020		Thorsen *et al.* (1996)
	Marine	Pelagic	"	1.0242			Sclafani *et al.* (1997)
	Marine	Pelagic	"	1.026			Riis-Vestergaard (2002)
	Marine	Pelagic	"			1.0306	Ådlandsvik *et al.* (2001)
	Marine	Pelagic	Atlantic mackerel	1.0205	1.0245		Coombs (1981)
	Marine	Pelagic	red porgy	1.0200	1.0242		Kitajima *et al.* (1993)
	Marine	Pelagic	black seabream	1.024			Kitajima *et al.* (1994a)
	Marine	Pelagic	Japanese flounder	1.022			Kitajima *et al.* (1994b)
Yolk-sac larva	Freshwater[d]	Pelagic[d]	ayu	1.038	1.032		Kitajima *et al.* (1998)
	Marine	Pelagic	black seabream	1.024	1.028		Kitajima *et al.* (1994a)
	Marine	Pelagic	blue whiting	1.03656			Ådlandsvik *et al.* (2001)
	Marine	Pelagic	Japanese flounder		1.028		Kitajima *et al.* (1994b)
Larva	Freshwater	Demersal	ayu			1.050	Kitajima *et al.* (1998)
	Marine	Pelagic	Atlantic menhaden			1.021	Power & Walsh (1992)
	Marine	Pelagic	red porgy			1.028	Kitajima *et al.* (1993)
	Marine	Pelagic	striped trumpeter	1.0245[e]	1.0260[e]	1.0265[f]	Trotter *et al.* (2005)
	Marine	Pelagic	black seabream			1.028	Kitajima *et al.* (1994a)
	Marine	Pelagic	Japanese flounder			1.060	Kitajima *et al.* (1994b)

[a] Measured at the beginning of the stage of development.
[b] Measured at the end of the stage of development.
[c] Single estimate measured within the stage of development.
[d] Anadromous species.
[e] Species lacking swimbladder or larva before swimbladder inflation.
[f] After swimbladder inflation.

PHYSIOLOGY OF BUOYANCY

Various terminologies have been proposed and employed for developing fishes (Balon, 1975; Kendall *et al.*, 1984; Urho, 2002). Here, the generic term egg refers to all egg and embryonic developmental stages until hatching. The term chorion, refers to the egg envelope (Laale, 1980). At hatching, a yolk-sac larva emerges and remains in this state until yolk and oil-globule (if present) are depleted through transformation of carbohydrates, lipids, and proteins via anabolic pathways (e.g. cartilage formation or protein synthesis), or through catabolism. After yolk and oil-globule depletion, a larva lives as a free swimming and feeding, yet developing form, until metamorphosis. Development proceeding throughout the embryonic, yolk-sac larval, and larval stages, through hatching and metamorphosis, are typically events that are concomitant with major changes in organ systems associated with changing ecological and physiological requirements.

Buoyancy of whole organisms varies among taxa and spawning habitat (Table 1). Almost all freshwater eggs and larvae are demersal, though a few taxa have pelagic propagules; most marine eggs and larvae have pelagic propagules, though, some are demersal (Balon, 1975; Løning *et al.*, 1988). Some of the eggs of the minnows (Cyprinidae) in large Asian Rivers may be neutrally buoyant and pelagic in non-laminar turbulent water flows (Balon, 1975) as are the eggs of the river herrings (Alosidae) and the anadromous basses (Moronidae), but the eggs of the freshwater drum *(Aplodinotus grunniens)* (Davis, 1959) and the goldeye, *(Hiodon alosoides)* (Battle & Sprules, 1960) are innately buoyant and pelagic. Pelagic eggs and larvae in freshwater are buoyant via the same physiological mechanisms used by larvae in estuarine and marine waters, but without the advantage of the higher relative specific gravity of ambient seawater (Craik & Harvey, 1987). Demersal eggs in freshwater, estuarine, and marine habitats are often adhesive, or have tendrils for attachment to the substrate, while the larvae that hatch from demersal eggs often have morphological holdfasts (Balon, 1975).

Oocytes, Eggs, and Embryos

Developmental events within the ovary that prepare eggs for release into ambient water influence egg buoyancy. In general, demersal eggs have lower water content than do pelagic eggs, whereas pelagic eggs have higher water and lipid contents, which provide buoyancy, depending on the specific gravity of the water in which they are immersed (Fig. 1). Some demersal eggs in both freshwater and seawater, have high levels of lipids, but have greater specific gravity overall than ambient water. In the ocean, eggs rise either to the surface, or to strata of neutral buoyancy depending upon salinity (Solemdal, 1967; Kendall & Kim, 1989; Sundby, 1991; Nissling *et al.*, 1994). Proteins and large, structural, carbohydrates with high specific gravity force eggs to sink. Sinking (Sundby, 1991; Ådlandsvik *et al.*, 2001), as well as ascension (Cambalik *et al.*, 1998), is reduced by drag. The high water content, with water of lower specific gravity than that of ambient seawater, provides most of the buoyancy of pelagic marine eggs (Craik & Harvey, 1987).

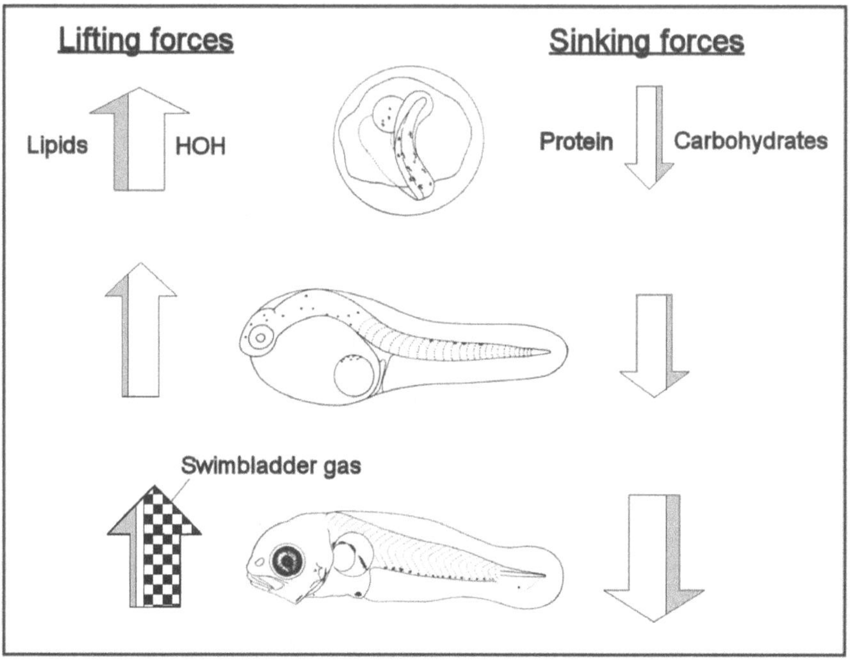

Fig. 1 Schematic diagram of the contributors to buoyancy of fish eggs, yolk-sac larvae, and larvae. Line drawings are of spot (*Leiostomus xanthurus*) from Powell & Gordy (1980). Relative sizes of arrow compartments are from Govoni *et al.* (1982) and Govoni and Hoss (2001). Size of arrow, and partitions within arrows, depict relative lift or sinking forces. For lifting forces, shaded areas depict lipids, clear areas water (HOH), and checkered areas gas; for sinking forces, clear areas indicate proteins and shaded areas carbohydrates.

The high water content of pelagic eggs arises from hydration of oocytes in the ovary. Hydration of oocytes is effected by an osmotic gradient set up by a high concentration of free amino acids (Fyhn *et al.*, 1999; Finn *et al.*, 2002a, b). The high free amino acid (FAA) concentration of eggs results from the hydrolysis of vitellogenin, a large lipo-glyco-phospho-protein, produced by the liver and released into the blood (Wallace, 1985; Greeley *et al.*, 1986; Mommsen & Walsh 1988). Following incorporation into the oocytes, vitellogenin is cleaved into a complex suite of yolk proteins (Patiño & Sullivan, 2002; LaFleur *et al.*, 2005; Finn, 2007a). During oocyte maturation in marine teleosts, the yolk proteins are differentially hydrolysed to render a pool of FAA (Thorsen & Fyhn, 1996; Matsubara *et al.*, 1999; Finn *et al.*, 2002a, b; Cerdà *et al.*, 2007; Finn, 2007b; Finn & Kristoffersen, 2007; Kolarevic *et al.*, 2008). The FAA cause a transient hyper-osmolality of the yolk in relation to the ovarian fluid, and thus drive the influx of oocyte water (Finn *et al.*, 2002a). Physiologically essential inorganic ions, e.g. K^+, Cl^- and P_i, are gathered by the oocytes across the plasma membrane before and during hydration (Alderdice, 1988; Finn *et al.*, 2002a). After hydration, the osmolality of fluid within eggs is roughly the same as that of maternal blood fluids, about

~350 m*Osm* (Finn *et al.*, 2002a). This osmolality is lower than the osmolality of most seawater (Riis-Vestergaard 2002). For example, the osmolality of seawater at 35 ppt is 1035 m*Osm*. Once released from the ovary into the water, i.e. after spawning, the vitelline membrane of the egg is relatively impermeable to water and inorganic ions (Thorsen *et al.*, 1996). After spawning, the chorion becomes less permeable to water and ions as well (Alderdice, 1988), owing largely to a change in chorion structure (Yamagami *et al.*, 1992; Mangor-Jensen & Waiwood, 1995). In seawater, the high FAA concentration precludes additional uptake of heavy, inorganic ions, principally Na^+, K^+, Ca^{2+}, and Mg^{2+} (Thorsen *et al.*, 1996; Finn *et al.*, 2002a). Demersal eggs may or may not exhibit a small buildup of FAAs prior to oocyte hydration and FAAs (Finn *et al.*, 2002b).

The overall permeability of pelagic and demersal oocytes and released eggs is related to the surface to volume ratio (Alderdice, 1988). Overall, egg buoyancy of pelagic eggs is related to the difference in osmolality between the egg and external medium (Nissling *et al.*, 1994).

Some oocyte lipids are derived from vitellogenin, after hydrolysis to lipovitellin (Mandich *et al.*, 2004) and incorporation into the oocyte, but most of the total lipid composition of eggs is derived from phospholipids that are transferred directly from the blood plasma during the previtellogenic stage (Lund *et al.*, 2000). Lipovitellin, phospholipids and neutral lipids are converted to the lipids of the yolk and oil globule, respectively when oocytes are released from ovarian follicules. The principal lipids that contribute to the buoyancy of fish eggs are the neutral lipids, principally the triacylglycerides, and their replacement molecules (50%), and the polar lipids, principally, the sterol esters, cholesterol, phosphatidyl choline, and the alcohols, phosphatidyl ethanol, and phosphatidyl ethanoline (Vetter *et al.*, 1983; Rainuzzo *et al.*, 1992; Finn *et al.*, 1995a, 1996; Lochmann *et al.*, 1995; Rønnestad *et al.*, 1995; Evans *et al.*, 1996).

The concentration of the components that cause eggs to rise (lipids) or sink (proteins and carbohydrates), changes with embryonic development (Blaxter, 1988; Kamler, 1992; Saborido-Rey *et al.*, 2003). Eggs, embryos, and yolk-sac larvae have few carbohydrates (Cetta & Capuzzo, 1982; Govoni *et al.*, 1982; Riis-Vestergaard, 2002), most of which is glycogen (Finn *et al.*, 1995a, 1996; Kunz, 2004). The small amount of carbohydrates, and the high amounts of lipids are catabolised (Riis-Vestergaard, 2002). Some FAAs are used for protein synthesis, to build muscle protein (Johnston & Hall, 2004), but most are catabolised (Finn *et al.*, 1995a, 1996; Ohkubo & Matsubara, 2002). FAAs are catabolised before lipids (Rønnestad *et al.*, 1994, 1995; Finn *et al.*, 1995a, 1996), and continue to remain an important source of energy during the larval stages (Finn *et al.*, 2002c). Most of the lipids reside in the yolk and are typically catabolised before the oil globule (Fyhn & Govoni, 1995). FAAs (~50%) and lipids fuel the catabolism of pelagic fish eggs, while carbohydrates, principally glycogen, exhibit a minor catabolic role during very early embryonic development (Finn *et al.*, 1995b, 1996).

Oil globules do not contribute greatly to buoyancy. The contribution of oil globules to buoyancy is greater among the few freshwater fishes with pelagic eggs than among many marine fishes with pelagic eggs (Craik & Harvey, 1987; Riss-Vestergaard 2002). Buoyancy imparted by the oil globule is indicated by the eggs of Atlantic menhaden as they rise toward the surface (Cambalik *et al.*, 1998) and accumulate there, with the oil-globule oriented upward (Fyhn & Govoni, 1995). Maintenance of an orientation with the oil globule toward the surface may be a gravitational factor in some developmental processes such as epiboly (Kunz, 2004).

Eggs and early embryos lack phylogenetic pattern in morphological features that contribute to buoyancy (Matarese & Sandknop, 1984). The perivitelline space between the plasma membrane and the integument of the developing embryos and the chorion is composed largely of water. Most eggs have narrow perivitelline spaces; whereas, eggs of both primitive (e.g. Clupeiformes, Anguilliformes, and Salmoniformes) and more derived taxa (e.g. cyprinids, moronids and Pleuronectiformes) have wide spaces. Most Gadiformes and Pleuronectiformes lack oil globules, but some species have one or more globules. Differences are apparent even within the same order, e.g. Clupeiformes. Most clupeiforms have oil globule(s), while round herring (*Etrumeus teres*) and the engraulids do not.

Yolk-sac Larvae

Catabolism of lipids continues in yolk-sac larvae (Rønnestad *et al.*, 1994, 1995; Finn *et al.*, 1995a, 1996). As a consequence, specific gravity increases toward the completion of larval development (Fig. 1). Overall, water content declines with development (Blaxter & Ehrlich, 1973). Yolk-sac larvae are, nonetheless, higher in water content than are larvae (Fig.1), but yolk-sac larvae lack full osmoregulatory capacities, because of the lack of fully functional kidneys responsible for water balance (Tytler *et al.*, 1996), and gills (Rombough, 2004), largely responsible for ionic balance. The integument of yolk-sac larvae is thin (Roberts *et al.*, 1973), but is relatively impermeable to water and ions (Tytler & Bell, 1989).

Yolk-sac larvae and larvae after yolk depletion are not in osmotic balance in either fresh- or seawater owing to physiological mediation. In freshwater, yolk-sac larvae are hyper-osmotic; in seawater they are hypo-osmotic. Yolk-sac larvae (and larvae) in freshwater gain water and lose ions passively, because of osmotic and concentration gradients, whereas larvae in seawater lose water and gain ions (Rombough, 2004; Kaneko & Hiroi, this volume). Water and ions pass through the epithelium of gill lamellae and the alimentary canal. Osmoregulation of larvae is integrated among many organs: the integument, gills, kidneys and alimentary canal (Rombough, 2004; Kaneko & Hiroi, this volume), and is dynamic, shifting from the integument to the visceral organs during yolk-sac depletion. Larvae appear to be capable of osmoregulation early-on (Rombough, 2004), with blood and body fluids of approximately 285-350 m*Osm*, three times lower than standard seawater (Tytler & Blaxter, 1988; Tytler *et al.*, 1996). Owing to the retention of physiologically essential, but heavy metal ions (though the titres of these ions are regulated), body fluids of freshwater larvae are less buoyant in freshwater than are the fluids of marine larvae in seawater.

Larvae

Without innate means of buoyancy compensation larvae sink. Drag is related to the surface to volume ratio and influences the rate of sinking, as it does with eggs and yolk-sac larvae. For larvae, drag changes because of changes in shape (Hare *et al.*, 2006), as well as the surface area to volume ratio. The tendency to sink increases as structural carbohydrates (chondromucin), metallic apatite (bone and calcified cartilage), and protein (muscle) are added as larvae feed and grow (Fig. 1). The principal organ that provides buoyancy for larval fishes, after yolk and oil-depletion, is the swimbladder. Larvae use residual lipids and water along with an inflated swimbladder, for buoyancy compensation, as do adult midwater fishes (Neighbors & Nafpaktitis, 1982). The development and inflation of a swimbladder meets the demand of countering the increase in specific gravity (Pelster, 2004).

ONTOGENY AND FUNCTION OF THE SWIMBLADDER

The pervasive pattern of development of the swimbladder throughout the phylogeny of fishes is this organ's origin as a dorsal evagination of the foregut, a process that involves the same germ cell lines that contribute to development of the alimentary canal. These same germ cell lines compose the swimbladder anlagen (Pelster, 2004). Endoderm contributes swimbladder mucosal epithelium (including gas gland tissue) and mesoderm contributes the muscularis and serosa (Kunz, 2004).

While an out-pocketing of the foregut appears to be the consistent mechanism for the swimbladder development, the timing of out-pocketing, i.e. evagination, varies greatly among fishes. Typically, evagination arises in the embryo or yolk-sac larvae of fishes that hatch from large demersal eggs, i.e. fishes with precocial development (Balon, 1981), while it arises at the depletion of the yolk and oil globule if present in fishes that hatch from small, pelagic eggs, i.e. fishes with altricial development (Balon, 1981). For example, evagination of the incipient swimbladder begins in yolk-sac larvae of the batracoidiform fishes, oyster toadfish (*Opsanus tau*) (Fig. 2) and plainfin midshipman (*Porichthys notatus*), fishes with precocial development that spawn large, adherent eggs in nests (Tracy, 1959). Evagination begins at the completion of yolk and oil globule absorption in the larvae of the perciform spot (*Leiostomus xanthurus*), and forms in the larva well after complete yolk and oil globule absorption in the clupeiform Atlantic menhaden (Govoni & Hoss, 2001), both fishes with small pelagic eggs and altricial development.

Following evagination, the swimbladder is typically filled with fluid (Pelster, 2004). The pervasive pattern of initial inflation with gas among fishes with pelagic larvae is by ascension to the surface and gulping of atmospheric gas, or gas bubbles near the surface (Pelster, 2004). The possible inflation of the swimbladder with gas from the digestion of organic material in the gut has been ruled out (Pelster, 2004). For larvae that are not pelagic (e.g. oyster toadfish) and do not have access to the surface, or species wherein the

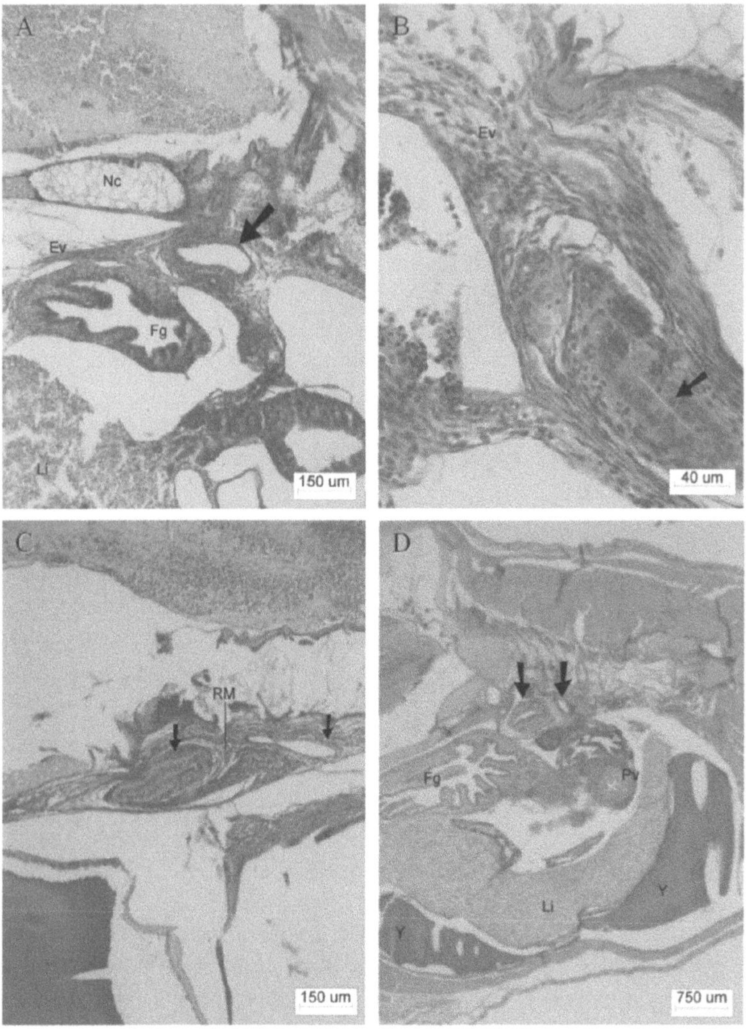

Fig. 2 Development of the swimbladder (arrows) in oyster toadfish (*Opsanus tau*), a batrachoidiform fish with demersal eggs and larvae and precocial with development and a two chambered, gas-inflated swimbladder with discrete gas gland and *retia mirabilia* develops in yolk-sac larvae, well before larvae are released from attachment to a nest. (A) histological section of a 10.0 mm Standard Length (SL) larva one day post hatching (dph) with a fluid-filled swimbladder (arrow) ventrad to the notochord (Nc) and dorsad to the foregut (Fg) and liver (Li), and a ligamentous remnant of connective tissue left from the evagination (Ev) of the swimbladder anlage; (B) section of a 11.5 mm SL larva, 4 dph with a more diminished remnant of evagination; (C) 14.5 mm SL larvae, 9 dph with a two chambered swimbladder and rete mirabilis (RM) with remnant yolk (Y); and (D) a 25.4 mm larvae, 10 dph with a two chambered swimbladder inflated with gas and a remnant yolk (Pv, pyloric valve separating the foregut from the stomach).

pneumatic duct has atrophied, internal gas secretion from the mucosa is evident (Chatain, 1986; Chatain & Ounais-Guschemann, 1990). Surfactants, composed of phosphatidylcholine, facilitate the forcing of a gas bolus to the swimbladder (Pelster, 2004). Surfactants are either produced by the pneumatic duct mucosa (Marty *et al.*, 1995) or membrane-bound laminar bodies in the swimbladder (Daniels *et al.*, 2004), or are secreted by the gall bladder and conveyed to the pneumatic duct by the bile duct (Trotter *et al.*, 2005).

Initial swimbladder inflation typically occurs at the completion of yolk-sac absorption, when inert biochemical means of achieving neutral buoyancy are near exhaustion, or when the density of larvae increases owing to the addition of heavy structural components (Power & Walsh, 1992), such as more dense muscle types (Johnston & Hall, 2004) or bone (Faustino & Power, 2001; Donoghue, 2002; Sire & Huysseune, 2003). Initial swimbladder inflation occurs relatively rapidly within an interval of hours or days (Trotter *et al.*, 2005).

The timing of initial swimbladder inflation during development differs widely among taxa; from the time of yolk-sac and oil globule depletion in spot (Govoni & Hoss, 2001) to a week or so after the depletion of yolk and oil in Atlantic menhaden (Govoni & Hoss, 2001), to several months in the clupeiform lake whitefish (*Coregonus clupeoformis*) (Tait, 1960), to a year or more in anguilliforms (Pelster, 2004). Failure to inflate the swimbladder initially has serious physiological and pathological consequences, such as increased energy expenditure owing to excess swimming for depth maintenance, deformation (most commonly lordosis or other vertebral column abnormities), or death (Pelster, 2004; Trotter *et al.*, 2005).

Though of common embryonic or larval origin, the swimbladder differs in function among larval fishes, and these differences can be partitioned between fishes that develop physostomous swimbladders and those that develop physoclistous swimbladders. All larvae are initially physostomous, insofar as the swimbladder develops as an evagination of the larval foregut (Pelster, 2004). Physostomous larvae maintain this connection with the alimentary canal throughout life; whereas, physoclistous larvae lose this connection as the pneumatic duct that connects to the larval forgut atrophies at some point after inflation of the swimbladder.

Physostomous fishes typically do not develop a discrete gas gland or rete mirabile, though some gas secretory structures do appear in ancillary pneumatic vessels of some clupieform fishes (O'Connell, 1955; Govoni & Hoss, 2001). Larvae of some physostomous fishes ascend to the surface, typically at night, to inflate their swimbladders with atmospheric gas (Hunter & Sanchez, 1976; Hoss *et al.*, 1989; Uotani *et al.*, 2000), then sink gradually as gas diffuses (Lapennas & Schmidt-Nielsen, 1977; Blaxter & Batty, 1984) through the thin-walled larval swimbladder (Govoni & Hoss, 2001), or is vented (Forward *et al.*, 1994). Guanine deposited in the developing swimbladder serosa serves as a diffusive barrier to gas (Denton *et al.*, 1972), in both physostomous and physoclistous larvae (Denton & Blaxter, 1976; Yamada *et al.*, 2004). Multiple diel ascensions are possible (Forward *et al.*, 1994). This mode of inflation and deflation does not then provide for the maintenance of neutral buoyancy.

The morphology of the swimbladder of adult fishes ranges from a simple single chambered vessel to a multi-chambered vessel with lobes or ramifying projections (Tracy, 1920; Harden-Jones & Marshall, 1953; Whitehead & Blaxter, 1989). Typically the adult condition is not developed until metamorphosis, though two chambers develop early-on within the yolk-sac larvae of batrachoidiform fishes, e.g., plainfin midshipman and oyster toadfish (Fig. 2). An organised rete mirabile develops in the larvae of some fishes before metamorphosis (e.g. Tracy, 1920; Fänge & Whittenberg, 1958; Yamashita, 1982), but does not appear in other fishes until after the onset of metamorphosis (Govoni & Hoss, 2001).

For most larval fishes the swimbladder mucosa functions as a gas gland secreting gas into the swimbladder lumen. Secretion is accomplished with the same mechanism as in the gas gland of adult fishes. Gas is salted out of the blood plasma or interstitial fluids of the mucosal epithelium by acidification via lactate and CO_2 production through aerobic and anaerobic metabolic pathways (Morris & Albright, 1984; Walsh & Milligan, 1993; Pelster, 2004). The principal pathway for lactic acid production is glycolysis. Whereas, some larval fishes do not show evidence of glycogen deposits in the mucosa of the swimbladder (Pelster, 2004), glycogen granules are evident in the cell of the mucosa of others (Marty *et al.*, 1995; Govoni & Hoss, 2001).

Inasmuch as most larval fish lack a rete miribile, multiplication of partial pressures of gases is not possible. Nonetheless, gas secretion against appreciable pressure occurs without benefit of a rete miribile (Govoni & Hoss, 2001). Moreover, many fishes with altricial development do not possess appreciable numbers of circulating erythrocytes (Govoni *et al.*, 2005), so the characteristic shifts in haemoglobin oxygen affinity and cooperative binding, the Bohr and Root effects, can not be employed to drive secretion; yet larvae can secrete O_2 into the swimbladder against appreciable pressure gradients (Govoni & Hoss, 2001).

The principal gases secreted into the swimbladder lumen of larval fishes are N_2 and O_2, with CO_2 and some inert gases also contributing minor quantities (Pelster, 2004). The gases that are secreted have proven difficult to assess, because $< \mu l$ quantities of gases can be collected. N_2, CO_2, and O_2 are produced by salting out from the blood plasma and interstitial tissue. Carbon monoxide and CO_2 are the only gases produced in quantities sufficient for swimbladder inflation by metabolic pathways (Pelster, 2004). While some CO_2 might enter the swimbladder lumen immediately following the stimulation of gas secretion by an increase in external pressure owing to a change in depth, CO_2 is rapidly re-absorbed by body fluids (Pelster, 2004). Carbon monoxide is not involved in swimbladder inflation (Pelster, 2004).

The role of NO in facilitating oxygen secretion into the swimbladder of larval fishes is unexplored. The possible role of NO may not simply involve it's influence on cardiac function (Eddy *et al.*, 1999), or its signalling of vasodilation (Haverkamp *et al.*, 1999), but it may involve the influence of NO on Root effect haemoglobins (Bonaventura *et al.*, 1976; Wink *et al.*, 1996) in larvae with abundant erythrocytes and haemoglobin (Govoni *et al.*, 2005). Whereas fish embryos and larvae have different haemoblobins than do adult fishes (e.g. Lancioni *et al.*, 2005), the oxygen affinity and cooperative binding of these haemoglobins changes with development.

Through its primary function of buoyancy compensation, the swimbladder provides a secondary function, energy economy. Energy is saved through the maintenance of hydrostatic equilibrium, while not having to swim to maintain depth (Hunt von Herbing *et al.*, 1996). A similar role is well demonstrated for adult fishes (Lefrançois *et al.*, 2001).

The specific gravity of larvae changes owing to nutrient deprivation (Blaxter & Erhlich, 1973) and consequent disruption of osmoregulation (Holliday & Blaxter, 1960; Sclafani *et al.*, 1997; Moustakas *et al.*, 2004). For example, a lipid deficient diet reduces growth, swimbladder size and survival of walleye pollock (*Theragra chalaogramma*) larvae. When fed lipid deficient prey, these larvae had to swim continuously in the water column to maintain their position (Davis & Olla, 1990). Apparent diel variation in specific gravity, not accounted for by nutrition or osmoregulation, has also been observed (Hare *et al.*, 2006).

Observations of diel variation in specific gravity and swimbladder volume, and attempts to relate this variation to function in vertical positioning in a water column, is plausible for physostomus larvae, but not for larvae that become physoclistus soon after initial swimbladder inflation. For both physostomous and physclistous larvae, vertical positioning is behaviourally mediated; buoyancy compensation to achieve hydrostatic equilibrium or neutral buoyancy at depth is physiologically mediated. Diel changes in specific gravity and in the volume of the swimbladder of larval fishes have been observed in density gradient columns in the laboratory (Kitajima *et al.*, 1993, 1994a, 1998; Trotter, 2005) and these changes have been loosely correlated with buoyancy. These changes imply that the swimbladder of larvae functions to adjust vertical positioning. Diel variation in the specific gravity of eggs and larvae might be attributed to diel variation in light-mediated metabolism that result in different biochemical reaction products (Mangor-Jensen & Waiwood 1995), but it cannot be attributed to changes in swimbladder volume, at least among physoclistous larvae. These diel changes in specific gravity will result in a shift in vertical position, which will in turn result in adjustments to the volume of the swimbladder to maintain neutral buoyancy as depth changes, but cause and effect are not directly linked. As an explicative example of the lack of cause and effect, diel variation in specific gravity of Japanese flounder (*Paralichthys olivaceus*) larvae, has also been observed (Kitajima *et al.*, 1994b), but cannot be explained by variation in swimbladder volume because the Japanese flounder larvae lack a swimbladder (Fukuhara, 1986). The swimbladder of adult fishes does not serve to accomplish vertical migration, and swimbladder volume changes do not underlie changes in vertical positions of larvae. Such a physiologically and morphologically complex organ, would probably not have evolved for a purpose that differs markedly in function between adults and larvae.

Fishes are extraordinarily diverse, and so is the development and function of their swimbladders in vertical positioning. The early ontogeny of herring-like fishes, Clupeiformes, and flatfishes, Pleuronectiformes, make good case studies of the presence and absence of a swimbladder, or the timing of swimbladder development, along with the function of the swimbladder in buoyancy compensation. All Clupeiformes are physostomous and all develop swimbladders as larvae. Yet, the clupeid Atlantic menhaden develops a swimbladder earlier (Govoni & Hoss, 2001) than does round herring (Chen *et al.*, 1992). On average, the larvae of Atlantic menhaden are found shallower in the water column than are round herring (Hare & Govoni, 2005). No systematic survey of the presence or absence of gas bladders in the larvae of pleuronectiform fishes exists, though some flatfishes develop swimbladders as larvae, while others do not. Most larvae of the left-eyed flatfishes, specifically larvae of the family Bothidae develop swimbladders that are resorbed (Ali-Maghazachi & Gibson, 1984) during metamorphosis, but larvae of the family Paralichthyidae reportedly develop small swimbladders that disappear before metamorphosis (Fahay, 1983), while some apparently do not develop swimbladders at all. Bisbal & Bengston (1995) report the development of a transient, vestigial swimbladder that does not inflate in the larval paralichthid, Japanese flounder, though such an organ has not been observed in the larvae of the congeners gulf flounder (*P. albigutta*) (Govoni, unpublished data) or Japanese flounder (Fukuhara, 1986). In a hierarchy scheme of vertical distribution in the ocean (Hare & Govoni, 2005), the eyed flounder (*Bothus* sp.) and dusky flounder larvae (*Syacium papillosum*), fish larvae with swimbladders, are found, on average, shallower in the water than are larvae that lack, or have small swimbladders, e.g. summer, gulf, and southern flounders (*Paralichthys* sp.), and shelf, small mouth, and gray flounders (*Etropus* sp.).

BEHAVIOURAL PHYSIOLOGY OF BUOYANCY

In a marine water column with environmental gradients, usually temperature decreases, salinity increases and hydrostatic pressures increase with depth. The effects on a larva with a swimbladder are that the water density increase with depth owing to the temperature decrease and salinity increase, will result in increased buoyancy as a larva descends. Alternatively, the pressure increase with depth will decrease buoyancy as the volume of the swimbladder in compressed. These considerations potentially lead to larval aggregation at depths in response to hydrostatic pressure.

Behavioural responses of larvae to changes in hydrostatic pressure are poorly studied. The mechanisms for actually detecting a change in hydrostatic pressure, however, are known for different species of larval fish. This sensitivity is gained through distortion of a dividing membrane within the otic bulla, a bulbous, gas-filled, anteriad extension of the swimbladder of the larvae of physostomous clupieforms (Denton & Blaxter, 1976; Hoss & Blaxter, 1979). In physoclistous larvae, this sensitivity may be achieved through priopriocеption owing to the distortion of the swimbladder serosa and consequent stimulation of proprioceptors (Tytler & Blaxter, 1977).

Alternatively, larvae could aggregate at the depth of density discontinuities (the pycnocline), which could result from temperature (thermocline) or salinity (halocline) changes. Field studies are mixed on vertical distributions relative to pycnoclines. Some species of fish larvae aggregate at pycnoclines (Frank *et al.*, 1989; Heath, 1992; Lough

& Potter, 1993); whereas, vertical movements of other species are restricted by thermoclines (Ahlstrom, 1959; Loeb, 1979, 1980; Kendall & Naplin, 1981; Sameoto, 1982; Röpke, 1993; Röpke *et al.*, 1993; Kendall *et al.*, 1994). Alternatively, vertical movements of some species are not restricted by pycnoclines (Sogard *et al.*, 1987; Gray 1993, 1996; Gray & Kingsford, 2003).

Distributions relative to pycnoclines could result from buoyancy changes with depth or from behavioural responses to temperature and salinity changes with depth. By comparison, invertebrate zooplankton have a negative feedback system in which the increase in salinity and decrease in temperature encountered upon descending in the water column evoke an ascent response owing to positive phototaxis, negative geotaxis, and increased activity (Forward, 1976, 1987; Sulkin, 1984). Similarly, they display a descent response upon exposure to the opposite changes in salinity and temperature, which occur upon moving up in the water column, as do fish larvae (Fig. 3). Fish larvae are capable of phototaxis (Woodhead, 1966; Blaxter 1969; 1973; Champalbert *et al.*, 1990, 1991, 1992; Masuda & Tsukamoto, 1996) and geotaxis (e.g. Champalbert *et al.*, 1992), and appear to exhibit a response similar to that of invertebrate larvae (Fig. 3).

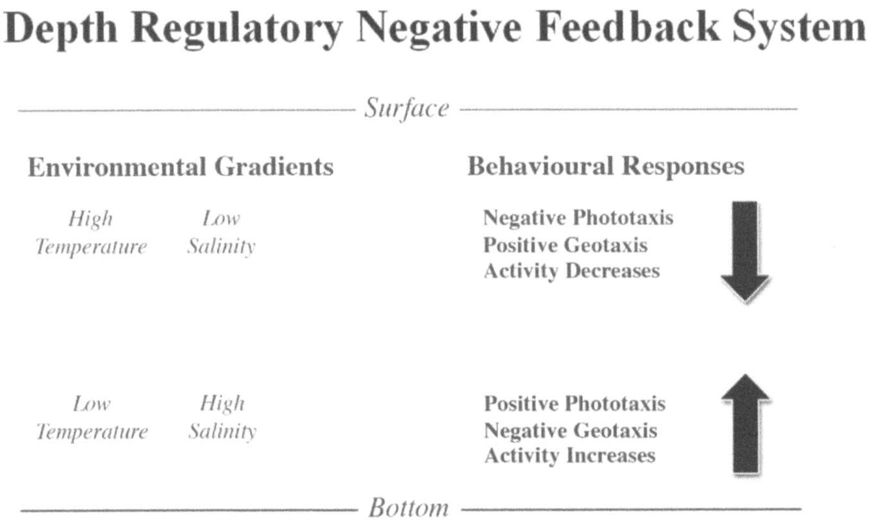

Fig. 3 A schematic illustration of depth regulation and the sensory feedback mechanisms of larval fishes which develop systems similar to those of larval invertebrates. Larval fishes are sensitive to changes in light and pressure; pressure sensitivity is gained through distortion of a dividing membrane within the otic bulla, a bulbous, gas-filled, anteriad extension of the swimbladder of the larvae of the physostomous clupeiforms (Denton & Blaxter, 1976; Hoss & Blaxter, 1979). In physoclistous larvae, this sensitivity may be achieved through priopriocеption, distortion of the swimbladder serosa (Tytler & Blaxter, 1977). Light and pressure sensitivity translates to phototaxis and geotaxis.

Thus, if fish larvae can perceive vertical salinity and temperature gradients as occur at a pycnocline and have this behavioural negative feedback system, they could regulate their depth. To sense a change in temperature and salinity, these factors should increase or decrease at detectable rates for detectable absolute amounts of change over short spatial scales (De Vries *et al.*, 1995a, b).

Considering behavioural responses to temperature, walleye pollock larvae swim up when exposed to an experimentally induced thermocline that decreases from 9 to 3°C (Olla & Davis, 1990; Olla *et al.*, 1996). This result agrees with field observations that walleye pollack larvae remain above or within a thermocline in the Gulf of Alaska (Kendall *et al.* 1994). Similarly responses to a laboratory thermocline were observed for older red hake (*Urophycis chuss*) larvae (Steiner & Olla, 1985). In the field, juvenile red hake descend from the planktonic to the benthic environment. When presented with a thermocline of 20°C in an upper section and 10-15°C in the lower section, older larvae avoided temperatures colder than 15°C. Thus, the descent to the bottom will be slowed by a thermocline (Steiner & Olla, 1985).

Atlantic menhaden larvae also respond to changes in temperature in the laboratory (De Vries *et al.*, 1995a). Temperature decreases from below caused an ascent response in both young (<2 weeks old) and old (2-4 weeks old) Atlantic menhaden larvae, but neither responded to this cue from above. The minimum (threshold) relative rates of decrease for initiating ascent were similar (7.9×10^{-2}, $10.7 \times 10^{-2}\,°C \cdot min^{-1}$) for both young and old larvae as were the minimum absolute amounts of decrease that must occurred before a response (0.1, 0.05°C). Young larvae did not respond to a temperature increase, while old larvae ascended regardless of whether the increase was presented from above or below. Threshold relative rates of increase were $8.59 \times 10^{-2}\,°C \cdot min^{-1}$ from below and $14.79 \times 10^{-2}\,°C \cdot min^{-1}$ from above for a temperature increase. The threshold rates and range of larval speed during vertical movements were used to calculate vertical temperature gradients that could be perceived (De Vries *et al.*, 1995a). On the continental shelf in areas where larvae occur, detectable temperature gradients appear common for temperature decreases that occur upon descending and temperature increases upon ascending.

A similar study with young and old Atlantic menhaden and spot larvae also found that larvae responded behaviourally to changes in salinity (De Vries *et al.*, 1995b). Both species and age groups showed a significant ascent response upon salinity increases. Rates of increase of 0.28 and 0.13×10^{-1} ppt $\cdot$ min^{-1} with absolute salinity increases of 0.8 and 0.3 ppt were thresholds for responses of young and old Atlantic menhaden larvae, respectively. For spot, rates of increase of 1.15 and 0.85×10^{-1} ppt $\cdot$ min^{-1} with absolute salinity increases of 1.4 and 2.0 ppt were the thresholds for responses by young and old larvae, respectively. Responses to decreasing salinity as occur upon ascending were absent in both species. Salinity gradients observed in areas where larvae occur were of sufficient magnitude for depth regulation by Atlantic menhaden, but not for spot. Collectively aspects of the negative feedback system for depth regulation (Fig. 3) apply to the few larval species that have been tested in that larvae ascend in response to temperature and salinity changes that would occur upon descending.

Although the light:dark cycle regulates depth of fish larvae that undergo diel vertical migration (Forward, 1976, 1988), it can also regulate buoyancy. Among larvae of clupeoid fishes there is a diel cycle in inflation and deflation of the swimbladder (Uotani, 1973; Hunter & Sanchez, 1976; Blaxter & Hunter, 1982). Both Atlantic menhaden (Hoss *et al.*, 1989) and gulf menhaden (*B. patronus)* (Hoss & Phonlor, 1984) inflate their swimbladders during the night and deflate them during the day. Inflation involves moving to the surface and swallowing air into the alimentary canal and moving it to the swimbladder through the pneumatic duct (Hoss & Phonlor, 1984). Although deflation could result from gas diffusion throughout the night (Tracy 1920; Hoss & Phonlor, 1984), it actually occurs by reversing the filling process (Forward *et al.*, 1994).

Factors controlling inflation and deflation have been studied in detail in Atlantic menhaden larvae (Forward *et al.*, 1993, 1994). The inflation and deflation cycle begins when larvae attain a total length of 10 mm. Larvae inflate their swimbladders when light levels fall below $6 \times 10^{13} - 6 \times 10^{14}$ photons $cm^{-2} \cdot s^{-1}$ at sunset by going to the surface where they ingest air, stop swimming and begin to sink (Forward *et al.*, 1993). The studies De Vries *et al.* (1995a, b) indicate temperature and salinity changes that will be encountered upon ascending will not impede upward movement. The ascent responses to temperature and salinity, however, changes upon descending and could regulate the depth to which they sink. Since, larvae have an activity rhythm in which they make frequent excursions to the upper part of a water column during the time of night (Forward *et al.*, 1996) and the swimbladder volume increases during the night in the laboratory, some larvae probably make a number of excursions to the surface at night to ingest air.

Vertical movement at night could function in predator avoidance by Atlantic menhaden larvae. Initially at sunset, there is a rapid ascent to the surface to ingest air to fill the swim bladder. Rapid upward swimming would reduce the probability of capture because the length of time that a larva is in the vicinity of a predator is small. Once the swim bladder is inflated, larvae are still negatively buoyant, but their sinking rates are reduced (Hoss *et al.*, 1989). Since an increase in hydrostatic pressure with depth will decrease the volume of gas in the swimbladder (Hoss & Blaxter, 1979), the sinking rate of larvae will increase with depth. Nevertheless, after swimbladder inflation, larvae sink more slowly, which would reduce the amount of energy required for position maintenance in the water column (Hunter & Sanchez, 1976). In addition, their movements for the greatest portion of the night would be minimised, thus reducing detection by predators (Uotani, 1973), such as chaetognaths that hunt by detection of vibrations.

At sunrise, once the light intensity reaches 9×10^{12} photons $cm^{-2} \cdot s^{-1}$, Atlantic menhaden larvae rapidly deflate their swimbladders by passing a bubble(s) into the alimentary canal where they are expelled through the mouth and anus (Forward *et al.*, 1994). Since the light intensity that induces deflation is above the intensity of full moonlight and starlight (10^{11} photons $cm^{-2} \cdot s^{-1}$, McFarland & Munz, 1975; Lythgoe, 1979), deflation will not occur at night. Considering functional significance, a fully inflated swimbladder may reduce the speed of movement (Forward *et al.*, 1999) and thereby, the effectiveness of prey capture by larvae, which capture prey by sight during the day (Blaxter & Hunter, 1982). In addition, an inflated swimbladder may increase detection of larvae during the day by visual predators. The difference in refractive index between air and water increased the contrast between an inflated swimbladder and the surrounding water for the relatively transparent larvae (Forward *et al.*, 1994). This increase in visibility could lead to increased predation. Thus, deflation at sunrise may aid feeding and predator avoidance.

SYNTHESIS

Some teleostean fishes have pelagic eggs and larvae that achieve positive, or near neutral buoyancy; others have demersal eggs with larvae that are not buoyant. For pelagic eggs and larvae, Archimedes' principle is at work in the buoyancy of developing fishes, whereby eggs and larvae displace an equivalent mass of water with a substance of lower specific gravity than ambient water. Incorporation of water of lower osmolality, and specific gravity, is the principal provider of buoyancy. Lipids secondarily contribute buoyancy to pelagic eggs and yolk-sac larvae. The proportional contribution to buoyancy of water and lipids varies phylogenetically and phenotypically depending on spawning habitat. A swimbladder, if present, is the principal organ that provides buoyancy in larvae, after yolk and oil globule absorption, though water content continues to contribute. Vertical distribution and neutral buoyancy is achieved by adjusting the volume of gas in the swim bladder of larvae; larvae do not ascend or descend by adjustments to swimbladder volume.

Behaviour is also important for buoyancy and depth maintenance of larvae. Phototaxis, geotaxis and activity changes in response to vertical gradients in hydrostatic pressure, temperature and salinity provide a negative feedback mechanism for aggregation at specific depths and at density discontinuity areas (pycnocline). In addition, diel changes in buoyancy of some larvae are accomplished by inflation and deflation of the swimbladder in response to the light:dark cycle.

Since depth regulation is important for transport, feeding, and predator avoidance, the amount of metabolic energy needed to maintain depth through swimming is reduced if a larva is neutrally buoyant. Subsequently, as older larvae rapidly increase their mass, buoyancy control shifts to the swimbladder, which allows more energy to be put into growth rather than position maintenance through swimming. Thus, eggs and young larvae use changes in specific gravity to regulate buoyancy, whereas older larvae use a swimbladder and behaviour for buoyancy and depth regulation.

References

Ådlandsvik B, Coombs S, Sundby S, Temple G (2001) Buoyancy and vertical distribution of eggs and larvae of blue whiting (*Micromesistius poutassou*): Observations and modeling. Fish Res 50: 59-72

Ahlstrom EH (1959) Vertical distribution of pelagic fish eggs and larvae off California and Baja California. Fish Bull 60: 107-146

Alderdice DG (1988) Osmotic and ionic regulation in teleost eggs and larvae. In: Hoar WS, Randall DJ (eds) Fish Physiology, Vol IIA, Academic Press, New York, p 163-251

Alexander RM (1966) Physical aspects of swimbladder function. Biol Rev 41: 347-355

Ali-Maghazachi SJ, Gibson R (1984) The developmental stages of larval turbot, *Scophthalmus maximus* (L.). J Exp Mar Biol Ecol 83: 35-51

Balon EK (1975) Reproductive guilds of fishes: a proposal and definition. J Fish Res Bd Can 32: 821-864

Balon EK (1981) Saltatory processes and altricial to precocial forms in the ontogeny of fishes. Am Zool 21: 573-596

Battle HI, Sprules WM (1960) A description of the semi-buoyant eggs and early developmental stages of the goldeye *Hiodon alosoides* (Rafinesdque). J Fish Res Bd Can 17: 245-266

Berenbrink M, Koldkjaer P, Kepp O, Cossins AR (2005) Evolution of oxygen secretion in fishes and the emergence of a complex physiological system. Science 307: 1752-1757

Bisbal GZ, Bengtson DA (1995) Development of the digestive tract in larval summer flounder. J Fish Biol 47: 277-291

Blaxter JHS (1969) Visual thresholds and spectral sensitivity of flatfish larvae. J Exp Biol 51: 221-230

Blaxter JHS (1973) Monitoring the vertical movements and light responses of herring and plaice larvae. J Mar Biol Assoc UK 53: 635-647

Blaxter JHS (1988) Pattern and variety in development. In: Hoar WS, Randall DJ (eds) Fish Physiology. Academic Press, San Diego, p. 1-58

Blaxter JHS, Ehrlich KF (1973) Changes in behavior during starvation of herring and plaice larvae. Blaxter JHS (ed) The Early Life History of Fish. Springer-Verlag, Berlin, p 575-588

Blaxter JHS, Hunter JR (1982) The biology of clupeoid fishes. Adv Mar Biol 20: 1-223

Blaxter JHS, Batty RS (1984) The herring swimbladder: loss and gain of gas. J Mar Biol Assoc UK 64: 441-459

Bonaventura C, Sullivan B, Bonaventura J (1976) Spot hemoglobin: studies on the root effect hemoglobin of a marine teleost. J Biol Chem 251: 1871-1876

Cambalik JJ, Checkley DM, Kamykowski D (1998) A new method to measure the terminal velocity of small particles: a demonstration using ascending eggs of the Atlantic menhaden (*Brevoortia tyrannus*). Limnol Oceanogr 43: 1722-1727

Cerdà J, Fabra M, Raldúa D (2007) Physiological and molecular basis of fish oocyte hydration. In: Babin PJ, Cerdà J, Lubzens E (eds) The Fish Oocyte: From Basic Studies to Biotechnological Applications. Springer, Dordrecht, The Netherlands, p 349-396

Cetta CM, Capuzzo JM (1982) Physiological and biochemical aspects of embryonic and larval development of the winter flounder *Pseudopleuronectes americanus*. Mar Biol 71: 327-337

Champalbert G, Castelbon C, Le Direach L, Macqart-Moulin C, Patriti G (1990) Experimental study of swimming activity and orientation of sole (*Solea solea* L.) larvae and juveniles: influence of endogenous rhythm, light, gravity, temperature and feeding. Le Mer 28: 248-254

Champalbert G, Macquart-Moulin C, Patriti G, Chiki D (1991) Ontogenic variations in the phototaxis of larval and juvenile sole (*Solea solea* L.). J Exp Mar Biol Ecol 149: 207-225

Champalbert G, Macquart-Moulin C, Patriti G, Le Direach-Boursier L (1992) Light control of vertical movements of larvae and juvenile sole (*Solea solea* L.). Mar Behav Physiol 19: 263-283

Chatain B (1986) La vessie natatoire chez *Dicentrarchus labrax* et *Sparus auratus*. I. Aspects morphologiques du developpement. Aquaculture 53: 303-311

Chatain B, Ounais-Guschemann N (1990) Improved rate of initial swimbladder inflation in intensively reared *Sparus auratus*. Aquaculture 84: 345-353

Chen W, Govoni JJ, Warlen SM (1992) Comparison of feeding and growth of larval round herring *Etrumeus teres* and gulf menhaden *Brevoortia patronus*. Fish Bull 90: 183-189

Coombs SH (1981) A density-gradient column for determining the specific gravity of fish eggs, with particular reference to eggs of the mackerel *Scomber scombrus*. Mar Biol 63: 101-106

Craik JCA, Harvey SM (1987) The causes of buoyancy in eggs of marine teleosts. J Mar Biol Assoc UK 67: 169-182

Daniels CB, Orgeig S, Sullivan LC, Ling N, Bennett MB, Schurch S, Val AL, Brauner CJ (2004) The origin and evolution of the surfactant system in fish: insights into the evolution of lungs and swim bladders. Physiol Biochem Zool 77: 732-749

Davis CC (1959) A planktonic fish egg from freshwater. Limnol Oceanogr 4: 353-355

Davis MW, Olla BL (1990) Comparison of growth, behavior and lipid concentration of walleye pollack *Theragra chalcogramma* larvae fed lipid-enriched, lipid deficient and field-collected prey. Mar Ecol Prog Ser 90: 23-30

De Vries MC, Forward RB Jr, Hettler WF (1995a) Behavioral responses of larval Atlantic menhaden *Brevoortia tyrannus* (Latrobe) and spot *Leiostomus xanthurus* (Lacepede) to rates of salinity change. J Exp Mar Biol Ecol 185: 93-93

De Vries MC, Forward RBJr, Hettler WF (1995b) Behavioural response of larval Atlantic menhaden to different rates of temperature change. J Fish Biol 47: 1081-1095

Denton EJ, Blaxter JHS (1976) The mechanical relationships between the clupeid swimbladder, inner ear, and lateral line. J Mar Biol Assoc UK 56: 787-807

Denton EJ, Liddicout JD, Taylor DW (1972) The permeability of gases of the swimbladder of the conger eel (*Conger conger*). J Mar Biol Assoc UK 52: 727-746

Eddy FB, McGovern L, Acock N, McGeer JC (1999) Cardiovascular responses of eggs, embryos and alevins of Atlantic salmon and rainbow trout to nitric oxide donors, sodium nitroprusside and isosorbide dinitrate, and inhibitors of nitric oxide synthase, N-nitro-L-arginine methyl ester and aminoguanidine, following short- and long-term exposure. J Fish Biol 55: 11-127

Evans RP, Parrish CC, Brown PJ (1996) Biochemical composition of eggs from repeat and first-time spawning captive Atlantic halibut (*Hippoglossus hippoglossus*). Aquaculture 139: 139-149

Fahay MP (1983) Guide to the early stages of marine fishes occurring in the western North Atlantic Ocean, Cape Hatteras to the southern Scotian Shelf. J NW Atlant Fish Sci 4: 1-423

Faustino M, Power DM (2001) Osteologic development of the viscerocranial skeleton in bream: alternative ossification strategies in teleost fish. J Fish Biol 58: 537-572

Finn RN (2007a) Vertebrate yolk complexes and the functional implications of phosvitins and other subdomains in vitellogenins. Biol Reprod 76: 936-948

Finn RN (2007b) The maturational disassembly and differential proteolysis of paralogous vitellogenins in a marine pelagophil teleost: A conserved mechanism of oocyte hydration. Biol Reprod 76: 936-948

Finn RN, Kristoffersen BA (2007) Vertebrate vitellogenin gene duplication in relation to the "3R Hypothesis": Correlation to the pelagic egg and the oceanic radiation of teleosts. PLoS ONE 2(1): e169

Finn RN, Henderson JR, Fyhn HJ (1995a) Physiological energetics of developing embryos and yolk-sac larvae of Atlantic cod (*Gadus morhua*). II. Lipid metabolism and enthalpy balance. Mar Biol 124: 371-379

Finn RN, Widdows J, Fyhn HJ (1995b) Calorespirometry of developing embryos and yolk-sac larvae of turbot (*Scophthalmus maximus*). Mar Biol 122: 157-163

Finn RN, Fyhn HJ, Henderson RJ, Evjen MS (1996) The sequence of catabolic substrate oxidation and enthalpy balance of developing embryos and yolk-sac larvae of turbot (*Scophthalmus maximus* L.). Comp Biochem Physiol 115A: 133-151

Finn RN, Østby GC, Norberg B, Fyhn HJ (2002a) *In vivo* oocyte hydration in Atlantic halibut (*Hippoglossus hippoglossus*); proteolytic liberation of free amino acids, and ion transport, are driving forces for osmotic water influx. J Exp Biol 205: 211-224

Finn RN, Wamboldt M, Fyhn HJ (2002b) Differential processing of yolk proteins during oocyte hydration in marine fishes (Labridae) that spawn benthic and pelagic eggs. Mar Ecol Prog Ser 237: 217-226

Finn RN, Rønnestad I, van der Meeren T, Fyhn HJ (2002c) Fuel and metabolic scaling during the early life stages of Atlantic cod *Gadus morhua*. Mar Ecol Prog Ser 243: 217-234

Forward RB Jr (1976) Light and diurnal vertical migration: photobehavior and photophysiology of plankton. In: Smith KC (ed) Photochemical and Photophysiological Reviews. Plenum, New York, p 157-209

Forward RB Jr (1987) Crustacean larval vertical migration: a perspective. In: Herrnkind WF, Thistle AB (eds) Signposts in the Sea. Florida State University, Tallahassee, Florida, p 29-44

Forward RB Jr (1988) Diel vertical migration: zooplankton photobiology and behavior. Oceanogr Mar Biol Ann Rev 26: 361-393

Forward RB Jr, McKelvey LM, Hettler WF, Hoss DE (1993) Swimbladder inflation of the Atlantic menhaden *Brevoortia tyrannus*. Fish Bull 91: 254-259

Forward RB Jr, Hettler WF, Hoss DE (1994) Swimbladder deflation in the Atlantic menhaden, *Brevoortia tyrannus*. Fish Bull 92: 641-646

Forward RB Jr, Tankersley RA, Burke JS (1996) Endogenous swimming rhythms of larval Atlantic menhaden, *Brevoortia tyrannus* Latrobe: implications for vertical migration. J Exp Mar Biol Ecol 204: 195-207

Forward RB Jr, De Vries MC, Tankersley RA, Rittschof D, Hettler WF, Burke JS, Welch JM, Hoss DE (1999) Behaviour and sensory physiology of Atlantic menhaden larvae, *Brevoortia tyrannus*, during horizontal transport. Fish Oceanogr 8: 37-56

Frank KT, Page FH, McRuer JR (1989) Hydrographic effects on the vertical distribution of haddock (*Melanogrammus aeglefinus*) eggs and larvae on the southwestern Scotian Shelf. Can J Fish Aquat Sci 46: 82-92

Fukuhara D (1986) Morphological and functional development of Japanese flounder in early life stage. Bull Jap Soc Sci Fish 52: 81-91

Fyhn HJ, Govoni JJ (1995) Endogenous nutrient mobilization during egg and larval development in two marine fishes – Atlantic menhaden and spot. ICES Mar Sci Symp 201: 64-69

Fyhn HJ, Finn RN, Reith M, Norberg B (1999) Yolk protein hydrolysis and oocyte free amino acids as key features in the adaptive evolution of teleost fishes to seawater. Sarsia 84: 451-456

Fänge R (1966) Physiology of the swimbladder. Physiol Rev 46: 299-322

Fänge R, Wittenberg JB (1958) The swimbladder of the toadfish (*Opsanus tau* L.). Biol Bull 115: 172-179

Govoni JJ (2004a) The development of form and function in fishes and the question of larval adaptation. In: Govoni JJ (ed) The Development of Form and Function in Fishes and the Question of Larval Adaptation. Am Fish Soc Symp 40, Bethesda, p 1-7

Govoni JJ (2004b) Epilogue. In: Govoni JJ (ed) The Development of Form and Function in Fishes and the Question of Larval Adaptation. Am Fish Soc Symp 40, Bethesda, p 195-198

Govoni JJ, Hoss DE (2001) Comparison of the development and function of the swimbladder of *Brevoortia tyrannus* (Clupeidae) and *Leiostomus xanthurus* (Sciaenidae). Copeia 2001: 430-442

Govoni JJ, Peters DS, Merriner JV (1982) Carbon assimilation during larval development of the marine teleost *Leiostomus xanthurus* Lacépède. J Exp Mar Biol Ecol 64: 287-299

Govoni JJ, West MA, Bonaventura J, Goddette G, Jenkins TE (2005) The ontogeny of haematopoiesis in the marine teleost *Leiostomus xanthurus* and a comparison of the site of initial haematopoiesis with *Opsanus tau*. J Fish Biol 67: 696-712

Gray CA (1993) Horizontal and vertical trends in the distribution of larval fishes in coastal waters off central New South Wales Australia. Mar Biol 116: 649-666

Gray CA (1996) Do thermoclines explain the vertical distribution of larval fishes in the dynamic coastal waters of south-eastern Australia? Mar Freshwater Res 47: 183-190

Gray CA, Kingsford MJ (2003) Variability in thermocline depth and strength, and relationships with vertical distributions of fish larvae and mesozooplankton in dynamic coastal waters. Mar Ecol Prog Ser 247: 211-244

Greeley MS Jr, Calder DR, Walace RA (1986) Changes in teleost yolk proteins during oocyte maturation: correlation of yolk proteolysis with oocyte hydration. Comp Biochem Physiol 84B: 1-9.

Harden-Jones FR, Marshall NB (1953) The structure and functions of the teleostean swimbladder. Biol Rev 28: 16-83

Hare JA, Govoni JJ (2005) Comparison of average larval fish vertical distributions among species exhibiting different transport pathways on the southeast United States continental shelf. Fish Bull 103: 728-736.

Hare JA, Walsh HJ, Wuenschel MJ (2006) Sinking rates of late-stage fish larvae: Implications for larval ingress into estuarine nursery habitats. J Exp Mar Biol Ecol 330: 493-504

Haverkamp S, Kolb H, Cuenca N (1999) Endothelial nitric oxide synthase (eNOS) is localized to Muller cells in all vertebrate retinas. Vision Res 39: 2299-2303

Heath MR (1992) Field investigations of the early life stages of marine fish. Adv Mar Biol 28: 1-174

Holliday FGT, Blaxter JHS (1960) The effects of salinity on the developing eggs and larvae of the herring. J Mar Biol Assoc UK 39: 591-603

Hoss DE, Blaxter JHS (1979) The effect of rapid changes in hydrostatic pressure on the Atlantic herring *Clupea harengus* L. I. Larval survival and behaviour. J Exp Mar Biol Ecol 41: 75-85

Hoss DE, Phonlor G (1984) Field and laboratory observation of diurnal swim bladder inflation-deflation in larvae of gulf menhaden, *Brevoortia patronus*. Fish Bull 82: 13-517

Hoss DE, Checkley DM Jr, Settle LR (1989) Diurnal buoyancy changes in larval Atlantic menhaden (*Brevoortia tyrannus*). Rapports et Proces-Verbaux des Reunions Conseil International pour L'Exploration de La Mer 191: 105-111

Hunt von Herbing I, Boutilier RG, Hall BK (1996) Effects of temperature on morphological landmarks critical to growth and survival in larval cod (*Gadus morhua*). Mar Biol 124: 593-606

Hunter JR, Sanchez C (1976) Diel changes in swim bladder inflation of the larvae of the northern anchovy, *Engraulis mordax*. Fish Bull 74: 847-855

Hyslop JRG, Bell MA (1987) Observations on the size, dry weight and energy content of the eggs of some demersal fish species from British marine waters. J Fish Biol 31: 1-20

Johnston IA, Hall TE (2004) Mechanisms of muscle development and responses to temperature change in fish larvae. In: Govoni JJ (ed) The Development of Form and Function in Fishes and the Question of Larval Adaptation. Am Fish Soc Symp 40, Bethesda, p 85-116

Kamler E (1992) Early Life History of Fish: An Energetic Approach. Chapman and Hall, London

Kendall AW Jr, Naplin NA (1981) Diel-depth distribution of summer ichthyoplankton in the middle Atlantic bight. Fish Bull 79: 705-726

Kendall AW Jr, Kim S (1989) Buoyancy of walleye pollock (*Theragra chalcogramma*) eggs in relation to water properties and movement in Shelikof Strait, Gulf of Alaska. Can Spec Publ Fish Aquat Sci 108: 169-180

Kendall AW Jr, Ahlstrom EH, Moser HG (1984) Early life history of fishes and their characters. In: Moser HG, Richards WJ, Cohen DM, Fahay MP, Kendall AWJ, Richardson SL (eds) Ontogeny and Systematics of Fishes. Allen Press, Inc, Lawrence, Kansas, USA, p 11-22

Kendall AW Jr, Incze LS, Ortner PB, Cummings SR, Brown PK (1994) Factors relating to the vertical distribution of eggs and larvae of walleye pollock *Theragra chalcogramma* in Shelikof Strait, Gulf of Alaska. Fish Bull 92: 540-554

Kitajima C, Yamane Y, Matsui S (1994a) Developmental changes in specific gravity in the early stages of the Japanese flounder *Paralichthys olivaceus*. Nippon Suisan Gakkaishi 60: 617-623

Kitajima C, Yamane Y, Matsui S (1994b) Ontogenetic changes in specific gravity during early development of black sea bream. Nippon Suisan Gakkaishi 60: 755-762

Kitajima C, Yamane Y, Matsui S, Yoshimatsu T (1998) Ontogenetic change in specific gravity in the early stages of the ayu *Plecoglossus altivelis*. Bull J Soc Sci Fish 64: 822-829

Kitajima C, Yamane Y, Matsui S, Kihara Y, Furuichi M (1993) Ontogenetic change in buoyancy in the early stage of red sea bream. Nippon Suisan Gakkaishi 59: 209-216

Kolarevic J, Nerland A, Nilsen F, Finn RN (2008) Goldsinny wrasse (*Ctenolabrus rupestris*) is an extreme *vtgAa-type* pelagophil teleost. Mol Reprod Dev DOI 10.1002/mrd.20845

Kunz YW (2004) Developmental Biology of Teleost Fishes. Springer, Dordrecht, Netherlands

Laale HW (1980) The perivitelline space and egg envelopes of bony fishes: a review. Copeia 1980: 210-226

LaFleur GJ Jr, Demetrio R, Fabra M, Carnevali O, Denslow N, Wallace RA, Cerdà J (2005) Derivation of major yolk proteins from parental vitellogenins and alternative processing during oocyte maturation in *Fundulus heteroclitus*. Biol Reprod 73: 815-824

Lancioni H, Lucentini L, Panara F (2005) Ontogenetic changes of circulating erythroid cells and haemoglobin components in *Esox lucius*. J Fish Biol 66: 1576-1588

Lapennas GN, Schmidt-Nielsen K (1977) Swimbladder permeability to oxygen. J Exp Biol 67: 175-196

Lefrançois C, Odion M, Claireaux G (2001) An experimental and theoretical analysis of the effect of added weight on the energetics and hydrostatic function of the swimbladder of European sea bass (*Dicentrarchus labrax*). Mar Biol 139: 13-17

Lochmann SE, Maillet GL, Frank KT, Taggart CT (1995) Lipid class composition as a measure of nutritional condition in individual larval Atlantic cod (*Gadus morhua*). Can J Fish Aquat Sci 52: 1294-1306

Loeb VJ (1979) Larval fishes in the zooplankton community of the North Pacific gyre. Mar Biol 53: 173-191

Loeb VJ (1980) Patterns of spatial and species abundance within the larval fish assemblages of the North Pacific Central Gyre during late summer. Mar Biol 60: 189-200

Lønning S, Kjorsvik E, Falk-Petersen I-B (1988) A comparative study of pelagic and demersal eggs from common marine fishes in northern Norway. Sarsia 73: 49-60.

Lough RG, Potter DC (1993) Vertical distribution patterns and diel migrations of larval and juvenile haddock, *Melanogrammus aeglefinus* and Atlantic cod, *Gadus morhua*, on Georges Bank. Fish Bull 91: 281-303

Lui WS (1993) Development of the respiratory gas bladder of *Pangasius sutchi*. J Fish Biol 42: 159-167

Lund ED, Sullivan CV, Place AR (2000) Annual cycle of plasma lipids in captive reared striped bass: effects of environmental conditions and reproductive cycle. Fish Physiol Biochem 22: 263-275

Lythgoe JV (1979) The Ecology of Vision. Clarendon Press, Oxford, England

Mandich A, Massari A, Bottero S, Pizzicori P, Goos H, Marino G (2004) Plasma sex steroid and vitellogenin profiles during gonad development in wild Mediterranean amberjack (*Seriola dumerilii*). Mar Biol 144: 127-138

Mangor-Jensen A, Waiwood KG (1995) The effect of light exposure on buoyancy of halibut eggs. J Fish Biol 47: 18-25

Marty GDE, Hinton E, Summerfelt RC (1995) Histopathology of swimbladder noninflation in walleye (*Stizostedion vitreum*). Aquaculture 138: 35-48

Masuda R, Tsukamoto K (1996) Morphological development in relation to phototaxis and rheotaxis in the striped jack, *Pseudocaranx dentex*. Mar Freshwater Behav Physiol 28: 75-90

Matarese AC, Sandrop EM (1984) Identification of fish eggs. In: Moser HG, Richards WJ, Cohen DM, Fahay MP, Kendall AWJ, Richardson SL (eds) Ontogeny and Systematics of Fishes. Allen Press, Inc, Lawrence, Kansas, USA, p 23-31

Matsubara T, Ohkubo N, Andoh T, Sullivan CV, Hara A (1999) Two forms of vitellogenin, yielding two distinct lipovitellins, play different roles during oocyte maturation and early development of barfin flounder, *Verasper moseri*, a marine teleost that spawns pelagic eggs. Dev Biol 213: 18-32

McFarland WN, Munz FW (1975) The evolution of photopic visual pigments in fish. Vision Res 15: 1071-1080

Mommsen TP, Walsh PJ (1988) Vitellogenesis and oocyte assembly. In: Hoar WS, Randall DJ (eds) Fish Physiology, Vol IIA, Academic Press, New York, p 347-406

Morris DM, Albright JT (1984) Catalase, glutathione peroxidase, and superoxide dimutase in the rete mirabile and gas gland epithelium of six species of marine fishes. J Exp Zool 232: 29-39.

Moustakas CT, Watanabe WO, Copeland KA (2004) Combined effects of photoperiod and salinity on growth, survival, and osmoregulatory ability of larval southern flounder *Paralichthys lethostigma*. Aquaculture 229: 159-179.

Neighbors MA, Nafpaktitis BG (1982) Lipid compositions, water contents, swimbladder morphologies and buoyancies of nineteen species of midwater fishes (18 Myctophids and 1 Neoscopelid). Mar Biol 66: 207-215

Nissling A, Kryvi H, Vallin L (1994) Variation in egg buoyancy of Baltic cod *Gadus morhua* and its implications for egg survival in prevailing conditions in the Baltic Sea. Mar Ecol Prog Ser 110: 67-74

O'Connell CP (1955) The gas bladder and its relation to the inner ear in *Sardinops caerulea* and *Engraulis mordax*. Fish Bull 56: 429-446

Ohkubo N, Matsubara T (2002) Sequential utilization of free amino acids, yolk proteins and lipids in developing eggs and yolk-sac larvae of barfin floounder *Verasper moseri*. Mar Biol 140: 187-196

Olla BL, Davis MW (1990) Effects of physical factors on the vertical distribution of larval walleye pollock *Theragra chalcogramma* under controlled laboratory conditions. Mar Ecol Prog Ser 63: 105-122

Olla BL, Davis MW, Ryes CH, Sogard SM (1996) Behavioural determinants of distribution and survival in early stages of walleye pollock, *Theragra Chalcogramma*: a synthesis of experimental studies. Fish Oceanogr 5: 167-178

Patiño R, Sullivan CV (2002) Ovarian follicle growth, maturation, and ovulation in teleost fish. Fish Physiol Biochem 26: 57-70

Pelster B (1997) Buoyancy. In: Evans DH (ed) The Physiology of Fishes. CRC Press, Boca Raton, Florida, p 25-42

Pelster B (2004) The development of the swimbladder: structure and performance. In: Govoni JJ (ed) The Development of Form and Function in Fishes and the Question of Larval Adaptation. Am Fish Soc Symp 40, Bethesda, p 37-46

Power JH (1989) Sink or swim: growth and zooplankton hydromechanisms. Am Nat 133: 706-721

Power JH (1996) Errors associated with using Archimedes, principle to determine mass and volume of small aquatic organisms. Hydrobiologia 335: 141-145

Power JH, Walsh PJ (1992) Metabolic scaling, buoyancy and growth in larval Atlantic menhaden, *Brevoortia tyrannus*. Mar Biol 112: 17-22

Power JH, Morrison WL, Zeringue J (1991) Determining the mass, volume, density and weight in water of small zooplankters. Mar Biol 110: 267-271

Rainuzzo JR, Reitan KJ, Jørgensen L (1992) Comparative study on the fatty acid and lipid composition of four marine fish larvae. Comp Biochem Physiol 103B: 21-26

Riis-Vestergaard J (2002) Energy density of marine pelagic fish eggs. J Fish Biol 60: 1511-1528

Roberts RJ, Bell M, Young H (1973) Studies on the skin of plaice (*Pleuronectes platessa* L). II. The development of larval plaice skin. J Fish Biol 5: 103-108

Rombough PJ (2004) Gas exchange, ionoregulation, and the functional development of the teleost gill. In: Govoni JJ (ed) The Development of Form and Function in Fishes and the Question of Larval Adaptation. Am Fish Soc Symp 40, Bethesda, p 47-83

Röpke A (1993) Do larvae of mesopleagic fishes in the Arabian Sea adjust their vertical distributions to physical and biological gradients? Mar Ecol Prog Ser 101: 223-235

Röpke A, Nellen W, Piatkowski U (1993) A comparative study of the influence of the pycnocline on the vertical distribution of fish larvae and cephalopod paralarvae in three ecologically different areas of the Arabian Sea. Deep-Sea Res II 40: 801-819

Rønnestad I, Koven WM, Tandler A, Harel M (1994) Energy metabolism during development of eggs and larvae of gilthead sea bream (*Sparus aurata*). Mar Biol 120: 187-196

Rønnestad I, Finn RN, Lein I, Lie O (1995) Compartmental changes in the contents of total lipid, lipid classes and their associated fatty acids in developing yolk-sac larvae of Atlantic halibut, *Hippoglossus hippoglossus* (L.). Aquacult Nutr 1: 119-130

Saborido-Rey F, Kjesbu OS, Thorsen A (2003) Buoyancy of Atlantic cod larvae in relation to developmental stage and maternal influences. J Plankton Res 25: 291-307

Sameoto D (1982) Vertical distribution and abundance of the Peruvian anchovy *Engraulis ringens,* and the sardine, *Sardinops sagax*, larvae during November 1977. J Fish Biol 21: 171-185

Sclafani M, Stirling G, Leggett WC (1997) Osmoregulation, nutritional effects and buoyancy of marine larval fish: a bioassay for assessing density changes during the earliest life-history stages. Mar Biol 129: 1-9

Sire JY, Huysseune A (2003) Formation of dermal skeletal and dental tissues in fish: a comparative and evolutionary approach. Biol Rev 78 : 219-249

Sogard SM, Hoss DE, Govoni JJ (1987) Density and depth distribution of larval Gulf menhaden, *Brevoortia patronus*, Atlantic croaker, *Micropogonias undulates*, spot, *Leiostomus xanthurus*, in the Northern Gulf of Mexico. Fish Bull 85: 601-609

Solemdal P (1967) The effect of salinity on buoyancy, size and development of flounder eggs. Sarsia 29: 431-442

Steen B (1970) The swimbladder as a hydrostatic organ. In: Hoar WS, Randall DJ (eds) Fish Physiology, Vol IV, Academic Press, New York, p 413-443

Steiner WW, Olla BL (1985) Behavioral responses of prejuvenile red hake, *Urophyclis chuss*, to experimental thermoclines. Environ Biol Fishes 14: 167-173

Sulkin SD (1984) Behavioral basis of depth regulation in larvae of brachyuran crabs. Mar Ecol Prog Ser 15: 181-205

Sundby S (1991) Factors affecting the vertical distribution of eggs. ICES Mar Sci Symp 192: 33-38

Tait JS (1960) The first filling of the swim bladder in salmonoids. Can J Zool 38: 179-187

Thorsen A, Fyhn HJ (1996) Final oocyte maturation *in vivo* and *in vitro* in marine fishes with pelagic eggs; yolk protein hydrolysis and free amino acid content. J Fish Biol 48: 1195-1209

Thorsen A, Kjesbu OS, Fyhn HJ, Solemdal P (1996) Physiological mechanisms of buoyancy in eggs from brackish water cod. J Fish Biol 48: 457-477

Tracy HC (1920) The membranous labyrinth and its relation to the precoelomic diverticulum of the swimbladder in clupeoids. J Comp Neurol 31: 219-257

Tracy HC (1959) Stages in the development of the anatomy of motility of the toadfish (*Opsanus tau*). J Comp Neurol 111: 219-257

Trotter AJ, Pankhurst PM, Battalglene SC (2005) A finite interval of initial swimbladder inflation in *Latris lineata* revealed by sequential removal of water-surface films. J Fish Biol 67: 730-741

Tytler P, Bell MV (1989) A study of diffusional permeability of water, sodium and chloride in yolk-sac larvae of cod (*Gadus morhua* L.). J Exp Biol 147: 125-132

Tytler P, Blaxter JHS (1977) The effect of swimbladder deflation on pressure sensitivity in the saithe *Pollachius virens*. J Mar Biol Assoc UK 57: 1057-1064

Tytler P, Blaxter JHS (1988) Drinking in yolk-sac stage larvae of the halibut, *Hippoglossus hippoglossus* (L.). J Fish Biol 32: 493-494.

Tytler P, Ireland J, Fitches E (1996) A study of the structure and function of the pronephros in the larvae of the turbot (*Scophthalmus maximus*) and the herring (*Clupea harengus*) . Mar Freshwater Behav Physiol 28: 3-18

Uotani L (1973) Diurnal changes in gas bladder and behavior of postlarval anchovy and other related species. Bull J Soc Sci Fish 39: 867-876

Uotani I, Fukui A, Osaki K, Ozawa T (2000) Experimental study on the inflation and deflation of gas bladder of Japanese anchovy, *Engraulis japonicus* larvae. Bull Mar Sci 66: 97-103

Urho L (2002) Characteristics of larvae – what are they? Folia Zoologica 51: 161-186

Vetter RD, Hodson RE, Arnold C (1983) Energy metabolism in a rapidly developing marine fish egg, the red drum (*Sciaenops ocellatus*). Can J Fish Aquat Sci 40: 627-634

Wallace RA (1985) Vitellogenesis and oocyte growth in nonmammalian vertebrates. In: Browder LW (ed) Developmental Biology. Plenum Press, New York, p 127-177

Walsh PJ, Milligan CL (1993) Roles of buffering capacity and pentose phosphate pathway activity in the gas gland of the Gulf toadfish *Opsanus beta*. J Exp Biol 176: 311-316

Whitehead PJP, Blaxter JHS (1989) Swimbladder form in clupeoid fishes. Zool J Linn Soc 97: 299-372

Wink DA, Grisham MB, Mtchell JB, Ford PC (1996) Direct and indirect effects of nitric oxide in chemical reactions relevant to biology. Methods Enzymol 268: 12-31

Woodhead PMJ (1966) The behavior of fish in relation to light in the sea. Oceanogr Mar Biol Ann Rev 4: 337-403

Yamada Y, Okamura A, Tanaka S, Mikawa N, Horie N, Utoh T, Akazawa A, Oka HP (2004) Monthly changes in the swim bladder morphology of the female Japanese eel *Anguilla japonicus* in the coastal waters of Mikawa Bay, Japan. Ichthyol Res 51: 52-56

Yamagami K, Hamazaki S, Yasumasu S, Masuda K, Luchi I (1992) Molecular and cellular basis of formation, hardening, and breakdown of the egg envelope in fish. Int Rev Cytol 136: 51-92

Yamashita K (1982) Differentiation of the swimbladder structure in larvae of the red seabream *Pagrus major*. Jap J Ichthyol 29: 193-202

Zheng WB, Liu WS (1988) Morphology and histology of the swimbladder and ultrastructure of respiratory epithelium in the air-breathing catfish, *Pangasius sutchi* (Pangasiidae). J Fish Biol 33: 147-154

CHAPTER

16

Swimming and Muscle

Ulrike K Müller

INTRODUCTION

Why Fish Swim: A Matter of Speed

Swimming is a fish's main weapon in its struggle for survival (Videler, 1993; Plaut, 2001). The same might apply to teleost fish larvae, which suffer considerable mortality, although the correlation between swimming performance and survival is weaker than in adult fish (Fuiman & Cowan, 2003). The possible causes for fish larval mortality – predation, starvation, and larval drift (e.g. Bailey and Yen, 1983; Houde, 1994; Paradis *et al.*, 1996; Castro & Hernandez, 2000; Fuiman & Werner, 2002; Sponaugle & Grorud-Covert, 2006) – are all related to larval swimming performance. Vulnerability to predation is related to maximum burst speed and routine swimming speed (Table 1), which co-determines encounter rate (MacKenzie & Leggett, 1991; Litvak & Leggett, 1992; Cowan *et al.*, 1996). Vulnerability to larval drift depends on the maximum sustained swimming speed (Keckeis & Schiemer, 2002). Fish larvae differ in their swimming behaviour and activity levels, depending on their ecology and developmental stage. Reef fish larvae steadily increase their critical swimming speed, but not their

Author's address: Department of Biology, California State University Fresno, 2555 E San Ramon Avenue, Fresno CA 93740, USA.
E-mail: umuller@csufresno.edu

Table 1 Different categories of swimming speed used to characterise fish swimming (Beamish, 1966; Plaut, 2001)

Type of speed	*Category*	*Powered*	*Behavioural context*
routine swimming speed	behaviour-based category of swimming; spontaneous	by aerobic muscles; no fatigue	spontaneous, e.g. cruising, station-holding, foraging
(maximum) sustained swimming speed	endurance-based category of swimming	by aerobic muscles; no fatigue after 240 minutes in adult fish (Beamish 1966)	spontaneous, e.g. cruising, station-holding, foraging
(maximum) prolonged swimming speed	endurance-based category of swimming	by aerobic muscles; fatigue after 20 minutes in adult fish (Beamish, 1966)	e.g. chasing; swimming against rapids or currents
critical swimming speed	measure of max. sustainable speed; determined experimentally – fish swims in a current that incrementally increases in speed	by aerobic muscles; refers to the speed increment prior to the onset of fatigue	often used to evaluate impact of environmental factors on condition (Plaut, 2001)
(maximum) burst swimming speed	endurance-based category of swimming	by anaerobic muscles; fatigue occurs after <20 s in adult fish (Beamish, 1966)	to catch prey, evade predator, respond to sudden disturbance, manoeuvre through currents (Reidy *et al.*, 2000)

sustained swimming speed during the transition period from their pelagic phase to settlement (Fisher *et al.*, 2000). Flatfish larvae even decrease routine swimming speed, but increase swimming activity during development beyond metamorphosis (Skiftesvik, 1992). In contrast, zebrafish larvae (*Danio rerio*) increase both routine swimming speed and activity (Fuiman & Webb, 1988).

Teleost fish larvae differ from adult fish in their size and morphology. Both aspects have profound consequences for larval swimming speed and the swimming techniques available to larval fish. Although larval morphology is arguably no less diverse than adult morphology (Webb, 1999), fish larvae hatch at a median body size of 2.9 mm (marine) to 5.4 mm (freshwater) (Fuiman, 2002). This small body size largely determines larval swimming behaviour. Swimming speed increases with body length: fish larvae are up to 4 orders of magnitude shorter than the largest adult fish and reach swimming speeds that are 3 orders of magnitude lower than those of the fastest and largest adult fish.

How to Study Fish Swimming – Some Brief Technical Notes

Experimental studies on larval fish have often been held back by technical problems associated with the larva's small body size (swimming energetics, muscle physiology) and fast movements (swimming kinematics). When fish larvae swim, their body movements are too fast to be captured by normal cine, video or digital cameras, which usually record at 25 or 30 frames per second. The first studies used high-speed cine film (Hunter, 1972; Batty, 1984; Eaton and DiDomenico, 1986; Fuiman, 1986), which is cumbersome because larval swimming bouts are brief and larval swimming performance is variable and not easily anticipated; cine cameras take time to reach a constant film speed, and a constant film speed is vital to extract quantitative information such as swimming speed from recordings. With the advent of high-speed digital video, studying larval swimming behaviour has become comparatively simple and affordable, as reflected in the recent literature on fish larval swimming (e.g. Budick & O'Malley, 2000; Müller & van Leeuwen, 2004; Thorson *et al.*, 2004; McHenry & Lauders, 2005).

Respiratory and endurance tests on larval fish have a considerable history (e.g. Kaufmann, 1990; Fisher *et al.*, 2000; Bagatto *et al.*, 2001). In most studies, fish larvae swim in small chambers or flow tunnels. Flow tunnels come with their own problems: fish larvae are so small and the flow speeds so slow that the larvae can 'hide' in the thick boundary layer of the flow tunnel, and therefore swim much slower than in the central so-called free-stream of the tunnel. This can lead to severe overestimations of larval swimming endurance and critical swimming speeds. Hence, compared with adult fish, experimenters have to go to much greater lengths to ensure a sufficient flow quality of their water tunnel (Fisher *et al.*, 2000; Bagatto *et al.*, 2001; for a more detailed discussion see Plaut, 2001).

Even more elusive than data on larval swimming behaviour are data on larval swimming muscles: few studies exist on the physiology of larval swimming muscles (e.g. Buss & Drapeau, 2000, 2001, 2002) and their mechanical behaviour (e.g. van der Meulen *et al.*, 2005; van Leeuwen *et al.*, 2005). There are good data on larval fibre composition (e.g. Waterman, 1969; Nag & Nursall, 1972) and muscle functional morphology (e.g. Koumans & Akster, 1995; van der Meulen *et al.*, 2005). We also have learnt much about embryonic development of muscle morphology (e.g. Devoto *et al.*, 1996) and neural control (e.g. Kimmel, 1972; Eaton *et al.*, 1986; Hale *et al.*, 2001). Yet, compared with the insights gathered from adult fish (Grillner *et al.*, 1991; Rome *et al.*, 1993; Wardle *et al.*, 1995; Shadwick & Gemballa, 2006; Syme, 2006), surprisingly little is known about how swimming muscles work in larval fish.

SWIMMING SPEED AND BODY SIZE: WHY FISH LARVAE ARE SLOW

How Swimming Speed Scales with Size

In fish, swimming speed is closely tied to body length. Across species, absolute swimming speed (measured in m s^{-1}) increases steadily with body length (Bailey & Batty, 1984;

Blaxter, 1986; Fuiman, 2002; Fisher & Wilson, 2004; Gibson, 2005). Maximum swimming speed increases almost linearly with body size when looking at inter-specific data (Bailey & Batty, 1984; Blaxter, 1986; Fuiman, 2002; Gibson, 2005). However, a slightly different picture emerges over the life history of a single species. In zebrafish, for example, routine swimming speed initially increases steeply, but does not continue to increase at this high rate in later development (Fuiman & Webb, 1988). Hatchlings are often weak swimmers, and routine swimming speed increases steadily during development (Fuiman & Webb, 1988). However, this trend can be obscured by switches in swimming style or by behavioural changes, such as the onset of feeding (Hunter, 1972; Skiftesvik, 1992). From hatchling to a length of ca. 20 mm, routine swimming speed increases four- to five-fold, and burst speed increases by a factor of two to three over the same four-fold increase in length (Fuiman & Webb, 1988; Gibson, 2005). Critical swimming speed increases curvilinearly with size (Flore & Keckeis, 1998) and age (Fisher *et al.*, 2000). Plaut (2001) cautions that critical swimming speed, although commonly used in ecological studies, cannot be applied directly to fish in the wild, but needs to be related carefully to ecologically relevant swimming behaviour (e.g. experiments on Danube freshwater species that explore the role of river flow speed: Flore & Keckeis, 1998; Kaufmann, 1990). All these aforementioned trends relate age and body length to *absolute* swimming speed (m s^{-1}).

Relative swimming speed – expressed in body lengths per second (BL s^{-1}) – scales very differently. This length-specific speed decreases drastically with increasing body size: adult fish have similar burst speeds around 10 BL s^{-1} (range 6 to 19 BL s^{-1}; 9.9±3.5 BL s^{-1}, n = 11 species) (Videler, 1993), whereas fish larvae can maintain speeds of up to 66 BL s^{-1} (range 10 to 66 BL s^{-1}) for many body lengths of travel (Miller et al., 1988; Williams *et al.*, 1996; Bellwood & Fisher, 2001). Although fish larvae excel when swimming speed is expressed relative to body length, their absolute swimming speeds are considerably lower than those of adult fish.

The Hydrodynamics of Scaling: Why many Fish Larvae Undulate their Body

The small size and low absolute swimming speeds of larval fish profoundly affect the way in which a swimming larva interacts with the surrounding water. Small, slow organisms, such as sperm cells or flagellates, experience water mainly as a viscous medium – water adheres to the surface of the swimming fish and exerts shearing forces that slow the fish down. Large, fast swimmers, such as whales, experience water mainly as an inertial medium. Water is roughly 800 times more dense than air, and it therefore exerts considerable inertial forces that resist acceleration and thereby resist the propulsive efforts of a swimming fish. While the water always exerts both viscous and inertial forces on a swimming organism, the relative importance of the two types of forces changes with the organism's body length L and swimming speed U. This relationship is expressed by the Reynolds number Re:

$$\mathrm{Re} = LU \rho \, v^{-1},$$

where ν is the kinematic viscosity of the water, which is the ratio of density (ρ) to viscosity of water. Kinematic viscosity depends strongly on temperature and to a lesser extent on salinity: it decreases with increasing temperature and increases slightly with salinity (Webb, 1975). Fish larvae have body sizes and swimming speeds that are intermediate between the two extremes of spermatozoa and whales. As fish larvae grow, they begin to experience water increasingly as an inertial rather than a viscous medium. And as the physical nature of the fish-water interaction changes, so do the available swimming techniques.

Rather than dividing swimming styles by swimming technique, fish propulsive movements (swimming modes) have traditionally been divided into different categories according to the body parts used to generate propulsion (e.g. Breder, 1926; Lindsey, 1978; Webb, 1984; Blake, 2004). Fish can beat or undulate their paired or unpaired fins – median and paired fin (MPF) mode (Webb, 1984); they can undulate a varying proportion of their body – body and caudal fin (BCF) mode. Yet, body part movement and propulsive technique are often related. A fish can exploit two main types of forces to propel themselves: lift forces, which act perpendicular to the direction of flow, and drag forces, which act parallel to the flow. These forces are generated by the fish's "propellers", that is its fins and its body. A beating fin, for example, generates lift forces perpendicular to the direction that water flows past the fin. It also generates drag forces that act parallel to the flow, in the opposite direction. An oscillating propeller accelerates and decelerates continually, so it can also exploit so-called unsteady effects, such as the acceleration reaction (Webb & Weihs, 1986). Fish use acceleration reaction when they bend their entire body to perform a fast start. The better-known examples for acceleration reaction-based swimming are squid and jellyfish, which expel a jet of water from their contracting mantle (McHenry & Jed, 2003; Anderson & Grosenbaugh, 2005). Pectoral fin swimmers such as bird wrasses (*Gomphosus varius*) use their fins like birds use their wings: they generate lift forces to swim forward (Ramamurti *et al.*, 2002). Other pectoral fin swimmers such as bluegill sunfish (*Lepomis macrochirus*) row with their fins: their fins work like oars, generating drag forces to push the fish forward (Drucker & Lauder, 1999). How well these mechanisms work strongly depends on the swimmer's size and speed.

Fish larvae are too small and too slow to effectively exploit lift forces and acceleration reaction with their pectoral fins and tail fin (Webb & Weihs, 1986). This considerably limits the range of swimming techniques and morphologies available to larval fish. Larvae mainly use BCF swimming: they undulate their body. This swimming style is used by sperm cells and whale sharks alike, proving its effectiveness across a five orders-of-magnitude length range. It might therefore be particularly suited for developing fish, which can traverse up to three orders of magnitude in body length during their life history. Small undulatory swimmers the size of fish larvae exploit mainly viscous drag forces to propel themselves (McHenry *et al.*, 2003). The finfold, which is quite ubiquitous in fish larvae, might help to exploit these friction forces more effectively to swim. As fish larvae grow, lift and inertial propulsion mechanisms are added to their behavioural repertoire (Weihs, 1980; Webb & Weihs,

1986). The increasing importance of inertial forces also enables fish larvae to exploit different swimming styles; older fish larvae begin to use intermittent or burst-and-coast swimming (Webb & Weihs, 1986; McHenry & Lauder, 2005), rather than relying mainly on continuous swimming (Hunter, 1972).

LARVAL SWIMMING STYLES: WAYS TO UNDULATE YOUR BODY

What is Undulatory Swimming: Some Terminology

Undulatory swimmers generate a transverse wave that travels down their body (Fig. 1). This transverse wave is characterised by a wave speed (speed with which the

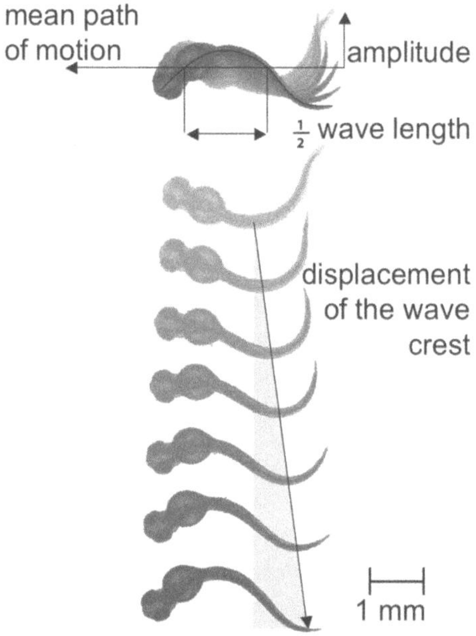

Fig. 1 Zebrafish (*Danio rerio*) larva (age 3 days post fertilisation), swimming cyclically at 8 body lengths (BL) per second along a straight mean path of motion for half a tail beat cycle, recorded with a high-speed video camera at 1000 frames per second; tail beat amplitude is 0.26 body lengths, the mean body wave length along the body is 1.0 BL. The top panel shows frames superimposed in an earth frame of reference (time between frames 2 ms). Amplitude at a given body section is defined as the maximum lateral excursion away from the mean path of motion; wavelength is defined by the distance between crossing points of the midline and the mean path of motion. The bottom panel shows the same frames, offset laterally; the body wave travels down the body and backwards in an earth frame of reference, as indicated by the arrow that crosses through the wave crest in successive frames. Body wave speed is 30 body length per second, almost four times higher than the swimming speed.

wave travels down the body), wavelength and amplitude (lateral excursion of the body). In larval fish, wave speed, wavelength and amplitude all change along the body (Batty,1981, 1984; Müller & van Leeuwen, 2004). The amplitude envelope is widest at the tail and is most narrow at the yolk-sac. Larval body wave shape also changes with the exact style of undulatory swimming (Osse & van den Boogaart, 1999; Budick & O'Malley, 2000; Müller & van Leeuwen, 2004).

Fish larvae use two styles of undulatory swimming, cyclic and intermittent (Hunter, 1972; Buss & Drapeau, 2001; Müller & van Leeuwen, 2004). During cyclic swimming, the larva executes many tail beats consecutively, approaching a steady mean swimming speed. During intermittent (burst-and-coast) swimming, the larva executes few tail beats, and does not reach a steady swimming speed, but begins to decelerate soon after the acceleration phase (Müller & van Leeuwen, 2004). Cyclic and intermittent swimming are used in different functional contexts, by different age groups and at different swimming speeds. Initially, fish larvae swim mainly cyclicly. As fish larvae grow, ontogenetic and hydrodynamic constraints change during larval development, and so do swimming speed and swimming style. Consequently, fish larvae switch from cyclic to intermittent swimming as their main swimming style (Hunter, 1972; Weihs, 1980). Furthermore, as the larva begins to feed externally, it spends more time swimming routinely to forage (Skiftesvik, 1992) executes "J" turns to track prey (McElligott & O'Malley, 2005) and uses intermittent bursts to capture prey (Budick & O'Malley, 2000).

The differences between the swimming styles are most apparent in the amplitude envelope (Fig. 2). During cyclic swimming fish larvae use a much wider body wave than during intermittent swimming: the tail beat amplitude can reach up to 0.24 BL during fast cyclic swimming versus 0.08 to 0.10 BL during a spontaneous burst (Osse & van den Boogaart, 1999; Müller & van Leeuwen, 2004). During such a slow burst, the larval body wave is narrow, resembling the body waves of many adult fish (Müller & van Leeuwen, 2004).

Undulatory Swimming: from Start to Finish

Fast start

Every swimming bout begins with a start. Both during escape and routine starts, the fish larvae usually execute a "C" start. A "C" start consists of the preparatory stroke, when the larva's body assumes a "C" shape, and the propulsive stroke, when the larva's centre of mass reaches peak acceleration (Fig. 3). During both the preparatory and the propulsive stroke, a wave of curvature travels along the body from head to tail. During the preparatory stroke, the body begins to bend behind the yolk-sac, and continues to curve, while the point of maximum curvature travels posteriorly. During the propulsive stroke, the anterior body begins to bend in the opposite direction of the initial "C" bend, and like the initial bend, this opposite-signed second bend travels down the body as maximum curvature along this bend increases (Müller & van Leeuwen, 2004). During

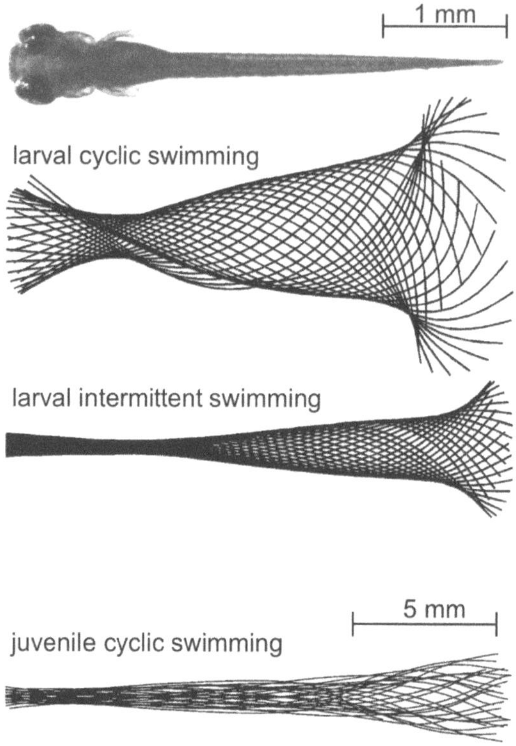

Fig. 2 Amplitude envelopes for cyclic and intermittent swimming of a zebrafish (*Danio rerio*) larva (4 days post fertilisation) versus those of a juvenile thicklip grey mullet (*Chelon labrosus*) (bottom panel). Top panel: dorsal view of a zebrafish larva. Larvae generate a body wave with a wide amplitude envelope during cyclic swimming and more narrow amplitude envelope during intermittent swimming. The larval amplitude envelope fans out at the tail compared with that of juvenile and adult fish.

strong escape responses, fish larvae can curl their body to form an "O", with their snout touching their tail, during the preparatory stroke. Then the curvature of the larval midline can reach up to 7 BL s^{-1}, which corresponds to a radius of curvature of 0.14 BL (0.5 mm in the case of a 3.5 mm zebrafish larva). During the propulsive stroke, zebrafish larvae in 27°C water can reach accelerations of 15 ms^{-2} and peak acceleration is reached within 10 to 15 ms from the onset of the "C" start. Peak performance and time to peak performance depend on temperature (Fuiman, 1986).

Zebrafish larvae reach peak acceleration at the transition from preparatory to propulsive phase. These peak accelerations can reach up to 60 m s^{-2} at the larva's the centre of mass. Speed peaks during the propulsive phase of the "C" start, reaching values of up to 60 BL s^{-1}. Adult fish for comparison reach accelerations of up to 250 m s^{-2} (Harper & Blake, 1989), but length-specific maximum speed is considerably lower at values below 20 BL s^{-1}.

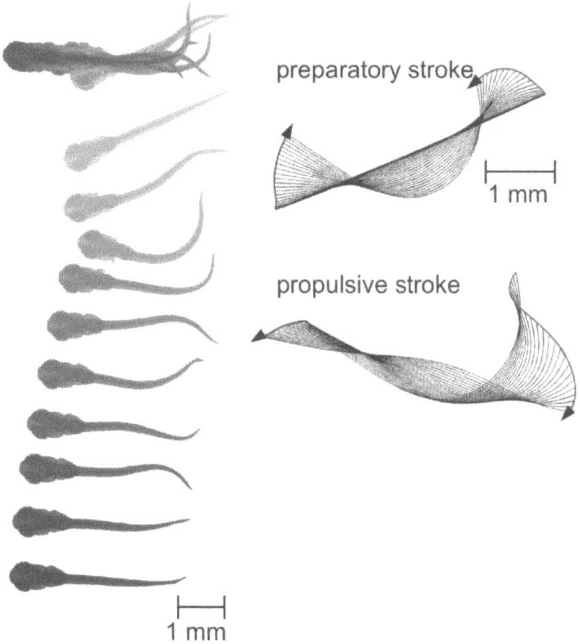

Fig. 3 Zebrafish (*Danio rerio*) larva (age 5 days post fertilisation, body length 4.2 mm) performing a spontaneous start, recorded with a high-speed video camera at 1000 frames per second. The left panel shows the larva at time steps of 10 ms. The right top panel shows the midlines (black) of the larva during the preparatory stroke, the right bottom panel shows the propulsive stroke (time step 1 ms). The grey lines in the right panels trace the path of the snout and the tail.

Fast start performance peaks early in ontogeny (Hale, 1999). Around the time of yolk-sac absorption, the time necessary to reach maximum speed and acceleration is lowest compared with younger and older larvae in three salmonid fishes: brown trout (*Salmo trutta*), Chinook salmon (*Oncorhynchus tshawytscha*), and coho salmon (*Oncorhynchus kisutch*) (Hale, 1999). Distance travelled, maximum speed, and maximum acceleration increase during ontogeny. Distance travelled more than doubles between hatchling and yolk-sac absorption and continues to increase throughout the juvenile period. Absolute maximum speed also increases ten-fold from 0.15 to 1.5 m s^{-1} between hatchling and juvenile over a fivefold increase in BL. Acceleration improves less strongly during ontogeny, with hatchlings reaching accelerations of up to 15 m s^{-2}, while juveniles reach values in excess of 100 m s^{-2}.

Cyclic swimming

Cyclicly swimming fish larvae undulate their body. Just as for start performance, the speeds reached during cyclic swimming strongly depend on temperature. Tropical freshwater and reef fish larvae can sustain swimming speeds of 10-17 BL s^{-1} at temperatures of 26-30°C (Fisher & Wilson, 2004), while freshwater and marine species in cooler waters (15 to 20°C) can sustain speeds of 3-10 BL s^{-1} (Keckeis & Schiemer, 2002). Swimming speed also depends on tail beat frequency. Atlantic herring (*Clupea harengus*) larvae in water of 5°C reach speeds of 5 BL s^{-1} at tail beat frequencies of 19 Hz (Batty & Blaxter, 1992), but at the higher temperature of 15°C, tail beat frequency increases to 36 Hz and swimming speed to 12 BL s^{-1}. Amongst the highest tail beat frequencies reported so far are for zebrafish larvae swimming at 27°C; at tail beat frequencies of 100 Hz, these larvae can reach speeds of up to 60 BL s^{-1} (Fuiman, 1986; Müller & van Leeuwen, 2004).

During fast cyclic swimming, larvae use a wider body wave amplitude than adult fish (Fig. 2) (Batty, 1984; Batty & Blaxter, 1992; Müller & van Leeuwen, 2004). The shape of the body wave amplitude differs most at the head and the tail. The larval head oscillates considerably with an amplitude comparable to the adult tail beat amplitude, while adult fish show hardly any such yaw (Tytell, 2004). At the tail, the larval amplitude envelope fans out, whereas the adult amplitude envelopes widens gradually along the body (Fig. 2). During fast cyclic swimming zebrafish tuck in their pectoral fins, hence they probably power and control fast cyclic swimming exclusively by their body wave (Thorson *et al.*, 2004).

Even when fish larvae swim at a constant mean speed ("cyclic swimming"), their instantaneous speed varies considerably during the tail beat cycle (Fig. 4). Thrust correlates with lateral tail speed, so instantaneous swimming speed is low when lateral tail speed is low and vice versa (Müller & van Leeuwen, 2004). By how much instantaneous speed oscillates depends on the swimming speed itself: in slowly swimming zebrafish larvae, instantaneous swimming speed can be up to 50% higher or lower than mean swimming speed, whereas at higher mean swimming speed these oscillations drop to less than 20% (Müller & van Leeuwen, 2004). These oscillations in speed in response to oscillations in thrust generation are much more pronounced in larval than adult fish. This difference is probably due to the relatively more important viscous forces versus the relatively less important inertial forces acting on the swimming larva.

Intermittent swimming

Pre-hatch exercises. Fish initiate brief bursts of body undulations already in the embryonic stage: zebrafish embryos can be observed to twitch and move around 18 hours after fertilisation, one day before they hatch (Saint-Amant & Drapeau, 1998; van der Meulen *et al.*, 2005).

Slow intermittent swimming. Hatchling and yolk-sac larvae regularly execute weak brief bursts of a few tail beats. These bursts consist of 2-6 tail beat cycles. Zebrafish larvae in water of 28°C reach maximum swimming speeds of 0 to 10 BL s^{-1} and move forward by less than half a body length (Müller & van Leeuwen, 2004). Tail beat frequency quickly drops during a slow burst from peak values of 30 Hz, and the amplitude envelope quickly narrows during a slow burst. During these slow bursts, zebrafish larvae beat their pectoral fins in conjunction with the body undulations to

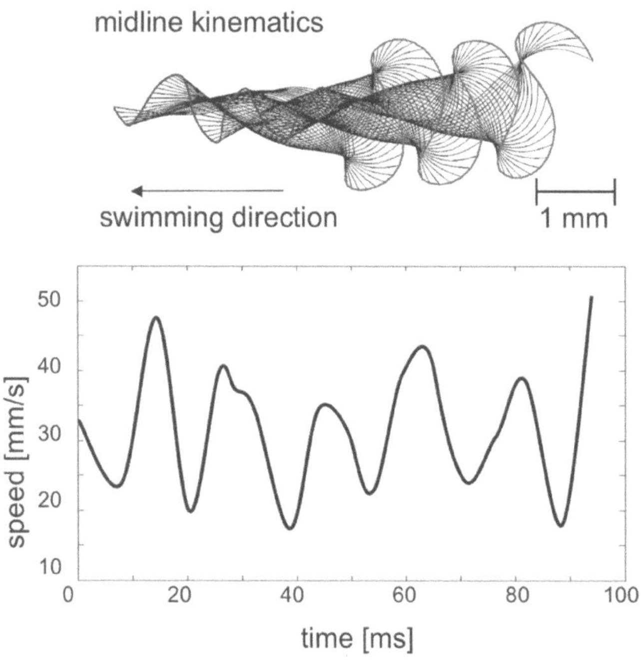

Fig. 4 A cyclicly swimming zebrafish (*Danio rerio*) larva (same sequence as in Figure 1). Top panel: midlines of the larva in an earth frame of reference (time between midlines 1 ms). Bottom panel: swimming speed (determined at the centre of mass) oscillates during the tail beat; values shown for the swimming sequence in top right panel.

generate extra thrust and possibly maintain control over their body posture (Thorson *et al.*, 2004). Common carp (*Cyprinus carpio*), swimming in 17°C water, use lower tail beat frequencies during slow bursts (Osse & van den Boogaart, 1999). The bout frequency of slow bursts depends on oxygen concentration, which gave rise to the hypothesis that slow bursts have a respiratory function (Osse & van den Boogaart, 1999). Even if the larva does not move forward, its undulating body movements and pectoral fin beats replace the water adjacent to its body with fresh, oxygen-richer water.

Fast intermittent swimming. Around the onset of feeding, swimming activity increases (Fuiman & Webb, 1988; Skiftesvik, 1992). While foraging, fish larvae use fast intermittent swimming, employing similar body kinematics to slow intermittent swimming, but at higher tail beat frequencies. In general, a higher burst swimming speed goes hand in hand with a higher tail beat frequency and tail beat amplitude. While during slow bursts, the larva only undulates the posterior portion of its body, during faster bursts, more and more anterior body sections get involved, which manifests as an increasing body wavelength with burst swimming speed (Müller & van Leeuwen, 2004). Concomitantly, the amplitude envelope of the body wave widens with speed.

LARVAL FLOW PATTERNS: HOW LARVAL SWIMMING WORKS

Flow Pattern and What They can Tell Us: Some Terminology

In order to understand how a fish swims, we need to know both where along the body and when during the body wave cycle the fish is generating thrust and drag. In terrestrial animals, it is possible to measure propulsive forces directly as ground reaction forces with force platforms. In water, such direct force measurements are impossible. Instead, the effect of these forces on the water are visualised as the flow patterns around and behind the fish.

The flow patterns around a swimming fish consist of vortices (areas of rotating and circulating flow) and jets (areas of fast flow). A swimming fish can generate thrust and drag at several locations: with its body, its tail fin, and its other unpaired and paired fins. Because body and fins oscillate, the generated forces also oscillate with the fin beat or body wave cycle. Hence, all these drag- and thrust producing surfaces generate each their own time-varying flow. These flows often interact because the flow-generating surfaces are arranged close to each other or one is downstream of the other, as e.g. dorsal and tail fin (Drucker & Lauder, 2001). The fish can control the phase relationship between these time-varying flow patterns (e.g. the interaction between body flow and tail flow) by changing its body wave (Ducker & Lauder, 2001; Müller *et al.*, 2002; Kern & Koumoutsakos, 2006; Tytell, 2006). Therefore, by altering the interaction between the flow patterns, the fish can control its swimming performance: it can for example maximise thrust or maximise swimming efficiency (Kern & Koumoutsakos, 2006). The interacting flows are ultimately shed into the wake behind the swimming fish (Drucker & Lauder, 2001; Tytell, 2006). From the patterns in the wake, which form into vortex rings, we can then glean information about the amount of net thrust generated and about the role of the various fins and body parts during swimming (Drucker & Lauder, 1999; Müller *et al.*, 2002; Tytell, 2004).

Wake shape indicates swimming performance, whether the fish is accelerating, swimming steadily or decelerating (Gopalkrishnan *et al.*, 1994; Müller *et al.*, 2002; Tytell, 2004), whether it is negatively buoyant (Wilga & Lauder, 2002), whether it is maximising thrust or efficiency (Kern & Koumoutsakos, 2006). If the wake contains simple vortex rings, then a simple measure of acceleration and negative buoyancy is the orientation of the vortex rings (Fig. 5). If the axis of the vortex ring is at a right angle to the mean path of motion, no net thrust or drag, upward or downward force has been generated (Müller *et al.*, 2001; Tytell & Lauder, 2004). If the vortex ring forms a downward angle, the fish generated an upward force, as seen in negatively buoyant sharks (Wilga & Lauder, 2002). If the vortex ring is angled backward, the fish is accelerating forward (Tytell, 2004). A brief introduction into the concepts and techniques involved in the hydromechanics of undulatory swimming is given in Lauder & Tytell (2006) and Müller & van Leeuwen (2006). Vortices are a manifestation of

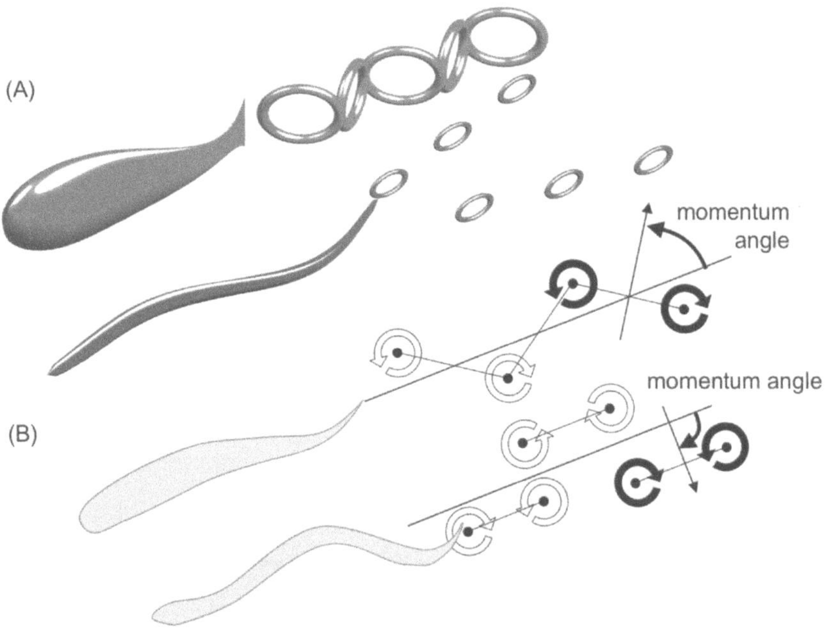

Fig. 5 Simplified vortex wake patterns observed behind juvenile fish. (A) Two three-dimensional impression of the vortex ring structure behind a juvenile thicklip grey mullet (*Chelon labrosus*) and a European eel (*Anguilla anguilla*) (Müller & van Leeuwen, 2006). Mullets shed a vortex chain wake, which consists of linked vortex rings (Müller *et al.*, 1997). European eels shed a vortex ring wake, which consists of individual vortex rings (Müller *et al.*, 2001). (B) Two cross sections through the vortex wakes of thicklip grey mullet and European eel. In thicklip grey mullet, vortex rings forms a momentum angle with the mean path of motion that is smaller than 90°. In European eel, vortex rings form a momentum angle of ca. 90° with the mean path of motion. Compare these wakes to a larval wake in fig. 7.

momentum transfer, and should only be considered as an indicator of the transferred momentum, not necessarily an honest and complete representation (Dabiri, 2005). New approaches are currently being developed to move towards more direct momentum estimates (Dabiri *et al.*, 2005).

Flow Pattern and Body Size Effects

Compared with adult fish, we know little about the flow patterns generated by fish larvae and about larval swimming mechanisms (Müller *et al.*, 2000; McHenry & Lauder, 2005). Fish larvae, due to their small size and lower swimming speed, swim in a more viscous flow regime and therefore generate different flow patterns from adult fish. With viscous forces relatively more important than inertial forces, fish larvae have for example a relatively thicker boundary layer. A boundary layer develops on any object moving relative to water: as a fish moves through water, water 'sticks' to its surface and is moved along with it. The thickness of the boundary layer depends on Reynolds number. As Reynolds number increases, the relative thickness compared with the size of the fish decreases. This means that larger and faster fish have a relatively thin boundary layer, so thin in adult fish that it only becomes visible when the experimenter zooms in on the local flow near the surface of the adult fish (Anderson *et al.*, 2001). In the much smaller and slower larval fish, the boundary layer is about as thick as the larval body is wide. It is therefore clearly visible around a swimming fish larva (Müller *et al.*, 2008).

The difference in boundary layer between larvae and adults becomes most obvious when the fish stops moving its body and begins to coast. A coasting larva drags along a rapidly growing layer of water, and the boundary layer grows thicker than the larva is wide within 100 ms. This quick growth is a manifestation of the considerable viscous drag that rapidly slows the larva down. This inability to coast for more than 0.5 mm might explain why fish larvae rely initially on cyclic swimming. While early larvae can burst, they cannot coast. Hence, they switch to intermittent swimming only once their coast performance has considerably improved (Weihs, 1980).

High viscous forces slow down not only the larva, they also cause the flow generated by the larva to slow down and die off quickly. Flow speeds in the wake of a cyclicly swimming larva drop to less than 10% of their peak value within 50 ms. This brief persistence of larval flow patterns means that larvae are less likely to be detected by predators or conspecifics due to their hydrodynamic signature (e.g. van Duren *et al.*, 1998; Dehnhardt *et al.*, 2001).

Flow Pattern during a Start

During the burst, the boundary layer has no time to build up and, therefore, larval and adult flow fields look quite similar during the burst phase, in particular during the start (Wolfgang *et al.*, 1999; Müller *et al.*, 2007) (Fig. 6). While assuming a "C" shape during the preparatory phase, the larva sucks water towards the increasingly concave side of its body, and a vortex ring is shed at the tail. During the propulsive stroke, another vortex ring is shed. These two rings form the wake of a larva starting to swim. The fact that the larva accelerates during the start manifests in the orientation of the vortex ring: it is angled backward relative to the larva's mean path of motion.

Flow Pattern during Cyclic Swimming

During steady swimming, larval and adult flow fields differ. These differences are caused not only by the differences in relative boundary layer thickness, but also by the differences in body shape. The larval flow pattern deviates most from that of adult fish along the anterior body (Fig. 7). Here, the thick larval yolk-sac generates a drag wake along the anterior body that is absent around the more streamlined adult fish. How important the contribution of the viscous flow regime is to this drag wake becomes obvious when comparing fish larvae to tadpoles, which have a similar body shape, but

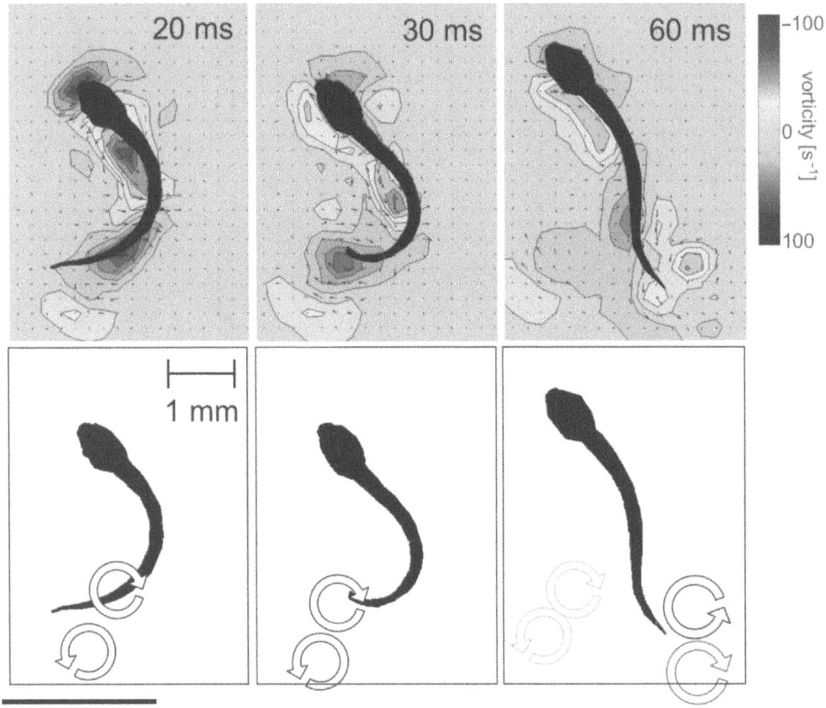

Fig. 6 Flow pattern of a larval zebrafish (*Danio rerio,* age 5 days post fertilisation) during a larval "C" start at the end of the preparatory phase (20 ms after initiation of "C" start), the beginning of the propulsive phase (30 ms) and the end of the propulsive phase (60 ms). The flow is visualised in the transverse plane through the larva. Therefore, any vortex rings in the wake manifest as two vortices of opposite sense of rotation. Top panel: The larva generates strong flows towards the concave side of its body (array of black arrows) (20 ms). These flows also cause elevated vorticity (gray shades), and this vorticity is shed at the tail (30 and 60 ms). The shed vorticity forms one vortex ring in the preparatory and one in the propulsive phase. Shed vortices are visible as areas of elevated vorticity (grey shades) behind the larva immediately after they were shed, but they die off quickly. Bottom panel: simplified sketch of the shed vortex wake.

are large enough to swim in the inertial flow regime (Liu *et al.*, 1997). A 4.7 cm long bullfrog (*Rana catesbeiana*) tadpole develops a relatively narrow drag wake compared with the much smaller 4 mm long zebrafish larva.

Flow fields of larval and adult fish resemble each other more closely along the posterior body, where body undulations prevent a thick boundary layer from developing and the flow is instead dominated by the interaction between the lateral body undulations and the adjacent water. The undulating body generates strong flow towards the concave side of the body, similar to flow patterns reported for juvenile and adult eel (Müller *et al.*, 2001; Tytell, 2004). When these body-generated flows reach the tail, they form vortex rings in the wake behind the fish. Zebrafish larvae and eels

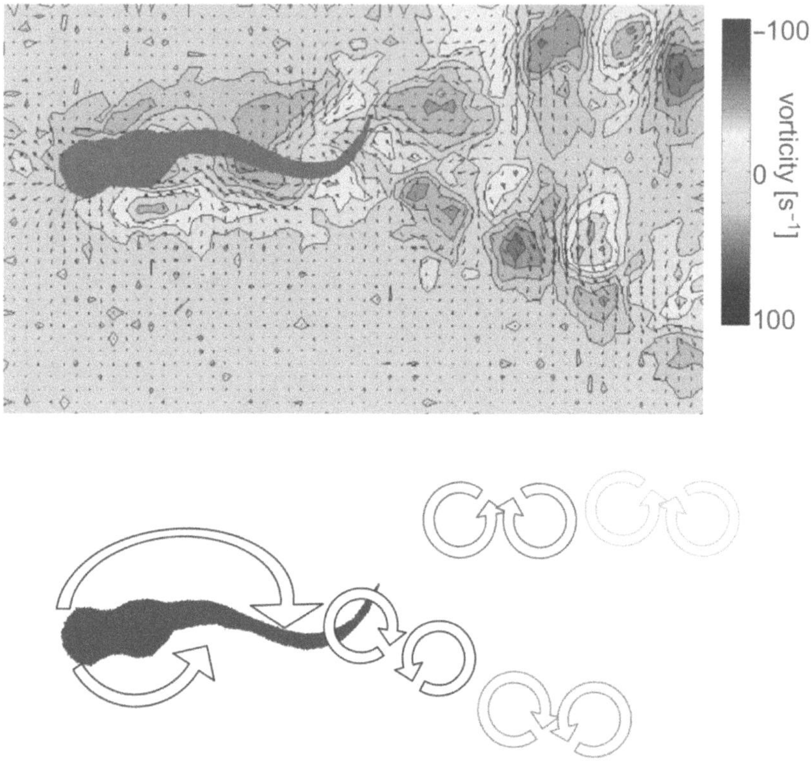

Fig. 7 Transverse cross-section through the flow pattern of a larval zebrafish (*Danio rerio,* age 3 days post fertilisation) during cyclic swimming. Top panel: the larva generates a strong drag wake along its anterior body, manifest as areas of elevated vorticity; along the posterior body, the larva generates a propulsive flow pattern with strong lateral flows and areas of elevated vorticity. The vorticity that is shed at the tail into the wake forms vortex rings. Bottom panel: sketch of a transverse cross-section through the wake of a fish larva; the wake comprises vortex rings, visible in the cross-section as counter-rotating vortex pairs, that are shed to the left and right of the mean path of motion. Compare to the wake shed behind a steadily swimming European eel in Fig. 5.

(European eel (*Anguilla anguilla*), American eel (*Anguilla rostrata*)) are all elongate fish that swim with a short body wave, and both shed a vortex pair wake – individual vortex rings shoot away left and right from the mean path of motion (Müller *et al.*, 2001; Tytell, 2004; Kern & Koumoutsakos, 2006). Whereas spindle-shaped fish with a narrow body wave, such as rainbow trout (*Oncorhynchus mykiss*), shed a chain of linked vortex rings (Blickhan *et al.*, 1992; Müller *et al.*, 1997, 2002; Nauen & Lauder, 2002). It is still unknown whether these differences in vortex arrangement in the wake are a result of differences in body shape (elongate *versus* spindle shaped), in body wave (wide *versus* narrow amplitude envelope), or in swimming performance (maximising thrust *versus* efficiency).

HOW FISH LARVAE POWER SWIMMING: AXIAL MUSCLE

Muscle Performance Required for Larval Swimming

Considering how little we know about how teleost fish larvae swim, even less is known about how fish larvae power swimming. We know that the demands on larval swimming muscles differ from those on most adult swimming muscles. Swimming fish larvae employ high tail-beat frequencies, up to 100 Hz (Budick & O'Malley, 2000; Müller & van Leeuwen, 2004). In combination with a body wave amplitude that is wider than in adult fish, these high frequencies might require the larval swimming muscles to operate at extreme strain rates, considerably higher than in adult fish (Müller & van Leeuwen, 2004). Muscles that contract cyclicly at frequencies approaching and exceeding 100 Hz are called "superfast" (Rome *et al.*, 1996). This muscle type usually operates at low strains (Young & Rome, 2001). Superfast muscles are also known for their poor power generation and are therefore considered unsuitable for locomotion (Young & Rome, 2001). However, given their much smaller body size, fish larvae need much less power to swim than adult fish. The scaling of muscle power generation *versus* propulsive power requirements with size might yet enable fish larvae to use muscle fibre types with superfast properties for locomotion.

To understand how larval swimming muscles meet locomotory requirements, it is necessary to characterise their morphology (shape of the myotomes, length and orientation of the muscle fibres), contractile behaviour (muscle fibre type, contractile properties), activation patterns (neural pathways and EMG patterns), and how these aspects change during larval ontogeny. While many of these aspects have been studied extensively in adult fish, much less is known about larval swimming muscles because many of the experimental techniques that worked so successfully in adults, are much more difficult or unfeasible in larval fish.

Muscle Fibres Required for Larval Swimming

Considerable work has been done on fibre typing in larval swimming muscles. Fish develop their muscle fibre composition in two main phases. First, during the embryonic and early larval stage, the two larval fibre types develop: red and white (e.g. Waterman, 1969; Nag & Nursall, 1972; van Ramsdonk *et al.*, 1978; Batty, 1984; Koumans & Akster, 1995; Hall & Johnston, 2003). During the embryonic phase, the red muscle fibres form at the midline near the notochord and migrate to the surface to form a single cell layer under the skin (Devoto *et al.*, 1996; Stickney *et al.*, 2000), so fish larvae hatch with a superficial layer of red muscle fibres and a deeper layer of white muscle fibres (Waterman, 1969; Nag & Nursall, 1972; Koumans & Akster, 1995; Rescan, 2005) (Fig. 8). One known exception is Northern anchovy *(Engraulis mordax)*, which hatches with two layers of oblique muscle fibres instead of one red and one white fibre

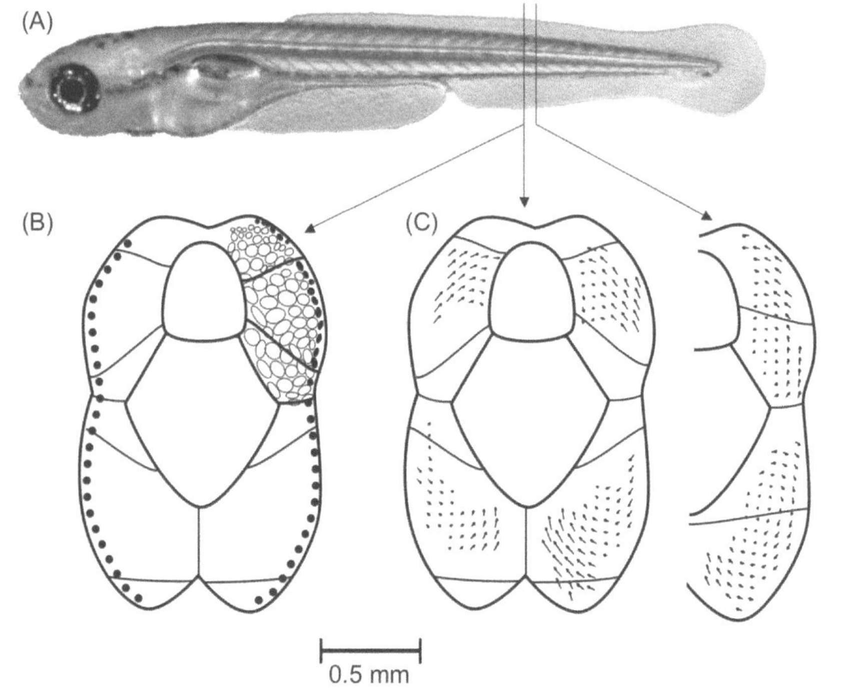

Fig. 8 Muscle fibre architecture of a zebrafish larva (*Danio rerio,* age 4 days post fertilisation). After van der Stelt *et al.* (1977) and van der Meulch *et al.* (2005). (A) A 4 day-old zebrafish larva has single chevron-shaped myotomes, whereas adult fish have a double chevron. (B) The cross-section through the 20th myotome shows the single layer of superficial red muscle fibres (black circles) and white muscle fibres (white circles). Fine lines indicate septa. (C) The white muscle fibres are arranged helicoidally (arrows).

layer (O'Connell, 1981). In the second phase, near yolk-sac absorption when exogenous feeding commences, the adult red and pink fibre types develop, and the white muscle layer starts growing by hyperplasia (Koumans & Akster, 1995; Alami-Durante *et al.*, 2006; Johnston, 2006).

The red muscle fibres of most adult fish would be too slow to power the high tail beat frequencies observed in fish larvae (Syme, 2006). This led to the hypothesis that the superficial red muscle fibres in fish larvae serve a respiratory rather than a locomotory function (El-Fiky & Wieser, 1988), and that swimming movements are powered by white muscle fibres (Osse & van den Boogaart, 1999). However, slow (red) muscle cells express genes for both slow and fast myosin heavy chains (Bryson-Richardson *et al.*, 2005). Hence, red muscle fibres might have slow and fast myosin heavy chains, while the fast (white) muscle fibres have fast myosin heavy chains. This expectation might explain the results of a recent study on the contractile behaviour of larval muscle fibres, which suggests that larval red muscle fibres are fast enough to power sustained swimming bouts in zebrafish (Buss & Drapeau, 2000). In these larvae, red muscle fibres are not only fast enough to power swimming. But surprisingly red and white fibres also seem to have similar contraction kinetics (Buss & Drapeau, 2000). In contrast, adult red and white muscle fibres differ substantially in their contraction kinetics, with slow red fibres powering slow, sustained swimming and fast white fibres powering fast, anaerobic bursts (Young & Rome, 2001; Syme, 2006). In this and in many other aspects of their contractile behaviour, larval muscle fibres differ significantly from adult muscle fibres (Buss & Drapeau, 2000).

The contractile behaviour of muscle fibres has been studied extensively in adult fish in order to measure muscle efficiency (Curtin & Woledge, 1993) and many other useful indicators of muscle performance. An important methodological breakthrough was the development of the workloop technique (Josephson, 1985), in which a single muscle fibre or a small bundle of fibres is mounted at one end on a force transducer and at the other end to a servomotor, which cyclically stretches the muscle fibres. The fibres are then stimulated at different phases of the stretching cycle. By varying the stimulus phase and duration, the stretch amplitude and frequency, and the shape of the stretch cycle (sine wave, ramp, etc.), it is possible to determine maximum power output, force development, maximum contractile speed, and titanic frequency (Altringham & Johnston, 1990; Rome *et al.*, 1993; Altringham & Ellerby, 1999; for a review see Syme, 2006). However, larval swimming muscles are too small to be mounted in a workloop-set-up. So, what little we know about muscle contractile properties of larval swimming muscles has mostly been gleaned from *in vivo* experiments (e.g. Buss & Drapeau, 2000).

Muscle Architecture Required for Larval Swimming

Muscle morphology and fibre composition is reasonably well-understood in larval fish, including ontogenetic changes. With the recent advances in molecular techniques, much has been learned about the development of muscles in fish embryos and larvae, how the location and number of muscle fibre types change during ontogeny. The functional context of myotome shape and muscle fibre arrangements are beginning to be well-understood in adult fish – we are beginning to understand how muscle morphology and physiology are related to the mechanical demands of swimming (van Leeuwen, 1999; Shadwick & Gemballa, 2006). Much less is known about the functional context of ontogenetic morphological changes. However, functional context is not synonymous with functional control: the shape of the larval locomotory apparatus seems to be mainly controlled genetically (Devoto *et al.*, 1996; van der Meulen *et al.*, 2005). Experiments with mutant zebrafish, whose mechanically intact muscle fibres do not contract due to defective neural signalling, show that myotome shape and muscle fibre arrangement are mainly controlled genetically rather than mechanically (van der Meulen *et al.*, 2005). But there is ontogenetic plasticity that enables the fish to respond to the environment. In particular temperature effects have been studied extensively (Hall & Johnston, 2003; Alami-Durante *et al.*, 2006).

Adult teleosts have a striking myotome shape (for a review see Shadwick & Gemballa, 2006) and helicoid fibre angles (Alexander, 1969; Mos & van der Stelt, 1982). This combination of double-coned shaped myotomes and helicoid fibre arrangement is hypothesised to ensure uniform strain. If the fibres were arranged not helicoidally, but parallel to the vertebral column, then the muscle fibres in a fish that undulates its body to swim, would experience different strains across the muscle cross section: the muscle fibres near the skin would experience the largest strains, and fibres near the vertebral column essentially no strain. Arranged along a suitably helicoidal path, the fibres in a given body cross section, from the skin to the midline, should experience the same strain (van der Stelt, 1968; Moss & van der Stelt, 1982). Uniform strain allows fibres of the same type to all operate under the same conditions, and therefore all deliver maximum power at their optimal strain and strain rate throughout the entire muscle cross-section. This helicoid fibre arrangement then requires cone-shaped myotomes, which ensure that muscle power is used mainly to bend the whole fish and to this end reduce wasteful internal deformations of the muscle (van Leeuwen, 1999).

In adult fish, this idea about uniform strain in particular and about the contractile behaviour of muscle fibres in general can be tested by measuring strain in a swimming fish. Adult fish have been implanted with electromyographic (EMG) electrodes to record muscle activity, and with sonomicrometry cystals to record the deformation of the muscles. Up to eight small crystals have been implanted into the muscle, and the distance between the crystals gives an estimate of muscle strain (Coughlin *et al.*, 1996; Katz *et al.*, 1999; Donley & Shadwick, 2003). The fish were then filmed and their swimming kinematics correlated with EMG and sonomicrometry recordings. There is still considerable debate about the results, but the emerging picture suggest a range of scenarios that corresponds to the range of swimming styles from anguilliform to thunniform swimmers (Rome *et al.*, 1993; Wardle *et al.*, 1995; Shadwick & Gemballa, 2006). Anguilliform swimmers seem to use their muscles rather uniformly along their body, carangiform swimmers might divide muscle function into a power-generating anterior zone and a force-transmitting posterior zone (Wardle *et al.*, 1995).

However, sonomicrometry cannot be applied to fish larvae – the smallest crystals are 1mm in diameter; neither is it feasible to insert EMG electrodes in free-swimming fish larvae. Hence, in fish larvae, muscle mechanical behaviour cannot yet be measured directly. Instead muscle contractile demands are estimated from the swimming movements, muscle fibre angles and myotome shape. Teleost larvae have myotomes with only one, not two cones such as adult teleosts (e.g. Buss & Drapeau, 2000; Gemballa & Vogel, 2002; van der Meulen *et al.*, 2005) (Fig. 8). In this respect, teleost larvae more closely resemble adults of lower chordates (Vogel & Gemballa, 2000; Gemballa & Röder, 2004). However, like adult teleosts, zebrafish larvae have helicoidal muscle fibre trajectories (van der Stelt *et al.*, 1977; van der Meulen *et al.*, 2005). Data on muscle fibre angles are scarce (adult fish: Alexander, 1969; larval fish: van der Meulen, 2005) (Fig. 8), and the mathematical tools for deriving muscle strain from muscle architecture are still being developed (van Leeuwen, 1999; van Leeuwen *et al.*,

2005). Hence, we do not know the *in vivo* contractile behaviour of larval swimming muscles. We also know little about the recruitment of different types of muscle fibres as a function of swimming speed or swimming behaviour.

To understand how muscle contractions are translated into propulsion, we need to know not only muscle contraction kinetics, muscle architecture, swimming kinematics and flow, we also need to know the mechanical pathways, which transfer muscle power to the propulsive structures. In adult fish, these data are gradually coming together (e.g. Cheng *et al.*, 1998; for a review see Summers & Long, 2006). However, in fish larvae, there are far fewer studies on a much narrower range of species, and many aspects, such as the mechanical behaviour of the larval swimming apparatus, have been barely touched upon.

FUTURE PERSPECTIVES

Research on larval fish is not only important to ichthyology, fishery and aquaculture. Fish larvae are becoming an increasingly important model to study the mechanics of swimming (McHenry & Lauder, 2006; Müller *et al.*, 2007), to study the motoneural control of locomotion (Kimmel, 1972; Eaton & DiDomenico, 1986; Hale *et al.*, 2001), and to understand the evolutionary aspects of vertebrate locomotion (Thorson *et al.*, 2004). In particular, innervation patterns and motoneural control of larval escape responses is receiving renewed attention (Fremont & Hale, 2005). By combining modern imaging techniques, genetic labelling and perturbation, by using mutants and knock-outs (Nüsslein-Volhard & Dahm, 2002), this line of research is becoming even more fruitful to study the evolution and ontogeny of neural control (Fetcho & Bhatt, 2004).

Acknowledgements

I would like to thank David Lentink, Otto Berg, Nigel Finn and an anonymous referee for encouraging comments on the manuscript. UKM was funded by NWO (project number 814.02.006) while working in the Experimental Zoology group of Johan van Leeuwen, Wageningen University, The Netherlands.

References

Alami-Durante H, Rouel M, Kentouri M (2006) New insights into temperature-induced white muscle growth plasticity during *Dicentrarchus labrax* early life: a developmental and allometric study. Mar Biol 149: 1551-1565

Alexander RM (1969) The orientation of muscle fibres in the myomeres of fishes. J Mar Biol Assoc UK 49: 263-290

Altringham JD, Johnston IA (1990) Scaling effects on muscle function: power output of isolated fish muscle fibres performing oscillatory work. J Exp Biol 151: 453-467

Altringham JD, Ellerby DJ (1999) Fish swimming: patterns in muscle function. J Exp Biol 202: 3397-3403

Anderson EJ, Grosenbaugh MA (2005) Jet flow in steadily swimming adult squid. J Exp Biol 208: 1125-1146

Anderson EJ, McGillis WR, Grosenbaugh MA (2001) The boundary layer of swimming fish. J Exp Biol 204: 81-201

Bagatto B, Pelster B, Burggren WW (2001) Growth and metabolism of larval zebrafish: effects of swim training. J Exp Biol 204: 4335-4343

Bailey KM, Yen J (1983) Predation by a carnivorous marine copepod, *Euchaeta elongata* Esterly, on eggs and larvae of the Pacific hake, *Merluccius productus*. J Plankton Res 5: 71-82

Bailey KM, Batty RS (1984) Laboratory study of predation by *Aurelia aurita* on larvae of cod, flounder, plaice and herring: development an vulnerability to capture. Mar Biol 83: 287-291

Batty RS (1981) Locomotion of plaice larvae. Symp Zool Soc Lond 48: 53-69

Batty RS (1984) Development of the swimming movements and musculature of larval herring (*Clupea harengus*). J Exp Biol 110: 217-229

Batty RS, Blaxter JHS (1992) The effect of temperature on the burst swimming performance of fish larvae. J Exp Biol 170: 187-201

Beamish FWH (1966) Swimming endurance of some northwest Atlantic fishes. J Fish Res Bd Can 23: 341-347

Bellwood DR, Fisher R (2001) Relative swimming speeds in reef fish larvae. Mar Ecol Prog Ser 211: 299-303

Blake RW (2004) Fish functional design and swimming performance. J Fish Biol 65: 1193-1222

Blaxter JHS (1986) Development of sense organs and behaviour in teleost larvae with special reference to feeding and predator avoidance. Trans Am Fish Soc 115: 98-114

Blickhan R, Krick C, Zehren D, Nachtigall W, Breithaupt T (1992) Generation of a vortex chain in the wake of a subundulatory swimmer. Naturwissenschaften 79: 220-221

Breder CM (1926) The locomotion of fishes. Zoologica 4: 159-297

Bryson-Richardson RJ, Daggett DF, Cortes F, Neyt C, Keenam DG, Currie PD (2005) Myosin heavy chain expression in zebrafish and slow muscle composition. Dev Dyn 233: 1018-1022

Budick SA, O'Malley DM (2000) Locomotor repertoire of the larval zebrafish: swimming, turning and prey capture. J Exp Biol 203: 2565-2579

Buss RR, Drapeau P (2000) Physiological properties of zebrafish embryonic red and white muscle fibers during early development. J Neurophysiol 84: 1545-1557

Buss RR, Drapeau P (2001) Synaptic drive to motorneurons during fictive swimming in the developing zebrafish. J Neurophysiol 86: 197-210

Buss RR, Drapeau P (2002) Activation of embryonic red and white muscle fibers during fictive swimming in the developing embryo. J Neurophysiol 87: 1244-1251

Castro LR, Hernandez EH (2000) Early life survival of the anchoveta *Engraulis ringens* off central Chile during the 1995 and 1996 winter spawning seasons. Trans Am Fish Soc 129: 1107-1117

Cheng J-Y, Pedley TJ, Altringham JD (1998) A continuous dynamic beam model for swimming fish. Phil Trans R Soc Lond B 353: 981-997

Coughlin DJ, Valdes L, Rome LC (1996) Muscle length changes during swimming in scup: sonomicrometry verifies the anatomical high-speed cine technique. J Exp Biol 199: 459-463

Cowan JH, Houde ED, Rose KA (1996) Size-dependent vulnerability of marine fish larvae to predation: an individual-based numerical experiment. ICES J Mar Sci 53: 23-37

Curtin NA, Woledge RC (1993) Efficiency of energy conversion during sinusoidal movement of white muscle fibres from the dogfish *Scyliorhinus canicula*. J Exp Biol 183: 137-147

Dabiri JO (2005) On the estimation of swimming and flying forces from wake measurements. J Exp Biol 208: 3519-3532

Dabiri JO, Colin SP, Costello JH, Gharib M (2005) Flow patterns generated by oblate medusan jellyfish: field measurements and laboratory analyses. J Exp Biol 208: 1257-1265

Dehnhardt G, Mauck B, Hanke W, Bleckmann H (2001) Hydrodynamic trail-following in harbor seals (*Phoca vitulina*). Science 293: 102-104

Devoto SH, Melancon E, Eisen JS, Westerfield M (1996) Identification of separate slow and fast muscle precursor cells *in vivo*, prior to somite formation. Development 122: 3371-3380

Donley JM, Shadwick RE (2003) Steady swimming muscle dynamics in the leopard shark *Triakis semifasciata*. J Exp Biol 206: 1117-1126.

Drucker EG, Lauder GV (1999) Locomotor forces on a swimming fish: Three-dimensional vortex wake dynamics quantified using digital particle image velocimetry. J Exp Biol 202: 2393-2412

Drucker EG, Lauder GV (2001) Locomotor function of the dorsal fin in teleost fishes: experimental analysis of wake forces in sunfish. J Exp Biol 204: 2943-2953

Eaton RC, DiDomenico R (1986) Role of the teleost escape response during development. Trans Am Fisher Soc 115: 128-142

El-Fiky N, Wieser W (1988) Life styles and patterns of development of gills and muscles in larval cyprinids (Cyprinidae, Teleostei). J Fish Biol 33: 135-145

Fisher R, Wilson SK (2004) Maximum sustainable swimming speeds of late-stage larvae of nine species of reef fishes. J Exp Mar Biol Ecol 312: 171-186

Fisher R, Bellwood DR, Job SD (2000) Development of swimming abilities in reef fish larvae. Mar Ecol Prog Ser 202: 163-173

Flore L, Keckeis H (1998) The effects of water current on foraging behaviour of the rheophilic cyprinid, *Chondrostoma nasus*, during early ontogeny: evidence of a trade-off between energetics benefit and swimming costs. Regul River 14: 141-154

Fremont RT, Hale ME (2005) Building neural networks: The response of the zebrafish startle circuit ectopic expression of mauthner cells. Integr Comp Biol 45: 1133-1133

Fuiman, LA (1986) Burst-swimming performance of larval zebra danios and the effects of diel temperature fluctuations. Trans Am Fish Soc 115: 143-148

Fuiman LA (2002) Special considerations of fish eggs and larvae. In: Fuiman, LA, Werner, RG (eds) Fishery Science: The Unique Contributions of Early Life Stages. Blackwell Publishing Ltd, Oxford, p 1-32

Fuiman LA, Webb PW (1988) Ontogeny of routine swimming activity and performance in zebra danios (Teleostei: Cyprinidae). Anim Behav 36: 250-261

Fuiman LA, Werner RG (2002) Fishery Science: The Unique Contributions of Early Life Stages. Blackwell Publishing Ltd, Oxford

Fuiman LA, Cowan JH Jr. (2003) Behavior and recruitment success in fish larvae: repeatability and covariation of survival skills. Ecology 84: 53-67

Fetcho JR, Bhatt DH (2004) Genes and photons: new avenues into the neural basis of behaviour. Curr Opin Neurobiol 14: 707-714

Gemballa S, Vogel F (2002) Spatial arrangement of white muscle fibres and myoseptal tendons in fishes. Comp Biochem Physiol 133A: 1013-1037

Gemballa S, Röder K (2004) From head to tail: the myoseptal system in basal Actinopterygians. J Morphol 259: 155-171

Gibson RN (2005) Flatfishes. Biology and Exploitation. Blackwell Publishing Ltd, Oxford

Gopalkrishnan R, Triantafyllou MS, Triantafylou GS, Barret DS (1994) Active vorticity control in a shear flow using a flapping foil. J Fluid Mech 274: 1-21

Grillner S, Wallen P, Brodin L, Lansner A (1991) Neuronal network generating locomotor behaviour in lamprey – circuitry, transmitters, membrane properties, and simulation. Ann Rev Neurosci 14: 169-199

Hale ME (1999) Locomotor mechanics during early life history: effects of size and ontogeny on fast-start performance of salmonid fishes. J Exp Biol 202: 1465-1479

Hale ME, Ritter DA, Fetcho JR (2001) A confocal study of spinal interneurons in living larval zebrafish. J Comp Neurol 437: 1-16

Hall EH, Johnston IA (2003) Temperature and developmental plasticity during embryogenesis in the Atlantic cod *Gadus morhua* L. Mar Biol 142: 833-840

Harper DG, Blake RW (1989) On the error involved in high-speed film when used to evaluate maximum acceleration in fish. Can J Zool 67: 1929-1936

Houde ED (1994) Differences between marine and freshwater fish larvae: implications for recruitment. ICES J Mar Sci 51: 91097

Hunter JR (1972) Swimming and feeding behaviour of larval anchovy *Engraulis mordax*. Fish B-NOAA 70: 821- 838

Johnston IA (2006) Environment and plasticity of myogenesis in teleost fish. J Exp Biol 209: 2249-2264

Josephson RK (1985) Mechanical power output from striated muscle during cyclic contraction. J Exp Biol 114: 493-512

Kaufmann R (1990) Respiratory cost of swimming in larval and juvenile cyprinids. J Exp Biol 150: 343-366

Katz SL, Shadwick RE, Rapoport HS (1999) Muscle strain histories in swimming milkfish in steady as well as sprinting gaits. J Exp Biol 202: 529-541

Keckeis H, Schiemer F (2002) Case studies. Understanding conservation issues of the Danube river. In: Fuiman LA, Werner RG (eds) Concepts in Fisheries Science: The Unique Contribution of Early Life Stages. Blackwell Publishing Ltd, Oxford, p 272-288

Kern S, Koumoutsakos P (2006) Simulations of optimized anguilliform swimming. J Exp Biol 209: 4841-4857

Kimmel CB (1972) Mauthner axons in living larvae. Dev Biol 27: 272-275

Koumans JTM, Akster HA (1995) Myogenic cells in development and growth of fish. Comp Biochem Physiol 110A: 3-20

Lauder GV, Tytell ED (2006) Hydrodynamics of undulatory propulsion. In: Shadwick RE, Lauder GV (eds) Fish Biomechanics. Fish Physiology Vol 23. Academic Press, Amsterdam, p 425-468

Lindsey CC (1978) Form, function and locomotory habits in fish. In: Hoar WS, Randall DJ (eds) Fish Physiology Vol 7, Academic Press, London, p 1-100

Litvak MK, Leggett WC (1992) Age and size-selective predation on larval fish: the bigger is better paradigm revisited. Mar Ecol Prog Ser 81: 1-12

Liu H, Wassersug R, Kawachi K (1997) The three-dimensional hydrodynamics of tadpole locomotion. J Exp Biol 200: 2807-2819

MacKenzie BR, Leggett WC (1991) Quantifying the contribution of small-scale turbulence to the encounter rates between larval fish and their zooplankton prey – effects of wind and tide. Mar Ecol Prog Ser 73: 149-160

McElligott MB, O'Malley DM (2005) Prey tracking by larval zebrafish: axial kinematics and visual control. Brain Behav Evol 66: 177-196

McHenry MJ, Jed J (2003) The ontogenetic scaling of hydrodynamics and swimming performance in jellyfish (*Aurelia aurita*). J Exp Biol 206: 4125-4137

McHenry MJ, Lauder GV (2005) The mechanical scaling of coasting in zebrafish (*Danio rerio*). J Exp Biol 208: 2289-2301

McHenry MJ, Lauder GV (2006) Ontogeny of form and function: locomotor morphology and drag in zebrafish. J Morphol 267: 1099-1109

McHenry MJ, Azizi E, Strother JA (2003) The hydrodynamics of locomotion at intermediate Reynolds numbers: undulatory swimming in ascidian larvae (*Botrylloides* sp.) J Exp Biol 206: 327-343

Miller TJ, Crowder LB, Rice JA, Marschall EA (1988) Larval size and recruitment mechanisms in fishes: towards a conceptual framework. Can J Fish Aquat Sci 45: 1657-1670

Mos W, van der Stelt A (1982) Efficiency in relation to the design of the segmented body musculature in *Brachydanio rerio*. Neth J Zool 32: 123-143

Müller UK, van Leeuwen JL (2004) Swimming of larval zebrafish: ontogeny of body waves and implications for locomotory development. J Exp Biol 207: 853-868

Müller UK, van Leeuwen JL (2006) Undulatory fish swimming: from muscles to flow. Fish Fisher 7: 84-103

Müller UK, van den Heuvel BLE, Stamhuis EJ, Videler JJ (1997) Fish foot prints: morphology and energetics of the wake behind a continuously swimming mullet (*Chelon labrosus*). J Exp Biol 200: 2893-2906

Müller UK, Stamhuis EJ, Videler JJ (2000) Hydrodynamics of unsteady fish swimming and the effects of body size: comparing fish larvae and adults. J Exp Biol 203: 193-206

Müller UK, Smit J, Stamhuis EJ, Videler JJ (2001) How the body contributes to the wake in undulatory fish swimming: flow fields of a swimming eel (*Anguilla anguilla*). J Exp Biol 204: 2751-2762

Müller UK, Stamhuis EJ, Videler JJ (2002) Riding the waves: the role of the body wave in undulatory fish swimming. Integr Comp Biol 42: 981-987

Müller UK, van den Boogaart JGM, van Leeuwen JL (2008) Flow patterns of larval fish: Undulatory swimming in the intermediate flow regime. J Exp Biol 211: 196-205

Nag AC, Nursall JR (1972) Histogenesis of white and red muscle fibres of trunk muscles of a fish *Salmo gairdneri*. Cytobios 6: 227-246

Nauen JC, Lauder GV (2002) Quantification of rainbow trout (*Oncorhynchus mykiss*) using three-dimensional stereoscopic particle image velocimetry. J Exp Biol 205: 3271-3279

Nüsslein C, Dahm R (2002) Zebrafish. Oxford University Press, Oxford

O'Connell (1981) Development of organ systems in the Northern anchovy *Engraulis mordax* and other teleosts. Am Zool 21: 429-446

Osse JWM, van den Boogaart JGM (1999) Dynamic morphology of fish larvae, structural implications of friction forces in swimming, feeding and ventilation. J Fish Biol 55: 156-174

Paradis AR, Pepin P, Brown JA (1996) Vulnerability of fish eggs and larvae to predation: Review of the influence of the relative size of prey and predator. Can J Fish Aquat Sci 53: 1226-1235

Plaut I (2001) Critical swimming speed: its ecological relevance. Comp Biochem Physiol 131A: 41-50

Ramamurti R, Sandberg WC, Lohner R, Walker JA, Westneat MW (2002) Fluid dynamics of flapping aquatic flight in the bird wrasse: three-dimensional unsteady computations with fin deformation. J Exp Biol 205: 2997-3008

Reidy SP, Kerr SR, Nelson JA (2000) Aerobic and anaerobic swimming performance of individual Atlantic cod. J Exp Biol 203: 347-357

Rescan PY (2005) Muscle growth patterns and regulation during fish ontogeny. Gen Comp Endocrinol 142: 111-116

Rome LC, Swank D, Corda D (1993) How fish power swimming. Science 261: 340-343

Rome LC, Syme DA, Hollingworth S, Lindstedt SL, Baylor SM (1996) The whistle and the rattle: the design of sound-producing muscles. Proc Natl Acad Sci USA 93: 8095-8100

Saint-Amant L, Drapeau P (1998) Time course of the development of motor behaviors in the zebrafish embryo. J Neurobiol 37: 622-632

Skiftesvik AB (1992) Changes in behavior at onset of exogenous feeding in marine fish larvae. Can J Fish Aquat Sci 49: 1570-1572

Sponaugle S, Grorud-Covert K (2006) Environmental variability, early life-history traits, and survival of new coral reef fish recruits. Integr Comp Biol 46: 623-633

Shadwick RE, Gemballa S (2006) Structure, kinematics and muscle dynamics in undulatory swimming. In: Shadwick RE, Lauder GV (eds) Fish Biomechanics. Fish Physiology Vol 23. Academic Press, Amsterdam, p 241-332

Stickney HL, Barressi MJF, Devoto SH (2000) Somite development in zebrafish. Dev Dyn 219: 287-303

Summers AP, Long JH Junior (2006) Skin and bones, sinew and gristle: the mechanical behaviour of fish skeletal tissue. In: Shadwick RE, Lauder GV (eds) Fish Biomechanics. Fish Physiology Vol 23, Academic Press, Amsterdam, p 141-177

Syme DA (2006) Functional properties of skeletal muscle. In: Shadwick RE, Lauder GV (eds) Fish Biomechanics. Fish Physiology Vol 23. Academic Press, Amsterdam, p 179-240

Thorson DH, Cassidy JJ, Hale ME (2004) Swimming of larval zebrafish: fin-axis coordination and implications for function and neural control. J Exp Biol 207: 4175-4183

Tytell ED (2004) Kinematics and hydrodynamics of linear acceleration in eels, *Anguilla rostrata*. Proc R Soc Lond B 271: 2535-2540

Tytell ED (2006) Median fin function in bluegill sunfish *Lepomis macrochirus*: streamwise vortex structure during steady swimming. J Exp Biol 209: 1516-1534

Tytell ED, Lauder GV (2004) The hydrodynamics of eel swimming. I. Wake structure. J Exp Biol 207: 1825-1841

van der Meulen T, Schipper H, van Leeuwen JL, Kranenbarg S (2005) Effects of decreased muscle activity on developing axial musclulature in nic^{b107} mutant zebrafish (*Danio rerio*). J Exp Biol 208: 3675-3687 van der Stelt A (1968) Spiermechanica en myotoombouw bij vissen. LJ Veen's Uitgeversmaatschappij N.V., University of Amsterdam, Amsterdam. PhD Thesis [in Dutch]

van der Stelt A, Mos W, Diegenbach PC (1977) Notes on reconstruction of muscle fibre orientation during the development of the teleost *Brachydanio rerio*. Bijdr Dierkd 46: 284-290

van Duren LA, Stamhuis EJ, Videler JJ (1998) Reading the copepod personal ads: increasing encounter probability with hydromechanical signals. Phil Trans R Soc Lond B 353: 691-700

van Leeuwen JL (1999) A mechanical analysis of myomere shape in fish. J Exp Biol 202: 3405-3414

van Raamsdonk W, Pool CW, te Kronnie G (1978) Differentiation of muscle fiber types in the teleost *Brachydanio rerio*. Anat Embryol 153: 137-155

Videler JJ (1993) Fish Swimming. Chapman and Hall, London

Vogel F, Gemballa S (2000) Locomotory design of cyclostome fishes: spatial arrangement and architecture of myosepta and lamellae. Acta Zool 81: 267-283

Wardle CS, Videler JJ, Altringham JD (1995) Tuning into fish swimming waves: body from, swimming mode and muscle function. J Exp Biol 198: 1629-1636

Waterman RE (1969) Development of the lateral musculature in the teleost, *Brachydanio rerio*: a fine structural study. Am J Anat 125: 457-494

Webb JF (1999) Larvae in fish development and evolution. In: Hall BK, Wake MH (eds) The Origin and Evolution of Larval Forms. Academic Press, San Diego p 109-158

Webb PW (1975) Hydrodynamics and energetics of fish propulsion. Bull Fish Res Bd Can 190: 1-159

Webb PW (1984) Body form, locomotion and foraging in aquatic vertebrates. Am Zool 24: 107-120

Webb PW, Weihs D (1986) Functional locomotor morphology of early-life-history stages of fishes. Trans Am Fish Soc 115: 115-127

Weihs D (1980) Energetic significance of changes in swimming modes during growth of anchovy larvae, *Engraulis mordax*. Fish. B-NOAA 77: 597-604

Wilga CD and Lauder GV (2002) Function of the heterocercal tail in sharks: quantitative wake dynamics during steady horizontal swimming and vertical maneuvering. J Exp Biol 205: 2365-2374

Williams PJ, Brown JA, Gotceitas V, Pepin P (1996) Developmental changes in escape response performance of five species of marine larval fish. Can J Fish Aquat Sci 53: 1246-1253

Wolfgang MJ, Anderson JM, Grosenbaugh MA, Yue DKP, Triantafyllou MS (1999) Near-body flow dynamics in swimming fish. J Exp Biol 202: 2303-2327

Young IS, Rome LC (2001) Mutually exclusive muscle designs: the power output of the locomotory and sonic muscles of the oyster toad fish (*Opsanus tau*). Proc R Soc Lond B268: 1965-2001

Part 6

Control and Defense

CHAPTER

17

Enteric Control

Anna Holmberg, Susanne Holmgren and Catharina Olsson*

INTRODUCTION

The enteric nervous system is one of the main actors involved in control of gut motility and secretion. It constitutes a part of the autonomic nervous system, but may act independently of central input. Nevertheless, extrinsic autonomic innervation is important for example for co-ordination of gut movements over longer distances as well as for different reflex patterns. Other key factors controlling gut activity are local hormones and the intrinsic electrical activity in the interstitial cells of Cajal (ICCs).

The importance of a functional enteric nervous system can be seen for example in people suffering from Hirschsprung's disease, a condition where enteric neurons fail to colonise the distal colon (Newgreen & Young, 2002a). It often causes severe constipation due to inability of the gut to carry out normal peristalsis. In order to reveal the causes behind this and other developmental abnormalities it is important to study the normal development of the enteric nervous system.

This chapter attempts to summarise the current knowledge about the development of the enteric nervous system in fish, with emphasis on its role in controlling gut motility. Teleosts, especially zebrafish (*Danio rerio*), have become increasingly important as model species regarding gut development (Shepherd *et al.*, 2001; Holmberg *et al.*,

Authors' address: Department of Zoology/Zoophysiology, Göteborg University, Box 463, SE 405 30 Göteborg, Sweden.

**Corresponding author:* E-mail: c.olsson@zool.gu.se

2003, 2004; Wallace & Pack, 2003; Wallace *et al.*, 2005). They have several advantages to mammals such as being easy to breed as well as having external fertilisation and development. The physiology of most organs is similar to other vertebrates and studying their formation in fish will give insight to normal development in other animal groups. Another important aspect is that fish can be used as a model for effects of environmental toxins on the development of the enteric nervous system. However, even if the number is increasing, there are still only a few studies on development of the enteric nervous system in fish. Hence, where appropriate, comparisons will be made to other vertebrate species.

GENERAL ANATOMY OF THE FISH GUT

Although the gross anatomy of the gut may vary between fish species, the general architecture is quite consistent and resembles the layout found in most vertebrate groups (Stevens & Hume, 1995) (Fig. 1). The oesophagus leads to a stomach, where mechanical and chemical digestion of the food begins. The stomach is followed by the intestine, which is the main site for digestion and absorption. The anatomical subdivisions of the intestine are often less clear-cut than in mammals, but usually one can distinguish an intestine (small intestine) and a rectum (large intestine). However, some fish like cyprinids (including zebrafish) and labrids, lack a proper stomach and in these species the intestine follows directly after the oesophagus (Koelz, 1992). In zebrafish, the proximal part of the intestine is enlarged to a bulb-like structure, commonly called the intestinal bulb, which possibly serves a transient storage purpose.

As in other vertebrates, the gut wall consists of several well-defined layers (Nilsson, 1983; Furness & Costa, 1987) (Fig. 1 B, C). The majority of the enteric nerve cell bodies are found in the myenteric plexus, situated between the outer longitudinal and inner circular smooth muscle layers. The submucosal plexus, between the circular muscle layer and the submucosa (a layer of connective tissue, vessels and glands below the mucosa), is usually less well-developed in fish than in mammals.

DEVELOPMENT OF THE GUT

A proper development of the enteric nervous system is dependent on the development of the gut itself. In amniotes, a primitive gut is formed from an endodermal tube at an early stage during development. This tube continues to develop in an oral-to-anal fashion, giving rise to the entire gastrointestinal tract (Wells & Melton, 1999). In contrast in fish embryos, at least in zebrafish, the digestive tube is formed comparatively late and from two separate endodermal precursors (Wallace & Pack, 2003). Between 34 and 42 hours post fertilisation (hpf) the endoderm forms a hollow tube, the primitive intestine, extending distally from the fin buds (Horne-Badovinac *et al.*, 2001; Wallace & Pack, 2003). At this stage, endoderm rostral to the gut is still relatively undifferentiated and does not develop into the oesophageal primordium until 50 hpf (Wallace & Pack, 2003). At 58 hpf the intestine and oesophagus have joined to form a continuous unit,

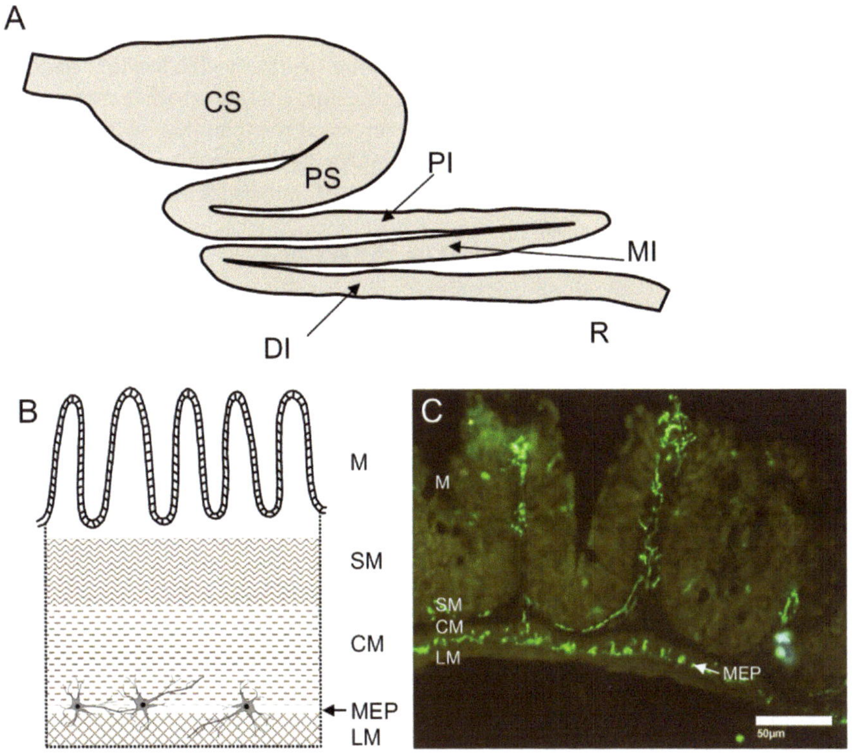

Fig. 1 **A.** Schematic drawing of the gastrointestinal tract of fish. CS, cardiac stomach; PS, pyloric stomach; PI, proximal intestine, MI, middle intestine; DI, distal intestine; R, rectum. **B.** Schematic drawing of the different layers of the gut wall in fish in general. **C.** Micrograph showing a cross section of the gut wall of an adult zebrafish (*Danio rerio*). M, mucosa; SM, submucosa; CM, circular muscle; MEP, myenteric plexus; LM, longitudinal muscle. (B. redrawn with permission from Olsson & Holmgren, 2001).

at 74 hpf the mouth is open and around 4 days post-fertilisation (dpf) the anus is open as well (Wallace & Pack, 2003).

The endoderm gives rise to the epithelial lining of the gut, while smooth muscle cells and connective tissues are formed from the mesoderm. The enteric nervous system that will be discussed in more detail later is derived from ectoderm. At 50 hpf (approx. 2 dpf), smooth muscle cells start to colonise the proximal intestine of the zebrafish, and at 70 hpf the entire gut is colonised by smooth muscle cells (Wallace *et al.*, 2005). Around 4 dpf, the circular smooth muscle forms a continuous sheet along and around the gut, thereby forming the foundation for motility.

The timing of development of different gastrointestinal regions varies between fish species (see Zambonino Infante & Cahu, 2001). For example, in sea bass (*Dicentrarchus labrax*) the stomach is anatomically almost fully developed at day 15, and it takes another ten days for pepsin secretion to start. In contrast, pepsin secretion is not detected in soles (*Solea solea*) younger than 200 days although the stomach is formed at day 10. As larvae grow, the gut is elongated and the epithelium continues to differentiate (see Zambonino Infante & Cahu, 2001). However, since the focus here will be on the enteric nervous system and motility, the development of other gastrointestinal functions like secretion and absorption is dealt with elsewhere (see Rønnestad & Morais, this volume).

THE AUTONOMIC NERVOUS SYSTEM

The autonomic nervous system controls numerous visceral functions including digestion and transport and has a direct and comparatively fast control of the gut. The teleost gut, like that of other vertebrates, is innervated by both extrinsic and intrinsic autonomic nerves (Nilsson, 1983; Jensen & Holmgren, 1994). Long, extrinsic gastrointestinal reflexes involve sensory pathways running from the gut to sympathetic ganglia, spinal cord or the brain and autonomic pathways going back to the gut. These generally have modulating and coordinating roles over gut functions, usually indirectly by affecting the intrinsic nerve signalling.

Intrinsic Innervation – the Enteric Division

The gut may function without input from the brain or spinal cord. This is due to the enteric nervous system – the intrinsic division of the autonomic nervous system (Furness & Costa, 1987). The enteric nervous system is located almost entirely within the gut wall, and gives rise to local reflexes including both sensory and motor components.

The enteric neurones are organised in networks, or plexa (Fig.1 B, C; Fig. 2). In teleosts, the myenteric plexus is well developed while the submucosal and deep muscular plexa contain, at most, a few ganglion cells. In comparison to mammals, where the myenteric nerve cells are organised in small groups or ganglia, the neurons are more scattered in teleosts (Kirtisinghe, 1940; Burnstock, 1959; Olsson & Karila, 1995). However, the overall nerve cell density is similar in teleosts and smaller mammals. The myenteric neurons include primary afferents (sensory receptors) and inhibitory as well as excitatory inter- and motor neurons.

Enteric reflexes underlie distinct motility patterns like peristalsis, the aboral propulsion of luminal content (Furness & Costa, 1987). Intrinsic primary afferent neurons (IPANs) are activated, mechanically or chemically, by contents in the lumen and stimulate interneurones in ascending (excitatory) and descending (inhibitory) neuronal pathways. These subsequently activate enteric motor neurones causing contractions and relaxation of the circular smooth muscle oral and anal to the site of stimuli, respectively (Fig. 2). Recent studies have also suggested that contractions occur synchronously in the longitudinal and circular muscle layers (Spencer *et al.*, 2003b).

Extrinsic Innervation

The extrinsic autonomic innervation of the gut consists of cranial and spinal outflows from the central nervous system. In teleosts, the vagal nerves contain sympathetic and

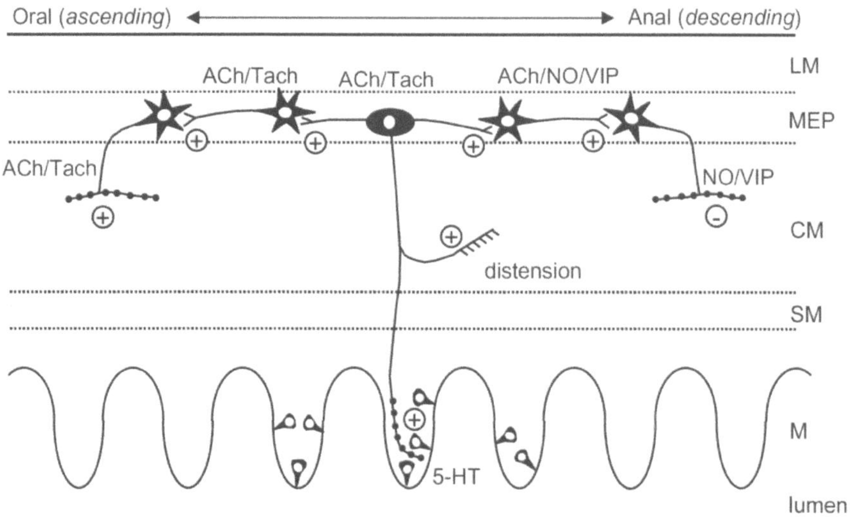

Fig. 2 Schematic drawing of the local enteric reflex underlying the control of peristalsis. Intrinsic sensory neurones are activated by chemical (luminal content via activation of endocrine cells that release serotonin (5-HT)) and mechanical (distension) stimulation. The sensory neurones subsequently activate both inhibitory descending and excitatory ascending nerve pathways, causing relaxation and contraction of the circular smooth muscle, respectively. Also included are examples of excitatory and inhibitory neurotransmitters (ACh, acetylcholine; Tach, tachykinin; VIP, vasaoactive intestinal polypeptide, NO, nitric oxide). M, mucosa; SM, submucosa; CM, circular muscle; MEP, myenteric plexus; LM, longitudinal muscle. After Olsson & Holmgren (2001).

parasympathetic efferent fibres as well as sensory, afferent fibres and generally innervate the upper part of the gastrointestinal tract, including oesophagus, stomach and the upper part of the intestine (see Nilsson, 1983) (Fig. 3). However, in stomach-less cyprinids, the vagal innervation usually includes the entire gut. The spinal outflows to the gut consist of the anterior splanchnic nerve(s) (arising from the sympathetic coeliac ganglion) innervating the stomach and intestine, and the posterior splanchnic nerve(s) (including sympathetic and parasympathetic fibres) innervating the distal part of the intestine. Similar to the vagus nerve, spinal nerves contain a large proportion of sensory fibres, conveying information about the state of the gut to higher centra or constituting part of autonomic reflexes involved in the control of motility patterns.

DEVELOPMENT OF THE ENTERIC NERVOUS SYSTEM

Enteric neurones in fish, as in other vertebrates, are derived from the neural crest (Le Douarin & Teillet, 1973; Lamers *et al.*, 1981; Sadaghiani & Vielkind, 1990; Raible *et al.*, 1992; Hall, this volume). This is a transient structure in the embryo, formed from the ectoderm along the dorsal margins of the developing neural tube.

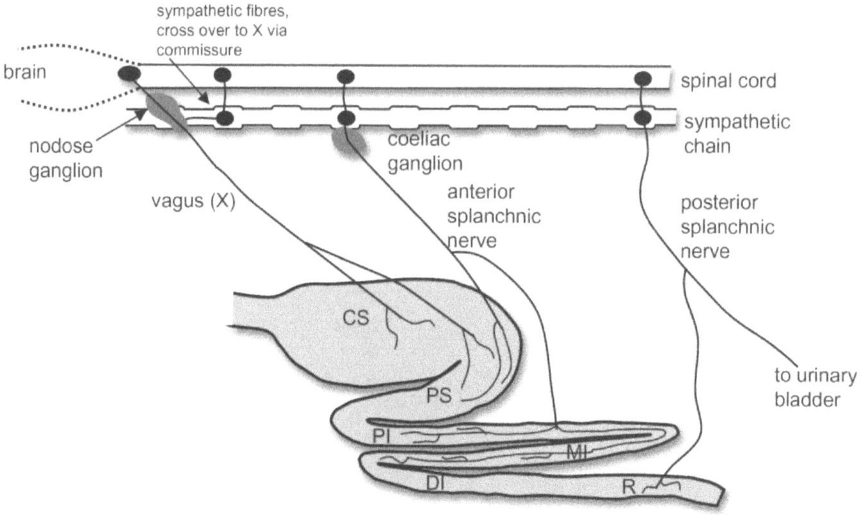

Fig. 3 Schematic drawing of the extrinsic innervation of the gastrointestinal tract. CS, cardiac stomach; PS, pyloric stomach; PI, proximal intestine, MI, middle intestine; DI, distal intestine; R, rectum. After Nilsson (1983).

Neural crest cells soon migrate and a subpopulation invades the gut. In birds and mammals, several series of experiments have been used to study the regional origin of the enteric nervous system. By removing defined regions of the chicken neural crest, Yntema and Hammond (1954) found that the majority of enteric neurons and glial cells is derived from the vagal (neck) part of the neural crest, corresponding to somites 1-7. The vagal source of enteric neurons was further confirmed by transplantation of quail vagal neural crest cells into chicken embryos in which the equivalent area had been removed (Le Douarin & Teillet, 1973). In the chimeras, quail nerve cells were observed throughout the gut. The exact origin of vagal crest cells entering the gut has been defined in chicken (mainly somites 3-6 for mid- and hindgut) and mouse (somites 1-4 for colonisation of the entire gut, 6-7 for the foregut) (Peters-van der Sanden *et al.*, 1993; Durbec *et al.*, 1996; Young & Newgreen, 2001).

In fish, however, this has not been described in detail although the anterior (vagal) crest region is suggested to give rise to enteric neurones (Sadaghiani & Vielkind, 1990; Bisgrove *et al.*, 1997). At 18 hpf, presumptive enteric neurone precursors have been found in zebrafish, adjacent to somites 3-4. These cells were detected using antisera against c-Ret (see below) and a few hours later (24 hpf) c-Ret positive cells were associated with the developing gut (Bisgrove *et al.*, 1997). Similarly, enteric nerve precursors have been localised, using antisera against the neuronal marker HNK-1, in swordtail (*Xiphophorus hellerii*). Immunoreactive cells were in close contact with the intestine from a stage corresponding to approximately 16 hpf (Sadaghiani & Vielkind, 1990; Kimmel *et al.*, 1995).

The sacral part of the neural crest is an additional source of enteric neurones in chicken and mouse (Le Douarin & Teillet, 1973; Serbedzija *et al.*, 1991). These sacral crest cells also develop into parasympathetic ganglion cells in the pelvic region while sympathetic and sensory (dorsal root) ganglia originate in the trunk region (Le Douarin *et al.*, 2004). Whether the sacral, in addition to the vagal, neural crest contributes to the enteric nervous system in fish is not known.

Guidance Cues

The fate of neural crest cells seems to be dependent on the migratory route they take and the environment they encounter. Two general pathways for cells leaving the neural crest have been observed in vertebrates: a lateral pathway between the somites and ectoderm, and a medial (or ventral) pathway between somites and the neural tube/notochord (Sadaghiani & Vielkind, 1990; Raible *et al.*, 1992; Jesuthasan, 1996; Newgreen & Young, 2002b) (Fig. 4). In fish as well as in birds and mammals, enteric neuronal precursor cells migrate via the medial pathway through the mesenchyme (i.e. loosely packed, undifferentiated mesodermal cells) to invade the primitive gut (Sadaghiani & Vielkind, 1990; Shepherd *et al.*, 2004). In mice, these enteric precursors are known to migrate in chains that are maintained by cell-to-cell contact (Young *et al.*, 2004b). The leading cells lay down the pathway that other migrating neural crest cells follow. There is a variation in speed and exact pathway of the migrating neural crest cells (Young *et al.*, 2004b). Various molecular signals are believed to direct the cells to their final destination in the gut, acting either as attractants or repellents. Signals not only stimulate neural crest cells to follow specific pathways to the gut but are also necessary to retain the cells within the gut, promote axon outgrowth and neuronal differentiation. Among those possible signals are neurotrophic factors, netrins,

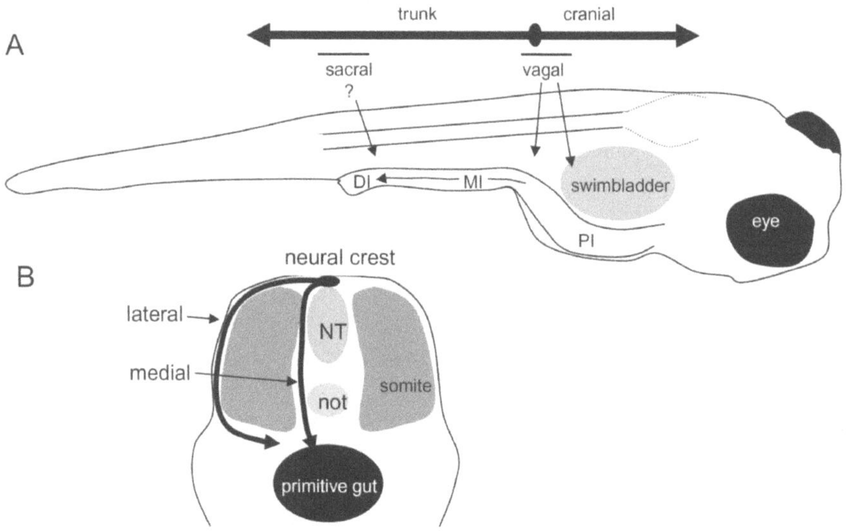

Fig. 4 **A.** Schematic drawing of a zebrafish (*Danio rerio*) larva. PI, proximal intestine, MI, middle intestine; DI, distal intestine. **B.** Migration pathways for neural crest cells in zebrafish embryos. NT, neural tube; not, notochord.

ephrins and extracellular matrix proteins some of which will be discussed in more detail below (Young & Newgreen, 2001; Gershon, 2002; Young *et al.*, 2004a).

During migration, most neural crest-derived cells are morphologically indistinguishable from the surrounding mesenchymal cells (but can be identified experimentally by their expression of specific markers) (Young & Newgreen, 2001). The vagal crest-derived cells enter the gut in the anterior region then spread in an oral-to-anal direction and differentiate into neurons or glial cells prompted by various stimulating or inhibiting signals (Gershon, 2002; Newgreen & Young, 2002b). Again, there is little information available from fish. In mammals, two of the most important signalling pathways for a normal development of the enteric nervous system seem to be the glial-derived neurotrophic factor (GDNF)/Ret-Gfrα1 and endothelin-3 (ET-3)/ endothelin B receptor systems. It is important to remember that as the neural crest cells migrate within the gut, the gut itself grows and differentiates, therefore providing different environments for the migrating cells depending on developmental stage.

GDNF is suggested to be involved in several important phases of a normal development of the enteric nervous system, including proliferation, differentiation, migration and neurite outgrowth (Chalazonitis *et al.*, 1998; Heuckeroth *et al.*, 1998; Taraviras *et al.*, 1999; Young *et al.*, 2001). It is secreted by the gut mesenchyme and Young and co-workers (2001) suggested that GDNF function as a chemoattractant to enteric precursor cells in mice. This might be one force that drives migration to the gut as well as the oral-to-anal migration inside the gut. GDNF and other members of the same family are present also in the gut of teleosts including zebrafish (Lucini *et al.*, 2005).

The receptor for GDNF is a complex consisting of a tyrosine kinase receptor (Ret) and a ligand-binding part (Gfrα1). These receptor units are normally expressed by enteric precursors before they colonise the gut in fish as well as in mammals (Bisgrove *et al.*, 1997; Young & Newgreen, 2001). In zebrafish, *c-ret* positive cells, suggested to be enteric neurone precursors, are detected in association with the developing gut (the gut primordium) at 24 hpf (Bisgrove *et al.*, 1997). These cells later spread anally to populate the entire length of the zebrafish gut at 72 hpf. At the same time, they express the neuronal marker Hu (Bisgrove *et al.*, 1997; Holmberg *et al.*, 2003; Wallace *et al.*, 2005). At 4 dpf, *c-ret* is no longer detected in the gut, indicating a transient dependence of the GDNF system. However, GDNF is expressed in endocrine cells in adult zebrafish (Lucini *et al.*, 2005).

Knockout mice lacking either the ligand (*gdnf -/-*) or the receptors (*ret -/-* or *gfrα1-/-*) have a markedly reduced number of enteric neurons caudal to the oesophagus (see Young *et al.*, 2001) indicating the importance of the GDNF/Ret-Gfrα1 system for the development of a large proportion of the enteric nervous system. Knock down experiments with *gdnf* as well as *ret* and *gfrα1* antisense morpholinos in zebrafish similarly disrupted the development of enteric neurons in the intestine (Shepherd *et al.*, 2001, 2004).

Belonging to the same superfamily as GDNF are the bone morphogenetic proteins (BMPs). BMPs are involved in induction of the neural crest but also in differentiation and apoptosis (Mehler *et al.*, 1997). In chicken, BMPs are essential for normal development of the enteric nervous system (Goldstein *et al.*, 2005). Further, inhibition of BMP reduced the response of the neural crest-derived cells to GDNF. In zebrafish, BMPs stimulate development of the epibranchial nerves suggesting that they have similar actions as in mammals although its involvement in the development of the enteric nervous system remains to be further studied (Holzschuh *et al.*, 2005).

Other neurotrophic factors are also important for development and survival of neurones but their exact roles in the early development of the enteric nervous system are not fully understood (Gershon, 2002; Newgreen & Young, 2002a). However, p75 that acts as a common receptor for the neurotrophins, is present on more or less all neural crest derived cells in the mammalian gut. It has been suggested that mainly neurotrophin-3 (NT-3) is involved in the development of the enteric nervous system (Gershon, 2002). In addition, TrkC, the NT-3-preferring neurotrophin tyrosine kinase receptor, is also found on enteric neurones. In some adult teleost species, only the related receptor TrkA has been seen on enteric neurones (Lucini *et al.*, 1999) while other studies have also demonstrated TrkC-like immunoreactivity in the teleost enteric nervous system (Hannestad *et al.*, 2000).

Another major signalling cue in mammals is endothelin 3. Endothelins regulate blood pressure in adults but are also known to be important in development. Lack of endothelin 3 or the endothelin B receptor causes absence of enteric neurones in the distal intestine in mice (Gershon, 2002; Newgreen & Young, 2002a). The reason why it only affects the colon, in contrast to the GDNF/Ret-Gfrα1 system, is not fully known. Endothelin 3 is shown to inhibit differentiation of most enteric precursors into neurones when in the gut (Gershon, 2002; Newgreen & Young, 2002a). It has thus been suggested that lack of endothelin would cause the enteric crest-derived cells to differentiate prematurely and stop migrating before they reach the colon. Another possible role for endothelin 3 is to stimulate proliferation of pre-migratory crest cells (Gershon, 2002). Endothelin plays a role in the development of at least some neural crest-derived cells also in zebrafish but the effect on enteric neurons is not described (Parichy *et al.*, 2000).

As mentioned earlier, extracellular matrix proteins also influence neural crest cells. For example, laminin promote either migration or differentiation of neural crest cells depending on the binding proteins the cells express (Gershon, 2002; Young *et al.*, 2004a). Laminin has also been demonstrated to stimulate axon outgrowth in the central nervous system of zebrafish, while peripheral axons do not seem to depend on laminin (Paulus & Halloran, 2006). Another extracellular component involved in axon development in zebrafish is tenascin that acts to repel the outgrowing axon (Becker *et al.*, 2004). The involvement of these molecules in the migration and differentiation of enteric neurones is, however, not known.

Differentiation

Neural crest cells develop into different cell types, including neurons and glial cells. As mentioned above, different regions in the neural crest give rise to neurons with different final destinations. What determines the fate of individual cells? Are the neural crest cells pluripotent and, if so, when do they become specified? (see Kelsh & Parichy, this volume). Replacement studies where neural crest cells from a region that normally do not migrate to the gut are transplanted into the vagal crest indicate that pre-migratory cells are pluripotent since they develop into normal enteric neurones (Le Douarin & Teillet, 1974; Le Douarin *et al.*, 1975; Rothman *et al.*, 1986).

On the other hand, there is evidence of pre-specification of at least a portion of the cells before they leave the neural crest to invade the gut. Le Douarin and Teillet (1974) created a chimera where the adrenomedullary level of the neural crest (from where enteric neurones are normally not derived) in chicken was replaced with quail vagal neural crest. Quail vagal neural crest cells were later observed in the gut of the chimera, suggesting that the neural crest cells are pre-specified and follow specific cues to reach the gut. For example, it is believed that only some neural crest cells, e.g. those that will develop into sensory, sympathetic or enteric neurones, express the GDNF receptor Ret (Gershon, 2002; Newgreen & Young, 2002a; Shepherd *et al.*, 2004).

The degree of pre-specification in fish is even less well documented. In zebrafish, trunk neural crest cells migrating at different times have different fates. Cells travelling along the lateral pathways start to migrate later than medially migrating cells, possibly due to different inhibitory cues produced by the somites (Raible *et al.*, 1992; Jesuthasan, 1996). Early migrating (medial pathway) trunk neural crest cells form sensory and sympathetic cells as well as glia, while late migrating (lateral pathway) neural crest cells become non-neural cell types (Raible & Eisen, 1994). By altering the time for migration for early and late migrating neural crest cells, Raible and Eisen (1996) showed that there was no change in the fate of individual cells in the zebrafish neural crest. This indicates that these cells have an intrinsic mechanism controlling their outcome and do not only depend on environmental factors. However, if early migrating neural crest cells are removed late migrating cells can differentiate into neuronal cell types, so the specification is not committed but can be altered, possibly due to interactions between individual neural crest cells (Raible & Eisen, 1994, 1996).

Once in the gut, enteric neurones continue to differentiate into different subclasses, showing different chemical coding and projections that will be discussed in the following section.

NEUROTRANSMITTERS IN THE DEVELOPING ENTERIC NERVOUS SYSTEM

Enteric neurones contain combinations of neurotransmitters of different types including acetylcholine, amines, neuropeptides and dissolved gaseous molecules (Gibbins, 1989; Costa *et al.*, 1996). Although the chemical coding is less well established in fish, most transmitters present in mammals appear to be present also in the fish gut. For an extensive review of the distribution and function of different neurotransmitters in adult fish see Jensen and Holmgren (1994). In general, in fish as in other vertebrates, acetylcholine and tachykinins (substance P and neurokinin A, NkA) usually have an excitatory effect on gut motility, while vasoactive intestinal peptide (VIP), pituitary adenylate cyclase-activating polypeptide (PACAP) and nitric oxide have an inhibitory action. Other transmitters, such as serotonin (5-HT) and ATP, may have either excitatory or inhibitory effects depending on pathways and receptors present (Lennard & Huddart, 1989; Jensen & Holmgren, 1994). Figure 2 summarises the roles of selected neurotransmitters in the control of peristalsis. In addition, many of the above mentioned neurotransmitters may also have neurotrophic roles as have been suggested for mammals (DiCicco-Bloom, 1996).

All of the above mentioned groups of transmitters have been detected in the developing mammalian enteric nervous system (Gintzler *et al.*, 1980; Rothman & Gershon, 1982; Larsson *et al.*, 1987; Pham *et al.*, 1991; Timmermans *et al.*, 1994; van Ginneken *et al.*, 1998). In general, cholinergic and serotonergic markers are found earlier than most of the peptides (Pham *et al.*, 1991). Similarly, nitric oxide synthetase (NOS), the enzyme responsible for nitric oxide production, is expressed early in the development (Timmermans *et al.*, 1994; van Ginneken *et al.*, 1998).

Several neurotransmitters have also been detected at early stages in the few studies on fish embryos and larvae (Fig. 5). NOS NkA and PACAP are present well before the onset of exogenous feeding in zebrafish (Poon *et al.*, 2003; Holmberg *et al.*, 2004, 2006; Holmqvist *et al.*, 2004) while in turbot (*Scophthalmus maximus*), only a few VIP-like immunoreactive enteric neurons were detected at this stage (Reinecke *et al.*, 1997). Later on, substance P-, somatostatin- and serotonin-immunoreactive nerve fibres were observed in the gut of turbot larvae (Reinecke *et al.*, 1997).

Little is so far known about the molecular cues that direct enteric neurones into different subclasses. It is believed that the cells belong to two main lineages (Gershon, 2002). Initially catecholaminergic cells are present in the gut, as detected by tyrosine hydroxylase and dopamine α-hydroxylase (DBH) immunoreactivities. However, these cells soon lose tyrosine hydroxylase expression and some develop into serotonergic neurons in the presence of Mash1. Another subset of enteric precursors never express tyrosine hydroxylase and are Mash1-independent. These differentiate later and express for example CGRP (calcitonin gene-related peptide).

The expression of different neurotransmitters in nerves is not enough to conclude that they actually affect the gut. The transmitters need to be released from the nerve cells, requiring functional synaptic machinery. Furthermore, to exert an effect, the appropriate receptors must be expressed on the effector cells and second messenger pathways have to be developed. For this reason, functional studies on gut motility or secretion serve a way to evaluate the significance of neurotransmitter expression at various developmental stages.

GUT MOTILITY

Gastrointestinal motility, together with secretion and absorption, is necessary to process food in an optimal way. To fulfil this task the functions need to be accurately controlled.

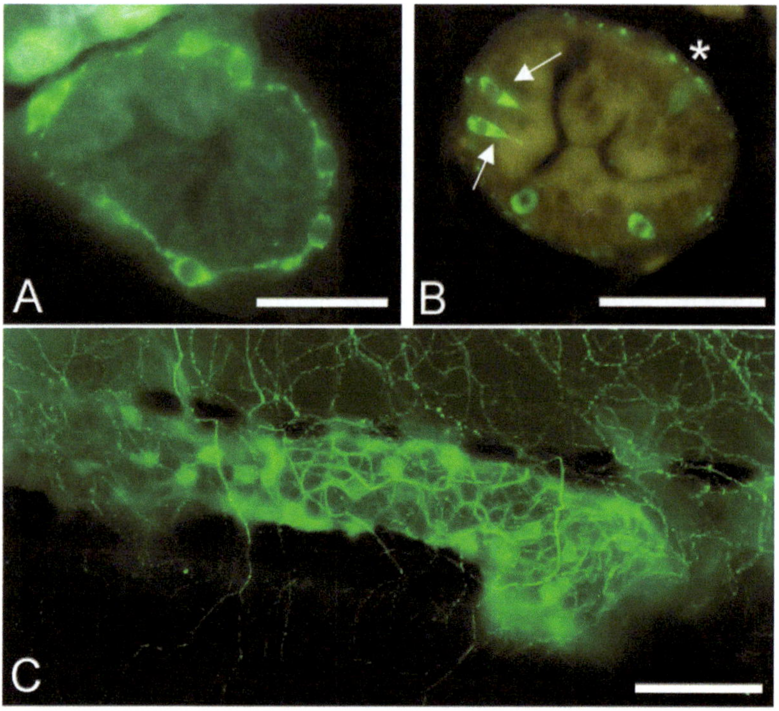

Fig. 5 Innervation of the gut of zebrafish (*Danio rerio*) larvae at 5 (**A, C**) and 7 (**B**) days post-fertilisation (dpf). Neurones and several known transmitters are present in the gut before the onset of feeding. **A**. Transverse section of the distal intestine showing enteric neurons expressing the enzyme nitric oxide synthase (NOS). NOS is involved in the production of nitric oxide. **B**. The photomicrograph shows NkA (neurokinin A) immunoreactive endocrine cells (arrows) and nerve fibres (*). **C.** The neuronal markers Hu (human neuronal protein) and AcT (acetylated tubulin) were used to identify neurones in whole-mount preparations of the zebrafish larvae. The photomicrograph shows the neuronal network innervating the distal intestine. Bars: 50 µm.

In this review we focus on the development of the enteric nervous system in relation to gut motility.

In fully developed smooth muscle, the muscle cells are electrically coupled to each other and as a consequence each muscle layer may contract more or less simultaneously over some distance. Electrical slow waves, originating in intramuscular ICCs, spread to muscles cells and underlie much of the contractile activity in the gut (Horowitz *et al.*, 1999). However, without the influence of neurones and other types of ICCs, gut motility will not be organised as is required for efficient digestion and absorption.

In many animals one has distinguished different motility patterns in fed and fasted states (Husebye, 1999). During feeding the main function of muscle activity is to ensure appropriate mixing and propulsion of gut contents. For example, peristalsis induced by chemical or mechanical activation of enteric reflexes, makes sure that food is transported along the intestine to allow efficient digestion and absorption. Spontaneous motility, described as migrating motor complexes (MMC), is known to occur between meals in adult vertebrates (Szurszewski, 1969). This activity might function as a housekeeper, to prevent bacterial overgrowth and remove waste products (Husebye, 1999). Motility patterns resembling phase III (i.e. a phase with rhythmic contractions) of the MMC have been observed in *in vitro* preparations from fasting adult teleosts (Karila & Holmgren, 1995; Olsson *et al.*, 1999).

Development of Gut Motility

There is little information about the development of smooth muscle activity of the gut in vertebrates in general. In some foetal mammals, gut motility develops and reaches a regular pattern already before birth and the first feeding (Gershon & Thompson, 1973; Bueno & Ruckebusch, 1979; Morriss *et al.*, 1986; Ruckebusch, 1986; Bisset *et al.*, 1988; Sase *et al.*, 2000; Acosta *et al.*, 2002; Anderson *et al.*, 2004). This can be compared to non-feeding fish larvae where regular gut contraction waves sweep the gut before the onset of exogenous feeding (Holmberg *et al.*, 2003). In contrast, in dogs the myoelectrical activity will not develop until after birth (Bueno & Ruckebusch, 1979; Ruckebusch, 1986). In lamb as in humans, the motor activity develops from an irregular pattern to well-defined MMCs during the last part of the gestation period (Bueno & Ruckebusch, 1979; Bisset *et al.*, 1988). This suggests that there is a difference between species in the degree of maturity of the gut motility at birth.

In zebrafish, sporadic irregular local contractions of the intestine appear at 3 dpf, while feeding usually starts around 5-6 dpf (at 28°C) (Holmberg *et al.*, 2003). One day later, at 4 dpf, regular contraction waves travel in anterograde (oral-to-anal) and retrograde (anal-to-oral) directions (Fig. 6). This regular motility pattern coincides in time with the formation of a continuous sheet of circular muscle, i.e. the contractions can travel circumferentially as well as along the gut (Holmberg *et al.*, 2003; Wallace *et al.*, 2005). Although at this stage the animals have not started to eat, the mouth has opened suggesting that bacteria and particles from the water might be present in the gut. The early gut motility could be necessary to prevent bacterial overgrowth and also to transport away dead mucosal cells and secreted matter and hence has a housekeeping function similar to MMCs in adult vertebrates.

As the fish starts to feed, anterograde peristaltic waves are necessary to propagate the food along the gut. Such peristalsis as well as retrograde motility has been observed in juvenile Atlantic halibut (*Hippoglossus hippoglossus*) (Rønnestad *et al.*, 2000). The retrograde waves originated in an area posterior to the pyloric caeca and are thought to enhance digestion by filling the pyloric caeca but also by ensuring proper mixing of the chyme and digestive enzymes, while the anterograde contraction waves sweeping the entire gut probably transport food along the intestine (Rønnestad *et al.*, 2000).

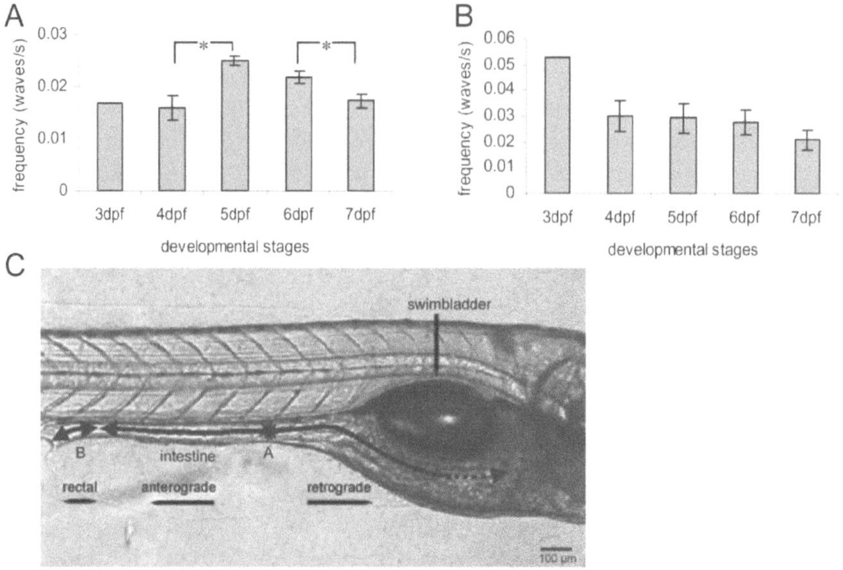

Fig. 6 **A, B**. The frequency of anterograde (**A**) and retrograde (**B**) contraction waves in zebrafish (*Danio rerio*) larvae at 3 to 7 days post fertilisation. Data are presented as mean ± SEM. * denotes a significant difference ($P<0.05$). **C.** A photomicrograph of an embedded anaesthetised zebrafish larva. Anterograde and retrograde contraction waves originated from an area in the middle intestine (A in picture) while the rectal activity originated in the distal part of the gut (B in picture). The arrows show the direction and approximately length of the contraction waves. After Holmgren *et al.* (2003).

Development of Nervous Control of Gut Motility

As mentioned above, an irregular contractile activity develops into a co-ordinated motility pattern before first feeding in zebrafish. In addition, there is also a pronounced development of neurons and expression of several known inhibitory and excitatory neurotransmitters in zebrafish (Holmberg *et al.*, 2003, 2004, 2006; Poon *et al.*, 2003; Holmqvist *et al.*, 2004). This suggests a parallel and presumably co-ordinated temporal development of gut motility and its nervous control before the fish starts actively to search for food. The importance of the enteric nervous system directly after birth is clearly seen in mutant mice that lack enteric neurons in the majority of the gut; these animals die soon after birth (Ward *et al.*, 1999). The absence of enteric neurons in parts of the gastrointestinal tract might not be lethal, but can cause major discomfort. In Hirschsprung's disease, for example, the absence of enteric neurones in the distal colon causes severe obstruction or constipation (see Newgreen & Young, 2002a). Furthermore, in adult mice the neurotoxin tetrodotoxin abolishes most of the MMC activity (Sarna *et al.*, 1981; Husebye, 1999; Spencer *et al.*, 2003a). However before birth, the absence of enteric neurones does seem to affect the propulsion of gut contents in foetal mice (Anderson *et al.*, 2004).

The role of enteric neurons in developing fish larvae is presently not known. However, our *in vivo* studies of zebrafish showed that acetylcholine initiates anterograde contraction waves at 4 dpf and from the same stage the cholinergic muscarinic antagonist atropine reduces anterograde contraction waves. This suggests that there is a cholinergic control system present and functional before the first feeding in zebrafish. Further, at the same stage as the cholinergic tonus developed (4 dpf), L-NAME (N^G-nitro-L-arginine methyl ester) increased the frequency of propagating anterograde waves. As L-NAME inhibits the production of NO, this indicates that there is an endogenous production of NO before the first feeding. There is probably a co-functionality between the excitatory (acetylcholine) and inhibitory (NO) pathways that balance the gut motility to a desired frequency. In addition, other transmitters are likely to play a supplementary role in regulation of gut motility. For example, application of the neuropeptides PACAP and NkA to the zebrafish larvae caused a decrease and increase in motility, respectively, indicating the presence of functional PACAP and NkA receptors from 5 dpf (Holmberg *et al.*, 2004).

CONCLUDING REMARKS

Studies of fish embryos and larvae show that fish, and in particular zebrafish, are excellent models when studying development of the enteric nervous system as well as gut motility. Clear advantages are the external fertilisation and development together with the transparency of the young embryos and larvae. This allows us to study the gut using non-invasive imaging methods.

Enteric neurones expressing several known inhibitory and excitatory neurotransmitters are present in the gut before the onset of exogenous feeding in the studied fish species. At the same time as regular gut motility develops in the (zebrafish) embryos and larvae, a functioning inhibitory nitrergic and excitatory cholinergic mechanisms act to balance the activity. This suggests a parallel development of gut motility and its nervous control before the fish starts to feed. In non-feeding larvae, gut motility might be important to control for example the bacterial level.

References

Acosta R, Lee JJ, Oyachi N, Buchmiller-Crair TL, Atkinson JB, Ross MG (2002) Anticholinergic suppression of fetal rabbit upper gastrointestinal motility. J Matern Fetal Neonatal Med 11: 153-157

Anderson RB, Enomoto H, Bornstein JC, Young HM (2004) The enteric nervous system is not essential for the propulsion of gut contents in fetal mice. Gut 53: 1546-1547

Becker CG, Schweitzer J, Feldner J, Schachner M, Becker T (2004) Tenascin-R as a repellent guidance molecule for newly growing and regenerating optic axons in adult zebrafish. Mol Cell Neurosci 26: 376-389

Bisgrove BW, Raible DW, Walter V, Eisen JS, Grunwald DJ (1997) Expression of c-ret in the zebrafish embryo: potential roles in motoneuronal development. J Neurobiol 33: 749-768

Bisset WM, Watt JB, Rivers RP, Milla PJ (1988) Ontogeny of fasting small intestinal motor activity in the human infant. Gut 29: 483-488

Bueno L, Ruckebusch Y (1979) Perinatal development of intestinal myoelectrical activity in dogs and sheep. Am J Physiol 237: E61-67

Burnstock G (1959) The innervation of the gut of the brown trout *Salmo trutta*. Quart J Microsc Sci 100: 199-220

Chalazonitis A, Rothman TP, Chen J, Gershon MD (1998) Age-dependent differences in the effects of GDNF and NT-3 on the development of neurons and glia from neural crest-derived precursors immunoselected from the fetal rat gut: expression of GFRα-1 *in vitro* and *in vivo*. Dev Biol 204: 385-406

Costa M, Brookes SJ, Steele PA, Gibbins I, Burcher E, Kandiah CJ (1996) Neurochemical classification of myenteric neurons in the guinea-pig ileum. Neuroscience 75: 949-967

DiCicco-Bloom E (1996) Region-specific regulation of neurogenesis by VIP and PACAP: direct and indirect modes of action. Ann N Y Acad Sci 805: 244-256; discussion 256-248

Durbec PL, Larsson-Blomberg LB, Schuchardt A, Costantini F, Pachnis V (1996) Common origin and developmental dependence on c-ret of subsets of enteric and sympathetic neuroblasts. Development 122: 349-358

Furness JB, Costa M (1987) The Enteric Nervous System. Churchill, Edingburgh

Gershon MD (2002) Development of the enteric nervous system. In: Brookes S, Costa M (eds) Innervation of the Gastrointestinal Tract. Taylor and Francis, London, p 469-515

Gershon MD, Thompson EB (1973) The maturation of neuromuscular function in a multiply innervated structure: development of the longitudinal smooth muscle of the foetal mammalian gut and its cholinergic excitatory, adrenergic inhibitory, and non-adrenergic inhibitory innervation. J Physiol 234: 257-277

Gibbins I (1989) Co-existence and co-function. In: Holmgren S (ed) The Comparative Physiology of Regulatory Peptides. Chapman and Hall, London, p 308-343

Gintzler AR, Rothman TP, Gershon MD (1980) Ontogeny of opiate mechanisms in relation to the sequential development of neurons known to be components of the guinea pig's enteric nervous system. Brain Res 189: 31-48

Goldstein AM, Brewer KC, Doyle AM, Nagy N, Roberts DJ (2005) BMP signaling is necessary for neural crest cell migration and ganglion formation in the enteric nervous system. Mech Dev 122: 821-833

Hannestad J, Marino F, Germana A, Catania S, Abbate F, Ciriaco E, Vega JA (2000) Trk neurotrophin receptor-like proteins in the teleost *Dicentrarchus labrax*. Cell Tissue Res 300: 1-9

Heuckeroth RO, Lampe PA, Johnson EM, Milbrandt J (1998) Neurturin and GDNF promote proliferation and survival of enteric neuron and glial progenitors *in vitro*. Dev Biol 200: 116-129

Holmberg A, Schwerte T, Fritsche R, Pelster B, Holmgren S (2003) Ontogeny of intestinal motility in correlation to neuronal development in zebrafish embryos and larvae. J Fish Biol 63: 318-331

Holmberg A, Schwerte T, Pelster B, Holmgren S (2004) Ontogeny of the gut motility control system in zebrafish *Danio rerio* embryos and larvae. J Exp Biol 207: 4085-4094

Holmberg A, Olsson C, Holmgren S (2006) The effects of endogenous and exogenous nitric oxide on gut motility in zebrafish, *Danio rerio*, embryos and larvae. J Exp Biol 209: 2472-2479

Holmqvist B, Ellingsen B, Forsell J, Zhdanova I, Alm P (2004) The early ontogeny of neuronal nitric oxide synthase systems in the zebrafish. J Exp Biol 207: 923-935

Holzschuh J, Wada N, Wada C, Schaffer A, Javidan Y, Tallafuss A, Bally-Cuif L, Schilling TF (2005) Requirements for endoderm and BMP signaling in sensory neurogenesis in zebrafish. Development 132: 3731-3742

Horne-Badovinac S, Lin D, Waldron S, Schwarz M, Mbamalu G, Pawson T, Jan Y, Stainier DY, Abdelilah-Seyfried S (2001) Positional cloning of heart and soul reveals multiple roles for PKC lambda in zebrafish organogenesis. Curr Biol 11: 1492-1502

Horowitz B, Ward SM, Sanders KM (1999) Cellular and molecular basis for electrical rhythmicity in gastrointestinal muscles. Annu Rev Physiol 61: 19-43

Husebye E (1999) The patterns of small bowel motility: physiology and implications in organic disease and functional disorders. Neurogastroenterol Motil 11: 141-161

Jensen J, Holmgren S (1994) The gastrointestinal canal. In: Nilsson S, S.Holmgren (eds) Comparative Physiology and Evolution of the Autonomic Nervous System. Harwood Academic Publisher, Chur, Switzerland, p 119-167

Jesuthasan S (1996) Contact inhibition/collapse and pathfinding of neural crest cells in the zebrafish trunk. Development 122: 381-389

Karila P, Holmgren S (1995) Enteric reflexes and nitric oxide in the fish intestine. J Exp Biol 198: 2405-2411

Kimmel CB, Ballard WW, Kimmel SR, Ullmann B, Schilling TF (1995) Stages of embryonic development of the zebrafish. Dev Dyn 203: 253-310

Kirtisinghe P (1940) The myenteric nerve plexus in some lower chordates. Quart J Microscop Sci 81: 521-539

Koelz HR (1992) Gastric acid in vertebrates. Scand J Gastroenterol Suppl 193: 2-6

Lamers CH, Rombout JW, Timmermans LP (1981) An experimental study on neural crest migration in *Barbus conchonius* (Cyprinidae, Teleostei), with special reference to the origin of the enteroendocrine cells. J Embryol Exp Morphol 62: 309-323

Larsson LT, Helm G, Malmfors G, Sundler F (1987) Ontogeny of peptide-containing neurons in human gut – an immunocytochemical study. Regul Pept 17: 243-256

Le Douarin NM, Teillet MA (1973) The migration of neural crest cells to the wall of the digestive tract in avian embryo. J Embryol Exp Morphol 30: 31-48

Le Douarin NM, Teillet MA (1974) Experimental analysis of the migration and differentiation of neuroblasts of the autonomic nervous system and of neurectodermal mesenchymal derivatives, using a biological cell marking technique. Dev Biol 41: 162-184

Le Douarin NM, Renaud D, Teillet MA, Le Douarin GH (1975) Cholinergic differentiation of presumptive adrenergic neuroblasts in interspecific chimeras after heterotopic transplantations. Proc Natl Acad Sci USA 72: 728-732

Le Douarin NM, Creuzet S, Couly G, Dupin E (2004) Neural crest cell plasticity and its limits. Development 131: 4637-4650

Lennard R, Huddart H (1989) Purinergic modulation in the flounder gut. Gen Pharmacol 20: 849-853

Lucini C, de Girolamo P, Maruccio L, Lamanna C, Castaldo L, Vega JA (1999) Trk-neurotrophin receptor-like immunoreactivity in the gut of teleost species. Cell Tissue Res 296: 323-330

Lucini C, Maruccio L, Tafuri S, Bevaqua M, Staiano N, Castaldo L (2005) GDNF family ligand immunoreactivity in the gut of teleostean fish. Anat Embryol 210: 265-274

Mehler MF, Mabie PC, Zhang D, Kessler JA (1997) Bone morphogenetic proteins in the nervous system. Trends Neurosci 20: 309-317

Morriss FH, Jr., Moore M, Weisbrodt NW, West MS (1986) Ontogenic development of gastrointestinal motility: IV. Duodenal contractions in preterm infants. Pediatrics 78: 1106-1113

Newgreen D, Young HM (2002a) Enteric nervous system: development and developmental disturbances - part 1. Pediatr Dev Pathol 5: 224-247

Newgreen D, Young HM (2002b) Enteric nervous system: development and developmental disturbances - part 2. Pediatr Dev Pathol 5: 329-349

Nilsson S (1983) Autonomic Nerve Function in the Vertebrates. Springer Verlag, Berlin

Olsson C, Karila P (1995) Coexistence of NADPH-diaphorase and vasoactive intestinal polypeptide in the enteric nervous system of the Atlantic cod (*Gadus morhua*) and the spiny dogfish (*Squalus acanthias*). Cell Tissue Res 280: 297-305

Olsson C, Holmgren S (2001) The control of gut motility. Comp Biochem Physiol 128A: 481-503

Olsson C, Aldman G, Larsson A, Holmgren S (1999) Cholecystokinin affects gastric emptying and stomach motility in the rainbow trout *Oncorhynchus mykiss*. J Exp Biol 202: 161-170

Parichy DM, Mellgren EM, Rawls JF, Lopes SS, Kelsh RN, Johnson SL (2000) Mutational analysis of endothelin receptor b1 (rose) during neural crest and pigment pattern development in the zebrafish *Danio rerio*. Dev Biol 227: 294-306

Paulus JD, Halloran MC (2006) Zebrafish *bashful/laminin-alpha1* mutants exhibit multiple axon guidance defects. Dev Dyn 235: 213-224

Peters-van der Sanden MJ, Kirby ML, Gittenberger-de Groot A, Tibboel D, Mulder MP, Meijers C (1993) Ablation of various regions within the avian vagal neural crest has differential effects on ganglion formation in the fore-, mid- and hindgut. Dev Dyn 196: 183-194

Pham TD, Gershon MD, Rothman TP (1991) Time of origin of neurons in the murine enteric nervous system: sequence in relation to phenotype. J Comp Neurol 314: 789-798

Poon KL, Richardson M, Lam CS, Khoo HE, Korzh V (2003) Expression pattern of neuronal nitric oxide synthase in embryonic zebrafish. Gene Expr Patterns 3: 463-466

Raible DW, Eisen JS (1994) Restriction of neural crest cell fate in the trunk of the embryonic zebrafish. Development 120: 495-503

Raible DW, Eisen JS (1996) Regulative interactions in zebrafish neural crest. Development 122: 501-507

Raible DW, Wood A, Hodsdon W, Henion PD, Weston JA, Eisen JS (1992) Segregation and early dispersal of neural crest cells in the embryonic zebrafish. Dev Dyn 195: 29-42

Reinecke M, Muller C, Segner H (1997) An immunohistochemical analysis of the ontogeny, distribution and coexistence of 12 regulatory peptides and serotonin in endocrine cells and nerve fibers of the digestive tract of the turbot, *Scophthalmus maximus* (Teleostei). Anat Embryol (Berl) 195: 87-101

Rothman TP, Gershon MD (1982) Phenotypic expression in the developing murine enteric nervous system. J Neurosci 2: 381-393

Rothman TP, Sherman D, Cochard P, Gershon MD (1986) Development of the monoaminergic innervation of the avian gut: transient and permanent expression of phenotypic markers. Dev Biol 116: 357-380

Ruckebusch Y (1986) Development of digestive motor patterns during perinatal life: mechanism and significance. J Pediatr Gastroenterol Nutr 5: 523-536

Rønnestad I, Rojas-Garcia CR, Skadal J (2000) Retrograde peristalsis, a possible mechanism for fillling the pyloric cecae? J Fish Biol 56: 216-218

Sadaghiani B, Vielkind JR (1990) Distribution and migration pathways of HNK-1-immunoreactive neural crest cells in teleost fish embryos. Development 110: 197-209

Sarna S, Stoddard C, Belbeck L, McWade D (1981) Intrinsic nervous control of migrating myoelectric complexes. Am J Physiol 241: G16-23

Sase M, Nakata M, Tashima R, Kato H (2000) Development of gastric emptying in the human fetus. Ultrasound Obstet Gynecol 16: 56-59

Serbedzija GN, Burgan S, Fraser SE, Bronner-Fraser M (1991) Vital dye labelling demonstrates a sacral neural crest contribution to the enteric nervous system of chick and mouse embryos. Development 111: 857-866

Shepherd IT, Beattie CE, Raible DW (2001) Functional analysis of zebrafish GDNF. Dev Biol 231: 420-435

Shepherd IT, Pietsch J, Elworthy S, Kelsh RN, Raible DW (2004) Roles for GFRα1 receptors in zebrafish enteric nervous system development. Development 131: 241-249

Spencer NJ, Sanders KM, Smith TK (2003a) Migrating motor complexes do not require electrical slow waves in the mouse small intestine. J Physiol 553: 881-893

Spencer NJ, Hennig GW, Smith TK (2003b) Stretch-activated neuronal pathways to longitudinal and circular muscle in guinea pig distal colon. Am J Physiol 284: G231-241

Stevens CE, Hume D (1995) Comparative Physiology of the Vertebrate Digestive System. Cambridge University Press, Cambridge

Szurszewski J (1969) A migrating electric complex of the canine small intestine. Am J Physiol 217: 1757-1763

Taraviras S, Marcos-Gutierrez CV, Durbec P, Jani H, Grigoriou M, Sukumaran M, Wang LC, Hynes M, Raisman G, Pachnis V (1999) Signalling by the RET receptor tyrosine kinase and its role in the development of the mammalian enteric nervous system. Development 126: 2785-2797

Timmermans JP, Barbiers M, Scheuermann DW, Bogers JJ, Adriaensen D, Fekete E, Mayer B, Van Marck EA, De Groodt-Lasseel MHA (1994) Nitric oxide synthase immunoreactivity in the enteric nervous system of the developing human digestive tract. Cell Tissue Res. 275: 235-245

Wallace KN, Pack M (2003) Unique and conserved aspects of gut development in zebrafish. Dev Biol 255: 12-29

Wallace KN, Akhter S, Smith EM, Lorent K, Pack M (2005) Intestinal growth and differentiation in zebrafish. Mech Dev 122: 157-173 van Ginneken C, van Meir F, Sommereyns G, Sys S, Weyns A (1998) Nitric oxide synthase expression in enteric neurons during development in the pig duodenum. Anat Embryol 198: 399-408

Ward SM, Ordog T, Bayguinov JR, Horowitz B, Epperson A, Shen L, Westphal H, Sanders KM (1999) Development of interstitial cells of Cajal and pacemaking in mice lacking enteric nerves. Gastroenterology 117: 584-594

Wells JM, Melton DA (1999) Vertebrate endoderm development. Annu Rev Cell Dev Biol 15: 393-410

Yntema CL, Hammond WS (1954) The origin of intrinsic ganglia of trunk viscera from vagal neural crest in the chick embryo. J Comp Neurol 101: 515-541

Young HM, Newgreen D (2001) Enteric neural crest-derived cells: origin, identification, migration, and differentiation. Anat Rec 262: 1-15

Young HM, Hearn CJ, Farlie PG, Canty AJ, Thomas PQ, Newgreen DF (2001) GDNF is a chemoattractant for enteric neural cells. Dev Biol 229: 503-516

Young HM, Anderson RB, Anderson CR (2004a) Guidance cues involved in the development of the peripheral autonomic nervous system. Auton Neurosci 112: 1-14

Young HM, Bergner AJ, Anderson RB, Enomoto H, Milbrandt J, Newgreen DF, Whitington PM (2004b) Dynamics of neural crest-derived cell migration in the embryonic mouse gut. Dev Biol 270: 455-473

Zambonino Infante JL, Cahu CL (2001) Ontogeny of the gastrointestinal tract of marine fish larvae. Comp Biochem Physiol 130C: 477-487

CHAPTER

18

Immunology

Agustín G Zapata* and Alfonso Cortés

INTRODUCTION

The immune system of fishes, as in other vertebrates, generally consists of two intercommunicating processes: innate and adaptive immunity. Cells and molecules involved in the non-specific recognition of foreign particles and organisms, largely pathogens, constitute the natural (innate) immune response, while specific mechanisms capable of discriminating between self and non-self via cell-mediated and humoral responses comprise the so-called adaptive immune system. Innate immunity is based on constitutively produced general receptors, expressed by different cell types that recognise conserved epitopes on distinct pathogens. Signalling through these types of receptors triggers a response that limits or stops the pathogenic invasion. By contrast, specific adaptive immunity depends upon the somatic diversification of antigen-receptor genes to generate a vast repertoire of lymphocytes, each of which expresses a different antigen receptor. These surface antigen receptors recognise antigen conformations (epitopes) of pathogens and respond to them by clonal amplification, cell differentiation and production of effective molecules, principally antibodies.

In the present chapter we review the appearance during fish ontogeny of both innate and adaptive immunity as well as of the cells and tissues involved in these

Authors' address: Department of Cell Biology, Faculty of Biology, Complutense University, 28040 Madrid.

**Corresponding author*: E-mail: zapata@bio.ucm.es

reactions. Although this is a growing issue to which many important contributions have been made over the past few years, results so far are unable to provide conclusive answers to many unresolved questions and are largely restricted to a few teleost species of commercial value or useful as experimental models, such as zebrafish (*Danio rerio*) or medaka (*Oryzias latipes*).

THE EARLY HAEMATOPOIETIC SITES OF TELEOSTS

Primitive and Definitive Haematopoiesis

Lymphocytes, the specialised cell type of adaptive immunity, and also natural killer (NK) cells, macrophages, granulocytes and other cell types that are involved in innate immune responses, are derived from multipotent haematopoietic stem cells (HSCs). During embryonic and larval development, HSCs appear in different locations. Studies have mainly been conducted in zebrafish and, more recently, in common carp (*Cyprinus carpio*). In all vertebrates, the first blood cells are derived from the mesoderm. In zebrafish, the first known expression of haematopoietic genes such as runt-related transcription factor (*runx-1, runx-3*), core-binding factor (*cbfb*), stem-cell leukemia

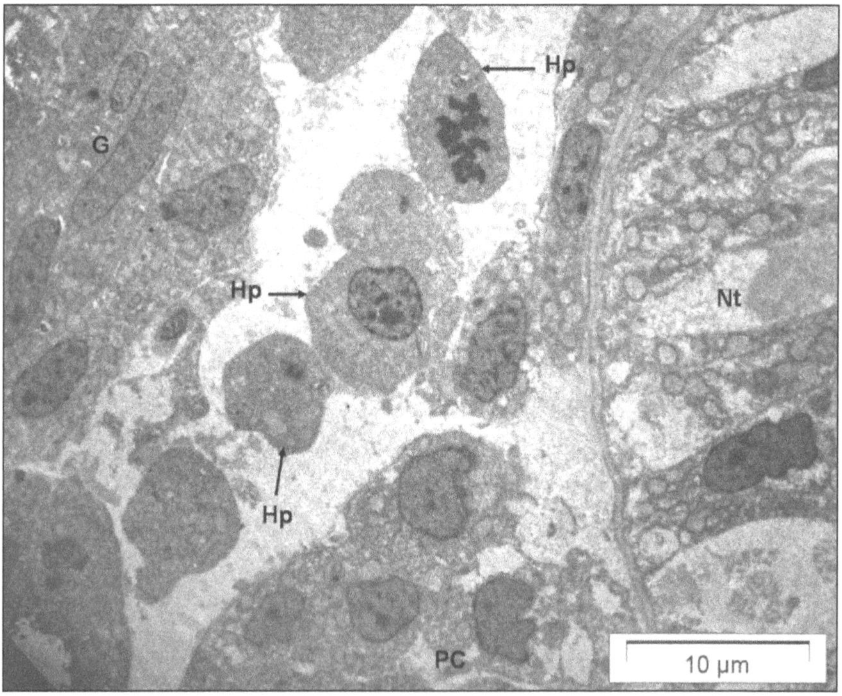

Fig. 1 Haematopoietic progenitors (Hp) in the intermediate cell mass (ICM) of a 20 hour-old zebrafish (*Danio rerio*). Notochord (Nt), pronephric duct (PC), gut (G).

(scl), LIM domain only *(lmo-2)*, Glutamyl-tRNA amidotransferase (GATA), subunit A *(gata-2)*, myeloblastosis *(c-myb)*, Friend leukemia integration *(fli-1)*, etc. occur in the lateral plate mesoderm (LPM) around 12 hours post fertilisation (hpf) (Thompson *et al.*, 1998; Trede *et al.*, 2001; Hall this volume). In fact, the lateral mesoderm is regionalised: the rostral region providing progenitors for both macrophages and granulocytes, and the caudal area containing the erythroid progenitors (Crowhurst *et al.*, 2002). Later, during migration of LPM cells, the so-called intra-embryonic intermediate cell mass (ICM) is formed. This structure gradually converges in a rostral to caudal direction to end just caudal to the future uro-genital pore (Al-Adhami *et al.*, 1977; Willett *et al.*, 1999). In some species, however, the first haematopoietic locus occurs extra-embryonically in the yolk-sac blood islands whereas in others, such as common mummichog (*Fundulus heteroclitus*) and rainbow trout (*Oncorhynchus mykiss*), early haematopoiesis occurs first in the yolk-sac and then in the ICM (Zapata *et al.*, 2006).

As early as 11-14 hpf cells present in the incipient ICM correspond to primitive haematopoietic progenitor cells (Fig. 1). Around 20 hpf, these cells have migrated medially and form two bilateral bands, which largely contain pro-erythroblasts and primitive endothelial cells (Willett *et al.*, 1999). In support of this, a subset of *scl* positive cells of the incipient ICM expresses the erythroid-specific transcription factor *gata-1* (Detrich *et al.*, 1995) whereas the remaining *gata-1* negative ICM cells are principally angioblasts capable of differentiating to *flk 1/vegfr 2* positive endothelial cells (Liao *et al.*, 1997; Davidson *et al.*, 2003). A similar situation has been recently described in common carp embryos (0-5 dpf) (Huttenthuis *et al.*, 2006a, b, c). Next, haematopoietic progenitors disappear from the ICM but the trunk blood vessels have already been formed and numerous developing erythroid cells appear in the circulation. Between 24-30 hpf (2 dpf onward in common carp) a few myelocytes appear between the yolk-sac and the body wall in an extravascular space that also seems to contain macrophages, which some authors consider to be functionally equivalent to the AGM region (an intra-embryonic haematopoietic locus based on the Aorta-Gonad-Mesonephros primordium) of other vertebrates (Thompson *et al.*, 1998). In support of this, *runx*, a transcription factor essential for murine AGM-derived HSC-formation (Okuda *et al.*, 1996), has been demonstrated to be expressed in the ventral wall of the dorsal aorta between 24 and 48 hpf (Burns *et al* 2002; Kalev-Zylinska *et al.*, 2002) as well as other HSC markers, including *c-myb* (Thompson *et al.*, 1998), *ikaros* (Willett *et al.*, 2001), *scl* and *lmo-2* (Davidson & Zon 2004). Also, in common carp an equivalent of the AGM area, capable of producing monocytes, macrophages and granulocytes, has been identified in the ventral part of the dorsal aorta (Huttenhuis *et al.*, 2006c). At this stage of development, neutrophils migrate into the circulation and can be seen in different tissues.

From 2 dpf onward a third haematopoietic locus has been recognised in zebrafish (Willett *et al.*, 1999; Lieschke *et al.*, 2001) and more recently in common carp (Huttenhuis *et al.*, 2006c). We first reported the occurrence of presumptive developing cells in the ventral vein region of zebrafish tail (Willett *et al.*, 1999). Some of them seemed to be circulating in the caudal vein but others are macrophages and developing granulocytes that house the connective tissue interposed between the caudal vein and the caudal artery (Willett *et al.*, 1999). More recently, we have ultrastructurally identified groups of cells intimately associated with the vessel walls or migrating throughout them, consisting of primitive multipotent blood cell progenitors but also of both myeloblasts and pro-erythroblasts (Fig. 2). Nevertheless whereas the occurrence of granulopoiesis is clearly supported by these results, the existence of *in situ* production of mature erythrocytes remains to be confirmed. The myelopoietic activity of this caudal area of zebrafish has been confirmed by the presence of numerous myeloperoxidase positive cells (Lieschke *et al.*, 2001). In common carp, a similar myelopoietic activity remains until 2 weeks pf, as demonstrated by both immunohistochemistry using specific monoclonal antibodies (mAbs) against leukocytes and electron microscopy (Rombout *et al.*, 2005; Huttenhuis *et al.*, 2006c).

The relationships between these haematopoietic regions are difficult to establish. Some authors consider this caudal haematopoietic organ to be merely a remnant of the ICM for producing blood cells longer (Willett *et al.*, 1999). However, another possibility

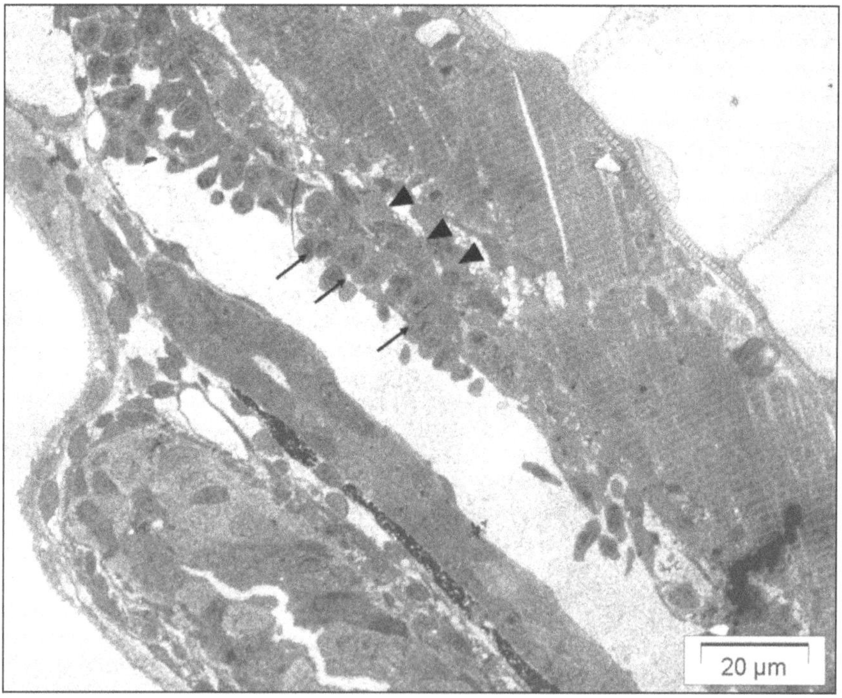

Fig. 2 Presence of developing blood cells into the caudal vein (arrows) and in the surrounding mesenchyme (arrowheads). 10 day-old zebrafish (*Danio rerio*).

is that the ICM only produces primitive blood cells for a short time and the caudal area represents a region for expanding and differentiating definitive haematopoietic progenitors generated in the dorsal aorta before colonising the definitive lympho-haematopoietic organs, i.e. thymus and kidney. Indeed, the origin of lymphoid and pluripotent haematopoietic progenitors that colonise these organ producers of lymphocytes and other blood cell lineages in adult teleosts remains to be conclusively demonstrated. This is a topic closely related to the occurrence of waves of "primitive" and "definitive" haematopoiesis in different anatomical locations of vertebrate embryos.

Indirect evidence suggests the existence of at least two different haematopoietic periods during zebrafish development. A first wave of maturation of primitive erythrocytes and macrophages, and presumably also granulocytes, occurs very early in zebrafish embryos. At the 4 somite-stage, *gata-1* positive *scl* positive cells from the rostral and later the trunk ICM give raise to circulating erythrocytes (Davidson & Zon, 2004). In addition, labelled pro-erythroblasts taken from 36 hpf embryos persist in the circulation of unlabelled host embryos for at least 10 days but they begin to decline at 6 dpf (Weinstein *et al.*, 1996). A "*bloodless*" zebrafish mutant with defects in ICM haematopoiesis starts to recover blood cell formation at day 5 when a second phase of haematopoiesis presumably begins to operate (Liao *et al.*, 2002).

Primitive macrophages, not derived from blood monocytes, appear in the zebrafish as early as the 3 somite-stage in the rostral area of LPM (Herbomel *et al.*, 1999; Bennett *et al.*, 2001; Lieschke *et al.*, 2002). These cells that later migrate toward the head midline, also appear in the Cuvier duct to finally accumulate in the ventral caudal vein (Fig. 2), an area that rapidly takes up latex beads or carbon particles injected into the bloodstream (Kimmel *et al.*, 1995; Lieschke *et al.*, 2001). In fact, at 30 hpf macrophages migrate to infection sites engulfing injected *Escherichia coli* or *Bacillus subtilis* (Herbomel *et al.*, 1999). Presumably these primitive macrophages are replaced later by monocyte-derived cells.

There is a general consensus that cells of the ICM are involved in the production of primitive blood cells, however the relationships between the haematopoietic site on the dorsal aorta walls and caudal haematopoiesis, as well as the origin of lympho-haematopoietic cells that colonise the definitive blood-forming organs, i.e. thymus and kidney, are not conclusively established. In zebrafish, it was proposed, but not tested, that lymphoid progenitors that seed the thymic primordium were descended from AGM cells produced around 24-48 hpf (Willett *et al.*, 1999). More recently, it has been reported that the elimination of *runx-1* expression or Notch signalling results in the absence of *c-myb* expression in the ventral area of dorsal aorta at 26 hpf and later of recombinase activating gene 1 (*rag-1*) expression in the thymus suggesting, but not demonstrating, a relationship between the two organs (Burns *et al.*, 2005; Gering & Patient, 2005). If we assume this to be the case, the next question is whether the seeding of AGM cells to the thymus and kidney is direct or AGM progenitors colonise other organs earlier, as occurs in murine fetal liver, in which they expand and differentiate. Could haematopoietic tissue from the caudal region be involved in these processes? Some results suggest that the caudal haematopoietic tissue could be a source for both the thymic lymphoid precursors and the multipotent cells that established the definitive haematopoiesis in the kidney (Murayama *et al.*, 2006).

HISTOGENESIS OF LYMPHOID ORGANS

The Origin of Immunocompetent Cells

After their appearance, blood cell progenitors migrate from the above described embryonic haematopoietic sites to the so-called primary lymphoid organs to produce the T and B lymphocytes. In all teleosts, thymus and kidney are the primary lymphoid organs whereas spleen, gut-associated lymphoid tisues (GALT) and again, the kidney are the peripheral lymphoid organs that house mature lymphocytes able to specifically react against antigens. Numerous studies have demonstrated that the thymus is the first organ to become lymphoid in freshwater teleosts, although earlier the kidney (pro- and mesonephros) can contain haematopoietic progenitors but not mature lymphocytes (reviewed by Bowden *et al.*, 2005; Rombout *et al.*, 2005; Zapata *et al.*, 2006). By contrast, in some studied marine teleosts the spleen appears earlier than the thymus, although in the first stages of development the splenic primordium contains more erythroid cells than developing or mature lymphocytes.

The thymus is a lympho-epithelial organ derived in all gnasthostomata from the pharyngeal pouches and involved in the phenotypical maturation and immunological "education" of T-cell progenitors. In mice, thymus organogenesis begins with migration of neural crest cells from the sixth hindbrain rhombomere to the pharyngeal region. In teleosts, the second, the third and fourth pharyngeal pouches are involved in thymus development (Bowden *et al.*, 2005). Apparently the interaction of neural crest-derived mesenchyme with endodermal epithelial cells of the pharyngeal pouches is crucial for thymus development. In this process, numerous genes, including *hoxa-3; pax 1, pax 3* and *pax* 9, are involved. In zebrafish, orthologues of these genes are involved in the development of pharyngeal pouches (Trede *et al.*, 2004) and their absence in distinct zebrafish mutants results in defective thymus organogenesis, although HSC formation is normal. One of these mutants is *van gogh* defective in the T-box 1 (*tbx-1*) gene, one of the earliest genes involved in the development of pharyngeal arches, which mimics the human Di George syndrome (Piotrowski *et al.*, 2003). Other mutants show defects in endoderm formation as a consequence of altered Nodal signalling, resulting in the lack of several organs including the thymus (Schier *et al.*, 1997; Alexander *et al.*, 1999; Reiter *et al.*, 2001). Also in the medaka, another interesting fish model due to its small genome size, mutants have been generated showing reduced *rag-1* expression in the thymic area in correlation with more or less acute defects in pharyngeal arch development (Iwanami *et al.*, 2004).

At 60 hpf the zebrafish thymus is an epithelial rudiment continuous with the pharyngeal epithelium a typical feature in all teleosts, which is colonised by the first lymphoid progenitors at 65-75 hpf. On day 4 pf, the organ is fully developed and the epithelium shows signs of maturation. Thymic epithelial cell development is dependent

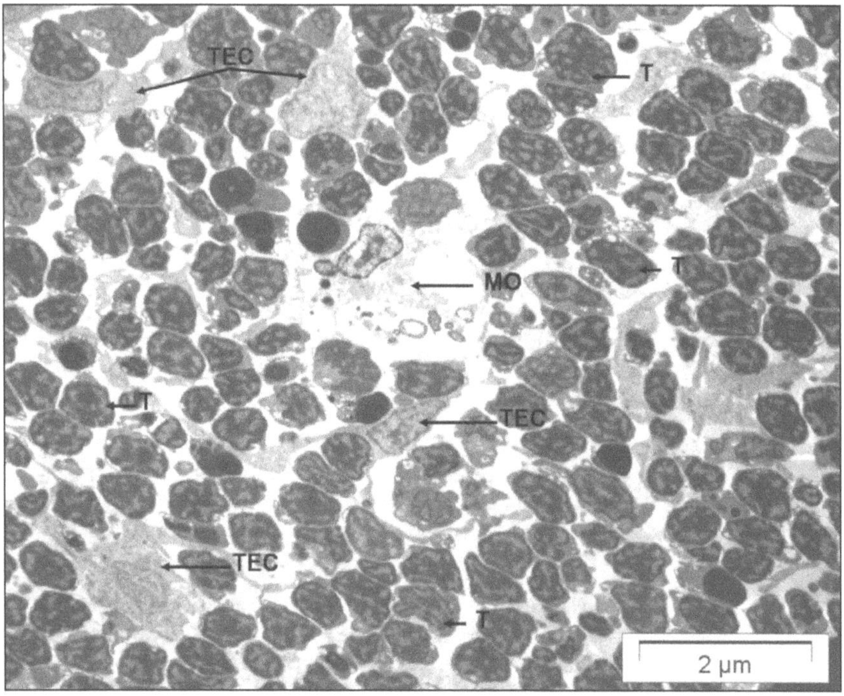

Fig. 3 Numerous small thymocytes (T) occur in the thymic parenchyma of a 3 month-old zebrafish (*Danio rerio*). Thymic epithelial cells (TECs), macrophages (MO).

on murine forkhead box N1 (*foxn1*) gene expression rather than on the arrival of lymphoid progenitors (Nehls *et al.*, 1994). Winged helix nude (*whn*) that was first reported as a *Foxn1* orthologue is expressed in thymic epithelial cells on day 3 pf (Schörpp *et al.*, 2002). In fact, there are two *whn*-like genes in zebrafish, *foxn1* and *foxn4* (Schlake *et al.*, 1997; Schörpp *et al.*, 2002). Whereas *foxn1* is expressed early in the zebrafish thymic epithelial cells as reported in mice (Schorpp *et al.*, 2002), *foxn4* expression is largely restricted to the nervous system. Expression in the thymus only occurs at 6 weeks and substantially increases in 6 month-old fish (Danilova *et al.*, 2004a). Remarkably, *cloche* mutants that do not contain HSCs and therefore have an alymphoid thymus, exhibit normal maturation of the thymic epithelium and retain *whn* expression (Schörpp *et al.*, 2002). Other zebrafish mutants, however, that do not express thymocyte markers in the thymic area but contain HSCs in the ICM, seem to show a good correlation between the disorganisation of thymic epithelial meshwork and the absence of thymocytes (Trede *et al.*, 2001).

Thymocyte maturation within the fish thymus correlates well with the expression of T-cell markers (Hansen 1997; Trede *et al.*, 2001, 2004). TcRα chain expression appears for the first time at 4 dpf (Danilova *et al.*, 2004b) two days after the appearance of lymphoblasts in the organ (Willett *et al.*, 1999). In other teleosts, as in the European seabass (*Dicentrarchus labrax*), presumptive T-cells appear in the thymus 3 days after the first lymphoid progenitors (Scapigliati *et al.*, 2002). Recently, *rag-1* and *rag-2* have been cloned in common carp and their expression studied during ontogeny (Huttenhuis *et al.*, 2005). Initial *rag* expression occurs on day 4 in the thymus and 3 days later is restricted to the lateral part of the organ, which contains (WCL 9) positive cortical thymocytes. It has been suggested, however, that lymphoid progenitors do not differentiate into immunocompetent T cells until the thymic stroma matures (Castillo *et al.*, 1990).

In this respect, in zebrafish, the organ grows rapidly reaching adult organisation by week 3. Then, some authors have reported the appearance of a thymic cortex close to the pharyngeal epithelium and a thymic medulla (Lam *et al.*, 2002; Schörpp *et al.*, 2002) but the thymus shape is not homogeneous and presumably changes with age (Romano *et al.*, 1999a) making it difficult to establish its demarcation. At least in both zebrafish and common carp, a functional zonation could exist because *rag-1* is only expressed under the pharyngeal epithelium (thymic cortex) but not in the inner region (thymic medulla) (Romano *et al.*, 1999b; Lam *et al.*, 2002), and in rainbow trout (Fischer *et al.*, 2005) the expression of major histocompatibility complex (MHC) class I and class II antigens also supports the occurrence of certain regionalisations in the teleost thymus despite the difficulties for determining its histological demarcation.

On the other hand, although an early involution of teleost thymus has been reported (Lam *et al.*, 2002), we and other authors (Danilova *et al.*, 2004b) have found lymphopoietic activity in zebrafish older than one year. Indeed, very little is known about the ageing of the lower vertebrate thymus (Torroba & Zapata, 2003) and the degree of involution is presumably variable in distinct species. For instance, *rag* expression has been reported in the thymus of 6 year-old common carp (Huttenhuis *et al.*, 2005) but no thymus was found in Japanese flounder (*Paralichthys olivaceus*) after 7 months of age (Liu *et al.*, 2004).

The origin of lymphoid precursors that colonise the thymus of teleost embryos is controversial, at least in species in which the thymus becomes lymphoid earlier than the pronephros. On the contrary, adult zebrafish kidney contains progenitors capable of long-term repopulation of irradiated host thymi (Langenau *et al.*, 2004). Injected thymocytes in irradiated recipient fish, however, resulted in short-term repopulation of the thymus. These results suggest that adult thymic progenitor cells arise from the kidney and are presumably derived from HSCs. Recently the occurrence of HSCs capable of colonising adult thymus has been confirmed in the ginbuna carp (*Carassius auratus langsdorfii*) (Kobayashi *et al.*, 2006). These authors have also demonstrated that trunk kidney (mesonephros) contains more haematopoietic precursors than head kidney (pronephros).

The kidney (both pro- and mesonephros) is also an early haematopoietic organ in teleosts, although the sequence of appearance of distinct blood cell lineages is still a matter of discussion that could reflect species-specific differences. On the other hand, published studies have used different technical approaches that make it difficult to compare the results obtained.

In zebrafish, we morphologically identified erythroblasts and myeloblasts in the embryonic kidney from day 4 onward (Willett *et al.*, 1999). At 5-6 dpf, mature and developing eosinophils also appeared but we were unable to conclusively identify lymphoblasts and small lymphocytes by electron microscopy until 2-3 weeks pf (Fig. 4). At that time, *rag-1* transcripts were clearly evidenced by *in situ* hybridisation on kidney histological sections (Zapata *et al.*, 2006). In fact, although some authors assume the site of origin and differentiation of teleost B lymphocytes there, their time of appearance remains controversial. We (Willett *et al.*, 1999) and other authors (Trede *et al.*, 2004) did not find *rag* expression in the embryonic kidney. However, Lam & colleagues (2004) by using RT-PCR found *rag-1* expression in whole 3 day-old fish, although these same authors only observed *rag-1* transcripts in a few cells of the pronephos at 2 weeks PF and clusters of *rag* positive cells at 3-4 weeks by *in situ* hybridisation on kidney sections. In addition, studies on larvae that expressed green fluorescent protein (GFP) under the control of *rag-2* expression suggested that the pronephros was the first extrathymic organ expressing *rag-2* in zebrafish at 8 dpf (Langenau *et al.*, 2004). In common carp, *rag-1* and *rag-2* expression evaluated by real time quantitative PCR was also detected for the first time around the first week pf (Huttenhuis, 2005). On the other hand, Danilova & Steiner (2002) reported *rag-1* expression in the pancreas of 4 day-old zebrafish by whole mount *in situ* hybridisation and a weak signal at 10 days pf on fish sections. Langenau and colleagues (2004) could not confirm this early presence

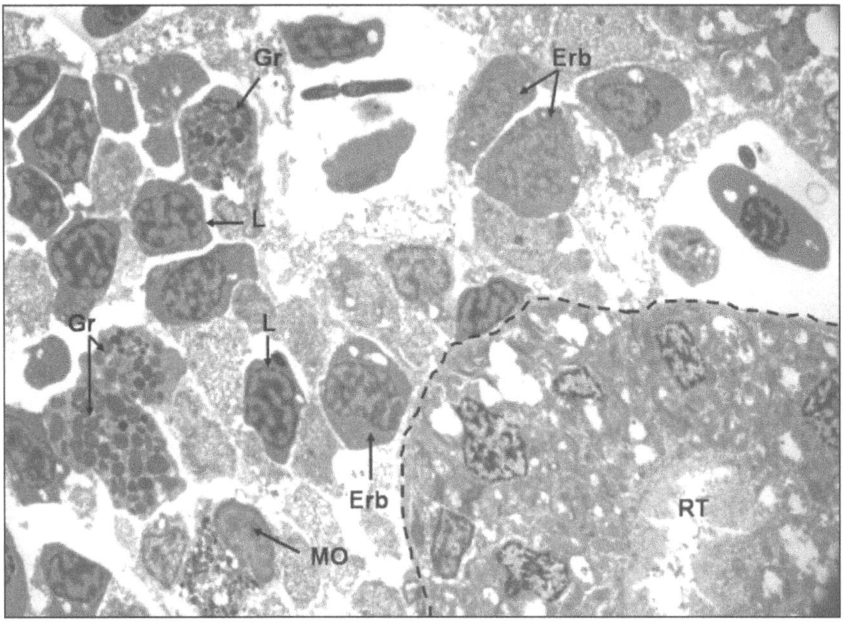

Fig. 4 Small lymphocytes (L), granulocytes (Gr), myeloblasts (MO) and erythroblasts (Erb) in the zebrafish (*Danio rerio*) renal haematopoietic tissue. Renal tubule (RT). 3 week-old zebrafish.

of *rag* positive cells in the pancreas of their transgenic fish and in common carp, neither immunoglobulin μ chain (Ig μ) expression nor monoclonal antibody specific for carp B lymphocytes (WCI 12) positive cells (B cells) have been found in the pancreas (Huttenhuis *et al.*, 2005). Each author concluded that the teleost pancreas is not involved in B-cell production.

Similar controversial results exist about the appearance of Ig-expressing cells in different teleosts. Danilova and Steiner (2002) reported rearrangements of genes encoding Ig μ in whole 4 dpf zebrafish and low levels of transcripts for surface Ig μ at 7-15 dpf but first Ig μ positive cells could be evidenced in the head kidney on day 19 pf. Likewise, Lam and colleagues (2004) detected Ig L isotypes by RT-PCR in whole zebrafish at 3 dpf but only 3 weeks pf by *in situ* hybridisation in kidney sections. In the zebrafish pronephros, Ig μ expression by RT-PCR was detected at 3 dpf but only 3 weeks pf by *in situ* hybridisation in kidney sections.

In the common carp pronephros, Ig μ expression has been detected by RT-PCR at 7 dpf (Huttenhuis *et al.*, 2005). One week later, the mAb WCL 12, specific to serum Ig-heavy (Ig-H) chain, detects surface Ig-positive cells in the head kidney (Romano *et al.*, 1997; Huttenhuis *et al.*, 2005) whereas the first cytoplasmic Ig-positive plasma cells appear one month pf, when the first serum specific Abs against T-independent antigens are produced (Huttenhuis *et al.*, 2005). A similar sequence of B-cell maturation occurs in the European seabass (Dos Santos *et al.*, 2000) and rainbow trout (Castillo *et al.*, 1993). In this last species, the kidney is a well-developed haematopoietic organ before hatching, principally concerned with erythroid and myeloid cell production. At hatching, lymphocytes can be ultrastructurally identified and both *rag* positive cells (Hansen & Zapata, 1998) and surface IgM positive lymphocytes (Razquin *et al.*, 1990) detected in the organ. However, other authors reported both *rag* expression (Hansen, 1999) and presence of cytoplasm IgM bearing lymphocytes (Castillo *et al.*, 1993) before hatching. A similar condition could occur in the Atlantic cod (*Gadus morhua*). In this species, IgM positive cells were detected in head kidney of 33 mm larvae (Schroeder *et al.*, 1998) and all VH families were present in 24-26 mm larvae, as evaluated by RT-PCR, but the VH-III family has been found in 13-14 mm larvae before the pronephros became lymphoid, suggesting the existence of other early B lymphopoietic sites in these species (Schroeder *et al.*, 1998).

Taking these results together we can conclude that: 1) the thymus and kidney are the first organs involved in T-cell and B-cell production, respectively in teleost, and 2) T lymphocytes appear earlier than B lymphocytes during ontogeny in correlation with the earlier appearance of the cell mediated immune responses of humoral immunity in fish (see later).

In most teleosts, the spleen becomes lymphocytic later (Ellis *et al.*, 1997; Dos Santos *et al.*, 2000; Petrie-Hanson & Ainsworth, 2000) although the splenic primordium appears early in development; i.e in zebrafish by 4 dpf (Langenau *et al.*, 2002). Even at 30 dpf, the zebrafish spleen remains a small organ that largely contains erythroblasts (Fig. 5). Two months later, when lymphocytes are histologically evident, the ellipsoids, anatomical elements crucial for antigen uptake, are just appearing. In this respect,

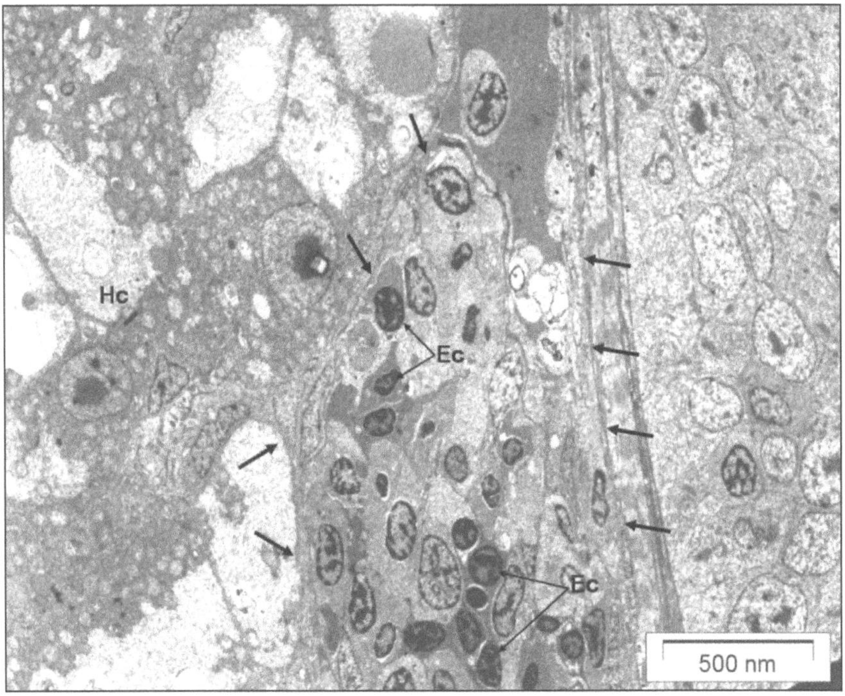

Fig. 5 Splenic primordium (limits are labelled by arrows) between the gut walls and the liver (Hc). Note the presence of numerous dark, erythroid cells (Ec) rather than lymphocytes. One-month-old zebrafish (*Danio rerio*).

rainbow trout splenocytes express MHC class I molecules for the first time on day 39 post-hatching (Fischer *et al.*, 2005). Nevertheless, IgM positive cells have been detected in channel catfish (*Ictalurus punctatus*) spleen by day 14, when there is apparently no organisation of the lymphoid compartment or ellipsoids (Petrie-Hanson & Ainsworth, 1999). The possibility that this delayed development of the peripheral lymphoid organs explains the late appearance of immune reaction in teleosts remains to be conclusively determined. In general, therefore, the spleen contains mature cells but not developing lymphocytes, although Hansen (1997) reported a moderate expression of both *rag-1* and *tdt* in the adult spleen of rainbow trout.

Adult teleosts do not contain important accumulations of lymphoid cells in the gut or other mucosae but instead lymphocytes, granulocytes and macrophages invade the intestinal epithelium whereas small aggregates of lymphocytes remain in the *lamina propria* (Fig. 6). In this respect, studies on teleost GALT have largely focused on the appearance and functional significance of intra-epithelial lymphocytes (IELs). Although IELs are largely considered to be T lymphocytes, whereas those accumulated in the gut *lamina propria* are principally B lymphocytes, they presumably include, as in mammals, numerous "non-conventional" T cells (Rombout *et al.*, 1993; Abelli *et al.*,

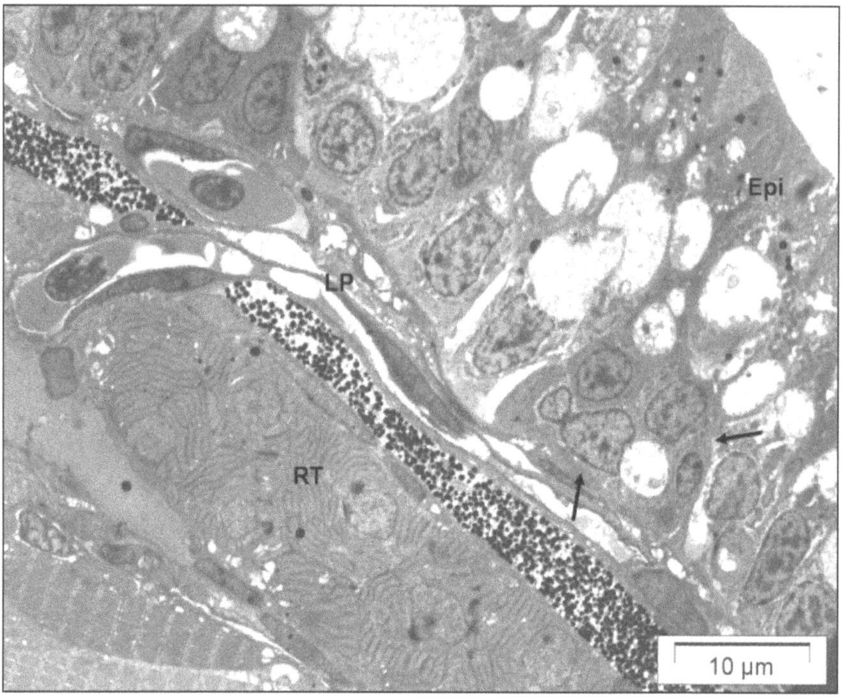

Fig. 6 Intra-epithelial lymphocytes (arrows) infiltrate the gut epithelium (Epi) of a 5 day-old zebrafish (*Danio rerio*). *Lamina propria* (LP); renal tubule (RT).

1997; Schroeder *et al.*, 1998; Fournier-Betz *et al.*, 2000). Unfortunately, the information on fish IELs is very limited. By electron microscopy, we found small lymphocytes and neutrophils in the intestine of 5 dpf zebrafish (Fig. 6) and Danilova & Steiner (2002) reported a few tcrα positive cells in the oesophagus and intestine of 9 dpf zebrafish. Other authors, however, were unable to detect T cells extrathymically in the first 3 weeks of life (Trede *et al.*, 2004).

Recently, Huttenhuis and colleagues (2006b) have extensively analysed the appearance of lymphoid cells in the common carp mucosae by combining flow cytometry and immunohistochemistry using mAbs specific to B cells (WCL 12), monocytes/macrophages (WCL 15) and WCL 38, an antibody that reacts with approximately 60% of lymphoid cells isolated from gills, gut and skin of adult carp (Rombout *et al.*, 1998). Their results demonstrate that WCL 38 positive cells are the first cells to appear in the intestine as early as 3 dpf. The nature of WCL 38 positive cells is, however, unclear. They are not B lymphocytes or NK cells (Rombout *et al.*, 1998) and exhibit no specific or non-specific cytotoxic responses (Huttenhuis *et al.*, 2006b). Rombout and colleagues (1998) have suggested that they are a type of T-cell found in all vertebrates characterised by using a second antigen T-cell receptor, so called $\gamma\delta$ cells, and more recently these authors pointed out that they are possibly regulatory T cells (Rombout *et al.*, 2005). In rainbow trout they exhibit non-specific cytotoxic activity (McMillan & Secombes, 1997). If we assume, therefore, that these cells are T lymphocytes, it is important to remark that they appear in the intestine before the thymus expresses *rag* and is histologically organised, suggesting that at least some of them have an extrathymic origin, presumably in the intestine itself. On the other hand, the numbers of gut WCL 38 positive cells really increase after the appearance of a functional thymus, indirectly suggesting that most of those cells must have originated in it. We could, therefore, speculate that WCL 38 positive intraepithelial (IE) cells could originate from the thymus but, as reported for mammalian IELs, do not undergo intrathymic selection or are a product of the *in situ* clonal expansion of a reduced number of differentiated IELs.

On the other hand, common carp monocytes/macrophages (WCL 15 positive cells) first appear in the intestinal *lamina propria* at 7 dpf (Romano *et al.*, 1998; Huttenhuis *et al.*, 2006b) and around 4 weeks pf in the intestinal epithelium. Their involvement in antigen trapping and presentation is recognised by all authors (Rombout *et al.*, 1989a, b; Joosten *et al.*, 1995). Finally, common carp WCL 12 positive cells (B lymphocytes) appear in both the intestine and the gills very late at 5-6 weeks pf, an observation also noted for the European seabass (Picchietti *et al.*, 1997) and spotted wolfish (*Anarhichas minor*) (Grontvedt & Espelid, 2003).

THE DEVELOPMENT OF IMMUNOLOGICAL CAPACITIES

Fish immunocompetence is determined by the appearance of immunological capacities rather than by the histological maturation of lymphoid tissues or the morphological evidence of immunocompetent cells. Indeed, there is an important delay between the early appearance of lymphoid cells and the late demonstration of immune responses in teleosts. In addition, only few data are available on the maturation of adaptive immunity based on a small number of "classical tests", such as allograft rejection, graft vs host reaction or antibody production against T-dependent and T-independent antigens. Only recently, some studies have correlated these responses with maturation of the molecular and cellular apparatus concerned with them (Nakanishi *et al.*, 2002; Shen *et al.*, 2002; Fischer *et al.*, 2006).

Cell-mediated immune responses develop a little earlier than humoral immunity. Thus, classical reports demonstrated that both rainbow trout and common carp are unable to reject scale allografts before 2-3 weeks of life (Manning & Grace, 1982; Tatner & Manning, 1983).

Although cells involved in fish-mediated cytotoxicity are difficult to characterise due to a lack of appropriate tools, we know that most molecules necessary for these reactions including MHC antigens, Tcr, CD8, β2-microglobulin, LPMs, PAP, CD3, etc. have been described in at least one teleost species (Fischer *et al.*, 2006). In addition, we know that cytotoxic responses against allogeneic cells are preformed by CD8-expressing lymphocytes (Fischer *et al.*, 2003), fish cytotoxic cells can be distinguished between virus-infected syngeneic and allogeneic target cells (Somamoto *et al.*, 2002) and the cytotoxic responses to virus-infected cells are MHC class I mediated (Fischer *et al.*, 2000).

There are unfortunately very few data concerning the ontogeny of these molecules or relating to fish cytotoxic cells. We have already reported the sequence of T and B lymphocyte appearance in distinct teleost species. Regarding the molecules involved in fish cytotoxicity, in rainbow trout CD8 and MHC class I transcripts are detected 1 week pf in the thymus, whereas mRNA encoding Tcr appears one week later (Fischer *et al.*, 2005). This early expression of MHC molecules in the thymus is probably related to T-cell education. However, MHC class I antigens are expressed in both gills and skin, earlier than in the thymus, presumably related to the trapping and processing of environmental antigens. In common carp, the results are similar (Rodrigues *et al.*, 1998). MHC expression is intimately associated with T-cell function. MHC class I was detected by RT-PCR on whole fish 1 dpf, while MHC class II antigens appeared 2 dpf and β2-microglobulin 7 dpf.

We have extensive information about the ability of adult fish to respond to antigens by producing specific antibodies (Bengten *et al.*, 2006; Marchalonis *et al.*, 2006; Solem & Stenvik, 2006) but, unfortunately, very little about its appearance during development despite the relevance of this information for vaccination and specific protection of eggs, embryos and larvae in an early pathogen confrontation. It is nevertheless reasonable to think that mechanisms used by larvae to acquire the antibody repertoire are similar to those of adult fish, which we summarise below.

In general, teleosts are high responders capable of producing relatively high numbers of specific antibodies to a variety of antigens (Solem & Stenvik, 2006). At the same time, they heavily modulate the response toward certain epitopes (Killie & Jorgensen, 1994). Some species, principally the Atlantic cod, however, seem to be unable to increase their antibody levels, or only to a minimal degree, after antigenic immunisation (Espelid *et al.*, 1991; Schroder *et al.*, 1992; Magnadóttir *et al.*, 2001).

The principal Ig present in teleost serum is an IgM tetramer with eight antigen-combining sites. In addition, some species have a monomeric form (H2 L2) in the serum with similar hapten binding affinity to the tetrameric IgM but less effective in complement activation (Wilson & Warr, 1992). A second Ig-H chain isotype, first identified in channel catfish, was designated Ig-HD due to certain sequence similarity, its location downstream of the Ig-HM gene and co-expression with the μ chain in some, but not all, B cells. The genomic structure and splicing pattern of the fish gene is, however, very variable among different species (Solem & Stenvik, 2006). Remarkably, certain constant domains of the teleost δ chain show a high similarity with non-μ isotypes, including those of cartilaginous fish (Hordvik *et al.*, 1999). In addition, a new Ig-H chain isotype (ω) recently reported in the marbled lungfish (*Protopterus aethiopicus*), which has been described as an orthologue of the cartilaginous fish non-μ chain gene (Ota *et al.*, 2003), also shows important similarities with the teleost δ1 and δ6 domains (Solem & Stenvik, 2006), supporting the notion of a common origin for the teleost δ and "ω family" of isotypes (Hordvik *et al.*, 1999). This fish IgD probably functions as a

B-cell receptor, although a secreted δ molecule has been found in the serum of channel catfish (Miller *et al.*, 1998).

A third teleost H chain isotype has been recently reported in two teleost species, zebrafish (IgZ) (Danilova *et al.*, 2005) and rainbow trout (IgT) (Hansen *et al.*, 2005). In addition, homologous cDNA sequences seem to exist in Atlantic salmon (*Salmo salar*) and common carp and a similar gene has been described in the Torafugu pufferfish (*Takifugu rubripes*) (Solem & Stenvik, 2006). According to bioinformatic data, all these new H chains seem to be restricted to teleosts. In both cases, the variable domain shares the same V gene segments as the μ chain but have their own D and J gene segments.

As indicated above, Igμ chain expression is weak at the start of development, but increases around 1 month pf (Danilova & Steiner, 2002). On the contrary, expression of the Igζ chain (Igζ) increased rapidly, reaching important levels at 2 weeks pf. Furthermore, expression of the Igζ chain gene in adult zebrafish is restricted to primary lymphoid organs, thymus and kidney, whereas the Igμ chain gene also appears in the peripheral lymphoid organs, such as spleen and intestine (Danilova *et al.*, 2005). Although phylogenetic analysis suggests an orthologous relationship for Igζ and IgT, they are differentially expressed, the former being restricted exclusively to primary lymphoid organs. On the contrary, IgT, apart from kidney and spleen, is also expressed in thymus, intestine (weakly), heart (presumably due to circulating B cells) and at low levels in peripheral blood lymphocytes (Hansen *et al.*, 2005). Furthermore, IgT transcripts appear before the presence of IgM positive B cells in the embryonic kidney.

Except for birds that seem to have only one Ig-light chain (Ig-L) isotype, in most vertebrates 2 or 3 isotypes have been described. In teleosts, 3 isotypes L1, L2 and L3 are recognised in most species, with a fourth in some species, such as the common carp (Ishikawa *et al.*, 2004) and one or two in other species (Solem & Stenvik, 2006). Obviously, the expression of several isotypes affects the diversity of the antibody repertoire although frequently expression is biased toward a given isotype. Nevertheless, in teleosts a large number of sterile Ig-L gene transcripts are expressed (Yoder *et al.*, 2002).

As mentioned above in relation to the development of B lymphocytes, data published on the appearance of Ig in fish serum or the onset of the capacity to mount humoral immune responses are controversial and difficult to compare owing to the different approaches used for their detection. Furthermore, the capacity to develop an antibody response does not necessarily correlate with the first appearance of IgM positive cells (Magnadóttir *et al.*, 2005).

On the other hand, during fish development, antibody production to T-independent antigens precedes that induced by T-dependent antigens. Thus, 2 week pf common carp responded to formalin-killed *Aeromonas salmonicida* but not to human IgG in Freund's complete adjuvant. Even 8 weeks post-immunisation no serum antibodies were detected (Manning & Grace, 1982). 7 dpf rainbow trout develop a primary immune response to *A. salmonicida* at 2 months with a memory element at 3 months (Tatner, 1986). Likewise, Lam and colleagues (2002) recently reported specific responses of zebrafish to *A. hydrophila* and HGG at 4 and 6 weeks pf, respectively.

THE INNATE SYSTEM IN EMBRYONIC AND LARVAL FISH

The innate immune system recognises generic classes of molecules produced by different pathogens that trigger an inflammatory response, in which several cell types and molecules secreted as a consequence of their activity, are involved. Innate and adaptive immunity are, in fact, elements of one physiological system, involved in recognising and eliminating pathogens. In ectotherms, innate immune responses are, however, particularly important due to the slow development of adaptive immunity and its limited immunological memory and capacity to produce an extensive antibody repertoire. Moreover, in embryonic and larval fish that are free-living organisms in an aquatic environment and have not fully developed their specific immunological capacity, the role of innate immunity is particularly relevant. This relevance of the innate immune system in teleosts is inferred from *in vivo* observations that *rag-1* deficient zebrafish that lack adaptive immunity have an apparently normal survival (Wienholds *et al.*, 2002). Despite this importance, many features of fish innate immune responses are, unfortunately, still unknown, particularly those related to their ontogenetic appearance.

Three large categories of components of the innate immune system have been recognised in fish: 1) physical parameters, 2) cells and 3) humoral/molecular parameters.

Physical Parameters

Fish scales and mucous surfaces of skin and gills constitute a first barrier against pathogens (Shephard, 1994). These elements constitute *per se* a physical barrier but the skin mucous is also an efficient agent to trap and slough pathogens, and also contains numerous antimicrobial molecules, such as lectins, complement proteins, antibacterial peptides, lysozyme, immunoglobulins, etc. (Magnadóttir, 2006).

Cells of the Innate Immune System

As in other vertebrates, granulocytes, macrophages and cytotoxic NK-like cells are the main cell types involved in fish innate immunity. Primitive macrophages, not derived from monocytes, appear very early in the embryonic zebrafish (Herbomel *et al.*, 1999; Bennett *et al.*, 2001; Lieschke *et al.*, 2002) and other teleost fish (Chantanachookin *et al.*, 1991; Romano *et al.*, 1998). These early macrophages show phagocytic activity-specific adherence to injected bacteria and form granulomas after mycobacterial infection (Herbomel *et al.*, 1999; Davis *et al.*, 2002; van der Sar *et al.*, 2003). Likewise, from 2 dpf onward neutrophils can be morphologically and immunohistochemically demonstrated in zebrafish (Willett *et al.*, 1999; Lieschke *et al.*, 2001; Crowhorst *et al.*, 2002) and carps (Huttenhuis *et al.*, 2006, c) and are able to migrate to sites of trauma and infection.

Non-specific cytotoxic cells have been identified in fish by different approaches. Presumably they are similar or equivalent to the mammalian NK cells capable of killing *in vitro* spontaneously tumour- or virus-infected cells. Cells with those capabilities were described in channel catfish as non-specific cytotoxic cells, morphologically different from large granular lymphocytes but functionally equivalent (Evans & Jaso-Friedmann, 1992). Furthermore, virus-infected autologous catfish lymphoid cell lines are killed by peripheral blood leukocytes from the same, non-immunised fish, but non-infected autologous target cells are not. The cells responsible for killing the viral-infected targets were different to those capable of allogeneic killing (Hogan *et al.*, 1996).

The best way to identify NK cells is to determine the occurrence of NK immune receptors in these primitive vertebrates, although even within mammals they exhibit major differences between distinct types. A non-specific cytotoxic cell receptor protein 1 (NCCRP-1), whose cross-linking results in receptor tyrosine and serine phosphorylation (Jaso-Friedman *et al.*, 2001), has been cloned in Nile tilapia (*Oreochromis niloticus*) (Jaso-Friedman *et al.*, 1997). NCCRP-1 positive cells were able to kill allogeneic, but also virus-infected fish (Yoshida *et al.*, 1995). In other fish species, another NK cell enhancing factor has been sequenced in rainbow trout (Zhang *et al.*, 2001) and common carp (Shin *et al.*, 2001) as well as group V C-type lectins homologous to the NKG2/CD94 family of mammalian NK receptors (Sato *et al.*, 2003; Kikuno *et al.*, 2004) and Fcμ receptors (Shen *et al.*, 2003). We do not know, however, about the existence of these molecules in fish embryos or larvae.

Apart from these receptors, Ig-type receptors, which share structural features with mammalian NK receptors, have recently been reported in various bony fish (Strong *et al.*, 1999; Litman *et al.*, 2001; Yoder, 2004). These receptors are encoded by the novel immune-type receptor (NITR) genes and consist of Ig-variable and intermediate ectodomains, transmembrane regions and cytoplasmic tails, largely containing immunoreceptor tyrosine-based inhibition motifs, although an activating cytoplasmic motif (NITR 5 family) has been described (Yoder *et al.*, 2004). In teleosts, 12 distinct families of NITRs have been described (Litman *et al.*, 2001), some of which are heavily expressed on NK cell lines that exhibit alloreactive specificity (Hawke *et al.*, 2001; Shen *et al.*, 2004). Furthermore, it is evident that different fish species possess distinct sets of NITR 5 families. On the other hand, C-type lectins are also encoded within the zebrafish NITR gene cluster (Yoder *et al.*, 2004). These receptors could, therefore, play a role in innate immunity during developmental stages in which adaptive immunity has not yet matured. They could be recognising a self-determinant similar to those recognised by NK receptors.

We will review also the occurrence of other receptors involved in innate immune responses the so-called Toll-like receptors (TLRs), an ancient family of proteins reported in recent years from invertebrates to humans. All family members of TLRs share the same structure and are largely expressed on macrophages and dendritic cells. They are composed of an extracellular domain containing leucine-rich repeats and a cysteine rich region, and a cytoplasmic tail that contains a highly conserved region named Toll/IL 1 receptor (TIR) domain (Akira, 2006).

TLRs have been identified in goldfish (*Carassius auratus*) (TLR 22) (Stanford *et al.*, 2003), Torafugu (TLR 1-5, TLR 5 soluble form, TLR 7-9, TLR 18, TLR 21,

TLR 22) (Oshiumi *et al.*, 2003) zebrafish, (TLR 1-5, TLR 5 soluble form, TLR 7-9, TLR 18-22) (Jault *et al.*, 2004; Meijer *et al.*, 2004), Japanese flounder (TLR 2, TLR 22) (Hirono *et al.*, 2004), the channel catfish (TLR 3, TLR 5) (Bilodeau & Waldbieser, 2005), and rainbow trout (TLR 5, TLR 5 soluble form) (Tsujita *et al.*, 2001; Tsukada *et al.*, 2005).

The zebrafish genome contains at least 17 expressed TLR genes that encode 24 putative variants, including orthologues of the 10 human TLR families, and homologous of IL 1R, IL 18R and S TIR-containing adaptor proteins e.g. myeloid differentiation factor 88 (MyD 88) myelin and lymphocyte protein (MAL) Toll/interleukin-1 receptor domain-containing adaptor protein inducing interferon β (TRIF) and selective androgen receptor antagonists/modulators (SARM), involved in the signalling pathway of these receptors (Jault *et al.*, 2004; Meijer *et al.*, 2004). The myD 88 cDNA and gene have also been identified in the Japanese flounder (Takano *et al.*, 2006).

The involvement of these receptors in fish innate immune responses has been demonstrated by distinct approaches. Bacterial flagellin, which is recognised by TLR 5, induces a TLR 5 soluble form (Tsujita *et al.*, 2004) and infection of zebrafish with *Mycobacterium marinus*, that induces granuloma formed by macrophages and lymphocytes, also up regulates the expression of several TLRs (Meijer *et al.*, 2004). Likewise, expression of a goldfish TLR was induced in macrophages exposed to *M. chelonei* (Stafford *et al.*, 2003). The zebrafish MyD 88 adaptor protein is also induced by *M. marinus* (Jault *et al.*, 2004; Meijer *et al.*, 2004). In the Japanese flounder, LPS (lipopolysaccharide ligand of TLR 4) and poly I:C (polyinosinic-polycytidylic acid ligand of TLR 3) treatment results in an immediately increased expression of *myD* 88 gene (Takano *et al.*, 2006). Recent results, however, question the relevance of myD 88 protein in the innate immune responses of teleosts. van der Sar and colleagues (2006) interfered with the myD 88 function by inhibiting its mRNA translation. Remarkably, *myD* 88 morphants apparently did not show myelopoietic defects, but were unable to effectively clear *Salmonella enterica*, although defective macrophages phagocytosed similarly to wild type (wt) cells.

At least 15 TLR genes are expressed in zebrafish at 1 dpf when macrophages and neutrophils begin to enter the blood circulation (van der Saar *et al.*, 2006). Most of these TLRs must also be maternally present, since their expression could be detected at the 4-cell stage which is prior to the onset of zygotic gene expression. Maternal transcripts of several adaptor genes including *myD* 88, mal, trif and sarm have also been identified (van de Sar *et al.*, 2006). Remarkably, different TLR genes display distinct expression patterns during fish ontogeny: TLR 1 transcript expression peaks during both blastula and gastrula stages and, in general, their levels are higher during embryogenesis than in adult fish. On the contrary, the expression of the *myD* 88 adaptor gene is highest in adult zebrafish. TLR 3 peaks during gastrulation and segmentation, is reduced during 1 to 5 dpf and again reaches high levels in the adult stage. In general, a peak in the expression of other TLRs, including TLR 5a, TLR 5b, TLR 7, TLR 8a, TLR 8b and TLR 18, coincides with the appearance of first macrophages 1 dpf.

Next, we wanted to emphasise the possible phylogenetic relevance of these receptors. Remarkably, lampreys have responses to pathogens based on variable lymphocyte receptors (VLRs) that contain, as do TLRs, arrays of leucine rich repeats (LRRs) (Pancer *et al.*, 2004). LRRs may bind different ligands with high affinities. This feature, together with the existence of a large number of LRR genes in cephalochordates (Pancer & Cooper, 2006), make the LRR units an interesting substrate to generate diversity in an agnathan ancestor.

On the other hand, apart from this cytotoxic activity, NK cells produce cytokines that affect the specific adaptive immune responses. Although over the past few years numerous cytokine gene sequences have been described in teleosts, their cellular origin and ontogenetic appearance are unfortunately poorly known (Secombes & Cunningham, 2004). Other cell types, apart from macrophages, granulocytes and NK cells, have been related to fish innate immune responses, largely mast cells and rodlet cells. However, no data on their occurrence and functional significance in embryos or larvae are available.

Most teleosts contain mast cells, principally in the intestine and gills but also in other tissues. They are morphologically and functionally equivalent to mammalian mast cells, except for the lack of histamine in their cytoplasmic granules (Reite & Evensen, 2006). Both increased numbers and recruitment of mast cells are frequent events in fish inflammatory reactions that result in degranulation and vasodilatation (Flaño *et al.*, 1996; Reite, 1997).

Rodlet cells are present in numerous teleost species. They are secretory cells, provided with remarkable cytoplasmic clublike inclusions that release their products to different body surfaces. Their origins are obscure but they have been involved in chronic inflammatory reactions in gills and intestine, as well as in responses to trematode and cestode parasites (Reite, 1997).

Humoral Parameters

Numerous humoral molecules, apart from the above mentioned cytokines, are involved in fish innate immune responses (Magnadóttir, 2006). They have been classified according to their pattern recognition specificities or effector functions, rather than their molecular features.

Transferrin

Transferrin inhibits bacterial growth by binding the available iron essential for their bacterial maintenance (Langston *et al.*, 1998).

Interferons

Two families of interferons (IFNs), type I and type II, are recognised on the basis of gene sequence, protein structure and functional properties. Type I IFNs include IFN α and IFN β, which are induced by viruses in most cells and play a major role in the defence against these pathogens. Type II IFNs comprise the IFN γ produced by NK cells and T lymphocytes in response to IL12, IL13, mitogens or antigens (Samuel, 2001).

IFN genes have been cloned in several teleost species, including zebrafish, Atlantic salmon, Torafugu, channel catfish and spotted green pufferfish (*Tetraodon nigroviridis*) (Altmann *et al.*, 2003; Lutfalla *et al.*, 2003; Robertsen *et al.*, 2003; Long *et al.*, 2004; Schultz *et al.*, 2004). Several fish viruses can induce IFN or IFN-like activity (de Kinkelin & Dorson, 1973; de Kinkelin *et al.*, 1982) and dsRNA poly I:C (an extensively used term for designing double stranded RNA rich in the bases I and C) is a strong inducer of IFN and IFN-inducible genes in several fish species as well as in mammals (Robertson, 2006). IFN production also occurs in very young rainbow trout fry after viral infection (Boudinot *et al.*, 1998) but the serum IFN levels following experimental infection of rainbow trout with viral haemorrhagic septicaemia virus (VHSH) were more a reflection of the levels of viraemia than resistance to virus (Ellis, 2001).

Protease inhibitors

Protease inhibitors are molecules present in fish serum and body fluids with an important role in their homeostasis, acute phase reactions and non-specific protection against pathogens, which secrete proteolytic enzymes (Magnadóttir, 2006). The best known is the α_2-macroglobulin that exhibits a broad specificity. In the eggs of a few fish species, other protease inhibitors have been reported but whether they play a role in innate immunity (see below) remains unknown (Magnadóttir *et al.*, 2005).

Lytic enzymes

Lytic enzymes are particularly important in the defense against bacteria. They have been classified as: hydrolases (i.e. lysozyme, chitinase, cathepsins), bacteriolytic and haemolytic enzymes. The lytic pathway of the complement system could also be included in this classification.

Lysozyme hydrolyses

Lysozyme hydrolyses β-(1,4)-linked glycoside bonds of bacterial cell wall proteoglycans resulting in the lysis of both Gram-positive and Gram-negative bacteria. In addition, it can also act as an opsonin activating the complement system and the phagocytes (Grinde, 1989). It is principally produced by leukocytes appearing in the mucus, lymphoid tissue, plasma and other body fluids of most teleosts except Atlantic cod and other marine species, which then exhibit a high chitinase activity, another lytic enzyme (Magnadóttir, 2006). Lysozyme has been reported to be present in oocytes, fertilised eggs and larvae of several fish. In the fertilised eggs of rainbow trout, it appears in the fertilisation envelope (chorion) that is involved in the physical protection of embryos and contains fungicidal and bactericidal enzymes (Kudo, 1991). In accordance with these results, lysozyme has been associated with the vertical protection from mother to offspring against some bacterial pathogens, such as *Aeromonas salmonicida*, *Renibacterium salmoninarum* and *Flavobacterium psychrophilum* (Magnadóttir *et al.*, 2005).

Cathepsins have also been reported in both eggs and larvae of several fish species as the major acid hydrolases responsible for yolk processing (Carnevali *et al.*, 2006; Tingaud-Sequeira & Cerdà, 2006). Cathepsins are also known to have bactericidal capacity having been demonstrated in the skin of Japanese eel (*Anguilla japonica*) and catfish (Aranishi, 1999; Cho *et al.*, 2002).

Fish agglutinins and precipitins present in both mucus and serum are C-type lectins and pentraxins. C-type lectins specifically bind different carbohydrates, such as mannose, N-acetyl glucosamine or fucose in the presence of calcium ions leading to opsonisation, phagocytosis and activation of complement (Arason, 1996). Like other lytic enzymes, C-type lectins, with different sugar specificities, have been found in oocytes, fertilised eggs and cells related to the immune system (Magnadóttir *et al.*, 2005).

Pentraxins, which have been found in fish eggs, can be divided into two large groups: CRP (C-reactive protein), pentraxins and SAP (serum amyloid protein) pentraxins. Some fish have only one of these two types, whereas others have both types, as occurs in higher vertebrates (Magnadóttir, 2006).

The complement system is a complex mechanism of defense comprising more than 30 plasma proteins as well as several membrane receptors. Activated complement components are active serine proteases that cleave other inactive components of the system and hence work in a sequential manner. This cascade of activation results in opsonisation and lysis of foreign micro-organisms and infected cells, and inflammation.

In fish, as in higher vertebrates, the complement system is activated through three different routes: the classical antibody-dependent pathway, the alternative pathway and the lectin pathway activated by binding lectin to carbohydrates expressed on micro-organisms (Magnadóttir *et al.*, 2005; Boshra *et al.*, 2006). In contrast to mammals, however, complement in teleosts is active at low temperatures and the titres generated through the alternative pathway are higher. Presumably, the teleost complement system could be capable of a wider recognition of molecules than mammals, a fact possibly related to the existence of multiple isoforms of some components of the fish complement system. Therefore, the isoforms of protein C3 bind with different efficiency to complement-activating surfaces (Boshra *et al.*, 2006).

On the other hand, the complement system is a phylogenetically old defence mechanism identified in primitive protostomes as well as in invertebrate deuterostomes, including echinoderms and tunicates. In cyclostomes, the most primitive group of vertebrates, complement components of both the alternative and lectin pathways have been identified, whereas elements of the classical route appear for the first time in the phylogeny in cartilaginous fish, the first vertebrates to possess Ig (Sunyer *et al.*, 2003; Nonaka & Yoshizaki, 2004).

Studies on the ontogeny of complement system are restricted to a few teleost species. The appearance of C3 was studied in Atlantic cod and Atlantic halibut

(*Hippoglossus hippoglossus*) by immunodetection, Western blotting and *in situ* hybridisation. In both species, the protein C3 could be detected at hatching in several organs (Magnodóttir *et al.*, 2005). Recently, the ontogeny of several components of the common carp complement system has been analysed. C3, α_2-macroglobulin, serum amyloid A (SAA) and C factor 1 r/s mannose binding lectin-associated serine protease-like molecule (C1/MASP2) transcripts are detected in carp embryos from 12 days pf onward. Moreover, LPS injection in the yolk-sac of 2 day-old common carp results in up-regulation of serum amyloid protein (SAA) expression and down-regulation of protein C3 and α_2-macroglobulin transcripts (Huttenhuis *et al.*, 2006, c). In rainbow trout, complement protein transcripts increase from 28 dpf to hatching and then decrease during yolk-sac resorption (Lovoll *et al.*, 2006). Nevertheless, no correlation between the appearance of transcripts and protein levels was found. Remarkably, both factor B and factor D transcripts, known o be involved in the alternative pathway, were expressed at much higher levels than protein C4, a key molecule of the classical pathway, emphasising again the relevance of the alternative pathway of complement system in fish defence mechanisms, particularly during embryonic and larval life.

Finally, just a few words to discuss the relevance of the presence of defence factors in teleost eggs. As already mentioned, eggs are rich in innate immune elements, such as lectins or lysozyme. In addition, maternal transfer of IgM has been repeatedly reported in numerous species of teleosts (Magnadóttir *et al.*, 2005). Most of this maternal IgM of teleost eggs is tetrameric and both structurally and antigenically identical to the maternal serum IgM, although monomeric and dimeric forms as well as truncated H chains have been reported. Apparently, however, the transference of maternal antibodies to the eggs has no protective value for the fry. Nevertheless, larvae of Atlantic salmon challenged with *Yersinia ruckeri* 2 weeks post hatch reached mortality levels of 3%, whereas these values rose to 13% and 60% when fry were challenged at 4 and 6 weeks post hatch, respectively (Lillehaug *et al.*, 1996). A similar greater resistance of young teleosts to pathogens has been observed in other species (Munro *et al.*, 1993). These results support the notion that, despite the lack of efficiency of maternal Ig for protecting fish larvae, other transferred molecules, presumably related to the innate immune system, provide important protection against micro-organisms in these early stages of development.

FUTURE PERSPECTIVES

Although data are very limited and numerous issues remain unresolved, over the past year the incorporation of some very useful new experimental models, such as medaka and zebrafish, to the "classical" ones, common carp, rainbow trout, European seabass and channel catfish has provided important information. In addition, the extensive use of the techniques of molecular biology and the combination of molecular and cell approaches have also contributed to the rapid development of our knowledge into fish immunology. Further studies in this area should attempt to resolve the unanswered questions and to apply emerging knowledge to improve methods for fish vaccination and the maintenance of fish fry of economic interest.

References

Abelli L, Picchietti S, Romano N, Mastrolia L, Scapigliati G (1997) Immunohistochemistry of gut-associated lymphoid tissue of the sea bass *Dicentrarchus labrax* (L.) Fish Shellfish Immunol 7: 235-245

Akira S (2006) TLR signalling. Curr Top Microbiol Immunol 311: 1-16

Al-Adhami M, Kunz YW (1977) Ontogenesis of haematopoietic sites in *Brachydanio rerio* (Hamilton-Buchanan) (Teleostei). Dev Growth Differ 19: 171-179

Alexander J, Rothenberg M, Henry GL, Stainier DY (1999) Casanova plays an early and essential role in endoderm formation in zebrafish. Dev Biol 215: 343-357

Altmann SM, Mellon MT, Distel DL, Kim CH (2003) Molecular and functional analysis of an interferon gene from the zebrafish, *Danio rerio*. J Virol 77: 1992-2002

Aranishi F (1999) Possible role for cathepsin B and L in bacteriolysis by Japanese eel skin. Fish Shellfish Immunol 9: 61-64

Arason GJ (1996) Lectins as defence molecules in vertebrates and invertebrates. Fish Shellfish Immunol 6: 277-289

Bengten E, Clem LW, Miller NW, Warr GW, Wilson M (2006) Channel catfish immunoglobulins: repertoire and expression. Dev Comp Immunol 30: 77-92

Bennett CM, Kanki JP, Rhodes J, Liu TX, Paw BH, Kieran MW, Langenau DM, Delahaye-Brown A, Zon LI, Fleming MD *et al.*, (2001) Myelopoiesis in the zebrafish, *Danio rerio*. Blood 98: 643-651

Bilodeau AL, Waldbieser GC (2005) Activation of TLR3 and TLR5 in channel catfish exposed to virulent *Edwardsiella ictaluri*. Dev Comp Immunol 29: 713-721

Boshra H, Li J, Sunyer O (2006) Recent advances on the complement system of teleost fish. Fish Shellfish Immunol 20: 239-262

Boudinot P, Blanco M, de Kinkelin P, Benmansour A (1998) Combined DNA immunization with the glycoprotein gene of viral hemorrhagic septicemia virus and infectious hematopoietic necrosis virus induces double-specific protective immunity and nonspecific response in rainbow trout. Virology 249: 297-306

Bowden TJ, Cook P, Rombout JH (2005) Development and function of the thymus in teleosts. Fish Shellfish Immunol 19: 413-427

Burns CE, DeBlasio T, Zhou Y, Zhang J, Zon L, Nimer SD (2002) Isolation and characterization of *runxa* and *runxb*, zebrafish members of the runt family of transcriptional regulators. Exp Hematol 30: 1381-1389

Burns CE, Traver D, Mayhall E, Shepard JL, Zon LI (2005) Hematopoietic stem cell fate is established by the Notch-Runx pathway. Genes Dev 19: 2331-2342

Carnevali O, Cionna C, Tosti L, Lubzens E, Maradonna F (2006) Role of cathepsins in ovarian follicle growth and maturation. Gen Comp Endocrinol 146: 195-203

Castillo A, Razquin E, Lopez-Fierro P, Alvarez F, Zapata A, Villena AJ (1990) Enzyme- and immuno-histochemical study of the thymic stroma in the rainbow trout, *Salmo gairdneri*, Richardson. Thymus 15: 153-166

Castillo A, Sanchez C, Dominguez J, Kaattari SL, Villena AJ (1993) Ontogeny of IgM and IgM-bearing cells in rainbow trout. Dev Comp Immunol 17: 419-424

Chantanachookhin CT, Seikai T, Tanaka M (1991) Comparative study of the lymphoid organs in three species of marine fish. Aquaculture 99: 143-155

Cho JH, Park IY, Kim MS, Kim SC (2002) Matrix metalloproteinase 2 is involved in the regulation of the antimicrobial peptide parasin I production in catfish skin mucosa. FEBS Letters p 459-463

Crowhurst MO, Layton JE, Lieschke GJ (2002) Developmental biology of zebrafish myeloid cells. Int J Dev Biol 46: 483-492

Danilova N, Steiner LA (2002) B cells develop in the zebrafish pancreas. Proc Natl Acad Sci USA 99: 13711-13716

Danilova N, Visel A, Willett CE, Steiner LA (2004a) Expression of the winged helix/forkhead gene, *foxn4*, during zebrafish development. Dev Brain Res 153: 115-119

Danilova N, Hohman VS, Sacher F, Ota T, Willett CE, Steiner LA (2004b) T cells and the thymus in developing zebrafish. Dev Comp Immunol 28: 755-767

Danilova N, Bussmann J, Jekosch K, Steiner LA (2005) The immunoglobulin heavy-chain locus in zebrafish: identification and expression of a previously unknown isotype, immunoglobulin Z. Nature Immunol 6: 295-302

Davidson AJ, Ernst P, Wang Y, Dekens MP, Kingsley PD, Palis J, Korsmeyer SJ, Daley GQ, Zon LI (2003) *cdx4* mutants fail to specify blood progenitors and can be rescued by multiple hox genes. Nature 425: 300-306

Davidson AJ, Zon LI (2004) The 'definitive' (and 'primitive') guide to zebrafish hematopoiesis. Oncogene 23: 7233-7246

Davis JM, Clay H, Lewis JL, Ghori N, Herbomel P, Ramakrishnan L (2002) Real-time visualization of mycobacterium-macrophage interactions leading to initiation of granuloma formation in zebrafish embryos. Immunity 17: 693-702

De Kinkelin P, Dorson M (1973) Interferon production in rainbow trout (*Salmo gairdneri*) experimentally infected with Egtved virus. J Gen Virol 19: 125-127

De Kinkelin P, Dorson M, Hattenberger-Baudouy AM (1982) Interferon synthesis on trout and carp after viral infection. Dev Comp Immunol 2: 167-174

Detrich III HW, Kieran MW, Chan FY, Barone LM, Yee K, Rundstadler JA, Pratt S, Ransom D, Zon LI (1995) Intraembryonic hematopoietic cell migration during vertebrate development. Proc Natl Acad Sci USA 92: 10713-10717

Dos Santos NM, Romano N, de Sousa M, Ellis AE, Rombout JH (2000). Ontogeny of B and T cells in sea bass (*Dicentrarchus labrax* L.). Fish Shellfish Immunol 10: 583-596

Ellis A (1977) Ontogeny of the immune response in *Salmo salar*. Histogenesis of the lymphoid organs and appearance of membrane immunoglobulin and mixed leucocyte reactivity. In: Solomon JB, Horton JD (eds). Dev Immunobiol. Elsevier/North-Holland Biomedical Press, London, p 225-231

Ellis AE (2001) Innate host defense mechanisms of fish against viruses and bacteria. Dev Comp Immunol 25: 827-839

Espelid S, Rødseth OM, Jørgensen TØ (1991) Vaccination experiments and studies of the humoral immune responses in cod, *Gadus morhua* L., to four strains of monoclonal-defined *Vibrio anguillarum*. J Fish Dis 14: 185-197

Evans DL, Jaso-Friedmann J (1992) Nonspecific cytotoxic cells as effectors of immunity in fish. Annu Rev Fish Dis 2: 109-121

Fischer U, Dijkstra JM, Aoyagi K, Xia C, Ototake M, Nakanishi T (2000) MHC class I restricted cytotoxicity in rainbow trout, *Oncorhynchus mykiss*. Dev Comp Immunol 24: 76-77

Fischer U, Utke K, Ototake M, Dijkstra JM, Kollner B (2003). Adaptive cell-mediated cytotoxicity against allogeneic targets by CD8-positive lymphocytes of rainbow trout (*Oncorhynchus mykiss*). Dev Comp Immunol 27: 323-337

Fischer U, Dijkstra JM, Kollner B, Kiryu I, Koppang EO, Hordvik I, Sawamoto Y, Ototake M (2005) The ontogeny of MHC class I expression in rainbow trout (*Oncorhynchus mykiss*). Fish Shellfish Immunol 18: 49-60

Fischer U, Utke K, Somamoto T, Kollner B, Ototake M, Nakanishi T (2006) Cytotoxic activities of fish leucocytes. Fish Shellfish Immunol 20: 209-226

Flaño E, Lopez-Fierro P, Razquin BE, Villena A (1996). *In vitro* differentiation of eosinophilic granular cells in *Renibacterium salmoninarum*-infected gill cultures from rainbow trout. Fish Shellfish Immunol 6: 173-184

Fournier-Betz V, Quentel C, Lamour F, LeVen A (2000) Immunocytochemical detection of Ig-positive cells in blood, lymphoid organs and the gut associated lymphoid tissue of the turbot (*Scophthalmus maximus*). Fish Shellfish Immunol 10: 187-202

Gering M, Patient R (2005) Hedgehog signaling is required for adult blood stem cell formation in zebrafish embryos. Dev Cell 8: 389-400

Grinde B (1989) Lysozyme from rainbow trout, *Salmo gairdneri* Richardson, as an antibacterial agent against fish pathogens. J Fish Dis 12: 95-104

Grontvedt RN, Espelid S (2003) Immunoglobulin producing cells in the spotted wolffish (*Anarhichas minor* Olafsen): localization in adults and during juvenile development. Dev Comp Immunol 27: 569-578

Hansen JD (1997) Characterization of rainbow trout terminal deoxynucleotidyl transferase structure and expression. TdT and RAG1 co-expression define the trout primary lymphoid tissues. Immunogenetics 46: 367-375

Hansen JD, Zapata AG (1998) Lymphocyte development in fish and amphibians. Immunol Rev 166: 199-220

Hansen JD, Landis ED, Phillips RB (2005) Discovery of a unique Ig heavy-chain isotype (IgT) in rainbow trout: Implications for a distinctive B cell developmental pathway in teleost fish. Proc Natl Acad Sci USA 102: 6919-6924

Hawke NA, Yoder JA, Haire RN, Mueller MG, Litman RT, Miracle AL, Stuge T, Shen L, Miller N, Litman GW (2001) Extraordinary variation in a diversified family of immune-type receptor genes. Proc Natl Acad Sci USA 98: 13832-13837

Herbomel P, Thisse B, Thisse C (1999) Ontogeny and behaviour of early macrophages in the zebrafish embryo. Development 126: 3735-3745

Hirono I, Takami M, Miyata M, Miyazaki T, Han HJ, Takano T, Endo M, Aoki T (2004) Characterization of gene structure and expression of two toll-like receptors from Japanese flounder, *Paralichthys olivaceus*. Immunogenetics 56: 38-46

Hogan RJ, Stuge TB, Clem LW, Miller NW, Chinchar VG (1996) Anti-viral cytotoxic cells in the channel catfish (*Ictalurus punctatus*). Dev Comp Immunol 20: 115-127

Hordvik I, Thevarajan J, Samdal I, Bastani N, Krossoy B (1999) Molecular cloning and phylogenetic analysis of the Atlantic salmon immunoglobulin D gene. Scand J Immunol 50: 202-210

Huttenhuis HB, Huising MO, van der Meulen T, van Oosterhoud CN, Sanchez NA, Taverne-Thiele AJ, Stroband HW, Rombout JH (2005) Rag expression identifies B and T cell lymphopoietic tissues during the development of common carp (*Cyprinus carpio*). Dev Comp Immunol 29: 1033-1047

Huttenhuis HB, Grou CPO, Taverne-Thiele AJ, Taverne N, Rombout JHWM (2006a) Carp (*Cyprinus carpio*) innate immune factors are present before hatching. Fish Shellfish Immunol 20: 586-596

Huttenhuis HB, Romano N, Van Oosterhoud CN, Taverne-Thiele AJ, Mastrolia L, Van Muiswinkel WB, Rombout JH (2006b) The ontogeny of mucosal immune cells in common carp (*Cyprinus carpio*). Anat Embryol (Berl) 211: 19-29

Huttenhuis HB, Taverne-Thiele AJ, Grou CP, Bergsma J, Saeij JP, Nakayasu C, Rombout JH (2006c) Ontogeny of the common carp (*Cyprinus carpio*) innate immune system. Dev Comp Immunol 30: 557-574

Ishikawa J, Imai E, Moritomo T, Nakao M, Yano T, Tomana M (2004) Characterisation of a fourth immunoglobulin light chain isotype in the common carp. Fish Shellfish Immunol 16: 369-379

Iwanami N, Takahama Y, Kunimatsu S, Li J, Takei R, Ishikura Y, Suwa H, Niwa K, Sasado T, Morinaga C *et al.* (2004) Mutations affecting thymus organogenesis in Medaka, *Oryzias latipes*. Mech Dev 121: 779-789

Jaso-Friedmann L, Leary JH, Evans DL (1997) NCCRP-1: a novel receptor protein sequenced from teleost nonspecific cytotoxic cells. Mol Immunol 34: 955-965

Jaso-Friedmann L, Leary JH, Evans DL (2001) The non-specific cytotoxic cell receptor (NCCRP-1): molecular organization and signaling properties. Dev Comp Immunol 25: 701-711

Jault C, Pichon L, Chluba J (2004) Toll-like receptor gene family and TIR-domain adapters in *Danio rerio*. Mol Immunol 40: 759-771

Joosten PHM, Avilés-Trigueros M, Sorgeloos P, Rombout JHWM (1995) Oral vaccination of juvenile carp (*Cyprinus carpio*) and gilthead seabream (*Sparus aurata*) with bioencapsulated bacterin. Fish Shellfish Immunol 5: 289-299

Kalev-Zylinska ML, Horsfield JA, Flores MV, Postlethwait JH, Vitas MR, Baas AM, Crosier PS, Crosier KE (2002) Runx1 is required for zebrafish blood and vessel development and expression of a human RUNX1-CBF2T1 transgene advances a model for studies of leukemogenesis. Development 129: 2015-2030

Kikuno R, Sato A, Mayer WE, Shintani S, Aoki T, Klein J (2004) Clustering of C-type lectin natural killer receptor-like loci in the bony fish *Oreochromis niloticus*. Scand J Immunol 59: 133-142

Killie JE, Jørgensen TO (1994) Immunoregulation in fish. I: Intramolecular-induced suppression of antibody responses to haptenated protein antigens studied in Atlantic salmon (*Salmo salar* L). Dev Comp Immunol 18: 123-136

Kimmel CB, Ballard WW, Kimmel SR, Ullmann B, Schilling TF (1995) Stages of embryonic development of the zebrafish. Dev Dyn 203: 253-310

Kobayashi I, Sekiya M, Moritomo T, Ototake M, Nakanishi T (2006) Demonstration of hematopoietic stem cells in ginbuna carp (*Carassius auratus langsdorfii*) kidney. Dev Comp Immunol 30: 1034-1046

Kudo H, Ueda H, Yamauchi K (1996) Immunocytochemical investigation of a salmonid olfactory system-specific protein in the kokanee salmon (*Oncorhynchus nerka*). Zool Sci 13: 647-653

Lam SH, Chua HL, Gong Z, Wen Z, Lam TJ, Sin YM (2002) Morphologic transformation of the thymus in developing zebrafish. Dev Dyn 225: 87-94

Lam SH, Chua HL, Gong Z, Lam TJ, Sin YM (2004) Development and maturation of the immune system in zebrafish, *Danio rerio*: a gene expression profiling, *in situ* hybridization and immunological study. Dev Comp Immunol 28: 9-28

Langston AL, Bricknell IR, Ellis AE (1998) Iron binding capacity of peripheral blood leucocyte lysates from Atlantic salmon (*Salmo salar* L.). In: Barnes AC, Davison GA, Hiney MP, McIntosh D (eds). Methodology in Fish Diseases Research. Fisheries Research Services. Aberdeen. p 111-116

Langenau DM, Palomero T, Kanki JP, Ferrando AA, Zhou Y, Zon LI, Look AT (2002) Molecular cloning and developmental expression of *Tlx* (*Hox11*) genes in zebrafish (*Danio rerio*). Mech Dev 117: 243-248

Langenau DM, Ferrando AA, Traver D, Kutok JL, Hezel JP, Kanki JP, Zon LI, Look AT, Trede NS (2004) *In vivo* tracking of T cell development, ablation, and engraftment in transgenic zebrafish. Proc Natl Acad Sci USA 101: 7369-7374

Liao W, Bisgrove BW, Sawyer H, Hug B, Bell B, Peters K, Grunwald DJ, Stainier DYR (1997) The zebrafish gene cloche acts upstream of a *flk*-1 homologue to regulate endothelial differentiation. Development 124: 381-389

Liao EC, Trede NS, Ransom D, Zapata A, Kieran M, Zon LI (2002) Non-cell autonomous requirement for the *bloodless* gene in primitive hematopoiesis of zebrafish. Development 129: 649-659

Lieschke GJ, Oates AC, Crowhurst MO, Ward AC, Layton JE (2001) Morphologic and functional characterization of granulocytes and macrophages in embryonic and adult zebrafish. Blood 98: 3087-3096

Lieschke GJ, Oates AC, Paw BH, Thompson MA, Hall NE, Ward AC, Ho RK, Zon LI, Layton JE (2002) Zebrafish SPI-1 (PU.1) marks a site of myeloid development independent of primitive erythropoiesis: implications for axial patterning. Dev Biol 246: 274-295

Lillehaug A, Sevatadal S, Endal T (1996) Passive transfer of specific maternal immunity does not protect Atlantic salmon (*Salmo salar* L.) fry against yersiniosis. Fish Shellfish Immunol 6: 521-535

Litman GW, Hawke NA, Yoder JA (2001) Novel immune-type receptor genes. Immunol Rev 181: 250-259

Liu Y, Zhang S, Jiang G, Yang D, Lian J, Yang Y (2004) The development of the lymphoid organs of flounder, *Paralichthys olivaceus,* from hatching to 13 months. Fish Shellfish Immunol 16: 621-632

Long S, Wilson M, Bengten E, Bryan L, Clem LW, Miller NW, Chinchar VG (2004) Identification of a cDNA encoding channel catfish interferon. Dev Comp Immunol 28: 97-111

Lovoll M, Kilvik T, Boshra H, Bogwald J, Sunyer JO, Dalmo RA (2006) Maternal transfer of complement components C3-1, C3-3, C3-4, C4, C5, C7, Bf, and Df to offspring in rainbow trout (*Oncorhynchus mykiss*). Immunogenetics 58: 168-179

Lutfalla G, Roest Crollius H, Stange-Thomann N, Jaillon O, Mogensen K, Monneron D (2003) Comparative genomic analysis reveals independent expansion of a lineage-specific gene family in vertebrates: the class II cytokine receptors and their ligands in mammals and fish. BMC Genomics 4: 29

Magnadóttir B (2006) Innate immunity of fish (overview). Fish Shellfish Immunol 20: 137-151

Magnadóttir B, Jonsdottir H, Helgason S, Bjornsson B, Solem ST, Pilstrom L (2001) Immune parameters of immunised cod (*Gadus morhua* L.). Fish Shellfish Immunol 11: 75-89

Magnadóttir B, Lange S, Gudmundsdottir S, Bogwald J, Dalmo RA (2005) Ontogeny of humoral immune parameters in fish. Fish Shellfish Immunol 19: 429-439

Manning MJ, Grace MF (1982). Developmental aspects of immunity and tolerance in fish. In: Roberts RJ (ed) Microbial Diseases of Fish. Academic Press, London, p 31-46

Marchalonis JJ, Adelman MK, Schluter SF, Ramsland PA (2006) The antibody repertoire in evolution: chance, selection, and continuity. Dev Comp Immunol 30: 223-247

Mcmillan DN, Secombes C J (1997) Isolation of rainbow trout (*Oncorhynchus mykiss*) intestinal intraepithelial lymphocytes (IEL) and measurement of their cytotoxic activity. Fish Shellfish Immunol 7: 527-541

Meijer AH, Gabby Krens SF, Medina Rodriguez IA, He S, Bitter W, Ewa Snaar-Jagalska B, Spaink HP (2004) Expression analysis of the Toll-like receptor and TIR domain adaptor families of zebrafish. Mol Immunol 40: 773-783

Miller N, Wilson M, Bengten E, Stuge T, Warr G, Clem W (1998) Functional and molecular characterization of teleost leukocytes. Immunol Rev 166: 187-197

Munro ALS, Hastings TS (1993) Furunculosis. In: Inglis V, Roberts RJ, Bromage NR (eds). Bacterial Diseases of Fish. Blackwell Scientific Publications. Oxford, p 122-144

Murayama E, Kissa K, Zapata A, Mordelet E, Briolat V, Lin HF, Handin RI, Herbomel P (2006) Tracing hematopoietic precursor migration to successive hematopoietic organs during zebrafish development. Immunity 25: 963-975

Nakanishi T, Fischer U, Dijkstra JM, Hasegawa S, Somamoto T, Okamoto N, Ototake M (2002) Cytotoxic T-cell function in fish. Dev Comp Immunol 26: 131-139

Nehls M, Pfeifer D, Schorpp M, Hedrich H, Boehm T (1994) New member of the winged-helix protein family disrupted in mouse and rat nude mutations. Nature 372: 103-107

Nonaka M, Yoshizaki F (2004) Evolution of the complement system. Mol Immunol 40: 897-902

Okuda T, van Deursen J, Hiebert SW, Grosveld G, Downing JR (1996) AML1, the target of multiple chromosomal translocations in human leukemia, is essential for normal fetal liver hematopoiesis. Cell 84: 321-330

Oshiumi H, Tsujita T, Shida K, Matsumoto M, Ikeo K, Seya T (2003) Prediction of the prototype of the human Toll-like receptor gene family from the pufferfish, *Fugu rubripes*, genome. Immunogenetics 54: 791-800

Ota T, Rast JP, Litman GW, Amemiya CT (2003) Lineage-restricted retention of a primitive immunoglobulin heavy chain isotype within the Dipnoi reveals an evolutionary paradox. Proc Natl Acad Sci USA 100: 2501-2506

Pancer Z, Amemiya CT, Ehrhardt GR, Ceitlin J, Gartland GL, Cooper MD (2004) Somatic diversification of variable lymphocyte receptors in the agnathan sea lamprey. Nature 430: 174-180

Pancer Z, Cooper MD (2006) The evolution of adaptive immunity. Annu Rev Immunol 24: 497-518

Petrie-Hanson L, Ainsworth AJ (1999) Humoral immune responses of channel catfish (*Ictalurus punctatus*) fry and fingerlings exposed to *Edwardsiella ictaluri*. Fish Shellfish Immunol 9: 579-589

Petrie-Hanson L, Ainsworth AJ (2000) Differential cytochemical staining characteristics of channel catfish leukocytes identify cell populations in lymphoid organs. Vet Immunol Immunopathol 73: 129-144

Picchietti S, Terribili FR, Mastrolia L, Scapigliati G, Abelli L (1997) Expression of lymphocyte antigenic determinants in developing gut-associated lymphoid tissue of the sea bass *Dicentrarchus labrax* (L.). Anat Embryol 196: 457-463

Piotrowski T, Ahn DG, Schilling TF, Nair S, Ruvinsky I, Geisler R, Rauch GJ, Haffter P, Zon LI, Zhou Y *et al.*, (2003) The zebrafish *van gogh* mutation disrupts *tbx1*, which is involved in the DiGeorge deletion syndrome in humans. Development 130: 5043-5052

Razquin BE, Castillo A, López-Fierro P, Álvarez F, Zapata A, Villena AJ (1990) Ontogeny of IgM-producing cells in the lymphoid organs of rainbow trout, *Salmo gairdneri* Richardson: an immuno and enzyme-histochemical study. J Fish Biol 36: 159-173

Reite OB (1997) Mast cells/eosinophilic granule cells of salmonids: staining properties and responses to noxious agents. Fish Shellfish Immunol 7: 567-584

Reite OB, Evensen O (2006) Inflammatory cells of teleostean fish: a review focusing on mast cells/eosinophilic granule cells and rodlet cells. Fish Shellfish Immunol 20: 192-208

Reiter JF, Kikuchi Y, Stainier DY (2001) Multiple roles for Gata5 in zebrafish endoderm formation. Development 128: 125-135

Robertsen B (2006) The interferon system of teleost fish. Fish Shellfish Immunol 20: 172-191

Robertsen B, Bergan V, Rokenes T, Larsen R, Albuquerque A (2003) Atlantic salmon interferon genes: cloning, sequence analysis, expression, and biological activity. J Interferon Cytokine Res 23: 601-612

Rodrigues PN, Hermsen TT, van Maanen A, Taverne-Thiele A J, Rombout JH, Dixon B, Stet R J (1998) Expression of MhcCyca class I and class II molecules in the early life history of the common carp (*Cyprinus carpio* L.). Dev Comp Immunol 22: 493-506

Romano N, Taverne-Thiele JJ, Van Maanen JC, Rombout JHMW (1997) Leucocyte subpopulations in developing carp (*Cyprinus carpio* L.): immunocytochemical studies. Fish Shellfish Immunol 7: 439-453

Romano N, Picchietti S, Taverne-Thiele JJ, Taverne N, Abelli L, Mastrolia L, Verburg-van Kemenade BM, Rombout J H (1998) Distribution of macrophages during fish development: an immunohistochemical study in carp (*Cyprinus carpio* L.). Anat Embr (Berl) 198: 31-41

Romano N, Fanelli M, Maria del Papa G, Scapigliati G, Mastrolia L (1999) Histological and cytological studies on the developing thymus of sharpsnout seabream, *Diplodus puntazzo*. J Anat 194: 39-50

Romano N, Taverne-Thiele A J, Fanelli M, Baldassini MR, Abelli L, Mastrolia L, Van Muiswinkel WB, Rombout JH (1999) Ontogeny of the thymus in a teleost fish, *Cyprinus carpio* L.: developing thymocytes in the epithelial microenvironment. Dev Comp Immunol 23: 123-137

Rombout JHWM, Bot HE, Taverne-Thiele AJ (1989a) Immunological importance of the second gut segment of carp. II. Characterisation of mucosal lymphocytes. J Fish Biol 35: 167-178

Rombout JHWM, van den Berg AA, van den Berg CTGA, Witte P, Egberts E (1989b) Immunological importance of the second gut segment of carp. III. Systemic and/or mucosal immune responses after immunization with soluble or particulate antigen. J Fish Biol 35: 179-186

Rombout JHWM, Taverne-Thiele AJ, Villena MI (1993) The gut-associated lymphoid tissue (GALT) of carp (*Cyprinus carpio* L.): an immunocytochemical analysis. Dev Comp Immunol 17: 55-66

Rombout JHWM, Joosten PH, Engelsma MY, Vos APTN, Taverne-Thiele J (1998) Indications for a distinct putative T cell population in mucosal tissue of carp (*Cyprinus carpio* L.). Dev Comp Immunol 22: 63-77

Rombout JHWM, Huttenhuis HBT, Picchietti S, Scapigliati G (2005) Phylogeny and ontogeny of fish leucocytes. Fish Shellfish Immunol 19: 441-455

Samuel CE (2001) Antiviral actions of interferons. Clin Microbiol Rev 14: 778-809

Sato A, Mayer WE, Overath P, Klein J (2003) Genes encoding putative natural killer cell C-type lectin receptors in teleostean fishes. Proc Natl Acad Sci USA 100: 7779-7784

Scapigliati G, Romano N, Buonocore F, Picchietti S, Baldassini MR, Prugnoli D, Galice A, Meloni S, Secombes CJ, Mazzini M *et al.* (2002) The immune system of sea bass, *Dicentrarchus labrax*, reared in aquaculture. Dev Comp Immunol 26: 151-160

Schier AF, Neuhauss SC, Helde KA, Talbot WS, Driever W (1997) The *one-eyed pinhead* gene functions in mesoderm and endoderm formation in zebrafish and interacts with *no tail*. Development 124: 327-342

Schlake T, Schörpp M, Nehls M, Boehm T (1997) The *nude* gene encodes a sequence-specific DNA binding protein with homologs in organisms that lack an anticipatory immune system. Proc Natl Acad Sci USA 94: 3842-3847

Schörpp M, Leicht M, Nold E, Hammerschmidt M, Haas-Assenbaum A, Wiest W, Boehm T (2002) A zebrafish orthologue (*whnb*) of the mouse *nude* gene is expressed in the epithelial compartment of the embryonic thymic rudiment. Mech Dev 118: 179-185

Schrøder MB, Espelid S, Jørgensen TØ (1992) Two serotypes of *Vibrio salmonicida* isolated from diseased cod (*Gadus morhua* L.); virulence, immunological studies and vaccination experiments. Fish Shellfish Immunol 2: 211-221

Schrøder MB, Villena A, Jørgensen TØ (1998) Ontogeny of lymphoid organs and immunoglobulin producing cells in Atlantic cod (*Gadus morhua* L.). Dev Comp Immunol 22: 507-517

Schrøder MB, Flaño E, Pilström L, Jørgensen TØ (1998) Localisation of Ig heavy chain mRNA positive cells in Atlantic cod (*Gadus morhua* L.) tissues; identified by *in situ* hybridisation. Fish Shellfish Immunol 8: 565-576

Schultz U, Kaspers B, Staeheli P (2004) The interferon system of non-mammalian vertebrates. Dev Comp Immunol 28: 499-508

Secombes C, Cunningham C (2004) Cytokines: an evolutionary perspective. Dev Comp Immunol 28: 373-374

Shen L, Stuge TB, Zhou H, Khayat M, Barker KS, Quiniou SM, Wilson M, Bengten E, Chinchar VG, Clem LW *et al.* (2002) Channel catfish cytotoxic cells: a mini-review. Dev Comp Immunol 26: 141-149

Shen L, Stuge TB, Evenhuis JP, Bengten E, Wilson M, Chinchar VG, Clem LW, Miller NW (2003) Channel catfish NK-like cells are armed with IgM via a putative FcmicroR. Dev Comp Immunol 27: 699-714

Shen L, Stuge TB, Bengten E, Wilson M, Chinchar VG, Naftel JP, Bernanke JM, Clem LW, Miller NW (2004). Identification and characterization of clonal NK-like cells from channel catfish (*Ictalurus punctatus*). Dev Comp Immunol 28: 139-152

Shephard KL (1994). Functions of fish mucus. Rev Fish Biol Fish 4: 401-429

Shin DH, Fujiki K, Nakao M, Yano T (2001) Organization of the NKEF gene and its expression in the common carp (*Cyprinus carpio*). Dev Comp Immunol 25: 597-606

Solem ST, Stenvik J (2006) Antibody repertoire development in teleosts – a review with emphasis on salmonids and *Gadus morhua* L. Dev Comp Immunol 30: 57-76

Somamoto T, Nakanishi T, Okamoto N (2002) Role of specific cell-mediated cytotoxicity in protecting fish from viral infections. Virology 297: 120-127

Stafford JL, Ellestad KK, Magor KE, Belosevic M, Magor BG (2003) A toll-like receptor (TLR) gene that is up-regulated in activated goldfish macrophages. Dev Comp Immunol 27: 685-698

Strong SJ, Mueller MG, Litman RT, Hawke NA, Haire RN, Miracle AL, Rast JP, Amemiya CT, Litman GW (1999) A novel multigene family encodes diversified variable regions. Proc Natl Acad Sci USA 96: 15080-15085

Sunyer JO, Boshra H, Lorenzo G, Parra D, Freedman B, Bosch N (2003) Evolution of complement as an effector system in innate and adaptive immunity. Res Immunol 27: 549-564

Takano T, Kondo H, Hirono I, Saito-Taki T, Endo M, Aoki T (2006) Identification and characterization of a myeloid differentiation factor 88 (MyD88) cDNA and gene in Japanese flounder, *Paralichthys olivaceus*. Dev Comp Immunol 30: 807-816

Tatner MF (1986) The ontogeny of humoral immunity in rainbow trout, *Salmo gairdneri*. Vet Immunol Immunopathol 12: 93-105

Tatner MF, Manning MJ (1983) The ontogeny of cellular immunity in the rainbow trout, *Salmo gairdneri* Richardson, in relation to the stage of development of the lymphoid organs. Dev Comp Immunol 7: 69-75

Thompson MA, Ransom DG, Pratt SJ, MacLennan H, Kieran MW, Detrich III HW, Vail B, Huber TL, Paw B, Brownlie AJ *et al.* (1998) The *cloche* and *spadetail* genes differentially affect hematopoiesis and vasculogenesis. Dev Biol 197: 248-269

Tingaud-Sequeira A, Cerdà J (2006) Phylogenetic relationships and gene expression pattern of three different cathepsin L (Ctsl) isoforms in zebrafish: Ctsla is the putative yolk processing enzyme. Gene 386: 98-106

Torroba M, Zapata AG (2003) Aging of the vertebrate immune system. Microsc Res Tech 62: 477-481

Trede NS, Zapata A, Zon LI (2001). Fishing for lymphoid genes. Trends Immunol 22: 302-307

Trede NS, Langenau DM, Traver D, Look AT, Zon LI (2004) The use of zebrafish to understand immunity. Immunity 20: 367-379

Tsujita T, Tsukada H, Nakao M, Oshiumi Y, Matsumoto M, Seya T (2004) Sensing bacterial flagellin by membrane and soluble orthologs of Toll-like receptor 5 in rainbow trout (*Onchorhynchus mykiss*). J Biol Chem 279: 48588-48597

Tsukada H, Fukui A, Tsujita T, Matsumoto M, Iida T, Seya T (2005) Fish soluble Toll-like receptor 5 (TLR5S) is an acute-phase protein with integral flagellin-recognition activity. Int J Mol Med 15: 519-525

Van der Sar AM, Musters RJ, van Eeden FJ, Appelmelk BJ, Vandenbroucke-Grauls CM, Bitter W (2003) Zebrafish embryos as a model host for the real time analysis of *Salmonella typhimurium* infections. Cell Microbiol 5: 601-611

Van der Sar AM, Stockhammer OW, van der Laan C, Spaink HP, Bitter W, Meijer AH (2006) MyD88 innate immune function in a zebrafish embryo infection model. Infect Immun 74: 2436-2441

Weinstein BM, Schier AF, Abdelilah S, Malicki J, Solnica-Krezel L, Stemple DL, Stainier DY, Zwartkruis F, Driever W, Fishman MC (1996) Hematopoietic mutations in the zebrafish. Development 123: 303-309

Wienholds E, Schulte-Merker S, Walderich B, Plasterk RH (2002) Target-selected inactivation of the zebrafish *rag1* gene. Science 297: 99-102

Willett CE, Cortes A, Zuasti A, Zapata AG (1999) Early hematopoiesis and developing lymphoid organs in the zebrafish. Dev Dyn 214: 323-336

Willett CE, Kawasaki H, Amemiya CT, Lin S, Steiner LA (2001) *Ikaros* expression as a marker for lymphoid progenitors during zebrafish development. Dev Dyn 222: 694-698

Wilson MR, Warr GW (1992) Fish immunoglobulins and genes that encode them. Annu Rev Fish Dis 2: 201-221

Yoder JA (2004) Investigating the morphology, function and genetics of cytotoxic cells in bony fish. Comp Biochem Physiol C 138: 271-280

Yoder JA, Nielsen ME, Amemiya CT, Litman GW (2002) Zebrafish as an immunological model system. Microbes Infect 4: 1469-1478

Yoder JA, Litman RT, Mueller MG, Desai S, Dobrinski KP, Montgomery JS, Buzzeo MP, Ota T, Amemiya CT, Trede NS *et al.* (2004) Resolution of the novel immune-type receptor gene cluster in zebrafish. Proc Natl Acad Sci USA 101: 15706-15711

Yoshida SH, Stuge TB, Miller NW, Clem LW (1995). Phylogeny of lymphocyte heterogeneity: cytotoxic activity of channel catfish peripheral blood leukocytes directed against allogeneic targets. Dev Comp Immunol 19: 71-77

Zapata A, Diez B, Cejalvo T, Gutierrez-de Frias C, Cortes A (2006) Ontogeny of the immune system of fish. Fish Shellfish Immunol 20: 126-136

Zhang H, Evenhuis JP, Thorgaard GH, Ristow SS (2001) Cloning, characterization and genomic structure of the natural killer cell enhancement factor (NKEF)-like gene from homozygous clones of rainbow trout (*Oncorhynchus mykiss*). Dev Comp Immunol 25: 25-35

PART 7

Functional Changes in Form

CHAPTER

19

Metamorphosis

DM Power*, N Silva and MA Campinho

INTRODUCTION

A unifying characteristic of all vertebrates is the change in form that frequently accompanies their development from an immature to mature state in which they are able to reproduce. Development is a continuous process composed of intervals of rapid and slower change. For convenience it is frequently subdivided into distinct stages which in fish most simply can be defined as embryo, larva (alevin), juvenile and adult, although some debate about terminology and the biological significance of stages exists (Blaxter, 1988; Balon, 1999). The embryologic development of fish normally occurs in a protected environment as the egg is encapsulated by a tough envelope, the chorion. The larval stage is a transitory stage in which organ and tissue systems further develop and the fish harbours nutrients and gains weight and size that will enable it to undertake the transition from larva to juvenile (Fig. 1). Fish in the juvenile stage, although smaller and sexually immature, are already similar in form, behaviour and physiology to adults. The processes controlling development are complex and are still far from resolved although considerable progress has been made in understanding fish embryo development through the use of the model organism zebrafish (*Danio rerio*) (as detailed

Authors' address: Comparative Molecular Endocrinology Group, Marine Science Centre (CCMAR), Universidade do Algarve, FERN, Campus de Gambelas, 8005-139 Faro, Portugal.
**Corresponding author:* E-mail: dpower@ualg.pt

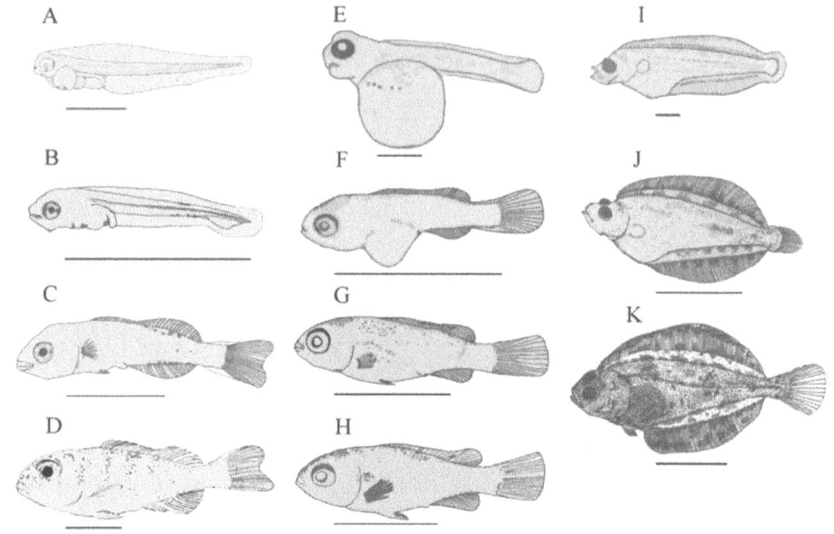

Fig. 1 Schematic drawing of the development of the gilthead seabream (*Sparus aurata*), the Mozambique tilapia (*Oreochromis mossambicus*) and the marine flatfish Atlantic halibut (*Hippoglossus hippoglossus*). The larval period in fish is characterised by the presence of embryonic organs such as the yolk-sac (A, E, F), the development of transitional structures such as the finfold and by a thin thread-like body shape (A, B, E, I). During ontogeny these structures disappear or develop into the adult form. Gilthead seabream – reared at 18°C. (A) 2 days post hatch (dph), the yolk-sac is still present, there are very few melanophores and the eyes are not developed; (B) 20 dph, notochord flexion, the caudal fin has started to differentiate and the larva is an active feeder; (C) 40 dph, the larva/juvenile transition starts around this age, all the fins are present and the gut is fully functional; (D) 60 dph transformation to a juvenile has occurred. Mozambique tilapia – reared at 27°C; (E) 4 dph, a large yolk-sac is evident; (F) 8 dph, a small yolk-sac persists, the fins are rapidly developing, pigmentation starts to appear and exogenous feeding starts; (G) 17 dph, the body shape is very similar to the juvenile although muscle and skeletal structures are not completely formed; (H) 26 dph, transformation to a juvenile has occurred. Atlantic halibut – reared at 8°C; (I) 44 dph, the pelagic larva is symmetrical, flexion of the notochord is noticeable and the larva is pre-metamorphic; (J) 58 dph, beginning of metamorphosis, eye migration has initiated, the body becomes oval and thicker as muscle develops, the pattern of pigmentation starts to resemble the juvenile; (K) 82 dph, a benthic asymmetrical juvenile with eyes on the same pigmented side of the body develops, which is similar in shape and pigmentation to the adult. Scale bar = 1 cm. A, E, I = 0.2 cm.

in (Wolpert *et al.*, 2002; Gilbert, 2006). In general developmental diversity occurs between vertebrates, although conserved embryonic processes exist, such as cleavage to form a blastula, followed by gastrulation and formation of the neural tube. A catalogue of genes and processes has been associated with specific embryonic stages and efforts to unravel the developmental program of vertebrates are intense.

Much less attention has been focused on regulatory factors of later developmental stages in fish. The present chapter will discuss metamorphosis from the Greek *metamorphoum*, which most simply can be defined as "transform" (Webster). Metamorphosis in fish is post-embryonic and involves profound morphological, biochemical and physiological transformations, which prepare the fish for life in a different habitat. This encompasses the relatively abrupt transformation of lamprey ammocoetes into juveniles; the protracted change of leptocephalus to glass eels; the remodelling of flatfish (Pleuronectiformes) from bilaterally symmetric pelagic larvae into an asymmetric benthic juveniles and smoltification of salmonid fish (for a review of smoltification, see Stefansson *et al.*, this volume). Although the larva to juvenile transition of roundfish is much less dramatic than that of flatfish, the underlying tissue-specific changes essential for survival and success are similar. Hence, despite the changes being more gradual and modest, roundfish also undergo metamorphosis. Each metamorphic process is commonly characterised by a period of rapid morphological/physiological change and then a stable period of slower change.

In teleost fishes a number of studies indicate that most elements of the endocrine system, the pineal, hypothalamus, pituitary, thyroid, ultimobranchial, endocrine pancreas, interrenal tissues, and corpuscles of Stannius, are already well developed in larvae and raise many questions which remain to be answered about their function in early stages but also during metamorphosis (Gorbman *et al.*, 1983; Lam, 1985; Bentley, 1998; Power *et al.*, 2001; Power, 2005). There is little question that metamorphosis in teleosts, in common with amphibian metamorphosis, is an endocrine driven process in which thyroid hormones (THs) play a central role (Leatherland, 1994; Yamano, 2005). Metamorphosis of the ammocoete into a juvenile lamprey is the exception and in both sea lamprey (*Petromyzon marinus*) and Southern hemisphere lamprey (*Geotria australis*), there is a dramatic decline in THs prior to this process (Wright & Youson, 1977; Lintlop & Youson, 1983; Leatherland *et al.*, 1990; Youson *et al.*, 1994). Moreover, administration of $KClO_4$, a goitrogenic chemical to sea lamprey promotes metamorphosis (Suzuki, 1986; Holmes & Youson, 1993), although its effectiveness is variable and it appears to have no effect in Southern hemisphere lamprey (Leatherland *et al.*, 1990). The regulation of metamorphosis in lamprey is still enigmatic and appears to be unlike that found in amphibians and flatfish. It will not be discussed further here and information can be obtained from recent reviews (Youson, 1997).

Only few species of fish have been studied in any detail and they are in no way representative of the evolutionary diversity of teleosts. This is an important consideration when working with new species. Some generalisations about molecular, biochemical and morphological alterations which accompany metamorphosis in different species may be possible. However, as has been reported in amphibians (see Denver *et al.*, 2002), is it likely that metamorphosis in fishes varies in response to environmental factors, and is a species-specific process.

THE ENDOCRINE SYSTEM AND METAMORPHOSIS IN FISH

Metamorphosis in urodeles (salamanders) and anurans (frogs and toads) is a dramatic process and as early as 1912 Gudernatch showed that the vertebrate thyroid gland contains a factor which can induce premature metamorphosis (see Gorbman & Bern, 1962). Subsequent studies revealed that thyroidectomy blocks metamorphosis and that thyroid hormones (THs) are the active principle in the gland (Shi *et al.*, 1986). Growth hormone and prolactin (PRL) produced by the pituitary gland and cortisol from the interrenal tissues (equivalent to the mammalian adrenal cortex) have also been implicated in metamorphosis, but far fewer studies of their role in this process exist (Kikuyama *et al.*, 1993). Prior to considering the experimental evidence demonstrating a role for THs in fish metamorphosis, an overview of this axis and the various elements which constitute it will be given. It will become evident during this section that the thyroid axis is conserved in vertebrates, and that there is much less molecular and functional information available in fish. For this reason the following overview will take a comparative approach presenting an outline of the state-of-the-art followed, when available, by a more specific consideration in fish.

Thyroid Hormones

THs are small hydrophobic molecules characterised by a diphenylether function which contains iodine. The THs are produced by thyroid follicles and circulate in the plasma bound to thyroid hormone-binding proteins. These include albumin, transthyretin and thyroxine-binding globulin (Schreiber & Richardson, 1997). Thyroxine (T4; 3-[4-(hydroxyl-3,5-diiodophenoxy)-3,5-diiodophenyl]-L-alanine) is the most abundant form of THs while triiodothyronine (T3; O-[4-(hydroxyl-3iodophenyl]-3,5-diiodo-L-tyrosine) is the most active form of the hormone. Iodine comprises 65% and 58% of the molecular weight of T4 and T3, respectively, and is an indispensable component of the hormones.

THs have been identified in the blood of representative species of both lower and higher vertebrates, from cyclostomes to humans, and are the only iodine-containing compounds of physiological significance in vertebrates. The metabolites 3,3′,5′-triiodothyronine (reverse-T3, rT3) and 3′,5′-, 3,3′-or 3,5- diiodotyronine (T2) are other forms of thyroid hormones, but they have little activity and their physiological role is still controversial. It has been reported that rT3 may have a direct action on plasma membrane proteins and T2 has a direct effect on the activation of the respiratory chain in mitochondria since a T2 binding site has been identified in a subunit of cytochrome-C-oxidase (Wrutniak-Cabello *et al.*, 2001).

Production of Thyroid Hormones

The thyroid follicles in cyclostomes and teleosts do not form an organised gland. In most fish (with the exception of parrotfish, swordfish and tuna fish) the follicles are scattered in the connective tissue of the pharynx around the ventral aorta (Fig. 2). Production of TH is controlled by the hypothalamus-pituitary-thyroid axis and ultimately by available iodine, hypothalamic-derived thyrotropin releasing hormone, and pituitary-derived thyrotropin, also called thyroid stimulating hormone (TSH). In marine fish, iodine is rarely limiting as it is available in the ambient water and is taken up through the gills, although in freshwater fish and larvae which lack fully functional gills, the diet may be an important source (Hunn & Fromm, 1966; Hamre *et al.*, 2005; Moren *et al.*, 2006). The biosynthesis of T4 in the thyrocytes of the thyroid follicles is fairly well characterised, although it is still not totally understood in vertebrates (www.thyroidmanager.org).

In simple terms, the thyrocytes sequester iodide, probably by means of a sodium-iodide-symporter in the basal membrane of the cell at the interface between the colloid with the thyrocyte apical membrane (Dai *et al.*, 1996; Eskandari *et al.*, 1997). Iodonium ions are incorporated into thyroglobulin via specific tyrosines giving rise to di- and mono-iodotyrosine. Thyroperoxidase and H_2O_2 via an oxidative process join the iodotyrosines and form T3, when di-iodotyrosine couples to mono-iodotyrosine, or T4 when two di-iodotyrosines couple together (Lamas *et al.*, 1974; Cahmann *et al.*, 1977; Deme *et al.*, 1978; Gavaret *et al.*, 1980, 1981; Virion *et al.*, 1981, 1985). Free hormone is liberated from thyroglobulin after it enters the thyrocyte by micro- or macro-pinocytosis and it is stored in the thyrocyte until it enters the circulation through an as yet

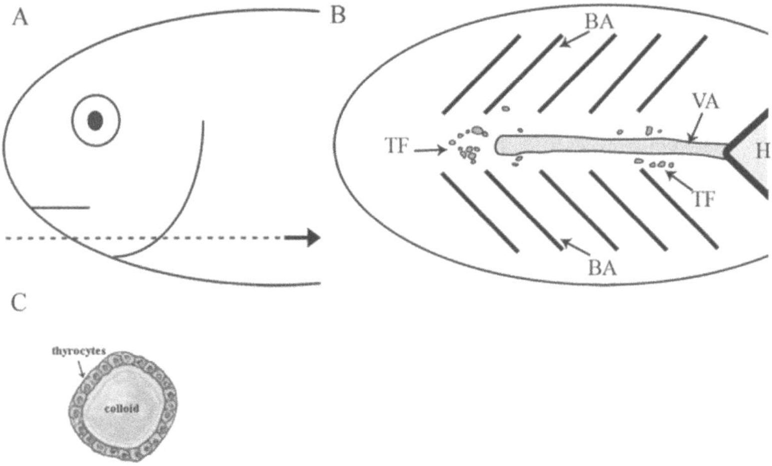

Fig. 2 Scheme showing the localisation of thyroid tissue in gilthead seabream and the organisation of a thyroid follicle. A – Lateral view of a fish head, a dashed line indicates the approximate position of thyroid tissue; B – Longitudinal section indicating the position of the thyroid follicles which are dispersed in the connective tissue at the insertion of the gill arches in the gill bar and close to the ventral aorta; C – Scheme of a thyroid follicle which consists of a unicellular layer of thyrocytes that surround a colloid filled lumen. BA: branchial arches; TF: thyroid follicles; VA: ventral aorta; H, heart.

undocumented mechanism (Tietze *et al.*, 1989; Andersson *et al.*, 1990; Dunn & Dunn, 2000).

TSH produced in the pituitary gland regulates TH synthesis and acts through a G protein-coupled receptor, the TSH receptor that transmits and amplifies the signal via cAMP. TSH stimulates the expression, production and post-translational modification of almost all factors involved in TH production. Transcription factors, such as thyroid transcription factor-1 and -2 and *pax8* are also involved, and are also regulated by TSH (Mascia *et al.*, 2002). Through a classical negative feedback mechanism THs control the release of TSH from thyrotrophs and this is the principal action of these hormones on the pituitary gland (Tannahill *et al.*, 2002).

Transport, Uptake, Conversion and Metabolism of Thyroid Hormones

TH Transport

Despite their hydrophobic nature, THs are soluble in aqueous media and can be transported directly in the blood stream (Rotzsch *et al.*, 1967). However, THs are normally transported in the circulation bound to thyroid hormone binding proteins and this limits their tendency to preferentially partition into lipid membranes. A number of proteins and lipoproteins are thought to be involved in TH serum transport, although thyroid binding globulin, transthyretin and serum albumin are the major serum proteins implicated in vertebrates. In terrestrial vertebrates thyroid binding globulin has the greatest affinity for T4 followed by transthyretin and albumin which have 50- and 7000-fold lower affinity for T4, respectively. In addition to TH transport, thyroid hormone binding proteins play an important physiological role in the maintenance of total TH levels in the serum, and thus safeguard cells against adverse abrupt changes in hormone secretion. They also prolong TH half-life by preventing their disposal in the kidneys (Bartalena & Robbins, 1993; Langsteger, 1997; Schussler, 2000). Relatively few studies exist of thyroid hormone binding proteins in fish, although transthyretin has been isolated in salmonids and cloned in gilthead seabream (*Sparus aurata*), and has been shown to bind THs, suggesting it plays a similar role in TH transport to that found in mammals (Santos & Power, 1999; Yamauchi *et al.*, 1999; Power *et al.*, 2000). The role and importance of thyroid hormone binding proteins during metamorphosis in fish and amphibians is unknown, although the importance of these proteins in T4 distribution in the brain of terrestrial vertebrates suggests they may play an essential role in brain development (Thomas *et al.*, 1988, 1989; Tu *et al.*, 1990; Thomas & Dziadek, 1993).

TH Uptake

Until recently it was assumed that because of their lypophilic nature TH uptake from serum by target cells occurred by diffusion through the cell membrane. However, new evidence suggests that several transmembrane transport systems are involved in TH uptake (Salvatore *et al.*, 1996a). This helps explain why the transport of THs from the serum to the interior of the cell is a rate-limiting step on their subsequent metabolism in rats (Salvatore *et al.*, 1996b). The transport systems identified include both Na^+-independent and Na^+-dependent transporters and also amino acid transporters (Abe *et al.*, 2002). The Na^+-independent TH transporters belong to the organic anion transporting polypeptide family which is also responsible for the transport of organic acids, bile acids, bromosulfophthalein and unconjugated steroid hormones by hepatocytes (Jacquemin *et al.*, 1994). The Na^+/taurocholate cotransporting polypeptide is the only reported Na^+-dependent transporter of THs in rats and humans (Abe *et al.*, 2002). L-type heterodimeric amino acid transporters are also involved in TH transport through the cell membrane and also act in a Na^+-independent manner (Friesema *et al.*, 2001). So far nothing is known about TH uptake by target cells in fish but it seems likely that it will occur by a similar mechanism and the genes for some of the transporters have already been identified.

TH Metabolism

The predominant form of circulating THs in vertebrates is T4, although in some teleosts circulating T3 may be more abundant (Eales, 1990). The bioactive form of TH is considered to be T3 which also has the greatest receptor affinity. Metabolism of T4 occurs by the action of three iodothyronine deiodinases, which have been identified in a number of vertebrates and in the tunicate *Halocynthia roretzi* and in phylogenetic analysis cluster independently in three groups (Fig. 3). The enzymes belong to the selenoprotein family and contain a selenocysteine in their catalytic centre and control local production and degradation of THs by deiodination (Köhrle, 2000; Bianco *et al.*, 2002; Shepherdley *et al.*, 2004). The deiodinases are transmembrane enzymes, either associated with the inner leaflet of the plasma membrane or with the cytosolic side of the endoplasmatic reticulum. The enzymes have a developmental and tissue-specific expression. Type 1 iodothyronine deiodinase (D1) and type 2 iodothyronine deiodinase (D2) catalyse activation (outer ring deionidination, ORD) and convert T4 to T3. Type 3 iodothyronine deiodinase (D3) catalyses inner ring deiodination (IRD) and generates the inactive metabolites rT3 or T2, respectively from T4 and T3 (Fig. 4A). All deiodinases are dependent on thiol cofactors for correct activity and in concert regulate the cellular concentration of T4 and all other iodothyronines independent of nuclear receptors.

Type 1 Deiodinase D1 is an enzyme with a high K_m, and V_{max} that prefers rT3 >> T4 > T3 as substrate. It catalyses the ORD of rT3 more effectively, while IRD of T4 and T3 is facilitated by sulfation (Visser, 1994). In the rat liver D1 is associated with the endoplasmatic reticulum and in the kidneys with the plasma membrane (Baqui *et al.*, 2000; Köhrle, 2000). The active enzyme is normally dimeric and composed of two units of 27 kDa although recent studies suggest the monomer has catalytic activity (Curcio-Morelli *et al.*, 2003). The selenocysteine in each subunit of the active enzyme is encoded by a UGA codon (which normally acts as a translation terminator) but

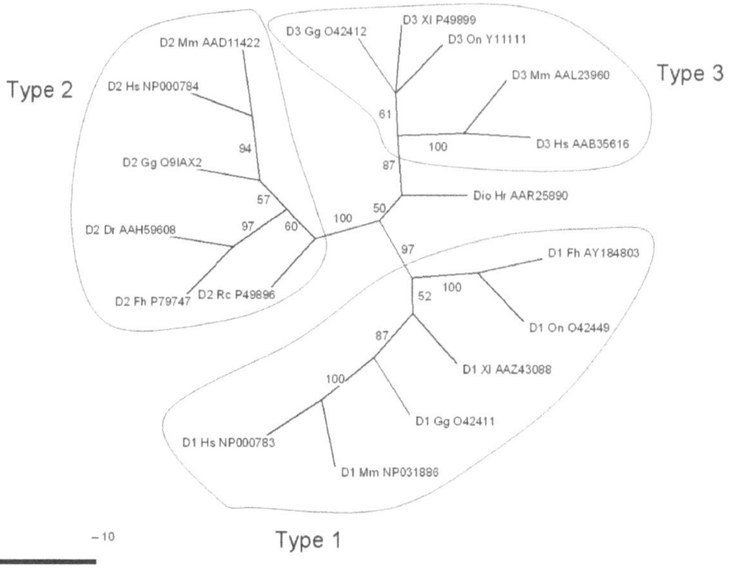

Fig. 3 Maximum parsimony phylogenetic tree of vertebrate protein deiodinases. The iodothyronine deiodinase from the urochordate *Halocynthia roretzi* was used as the outgroup. The bar on the bottom left represents 10% sequence divergence. Hs: *Homo sapiens*; Mm: *Mus musculus*; Gg: *Gallus gallus*; Xl: *Xenopus laevis*; On: *Oreochromis niloticus*; Dr: *Danio rerio*; Fh: *Fundulus heteroclitus*; Rc: *Rana catesbeiana*; Hr: *Halocynthia roretzi*. The GenBank accession number of each sequence is shown after the species name.

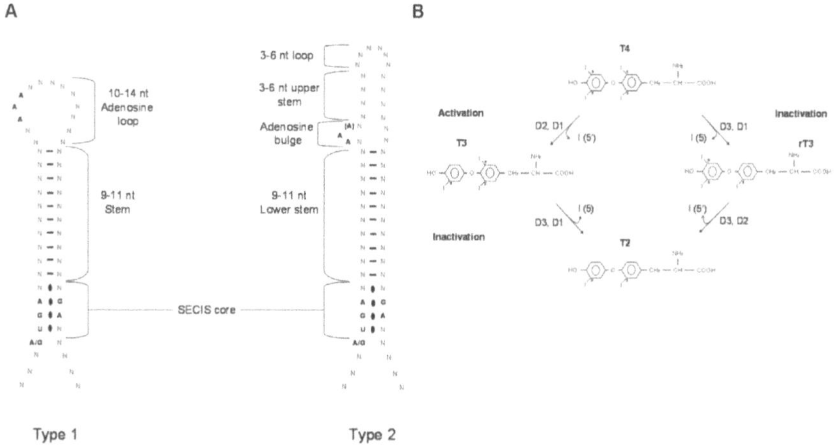

Fig. 4 A) Diagram depicting the RNA secondary structure of the two known types of vertebrate SElenoCysteine Insertion Sequence (SECIS), which are present in the 3'UTR of mRNA encoding selenoproteins and are responsible for the insertion of the selenocysteine amino acid encoded by a TGA codon. D1 and D3 are characteristic of Type 1 enzymes and D2 is characteristic of type 2 enzymes. B) Metabolism of thyroid hormones by the three known vertebrate deiodinases.

which when followed by a selenocysteine-insertion sequence (SECIS) in the 3'UTR of the mRNA leads to insertion of a selenocysteine residue (Fig. 4B) (Berry *et al.*, 1991a, b). The 3'UTR SECIS motif is also present in common mummichog (*Fundulus heteroclitus*) D1 cDNA, suggesting that in fish D1 is also a selenoenzyme (Orozco *et al.*, 2003).

The selenocysteine residue seems to be essential for correct enzymatic activity (Berry *et al.*, 1991b), and the thiol cofactor is highly reactive with sulfenyl iodide groups of proteins and is thought to reduce the Se-I group of the intermediate and restore the enzyme to its native form (Visser, 1994; Leonard & Köhrle, 2000).

In adult mammals, D1 is principally expressed in the liver, kidney, thyroid gland and intestine (Bates *et al.*, 1999) and hepatic D1 is thought to be responsible for the majority of T3 production in the peripheral tissues and T3 levels in plasma of mammals (Fig. 4A). The liver and kidney are also considered to be responsible for the metabolism of rT3 to T2 (Köhrle, 2000). In the thyroid gland D1 is implicated in the recycling of iodine by removal of the ion from mono-iodotyrosine and di-iodotyrosine residues after thyroglobulin digestion and conversion of ~10% of T4 to T3 in the gland (Toyoda *et al.*, 1992).

In vitro enzyme activity studies and gene cloning indicate D1 is present in the liver and kidney of Nile tilapia (*Oreochromis niloticus*), rainbow trout (*Oncorhynchus mykiss*), African catfish (*Clarias gariepinus*), common mummichog, and gilthead seabream (Orozco *et al.*, 1997; 2003; Sanders *et al.*, 1997; Klaren *et al.*, 2005). Moreover, the catalytic properties and activity of D1 in fish resemble those of mammals, with similar maximal velocities, affinities for the substrates, and pH and temperature optimum (Mol, 1998). The major difference of D1 enzymes between fish and mammals appears to be the insensitivity of this enzyme in fish to 6-propyl-2-thiouracil, which in mammals inhibits its activity. The response of D1 to physiological challenges in fish varies according to species and is a reminder that alternative mechanisms of thyroid axis adaptation may exist in different species (Mol, 1998; Köhrle, 2000; Van der Geyten *et al.*, 2001a, b; Orozco *et al.*, 2003). So far, no reports exist about D1 ontogeny in teleosts so its importance in development and metamorphosis has yet to be established. The more restricted distribution of D1 in teleosts suggests, in contrast to mammals, it may not be the major enzyme responsible for T3 plasma levels.

Type 2 Deiodinase D2 acts by ORD at the 5'position of the phenolic ring but it lacks IRD activity (Fig. 4A). Unlike D1, D2 is a low K_m and V_{max} enzyme that prefers T4 > rT3 as substrate and does not deiodinate sulphated T4 or rT3. In common with other deiodinases, D2 is located in cellular and intracellular membranes (Baqui *et al.*, 2000; Köhrle, 2000). The active enzyme complex seems to be composed of several different subunits of ~30 kDa protein encoded by D2 mRNA and by a 29 kDa subunit that is encoded by a different gene (Leonard *et al.*, 2000). Recent reports indicate that *in vivo* monomers of D2 also have a catalytic effect (Curcio-Morelli *et al.*, 2003).

The D2 cDNA in vertebrates is also characterised by the presence of a 3'UTR SECIS motif (Davey *et al.*, 1995, 1999; Celi *et al.*, 2000). In fish the cDNA for D2 has been isolated in common mummichog and zebrafish and has a well conserved organisation when compared to the human D2 sequence (Orozco *et al.*, 2002). The mature active enzyme appears to be a complex of selenoproteins and non-selenoproteins in tetrapods, making D2 a very unusual deiodinase. Despite the central importance of the selenocysteine residue for enzyme activity of D2 and D1, they are proposed to act by different mechanisms (Mol, 1998; Köhrle, 2000; Kuiper *et al.*, 2002).

In mammals D2 has been implicated in the fine-tuning of locally expressed T3-regulated genes through control of intracellular T4 and T3 levels. It is expressed in brain cell populations, the anterior pituitary gland, thyroid gland, heart and skeletal muscle and brown adipose tissue (Croteau *et al.*, 1996; Bates *et al.*, 1999). However, in fish D2 is the main selenoenzyme expressed in liver and kidney and it is undetectable in the brain (Mol, 1998; Orozco *et al.*, 2000, 2002). The high level of D2 activity in the liver and kidney and the absence of D1 activity led to the proposal that in fish D2 is the major contributor to T3 plasma levels. Although the tissue distribution of D2 and mRNA expression in fish is quite different from terrestrial vertebrates, the enzymatic and physical properties of fish D2 are similar to those in higher vertebrates (Mol, 1998). For example, in both rat and fish T4 and rT3 have an inhibitory effect on D2 enzymatic activity, but not mRNA levels. Overall the rapid inactivation of D2 by substrate and the relatively short half-life of the protein and mRNA suggest a role for D2 in maintaining local tissue concentrations of T3 despite rapid T4 serum variations (Burmeister *et al.*, 1997; Mol, 1998; Köhrle, 2000; Van der Geyten *et al.*, 2001b).

Type 3 Deiodinase Inactivation of THs occurs by the action of D3, a low K_m, low V_{max} enzyme that prefers T3 > T4 as substrate. It acts exclusively as an IRD in mammals and fishes (Croteau *et al.*, 1995; Salvatore *et al.*, 1995; Mol, 1998; Hernandez *et al.*, 1999; Sanders *et al.*, 1999). Generally, D3 inactivates THs by either converting T3 to T2 or by converting T4 to rT3 through IRD of the 5-position of the tyrosine ring (Fig. 4A). D3 is thought to be mainly involved in protecting tissue locally from excessive T3 concentrations thus controlling the expression of T3-dependent genes. This is particularly important in the developing mammalian foetus and it is localised in human maternal-foetal interfaces (St Germain *et al.*, 1994; Becker *et al.*, 1997; Richard *et al.*, 1998; Huang *et al.*, 1999, 2003; Köhrle, 2000).

The putative vertebrate D3 protein sequences have a predicted molecular weight of 20 to 31 kDa. Evidence suggests the enzyme acts exclusively as a homodimer (Curcio-Morelli *et al.*, 2003). D3 from human (Salvatore *et al.*, 1995), rat (Croteau *et al.*, 1995), mouse (Hernandez *et al.*, 1999), African clawed frog (*Xenopus laevis*) (St Germain *et al.*, 1994) and Nile tilapia (Sanders *et al.*, 1999) has a highly conserved selenocysteine in its active site and also the SECIS motif in the 3'UTR of the mRNA (Fig. 4B). D3 from different vertebrates in common with D2 and D1 in fish is relatively insensitive to 6-propyl-2-thiouracil and thiol cofactors restore the enzyme to its native form, suggesting it may act by a similar mechanism to D2 (Köhrle, 2000).

In the adult rat D3 mRNA expression and enzymatic activity is detected in placenta, cerebral cortex and skeletal muscle (Croteau *et al.*, 1995; Bates *et al.*, 1999). In salmonids, D3 activity is found in liver, gill and brain (Leatherland & Farbridge, 1992; MacLatchy

& Eales, 1992; Morin *et al.*, 1993), although in rainbow trout (Mol, 1998) D3 activity is absent from the gills. In contrast, D3 activity is principally detected in gill, brain and skin of Nile tilapia and gill and kidney of turbot (*Scophthalmus maximus*) (Mol, 1998).

In general D3 appears to be responsive to THs and a reduction in THs (hypothyroidism) causes a reduction in gene expression and enzyme activity, while an increase in THs (hyperthyroidism) has the opposite effect (MacLatchy *et al.*, 1992; Mol, 1998; Finnson & Eales, 1999). Overall, fish D3 expression and activity pattern in either eu-, hypo- or hyper-thyroid animals resembles that observed in mammals (MacLatchy *et al.*, 1992; Mol, 1998; Finnson & Eales, 1999; Van der Geyten *et al.*, 2001a). Taken together the evidence suggests D3 regulation and function in fish is well conserved when compared to other vertebrates and the principal role of D3 is to protect tissue locally from excessive T3 concentration.

Mechanism of Action of Thyroid Hormones

THs mediate their action by binding to nuclear receptors that act directly on target genes. TH receptors (THRs) are members of the steroid-thyroid receptor superfamily that also includes receptors for ligands such as steroid hormones, retinoids, melatonin and vitamin D3. The structure of THR proteins is well conserved in vertebrates. They contain an N-terminal A/B domain which interacts with transcription factors, a DNA binding domain (C-domain) comprising two zinc fingers, an E-domain, also called the ligand binding domain, is involved in corepressor interaction, ligand binding, dimerisation and transactivation of gene expression and at the C-terminal end, the F-domain is responsible for corepressor and coactivator interaction (Fig. 5) (Nilsson *et al.*, 2001).

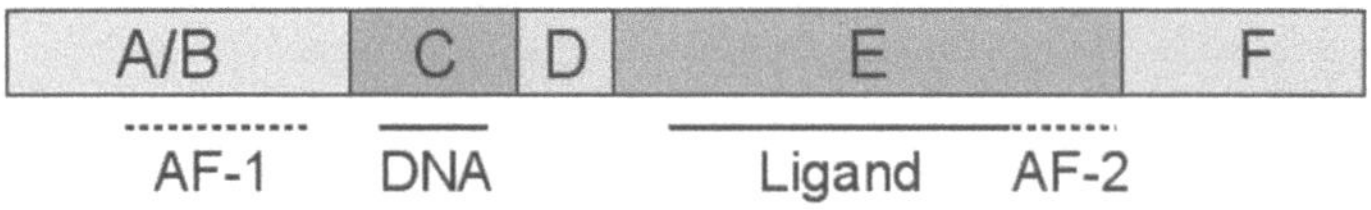

Fig. 5 Domain structure of Thyroid Hormone Receptors. The N-terminal A/B domain contains the AF-1 site where other transcription factors interact, the C domain contains the two- zinc finger structures that bind DNA, and the E domain contains the ligand binding domain as well as the AF-2 region that directly contacts coactivator peptides.

THRs bind as heterodimers to specific DNA sequences (Thyroid Response Elements - TRE), and in the presence or absence of ligand, activate or repress gene expression, respectively, and bring about the appropriate cellular response (Yen & Chin, 1994). The TRE sequences are well conserved between vertebrates and are comprised of two half-sites containing 6-base pair sequences separated by a four nucleotide linker sequence (Fig. 6) or by the higher affinity octamer sequence TNAGGTCA which may be present as direct repeats, palindromic sequences or inverted palindrome sequences. Evidence suggests that the different types of TRE may be of differing

Fig. 6 Thyroid Response Element (TRE) consensus sequence. The sequence depicted represents the most common direct repeat TRE. Other TREs have the same sequence and can be present as a palindrome or an inverted sequence. *n* represents any base.

biological relevance although their precise role remains to be established (Katz & Lazar, 1993; Nagaya & Jameson, 1993; Reginato *et al.*, 1996; Yang *et al.*, 1996).

In mammals, two highly homologous TR genes have been identified, *thrα (nr1a1)* and *thrβ (nr1a2)* (Sap *et al.*, 1986; Weinberger *et al.*, 1986) which encode proteins that act frequently and efficiently as heterodimers with Retinoid X Receptors (Shi *et al.*, 1992). The presence of coactivaters and corepressors are essential for the activating or repressing the action of THRs (Chin & Yen, 1997). In general, the nature of THRs means they are widespread and present in low abundance in most tissues and this represents a considerable barrier to their identification and for studies of their action.

In fish, the first THRs were cloned and isolated in Japanese flounder (*Paralichthys olivaceus*) (Yamano & Inui, 1995; Yamano *et al.*, 1994a). Subsequently, an Atlantic halibut (*Hippoglossus hippoglossus*) *thrα* cDNA (Llewellyn *et al.*, 1999) and a zebrafish *thrβ* cDNA was cloned (Liu *et al.*, 2000). In gilthead seabream, Japanese flounder, Atlantic salmon (*Salmo salar*), and conger eel (*Conger myriaster*) two different genes for THR were isolated which correspond to *thrα* and *thrβ* (Yamano *et al.*, 1994a; Llewellyn *et al.*, 1998; Yamano & Miwa, 1998; Nowell *et al.*, 2001; Jones *et al.*, 2002; Kawakami *et al.*, 2003a, b) and a number of THR splice variants have also been identified (Marchand *et al.*, 2001). THRs in fish have a widespread tissue distribution. THRα in fish binds T3 and stimulates gene transcription (Essner *et al.*, 1997; Liu *et al.*, 2000), although the way in which THRα influences transcription in the absence of ligand remains to be established. It would appear that in common with the mammalian *thrα* gene, a number of mechanisms may generate isoform diversity and increase receptor diversity and in this way THR action in adult fish and during development and metamorphosis.

Fish THRβ shares 85% amino acid identity with the human orthologue and in all fish studied to date, with the exception of zebrafish, two THRβ isoforms occur, a short and a long form which contains an additional nine amino acids at the beginning of the ligand binding domain (Yamano & Inui, 1995; Liu *et al.*, 2000; Marchand *et al.*, 2001; Nowell *et al.*, 2001). It has been proposed that this insertion could be due to an alternative splice exon or alternative splice acceptor sites in exon 6. In gilthead seabream and Atlantic salmon the longer THRβ isoform displayed affinity for T3 comparable with amphibian and mammalian receptors (Marchand *et al.*, 2001; Nowell *et al.*, 2001).

Considerable conservation exists between teleost and terrestrial vertebrate TRs; the A/B domain is the least conserved region of the protein (60-69% similarity) whereas the C-, D-, E- and F-domains are highly conserved. The amino acid sequence conservation observed in the C-domain of THRs from teleosts, mammals and birds suggests the same sequences are likely to function as TREs in all vertebrates (Umesono & Evans, 1989). The high level of conservation for the remaining domains, E-, D- and F- in all vertebrates suggests that TH binding, receptor dimerisation and recruitment of co-activators and repressors is conserved and similar to that described for the mammalian receptors. This is confirmed by studies with Atlantic salmon THRβ which binds to DNA in a specific manner in the presence of different TREs and transactivates gene expression in the presence of T3 and represses basal transcription in its absence (Marchand *et al.*, 2001).

THYROID HORMONES AND METAMORPHOSIS OF FISH

The following section will aim to integrate information about changes during metamorphosis with experimental evidence indicating when THs intervene and the consequence of this intervention. The role of THs prior to and during metamorphosis appears to be maturational as they promote changes which prepare larvae for the successful transition into a juvenile.

Deiodinases

There is no question about the importance of the deiodinases during development as by a co-ordinate action they regulate the availability of bioactive hormone and confer together with THRs specific tissue and cellular actions. Moreover, a number of studies in amphibians, birds and mammals demonstrate that normal embryonic and subsequent development requires the correct spatial and temporal expression of the iodothyronine deiodinases (Croteau *et al.*, 1995; Becker *et al.*, 1997; Van der Geyten *et al.*, 1997; 2002; Richard *et al.*, 1998; Bates *et al.*, 1999; Köhrle, 2000; Forrest *et al.*, 2002; Shintani *et al.*, 2002; Huang *et al.*, 2003). In fish despite cloning and characterisation of the tissue distribution of the three deiodinases in adults, few studies exist that describe the role of these key enzymes in any early life stage of teleosts. In turbot D2 has a comparable expression pattern to *thrα*, which is up-regulated during metamorphosis, although tissue localisation and enzyme activity have not been established (Marchand *et al.*, 2004).

The potential scope of deiodinase actions in fish metamorphosis is evident from their essential role in amphibian metamorphosis. Tail regression and progressive development of the limbs in amphibians provide a good example. In all tissues D2 is up-regulated at the climax of remodelling and D3 decreases slightly, suggesting a dynamic interaction that tightly controls intracellular T3 levels and hence its action on gene transcription and morphogenetic processes. In anuran metamorphosis D1 has a discrete or negligible role whereas D2 and D3 are involved in the co-ordination of TH action during this developmental period (Becker *et al.*, 1997; Huang *et al.*, 1999; Shintani *et al.*, 2002).

External Morphology

The probable role of THs in flatfish metamorphosis was shown in the 1980's in the Japanese flounder when a peak of THs was correlated with the metamorphic climax and transformation to an asymmetric benthic juvenile (Miwa *et al.*, 1988). Subsequent studies in which THs or TSH was administered directly to larvae caused precocious metamorphosis confirming the role of the TH axis in this process (Inui & Miwa, 1985; Miwa & Inui, 1987; Inui *et al.*, 1989; de Jesus *et al.*, 1993). More recent studies in other flatfish, such as, the summer flounder (*Paralichthys dentatus*) and Atlantic halibut confirm the results from Japanese flounder and TH administration to both species accelerates metamorphosis (Schreiber & Specker, 1998; Solbakken *et al.*, 1999). In the Atlantic halibut, the visible morphological effects of exogenous T4 are stimulated muscle and skeletal growth, pigmentation and eye migration (Solbakken *et al.*, 1999). A recent study with the spotted halibut (*Verasper variegatus*) suggests the response of flatfish to exogenous T4 probably depends on the time that it is administered prior to spontaneous metamorphosis (Tagawa & Aritaki, 2005). The results of the study are intriguing and may indicate that the outcome of TH administration depends on tissue responsiveness/sensitivity which is probably dependent on deiodinase and THR expression.

The molecular basis of asymmetry in flatfish remains to be established and recently one potential candidate, a bicoid-type homeobox gene *pitx2* which changes sidedness of visceral organs and eye migration in the Japanese flounder, was identified (Hashimoto *et al.*, 2002). Suppression subtraction hybridisation also in Japanese flounder has led to the identification of several upregulated transcripts during metamorphosis one of which is splicing factor arginine/serine rich-3 and is localised in rapidly proliferating tissue in the head region (Bao *et al.*, 2005). TH target genes identified in African clawed frog metamorphosis have been cloned by reverse transcriptase-polymerase chain reaction in turbot and their expression analysed during metamorphosis. Arginase, cytosolic thyroid hormone binding protein, elongation factor-1α and D2 have a comparable expression profile to *thr*α which is upregulated in this species during metamorphosis (Marchand *et al.*, 2004). The localisation and functional significance of the genes in turbot was not shown. Clearly much more work will be required to understand the changes at a molecular level which lead to the switch from bilaterally symmetrical larva to an asymmetrical juvenile.

Relatively few studies exist about the endocrine regulation of the dramatic metamorphosis of leptocephalus to elver, probably as a consequence of their enigmatic lifecycle. However, in the conger eel the role of THs on metamorphosis has been demonstrated (Kitajima *et al.*, 1967; Yamano *et al.*, 1991b). Metamorphosis in the conger eel is associated with a change from planktonic to benthic ecology and changes in its external morphology have been reported, although morphological and molecular changes in tissue and organs have yet to be described. The changing pattern of expression of *thr*α*s* and *thr*β*s* during metamorphosis provides further evidence supporting a role for these hormones in this species (Kawakami *et al.*, 2003a). Further studies will be required to determine if the maturational effects of THs during metamorphosis in Anguilliformes, a more ancient order of teleosts, is comparable with that in flatfish and roundfish.

In the zebrafish, a roundfish, the effect of exogenous THs and inhibitors of the thyroid gland are comparable to that seen in the Japanese flounder. The principal modifications associated with the larva to juvenile transition in zebrafish is the differentiation of the paired pectoral fins into a pair of fan-like paddles and the development of the pelvic fins and striped pattern of the adult fish. Exogenous THs were found to accelerate the differentiation of the paired pectoral fins, but failed to accelerate pelvic fin development, while goitrogens inhibit the transition from larva to juvenile (Brown, 1997). In the red seabream (*Pagrus major*) and orange-spotted grouper (*Epinephelus coioides*), TH administration also accelerates the metamorphic process (Hirata *et al.*, 1989; de Jesus *et al.*, 1998). In the orange-spotted grouper it accelerated onset of pigmentation and changes in swimming behaviour and also accelerated resorption of the dorsal fin ray (de Jesus *et al.*, 1998). The preceeding results suggest that THs may also modulate the less dramatic larva to juvenile transition in several roundfish. Moreover, the differing responsiveness of tissues to exogenous THs probably marks their differing sensitivity, probably associated with deiodinase activity and THR distribution that is a hallmark of metamorphosis.

It appears likely that THs are obligatory for the larva to juvenile transformation in both roundfish and flatfish, although studies of a wider range of species are required to substantiate this idea. Moreover, the pivotal importance of deiodinases in regulating tissue availability of THs and the presence of THRs to activate or repress gene transcription will influence the timing and character of morphological changes. Studies of these factors will be essential, if a better understanding of metamorphosis is to be achieved in teleosts.

Skin and Pigmentation

The skin is a complex epithelium, which in teleosts contains sensory systems and helps maintain body shape and protect against shock and infection. In embryonic and larval teleosts the skin is composed principally of epidermis and during maturation to a juvenile it becomes multi-layered and composed of an outermost epidermal layer, the epidermis and the innermost hypodermis. Relatively few studies of the structure and ontogeny of skin exist in teleosts, although those that do suggest it is fairly well conserved (Liu & Chan, 2002; Le Guellec *et al.*, 2004). The way in which THs regulate skin maturation has not been described, although studies in Atlantic halibut reveal a close link between differentiation of skin cells and gene expression with the surge of THs which occur at the metamorphic climax (Campinho *et al.*, 2007). The transformation of Atlantic halibut skin during metamorphosis resembles that reported in anurans (Miyatani *et al.*, 1986; Nishikawa *et al.*, 1992) and may indicate that it is a general phenomenon in other teleost fish, although confirmation will require studies in other species.

In addition to structural modifications in skin, the embryonic/larval pigment pattern changes to the adult pattern during the larva to juvenile transition (see Kelsh & Parichy, this volume). In zebrafish embryos and larvae the change from a simple pattern of stripes to the complex adult pattern is a consequence of the shift to an adult population of black melanophores, yellow xanthophores and iridescent iridophores, although some embryonic/larval melanophores still persist (Parichy & Turner, 2003). It seems likely that the changes in skin pigmentation noted during the spontaneous larva to juvenile transition in other roundfish and flatfish are also the consequence of a change in melanophores (Inui & Miwa, 1985; Hirata *et al.*, 1989; Brown & Kim, 1995; Brown, 1997; de Jesus *et al.*, 1998; Huang *et al.*, 1998). Exposure to T3 advances adult pigmentation in striped bass (*Morone saxatilis*) and the sixfinger threadfin (*Polydactylus sexfilis*) suggesting changes in the population of melanophores may be regulated by THs (Brown & Kim, 1995; Huang *et al.*, 1998). Currently the molecular mechanisms responsible for the change in pigmentation during spontaneous metamorphosis and in response to THs are unknown. However, the recently identified SLC24A5, a cation exchanger with an important role in pigmentation in zebrafish and humans (Lamason *et al.*, 2005), may be an interesting candidate to study in this respect.

Musculo-skeletal System

The influence of THs on muscle development is well documented in terrestrial vertebrates (Larsson *et al.*, 1994, 1995; Adams *et al.*, 1999; Soukup & Jirmanova, 2000; Chanoine & Hardy, 2003). In rats full differentiation of the skeletal muscle phenotype is only achieved with appropriate TH levels (Vadaszova *et al.*, 2004).

In teleosts a number of studies have described biochemical changes of muscle proteins in response to THs during metamorphosis. The manner in which changes in protein isoform affects muscle function during ontogeny remains to be established but is probably a response to changes in hydrodynamic environment, locomotive strategies, respiration and intracellular environment (Osse, 1990; Johnston, 1994; Koumans & Akster, 1995; Johnston *et al.*, 1997; Patruno *et al.*, 1998; Osse & Boogaart, 1999; Watabe, 1999; Verhagen, 2004). In fact in developing larvae of orange-spotted grouper administration of THs led to differences in locomotion between treated and control larvae (de Jesus *et al.*, 1998), supporting the idea that the TH driven changes in muscle may lead to functional modifications.

Significant behavioural and morphological changes during metamorphosis of the Japanese flounder are accompanied by biochemical changes in muscle proteins. A change in myosin light chain proteins and fast troponin T (Ftnt) has been reported during spontaneous metamorphosis. In particular, different protein isoforms of Ftnt are present in muscle from pre- and post-metamorphic larvae (Yamano *et al.*, 1991a). Myosin light chain proteins (MLC), alkali light chains (MLC1 and 3) and 5,5′-dithio-bis-nitrobenzoic acid light chain (MLC2) that form a complex with myosin heavy chain to form myosin, the major component of striated muscle, undergo a change during spontaneous metamorphosis. MLC2 in premetamorphic larvae undergoes isoform switching so that after metamorphic climax it is replaced by an adult specific isoform of MLC2 (Yamano *et al.*, 1994b). This process can be induced by T4 and inhibited by thiourea. It would appear therefore, that the spontaneous change in contractile protein isoforms during Japanese flounder metamorphosis is regulated by THs, although the mechanism by which this occurs remains to be elucidated. It will be of interest to further explore the molecular changes of muscle in response to THs during metamorphosis in order to understand the way in which function is modified.

Relatively few studies of the biochemical composition of muscle during the larval to juvenile transition of roundfish exist. However, in a recent study of slow troponin T (*Stnt*) in the gilthead seabream, a change in isoform abundance was shown to occur at the larva to juvenile transition (Campinho *et al.*, 2005). Moreover, analysis of *Stnt* and *Ftnt* isoforms in gilthead seabream treated with THs revealed that the change in isoform expression was associated with changes in THs, suggesting that these hormones may act as regulatory factors (Campinho *et al.*, 2006b). Comparison of ftnt gene expression in gilthead seabream and the protein in Japanese flounder reveals similar isoforms exist but there is a species specific pattern of expression. For example, in gilthead seabream an embryonic/larval ftnt transcript is down-regulated to undetectable levels at hatch while in flounder the 41.5 kDa embryonic/larval *Ftnt* protein isoform persists until metamorphosis when it is substituted by a 34 kDa adult ftnt protein (Yamano *et al.*, 1991a). Nonetheless, although further studies are required, it would appear that in fish that undergo a dramatic metamorphosis and those that undergo a less dramatic transition, THs may regulate molecular changes in muscle to better adapt the larva to the functional demands of the juvenile.

It seems likely, taking into consideration the profound alteration which occurs in the viscerocranial skeleton during flatfish metamorphosis, that this process may be regulated by THs (Brewster, 1987; Wagemans *et al.*, 1998; Okada *et al.*, 2003; Sæle & Pittman, 2006). However, so far the mechanisms which lead to the complex remodelling of the viscerocranial skeleton have not been identified and make establishing the influence of THs difficult to demonstrate. That THs can act on the flatfish skeleton is evident from *in situ* hybridisation studies which demonstrate the presence of *thr*β in putative chondrocytes in cartilage and presumptive osteoblasts in the periosteum of Japanese flounder larvae (Yamano & Miwa, 1998). However, more studies will be required to establish how THs influence skeletal development and if they are obligatory for the skeletal changes seen during metamorphosis in flatfish.

Interestingly, in roundfish, exogenous THs have been shown to stimulate scale formation which is generally the last skeletal element to form, and marks the larval to juvenile transition (Eales, 1979; Reddy *et al.*, 1992). However, although studies about the role of THs on skeletal transformation during the larva to juvenile transition are missing, in the embryo to larva transition, in the zebrafish, THs appear to be important for normal cartilage development in the jaw (Liu & Chan, 2002). This indicates that such tissues can respond to THs and raises the possibility of a role in subsequent processes. Further studies will be required to clarify the importance of THs in this process and may contribute to the ongoing discussion about the relatively high incidence of certain skeletal abnormalities during aquaculture production of fish.

Digestive System

The digestive system of fish undergoes a prodigious process of growth and differentiation during ontogeny in order for the larvae to be able to extract nutrients from its exogenous food supply (see Rønnestad & Morais, this volume). Associated with the change in diet that the pelagic to benthic transition implies during flatfish metamorphosis a number of changes to the digestive system occur. Most notable of the changes is the appearance of a functional stomach (gastric glands). In both Japanese and summer flounder associated with the spontaneous TH driven metamorphosis, the gastric glands develop and pepsinogen-like immunoreactivity is detected (Miwa *et al.*, 1992; Huang *et al.*, 1998; Soffientino & Specker, 2003). Modification of the digestive tract in Japanese flounder and turbot during metamorphosis is also associated with HCl secretion suggesting it also becomes functionally active (Rønnestad, 1993; Hoehne-Reitan *et al.*, 2001). Moreover, administration of THs to premetamorphic Japanese and summer flounder causes the precocious appearance of gastric glands and pepsinogen (Miwa *et al.*, 1992; Inui *et al.*, 1995; Huang *et al.*, 1998).

A number of studies of the ontogeny of the digestive system in roundfish exist generally in the context of nutritional studies. Relatively few studies have linked the larva to juvenile transition to gastric gland differentiation and the appearance of a functional stomach. The direct stimulatory effect of THs on gastric gland development has only been shown in the striped bass (Huang *et al.*, 1998), suggesting THs may be expected to have a similar effect on the digestive system in flat- and roundfish.

Haematopoietic System

The expression of haemoglobin (Hb) genes is regulated in a tissue- and stage-specific manner in vertebrates (Weber, 1996). In the spontaneously metamorphosing Japanese flounder, larval erythrocytes, which are large round cells with a small compact nucleus, are replaced by smaller elliptical adult erythrocytes. Concurrent with the change in erythrocytes, is the change in Hb from larval to juvemile isoforms, a phenomenon also seen in amphibians (Jurd, 1985; Hourdry, 1993; Yamano *et al.*, 1994a; Sæle *et al.*, 2003). Moreover, TH treatment of premetamophic Japanese flounder causes a precocious switch in erythrocyte character and Hb, suggesting that THs suppress larval, but stimulate adult erythropoiesis (Miwa & Inui, 1991; Inui *et al.*, 1995). Presumably the change in Hb and erythrocyte character is related to the change in respiration accompanying metamorphosis and the development of fully functional gills for gas exchange.

In Angelfish (*Pterophyllum scalare*) a change in erythrocyte character has been observed during development (Al-Adhami & Kunz, 1976), while in the rainbow trout both erythrocytes and Hb are changed (Iuchi, 1973; Iuchi *et al.*, 1975; Yamamoto & Iuchi, 1975), such changes have not been related to the larva to juvenile transition. However, the identification of nuclear thyroid receptors in trout erythrocytes (Sullivan *et al.*, 1987) suggests that THs may act directly on erythrocyte maturation in both roundfish and flatfish and further studies will be required to establish at what stage it occurs.

The implication of THs in immune system maturation in African clawed frog (Rollins-Smith, 1998) suggests they may also play a similar role in teleosts. At the level of the innate immune response (see Zapata & Cortés, this volume), the larva to juvenile transition in teleosts involves maturation of the skin and stomach pH. Studies of the ontogeny of the acquired immune response in teleost fish are still rudimentary. Whether the immune system undergoes maturation during flatfish metamorphosis and if THs are involved is unstudied. In the zebrafish, a recent study indicated that treatment with THs prior to and during the larval/juvenile transition stimulated thymus development and lymphopoiesis (Lam *et al.*, 2005). However, more studies are required to clearly demonstrate a role for THs in immune system maturation in fish and to establish if, during spontaneous TH driven larval/juvenile transition of roundfish and flatfish, similar changes occur.

Miscellaneous

Accompanying the change in external morphology and tissue maturation during the larval/juvenile transition in fish is a change in behaviour. For example, juveniles may change from a pelagic to a benthic ecology or migrate from an offshore to an estuarine environment. Associated with a change in environment are a number of physiological challenges to which the juvenile must adapt. The response to the osmoregulatory challenge implied by the movement from the sea to estuaries is in part mediated by THs in both juvenile Japanese and summer flounder and in the latter species THs also influence the development of a juvenile form of mitochondria-rich cells (putative chloride cells) in the gill, (Hiroi *et al.*, 1997; Schreiber & Specker, 1999, 2000). A role for the "classical" osmoregulatory pituitary hormone, PRL (Loretz & Bern, 1982) has also been observed in Japanese flounder in this migration.

Studies in amphibians indicate that both PRL and cortisol may play a role in metamorphosis (Denver *et al.*, 2002) and until recently PRL was considered to correspond to a "juvenile" hormone. However, in transgenic African clawed frog overexpressing PRL, the timing of metamorphosis is unaffected, although tail resorption is blocked (Huang & Brown, 2000), and such observations have raised questions about the role of PRL during amphibian metamorphosis (see Denver *et al.*, 2002). Evidence indicating a role for PRL in the tissue and molecular changes during metamorphosis in teleosts is scarce and a single study exists indicating PRL antagonises the stimulatory effect of T3 on resorption of dorsal fin rays in Japanese flounder (de Jesus *et al.*, 1994).

In amphibians the function of corticosteroids is enigmatic and depends on the developmental stage at which they are administered and in pre-metamorphic and pro-metamorphic larvae they accelerate or decelerate metamorphosis respectively (Denver *et al.*, 2002). In teleost fish during spontaneous metamorphosis the profile of THs and cortisol is similar (de Jesus *et al.*, 1991, 1993; Yamano *et al.*, 1991b). Moreover, in a number of different teleosts the administration of cortisol enhanced the effect of THs on metamorphosis (de Jesus *et al.*, 1990, 1991; Brown & Kim, 1995). Despite the relatively limited number of studies it seems probable that PRL and cortisol play a role in fish metamorphosis and this should be a fruitful area of research in the future.

FUTURE PERSPECTIVES

THs are principally responsible for the myriad changes that are linked to the larva to juvenile transition in teleost fish. Although flatfish undergo a more dramatic change in external morphology and behaviour, the biochemical and tissue changes studied so far appear to be surprisingly similar between flatfish and roundfish. A better understanding of the relative importance and role of deiodinases and their interplay during the larva to juvenile transition is needed in order to understand how tissue specific TH metabolism regulates, along with THR availability, tissue responsiveness. At the molecular level much more work is required to understand THR function and identify target genes and interacting factors which modulate gene transcription. The increasing availability of molecular resources for some teleost fish coupled to existing biochemical resources and the strong research background in fish metamorphosis is providing exciting opportunities for development of this area. THs have a plethora of different actions either on their own or with other hormones and an improved understanding of the ontogeny, biology and physiology of teleost fish should help distinguish true metamorphic effects.

Understanding metamorphosis or endocrine regulated processes in general represents an important scientific goal and may also have important socio-economic and ecological impacts. For example, the failing wild stocks of fish make fish aquaculture a viable alternative source, although success and sustainability require that current bottlenecks in larval production be overcome and production of good quality juveniles assured. Further, the fundamental importance of endocrine factors in regulating physiological processes is highlighted by the effects of endocrine disrupting chemicals on fish development, growth, reproduction and survival. The vunerability of fish to aquatic pollution and the importance of THs for the larva to juvenile transition represent important areas for future investigation from an aquaculture, ecological and endocrinological perspective.

Acknowledgements

The authors acknowledge financial support from the Commission of the European Communities, ENDCONAQUA (Q4RS-1996-1442) and ARRDE (Q5RS-2002-01192); funding from POCI 2010 and the European Social Funds attributed by the Portuguese National Science Foundation (FCT) to CCMAR and a PhD fellowship to MAC (SFRH/BD/6133/2001). The contents of the present chapter do not necessarily reflect the Commission's views and in no way anticipates its future policy in this area. Thanks to all past and present collaborators in this research area: Thrandur Björnsson, Inga Einarsdóttir, Glen Sweeny, Malyka Galay-Burgos, Karin Pittmann, Pat Ingleton, Heiddis Smáradóttir (Fiskey, Iceland), Joana Amaral (Viveiros Vila Nova de Millfontes, Portugal) and Vitor Baptista.

References

Abe T, Suzuki T, Unno M, Tokui T, Ito S (2002) Thyroid hormone transporters: recent advances. Trends Endocrinol Metab 13: 215-220

Adams GR, McCue SA, Zeng M, Baldwin KM (1999) Time course of myosin heavy chain transitions in neonatal rats: importance of innervation and thyroid state. Am J Physiol 276: R954-R961

Al-Adhami MA, Kunz YW (1976) Haemopoietic centres in the developing angelfish, *Pterophyllum scalare* (Cuvier and Valenciennes). Dev Genes Evol 179: 393-401

Andersson H, Kohn L, Bernardini I, Blom H, Tietze F, Gahl W (1990) Characterization of lysosomal monoiodotyrosine transport in rat thyroid cells. Evidence for transport by system H. J Biol Chem 265: 10950-10954

Balon EK (1999) Alternative ways to become a juvenile or a definitive phenotype (and on some persisting linguistic offenses). Env Biol Fishes 56: 17-38

Bao B, Yang G, Liu Z, Li S, Wang Z, Ren D (2005) Isolation of SFRS3 gene and its differential expression during metamorphosis involving eye migration of Japanese flounder *Paralichthys olivaceus*. Biochim Biophys Acta 1725: 64-70

Baqui MM, Gereben B, Harney JW, Larsen PR, Bianco AC (2000) Distinct subcellular localization of transiently expressed types 1 and 2 iodothyronine deiodinases as determined by immunofluorescence confocal microscopy. Endocrinology 141: 4309-4312

Bartalena L, Robbins J (1993) Thyroid hormone transport proteins. Clin Lab Med 13: 583-598

Bates JM, St Germain DL, Galton VA (1999) Expression profiles of the three iodothyronine deiodinases, D1, D2, and D3, in the developing rat. Endocrinology 140: 844-851

Becker KB, Stephens KC, Davey JC, Schneider MJ, Galton VA (1997) The type 2 and type 3 iodothyronine deiodinases play important roles in coordinating development in *Rana catesbeiana* tadpoles. Endocrinology 138: 2989-2997

Bentley PJ (1998) Comparative Vertebrate Endocrinology. Cambridge University Press, Cambridge

Berry MJ, Banu L, Larsen PR (1991a) Type I iodothyronine deiodinase is a selenocysteine-containing enzyme. Nature 349: 438-440

Berry MJ, Kieffer JD, Harney JW, Larsen PR (1991b) Selenocysteine confers the biochemical properties characteristic of the type I iodothyronine deiodinase. J Biol Chem 266: 14155-14158

Bianco AC, Salvatore D, Gereben B, Berry MJ, Larsen PR (2002) Biochemistry, cellular and molecular biology, and physiological roles of the iodothyronine selenodeiodinases. Endocr Rev 23: 38-89

Blaxter JHS (1988) Pattern and variety in development. In: Hoar WS, Randall DJ (eds) The Physiology of Developing Fish. Vol XI, Part A, Eggs and Larvae. Academic Press, San Diego, p 1-58

Brewster B (1987) Eye migration and cranial development during flatfish metamorphosis: a reappraisal (Teleostei: Pleuronectiformes). J Fish Biol 31: 805-833

Brown C, Kim B (1995) Combined application of cortisol and triiodothyronine in the culture of larval marine finfish. Aquaculture 135: 79-86

Brown DD (1997) The role of thyroid hormone in zebrafish and axolotl development. Proc Natl Acad Sci USA 94: 13011-13016.

Burmeister LA, Pachucki J, St Germain DL (1997) Thyroid hormones inhibit type 2 iodothyronine deiodinase in the rat cerebral cortex by both pre- and posttranslational mechanisms. Endocrinology 138: 5231-5237

Cahmann H, Pommier J, Nunez J (1977) Spatial requirement for coupling of iodotyrosine residues to form thyroid hormone. Proc Natl Acad Sci USA 74: 5333-5335

Campinho MA, Power DM, Sweeney GE (2005) Identification and analysis of teleost slow muscle troponin T (sTnT) and intronless TnT genes. Gene 361: 67-79

Campinho M, Sweeney GE, Power DM (2006) Regulation of troponin T expression by thyroid hormones during muscle development in sea bream (*Sparus auratus* Linnaeus). J Exp Biol 209: 4751-4767

Campinho M, Silva N, Sweeney G, Power DM (2007) Molecular, cellular and histological changes in skin from a larval to an adult phenotype during bony fish metamorphosis. Cell Tissue Res 327: 267-284

Celi FS, Canettieri G, Mentuccia D, Proietti-Pannunzi L, Fumarola A, Sibilla R, Predazzi V, Ferraro M, Andreoli M, Centanni M (2000) Structural organization and chromosomal localization of the human type II deiodinase gene. Eur J Endocrinol 143: 267-271

Chanoine C, Hardy S (2003) *Xenopus* muscle development: from primary to secondary myogenesis. Dev Dyn 226: 12-23.

Chin WW, Yen PM (1997) Molecular mechanisms of nuclear thyroid hormone action. In: Braverman LR (ed) Diseases of the Thyroid. Humana Press Inc, Totawa, NJ, p 1-15

Croteau W, Whittemore SL, Schneider MJ, St Germain DL (1995) Cloning and expression of a cDNA for a mammalian type III iodothyronine deiodinase. J Biol Chem 270: 16569-16575

Croteau W, Davey JC, Galton VA, St Germain DL (1996) Cloning of the mammalian type II iodothyronine deiodinase. A selenoprotein differentially expressed and regulated in human and rat brain and other tissues. J Clin Invest 98: 405-417

Curcio-Morelli C, Gereben B, Zavacki AM, Kim BW, Huang S, Harney JW, Larsen PR, Bianco AC (2003) *In vivo* dimerization of types 1, 2, and 3 iodothyronine selenodeiodinases. Endocrinology 144: 937-946

Dai G, Levy O, Carrasco N (1996) Cloning and characterization of the thyroid iodide transporter. Nature 379: 458-460

Davey JC, Becker KB, Schneider MJ, St Germain DL, Galton VA (1995) Cloning of a cDNA for the type II iodothyronine deiodinase. J Biol Chem 270: 26786-26789

Davey JC, Schneider MJ, Becker KB, Galton VA (1999) Cloning of a 5.8 kb cDNA for a mouse type 2 deiodinase. Endocrinology 140: 1022-1025 de Jesus EG, Inui Y, Hirano T (1990) Cortisol enhances the stimulating action of thyroid hormones on dorsal fin-ray resorption of flounder larvae *in vitro*. Gen Comp Endocrinol 79: 167-173

de Jesus EG, Hirano T, Inui Y (1991) Changes in cortisol and thyroid hormone concentrations during early development and metamorphosis in the Japanese flounder, *Paralichthys olivaceus*. Gen Comp Endocrinol 82: 369-376

de Jesus EG, Hirano T, Inui Y (1993) Flounder metamorphosis: Its regulation by various hormones. Fish Physiol Biochem 11: 323-328

de Jesus EG, Hirano T, Inui Y (1994) The antimetamorphic effect of prolactin in the Japanese flounder. Gen Comp Endocrinol 93: 44-50

de Jesus EG, Toledo JD, Simpas MS (1998) Thyroid hormones promote early metamorphosis in grouper (*Epinephelus coioides*) larvae. Gen Comp Endocrinol 112: 10-16

Deme D, Pommier J, Nunez J (1978) Specificity of thyroid hormone synthesis. The role of thyroid peroxidase. Biochim Biophys Acta 540: 73-82

Denver RJ, Boorse GC, Glennemeier KA (2002) Endocrinology of complex life cycles: Amphibians. In: Pfaff D, Arnold A, Etgen A, Fahrbach S, Moss R, Rubin R (eds) Hormones, Brain and Behavior, Vol 2, Academic Press, San Diego, p 469-513

Dunn J, Dunn A (2000) Thyroglobulin: chemistry, biosynthesis, and proteolysis. In: Braverman LE (ed) The Thyroid, 8th Edition. Utiger R. Lippincott Williams & Wilkins, Philadelphia, p 91-104.

Eales JG (1979) Thyroid Function in cyclostomes and fishes. In: Barrington EJW (ed) Hormones and Evolution, Vol 1, Academic Press, London, p 342-436

Eales JG (1990) Thyroid function in poikilotherms. Prog Clin Biol Res 342: 415-420

Eskandari S, Loo DD, Dai G, Levy O, Wright EM, Carrasco N (1997) Thyroid Na^+/I^- symporter. Mechanism, stoichiometry, and specificity. J Biol Chem 272: 27230-27238

Essner JJ, Breuer JJ, Essner RD, Fahrenkrug SC, Hackett PB (1997) The zebrafish thyroid hormone receptor $\alpha 1$ is expressed during early embryogenesis and can function in transcriptional repression. Differentiation 62: 107-117

Finnson KW, Eales JG (1999) Effect of T3 treatment and food ration on hepatic deiodination and conjugation of thyroid hormones in rainbow trout, *Oncorhynchus mykiss*. Gen Comp Endocrinol 115: 379-386

Forrest D, Reh TA, Rusch A (2002) Neurodevelopmental control by thyroid hormone receptors. Curr Opin Neurobiol 12: 49-56

Friesema EC, Docter R, Moerings EP, Verrey F, Krenning EP, Hennemann G, Visser TJ (2001) Thyroid hormone transport by the heterodimeric human system L amino acid transporter. Endocrinology 142: 4339-4348

Gavaret J-M, Nunez J, Cahnmann H (1980) Formation of dehydroalanine residues during thyroid hormone synthesis in thyroglobulin. J Biol Chem 255: 5281-5285

Gavaret J-M, Cahnmann H, Nunez J (1981) Thyroid hormone synthesis in thyroglobulin; the mechanism of the coupling reaction. J Biol Chem 256: 9167-9172

Gilbert SF (2006) Developmental Biology. Sinauer Associates, Inc, Sunderland, Massachusetts

Gorbman A, Bern HA (1962) A Textbook of Comparative Endocrinology. John Wiley & Sons, New York

Gorbman A, Dickhoff WW, Vigna SR, Clark NB, Ralph CL (1983) Comparative Endocrinology. John Wiley & Sons, New York

Hamre K, Moren M, Solbakken J, Opstad I, Pittman K (2005) The impact of nutrition on metamorphosis in Atlantic halibut (*Hippoglossus hippoglossus* L.). Aquaculture 250: 555-565

Hashimoto H, Mizuta A, Okada N, Suzuki T, Tagawa M, Tabata K, Yokoyama Y, Sakaguchi M, Tanaka M, Toyohara H (2002) Isolation and characterization of a Japanese flounder clonal line, *reversed*, which exhibits reversal of metamorphic left-right asymmetry. Mech Dev 111: 17-24

Hernandez A, Lyon GJ, Schneider MJ, St Germain DL (1999) Isolation and characterization of the mouse gene for the type 3 iodothyronine deiodinase. Endocrinology 140: 124-130

Hirata Y, Kurokura H, Kasahara S (1989) Effects of thyroxine and thiourea on the development of larval red sea bream *Pagrus major*. Nippon Suisan Gakkaishi 55: 1189-1195

Hiroi J, Sakakura Y, Tagawa M, Seikai T, Tanaka M (1997) Developmental changes in low-salinity tolerance and responses of prolactin, cortisol and thyroid hormones to low-salinity environment in larvae and juveniles of Japanese flounder (*Paralichthys olivaceus*). Zool Sci 14: 987-992

Hoehne-Reitan K, Kjørsvik E, Reitan K (2001) Development of the pH in the intestinal tract of larval turbot. Mar Biol 139: 1159-1164

Holmes JA, Youson JH (1993) Induction of metamorphosis in landlocked sea lampreys, *Petromyzon marinus*. J Exp Zool 267: 598-604

Hourdry J (1993) Passage to the terrestrial life in amphibians: II. Endocrine determinism. Zool Sci 10: 887-902

Huang H, Brown DD (2000) Prolactin is not a juvenile hormone in *Xenopus laevis* metamorphosis. Proc Natl Acad Sci USA 97: 195-199

Huang LY, Miwa S, Bengtson DA, Specker JL (1998) Effect of triiodothyronine on stomach formation and pigmentation in larval striped bass (*Morone saxatilis*). J Exp Zool 280: 231-237

Huang H, Marsh Armstrong N, Brown DD (1999) Metamorphosis is inhibited in transgenic *Xenopus laevis* tadpoles that overexpress type III deiodinase. Proc Natl Acad Sci USA 96: 962-967

Huang SA, Dorfman DM, Genest DR, Salvatore D, Larsen PR (2003) Type 3 iodothyronine deiodinase is highly expressed in the human uteroplacental unit and in fetal epithelium. J Clin Endocrinol Metab 88: 1384-1388

Hunn JB, Fromm PO (1966) *In vivo* uptake of radioiodide by rainbow trout. J Water Pollut Control Fed 38: 1981

Inui Y, Miwa S (1985) Thyroid hormone induces metamorphosis of flounder larvae. Gen Comp Endocrinol 60: 450-454

Inui Y, Tagawa M, Miwa S, Hirano T (1989) Effects of bovine TSH on the tissue thyroxine level and metamorphosis in prometamorphic flounder larvae. Gen Comp Endocrinol 74: 406-410

Inui Y, Yamano K, Miwa S (1995) The role of thyroid hormone in tissue development in metamorphosing flounder. Aquaculture 135: 87-98

Iuchi I (1973) The post-hatching transition of erythrocytes from larval to adult type in the rainbow trout, *Salmo gairdnerii irideus*. J Exp Zool 184: 383-396

Iuchi I, Suzuki R, Yamagami K (1975) Ontogenetic expression of larval and adult hemoglobin phenotypes in the intergeneric salmonid hybrids. J Exp Zool 192: 57-64

Jacquemin E, Hagenbuch B, Stieger B, Wolkoff AW, Meier PJ (1994) Expression cloning of a rat liver Na^+-independent organic anion transporter. Proc Natl Acad Sci USA 91: 133-137

Johnston IA (1994) Development and plasticity of fish muscle with growth. Basic Appl Myol 4: 353-368

Johnston IA, Cole NJ, Vieira VLA, Davidson I (1997) Temperature and developmental plasticity of muscle phenotype in herring larvae. J Exp Biol 200: 849-868

Jones I, Rogers SA, Kille P, Sweeney GE (2002) Molecular cloning and expression of thyroid hormone receptor α during salmonid development. Gen Comp Endocrinol 125: 226-235

Jurd RD (1985) Heamatological and immunological metamorphosis in neotenous urodeles. In: Balls M, Brownes M (eds) Metamorphosis. Clarendon Press, Oxford, p 313-331

Katz D, Lazar MA (1993) Dominant negative activity of an endogenous thyroid hormone receptor variant (α2) is due to competition for binding sites on target genes. J Biol Chem 268: 20904-20910

Kawakami Y, Tanda M, Adachi S, Yamauchi K (2003a) Characterization of thyroid hormone receptor α and β in the metamorphosing Japanese conger eel, *Conger myriaster*. Gen Comp Endocrinol 132: 321-332

Kawakami Y, Tanda M, Adachi S, Yamauchi K (2003b) cDNA cloning of thyroid hormone receptor betas from the conger eel, *Conger myriaster*. Gen Comp Endocrinol 131: 232-240

Kikuyama S, Kawamura K, Tanaka S, Yamamoto K (1993) Aspects of amphibian metamorphosis: hormonal control. Int Rev Cytol 145: 105-148

Kitajima C, Sato T, Kawanishi M (1967) On the effect of thryoxine to promote the metamorphosis of a conger eel. Bull Jap Soc Sci Fish 33: 919-922

Klaren PH, Haasdijk R, Metz JR, Nitsch LM, Darras VM, Van der Geyten S, Flik G (2005) Characterization of an iodothyronine 5'-deiodinase in gilthead seabream (*Sparus auratus*) that is inhibited by dithiothreitol. Endocrinology 146: 5621-5630

Köhrle J (2000) The selenoenzyme family of deiodinase isozymes controls local thyroid hormone availability. Rev Endocr Metab Disord 1: 49-58

Koumans JTM, Akster HA (1995) Myogenic cells in development and growth of fish. Comp Biochem Physiol 110A: 3-20

Kuiper GG, Klootwijk W, Visser TJ (2002) Substitution of cysteine for a conserved alanine residue in the catalytic center of type II iodothyronine deiodinase alters interaction with reducing cofactor. Endocrinology 143: 1190-1198

Lam SH, Sin YM, Gong Z, Lam TJ (2005) Effects of thyroid hormone on the development of immune system in zebrafish. Gen Comp Endocrinol 142: 325-335

Lam T (1985) Role of thyroid hormone on larval growth and development in fish. In: Lofts B, Holms WN (eds) Current Trends in Comparative Endocrinology, Hong Kong University Press, Hong Kong, p 481-485

Lamas L, Taurog A, Salvatore G, Edelhoch H (1974) The importance of thyroglobulin structure in thyroid peroxidase-catalyzed conversion of diiodotyrosine to thyroxine. J Biol Chem 249: 2732-2737

Lamason RL, *et al* (2005) SLC24A5, a putative cation exchanger, affects pigmentation in zebrafish and humans. Science 310: 1782-1786

Langsteger W (1997) Clinical aspects and diagnosis of thyroid hormone transport protein anomalies. Curr Top Pathol 91: 129-161

Larsson L, Li X, Teresi A, Salviati G (1994) Effects of thyroid hormone on fast- and slow-twitch skeletal muscles in young and old rats. J Physiol 481: 149-161

Larsson L, Muller U, Li X, Schiaffino S (1995) Thyroid hormone regulation of myosin heavy chain isoform composition in young and old rats, with special reference to IIX myosin. Acta Physiol Scand 153: 109-116

Le Guellec D, Morvan-Dubois G, Sire JY (2004) Skin development in bony fish with particular emphasis on collagen deposition in the dermis of the zebrafish (*Danio rerio*). Int J Dev Biol 48: 217-231

Leatherland JF (1994) Reflections on the thyroidology of fishes: from molecules to humankind. Guelph Ichthyol Rev 2: 2-67

Leatherland JF, Farbridge KJ (1992) Chronic fasting reduces the response of the thyroid to growth hormone and TSH, and alters the growth hormone-related changes in hepatic 5'-monodeiodinase activity in rainbow trout, *Oncorhynchus mykiss*. Gen Comp Endocrinol 87: 342-353

Leatherland JF, Hilliard RW, Macey DJ, Potter IC (1990) Changes in serum thyroxine and triiodothyronine concentrations during metamorphosis of the southern hemisphere lamprey *Geotria australis*, and the effect of propylthiouracil, triiodothyronine and environmental temperature on serum thyroid hormone concentrations of ammocoetes. Fish Physiol Biochem 8: 167-177

Leonard J, Köhrle J (2000) Intracellular pathways of iodothyronine metabolism. In: Braverman L, Utiger R (eds) The Thyroid. Lippincott-Raven, Philadelphia, p 136-173

Leonard DM, Stachelek SJ, Safran M, Farwell AP, Kowalik TF, Leonard JL (2000) Cloning, expression, and functional characterization of the substrate binding subunit of rat type II iodothyronine 5'-deiodinase. J Biol Chem 275: 25194-25201

Lintlop SP, Youson JH (1983) Concentration of triiodothyronine in the sera of the sea lamprey, *Petromyzon marinus*, and the brook lamprey, *Lampetra lamottenii*, at various phases of the life cycle. Gen Comp Endocrinol 49: 187-194

Liu YW, Chan WK (2002) Thyroid hormones are important for embryonic to larval transitory phase in zebrafish. Differentiation 70: 36-45

Liu YW, Lo LJ, Chan WK (2000) Temporal expression and T3 induction of thyroid hormone receptors $\alpha 1$ and $\beta 1$ during early embryonic and larval development in zebrafish, *Danio rerio*. Mol Cell Endocrinol 159: 187-195

Llewellyn L, Ramsurn VP, Sweeney GE, Wigham T, Power DM (1998) Expression of thyroid hormone receptor during early development of the sea bream (*Sparus aurata*). Ann N Y Acad Sci 839: 610-611

Llewellyn L, Nowell MA, Ramsurn VP, Wigham T, Sweeney GE, Kristjánsson B, Halldórsson O (1999) Molecular cloning and developmental expression of the halibut thyroid hormone receptor-α. J Fish Biol 55: 148-155

Loretz CA, Bern HA (1982) Prolactin and osmoregulation in vertebrates. An update. Neuroendocrinology 35: 292-304

MacLatchy DL, Eales JG (1992) Properties of T4 5'-deiodinating systems in various tissues of the rainbow trout, *Oncorhynchus mykiss*. Gen Comp Endocrinol 86: 313-322

MacLatchy DL, Kawauchi H, Eales JG (1992) Stimulation of hepatic thyroxine 5'-deiodinase activity in rainbow trout (*Oncorhynchus mykiss*) by Pacific salmon growth hormone. Comp Biochem Physiol 101A: 689-691

Marchand O, Safi R, Escriva H, Van Rompaey E, Prunet P, Laudet V (2001) Molecular cloning and characterization of thyroid hormone receptors in teleost fish. J Mol Endocrinol 26: 51-65

Marchand O, Duffraisse M, Triqueneaux G, Safi R, Laudet V (2004) Molecular cloning and developmental expression patterns of thyroid hormone receptors and T3 target genes in the turbot (*Scophthalmus maximus*) during post-embryonic development. Gen Comp Endocrinol 135: 345-357

Mascia A, Nitsch L, Di_Lauro R, Zannini M (2002) Hormonal control of the transcription factor Pax8 and its role in the regulation of thyroglobulin gene expression in thyroid cells. J Endocrinol 172: 163-176

Miwa S, Inui Y (1987) Effects of various doses of thyroxine and triiodothyronine on the metamorphosis of flounder (*Paralichthys olivaceus*). Gen Comp Endocrinol 67: 356-363

Miwa S, Inui Y (1991) Thyroid hormone stimulates the shift of erythrocyte populations during metamorphosis of the flounder. J Exp Zool 259: 222-228

Miwa S, Tagawa M, Inui Y, Hirano T (1988) Thyroxine surge in metamorphosing flounder larvae. Gen Comp Endocrinol 70: 158-163

Miwa S, Yamano K, Inui Y (1992) Thyroid hormone stimulates gastric development in flounder larvae during metamorphosis. J Exp Zool 261: 424-430

Miyatani S, Winkles JA, Sargent TD, Dawid IB (1986) Stage-specific keratins in *Xenopus laevis* embryos and tadpoles: the XK81 gene family. J Cell Biol 103: 1957-1965

Mol KA, van der Geyten S, Burel C, Kuhn ER, Boujard T, Darras VM (1998) Comparative study of iodothyronine outer ring and inner ring deiodination activities in five teleostean fishes. Fish Physiol Biochem 18: 253-266

Moren M, Opstad I, Van der Meeren T, Hamre K (2006) Iodine enrichment of *Artemia* and enhanced levels of iodine in Atlantic halibut larvae (*Hippoglossus hippoglossus* L.) fed the enriched *Artemia*. Aquacult Nutr 12: 97-102

Morin PP, Hara TJ, Eales JG (1993) Thyroid hormone deiodination in brain, liver, gill, heart and muscle of Atlantic salmon (*Salmo salar*) during photoperiodically-induced parr-smolt transformation. I. Outer- and inner-ring thyroxine deiodination. Gen Comp Endocrinol 90: 142-156

Nagaya T, Jameson JL (1993) Distinct dimerization domains provide antagonist pathways for thyroid hormone receptor action. J Biol Chem 268: 24278-24282

Nelson JS (1994) Fishes of the World. John Wiley and Sons, New York

Nilsson S, Makela S, Treuter E, Tujague M, Thomsen J, Andersson G, Enmark E, Pettersson K, Warner M, Gustafsson JA (2001) Mechanisms of estrogen action. Physiol Rev 81: 1535-1565

Nishikawa A, Shimizu-Nishikawa K, Miller L (1992) Spatial, temporal, and hormonal regulation of epidermal keratin expression during development of the frog, *Xenopus laevis*. Dev Biol 151: 145-153

Nowell MA, Power DM, Canario AV, Llewellyn L, Sweeney GE (2001) Characterization of a sea bream (*Sparus aurata*) thyroid hormone receptor-beta clone expressed during embryonic and larval development. Gen Comp Endocrinol 123: 80-89

Okada N, Takagi Y, Tanaka M, Tagawa M (2003) Fine structure of soft and hard tissues involved in eye migration in metamorphosing Japanese flounder (*Paralichthys olivaceus*). The Anatomical Record Part A: Discoveries in Molecular, Cellular, and Evolutionary Biology 273A: 663-668

Orozco A, Silva JE, Valverde RC (1997) Rainbow trout liver expresses two iodothyronine phenolic ring deiodinase pathways with the characteristics of mammalian types I and II 5'-deiodinases. Endocrinology 138: 254-258

Orozco A, Linser P, Valverde RC (2000) Kinetic characterization of outer-ring deiodinase activity (ORD) in the liver, gill and retina of the killifish *Fundulus heteroclitus*. Comp Biochem Physiol 126B: 283-290

Orozco A, Jeziorski MC, Linser PJ, Greenberg RM, Valverde RC (2002) Cloning of the gene and complete cDNA encoding a type 2 deiodinase from *Fundulus heteroclitus*. Gen Comp Endocrinol 128: 162-167

Orozco A, Villalobos P, Jeziorski MC, Valverde RC (2003) The liver of *Fundulus heteroclitus* expresses deiodinase type 1 mRNA. Gen Comp Endocrinol 130: 84-91

Osse JW (1990) Form changes in fish larvae in relation to changing demands of function. Neth J Zool 40: 362-385

Osse JW, Boogaart vd (1999) Dynamic morphology of fish larvae, strutural implications of friction forces in swimming, feeding and ventilation. J Fish Biol 55: 156-174

Parichy DM, Turner JM (2003) Zebrafish *puma* mutant decouples pigment pattern and somatic metamorphosis. Dev Biol 256: 242-257

Patruno M, Radaelli G, Mascarello F, Carnevali MDC (1998) Muscle growth in response to changing demands of functions in the teleost *Sparus aurata* (L.) during development from hatching to juvenile. Anat Embryol 198: 487-504

Power DM (2005) Developmental ontogeny of prolactin and its receptor in fish. Gen Comp Endocrinol 142: 25-33

Power DM, Elias NP, Richardson SJ, Mendes J, Soares CM, Santos CRA (2000) Evolution of the thyroid hormone binding protein, transthyretin. Gen Comp Endocrinol 119: 241-255

Power DM, Llewellyn L, Faustino M, Nowell MA, Bjornsson BT, Einarsdottir IE, Canario AVM, Sweeney GE (2001) Thyroid hormones in growth and development of fish. Comp Biochem Physiol 130C: 447-459

Reddy PK, Brown CL, Leatherland JF, Lam TJ (1992) Role of thyroid hormones in tilapia larvae (*Oreochromis mossambicus*). 2. Changes in the hormones and 5'-monodeiodinase activity during development. Fish Physiol Biochem 9: 487-496

Reginato MJ, Zhang J, Lazar MA (1996) DNA-independent and DNA-dependent mechanisms regulate the differential heterodimerization of the isoforms of the thyroid hormone receptor with retinoid X receptor. J Biol Chem 271: 28199-28205

Richard K, Hume R, Kaptein E, Sanders JP, van Toor H, De Herder WW, den Hollander JC, Krenning EP, Visser TJ (1998) Ontogeny of iodothyronine deiodinases in human liver. J Clin Endocrinol Metab 83: 2868-2874

Rollins-Smith LA (1998) Metamorphosis and the amphibian immune system. Immunol Rev 166: 221-230

Rønnestad I (1993) No efflux of free amino acids from yolk-sac larvae of Atlantic halibut (*Hippoglossus hippoglossus* L). J Exp Mar Biol Ecol 167: 39-45

Rotzsch W, Kohler H, Martin H (1967) Zur loslichkeit von thyroxin. Hoppe-Seylers Zeitschrift Fur Physiologische Chemie 348: 939

Sæle Ø, Nadia S, Pittman KA (2006) Post embryonic remodelling of neurocranial elements, a comparative study of normal versus abnormal eye migration in the flatfish Atlantic halibut. J Anat 209: 31-41

Sæle Ø, Solbakken JS, Watanabe K, Hamre K, Pittman KA (2003) The effect of diet on ossification and eye migration in Atlantic halibut larvae (*Hippoglossus hippoglossus* L.). Aquaculture 220: 683-696

Salvatore D, Low SC, Berry M, Maia AL, Harney JW, Croteau W, St Germain DL, Larsen PR (1995) Type 3 lodothyronine deiodinase: cloning, *in vitro* expression, and functional analysis of the placental selenoenzyme. J Clin Invest 96: 2421-2430

Salvatore D, Bartha T, Harney JW, Larsen PR (1996a) Molecular biological and biochemical characterization of the human type 2 selenodeiodinase. Endocrinology 137: 3308-3315

Salvatore D, Tu H, Harney JW, Larsen PR (1996b) Type 2 iodothyronine deiodinase is highly expressed in human thyroid. J Clin Invest 98: 962-968

Sanders JP, Van der Geyten S, Kaptein E, Darras VM, Kuhn ER, Leonard JL, Visser TJ (1997) Characterization of a propylthiouracil-insensitive type I iodothyronine deiodinase. Endocrinology 138: 5153-5160

Sanders JP, Van der Geyten S, Kaptein E, Darras VM, Kuhn ER, Leonard JL, Visser TJ (1999) Cloning and characterization of type III iodothyronine deiodinase from the fish *Oreochromis niloticus*. Endocrinology 140: 3666-3673

Santos CR, Power DM (1999) Identification of transthyretin in fish (*Sparus aurata*): cDNA cloning and characterisation. Endocrinology 140: 2430-2433

Sap J, Munoz A, Damm K, Goldberg Y, Ghysdael J, Leutz A, Beug H, Vennstrom B (1986) The c-erb-A protein is a high-affinity receptor for thyroid hormone. Nature 324: 635-640

Schreiber AM, Specker JL (1998) Metamorphosis in the summer flounder (*Paralichthys dentatus*): stage-specific developmental response to altered thyroid status. Gen Comp Endocrinol 111: 156-166

Schreiber AM, Specker JL (1999) Metamorphosis in the summer flounder, *Paralichthys dentatus*: thyroidal status influences salinity tolerance. J Exp Zool 284: 414-424

Schreiber AM, Specker JL (2000) Metamorphosis in the summer flounder, *Paralichthys dentatus*: thyroidal status influences gill mitochondria-rich cells. Gen Comp Endocrinol 117: 238-250

Schreiber G, Richardson SJ (1997) The evolution of gene expression, structure and function of transthyretin. Comp Biochem Physiol 116B: 137-160

Schussler G (2000) The thyroxine-binding proteins. Thyroid 10: 141-149

Shepherdley CA, Klootwijk W, Makabe KW, Visser TJ, Kuiper GG (2004) An ascidian homolog of vertebrate iodothyronine deiodinases. Endocrinology 145: 1255-1268

Shi YB, Wong J, Puzianowska-Kuznicka M, Stolow MA (1986) Tadpole competence and tissue specific temporal regulation of amphibian metamorphosis: roles of thyroid hormone and its receptors. BioEssays 18: 391-399

Shi YB, Yaoita Y, Brown DD (1992) Genomic organization and alternative promoter usage of the two thyroid hormone receptor β genes in *Xenopus laevis*. J Biol Chem 267: 733-738

Shintani N, Nohira T, Hikosaka A, Kawahara A (2002) Tissue-specific regulation of type III iodothyronine 5-deiodinase gene expression mediates the effects of prolactin and growth hormone in *Xenopus* metamorphosis. Develop Growth Differ 44: 327-335

Soffientino B, Specker JL (2003) Age-dependent changes in the response of the stomach to thyroidal signaling in developmentally arrested larval summer flounder. Gen Comp Endocrinol 134: 237-243

Solbakken JS, Norberg B, Watanabe K, Pittman K (1999) Thyroxine as a mediator of metamorphosis of Atlantic halibut, *Hippoglossus Hippoglossus*. Env Biol Fishes 56: 53-65

Soukup T, Jirmanova I (2000) Regulation of myosin expression in developing and regenerating extrafusal and intrafusal muscle fibers with special emphasis on the role of thyroid hormones. Physiol Res 49: 617-633

St Germain DL, Schwartzman RA, Croteau W, Kanamori A, Wang Z, Brown DD, Galton VA (1994) A thyroid hormone-regulated gene in *Xenopus laevis* encodes a type III iodothyronine 5-deiodinase. Proc Natl Acad Sci USA 91: 7767-7771

Sullivan CV, Darling DS, Dickhoff WW (1987) Nuclear receptors for L-triiodothyronine in trout erythrocytes. Gen Comp Endocrinol 65: 149-160

Suzuki S (1986) Induction of metamorphosis and thyroid function in the larval lamprey. In: Mederios-Neto G, Gaitan E (eds) Frontiers in Thyroidology, Vol I, Plenum Press, New York, p 667-670

Tagawa M, Aritaki M (2005) Production of symmetrical flatfish by controlling the timing of thyroid hormone treatment in spotted halibut *Verasper variegatus*. Gen Comp Endocrinol 141: 184-189

Tannahill LA, Visser TJ, McCabe CJ, Kachilele S, Boelaert K, Sheppard MC, Franklyn JA, Gittoes NJ (2002) Dysregulation of iodothyronine deiodinase enzyme expression and function in human pituitary tumours. Clin Endocrinol 56: 735-743

Thomas T, Dziadek M (1993) Capacity to form choroid plexus-like cells *in vitro* is restricted to specific regions of the mouse neural ectoderm. Development 117: 253-262

Thomas T, Schreiber G, Jaworowski A (1989) Developmental patterns of gene expression of secreted proteins in brain and choroid plexus. Dev Biol 134: 38-47

Thomas T, Power B, Hudson P, Schreiber G, Dziadek M (1988) The expression of transthyretin messenger RNA in the developing rat brain. Dev Biol 128: 415-427

Tietze F, Kohn L, Kohn A, Bernardini I, Andersson H, Adamson M, Harper G, Gahl W (1989) Carrier-mediated transport of monoiodotyrosine out of thyroid cell lysosomes. J Biol Chem 264: 4762-4765

Toyoda N, Nishikawa M, Mori Y, Yoshimura M, Masaki H, Gondou A, Yonemoto T, Inada M (1992) Identification of a 27-kilodalton protein with the properties of type I iodothyronine 5'-deiodinase in human thyroid gland. J Clin Endocrinol Metab 74: 533-538

Tu GF, Cole T, Southwell BR, Schreiber G (1990) Expression of the genes for transthyretin, cystatin-C and β-A4 amyloid precursor protein in sheep choroid plexus during development. Dev Brain Res 55: 203-208

Umesono K, Evans RM (1989) Determinants of target gene specificity for steroid/thyroid hormone receptors. Cell 57: 1139-1146

Vadaszova A, Zacharova G, Machacova K, Jirmanova I, Soukup T (2004) Influence of thyroid status on the differentiation of slow and fast muscle phenotypes. Physiol Res 53 Suppl 1: S57-61

Van der Geyten S, Sanders JP, Kaptein E, Darras VM, Kuhn ER, Leonard JL, Visser TJ (1997) Expression of chicken hepatic type I and type III iodothyronine deiodinases during embryonic development. Endocrinology 138: 5144-5152

Van der Geyten S, Segers I, Gereben B, Bartha T, Rudas P, Larsen PR, Kuhn ER, Darras VM (2001a) Transcriptional regulation of iodothyronine deiodinases during embryonic development. Mol Cell Endocrinol 183: 1-9

Van der Geyten S, Toguyeni A, Baroiller J-F, Fauconneau B, Fostier A, Sanders JP, Visser TJ, Kuhn ER, Darras VM (2001b) Hypothyroidism induces type I iodothyronine deiodinase expression in tilapia liver. Gen Comp Endocrinol 124: 333-342

Van der Geyten S, Van den Eynde I, Segers IB, Kuhn ER, Darras VM (2002) Differential expression of iodothyronine deiodinases in chicken tissues during the last week of embryonic development. Gen Comp Endocrinol 128: 65-73

Verhagen JHG (2004) Hydrodynamics of burst swimming fish larvae; a conceptual model approach. J Theor Biol 229: 235-248

Virion A, Pommier J, Deme D, Nunez J (1981) Kinetics of thyroglobulin iodination and thyroid hormone synthesis catalyzed by peroxidase:therole of H_2O_2. Eur J Biochem 117: 103-109

Virion A, Courtin F, Deme D, Michot J, Kaniewski J, Pommier J (1985) Spectral characteristics and catalytic properties of thyroid peroxidase-H_2O_2 compounds in the iodination and coupling reactions. Arch Biochem Biophys 242: 41-47

Visser TJ (1994) Role of sulfation in thyroid hormone metabolism. Chem Biol Interact 92: 293-303

Wagemans F, Focant B, Vandewalle P (1998) Early development of the cephalic skeleton in the turbot. J Fish Biol 52: 166-204

Watabe S (1999) Myogenic regulatory factors and muscle differentiation during ontogeny in fish. J Fish Biol 55: 1-18

Weber R (1996) Switching of globin genes during anuran metamorphosis. In: Gilbert LI, Tata JR, Atkinson B (eds) Metamorphosis. Academic press, New York, p 567-597

Weinberger C, Thompson CC, Ong ES, Lebo R, Gruol DJ, Evans RM (1986) The c-erb-A gene encodes a thyroid hormone receptor. Nature 324: 641-646

Wolpert L, Beddington R, Jessell T, Lawrence P, Meyerowitz EM, Smith J (2002) Principles of Development. Oxford University Press, Oxford

Wright GM, Youson JH (1977) Serum thyroxine concentrations in larval and metamorphosing anadromous sea lamprey, *Petromyzon marinus* L. J Exp Zool 202: 27-32

Wrutniak-Cabello C, Casas F, Cabello G (2001) Thyroid hormone action in mitochondria. J Mol Endocrinol 26: 67-77

Yamamoto M, Iuchi I (1975) Electron microscopic study of erythrocytes in developing rainbow trouts, *Salmo gairdnerii irideus*, with particular reference to changes in the cell line. J Exp Zool 191: 407-426

Yamano K (2005) The role of thyroid hormone in fish development with reference to aquaculture. Jarq-Jpn Agr Res Q 39: 161-168

Yamano K, Inui Y (1995) cDNA cloning of thyroid hormone receptor beta for the Japanese flounder. Gen Comp Endocrinol 99: 197-203

Yamano K, Miwa S (1998) Differential gene expression of thyroid hormone receptor alpha and beta in. Gen Comp Endocrinol 109: 75-85

Yamano K, Miwa S, Obinata T, Inui Y (1991a) Thyroid hormone regulates developmental changes in muscle during flounder metamorphosis. Gen Comp Endocrinol 81: 464-472

Yamano K, Tagawa M, Dejesus EG, Hirano T, Miwa S, Inui Y (1991b) Changes in whole body concentrations of thyroid hormones and cortisol in metamorphosing conger eel. J Comp Physiol B 161: 371-375

Yamano K, Araki K, Sekikawa K, Inui Y (1994a) Cloning of thyroid hormone receptor genes expressed in metamorphosing flounder. Dev Genet 15: 378-382

Yamano K, Takano-Ohmuro H, Obinata T, Inui Y (1994b) Effect of thyroid hormone on developmental transition of myosin light chains during flounder metamorphosis. Gen Comp Endocrinol 93: 321-326

Yamauchi K, Nakajima J, Hayashi H, Hara A, Yamauchi K (1999) Purification and characterization of thyroid-hormone-binding protein from masu salmon serum – A homolog of higher-vertebrate transthyretin. Eur J Biochem 265: 944-949

Yang YZ, Burgos-Trinidad M, Wu Y, Koenig RJ (1996) Thyroid hormone receptor variant α2. Role of the ninth heptad in DNA binding, heterodimerization with retinoid X receptors, and dominant negative activity. J Biol Chem 271: 28235-28242

Yen PM, Chin WW (1994) New advances in understanding the molecular mechanisms of thyroid hormone action. Trends Endocrinol Metabol 5: 65-72

Youson JH (1997) Is lamprey metamorphosis regulated by thyroid hormones? Am Zool 37: 441-460

Youson JH, Plisetskaya EM, Leatherland JF (1994) Concentrations of insulin and thyroid hormones in the serum of landlocked sea-lampreys (*Petromyzon marinus*) of 3 larval year classes, in larvae exposed to 2 temperature regimes, and in individuals during and after metamorphosis. Gen Comp Endocrinol 94: 294-304

CHAPTER

20

Smoltification

Sigurd O Stefansson[1,*], Björn Th Björnsson[2], Lars OE Ebbesson[1] and Stephen D McCormick[3]

INTRODUCTION

The Anadromous Life History Strategy

Several salmonid species, including members of the genera *Oncorhynchus*, *Salmo* and *Salvelinus*, are anadromous and undergo a significant structural and functional transformation prior to migration to seawater. This has been termed parr-smolt transformation herein referred to as smoltification, and has previously been reviewed by Hoar (1976, 1988), Wedemeyer *et al.* (1980), Barron (1986), McCormick & Saunders (1987) and McCormick *et al.* (1998). Early biologists believed that the downstream migrating smolt was a distinct species of fish, but it was later discovered that it was the round, cryptically coloured, bottom-dwelling parr which, in the spring, transformed into a streamlined, silvery, pelagic smolt. During smoltification, the behaviour of the juvenile salmon changes from territorial and bottom-dwelling to pelagic and schooling, a strategy which reduces the risk of predation in open waters

Authors' addresses: [1]Department of Biology, University of Bergen, Bergen High Technology Centre, N-5008 Bergen, Norway.
[2]Fish Endocrinology Laboratory, Department of Zoology/Zoophysiology, Göteborg University, Box 463, S40530, Göteborg, Sweden.
[3]USGS, Conte Anadromous Fish Research Center, Turners Falls, MA 01376, USA.
**Corresponding author:* E-mail: sigurd.stefansson@bio.uib.no

(Handeland *et al.*, 1996). Despite contrasting evidence of changes in swimming performance of parr and smolts (Virtanen & Forsman, 1987; Peake & McKinley, 1998), parr actively swim against the water current to keep position, while smolts develop increased buoyancy and a preference for moving downstream out of the river system. The biological significance of this behaviour is to enter the ocean at an appropriate time, avoid predation in the estuary, and migrate from coastal waters to marine feeding grounds offshore.

Smoltification consists of a number of independent, but co-ordinated developmental changes in the biochemistry, physiology, morphology and behaviour of the young salmon. Alterations in lipid metabolism, osmoregulation, oxygen transport, buoyancy, growth, colour, shape, rheotaxis, and schooling behaviour can be understood as being preparatory to maximise the success of a life history strategy which involves downstream migration, ocean entry and subsequent long-distance feeding migrations in the marine environment. At the same time, many of these changes would be detrimental for continued life in freshwater, and it is thus not surprising that smoltification-related changes revert relatively quickly (desmoltification), if the fish are unable to enter the marine environment during the period of complete smoltification (the smolt 'window').

The anadromous strategy allows the different species to take advantage of the two very different ecosystems; the limnic, including rivers and lakes, and the marine, including both coastal areas and the open ocean. Reproduction is limited to rivers, an ecosystem which in temperate and arctic regions holds relatively few predators, thereby increasing the chances of survival during the early stages. The ecosystem is also characterised by low productivity, supporting only a limited growth rate and size of the fish. By leaving the river at a certain life stage, and entering the ocean to exploit the rich marine food supply, salmonids can increase their reproductive success ('fitness'). Models predicting the probability of an individual adopting an anadromous versus a resident strategy suggest that several factors including growth conditions, cost of migration, mortality rate and fecundity are involved (Gross, 1987). As will be discussed later in this chapter, residency is an evolutionary viable reproductive strategy for males, and in many anadromous populations a portion of the males do not migrate to the sea. On a global scale, anadromy is more common in temperate and sub-arctic regions, and is more widespread in the Northern than Southern hemisphere (McDowall, 1987). In contrast, catadromy is more common in the tropics and sub-tropic/warm temperate regions. Hence, in general terms, the frequency of anadromous and catadromous species reflect the relative productivity of the freshwater and marine environments.

Anadromy is not unique to salmonids. It is a strategy found among several fish groups both in the Northern and Southern hemispheres, including lampreys, sturgeons, osmerids, clupeids (shads) and basses (e.g. striped bass, *Morone saxatilis*). What sets the salmonids apart from other anadromous species is the smolt stage. Smoltification is a 'preparatory adaptation' occurring in fresh water before the juvenile salmon experience saltwater. This contrasts with acclimation that is initiated due to exposure to a new or variable environment. Under natural conditions in rivers and lakes, smoltification is a seasonal process usually occurring in late spring. Due to its similarity to metamorphosis (see Power *et al.*, this volume), smoltification was thought to be a single process, where the regulatory mechanisms for changes in morphology, physiology and behaviour were coupled. However, detailed studies have revealed that smoltification is a synchronisation of several independent developmental processes, and further, that certain processes may become uncoupled, e.g. the lack of correlation between salinity tolerance and body silvering (Bern, 1978). Manipulations which affect only certain components of smoltification can result in incomplete transformation and a reduced potential to exploit or survive in the marine environment. Such reduced adaptive potential may explain the limited success of hatchery-released smolts noted for different species during the last decades. There is evidence that the seasonal nature of smoltification has a basis in an endogenous rhythm, yet environmental cues are of major importance as 'zeitgebers' for complete smoltification (Stefansson *et al.*, 1991; Berge *et al.*, 1995; Björnsson *et al.*, 1998; McCormick *et al.*, 2000). As will be discussed later, this interplay between endogenous and exogenous control of smoltification is mediated by the neuroendocrine system.

The age and size at which smoltification occurs and its relative intensity differs among salmonid species. In pink salmon (*Oncorhynchus gorbuscha*) and chum salmon (*Oncorhynchus keta*), seaward migration takes place soon after emergence from the gravel, and almost the entire life cycle and all growth takes place in the sea. At the other end of the scale are the charr (*Salvelinus* spp.), which are entirely freshwater resident or spend only a few weeks feeding in seawater during the summer. Smoltification in Atlantic salmon (*Salmo salar*), coho salmon (*Oncorhynchus kisutch*) and steelhead trout (*Oncorhynchus mykiss*) occurs after at least one year residence in fresh water. A possible explanation for the heterochrony of salmonid smoltification is discussed by McCormick (1994). In this review we will focus on smoltification in Atlantic salmon, which is the most widely used salmon in aquaculture and has the largest body of scientific literature related to smolt development.

ALTERNATIVE LIFE HISTORIES

Desmoltification

Entry into seawater can be regarded as the manifestation of smoltification, a commitment to an oceanic life. However, if smolts are prevented from exposure to seawater, several of the preparatory changes associated with marine life are abandoned, a process known as desmoltification in Atlantic salmon or parr-reversion in Pacific salmonids (Hoar, 1988; Duston *et al.*, 1991). Accordingly, a 'smolt-window' exists, during which smolts can enter, and quickly adapt to, seawater. Desmoltification is accelerated by short photoperiod (Kurokawa, 1990) and by high and increasing freshwater temperatures (the smolt window closes again, Soivio *et al.*, 1988).The duration of the smolt window is suggested to be in the range of 300-400 degree-days (d°C; Stefansson *et al.*, 1998; McCormick *et al.*, 1999, see also Temperature below).

Desmoltification is generally viewed as a loss of important preparatory adaptations to marine life, including loss of hypo-osmoregulatory ability, critical metabolic adaptations, behavioural changes (e.g. re-establishment of positive rheotaxis) and major endocrine changes. Fish which enter seawater around or after the time of the closing of the smolt window generally show lower feed intake and growth, higher plasma ion levels, and reduced growth hormone (GH) levels, all indications of loss of seawater adaptability (Arnesen *et al.*, 2003). The desmoltification process is generally characterised by only a partial loss of morphological smolt characters, the fish again taking on a darker appearance, although not reverting to the morphology of parr, despite the loss of critical physiological functions. Evidence from culture of Atlantic salmon demonstrate that seawater growth of post-smolts is poor, disease resistance is lower and the fish often fail to adapt to harsh conditions (low temperature, high salinity) during winter.

Parr Maturation

According to Thorpe (1994) smoltification can be regarded as an alternative to the strategy of sexual maturation at the earliest opportunity (parr). Hence, some juvenile salmon become sexually mature in freshwater, without prior smoltification and seawater migration. Parr maturation is a viable strategy among male salmonids in many species and populations, and is generally observed when conditions for freshwater growth and development are favourable. On the other hand, maturation among female parr is normally not a viable strategy, as fecundity is drastically reduced (Hutchings & Jones, 1998). However, in cases of extremely low adult returns, due to high marine mortality or prevention of upstream passage, female parr maturation may be observed in salmonids as a response to the altered conditions and suggest the development of a landlocked life history. The basic mechanisms controlling the onset of puberty in fish, or indeed in any vertebrate, are still unresolved, although evidence points to a complex interaction between energy status, growth rate, size and genetics.

The process of sexual maturation involves major endocrine changes along the brain-pituitary-gonad axis. At the brain-pituitary level an increase in gonadotropic releasing hormone (GnRH) and a reduction in dopamine (DA) stimulate the synthesis and release of follicle stimulating hormone (FSH) and luteinising hormone (LH) from pituitary gonadotropic cells which in turn stimulate the production and release of sex steroids from gonadal tissues. Several points of interaction between the endocrine control of sexual maturation and smoltification can be identified, including a wide range of neuroendocrine factors, pituitary hormones and peripheral endocrine systems. Androgens have been shown to impair seawater adaptability (Lundqvist *et al.*, 1989), and downstream migratory activity (Berglund *et al.*, 1994) in smolting Atlantic salmon. Jakobsson *et al.* (1997) demonstrated receptor-like binding of androgens in Atlantic salmon gills, suggesting that the effect of androgens on osmoregulation may be exerted directly at the gill level, in addition to subsequent cascade systems. Madsen *et al.* (1997, 2004) and McCormick *et al.* (2005) found that estradiol (E2) and the synthetic estrogenic compound 4-nonylphenol (4-NP) inhibited the progress of smoltification, as judged by an elevated condition factor and a significantly poorer hypo-osmoregulatory performance, that was correlated with a reduction of gill Na^+,K^+-ATPase (NKA) activity, relative α-subunit NKA mRNA expression and gill chloride cell density (see further discussion of ion/osmoregulation later in this chapter). McCormick *et al.* (2005) further reported decreased plasma insulin-like growth factor-I (IGF-I) and thyroid hormone (TH) levels, suggesting complex endocrine pathways for the actions of sex steroids and endocrine disruptors. Negative effects of E2 on salinity tolerance appears to be a general response among several species of teleost, as Vijayan *et al.* (2001) found reduced hypo-osmoregulatory capacity in E2 treated Mozambique tilapia (*Oreochromis mossambicus*).

In a recent study on maturing adult sockeye salmon (*Oncorhynchus nerka*) Shrimpton *et al.* (2005) described an overall decline in gill NKA activity and changes in NKA isoform mRNA expression from open ocean through coastal, estuarine and river migration. Although no endocrine data were presented, studies by Cooke *et al.* (2006) documented a concurrent increase in levels of sex steroids in maturing sockeye. Taken together, these data suggest one mechanism by which sex steroids may reduce seawater tolerance in maturing salmonids.

Landlocked Atlantic Salmon

It is argued that landlocked forms of Atlantic salmon are derived independently from various anadromous founder populations, which were later prevented from reaching the upper regions of watersheds due to the post-glaciation elevation of land (Power, 1958; Behnke, 1972). Populations of landlocked Atlantic salmon from North America and Europe differ in their capacity to adapt to seawater (Barbour & Garside, 1983; Chernitsky & Loenko, 1983; Burton & Idler, 1984; Birt & Green, 1986, 1993; Birt *et al.*, 1991; Staurnes *et al.*, 1992; Schmitz, 1995), although they seem to have retained a certain degree of silvering enabling them to adapt to a pelagic life in lakes (Dahl, 1928; Berg, 1953; Birt & Green, 1986; Nilsen *et al.*, 2003). Non-anadromous salmon from Five Mile Pond East stock (Avalon Peninsula, Newfoundland) show no smolt-related increase in gill NKA activity (Birt & Green, 1993) and are unable to maintain normal hydro-mineral balance when challenged with seawater (Burton & Idler, 1984). In contrast, landlocked 'Gullspång' Atlantic salmon from lake Vänern in south-western Sweden show a gradual increase in hypo-osmoregulatory ability during spring (Schmitz, 1995), although no information is available on changes in gill NKA activity nor long-term seawater adaptability. Studies of landlocked strains of Atlantic salmon in eastern North America have yielded somewhat contrasting results. The general pattern seems to be a gradual improvement in seawater tolerance during spring, despite lower levels of gill NKA activity than those found in anadromous populations (Birt & Green, 1986, 1993; Birt *et al.*, 1991).

Recent findings by Nilsen *et al.* (2003) show that the 'Bleke' population from lake Byglandsfjord, south-central Norway, having been isolated from the sea for more than

9000 years (Berg, 1985), have abandoned key elements of smoltification associated with long-term adaptation to seawater (Fig. 1). Despite still showing some euryhaline capacity and ability for short-term hypo-osmoregulation, Bleke show peak freshwater NKA activity at only 50% of levels observed in anadromous smolts and no preparatory changes in gill *nka* isoform expression nor $Na^+,K^+,2Cl^-$ co-transporter (*nkcc*) or cystic fibrosis transmembrane conductance regulator I (*cftr1*) mRNA levels (Nilsen *et al.*, 2007). In a similar fashion, smolt-related development of critical brain regions and major endocrine changes are significantly attenuated in Bleke (Nilsen *et al.*, 2007; Ebbesson *et al.*, unpublished data). In line with the lack of preparatory changes in ion regulation, no smolt-related reduction in condition factor can be observed among landlocked populations, suggesting the absence of preparatory metabolic changes (Birt & Green, 1986; Birt *et al.*, 1991; Nilsen *et al.*, 2003). These differences may be viewed as adaptations to the generally poorer growing conditions of freshwater lakes compared with the North Atlantic ocean. Smoltification is an energetically demanding process (Hoar, 1976; 1988), and in the case of landlocked populations aspects of smoltification associated with marine life may thus be considered maladaptive. The period that populations have been released from selection pressure on maintenance of physiological capacity for seawater tolerance may have resulted in the observed differences in salinity tolerance and other aspects of smolt development seen among landlocked populations of Atlantic salmon.

PHYSIOLOGY – PREPARATORY CHANGES FOR MARINE LIFE

Growth and Metabolism

Through the evolution of each species, the anadromous strategy has been favoured in regions where conditions have enabled salmon to utilise freshwater for reproduction, with juveniles migrating into the ocean to feed on the rich supply of fish, squid and crustaceans. Concurrent with the increase in hypo-osmoregulatory ability, other changes occur which are related to growth and metabolism and which prepare the young salmon for the changes in availability and composition of prey organisms.

Growth pattern in freshwater

A major component of the anadromous strategy is for the smolts to enter the ocean at a size where risk of predation is significantly reduced. The schooling behaviour of smolts in the ocean is also advantageous for predator evasion. However, these strategies require that the juvenile salmon achieve a minimum threshold size, and that the smolts are fairly uniform in size. A primary observation of these relations is the occurrence of a bimodal size–frequency distribution that is related to smolt development (Kristinsson *et al.*, 1985). Depending on the overall conditions for growth and development, bimodality can be observed at the end of the first growing season in southern populations (0+, Utrilla & Lobon-Cervia, 1999), whereas the same bimodal distributions may take

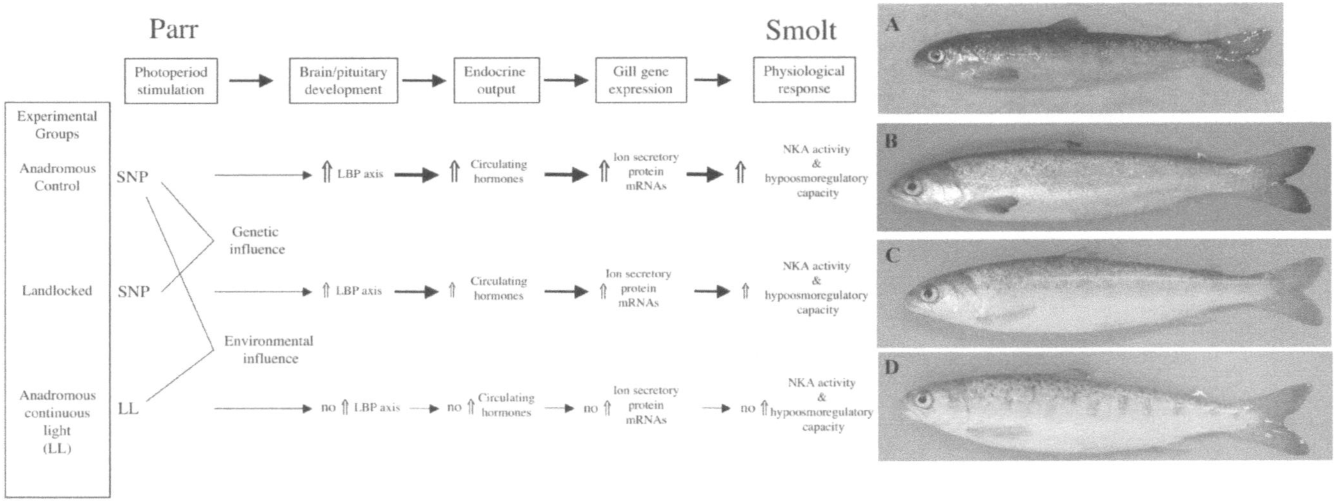

Fig. 1 This schematic representation illustrates a sequence of events that occurs during smoltification in Atlantic salmon (*Salmo salar*) leading to hypo-osmoregulatory development. Here three groups of Atlantic salmon are presented (anadromous control, parr in February (A) and smolt in May (B) reared under simulated natural photoperiod (SNP); juvenile landlocked salmon in May (C) reared under SNP; and anadromous juveniles in May reared under continuous light (D)) to demonstrate the importance of development of the light-brain-pituitary axis (LBP) early in smoltification on the downstream endocrine output, gill gene expression and hypo-osmoregulatory capacity. The degree of LBP activation is reflected through all downstream processes including morphological and physiological development, here represented by Na^+,K^+-ATPase (NKA) activity. After Ebbesson *et al.* (2007); Nilsen *et al.* (2007) and Stefansson *et al.* (2007).

years to become manifested under the poorer growth conditions of northern populations (Heggenes & Metcalfe, 1991; Skilbrei, 1991; Nicieza *et al.*, 1994). Under the exceptionally favourable growth conditions (temperature, photoperiod, food availability etc.) in Atlantic salmon culture, bimodality is normally observed in 0+ parr. Much of what we know about bimodal growth comes from laboratory studies, under natural conditions, factors such as overlapping year- and size-classes, migration and predation may obscure or override an underlying bimodal growth pattern.

Bimodality is the result of differential growth responses within a cohort of juvenile Atlantic salmon, where potential smolts of the upper mode (UM) continue to feed and maintain a relatively high growth rate through the autumn and winter (Metcalfe *et al.*, 1988), while the lower mode (LM) fish cease growth and development. Whether or not to continue growth through winter is a response based on an individual threshold of growth rate and/or surplus energy by summer or early autumn (Thorpe, 1989). Although the physiological mechanisms for size/energy-related developmental phenomena in fish are not fully understood (see below), their relationship to body size is fairly well described. When juvenile Atlantic salmon reach the critical size threshold, approximately 7.5-8.5 cm fork length in autumn among most European stocks, growth rate increases transiently during the 'fast growing phase' (Skilbrei, 1988; Skilbrei *et al.*, 1997). Laboratory studies have demonstrated that specific growth rate during this phase may increase to levels 4-5 times the typical growth rate of salmon parr of this size range, and length growth of as much as 1 mm day^{-1} has been recorded (Benfey *et al.*, 1994). In contrast, the LM fish enter a phase of anorexia during winter (Metcalfe *et al.*, 1986; Metcalfe & Thorpe, 1992), with growth resuming in spring when conditions again become favourable. The LM fish can then enter the UM or remain as LM parr based on their size the following summer. These different life-history strategies are gradually manifested as the UM fish smoltify during the spring, whereas the smaller LM fish delay smoltification and migration for at least one more year. A bimodal length distribution develops even among full-siblings, influenced both by genetics (Stefansson *et al.*, 1990; Nicieza *et al.*, 1994) and environmental conditions (Stefansson *et al.*, 1989; Skilbrei *et al.*, 1997) including food supply to the individual fish.

The underlying mechanisms that control the size-related development of smolts are still largely unknown. It has been proposed that they involve neuroendocrine changes in the brain (McCormick *et al.*, 1998, see also Neuroendocrine development below), and studies have shown plasma thyroxine (T_4) levels of UM fish to be several fold higher than those of LM fish during winter (Kristinsson *et al.*, 1985), implicating the involvement of the brain-pituitary-thyroid axis. In a recent study, McCormick *et al.* (2007) noted significant differences in the endocrine status of LM (parr) and UM (pre-smolt) Atlantic salmon throughout smolt development. Plasma levels of GH (secreted by somatotropes in the adenohypophysis) of UM fish increased steadily through spring, whereas plasma GH levels of parr remained low and did not respond to increased daylength. Plasma IGF-I levels were substantially lower in LM compared to UM in January and remained below levels of pre-smolts and smolts throughout spring. Similarly, plasma cortisol, which is secreted by the interrenal cells of the head kidney in response to adrenocorticotropic hormone (ACTH) from the pituitary, was low in LM fish, whereas cortisol levels in UM fish increased through late winter and spring. In line with the findings of Kristinsson *et al.* (1985), plasma thyroid hormones (T_4 and T_3) were generally lower in LM compared with UM fish. The key importance of a functional light-pituitary axis (resulting in elevated secretion of GH) for the initiation of the smoltification process is further demonstrated in a recent study in which Atlantic salmon parr were exposed to increased daylength at different body sizes. At 5 cm fork length, no fish responded to increased daylength. At 8 cm, however, fish destined to smoltify responded by a significant elevation of GH levels, while parr <8 cm did not (Nordgarden *et al.*, 2007). The differential activation of these endocrine axes during winter and spring points to fundamental physiological differences between parr and smolt in response to increased daylength, and also emphasises the importance of photoperiod as a is critical factor underlying smolt development. These differences may be explained by divergent development of the light-brain-pituitary axis (LBP; Ebbesson *et al.*, 2003, 2007) resulting in greater photoperiod response of smolt, though the precise pathways involved have not been elucidated (see Neuroendocrine development below).

Key role of growth hormone

Concurrent with the increase in hypo-osmoregulatory ability, other changes occur which are related to growth and metabolism. There is evidence that metabolism of proteins, lipids and carbohydrates is altered, which may relate both to the increased metabolic demands of smoltification and changes in availability and composition of marine prey organisms. Numerous studies have shown that GH has a strong growth-promoting action in salmonids (reviewed by McLean & Donaldson, 1993; Björnsson, 1997). This effect is either direct, or indirect through the subsequent GH-induced action of IGF-I (reviewed by Moriyama *et al.*, 1994; Moriyama, 1995; Björnsson, 1997, Björnsson *et al.*, 2002). The growth-promoting effects of GH concern both weight (muscle) growth and length (skeletal) growth. The smoltification-associated changes in growth pattern, which typically includes faster length than weight growth and consequent reduction in condition factor, are fully consistent of the growth-promoting effect of GH in salmonids, and the elevated plasma GH levels during smoltification. Thus, GH treatment of juvenile salmonids usually stimulates length growth more than weight growth, resulting in leaner fish (Higgs *et al.*, 1975, Johnsson & Björnsson, 1994; Neregård *et al.*, 2007). Recently, Wargelius *et al.* (2005) demonstrated that in Atlantic salmon, a peak in GH-receptor expression in vertebral bone tissue is associated with growth activation, while an upregulation of the IGF-I-receptor in vertebral bone appears to be related to increased bone density.

The stimulating effect of GH on body growth is further associated with increased food intake (Markert *et al.*, 1977; Johnsson & Björnsson, 1994) and increased feed conversion efficiency (Markert *et al.*, 1977; Gill *et al.*, 1985; Garber *et al.*, 1995). The effects of GH on food intake are not fully explained, but higher GH levels are associated with behavioural changes such as increased swimming activity, aggression and dominance, as well as decreased predator-avoidance (Johnsson *et al.*, 1996a, b, 1999; 2000; Jönsson *et al.*, 1996, 1998), consistent with increased opportunity to obtain food. The increased growth of GH-treated salmon parr further suggests an elevated foraging activity within the territory (Martin-Smith *et al.*, 2004). Recent studies indicate that in part, the behavioural effects of GH in salmonids may be direct, at the level of the central nervous system (Jönsson *et al.*, 2003; Johansson *et al.*, 2004, 2005).

In terms of tissue metabolism, the two major effects of GH in salmonids appear to be the stimulation of lipid mobilisation and protein accretion (Björnsson, 1997), but few studies have been carried out in relation to smoltification. The anabolic effects of GH on protein metabolism in salmonids have been studied most closely in rainbow trout (*Oncorhynchus mykiss*) (Foster *et al.*, 1991), where GH is found primarily to stimulate protein synthesis, and to a lesser extent to decrease protein degradation. In Atlantic salmon, standard and active metabolic rates are about 50% higher in smolts than in parr. The reasons for this are suggested to be the elevated respiratory enzyme activity and mitochondrial proliferation (Maxime *et al.*, 1989) and the metabolic demands of development and differentiation (McCormick & Saunders, 1987). Oxygen consumption is also seen to increase in smolts, and this increase is stimulated by several hormones, including GH (see above; Seddiki *et al.*, 1996). However, the increased metabolism/oxygen consumption may partly be due to increased physical activity such as seen following GH treatment (Herbert *et al.*, 2001). Smolts typically show higher cardiosomatic indices (CSI) than parr (Poupa *et al.*, 1974; Leonard & McCormick, 2001). Heart growth is accelerated (relative to body growth) during smoltification, probably as an adaptation to the increased cardiac output required for the pelagic schooling of the marine salmonids. Another developmental event which is worth noting in this context is the changes in the oxygen-binding proteins (haemoglobins) of the blood. There is conflicting evidence whether these changes are specific to smoltification or whether they are more related to the continuous increase in size and hence not strictly an aspect of smoltification (Hoar, 1988; Fyhn *et al.*, 1991). An increasing complexity of haemoglobins has been described in parr, smolts and post-smolts of *Oncorhynchus* and *Salmo* species, and is generally considered to be an adaptation to the differences in availability of oxygen between fresh water and seawater and the changes in behaviour, feeding and growth.

There is also some evidence of a reorganisation of amino acid distribution during smoltification; however, previous studies have reported no net change in the total amino acid content of smolts (Fontaine & Marchelidon, 1971). It has been suggested that this reorganisation is linked to guanine and hypoxanthine deposition in the skin and scales (smolt silvering, Fontaine & Marchelidon, 1971) as well as possible changes in cellular osmoregulatory gradients (McCormick & Saunders, 1987). There is also evidence of an increased activity of respiratory enzymes in the gill (Langdon & Thorpe, 1985; Chernitsky, 1986; McCormick & Saunders, 1987) and liver (Blake *et al.*, 1984) of Atlantic salmon smolts. From a growth and energetics perspective, smolts appear to prioritise protein deposition and growth of muscle and skeleton, at the expense of reduced lipid and carbohydrate stores (see Stefansson *et al.*, 2003a and discussion below). Studies under controlled conditions demonstrate an increase in growth rate following seawater transfer, despite no significant change in temperature or feeding (Handeland & Stefansson, 2001, 2002; Arnesen *et al.*, 2003), suggesting an influence of salinity itself on growth performance. The physiological basis of this growth increase may be related to the increase in GH observed as a response to salinity in salmonids (Björnsson, 1997; Björnsson *et al.*, 2000; Arnesen *et al.*, 2003; Handeland *et al.*, 2003).

Smolt development is generally accompanied by decreases in lipid and glycogen reserves in liver and muscle, even when food is present in excess (Wendt & Saunders, 1973; Sheridan, 1989). In coho salmon the reduction in liver glycogen during smoltification is caused by a combination of decreased glycogen synthesis and increased glycogenolysis (Sheridan *et al.*, 1985). Further studies of the biochemical basis for the changes in energy status during smoltification in Atlantic salmon have revealed both a seasonal, environmental and developmental tissue-specific regulation of several key metabolic enzymes (Leonard & McCormick, 2001). Smolts consistently showed lower liver lactate dehydrogenase (LDH), suggesting a decreased reliance on lactate–pyruvate conversion in line with findings by Plisetskaya *et al.* (1994) and Ji *et al.* (1996) who found low concentrations of blood glucose and liver glycogen due to decreased gluconeogenesis. Of potentially particular significance were changes in white muscle phosphofructokinase (PFK), which increased in both wild and hatchery-reared smolts but not in hatchery-reared parr. The same metabolic phenomenon was observed in white muscle LDH which showed an earlier increase in hatchery-reared than wild smolts and wild parr. This suggests that both habitat and developmental stage may influence metabolic parameters (Leonard & McCormick, 2001). Recent evidence from anadromous Arctic charr (*Salvelinus alpinus*) demonstrates an increase in activity of several hepatic enzymes concurrent with a reduction in liver glucose and glycogen (Aas-Hansen *et al.*, 2005), supporting the concept of smolts as glucose deficient. The results of Stefansson *et al.* (2003a, see below) support this view, and demonstrate that smolts have low energy reserves during the early marine phase.

Lipid metabolism is altered during smoltification resulting in reduced body and muscle lipid (Saunders & Henderson, 1970, 1978), often observed as a reduction in condition factor. Studies of chinook salmon (*Oncorhynchus tshawytscha*, Cowley *et al.*, 1994) and coho salmon (Sheridan *et al.*, 1985; Sheridan, 1986) suggest that this depletion of lipid stores is caused by increased lipolysis and decreased fatty acid synthesis. In coho salmon parr, Sheridan (1986, 1994) found GH to increase triacylglycerol (TAG) lipase activity, whereas GH treatment of Atlantic salmon smolts failed to raise plasma FA levels (McKeown *et al.*, 1975) or hepatic lipid mobilisation (Sheridan, 1986). Possibly, elevated endogenous GH levels during smoltification have already increased the rate of lipolysis to a level at which GH treatment is ineffective. The increased activity of several lipases during smoltification (Sheridan, 1994) suggests that the lipid depletion is not a function of reduced feeding, but rather an integral part of the changes preparing the smolt for migration and marine life. Changes in specific lipid classes may also be an important part of smolt development. Li & Yamada (1992) observed a proportional decrease of TAG and an increase of phospholipid (PL) in the gut and gill of masu salmon (*Oncorhynchus masou*) smolts in seawater and concluded that the increased proportion of (n-3) polyunsaturated fatty acids (PUFA), particularly in polar lipids of osmoregulatory organs (gut, gill) was critical for seawater adaptation.

As for other preparatory changes associated with smoltification, the changes in lipid metabolism are reversible, and qualitative and quantitative reversions to a pre-smolt stage are observed if smolts are maintained in fresh water. In a study of tissue lipid composition of smolts, seawater post-smolts and desmoltified masu salmon, Li & Yamada (1992) concluded that lipid contents of muscle, liver, gut and gills of desmolts increase compared with smolt levels, and further, that the proportion of TAG increases while PL decreases in tissue lipids of desmolts compared with smolts. As would be expected based on the discussions above, tissue lipids of desmolts contain more mono-unsaturated fatty acids and saturated fatty acids and less PUFA, especially (n-3) PUFA such as 22:6(n-3), than those of smolts in fresh water and post-smolts in seawater.

Freshwater prey organisms are rich in 18:2(n-6) and 18:3(n-3) fatty acids in contrast with the marine diet which is rich in the longer chain highly unsaturated fatty acids (HUFA) such as 20:5(n-3) and 22:6(n-3), hence marine fish differ from freshwater fish in their fatty acid composition, and are generally characterised by higher proportions of PUFA (Sargent *et al.*, 1999). It is important to note that smolts initiate these preparatory changes in both fatty acid metabolism and tissue fatty acid composition while still in fresh water. This includes an increase in fatty acid elongase and desaturase activity of isolated hepatocytes as well as increases in the ratio of C20 and C22: C18 and n-3: n-6 fatty acids in salmonid tissue (Tocher *et al.*, 2000, 2003). HUFA biosynthesis in Atlantic salmon reaches a transient peak activity at the time of completion of smoltification and seawater transfer (Tocher *et al.*, 2003; Zheng *et al.*, 2004). Consistent with this, the gene expression of $\Delta 6$ desaturase, the rate-limiting step in the HUFA biosynthetic pathway, is highest around the point of seawater transfer.

In addition to the changes occurring in the main lipid stores (muscle, mesenteric fat), changes have been described in other tissues, such as gills, during the early seawater phase (Handeland *et al.*, 2000). This occurs even without a change in diet, suggesting that these changes in lipid composition are related to functional changes in the gill tissue (e.g. gas-transfer, ion regulation, acid-base balance). An important smolt-related feature of gill lipid metabolism is the transient reduction in prostaglandin F (PGF) production at seawater transfer (Bell *et al.*, 1997; Tocher *et al.*, 2003), with peaks of activity before and after transfer (Bell *et al.*, 1997). Whether this is a developmental (Bell *et al.*, 1997) and/or seasonal (Tocher *et al.*, 2003) phenomenon remains to be fully elucidated. Prostaglandins are known to modulate adaptation to salinity alteration in ectothermic animals (Mustafa & Srivastava, 1989). Gill prostaglandin F2α (PGF2α) production increases in Atlantic salmon post-smolts with increased gill polar lipid arachidonic acid (ARA) and decreases eichosapentinoic acid, EPA/ARA ratio (Bell *et al.*, 1997a; Tocher *et al.*, 2000; Bell & Sargent, 2003). In addition, fish with higher ARA levels in the gill PL tend to adapt better to seawater, implicating the importance of feed composition in stimulating preparatory changes in smolts and improving their capacity to respond appropriately to salinity change (reviewed by Bell & Sargent, 2003). Changes in PGF production may also have consequences for disease susceptibility of smolts, which are characterised by changes in serum proteins, immunoglobulin M (IgM) levels and leucocyte populations, which may affect cellular and humoral immunity (Maule *et al.*, 1989; Melingen & Wergeland, 2000; Melingen *et al.*, 2002; see also Zapata & Cortés, this volume).

The results of Stefansson *et al.* (2003a) on wild migrating smolts in river and coastal waters support the models outlined above, and further demonstrate that wild smolts have low energy reserves also during the early marine phase. Lipid and glycogen content decreased towards depletion from the river, through the fjord and into the ocean. There was, however, no significant change in protein content. These results confirm that smolts are naturally 'energy deficient' during downstream migration, and suggest that post-smolts also mobilise energy reserves during their early marine phase. Plasma GH levels increased transiently during passage through the estuary and fjord, with lower levels observed in post-smolts caught off-shore, i.e. in fish which were feeding on marine prey and had adapted to the marine environment. Taken together, these findings suggest that post-smolts maintain a positive protein balance, and that there is a high scope for growth during this period, at the expense of energy stores. This model is consistent with increasing RNA:DNA ratios in wild post-smolts from estuary through fjords and into the ocean (Haugland, Stefansson, Bjørn, Finstad & Hvidsten, unpublished data).

Ion/Osmoregulation

Perhaps the single most critical element of smoltification is the development of hypo-osmoregulatory ability (Fig. 1). Prior to smolt development parr have relatively poor seawater tolerance; and direct transfer of parr to seawater results in high levels of ionic and osmotic perturbation, increased mortality and reduced growth. These responses are only partly due to the size difference between parr and smolt, since fish large enough to become smolts but prevented from doing so by photoperiod manipulation have reduced hypo-osmoregulatory ability and growth in seawater (McCormick *et al.*, 1987, Stefansson *et al.*, 1991, 2007). The higher salinity tolerance of smolts permits rapid movements through estuaries and into coastal waters and reduces predation (Handeland *et al.*, 1996).

All teleosts maintain a relatively narrow range of osmotic pressure in their extracellular fluid, generally 260-360 mOsm kg^{-1}. Where salmonids are found, fresh water is normally 1–10 mOsm kg^{-1}, and fish counteract the passive gain of water and loss of ions by having the kidney produce an abundant and dilute urine, and by actively taking up salts across the gill. Salts may also be absorbed from food in the intestine. In seawater (~1000 mOsm kg^{-1}), fish reverse the passive loss of water and gain of ions by drinking seawater, absorbing water and monovalent ions across the gut and excreting Na^+ and Cl^- over the gills and divalent ions by the gut and kidney.

The role of chloride cells in ion transport

In juvenile and adult teleosts, the gill is a critical site for ion regulation. It is also the site for a wide range of other critical physiological functions for maintaining homeostasis. It is the site for gas exchange, ammonia excretion, and plays a critical role in acid/base balance (see Brauner, this volume; Pelster, this volume; Terjesen, this volume). In the context of ion regulation and acid base balance, the mitochondrion-rich chloride cells (CC, Fig. 2) play a key role in uptake and excretion of Na^+, Cl^-, HCO_3^- and H^+. A detailed discussion of ion regulation in fish larvae can be found in Kaneko & Hiroi (this volume), however, we have included a brief summary, based on McCormick (2001), Marshall (2002), Evans *et al.* (2005), of current models for transport of monovalent ions.

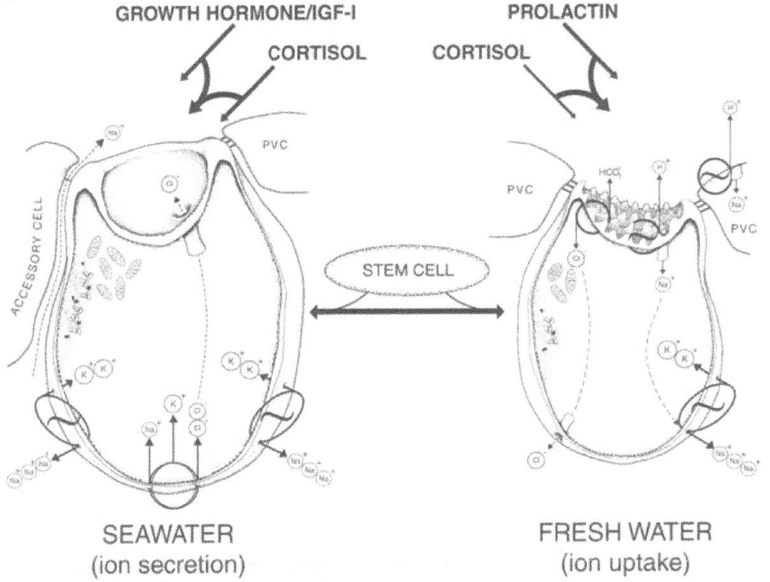

Fig. 2 Morphology and transport mechanisms of gill chloride cells in seawater and fresh water. See text for details of transport mechanisms. Chloride cells are characterised by numerous mitochondria and an extensive tubular system that is continuous with the basolateral membrane. In seawater, chloride cells are generally larger and contain a deep apical crypt, whereas in freshwater the apical surface is broad and contains numerous microvilli. In some species, such as tilapia the H^+-ATPase and apical sodium channel may be present in pavement cells rather than chloride cells. Recent evidence suggests that individual chloride cells can move between these two morphological states (Hiroi *et al.*, 1999), and also arise from undifferentiated stem cells (Wong & Chan, 1999; Kaneko & Hiroi, this volume). Growth hormone and cortisol can individually promote the differentiation of the seawater chloride cell, and also interact positively to control epithelial transport capacity. Prolactin inhibits the formation of seawater chloride cells and promotes the development of fresh water chloride cells. Cortisol also promotes acclimation to fresh water by maintaining ion transporters and chloride cells, and by interacting to some degree with prolactin. PVC = pavement cell. After McCormick (2001).

Mechanisms of Ion uptake in freshwater

To remain hyper-osmotic in fresh water, fish take up ions (Na^+, Cl^-, Ca^{2+}) from the water by active and secondary active transport mechanisms to compensate for passive ion loss, and produce large volumes of dilute urine to balance the osmotic gain of water. Ion uptake across the gill is an active area of research, and there remains some uncertainty as to the mechanisms involved among teleost groups and in different fresh water environments. Current consensus suggests that Na^+ moves across the apical membrane either through exchange with H^+ (NHE) or through an epithelial Na^+-channel (ENaC) energised by H^+-ATPase. Na^+ is then moved across the basolateral membrane by NKA. Apical Cl^-/HCO_3^- exchange mechanisms have been found in the apical membrane of CCs, implicating HCO_3^- excretion as a driving force for Cl^- uptake. The channels or transporters involved in movement of Cl^- across the basolateral membrane and into the blood have yet to be characterised, but probably involve the CFTR channel.

Mechanisms of Ion excretion in seawater

Marine teleosts drink seawater to replace osmotic water loss, and excrete excess Na^+ and Cl^- across the gills through CCs. Basolateral NKA provides both chemical and electrical gradients for Na^+ and Cl^- secretion. By transporting 3 Na^+ for 2 K^+, NKA generates an intracellular electronegative gradient (approx. –100 mV) low in Na^+, whereas the blood becomes high in Na^+ and positive. The large Na^+ gradient created by NKA allows a basolateral NKCC to shuttle Cl^- into the CC. Cl^- can then leave the highly negative CC on a downhill electrical gradient through and an apically located CFTR anion channel. Na^+ exits through a paracellular pathway between the CC and its accessory cells on a downhill electrical gradient (also created by NKA). The mechanistic basis of the paracellular Na^+ pathway has yet to be investigated in fish (for review, see Evans *et al.*, 2005).

Development of salinity tolerance: cellular and biochemical basis

The development of salinity tolerance is functionally linked to increases in transport proteins and the number and size of CC. Seawater CCs are characterised by an extensive tubular system, continuous with the basolateral membrane, creating a large surface area for transport proteins, and the presence of a large apical crypt (McCormick, 2001; Marshall, 2002, Fig. 2). The appearance of these morphological features of seawater CCs have been shown to develop in spring in smolts while still in fresh water, and to be absent in parr (Payan *et al.*, 1984; Pisam *et al.*, 1988; Lubin *et al.*, 1989). Seawater CCs are also larger than those in fresh water, and numerous studies have shown that the number and size of CCs increase during smoltification but subsequently decrease (along with salinity tolerance) if fish remain in fresh water after smoltification (McCormick, 2001).

Gill NKA activity and protein abundance has been shown to increase during the spring and to be much higher in smolts compared to parr. There is a strong genetic component to NKA expression that can explain performance difference between strains (Nilsen *et al.*, 2003, Mackie *et al.*, 2005). NKA consists of three subunits (α, β and γ), and several isoforms of each of these subunits have been found in vertebrates. Richards *et al.* (2003) found that the α1a subunit mRNA levels decrease and the α1b subunit mRNA levels increase during seawater acclimation of rainbow trout. In Atlantic salmon, α1a mRNA levels decrease and α1b mRNA levels increase during smolt development (Nilsen *et al.*, 2007). These data suggest a possible differential role of these two isoforms in fresh water and seawater, though to date the protein levels, localisation and physiological significance of these isoforms have yet to be characterised.

Other ion transporters

In salmonids, NKCC has been localised to the basolateral membrane of CCs (Fig. 2), and appears to be regulated by the same developmental program of CC differentiation as NKA. The abundance of NKCC protein shows a clear increase during smoltification of Atlantic salmon (Pelis *et al.*, 2001). Similarly, the number of NKCC-immunoreactive chloride cells also increases during smolt development and decreases coincident with loss of salinity tolerance. Photoperiod manipulation results in altered timing of NKCC levels, and correlates with coincident changes in NKA and salinity tolerance (Stefansson *et al.*, 2007). Nilsen *et al.* (2007) found that NKCC abundance and mRNA levels were higher in anadromous relative to land-locked populations of Atlantic salmon.

Atlantic salmon express at least two isoforms of the CFTR anion channel, CFTR I and II (Singer *et al.*, 2002). CFTR I and II mRNA levels increase in Atlantic salmon smolts in response to seawater, though the latter only transiently (Singer *et al.*, 2002). Increased CFTR mRNA levels have also been found during smoltification, and are higher in anadromous relative to landlocked strains of Atlantic salmon during spring (Nilsen *et al.*, 2007). Gill CFTR I mRNA levels also decrease during disruption of normal smoltification by exposure to acid/aluminum (Monette, McCormick and Yada, unpublished data) or estrogenic compounds (Lerner, McCormick and Yada, unpublished data).

The role of the intestine in hydro-mineral balance

The increase in ionic concentration, prey availability and composition, and scope for growth associated with marine life require alterations in uptake and transport mechanisms in the intestine. A functional regionalisation of the intestine appears to be part of this preparatory adaptation for life in the marine environment. In fresh water, drinking rate is low and ions are absorbed from dietary sources, in addition to an active uptake across the gills, to replace losses to the hypotonic medium. During smoltification and seawater acclimation, changes take place which are characteristic of marine teleosts. The fluid transport rate (J_v) of the anterior and mid-sections of the intestine decrease from parr levels of 15-20 μl cm^{-2} h^{-1} to below 10 μl cm^{-2} h^{-1}, while the posterior sections show a distinct increase from typical parr levels of 10-15 μl cm^{-2} h^{-1} to 20-25 μl cm^{-2} h^{-1} in smolts, causing an overall 2-3 fold increase in intestinal fluid absorption in smolts (Collie & Bern, 1982; Veillette *et al.*, 1993, 1995, Sundell *et al.*, 2003). This increase in fluid transport is functionally linked to increased NKA activity (Veilette *et al.*, 1993) and epithelial paracellular permeability (Sundell *et al.*, 2003). Acclimation to SW involves increased drinking rate (Usher *et al.*, 1988; Nielsen *et al.*, 1999), a further increase in intestinal NKA activity, generally concomitant with a further increase in fluid absorption of the posterior intestine beyond the freshwater preparatory levels (Veillette *et al.*, 1993), and a shift in intestinal water uptake from a paracellular to a transcellular pathway (Sundell *et al.*, 2003). The discovery that aquaporins are recruited in the intestine of marine adapted eels (Lignot *et al.*, 2002), suggests that a similar mechanism of transcellular water uptake may be present in salmonids. Cortisol appears to be a major endocrine factor controlling these changes in intestinal fluid transport (Veillette *et al.*, 1995, 2004; Sundell *et al.*, 2003). In addition to their role in nutrient uptake, recent evidence also indicates that the pyloric caeca are important sites for hydro-mineral balance in smolts (Nielsen *et al.*, 1999; Veilette & Young, 2004), since they show similar changes in NKA activity and responsiveness to cortisol to that of the posterior intestine.

NEUROENDOCRINE CONTROL

The endocrine system and how it regulates smoltification processes has been studied for more than six decades and has been the subject of numerous reviews on growth, osmoregulation, imprinting, metabolism and its regulation (Hoar, 1988; McCormick *et al.*, 1998). What has been less studied and is emerging now as a key element in the smoltification process is the development of the regulator of the endocrine system, the brain (Ebbesson *et al.*, 2003; Ebbesson *et al.*, 2007). Here we briefly present the LBP and illustrate how structural changes in the LBP occur prior to the characteristic endocrine surges, indicating that these changes in the LBP may make the neuroendocrine system more photosensitive. Further, we discuss brain changes during smoltification and where possible relate their individual roles to the smoltification processes.

Light-brain-pituitary Axis and Endocrine Influence

Environmental information is transduced through sensory processes to neuroendocrine responses in the brain, that subsequently control pituitary function and ultimately physiological responses through several endocrine axes. The LBP is the pathway by which photoperiod information is conveyed to the endocrine system (Fig. 3). Light is detected by the retina and pineal organ, which transmits this information through neurons projecting to central brain regions, one of which is the preoptic area (POA). Here the information is integrated with other neural inputs and relayed to the pituitary, regulating hormone release into circulation (Holmqvist *et al.*, 1992). Of the wide range of hormones known to be involved in smoltification, cortisol, GH and TH are key regulators that display sequential increases in circulating levels that promote their transformational roles (Hoar, 1988; McCormick *et al.*, 1998; Björnsson *et al.*, 2002).

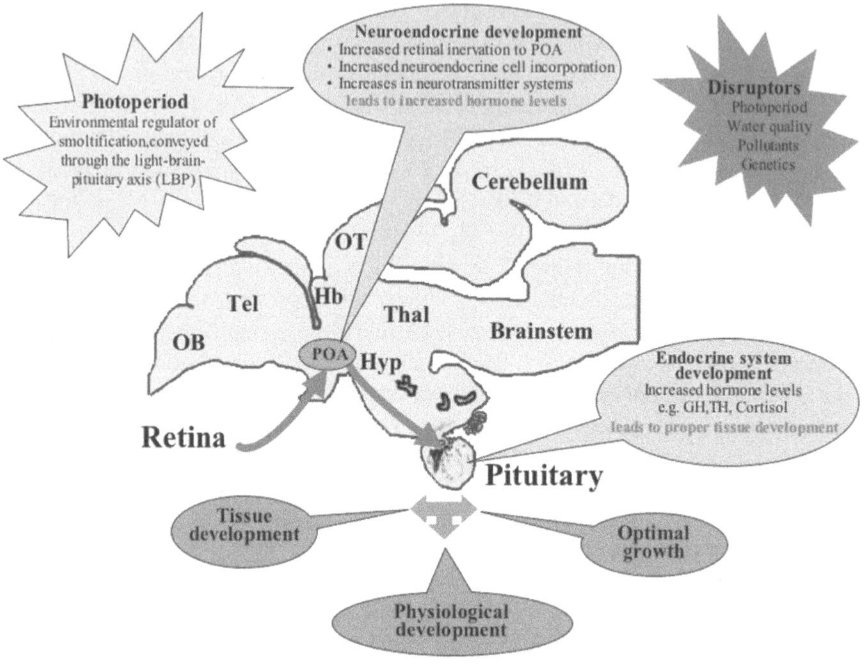

Fig. 3 An overview of the light-brain-pituitary axis that illustrates the role of brain changes during smoltification on endocrine and physiological development. Abbreviations: OB, olfactory bulb; Tel, telencephalon; Hb, habenula; POA, preoptic area; OT, optic tectum; Hyp, hypothalamus; Thal, thalamus; GH, growth hormone; TH, thyroid hormone.

Thyroid stimulating hormone (TSH) is released from the pituitary and stimulates the production and release of TH. GH is released from the pituitary into the circulation and ACTH regulates the production and release of cortisol from the head kidney (Holmqvist *et al.*, 1994; Holmqvist & Ekström, 1995; Baker *et al.*, 1996; Holloway & Leatherland, 1998; Ágústsson *et al.*, 2000). The releases of TH, GH and cortisol are directly or indirectly regulated by complex negative and positive feedback systems, including the actions of DA, GnRH, serotonin (5-HT), corticotropin releasing factor (CRF), thyrotropin releasing hormone (TRH), opiate peptides, or arginine vasotocin (AVT, Eales, 1995; Baker *et al.*, 1996; Björnsson *et al.*, 2002; Canosa *et al.*, 2007).

Smolt-related increases in circulating hormones have shown that they often act synergistically to affect hypo-osmoregulatory capacity, morphological changes (silvering of scales and skin), metabolic changes and schooling behaviour. For example, cortisol, in addition to its overall role in energy metabolism, regulation of protein turnover and lipid mobilisation, has important roles in the proliferation and differentiation of gill CC, gill NKA, NKCC and salinity tolerance. These functions are mediated through the interactions of cortisol with GH, IGF-I and corticosteroid receptors (Björnsson *et al.*, 1995; Björnsson, 1997; Seidelin *et al.*, 1999; McCormick, 2001). THs have also been shown to be essential in many smoltification related processes including silvering, olfactory imprinting, behaviour, and changes in visual sensitivity (Morin *et al.*, 1995; Hutchison & Iwata, 1998; Specker *et al.*, 2000; Power *et al.*, 2001; Lema & Nevitt, 2004; Allison *et al.*, 2006). Despite these studies, mechanisms of TH action during smoltification are still largely unresolved, most likely due to the complex TH regulation (Eales *et al.*, 1993; Morin *et al.*, 1993; Eales, 1995), changes in the system (Richardson *et al.*, 2005) and their often synergistic role in many of these key processes. THs are known to act synergistically with a range of other hormones, including cortisol, GH and prolactin (PRL), likely through modulation of receptor characteristics, as is shown for the corticosteroid receptors (McCormick, 2001). The changes in hormone levels have been studied extensively, yet only recently has the development of the neuroendocrine system become a focus and revealed its critical role in opening a 'gate' that increases neuroendocrine responsiveness to environmental photic inputs (see Fig. 5; Ebbesson *et al.*, 2003, 2007; McCormick *et al.*, 2007).

Neuroendocrine Development

Our recent data emphasise that the structural reorganisation of specific brain regions occurs prior to the major hormone surges that regulate the smoltification processes (Ebbesson *et al.*, 2003), while changes in neurotransmitter systems occur throughout smoltification regulating neuroendocrine and behavioural processes (Ebbesson *et al.*, 1996a, b; Morin *et al.*, 1997). Neural tract tracing has shown that some of these changes involve increased innervation of retinal and pineal fibres into POA and other brain regions (Holmqvist *et al.*, 1994; Ebbesson *et al.*, 2003, 2007), and a specific period of cell differentiation, as shown by transient growth-associated protein-43 (GAP-43) immunoreactive cells and fibres (Ebbesson, 2000; Ebbesson *et al.*, 2003). These structural reorganisations of the brain occur prior to the major increases in circulating TH, GH and cortisol levels, pivotal hormones in the physiological processes of smoltification (Ebbesson *et al.*, 2003). These data support the model that increased retinal innervation of the POA facilitates the increased endocrine response to changes in photoperiod (Fig. 4).

In addition to structural reorganisation during smoltification, there are changes in neurotransmitter systems including the transient appearance of serotonergic neurons in the habenula and the lateral POA (Ebbesson *et al.*, 1992) and the appearance of GnRH in the basal forebrain and midbrain of chum salmon (Lewis *et al.*, 1992; Parhar & Iwata, 1996). There are also sequential changes of certain neurotransmitter levels, e.g. GnRH, 5-HT, DA and norepinephrine (Lewis *et al.*, 1992; Ebbesson *et al.*, 1996b) in specific regions of the brain. Some of these systems are affected by THs (Morin *et al.*, 1997) and environmental/rearing conditions (Ebbesson *et al.*, 1994). There are also temporal changes in opiate receptors (Ebbesson *et al.*, 1996a) and TH receptors (Kudo *et al.*, 1994) in specific brain regions. It is hypothesised that thyroidal status in the brain affects monodeiodinase activities (Eales *et al.*, 1993; Morin *et al.*, 1993) that appears to be related to olfactory sensitivity (Morin *et al.*, 1994). Additional changes of the LBP axis include changes in retinal UV photosensitivity and visually-evoked electrophysiological responses in the telencephalon and *torus semicircularis* (Hofmann & Meyer 1993; Browman & Hawryshyn, 1994). Such data suggest that neural development during critical periods of development precede the physiological changes that direct life histories and fitness. At this point our understanding of larval and juvenile brain development in salmon is well underway, but the effects of undesirable environmental influences on brain development and smoltification during these critical periods in salmon are only partly known.

In salmon, as in other vertebrates, the life cycle includes specific critical periods of neural development (e.g. early development, smoltification, and sexual maturation). Disruption of the developmental program profoundly impacts the future outcome. These life history stages are finely tuned and sensitive to environmental stimuli that synchronise the proper timing of the development processes to maximise survival (McCormick *et al.*, 1998). Genetic specification determines much of the basic structure and function of the nervous system, however, the environment to a great extent affects the timing of development of neural pathways and functions. These developmental processes are guided by evolutionary and ontogenetic principles, partly explained by 'the parcellation theory' (Ebbesson, 1980). Thus, once connections are established, the survival of the new connections are dependent on the subsequent environmental stimuli, moulding the brain, by maintaining stimulated synapses, and losing unused connections to establish functional circuits in order for the animal to

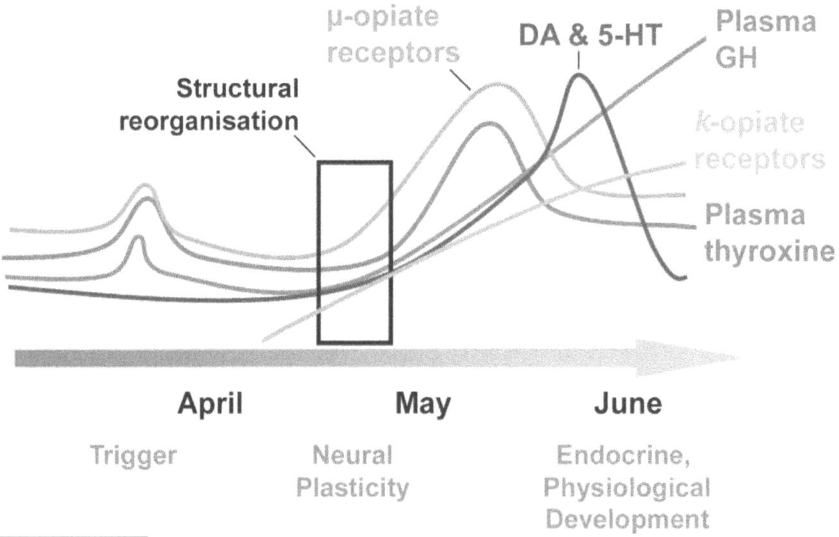

Fig. 4 Schematic diagram showing temporal changes in brain and endocrine systems during salmon smoltification. Early structural brain changes (box; e.g. retinal tract tracing) are followed by changes in neurotransmitter systems (e.g. opiate receptors, dopamine, DA and serotonin, 5-HT) and surges in circulating hormone levels (e.g. thyroxine, growth hormone, GH). After Ebbesson *et al.* (1992), (1996a, b), (2000) and (2003).

progress into the next stage of life. Indeed, the tailoring of neuronal circuits to the requirements of each individual may be considered to be the basis of critical periods and in salmon can lead to a variety of life strategies. Recent evidence suggests the importance of normal brain development during smoltification to attain the necessary endocrine and physiological changes (Ebbesson *et al.*, 2007).

PHYSIOLOGICAL BASIS FOR MORPHOLOGICAL CHANGES

Increased silvering is probably the best known external characteristic of salmonid smoltification (Fig. 1). The overall silvery colour characteristic to the smolt is caused by the deposition of purines (guanine and hypoxanthine) in the scales, scale layer and *stratum argentum* layer of the skin (Johnston & Eales, 1967; Staley & Ewing, 1992). Both layers are thicker in smolts than in parr. Consequently, the characteristic parr-marks ('finger marks') are obscured by layers of silvery skin and scales, however, they remain in the deeper skin layers and may become visible if the outer skin layers are eroded. Concurrent with the increased silvering, the juvenile salmon develops black fin margins, particularly on the dorsal, caudal and pectoral fins, while the fins become lighter, almost translucent. These changes are easily observed when schools of smolts are viewed either in nature or in culture. There is no firm evidence for the adaptive significance of changes in fin colouration, however, it can be hypothesised that the contrasting fin margin, especially of the caudal fin, contributes to visual signalling within a school of smolts.

The scales of smolts are looser than in parr, and are therefore easily dislodged during handling. It is difficult to propose any adaptive significance of loose scales, rather, this phenomenon should perhaps be viewed as a consequence of the increased growth rate observed in smolts. Evidence from the culture of Atlantic salmon suggests that loose scales are observed during periods of rapid growth, hence it is suggested that this may be related to the turnover of connective tissue associated with growth processes. From a practical point of view, scale losses from handling, netting and transportation of smolts will reduce the barrier towards osmotic disturbances and increase the osmoregulatory challenge associated with seawater entry.

The increased deposition of purines in the body may represent an alternative biochemical route for nitrogen metabolism which serves a dual purpose (Hoar, 1988; see also Terjesen, this volume). Firstly, by permanently storing purines in the body, no further energy consuming enzymatic oxidation to soluble products is required, hence water is not wasted in excreting such compounds. This may be regarded as an integral part of adapting to the hyper-osmotic conditions of seawater. Secondly, purine storage in the skin contributes to the adaptive counter-shading camouflage characteristic of pelagic schooling fish species. The endocrine basis for the changes in protein metabolism which cause the increased silvering is still not fully resolved, however, activation of the TH axis is known to increase purine deposition in teleosts. TH-induced changes are generally not as intense as the silvering observed in normal smolts, suggesting synergy with other hormones. GH has been implicated in stimulating silvering as has been seen in eels (Durif *et al.*, 2005). However, in a recent study using long-term GH-implants in Atlantic salmon parr, the silvering effect was found to be size-dependent rather than GH-dependent (Neregård *et al.*, unpublished data). The importance of these synergies are further supported by the observations that THs are ineffective in inducing the development of black fin margins, rather increased ACTH (but not cortisol) has been shown to cause the darkening of the dorsal surface and fin margins of juvenile Atlantic salmon (Langdon *et al.*, 1984).

ENVIRONMENTAL REGULATION

Photoperiod

From the late 1960s extensive research has been carried out to define the environmental control of smoltification. In several salmonids, including the Atlantic salmon, the most important environmental cue for the synchronisation of these processes is the seasonal changes in photoperiod. Concurrent with these findings, studies have suggested that important physiological changes related to smoltification, e.g. increased growth rate, seawater tolerance, silvering and reduced condition factor may occur independently and in the absence of environmental signals (zeitgeber; Eriksson and Lundqvist, 1982). Such endogenously driven circannual rhythms will, however, generally be unsynchronised, as the circannual rhythms are not sufficiently precise in the absence of a zeitgeber. The different aspects of smoltification may occur out of phase and out of step with the time of year. Further, the intensity of such changes is lower than in smolts, and despite a certain degree of silvering and somewhat improved seawater tolerance (Fig. 1, see further below), such incompletely smoltified juveniles often fail to adapt normally to seawater, hence survival and growth may be highly variable and disease outbreaks and mortalities often occur. Signals from the environment are needed to synchronise the processes, and juvenile Atlantic salmon detect and respond to changes in daylength. There is a large literature on the impact of normal and artificial photoperiods on smolt development (see references in Stefansson *et al.*, 2007).

Aspects of the biological clock have been studied in model teleosts such as zebrafish, and evidence points to conserved elements in the organisation of biological time-keeping in teleosts as in other vertebrates. In the context of smoltification, postulated circannual clocks are likely the most important as they translate changes in daylength into a physiological response. In contrast to the circadian system, hardly any information is available on the circannual time-keeping system in lower vertebrates. The current vertebrate model predicts the presence of integrative zones, calendar cells, which translate seasonal and diel changes in melatonin levels into long-term changes in physiology. Although very limited information is available for salmonids, studies of the rainbow trout have shown the expression of *clock*, a key time-keeping gene, in γ-aminobutyric acid (GABA) neurons which also express the melatonin receptor, localised to specific regions of the brain. These regions are suggested to be the sites at which environmental information is integrated into the biological clock to influence behavioural and endocrine responses (Mazurais *et al.*, 2000).

Growth rate is also significantly influenced by photoperiod, being high in spring and early summer as daylength is increasing, with a subsequent reduction later in summer and autumn as daylength decreases. Growth rate often shows better correlation with photoperiod than with temperature. In Atlantic salmon culture, long photoperiod and continuous light may be used to enhance growth rate of juvenile salmon in freshwater, and is routinely used during first-feeding, and to maintain juvenile growth during autumn and early winter. Correct use of photoperiod manipulation is used to improve survival during first-feeding and to increase subsequent growth rate, number of potential 1+ smolts and smolt size. Extended use of continuous light will, however, interfere with the completion of smoltification, as such unnatural photoperiods through winter and spring desynchronise the smoltification process (Fig. 1, McCormick *et al.*, 1987; Stefansson *et al.*, 1991). Such 'pseudo-smolts' will differ in smolt status at the time of release, and seawater performance will, at best, be variable. Recent evidence on the importance of a natural photoperiod for the completion of smoltification, and the failure of juvenile salmon to develop under constant light conditions demonstrate that smoltification is disrupted at a fundamental level (Ebbesson *et al.*, 2007; Stefansson *et al.*, 2007).

Parr exposed to continuous light, LL, do not develop the same extensive retinal innervation to the preoptic nucleus (NPO, Ebbesson *et al.*, 2007) that is seen in control Atlantic salmon smolts raised under natural or simulated natural photoperiod (Fig. 5, Ebbesson *et al.*, 2003, 2007). These retinal fibres, in part making up the light-brain section of the LBP, terminate on neurohypophysiotropic cells (e.g. CRF and AVT) in the NPO that either directly or indirectly regulate the release of pituitary hormones (e.g. GH, TSH, ACTH) or directly release the neurohormones into the circulation via the pituitary (e.g. AVT and oxytocin; Holmqvist *et al.*, 1994; Ebbesson *et al.*, 2003). Recent data show that, in addition to the lack of increased retinal innervation in the LL parr, they also do not increase the number of CRF producing neurones to the same extent as controls (Ebbesson, unpublished observations). These central changes provide potential pathways for increased signalling through the LBP that increase the responsiveness to photic cues resulting in the elevation of circulating hormone levels essential for smoltification and seawater adaptation. Accordingly, juveniles reared on LL largely failed to show the smolt-related physiological changes (Fig. 1, Stefansson *et al.*, 2007). Gill mRNA levels of *nka* isoforms $\alpha 1a$ and $\alpha 1b$ remain unchanged, as does NKCC. Consequently, gill NKA activity and NKAα and NKCC protein abundance are significantly lower in LL pseudo-smolts than in true smolts. The endocrine system also remains under-developed, with lower circulating levels of THs, GH and cortisol and lower expression of gill GH receptor. Taken together, deprivation of photoperiod information, e.g. continuous exposure to LL, disrupts smoltification by preventing the natural developmental program. Such photoperiod disruption has wide-ranging physiological consequences, suggesting that the photoperiod acts on key elements of the central regulatory system through its key role in the LBP.

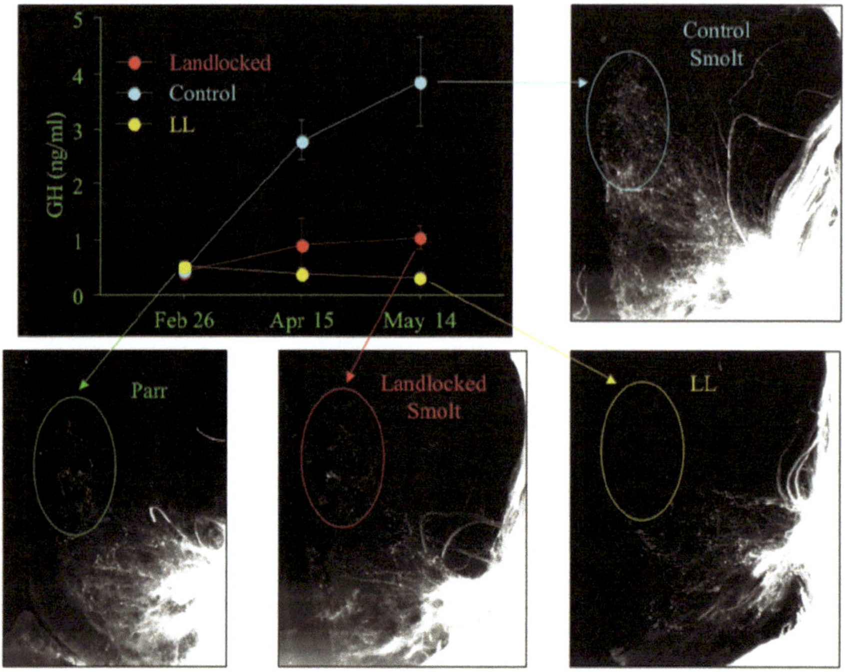

Fig. 5 Circulating growth hormone (GH) levels increase from parr in February to smolt in May in anadromous Atlantic salmon (*Salmo salar*), while they do not increase in either the continuous light (LL) or landlocked groups (A). The differences in hormone levels may be explained by degrees of brain development during this period. Retinal innervation into the preoptic nucleus (oval) in Atlantic salmon increases from parr in February (C) to control smolt in May (B). This increase is not observed in May in the landlocked salmon (D) or juveniles exposed to continuous light (LL, E). After Ebbesson *et al.* (2007) and Stefansson *et al.* (2007).

In Atlantic salmon culture, special attention has been given to the prospects of reducing the duration of the freshwater phase and transferring fish to sea out of season. Under standard farming conditions, smolts will be available only in spring (as either 1+ or 2+). However, it is possible to control and manipulate environmental conditions (temperature and specifically photoperiod) in order to produce off-season smolts, which may be transferred to seawater in the autumn of their first year. These smolts are referred to variously as 'half-year', 'underyearling' or 0+ smolts. The principle of off-season smolt production is to provide the juvenile salmon with a phase-advanced photoperiod signal of sufficient duration and amplitude to stimulate the completion of smoltification. Two strategies have been successful, one being the use of a segment of a simulated natural increase from approximately LD12:12 to LD24:0 with the natural rate of change (Berge *et al.*, 1995), the other strategy involves a 'square wave' photoperiod with a reduction in daylength from LD24:0 to LD12:12, keeping the fish on LD12:12

'winter photoperiod' for several weeks, followed by a return to LD24:0, with monitoring of smolt development during the next few weeks (Arnesen *et al.*, 2003). The physiological changes associated with smoltification in 0+ smolts parallel those of ordinary 1+ smolts reared under commercial culture conditions (Sundell *et al.*, 2003), with improved seawater tolerance co-occurring with increased NKA activity (Berge *et al.*, 1995), elevated plasma GH (Björnsson *et al.*, 2000; Arnesen *et al.*, 2003), cortisol (Sundell *et al.*, 2003) and fully developed silvering and reduced condition factor.

Indeed, through judicious use of the two environmental control factors, photoperiod and temperature (see below), it is theoretically possible to produce successive cohorts of smolts throughout most of the year. In this scenario, specific temperature regimes can be used to either accelerate or delay growth while photoperiod can be used at the appropriate time to provide the necessary winter and spring signals to initiate and complete smoltification. The result will be a fully controlled, season-independent smolt production.

Temperature

Temperature is, together with daylength, the most important environmental factor controlling growth and developmental rates of juvenile salmonids. In natural environments, the annual temperature cycle defines the growth potential of the fish, and thereby the age at which juvenile salmonids reach the minimum size for initiation of smoltification. Accordingly, in European Atlantic salmon, smolt ages increases from 1+ in the southern part of their distribution range (northern Spain) to 4+ or even 5+ in the rivers of northern Norway and Russia. A similar picture is seen among Atlantic salmon populations in North America.

Temperature also influences the timing of smolt migration, directly and/or indirectly. In some rivers, smolt migration takes place as water temperature reaches a certain threshold temperature (e.g. River Ims, Jonsson & Ruud-Hansen, 1985). However, in contrast to the results suggesting a temperature threshold for initiation of downstream movement, recent findings by Zydlewski *et al.* (2005) suggest that d°C are a better indicator of the initiation and termination of downstream movement than is absolute temperature. Thermal cues are, however, probably not universal among salmon populations, as other river stocks appear to cue on to water flow and probably other environmental signals for the timing of their migration (see below). In rivers of northern Sweden, with ice cover and temperatures close to zero until late spring (end of April), it has been suggested that the time window for smoltification and downstream migration is compressed, with full hypo-osmoregulatory ability still developing after the fish have entered the low-salinity Baltic sea (Schmitz *et al.*, 1994). The sea temperature that smolts experience at the time they reach salt water may be a strong selective force that has operated to affect the timing of emigration from seawater. Moving from south to north in Norway, smolt migration co-occurs with a sea temperature of approximately 9°C, which normally happens in April in the south and in June in the north

(NA Hvidsten, personal communication). A temperature of approximately 9°C is also the temperature suggested by Handeland *et al.* (2000) to cause the least osmotic disturbance in Atlantic salmon smolts during acclimation to full-strength seawater.

Under controlled conditions, through its influence on overall scope for growth, temperature controls the distribution of fish between the two growth modes (see discussion on bimodality above). Higher growth rate stimulates recruitment to the UM, and increases the number and size of potential smolts. Temperature also influences smoltification directly, by controlling the rate of physiological, morphological and behavioural development (Fig. 6, Björnsson *et al.*, 1989; McCormick *et al.*, 2000). Although temperature regulates the rate of development, it cannot act by itself as an independent zeitgeber, but rather appears to control the rate of response to photoperiodic cues (Solbakken *et al.*, 1994; McCormick *et al.*, 2002). Handeland *et al.* (2004) estimated a period of approximately 350 d°C between the onset of the smolt-related increase in gill NKA activity of Atlantic salmon and the peak in enzyme activity, providing a useful model for the prediction of completion of smoltification under various rearing temperatures. In a similar manner, temperature influences the duration of the period of peak smolt characters, e.g. high seawater tolerance, often referred to as the 'smolt window' (Fig. 6). The smolt window may ultimately be regulated by environmental influences on the neuroendocrine system (see above). Smolt characters are lost sooner when smolts are held in freshwater at higher temperatures, and this de-smoltification eventually prevents the smolts from acclimating to seawater. In the study of Handeland *et al.* (2004) the duration of the smolt window, defined as the period when gill NKA activity was > 90% of peak value, was estimated at 350 d°C. Similar models have been presented by Sigholt *et al.* (1998), Stefansson *et al.* (1998) and McCormick *et al.* (1999) who have proposed the duration of the smolt window in the range of 300-400 d°C, depending on the definition of the threshold level. Interestingly, recent findings by Zydlewski *et al.* (2005) on the behavioural aspects of smoltification seem to concur with physiological observations, with the opening and closing of the migratory window occurring at approximately the same number of d°C, independently of absolute temperature.

At the extremes, temperature may limit or even inhibit the completion of smoltification. Low temperatures may prevent the fish from responding to the increasing photoperiod by limiting the endocrine responsiveness of key endocrine tissues and target organs (McCormick *et al.*, 2000). Using the models outlined above, a temperature from April through June around 4°C would provide the juvenile salmon with approximately 350 d°C, and may be hypothesised as a lower temperature limit for smoltification. However, Handeland *et al.* (2004, see Fig. 6) and our own unpublished results suggest that different salmon strains are adapted to different temperature conditions, and smolt development differs among strains held at the same temperature, hence absolute limits may be difficult to define.

Studies on steelhead trout have suggested an upper limit for the completion of smoltification in the range of 13-15°C (Zaugg *et al.*, 1972; Zaugg & Wagner, 1973; Zaugg, 1981), based on the suppression of gill NKA activity above these temperatures.

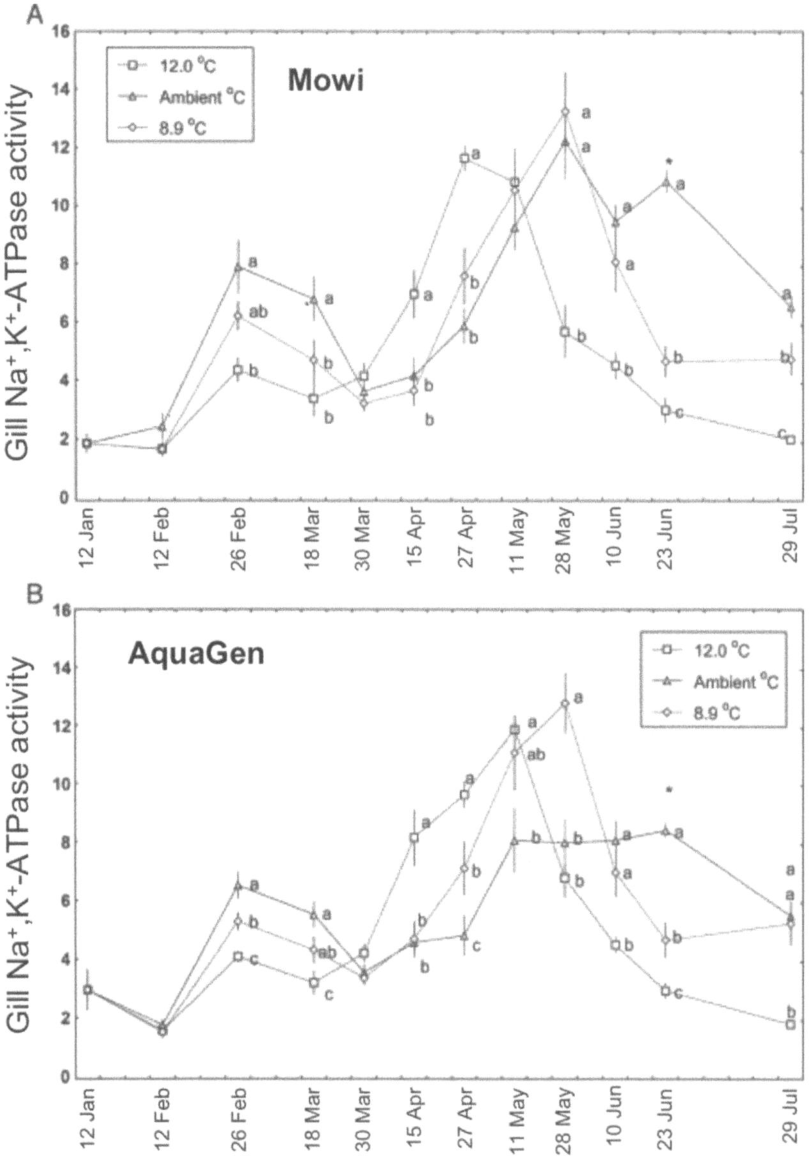

Fig. 6 Gill Na^+,K^+-ATPase activity in two farmed Norwegian strains of Atlantic salmon (*Salmo salar*) smolts (A: Mowi and B: AquaGen) at three temperatures (ambient, 8.9 and 12.0°C) during the period 12 January-29 July. Ambient temperature increased from 3°C in January to 10°C in June. Values are given as means (sem, n = 12). Significant differences among temperatures are indicated by different letters ($p < 0.05$). The asterisk indicates significant differences among strains ($p < 0.05$). After Handeland *et al.* (2004).

Other studies have failed to show such inhibition in other salmonid species. Atlantic salmon are able to adapt to seawater at temperatures as high as 19°C, although physiological stress and poor growth are observed at such high temperatures (Handeland *et al.*, 2000). Recent evidence has further shown that chinook salmon reared at temperatures above 17°C show impaired smoltification (Marine & Cech, 2004), suggesting that these temperatures are also restrictive for smoltification in this species. Taken together, studies to date indicate that temperature has a role in the timing of smolt development primarily through control of developmental rate, but that species-specific temperature limits may also exist.

Flow and Water Quality

The impact of flow on behaviour of Pacific and Atlantic salmon smolts has been reviewed previously (Hoar, 1988; McCormick *et al.*, 1998). Several studies have associated flow (water velocity) with smolt migration. Since smolts can move downstream both passively and actively, increased flow is likely to be detected as a factor in field studies of smolt migration simply through its effects on passive migration speed. The impact of flow on migration may relate to the initiation of migration and to the speed of migration. Fish may use other non-flow related signals such as water level, turbidity or novel chemical cues associated with flow events to initiate or increase their speed of migration.

Smolts are inherently more responsive than other life history stages of salmonids to a variety of external stressors, including poor water quality. Acid and associated aluminum toxicity have been shown to have severe impacts on smolts (Rosseland *et al.*, 2001). Direct mortality related to osmoregulatory failure in fresh water (loss of plasma ions) has been found at pH 5.0-5.5 with moderate to high aluminum levels (Staurnes *et al.*, 1996). At more moderate pH and aluminum conditions smolts survive in fresh water but incur a loss in gill NKA activity and salinity tolerance (Saunders *et al.*, 1983; Staurnes *et al.*, 1993). Reductions in salinity tolerance can occur within two days (Monette and McCormick, unpublished data) and are associated with loss of NKA activity and enzyme subunit disturbances (Kverneland, Nilsen and Stefansson, unpublished data). Although it remains to be demonstrated conclusively, aluminum exposure under acidic conditions may induce compensatory mechanisms for ion uptake that are in direct conflict with normal smoltification. Heavy metals also exert their toxic influence through impacts on ion regulation, and it stands to reason that smolts may be more sensitive than other life stages (or other species) due to the osmoregulatory change they undergo during smoltification, though to our knowledge this has not been investigated.

As an endocrine-driven developmental event, smoltification is susceptible to endocrine disruption by contaminants that act through hormone signalling pathways. It has been known for some time that androgens are detrimental to development of normal smolt behaviour and physiology (Berglund *et al.*, 1994). In a retrospective study, Fairchild *et al.* (1999) found that areas of New Brunswick, Canada that had been heavily sprayed for spruce budworm with a formulation that contained the xenoestrogen nonylphenol experienced reduced marine survival of smolts. Madsen *et al.* (1997) subsequently found that nonylphenol exposure can reverse the normal increases in gill CCs, NKA activity and salinity tolerance that occur during smoltification. These reductions in salinity tolerance are likely due to reductions in circulating levels of IGF-I (known to be important development mechanisms for salt secretion) that occurs after nonylphenol exposure (McCormick *et al.*, 2005). However, relatively high levels of aqueous nonylphenol are required to impact smolt development, levels that seem unlikely to occur in many river systems (Moore *et al.*, 2003). Lerner *et al.* (2007) recently found that yolk-sac larvae are much more sensitive to nonylphenol than other life stages of Atlantic salmon, and that exposure to moderate levels (6 µg/L) can result in reduced salinity tolerance and salinity preference of smolts a full year after exposure. For a recent review of the influence of xenobiotics on salmonid eggs and larvae, see Finn (2007).

Other organic pollutants may also impact smolts, either through endocrine disruption or through more 'classic' toxic actions. Polychlorinated biphenyl (PCB) exposure reduces salinity tolerance, salinity preference and plasma TH (Lerner *et al.*, 2006). Atrazine, a commonly used herbicide, has been found to reduce salinity tolerance of Atlantic salmon smolts (Waring & Moore 2004; Nieves-Puigdoller and McCormick, unpublished data). Contaminants may also interact to affect smolt development. Moore *et al.* (2003) found that atrazine at 1 µg/L or nonylphenol at 5 µg/L individually had no impact on smolts, but in combination the two compounds resulted in loss of salinity tolerance. A variety of contaminants interfere with olfaction in fishes (see Døving & Kasumyan, this volume), and compounds that affect olfaction may interfere both with the imprinting process that occurs primarily during smoltification, as well as the homing process during return of adults to their natal streams for spawning (Scholz *et al.*, 2000; Moore & Lower 2001; Jarrard *et al.*, 2004).

FUTURE PERSPECTIVES

Recent scientific progress concerning other anadromous species, e.g. American shad and striped bass (see Dadswell *et al.*, 1987, for a discussion of anadromous species), has raised the question of whether smoltification is unique to salmonids, or whether similar processes can be found among other families of teleosts. The preparatory nature of the smoltification process sets it apart from the widespread capacity among euryhaline teleosts to adapt to gradual changes in salinity. Recent evidence obtained from American shad, however, suggests that these fish develop salinity tolerance while still in freshwater, with a loss of freshwater adaptations during migration (Zydlewski & McCormick, 2001), akin in many ways to the osmoregulatory changes occurring in smolts. Further studies comparing preparatory adaptations to seawater in various anadromous and euryhaline species are needed to determine whether smoltification is indeed unique to salmonids and to provide a better understanding of the various physiological mechanisms used among teleosts in seawater adaptation.

Stefansson *et al.* (2003b) suggested specific research strategies that would be particularly fruitful for increasing our understanding of the smoltification process, including a better understanding of the variability in smoltification between and within strains, the importance of local adaptation for smolt quality and migration, challenges and actions under a global warming scenario, parameters defining and predicting smolt quality, and perspectives for improving hatchery smolt production. Further, functional genomics is another approach quickly making its way into smolt research. Large-scale gene expression/identification studies are likely to describe, on a very basic level of biological organisation, the control of smoltification. As more genes are identified in salmonids and other species and their products tied to specific functions (proteomics), an understanding of the biochemical and physiological processes involved in smoltification should emerge. Indeed, as we have pointed out, integrating information from other teleost models or target species will allow us to make rapid progress in traditional behavioural and physiological studies, as well as in functional genomic approaches to smoltification. Such work will hopefully also lead to development of useful tools such as microarrays for use in staging smolt development.

We would like to emphasise the importance of maintaining and strengthening the 'traditional' ecological, physiological, behavioural and environment-fish interaction studies at the organismal level. More work is required to understand the physiological and behavioural changes that occur in smolts in the wild and how these relate to survival and population dynamics. Such studies are critical for interpretation of findings at the molecular, cellular and organ levels, and are vital as input for the management of wild and reared salmon stocks.

References

Aas-Hansen Ø, Vijayan MM, Johnsen HK, Cameron C, Jørgensen EH (2005) Resmoltification in wild, anadromous Arctic char (*Salvelinus alpinus*): a survey of osmoregulatory, metabolic, and endocrine changes preceding annual seawater migration. Can J Fish Aquat Sci 62: 195-204

Ágústsson T, Ebbesson LOE, Björnsson BT (2000) Dopaminergic innervation of the rainbow trout pituitary and stimulatory effect of dopamine on growth hormone secretion in vitro. Comp Biochem Physiol 127A: 355-364

Allison WT, Veldhoen KM, Hawryshyn CW (2006) Proteomic analysis of opsins and thyroid hormone-induced retinal development using isotope-coded affinity tags (ICAT) and mass spectrometry. Mol Vision 12: 655-672

Arnesen AM, Toften H, Ágústsson T, Stefansson SO, Handeland SO, Bjömsson BT (2003) Osmoregulation, feed intake, growth and growth hormone levels in 0+ Atlantic salmon (*Salmo salar* L.) transferred to seawater at different stages of smolt development. Aquaculture 222: 167-187

Baker BI, Bird DI, Buckingham JC (1996) In the trout, CRH and AVT synergize to stimulate ACTH release. Regulatory Peptides 67: 207-210

Barbour SE, Garside ET (1983) Some physiologic distinctions between freshwater and diadromous forms of the Atlantic salmon, *Salmo salar* L. Can J Zool 61: 1165-1170

Barron, MG (1986) Endocrine control of smoltification in anadromous salmonids. J Endocrinol 108: 313-319

Behnke RJ (1972) The systematics of salmonid fishes of recently glaciated lakes. J Fish Res Bd Can 29: 639-671

Bell JG, Sargent JR (2003) Arachidonic acid in aquaculture feeds: current status and future opportunities. Aquaculture 218: 491-499

Bell JG, Tocher DR, Farndale BM, Cox DI, McKinney RW, Sargent JR (1997) The effect of dietary lipid on polyunsaturated fatty acid metabolism in Atlantic salmon (*Salmo salar*) undergoing parr–smolt transformation. Lipids 32: 515-525

Benfey TJ, Saunders RL, Knox DE, Harmon PR (1994) Muscle ornithine decarboxylase activity as an indication of recent growth in pre-smolt Atlantic salmon, *Salmo salar*. Aquaculture 121: 125-135

Berg M (1953) A relict salmon, *Salmo salar* L., called "småblank" from the river Namsen, North Tröndelag. Acta Boreal Sci 6: 1-19

Berg OK (1985) The formation of non-anadromous populations of Atlantic salmon, *Salmo salar* L., in Europe. J Fish Biol 27: 805-815

Berge ÅI, Berg A, Fyhn HJ, Barnung T, Hansen T, Stefansson SO (1995) Development of salinity tolerance in underyearling smolts of Atlantic salmon (*Salmo salar*) reared under different photoperiods. Can J Fish Aquat Sci 52: 243-251

Berglund I, Lundqvist H, Fängstam H (1994) Downstream migration of immature salmon (*Salmo salar*) smolts blocked by implantation of the androgen 11-ketoandrostenedione. Aquaculture 121: 269-276

Bern HA (1978) Endocrinological studies on normal and abnormal salmon smoltification. In: Gaillard PJ, Boer HH (eds) Comparative Endocrinology. Elsevier/North Holland Biomedical Press, Amsterdam, p 77-100

Birt TP, Green JM (1986) Parr–smolt transformation in female and sexually mature male anadromous and nonanadromous Atlantic salmon, *Salmo salar*. Can J Fish Aquat Sci 43: 680-686

Birt TP, Green JM (1993) Acclimation to seawater of dwarf nonanadromous Atlantic salmon, *Salmo salar*. Can J Zool 71: 1912-1916

Birt TP, Green JM, Davidson WS (1991) Contrasts in development and smolting of genetically distinct sympatric anadromous and nonanadromous Atlantic salmon, *Salmo salar*. Can J Zool 69: 2075-2084

Björnsson BT (1997) The biology of salmon growth hormone: from daylight to dominance. Fish Physiol Biochem 17: 9-24

Björnsson BT, Thorarensen H, Hirano T, Ogasawara T, Kristinsson JB (1989) Photoperiod and temperature affect plasma growth hormone levels, growth, condition factor and hypoosmoregulatory ability of juvenile Atlantic salmon (*Salmo salar*) during parr-smolt transformation. Aquaculture 82: 77-91

Björnsson BT, Stefansson SO, Hansen T (1995) Photoperiod regulation of plasma growth hormone levels during parr-smolt transformation of Atlantic salmon: implications for hypoosmoregulatory ability and growth. Gen Comp Endocrinol 100: 73-82

Björnsson BT, Stefansson GV, Berge AI, Hansen T, Stefansson SO (1998) Circulating growth hormone levels in Atlantic salmon smolts following seawater transfer: effects of photoperiod regime, salinity, duration of exposure and season. Aquaculture 168: 121-137

Björnsson BT, Hemre GI, Bjornevik M, Hansen T (2000) Photoperiod regulation of plasma growth hormone levels during induced smoltification of underyearling Atlantic salmon. Gen Comp Endocrinol 119: 17-25

Björnsson BT, Johansson V, Benedet S, Einarsdottir IE, Hildahl J, Agustsson T, Jonsson E (2002) Growth hormone endocrinology of salmonids: regulatory mechanisms and mode of action. Fish Physiol Biochem 27: 227-242

Blake RL, Roberts FL, Saunders RL (1984) Parr–smolt transformation of Atlantic salmon (*Salmo salar*): activities of two respiratory enzymes and concentrations of mitochondria in the liver. Can J Fish Aquat Sci 41: 199-203

Browman HI, Hawryshyn CW (1994) The developmental trajectory of ultraviolet photosensitivity in rainbow trout is altered by thyroxine. Vision Res 34: 1397-1406

Burton MP, Idler DR (1984) Can Newfoundland landlocked salmon, *Salmo salar* L., adapt to sea water? J Fish Biol 24: 59-64

Canosa LF, Chang JP, Peter RE (2007) Neuroendocrine control of growth hormone in fish. Gen Comp Endocrinol 151: 1-26

Chernitsky AG (1986) Quantitative evaluation of the degree of parr-smolt transformation in wild smolts and hatchery juveniles of Atlantic salmon (*Salmo salar* L.) by SDH activity of chloride cells. Aquaculture 59: 287-297

Chernitsky AG, Loenko AA (1983) The osmoregulatory system and possible ways of differentiation in ecological forms of Atlantic salmon, *Salmo salar* (Salmonidae). J Ichthyol 23: 84-94

Collie NL, Bern HA (1982) Changes in intestinal fluid transport associated with smoltification and seawater adaptation in Coho salmon *Oncorhynchus kisutch* (Walbaum). J Fish Biol 21: 337-348

Cooke SJ, Hinch SG, Crossin GT, Patterson DA, English KK, Shrimpton JM, Van der Kraak G, Farrell AP (2006) Physiology of individual late-run Fraser River sockeye salmon (*Oncorhynchus nerka*) sampled in the ocean correlates with fate during spawning migration. Can J Fish Aquat Sci 63: 1469-1480

Cowley DJ, Sheridan MA, Hoffnagle TL, Fivizzani AJ, Barton BA, Eilertson CD (1994) Changes in lipid metabolism and plasma concentrations of thyroxine, cortisol, and somatostatin in land-locked chinook salmon, *Oncorhynchus tshawytscha*, during smoltification. Aquaculture 121: 147-155

Dadswell MJ, Melvin GD, Williams PJ, Themelis DE (1987) Influences of origin, life history, and chance on the Atlantic coast migration of American shad. Am Fish Soc Symp 1: 313-330

Dahl K (1928) The Blege or dwarf-salmon. Skr Utg Det Norske Videnskabsakad 1927, Oslo (in Norwegian)

Durif C, Dufour S, Elie P (2005) The silvering process of *Anguilla anguilla*: a new classification from the yellow resident to the silver migrating stage J Fish Biol 66: 1025-1043

Duston J, Saunders RL, Knox DE (1991) Effects of increases in fresh-water temperature on loss of smolt characteristics in Atlantic salmon (*Salmo salar*). Can J Fish Aquat Sci 48: 164-169

Eales JG (1995) Regulation and measurement of thyroidal status in fish. Neth J Zool 45:175-180

Eales JG, Morin PP, Tsang P, Hara TJ (1993) Thyroid hormone deiodination in brain, liver, gill, heart, and muscle of Atlantic salmon (*Salmo salar*) during photoperiodically-induced parr–smolt transformation, II. Outer and inner-ring 3,5,3'-triiodo-L-thyronine and 3,3',5'-triiodo-L-thyronine reverse T3 deiodination. Gen Comp Endocrinol 90: 157-167

Ebbesson LOE (2000) Temporal dynamics of brain and hormone changes during parr-smolt transformation in salmon. PhD thesis, Lund University, Sweden

Ebbesson LOE, Holmqvist B, Östholm T, Ekström P (1992) Transient serotonin-immunoreactive neurons coincide with a critical period of neural development in coho salmon (*Oncorhynchus kisutch*). Cell Tissue Res 268: 389-392

Ebbesson LOE, Drew K, Ebbesson SOE (1994) Brain development during smolt transformation in salmon may be affected by environmental conditions. Soc Neurosci Abstracts 20: 1301

Ebbesson LOE, Deviche P, Ebbesson SOE (1996a) Distribution and changes in mu- and kappa-opiate receptors during the midlife neurodevelopmental period of coho salmon, *Oncorhynchus kisutch*. J Comp Neurol 366: 448-464

Ebbesson LOE, Ekström P, Ebbesson SOE, Stefansson SO, Holmqvist B (2003) Neural circuits and their structural and chemical reorganization in the light-brain-pituitary axis during parr-smolt transformation in salmon. Aquaculture 222: 59-70

Ebbesson LOE, Ebbesson SOE, Nilsen TO, Stefansson SO, Holmqvist B (2007) Exposure to continuous light disrupts retinal innervation of the preoptic nucleus during parr-smolt transformation in Atlantic salmon. Aquaculture (in press)

Ebbesson SOE (1980) The parcellation theory and its relation to interspecific variability in brain organization evolutionary and ontogenetic development and neuronal plasticity. Cell Tissue Res 213: 179-212

Ebbesson SOE, Smith J, Co C, Ebbesson LOE (1996b) Transient alterations in neurotransmitter levels during a critical period of neural development in coho salmon (*Oncorhynchus kisutch*). Brain Res 742: 339-342

Eriksson LO, Lundqvist H (1982) Circannual rhythms and photoperiod regulation of growth and smolting in Baltic salmon (*Salmo salar* L.). Aquaculture 28: 113-121.

Evans DH, Piermarini PM, Choe KP (2005) The multifunctional fish gill: Dominant site of gas exchange, osmoregulation, acid-base regulation, and excretion of nitrogenous waste. Physiol Rev 85: 97-177

Fairchild WL, Swansburg EO, Arsenault JT, Brown SB (1999) Does an association between pesticide use and subsequent declines in catch of Atlantic salmon (*Salmo salar*) represent a case of endocrine disruption? Env Health Perspectives 107: 349-357

Finn RN (2007) The physiology and toxicology of salmonid embryos and larvae in relation to water quality criteria. Aquat Toxicol 81: 337-354

Fontaine M, Marchelidon J (1971) Amino acid contents of the brain and the muscle of young salmon (*Salmo salar* L.) at parr and smolt stages. Comp Biochem Physiol 40: 127-134

Foster AR, Houlihan DF, Gray C, Medale F, Fauconneau B, Kaushikj SJ, Le Bail PY (1991) The effects of ovine growth hormone on protein turnover in rainbow trout. Gen Comp Endocrinol 82: 111-120

Fyhn UEH, Clarke WC, Whitler RE (1991) Hemoglobins in smoltifying Chinook salmon, *Oncorhynchus tshawytscha*, subjected to photoperiod control. Aquaculture 95: 359-372

Garber MJ, DeYonge KG, Byatt JC, Lellis WA, Honeyfield DC, Bull RC, Schelling GT, Roeder RA (1995) Dose-response effects of recombinant bovine somatotropin (Posilac) on growth performance and body composition of two-year-old rainbow trout (*Oncorhynchus mykiss*). J Anim Sci 73: 3216-3222

Gill JA, Sumpter JP, Donaldson EM, Dye HM, Souza L, Berg T, Wypych J, Langley K (1985) Recombinant chicken and bovine growth-hormone accelerate growth in aquacultured juvenile Pacific salmon, *Oncorhynchus kisutch*. Biotechnology 3: 643-646

Gross M (1987) Evolution of diadromy in fishes. Am Fish Soc Symp 1: 14-25

Handeland SO, Stefansson SO (2001) Photoperiod control and influence of body size on off-season parr-smolt transformation and post-smolt growth. Aquaculture 192: 291-307

Handeland SO, Stefansson SO (2002) Effects of salinity acclimation on pre-smolt growth, smolting and post-smolt performance in off-season Atlantic salmon smolts (*Salmo salar* L.). Aquaculture 209: 125-137

Handeland SO, Järvi T, Fernö A, Stefansson SO (1996) Osmotic stress, antipredator behaviour, and mortality of Atlantic salmon (*Salmo salar*) smolts. Can J Fish Aquat Sci 53: 2673-2680

Handeland SO, Berge A, Björnsson BT, Lie Ø, Stefansson SO (2000) Seawater adaptation by out-of-season Atlantic salmon (*Salmo salar* L.) smolts at different temperatures. Aquaculture 181: 377-396

Handeland SO, Björnsson BT, Arnesen AM, Stefansson SO (2003) Seawater adaptation and growth of post-smolt Atlantic salmon (*Salmo salar*) of wild and farmed strains. Aquaculture 233: 367-384

Handeland SO, Wilkinson E, Sveinsbø B, McCormick SD, Stefansson SO (2004) Temperature influence on the development and loss of seawater tolerance in two fast-growing strains of Atlantic salmon. Aquaculture 233: 513-529

Heggenes J, Metcalfe NB (1991) Bimodal size distributions in wild juvenile Atlantic salmon populations and their relationship with age at smolt migration. J Fish Biol 39: 905-907

Herbert NA, Armstrong JD, Björnsson BT (2001) Evidence that growth hormone-induced elevation in routine metabolism of juvenile Atlantic salmon is a result of increased spontaneous activity. J Fish Biol 59: 754-757

Higgs DA, Donaldson EM, Dye HM, McBride JR (1975) Preliminary investigation of effect of bovine growth-hormone on growth and muscle composition of coho salmon (*Oncorhynchus kisutch*). Gen Comp Endocrinol 27: 240-253

Hiroi J, Kaneko T, Tanaka M (1999) *In vivo* sequential changes in chloride cell morphology in the yolk-sac membrane of Mozambique tilapia (*Oreochromis mossambicus*) embyros and larvae during seawater adaptation. J Exp Biol 202: 3485-3495

Hoar WS (1976) Smolt transformation: evolution, behaviour, and physiology. J Fish Res Bd Can 33: 1234-1252

Hoar WS (1988) The physiology of smolting salmonids. In: Hoar WS, Randall DJ (eds) Fish Physiology, Vol 11B, Academic Press, London, p 275-343

Hofmann MH, Meyer DL (1993) Visual-evoked responses in the salmon telencephalon change during smolt transformation. Neurosci Lett 155: 234-236

Holmqvist BI, Östholm T, Ekström P (1992) Retinohypothalamic projections and suprachiasmatic nucleus of the teleost brain. In: Ali MA (ed) Rhythms in Fish. Plenum Press, New York, p 293-318

Holmqvist BI, Östholm T, Ekström P (1994) Neuroanatomical analysis of the visual and hypophysiotrophic systems in Atlantic salmon (*Salmo salar*) with emphasis on possible mediators of photoperiodic cues during parr-smolt transformation. Aquaculture 121: 1-12

Hutchings JA, Jones MEB (1998) Life history variation and growth rate thresholds for maturity in Atlantic salmon, *Salmo salar*. Can J Fish Aquat Sci 55: 22-47

Hutchison MJ, Iwata M (1998) Effect of thyroxine on the decrease of aggressive behaviour of four salmonids during the parr-smolt transformation. Aquaculture 168: 169-175

Jakobsson S, Mayer I, Borg B (1997) Androgen binding in the gills of Atlantic salmon, *Salmo salar*, mature male parr, and immature male smolts. J Exp Biol 278: 391-394

Jarrard HE, Delaney KR, Kennedy CJ (2004) Impacts of carbamate pesticides on olfactory neurophysiology and cholinesterase activity in coho salmon (*Oncorhynchus kisutch*). Aquat Toxicol 69: 133-148

Ji H, Bradley TM, Tremblay GC (1996) Lactate-dependent gluconeogenesis and atractyloside-sensitive flux through pyruvate carboxylase are reduced during smoltification of Atlantic salmon (*Salmo salar*). J Exp Zool 276: 375-386

Johansson V, Winberg S, Hall D, Jönsson E, Björnsson BT (2004) Peripherally administrated growth hormone affects behaviour and brain dopaminergic activity in rainbow trout. Horm Behav 46: 436-443

Johansson V, Winberg S, Björnsson BT (2005) Growth hormone-induced stimulation of swimming and feeding behaviour of rainbow trout is abolished by the D1 dopamine antagonist SCH23390. Gen Comp Endocrinol 141: 58-65

Johnsson JI, Björnsson BT (1994) Growth hormone increases growth rate, appetite and dominance in juvenile rainbow trout, *Oncorhynchus mykiss*. Animal Behav 48: 177-186

Johnsson JI, Jönsson E, Björnsson BT (1996a) Dominance, nutritional status and growth hormone levels in rainbow trout (*Oncorhynchus mykiss*). Horm Behav 30: 13-21

Johnsson JI, Petersson E, Jönsson E, Björnsson BT, Järvi T (1996b) Domestication and growth hormone alter growth patterns and anti-predator behaviour in juvenile brown trout, *Salmo trutta*. Can J Fish Aquat Sci 53: 1546-1554

Johnsson JI, Petersson E, Jönsson E, Järvi T, Björnsson BT (1999) Growth hormone induced effects on mortality, energy status and growth: a field study on brown trout (*Salmo trutta*). Funct Ecol 13: 514-522

Johnsson JI, Jönsson E, Petersson E, Järvi T, Björnsson BT (2000) Fitness-related effects of growth investment in brown trout under field and hatchery conditions. J Fish Biol 57: 326-336

Johnston CE, Eales JG (1967) Purines in the integument of Atlantic salmon (*Salmo salar*) during parr-smolt transformation. J Fish Res Bd Can 24: 955-964

Jonsson B, Ruud-Hansen J (1985) Water temperature as the primary influence on timing of seaward migrations of Atlantic salmon (*Salmo salar*) smolts. Can J Fish Aquat Sci 42: 593-595

Jönsson E, Johnsson JI, Björnsson BT (1996) Growth hormone increases predation exposure of rainbow trout. Proc R Soc Lond B 263: 647-651

Jönsson E, Johnsson JI, Björnsson BT (1998) Growth hormone increases aggressive behaviour in juvenile rainbow trout. Horm Behav 33: 9-15

Jönsson E, Johansson V, Björnsson BT, Winberg S (2003) Central nervous system actions of growth hormone on brain monoamine levels and behaviour of juvenile rainbow trout. Horm Behav 43: 367-374

Kristinsson JB, Saunders RL, Wiggs AJ (1985) Growth dynamics during the development of bimodal length-frequency distribution in juvenile Atlantic salmon (*Salmo salar* L.). Aquaculture 45: 1-20

Kudo H, Tsuneyoshi Y, Nagae M, Adachi S, Yamauchi K, Ueda U, Kawamura H (1994) Detection of thyroid hormone receptors in the olfactory system and brain of wild masu salmon, *Oncorhynchus masou* (Brevoort), during smolting by *in vitro* autoradiography. Aquaculture and Fisheries Management 25: 171-182

Kurokawa T (1990) Influence of the date and body size at smoltification and subsequent growth rate and photoperiod on desmoltification in underyearling masu salmon (*Oncorhynchus masou*). Aquaculture 86: 209-218

Langdon JS, Thorpe JE, Roberts RJ (1984) Effects of cortisol and ACTH on gill Na^+/K^+-ATPase, SDH and chloride cells in juvenile Atlantic salmon *Salmo salar* L. Comp Biochem Physiol 77: 9-12

Langdon JS, Thorpe JE (1985) The ontogeny of smoltification: Developmental patterns of gill Na^+/K^+-ATPase, SDH, and chloride cells in juvenile Atlantic salmon, *Salmo salar* L. Aquaculture 45: 83-95

Lema SC, Nevitt GA (2004) Evidence that thyroid hormone induces olfactory cellular proliferation in salmon during a sensitive period for imprinting. J Exp Biol 207: 3317-3327

Leonard JBK, McCormick SD (2001) Metabolic enzyme activity during smolting in stream- and hatchery-reared Atlantic salmon (*Salmo salar*). Can J Fish Aquat Sci 58: 1585-1593

Lerner DT, Björnsson BT, McCormick SD (2006) Effects of aqueous exposure to polychlorinated biphenyls (Arocolor 1254) on physiology and behaviour of smolt development of Atlantic salmon. Aquat Toxicol 81: 329-336

Lerner DT, Björnsson BTh, McCormick SD (2007) Larval exposure to 4-nonylphenol and 17β-estradiol affects physiological and behavioral development of seawater adaptation in Atlantic salmon smolts. Environ Sci Technol 41: 4479-4485

Lewis KA, Swanson P, Sower SA (1992) Changes in brain gonadotropin-releasing hormone pituitary and plasma gonadotropins and plasma thyroxine during smoltification in chinook salmon (*Oncorhynchus tshawytscha*). Gen Comp Endocrinol 87: 461-470

Li HO, Yamada, J (1992) Changes of the fatty acid composition in smolts of masu salmon (*Oncorhynchus masou*), associated with desmoltification and sea-water transfer. Comp Biochem Physiol 103: 221-226

Lignot JH, Cutler CP, Hazon N, Cramb G (2002) Immunolocalisation of aquaporin 3 in the gill and the gastrointestinal tract of the European eel (*Anguilla anguilla* L.). J Exp Biol 205: 2653-2663

Lubin RT, Rourke AW, Bradley TM (1989) Ultrastructural alteration in branchial chloride cells of Atlantic salmon, *Salmo salar*, during the parr-smolt transformation and early development in sea water. J Fish Biol 34: 259-272

Lundqvist H, Borg B, Berglund I (1989) Androgens impair seawater adaptability in smolting Baltic salmon (*Salmo salar*). Can J Zool 67: 1733-1736

Mackie P, Wright PA, Glebe BD, Ballantyne JS (2005) Osmoregulation and gene expression of Na^+/K^+ ATPase in families of Atlantic salmon (*Salmo salar*) smolts. Can J Fish Aquat Sci 62: 2661-2672

Madsen SS, Mathiesen AB, Korsgaard B (1997) Effects of 17 beta-estradiol and 4-nonylphenol on smoltification and vitellogenesis in Atlantic salmon (*Salmo salar*). Fish Physiol Biochem 17: 303-312

Madsen SS, Skovbolling S, Nielsen C, Korsgaard B (2004) 17-beta estradiol and 4-nonylphenol delay smolt development and downstream migration in Atlantic salmon, *Salmo salar*. Aquat Toxicol 68: 109-120

Marine KR, Cech JJ (2004) Effects of high water temperature on growth, smoltification, and predator avoidance in Juvenile Sacramento River Chinook salmon. North Am J Fish Management 24: 198-210

Markert JR, Higgs DA, Dye HM, MacQuarrie DW (1977) Influence of bovine growth-hormone on growth-rate, appetite, and food conversion of yearling coho salmon (*Oncorhynchus kisutch*) fed two diets of different composition. Can J Zool 55: 74-83

Marshall WS (2002) Na^+, Cl^-, Ca^{2+} and Zn^{2+} transport by fish gills: Retrospective review and prospective synthesis. J Exp Zool 293: 264-283

Martin-Smith KM, Armstrong JD, Johnsson JI, Björnsson BT (2004) Growth hormone increases growth and dominance of wild juvenile Atlantic salmon without affecting space use. J Fish Biol 65: 156-172

Maule AG, Tripp RA, Kaattari SL, Schreck CB (1989) Stress alters immune function and disease resistance in Chinook salmon (*Oncorhynchus tshawytscha*). J Endocrinol 120: 135-142

Maxime V, Boeuf G, Pennec JP, Peyraud C (1989) Comparative study of the energetic metabolism of Atlantic salmon (*Salmo salar*) parr and smolts. Aquaculture 82: 163-171

Mazurais D, Le Drean G, Brierley I, Anglade I, Bromage N, Williams LM, Kah O (2000) Expression of *clock* gene in the brain of rainbow trout: Comparison with the distribution of melatonin receptors. J Comp Neurol 422: 612-620

McCormick SD (1994) Ontogeny and evolution of salinity tolerance in anadromous salmonids: Hormones and heterochrony. Estuaries 17: 26-33

McCormick SD (2001) Endocrine control of osmoregulation in teleost fish. Am Zool 41: 781-794

McCormick SD, Saunders RL (1987) Preparatory physiological adaptations for marine life of salmonids: osmoregulation, growth, and metabolism. Am Fish Soc Symp 1: 211-299

McCormick SD, Saunders RL, Henderson EB, Harmon PR (1987) Photoperiod control of parr-smolt transformation in Atlantic salmon (*Salmo salar*): changes in salinity tolerance, gill Na^+,K^+-ATPase activity and plasma thryoid hormones. Can J Fish Aquat Sci 45: 1462-1468

McCormick SD, Hansen LP, Quinn TP, Saunders RL (1998) Movement, migration, and smolting of Atlantic salmon (*Salmo salar*). Can J Fish Aquat Sci 55: 77-92

McCormick SD, Cunjak RA, Dempson B, O'Dea MF, Carey JB (1999) Temperature-related loss of smolt characteristics in Atlantic salmon (*Salmo salar*) in the wild. Can J Fish Aquat Sci 56: 1649-1658

McCormick SD, Moriyama S, Björnsson BT (2000) Low temperature limits photoperiod control of smolting in Atlantic salmon through endocrine mechanisms. Am J Physiol 278: R1352-R1361

McCormick SD, Björnsson BT, Moeckel AM, Lerner DT, O'Dea MF (2001) Endocrine pathways for the disruption of the parr-smolt transformation by estradiol and nonylphenol. Am Zool 41: 1518-1518

McCormick SD, Shrimpton JM, Moriyama S, Björnsson BT (2002) Effects of an advanced temperature cycle on smolt development and endocrinology indicate that temperature is not a zeitgeber for smolting in Atlantic salmon. J Exp Biol 205: 3553-3560

McCormick SD, O'Dea MF, Moeckel AM, Lerner DT, Björnsson BT (2005) Endocrine disruption of parr-smolt transformation and seawater tolerance of Atlantic salmon by 4-nonylphenol and 17 beta-estradiol. Gen Comp Endocrinol 142: 280-288

McCormick SD, Shrimpton JM, Moriyama S, Björnsson BT (2007) Differential hormonal responses of Atlantic salmon parr and smolt to increased daylength: a possible developmental basis for smolting. Aquaculture 273: 337-344

McDowall RM (1987) The occurrence and distribution of diadromy among fishes. Am Fish Soc Symp 1: 1-13

McKeown BA, Leatherland JF, John TM (1975) The effect of growth hormone and prolactin on the mobilization of free fatty acids and glucose in the kokanee salmon, *Oncorhynchus nerka*. Comp Biochem Physiol 50: 425-430

McLean E, Donaldson EM (1993) The role of growth hormone in growth of poikilotherms. In: Schreibman MP, Scanes CG, Pang PKT (eds) The Endocrinology of Growth, Development and Metabolism in Vertebrates. Academic Press, New York, p 43-71

Melingen GO, Wergeland HI (2000) Serum protein and IgM profiles in connection with the smolting and vaccination of out-of-season Atlantic salmon (*Salmo salar* L.). Aquaculture 188: 189-201

Melingen GO, Pettersen EF, Wergeland HI (2002) Leucocyte populations and responses to immunization and photoperiod manipulation in Atlantic salmon (*Salmo salar* L.) 0+ smolt. Aquaculture 214: 381-396

Metcalfe NB, Thorpe JE (1992) Anorexia and defended energy levels in over-wintering juvenile salmon. J Animal Ecol 61: 175-181

Metcalfe NB, Huntingford FA, Thorpe JE (1986) Seasonal changes in feeding motivation of juvenile Atlantic salmon *Salmo salar*. Can J Zool 64: 2439-2446

Metcalfe NB, Huntingford FA, Thorpe JE (1988) Feeding intensity, growth rates, and the establishment of life-history patterns in juvenile Atlantic salmon *Salmo salar*. J Animal Ecol 57: 463-474

Moore A, Lower N (2001) The impact of two pesticides on olfactory-mediated endocrine function in mature male Atlantic salmon (*Salmo salar* L.) parr. Comp Biochem Physiol 129: 269-276

Moore A, Scott AP, Lower N, Katsiadaki I, Greenwood L (2003) The effects of 4-nonylphenol and atrazine on Atlantic salmon (*Salmo salar* L.) smolts. Aquaculture 222: 253-263

Morin PP, Hara TJ, Eales JG (1993) Thyroid hormone deiodination in brain, liver, gill, heart, and muscle of Atlantic salmon (*Salmo salar*) during photoperiodically-induced parr-smolt transformation I. Outer and inner-ring thyroxine deiodination. Gen Comp Endocrinol 90: 142-156

Morin PP, Andersen Ø, Haug E, Døving KB (1994) Changes in serum free thyroxine, prolactin, and olfactory activity during induced smoltification in Atlantic salmon (*Salmo salar*). Can J Fish Aquat Sci 51: 1985-1992

Morin PP, Hara TJ, Eales JG (1995) T-4 depresses olfactory responses to L-alanine and plasma T-3 and T-3 production in smoltifying Atlantic salmon. Am J Physiol 38: R1434-R1440

Morin PP, Winberg S, Nilsson GE, Hara TJ, Eales JG (1997) Effects of L-thyroxine on brain monoamines during parr-smolt transformation of Atlantic salmon (*Salmo salar* L). Neurosci Lett 224: 216-218

Moriyama S (1995) Increased plasma insulin-like growth-factor-I (IGF-I) following oral and intraperitoneal administration of growth hormone to rainbow trout, *Oncorhynchus mykiss*. Growth Regulation 5: 164-167

Moriyama S, Swanson P, Nishii M, Takahashi A, Kawauchi H, Dickhoff WW, Plisetskaya EM (1994) Development of a homologous radioimmunoassay for coho salmon insulin-like growth-factor-I. Gen Comp Endocrinol 96: 149-161

Mustafa T, Srivastava KC (1989) Prostaglandins (eicosanoids) and their role in ectothermic organisms. Adv Comp Environ Physiol 5: 157-207

Neregård L, Sundt-Hansen L, Hindar K, Einum S, Johnsson JI, Devlin RH, Fleming I, Björnsson BT (2007) Selective breeding of Atlantic salmon affects growth hormone action on growth and lipid metabolism. Can J Aquat Fish Sci (in press)

Nicieza AG, Reyesgavilan FG, Brana F (1994) Differentiation in juvenile growth and bimodality patterns between northern and southern populations of Atlantic salmon (*Salmo salar* L.). Can J Zool 72: 1603-1610

Nielsen C, Madsen SS, Björnsson BT (1999) Changes in branchial and intestinal osmoregulatory mechanisms and growth hormone levels during smolting in hatchery-reared and wild brown trout. J Fish Biol 54: 799-818

Nilsen TO, Ebbesson LOE, Stefansson SO (2003) Smolting in anadromous and landlocked strains of Atlantic salmon (*Salmo salar*). Aquaculture 222: 71-82

Nilsen TO, Ebbesson LOE, Madsen SS, McCormick SD, Andersson E, Björnsson BT, Prunet P, Stefansson SO (2007) Differential expression of gill Na^+, K^+-ATPase α and β subunits, Na^+,K^+,$2Cl^-$ cotransporter and CFTR anion channel in Juvenile anadromous and landlocked Atlantic salmon (*Salmo salar*). J Exp Biol 210: 2885-2896

Nordgarden U, Björnsson BT, Hansen T (2007) Developmental stage of Atlantic salmon parr regulates pituitary GH secretion and parr–smolt transformation Aquaculture 264: 441–448

Parhar IS, Iwata M (1996) Intracerebral expression of gonadotropin-releasing hormone and growth hormone-releasing hormone is delayed until smoltification in the salmon. Neurosci Res 26: 299-308

Payan P, Girard JP, Mayergostan N (1984) Branchial ion movements in teleosts – the roles of respiratory and chloride cells. In: Hoar WS, Randall DJ (eds) Fish Physiology, Vol 10B, Academic Press, London, p 39-63

Peake S, McKinley RS (1998) A re-evaluation of swimming performance in juvenile salmonids relative to downstream migration. Can J Fish Aquat Sci 55: 682-687

Pelis RM, Zydlewski J, McCormick SD (2001) Gill Na^+-K^+-$2Cl^-$ cotransporter abundance and location in Atlantic salmon: effects of seawater and smolting. Am J Physiol 280: R1844-R1852

Pisam M, Prunet P, Boeuf G, Rambourg A (1988) Ultrastructural features of chloride cells in the gill epithelium of the Atlantic salmon, *Salmo salar*, and their modifications during smoltification. Am J Anat 183: 235-244

Plisetskaya EM, Moon TW, Larsen DA, Foster GD, Dickhoff WW (1994) Liver glycogen, enzyme activities, and pancreatic hormones in juvenile Atlantic salmon (*Salmo salar*) during their first summer in seawater. Can J Fish Aquat Sci 51: 567-576

Poupa O, Gesser H, Jonsson S, Sullivan L (1974) Coronary-supplied compact shell of ventricular myocardium in salmonids: Growth and enzyme pattern. Comp Biochem Physiol 48: 85-88

Power DM, Llewellyn L, Faustino M, Nowell MA, Björnsson BT, Einarsdottir IE, Canario AVM, Sweeney GE (2001) Thyroid hormones in growth and development of fish. Comp Biochem Physiol 130: 447-459

Power G (1958) The evolution of the freshwater races of the Atlantic salmon (*Salmo salar* L.). Arctic 11: 86-92

Richards JG, Semple JW, Bystriansky JS, Schulte PM (2003) Na^+/K^+-ATPase α-isoform switching in gills of rainbow trout (*Oncorhynchus mykiss*) during salinity transfer. J Exp Biol 206: 4475-4486

Richardson SJ, Monk JA, Shepherdley CA, Ebbesson LOE, Sin F, Power DM, Frappell PB, Kohrle J, Renfree MB (2005) Developmentally regulated thyroid hormone distributor proteins in marsupials, a reptile, and fish. Am J Physiol 288: R1264-R1272

Rosseland BO, Kroglund F, Staurnes M, Hindar K, Kvellestad A (2001) Tolerance to acid water among strains and life stages of Atlantic salmon (*Salmo salar* L.). Water, Air and Soil Pollution 130: 899-904

Sargent JR, Bell JG, McEvoy L, Tocher DR, Estevez A (1999) Recent developments in the essential fatty acid nutrition of fish. Aquaculture 177: 191-199

Saunders RL, Henderson EB (1970) Influence of photoperiod on smolt development and growth of Atlantic salmon (*Salmo salar*). J Fish Res Bd Can 27: 1295-1311

Saunders RL, Henderson EB (1978) Changes in gill ATPase activity and smolt status of Atlantic salmon (*Salmo salar*). J Fish Res Bd Can 35: 1542-1546

Saunders RL, Henderson EB, Harmon PR, Johnston CE, Eales JG (1983) Effects of low environmental pH on smolting of Atlantic salmon (*Salmo salar*). Can J Fish Aquat Sci 40: 1203-1211

Schmitz M (1995) Seasonal changes in hypoosmoregulatory ability in landlocked and anadromous populations of Arctic charr, *Salvelinus alpinus*, and Atlantic salmon, *Salmo salar*. Environ Biol Fishes 42: 401-412

Schmitz M, Berglund I, Lundqvist H, Björnsson BT (1994) Growth hormone response to seawater challenge in Atlantic salmon, *Salmo salar*, during parr-smolt transformation. Aquaculture 121: 209-221.

Scholz NL, Truelove NK, French BL, Berejikian BA, Quinn TP, Casillas E, Collier TK (2000) Diazinon disrupts antipredator and homing behaviours in chinook salmon (*Oncorhynchus tshawytscha*). Can J Fish Aquat Sci 57: 1911-1918

Seddiki H, Boeuf G, Maxime V, Peyraud C (1996) Effects of growth hormone treatment on oxygen consumption and sea water adaptability in Atlantic salmon parr and pre-smolts Aquaculture 148: 49-62

Seidelin M, Madsen SS, Byrialsen A, Kristiansen K (1999) Effects of insulin-like growth factor-I and cortisol on Na^+,K^+-ATPase expression in osmoregulatory tissues of brown trout (*Salmo trutta*). Gen Comp Endocrinol 113: 331-342

Sheridan MA (1986) Effects of thyroxin, cortisol, growth hormone, and prolactin on lipid metabolism of coho salmon, *Oncorhynchus kisutch*, during smoltification. Gen Comp Endocrinol 64: 220-238

Sheridan MA (1989) Alterations in lipid metabolism accompanying smoltification and seawater adaptation of salmonid fish. Aquaculture 82: 191-203

Sheridan MA (1994) Regulation of lipid metabolism in poikilothermic vertebrates. Comp Biochem Physiol 107: 495-508

Sheridan MA, Woo NYS, Bern HA (1985) Changes in the rates of glycogenesis, glycogenolysis, lipogenesis, and lipolysis in selected tissues of the coho salmon (*Oncorhynchus kisutch*) associated with parr-smolt transformation. J Exp Zool 236: 35-44

Shrimpton JM, Patterson DA, Richards JG, Cooke SJ, Schulte PM, Hinch SG, Farrell AP (2005) Ionoregulatory changes in different populations of maturing sockeye salmon *Oncorhynchus nerka* during ocean and river migration. J Exp Biol 208: 4069-4078

Sigholt T, Asgard T, Staurnes M (1998) Timing of parr-smolt transformation in Atlantic salmon (*Salmo salar*): effects of changes in temperature and photoperiod. Aquaculture 160: 129-144

Singer TD, Koreen M, Clements KM, Semple JW, Schulte PM, Bystriansky JS, Finstad B, Fleming IA, McKinley RS (2002) Seawater tolerance and gene expression in two strains of Atlantic salmon smolts. Can J Fish Aquat Sci 59: 125-135

Skilbrei OT (1988) Growth pattern of pre-smolt Atlantic salmon (*Salmo salar* L.) – The percentile increment method (PIM) as a new method to estimate length-dependent growth. Aquaculture 69: 129-143

Skilbrei OT (1991) Importance of threshold length and photoperiod for the development of bimodal length frequency-distribution in Atlantic salmon (*Salmo salar*). Can J Fish Aquat Sci 48: 2163-2172

Skilbrei OT, Hansen T, Stefansson SO (1997) Effects of decreases in photoperiod on growth and bimodality in Atlantic salmon *Salmo salar* L. Aquaculture Research 28: 43-49

Soivio A, Virtanen E, Muona M (1988) Desmoltification of heat-accelerated Baltic salmon (*Salmo salar*) in brackish water. Aquaculture 71: 89-97

Solbakken VA, Hansen T, Stefansson SO (1994) Effects of photoperiod and temperature on growth and parr smolt transformation in Atlantic salmon (*Salmo salar*) and subsequent performance in seawater. Aquaculture 121: 13-27

Specker JL, Eales JG, Tagawa M, Tyler WA (2000) Parr-smolt transformation in Atlantic salmon: thyroid hormone deiodination in liver and brain and endocrine correlates of change in rheotactic behaviour. Can J Zool 78: 696-705

Staley KB, Ewing RD (1992) Purine levels in the skin of juvenile coho salmon (*Oncorhynchus kisutch*) during parr-smolt transformation and adaptation to seawater. Comp Biochem Physiol 101: 447-452

Staurnes M, Lysfjord G, Berg OK (1992) Parr-smolt transformation of a nonanadromous population of Atlantic salmon (*Salmo salar*) in Norway. Can J Zool 70: 197-199

Staurnes M, Blix P, Reite OB (1993) Effects of acid water and aluminium on parr-smolt transformation and seawater tolerance in Atlantic salmon, *Salmo salar*. Can J Fish Aquat Sci 50: 1816-1827

Staurnes M, Hansen LP, Fugelli K, Haraldstad O (1996) Short-term exposure to acid water impairs osmoregulation, seawater tolerance, and subsequent marine survival of smolts of Atlantic salmon (*Salmo salar* L). Can J Fish Aquat Sci 53: 1695-1704

Stefansson SO, Nævdal G, Hansen T (1989) The influence of three unchanging photoperiods on growth and parr-smolt transformation in Atlantic salmon, *Salmo salar* L. J Fish Biol 35: 237-247

Stefansson SO, Nævdal G, Hansen T (1990) Growth of different families of Atlantic salmon (*Salmo salar*) under three experimental photoperiods. Aquaculture 86: 271-281

Stefansson SO, Björnsson BT, Hansen T, Haux C, Taranger GL, Saunders RL (1991) Growth, parr-smolt transformation, and changes in growth hormone of Atlantic salmon reared under different photoperiods. Can J Fish Aquat Sci 48: 2100-2108

Stefansson SO, Berge AI, Gunnarsson GS (1998) Changes in seawater tolerance and gill Na^+,K^+-ATPase activity during desmoltification in Atlantic salmon kept in freshwater at different temperatures. Aquaculture 168: 271-277

Stefansson SO, Björnsson BT, Sundell K, Nyhammer G, McCormick SD (2003a) Physiological characteristics of wild Atlantic salmon post-smolts during estuarine and coastal migration. J Fish Biol 63: 942-955

Stefansson SO, McGinnity P, Bjornsson BT, Schreck CB, McCormick SD (2003b) The importance of smolt development to salmon conservation, culture, and management: perspectives from the 6th International Workshop on Salmonid Smoltification. Aquaculture 222: 1-14

Stefansson SO, Nilsen TO, Ebbesson LOE, Wargelius A, Madsen SS, Björnsson BT, McCormick SD (2007) Molecular mechanisms of continuous light inhibition of Atlantic salmon parr-smolt transformation. Aquaculture 273: 235-245

Sundell K, Jutfelt F, Ágústsson T, Olsen RE, Sandblom E, Hansen T, Björnsson BT (2003) Intestinal transport mechanisms and plasma cortisol levels during normal and out-of-season parr-smolt transformation of Atlantic salmon, *Salmo salar*. Aquaculture 222: 265-285

Thorpe JE (1989) Developmental variation in salmonid populations. J Fish Biol 35: 295-303

Thorpe JE (1994) Performance thresholds and life-history flexibility in salmonids. Conservation Biol 8: 877-879

Tocher DR, Bell JG, Dick JR, Henderson RJ, McGhee F, Mitchell D, Morris PC (2000) Polyunsaturated fatty acid metabolism in Atlantic salmon (*Salmo salar*) undergoing parr-smolt transformation and the effects of dietary linseed and rapeseed oils. Fish Physiol Biochem 23: 59-73

Tocher DR, Bell JG, McGhee F, Dick JR, Fonseca-Madrigal J (2003) Effects of dietary lipid level and vegetable oil on fatty acid metabolism in Atlantic salmon (*Salmo salar* L.) over the whole production cycle. Fish Physiol Biochem 29: 193-209

Usher ML, Talbot C, Eddy FB (1988) Drinking in Atlantic salmon smolts transferred to seawater and the relationship between drinking and feeding. Aquaculture 73: 237-246

Utrilla CG, Lobon-Cervia J (1999) Life-history patterns in a southern population of Atlantic salmon. J Fish Biol 55: 68-83

Veillette PA, Young G (2004) Temporal changes in intestinal Na^+, K^+-ATPase activity and *in vitro* responsiveness to cortisol in juvenile chinook salmon. Comp Biochem Physiol 138: 297-303

Veillette PA, White RJ, Specker JL (1993) Changes in intestinal fluid transport in Atlantic salmon (*Salmo salar* L.) during parr-smolt transformation. Fish Physiol Biochem 12: 193-202

Veillette PA, Sundell K, Specker JL (1995) Cortisol mediates the increase in intestinal fluid absorption in Atlantic salmon during parr-smolt transformation. Gen Comp Endocrinol 97: 250-258

Vijayan MM, Takemura A, Mommsen TP (2001) Estradiol impairs hyposmoregulatory capacity in the euryhaline tilapia, *Oreochromis mossambicus*. Am J Physiol 281: R1161-R1168

Virtanen E, Forsman L (1987) Physiological responses to continuous swimming in wild salmon (*Salmo salar* L.) parr and smolt. Fish Physiol Biochem 4: 157-163

Wargelius A, Fjelldal P-G, Benedet S, Hansen T, Björnsson BT, Nordgarden U (2005) A peak in GH-receptor expression is associated with growth activation in Atlantic salmon vertebrae, while upregulation of IGF-I receptor expression is related to increased bone density. Gen Comp Endocrinol 142: 163-168

Waring CP, Moore A (2004) The effect of atrazine on Atlantic salmon (*Salmo salar*) smolts in fresh water and after sea water transfer. Aquat Toxicol 66: 93-104

Wedemeyer GA, Saunders RL, Clarke WC (1980) Environmental factors affecting smoltification and early marine survival of anadromous salmonids. Mar Fish Rev 42: 1-14

Wendt CAG, Saunders RL (1973) Changes in carbohydrate metabolism in young Atlantic salmon in response to various forms of stress. Int Atl Salmon Found, Spec Publ Ser 4: 55-82

Wong CKC, Chan DKO (1999) Chloride cell subtypes in the gill epithelium of Japanese eel *Anguilla japonica*. Am J Physiol 46: R517-R522

Zaugg WS (1981) Advanced photoperiod and water temperature effects on gill Na^+,K^+-adenosine-triphosphatase activity and migration of juvenile steelhead (*Salmo gairdneri*). Can J Fish Aquat Sci 38: 758-764

Zaugg WS, Wagner HH (1973) Gill ATPase activity related to parr-smolt transformation and migration in Steelhead trout (*Salmo gairdneri*) – influence of photoperiod and temperature. Comp Biochem Physiol 45: 955-965

Zaugg WS, Adams BL, McClain LR (1972) Steelhead migration – potential temperature effects as indicated by adenosine-triphosphatase activities. Science 176: 415-416

Zheng XZ, Tocher DR, Dickson CA, Bell JG, Teale AJ (2004) Effects of diets containing vegetable oil on expression of genes involved in highly unsaturated fatty acid biosynthesis in liver of Atlantic salmon (*Salmo salar*). Aquaculture 236: 467-483

Zydlewski GB, Haro A, McCormick SD (2005) Evidence for cumulative temperature as an initiating and terminating factor in downstream migratory behaviour of Atlantic salmon (*Salmo salar*) smolts. Can J Fish Aquat Sci 62: 68-78

Zydlewski J, McCormick SD (2001) Developmental and environmental regulation of chloride cells in young American shad, *Alosa sapidissima*. J Exp Zool 290: 73-87

Glossary

Note: gene names are italicised.

A

AA	amino acids
ABR	auditory brainstem response
ACh	acetylcholine
AcT	acetylated tubulin
ACTH	adrenocorticotropic hormone
ADF	anatomical diffusion factor
AGA	N-acetyl-L-glutamate
AGM	Aorta-Gonad-Mesonephros primordium
Alevin	yolk-sac larva (salmonid)
alk	*anaplastic lymphoma receptor tyrosine kinase*
ALL	anterior lateral line system
AP	alkaline phosphatase
APC	active phase coupling
ARA	arachidonic acid
ASL	argininosuccinate lyase
ATP	adenosine triphosphate
ATU	accumulated thermal unit
AVT	arginine vasotocin

B

BCF	body and caudal fin
BL	body length
bls	*bloodless* mutant
bon	*bonnie & clyde*

BMP	bone morphogenetic protein
BSDL	bile salt-dependent lipase

C

C1/MASP2	C factor 1r/s mannose binding lectin-associated serine protease-like molecule
C3	complement component 3
C4	complement component 4
CaP	caudal primary neurones
cas	*casanova*
cbfb	*core-binding factor, beta subunit*
CC	mitochondrion-rich chloride cells
CCH	carbamylcholine chloride
CCK	cholecystokinin
CD8	T-cell receptor co-molecule
CFTR	cystic fibrosis transmembrane conductance regulator
CGRP	calcitonin gene related peptide
c-myb	*myeloblastosis* (alias for *v-myb*: *avian myeloblastosis viral oncogene homolog*)
CNS	central nervous system
CPSase	carbamoyl phosphate synthetase
c-ret	*c-(rearranged during transfection) proto-oncogene*
CRF	corticotropin releasing factor
CRP	C-reactive protein
csf1	*colony stimulating factor 1*
CSI	cardiosomatic index
CYA	complete yolk absorption
cyc	*cyclops*
CXCR4	chemokine (C-X-C motif) receptor 4

D

d°C	degree-days
D1	type I iodothyronine deiodinase
D2	type II iodothyronine deiodinase
D3	type III iodothyronine deiodinase
DA	dopamine
DAG	diacylglycerol
DBH	dopamine α-hydroxylase
Df	frequency difference
dpf	days post fertilisation
dpff	days post first-feeding
dph	days post hatch
DPIV	digital particle image velocimetry
dsRNA	double-stranded RNA

E

E_2	estradiol
ednrb1	*endothelin receptor type B1*
EFA	essential fatty acids
EMG	electromyographic
ENaC	epithelial Na^+-channel
EOD	electric organ discharge
EOG	electro-olfactogram
EPA	eichosapentinoic acid
ER	endoplasmatic reticulum
erbb3	*v-erb-b2 erythroblastic leukemia viral oncogene homolog 3*
ET-3	endothelin-3
eUT	eel urea transporter
EVL	enveloping layer

F

FA	fatty acid
FAA	free amino acid
FABP	fatty acid binding protein
fau	*faust (Finkel-Biskis-Reilly murine sarcoma virus)*
FFA	free fatty acids
FGF	fibroblast growth factor
flh	*floating head*
fli-1	*friend leukemia integration*
foxd3	*forkhead box D3*
foxn1	*forkhead box N1*
foxn4	*forkhead box N4*
FSH	follicle stimulating hormone
Ftnt	fast troponin T
FW	freshwater

G

GABA	γ-aminobutyric acid
GADPH	glyceraldehyde-3-phosphate dehydrogenase
GALT	gut-associated lymphoid tissue
GAP-43	growth-associated protein-43
GATA	glutamyl-tRNA amidotransferase
gata5	GATA-*binding protein 5*
GDH	glutamate dehydrogenase
GDNF	glial cell line-derived neurotrophic factor
GFP	green fluorescent protein
Gfrα1	glial cell line-derived neurotrophic factor receptor
GH	growth hormone

GnRH	gonadotropic releasing hormone
GSase	glutamine synthetase
gsc	*goosecoid*

H

HGF	hepatocyte growth factor
HGG	human gamma-globulin
HL	head length
hpf	hours post fertilisation
HNK-1	human natural killer-1
hox	*homeobox*
HSC	haematopoietic stem cell
5-HT	5-hydroxytryptamine (serotonin)
HUFA	highly unsaturated fatty acid

I

ICC	interstitial cell of Cajal
ICM	intermediate cell mass
IE	intra-epithelial
IEL	intra-epithelial lymphocyte
IFN	interferon
Ig μ	immunoglobulin μ chain
IGF-I	insulin-like growth factor-I
Ig-H	immunoglobulin heavy chain
Ig-L	immunoglobulin light chain
IgM	immunoglobulin M
IL	interleukin
IRD	inner ring deiodinase

J

JAR	jamming avoidance response
$J_{in}Cl$	unidirectional Cl^- uptake
$J_{in}Na^+$	unidirectional Na^+ uptake
J_v	fluid transport rate

K

Kir7.1	*(alias for kcnj13) potassium inwardly-rectifying channel, subfamily J, member 13*
kit	*v-kit Hardy-Zuckerman 4 feline sarcoma viral oncogene homolog*

L

LBP	light-brain-pituitary axis
LC_{50}	lethal concentration resulting in 50% mortality
LDH	lactate dehydrogenase

LH	luteinising hormone
LL	continuous light
LM	lower mode
lmo-2	*LIM domain only 2*
L-NAME	N^G-nitro-L-arginine methyl ester
LPM	lateral plate mesoderm
LPS	lipopolysaccharide
LRR	leucine rich repeat
LWS	long-to middle-wave class cone pigment sensitive to red-green

M

mAb	monoclonal antibody
MAG	monoacylglycerols
MAL	myelin and lymphocyte protein
mash1	*mammalian achaete-scute homologue 1*
MBT	mid-blastula transition
MC1R	melanocortin 1 receptor
MCH	melanin-concentrating hormone
MHC	major histocompatibility complex
mil	*miles apart*
MiP	middle primary neurones
mitf	*microphthalmia-associated transcription factor*
MLC1	myosin light chain 1
MLC2	myosin light chain 2
MLC3	myosin light chain 3
MMC	migrating motor complexes
MPF	median and paired fin
mpf	months post fertilisation
MRC	mitochondrial rich cell (chloride cell)
MSA	minimum separable angle
α-MSH	α-melanocyte-stimulating hormone
MSO	methionine sulfoximine
MSP	microspectrophotometry
mT	millitesla ($^1/_{1000}$ unit of magnetic flux density)
MTP	microsomal triglyceride transfer protein
MUFA	monounsaturated fatty acids
myc	*v-myc myelocytomatosis viral oncogene homolog*
MyD	myeloid differentiation factor

N

NA	numerical aperture
NCCRP-1	non-specific cytotoxic cell receptor protein 1
NEC	neuroepithelial cells

NHE	Na^+,H^+ exchanger
NITR	novel immune-type receptor
NK	natural killer
NKA	Na^+,K^+-ATPase
NkA	neurokinin A
NKCC	$Na^+/K^+/2Cl^-$ cotransporter
NKG2	killer cell lectin-like receptor subfamily C, member 1
NMDA	N-methyl-D-aspartate
NO	nitric oxide
NOS	nitric oxide synthetase
4-NP	4-nonylphenol
NPO	preoptic nucleus
nr1a1	homologue of human THRA: *thyroid hormone receptor, alpha*
nr1a2	homologue of human THRB: *thyroid hormone receptor, beta*
NRP-1	neuropilin-1
NT-3	neurotrophin-3
ntl	*no tail*

O

OA	oleic acid
OCT	ornithine carbamoyl transferase
OR	odorant receptor
ORD	outer ring deiodinase
OUC	ornithine-urea cycle

P

PACAP	pituitary adenylate cyclase-activating polypeptide
PAP	pancreatitis-associated protein
pax	*pair box*
PC	phosphatidylcholine
PCB	polychlorinated biphenyl
pCO_2	partial pressure of carbon dioxide
PCR	polymerase chain reaction
PDGF	platelet-derived growth factor
PF	post fertilisation
PFK	phosphofructokinase
PGF	prostaglandin F
PGF2α	prostaglandin F2α
ph	post hatch
pHe	extracellular pH
pHi	intracellular pH
pitx2	*paired-like homeodomain 2*
pK_a	acid ionisation constant

PL	phospholipid
PLA_2	phospholipase A_2
PLL	posterior lateral line system
pO_2	partial pressure of oxygen
POA	preoptic area
Poly I:C	polyinosinic-polycytidylic acid
PPE	pre-placodal ectoderm
prdm1	*PR domain containing 1, with ZNF domain*
PRL	prolactin
Prolarva	yolk-sac larva (eleutheroembryo)
PUFA	polyunsaturated fatty acid
PVF	perivitelline fluid
PVS	perivitelline space

R

rag1	*recombination-activating gene-1*
RBI	rostral blood island
RH1	rod class of pigment sensitive to green
RH2	middle-wave class cone pigment sensitive to green
rhcg1	*rhesus monkey glycoprotein C1*
RM	retinomotor movement
RoP	rostral primary neurones
RPE	retinal pigment epithelium
rT3	reverse T_3
RT-PCR	real-time polymerase chain reaction
runx	runt-related transcription factor

S

SAA	serum amyloid A
SAP	serum amyloid protein
SARM	selective androgen receptor antagonists/modulators
SCC	solitary chemosensory cells
scl	stem-cell leukemialeukaemia
SDF1a	stromal cell-derived factor 1a
SECIS	selenocysteine-insertion sequence
SEM	scanning electron micrograph
SFA	saturated fatty acids
ShUT	shark urea transporter
SL	standard length
SLC24A5	solute carrier family 24, member 5
SNP	simulated natural photoperiod
sox9	*SRY (sex determining region Y)-box 9*
sox10	*SRY (sex determining region Y)-box 10*

spt	*spade tail*
SPV	*stratum periventricular*
sqt	*squint*
SSP	salmon serum protein
sTnT	slow Troponin T
SW	seawater
SWS	short wave class cone pigment
SWS1	SWS sensitive to blue-violet
SWS2	SWS sensitive to violet-ultraviolet
SZ	*stratum zonale*

T

T_2	di-iodothyronine
T_3	tri-iodothyronine
T_4	thyroxine
Tach	tachykinin
TAG	triacylglycerol
Tcr	T-cell receptor
tcrα	*T-cell receptor subunit* α
tbx-1	*T-box 1*
tdt	*terminal deoxynucleotidyl transferase*
TEM	transmission electron micrograph
tfap2a	*transcription factor AP-2 alpha (activating enhancer binding protein 2 alpha)*
TGF-β	transforming growth factor-β
TH	thyroid hormone
THR	thyroid hormone receptor
trhα	see *nr1a1*
trhβ	see *nr1a2*
TIR	Toll/IL-1 receptor
TL	*torus longitudinalis*
TL	total length
TLR	Toll-like receptor
TRH	thyrotropin releasing hormone
TRIF	Toll/interlukin-1 receptor domain-containing adaptor protein inducing interferon β
TrkC	tropomyosin receptor kinase C
trpm7	*transient receptor potential cation channel, subfamily M, member 7*
TSH	thyroid stimulating hormone

U

UBL	unstirred boundary layer
UM	upper mode
Urea-N	moles of nitrogen contained in urea
UT	urea transporter

V

VEGF	vascular endothelial growth factor
VEGFR	vascular endothelial growth factor receptor
VH	V gene segments of H immunoglobulin chain
VHSH	viral haemorrhagic septicaemia virus
VIP	vasoactive intestinal peptide
VLDL	very low density lipoprotein
VLR	variable lymphocyte receptor
VR	vomeronasal receptor probe

W

WCL	Wageningen cell lines
WE	wax ester
whn	*winged helix nude*

Y

YCL	yolk cytoplasmic layer
YSL	yolk syncytial layer
ZOR	zebrafish odorant receptor
α-MSH	α-melanocyte stimulating hormone

Species Index

A

T

Common Name Index

Subject Index

B

C

Q

R

U

V

W

X

Y